Advances in Intelligent Systems and Computing

Volume 1200

The series "Advances in Intelligent Systems and Computing" contains publications on theory, applications, and design methods of Intelligent Systems and Intelligent Computing. Virtually all disciplines such as engineering, natural sciences, computer and information science, ICT, economics, business, e-commerce, environment, healthcare, life science are covered. The list of topics spans all the areas of modern intelligent systems and computing such as: computational intelligence, soft computing including neural networks, fuzzy systems, evolutionary computing and the fusion of these paradigms, social intelligence, ambient intelligence, computational neuroscience, artificial life, virtual worlds and society, cognitive science and systems, Perception and Vision, DNA and immune based systems, self-organizing and adaptive systems, e-Learning and teaching, human-centered and human-centric computing, recommender systems, intelligent control, robotics and mechatronics including human-machine teaming, knowledge-based paradigms, learning paradigms, machine ethics, intelligent data analysis, knowledge management, intelligent agents, intelligent decision making and support, intelligent network security, trust management, interactive entertainment, Web intelligence and multimedia.

The publications within "Advances in Intelligent Systems and Computing" are primarily proceedings of important conferences, symposia and congresses. They cover significant recent developments in the field, both of a foundational and applicable character. An important characteristic feature of the series is the short publication time and world-wide distribution. This permits a rapid and broad dissemination of research results.

**** Indexing: The books of this series are submitted to ISI Proceedings, EI-Compendex, DBLP, SCOPUS, Google Scholar and Springerlink ****

More information about this series at http://www.springer.com/series/11156

Joy Iong-Zong Chen · João Manuel R. S. Tavares ·
Subarna Shakya · Abdullah M. Iliyasu
Editors

Image Processing and Capsule Networks

ICIPCN 2020

Springer

Editors
Joy Iong-Zong Chen
Department of Electrical Engineering
Dayeh University
Changhua, Taiwan

Subarna Shakya
Department of Electronics
and Computer Engineering
Tribhuvan University
Lalitpur, Nepal

João Manuel R. S. Tavares
Instituto de Ciência e Inovação em
Engenharia Mecânica e Engenharia
Industrial
Departamento de Engenharia Mecânica
Faculdade de Engenharia
Universidade do Porto
Porto, Portugal

Abdullah M. Iliyasu
College of Engineering
Prince Sattam Bin Abdulaziz University
Al-Kharj, Saudi Arabia

ISSN 2194-5357 ISSN 2194-5365 (electronic)
Advances in Intelligent Systems and Computing
ISBN 978-3-030-51858-5 ISBN 978-3-030-51859-2 (eBook)
https://doi.org/10.1007/978-3-030-51859-2

This Springer imprint is published by the registered company Springer Nature Switzerland AG
The registered company address is: Gewerbestrasse 11, 6330 Cham, Switzerland

We are honored to dedicate the ICIPCN 2020 proceedings to all the organizers, editors, and authors of ICIPCN.

Foreword

It is our great pleasure to warmly invite you to the proceedings of the International Conference on Image Processing and Capsule Networks (ICIPCN 2020), which was held during May 6–7, 2020, at Bangkok, Thailand.

The primary theme of the conference is "Imaging Science and Capsule Networks". This two days conference event is designed to establish, share, and exchange the state-of-the-art research information and developments to all the researchers, academicians, and industrialists, who are actively participating in the computational intelligence and imaging science research. The proceedings of the conference highlight numerous research discoveries and significant breakthrough of capsule networks in the field of imaging science. It is always verified that the conference goes to the heart of all the research matters relating to image processing.

ICIPCN 2020 brings together the computational intelligence and image processing researchers from around the world by allowing all the participants to hear and meet the research experts at the forefront of our conference event and is being held at a enjoyable and lively location. In this way, we the conference organizers are confident that you will enjoy the stimulating conference with state-of-the-art research contributions, which has the potential to enhance the vibrancy of the research discussions around the conference theme, knowledge insights, and research collaborations.

I sincerely hope that the proceedings of the conference event will deliberate on the issues that need to be addressed when the artificial intelligence (AI) has started to leave its footprints in almost all the domains with a considerable impact in the imaging science domain.

Joy Iong-Zong Chen

Preface

It is our pleasure to welcome you to the International Conference on Image Processing and Capsule Networks (ICIPCN 2020) in Bangkok, Thailand. A major goal and feature of the conference is to bring academia and industries together to share and exchange their significant research experiences and results in the field of imaging science, with a particular interest on the capsule network algorithms and models by discussing the practical challenges encountered and solutions adopted to it. This conference will deliver a technically productive experience to the budding researchers in the field of image processing and capsule networks by stimulating a good awareness on this emerging research field. ICIPCN promises to provide a produce a bright landscape for the image processing research works, while the response received, and research eagerness witnessed, have truly exceeded our expectations. At the end of the conference event, we are overwhelmed with the high level of satisfaction.

The response for the conference event has been increasing at an unprecedented rate both from Thailand and overseas. Due to the professional expertise of both internal and external reviewers, the papers have been selectively accepted based on their extensive research and publication quality. Some state-of-the-art research works could not be accepted due to the capacity constraints of the conference proceedings. We have received a total submission of 223 out of which only 72 papers were accepted for publication based on its research effectiveness and applicability.

We would like to thank the guest editors **Dr. João Manuel R.S. Tavares,** Professor, Instituto de Ciência e Inovação em Engenharia Mecânica e Engenharia Industrial, Departamento de Engenharia Mecânica, Faculdade de Engenharia, Universidade do Porto, Portugal, **Dr. Subarna Shakya,** Professor, Department of Electronics and Computer Engineering, Central Campus, Institute of Engineering, Pulchowk, Tribhuvan University, Nepal, and **Dr. Abdullah M. Iliyasu,** Professor, concurrently with the College of Engineering, Prince Sattam Bin Abdulaziz University, Saudi Arabia and School of Computing, Tokyo Institute of Technology, Japan, for their valuable guidance and technical support for the selection of articles. We would also like to extend our sincere thanks to the members of the organizing committee for their hard work in delivering spontaneous response to all the conference participants.

We are now enthusiastic to get the proceedings of the ICIPCN conference event covered by Springer. We also appreciate all the authors of ICIPCN 2020 for their timely response to all the queries raised from the conference.

Finally, we would like to thank the Springer publications for producing this volume.

Joy Iong-Zong Chen
Conference Chair, ICIPCN 2020

Acknowledgments

We ICIPCN 2020 are grateful for the support of all the researchers, academicians, and industrialists, who worked hard to make this conference a successful event. Nevertheless, we thank King Mongkut's University of Technology Thonburi, Thailand, Dayeh University, Taiwan, and Tribhuvan University, Nepal, for their immense support and a very descriptive assistance during the conference event.

The conference organizers are particularly grateful to all the reviewers and advisory board members, who contributed their state-of-the-art research knowledge with the conference by offering significant impact on the received research manuscripts. Furthermore, the ICIPCN 2020 heartily acknowledges the efforts made by **Dr. Joy Iong-Zong Chen**, who helped out with the successful organization of the conference event at various instances. We are very pleased to thank our conference keynote speaker **Dr. João Manuel R. S. Tavares,** who delivered state-of-the-art and significant research insights and expertise, which has the potential to assist the next-generation research works in all the areas of image processing and capsule networks to the conference attendees. We thank all our technical and non-technical faculty members for their impeccable support toward the organization of the conference event.

Further, we wish to extend our warm compliments to all the authors, who contributed their significant and timely research works to enhance the research quality of ICIPCN 2020. Finally, we would like to extend our gratitude to all the session chairs, technical program committee members, and organizing committee members for contributing their tireless efforts to complete this successful conference event.

Contents

Short-Bio

João Manuel R. S. Tavares Faculdade de Engenharia da Universidade do Porto, Portugal graduated in Mechanical Engineering at the Universidade do Porto, Portugal in 1992. He also earned his M.Sc. degree and Ph.D. degree in Electrical and Computer Engineering from the Universidade do Porto in 1995 and 2001, and attained his Habilitation in Mechanical Engineering in 2015. He is a senior researcher at the Instituto de Ciência e Inovação em Engenharia Mecânica e Engenharia Industrial (INEGI) and Associate Professor at the Department of Mechanical Engineering (DEMec) of the Faculdade de Engenharia da Universidade do Porto (FEUP). João Tavares is co-editor of more than 55 books, co-author of more than 50 book chapters, 650 articles in international and national journals and conferences, and 3 international and 3 national patents. He has been a committee member of several international and national journals and conferences, is co-founder and co-editor of the book series "Lecture Notes in Computational Vision and Biomechanics" published by Springer, founder and Editor-in-Chief of the journal "Computer Methods in Biomechanics and Biomedical Engineering: Imaging & Visualization" published by Taylor & Francis, Editor-in-Chief of the journal "Computer Methods in Biomechanics and Biomedical Engineering" published by Taylor & Francis, and co-founder and co-chair of the international conference series: CompIMAGE, ECCOMAS VipIMAGE, ICCEBS and BioDental. Additionally, he has been (co-) supervisor of several MSc and PhD thesis and supervisor of several post-doc projects, and has participated in many scientific projects both as researcher and as scientific coordinator.

His main research areas include computational vision, medical imaging, computational mechanics, scientific visualization, human-computer interaction and new product development. (More information can be found at: www.fe.up.pt/~tavares)

Efficient GAN-Based Remote Sensing Image Change Detection Under Noise Conditions

Wenzhun Huang[1], Shanwen Zhang[1], and Harry Haoxiang Wang[2(✉)]

[1] School of Information Engineering, Xijing University, Xi'an 710123, China
[2] GoPerception Laboratory, Ithaca, NY 14850, USA
Hw496@goperception.com

Abstract. Efficient GAN-based remote sensing image change detection model under noise conditions is studied in this research work. Based on the multi-scale segmentation remote sensing change detection, this research work proposes an optimal method for the remote sensing image change detection with stable features, which provides an active way to increase the accuracy of the change detection. The target object selected in this article is relatively flat, but most target objects are different in practical applications. Therefore, the change detection method based on stable feature point removal based on multi-scale change detection is closely related to the multi-scale change detection, so the result is not avoiding the presence of "salt and pepper" noise. The GANs are integrated to pre-process the images, and the de-noising work is enhanced for the high-resolution images. The experiment has proved its effectiveness.

Keywords: GAN-based · Remote sensing · Image processing · Change detection · Noise condition · Fuzzy clustering

1 Introduction

The generative adversarial network is a generative deep learning model proposed in 2014. Once proposed, this model has become one of the hot research directions in the field of computer vision research. In recent years, with the rapid development of deep learning and mobile devices, the fields of image processing, image style migration, image content-based retrieval and classification, and image generation have become a topic of the great application value. The generative adversarial network contains a generative model and a discriminative model. Among them, the generation model is responsible for capturing the distribution of the sample data, and the discrimination model is generally a two-classifier to determine whether the input is real data or generated samples. The optimization process of this model is a "binary minimax game" problem. During training, one of the parties is fixed (discrimination network or generative network), the parameters of the other model are updated, and iterative iteration is performed. Finally, the generated model can estimate the sample data distribution. The emergence of the generation antagonism network greatly promotes the study of unsupervised learning and image generation [1–5].

Based on the literature review, the GANs used in the computer vision can be summarized in the following aspects. (1) The original application of GAN is in image

J. I.-Z. Chen et al. (Eds.): ICIPCN 2020, AISC 1200, pp. 1–8, 2021.
https://doi.org/10.1007/978-3-030-51859-2_1

generation and modeling. Regardless of the way of supervised learning or unsupervised learning, GAN can learn the distribution of real data. GAN uses this principle to first generate low-resolution images from low-resolution samples, and then use the generated low-resolution images as part of the next stage input and corresponding high-resolution samples to generate corresponding high-resolution images. The generator of a stage corresponds to a discriminator, to then determine whether the image at this stage is generated or real. (2) An interesting application in GAN is style transfer, that is, transforming images from one style to another. It is an improved CNN network. In style transfer, many pixel arrangements are not changed. Therefore, in addition to the normal convolution pooling operation, a part of the GAN structure is directly passed to the next layer, which ensures that the image content is unchanged. The classifier uses a convolutional GAN classifier. After experiments, the classification results using local classifier are better than the global one, and the number of parameters is greatly reduced, which improves the speed and efficiency of training [6–9].

As the demonstration, Fig. 1 gives the organization of GAN. Inspired by the satisfactory performance of the GAN, this paper will apply this model to the change detection task. The framework of the proposed model contains two major pipelines, namely the image de-noising and difference detection. When additive white Gaussian or impulse noise is severe, such as the two-stage strategy becomes less effective. Hence, the GAN will be applied for solving this task. Remote sensing change detection is one of the hot topics in the field of remote sensing. It has important commercial and application values in land cover change monitoring, environmental change dynamic monitoring, natural disaster monitoring, and land and resources survey. Feature-based image change detection is a feature that is generated based on the original data. During the feature extraction process, some parts of information may be lost, and there are

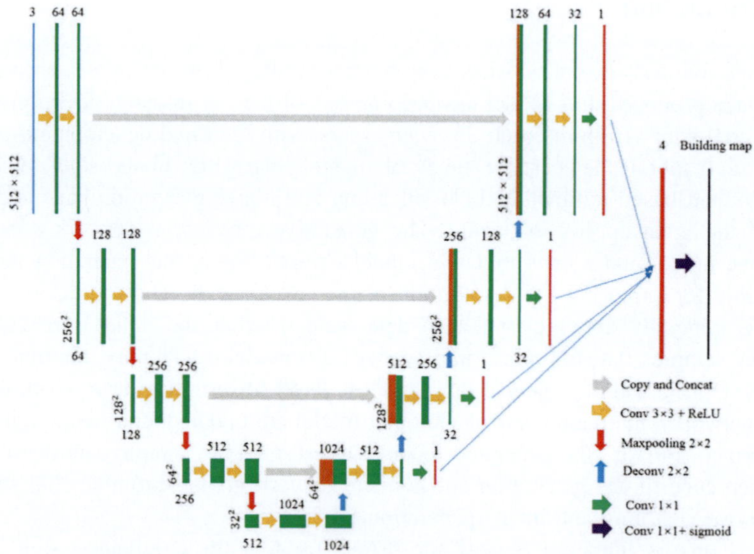

Fig. 1. The organization of the generative adversarial network

disadvantages that it is difficult to provide fine information. To begin with, in Fig. 2, the data set is demonstrated for analyzing the change detection tasks [10–12].

Fig. 2. The change detection data set

2 The Proposed Methodology

2.1 Image de-Noising Framework

In the process of collecting, transmitting and storing, images are usually disturbed by various noises, which leads to the degradation of the image quality, therefore, image de-noising is the most common and extremely important issue in digital image processing. Traditional image de-noising algorithms are mostly aimed at grayscale images, but color images are more common in real life [13, 14].

A series of breakthroughs of image processing algorithms were initiated by deep convolution neural network [15, 16].

$$\begin{pmatrix} f \\ g \end{pmatrix} \mapsto \begin{pmatrix} R_0 f \\ zg - S^*(z^*)f(0) \end{pmatrix} \tag{1}$$

In formula 1, the model details are presented. The receptive field is an important concept in convolutional neural networks, which is used to represent the size of the range of the perception of the original image by neurons at different locations within the network. the larger the value of the neuron receptive field indicates that the larger the range of the original image it can reach, the more global and also semantically higher features can be extracted; and the smaller the value indicates that the more features it contains, the more local and detail. So the value of the receptive field can be used to judge each layer of abstraction. Since the currently collected remote sensing images are usually high-resolution images, they usually contain huge amounts of data information. By comparison, the first two de-noising methods need to process all data information during the reconstruction process, which not only consumes a lot of time but also is difficult to reconstruct the original image in some cases [17–19].

The CS de-noising reconstruction method can then achieve the high-quality reconstruction of the signal using only a small amount of observation data, and can effectively solve the problem of reconstruction of high-dimensional signals.

2.2 GANs for Visual Analysis

When faced with complex tasks, the reward function of reinforcement learning is difficult to specify. On the one hand, the environment can only obtain less return information before the final result is obtained; on the other hand, it is difficult to accurately design the real-time reward function, and it is subjective and empirical. More importantly, different reward functions will then bring different optimal strategies. If the instant reward setting is not appropriate, it will make it difficult for the reinforcement learning algorithm to converge. Through empirical research, it is found that when examining the behavior of humans and animals, the reward function must be considered as an unknown condition. However, in the real world, due to the influence of natural factors and the limitations of data recording conditions, it is often unrealistic to obtain large-scale labeled data sets, with only a small number of labeled data samples. For example, to build an intelligent forecast model based on the dense fog weather situation field in a certain area, since the weather situation field is a texture map composed of contour lines, the fog type and the texture have a high degree of correlation is the best choice to solve this problem. This paper uses a generative adversarial network to solve the problem of zero-sample image classification. The main idea is to use random noise and semantic description of unknown classes to generate features of unknown classes of images, and then train a classifier to classify the generated image features. Although the above method indirectly solves the problem of zero-sample image classification, this method introduces a new problem, namely how to train a generative adversarial network that can accurately generate image features, that is to find the solution for the following formula.

$$\begin{aligned} &\min_{U,V} \sum_{m=1}^{M} \|\boldsymbol{x}_m - \boldsymbol{u}_m \boldsymbol{V}\|^2 + \lambda|\boldsymbol{u}_m| \\ &s.t \ \ \|\boldsymbol{v}_k\| \leq 1 \quad k = 1, 2, \ldots, K \end{aligned} \tag{2}$$

Due to the loss of the certain amount of information in the process of image down-sampling, there is bound to be some generation bias in the reconstruction of the corresponding high-resolution image from the low-resolution image that has lost some features, which will increase with the increase of the complexity of image content, category and texture [20].

The classifier weights are fixed and import the network structure into the SRGAN model, so the weights of the classifier parameters are no longer updated during model training. Finally, replace the nodes in the classifier and connect the existing SRGAN structure diagram with the classifier structure diagram to use additional category information in the SRGAN model as follows.

Figure 3 gives the pipeline for reference. By adding the classifier category loss term to the generator loss, the generator is optimized for the correct prediction on the classifier, which explicitly indicates that the result is close to the original image, and the category loss is reversed to the parameter weight of the generator, which makes the classifier present the additional information learned by the generator, such as some gender characteristics or the appearance of the glasses, etc.

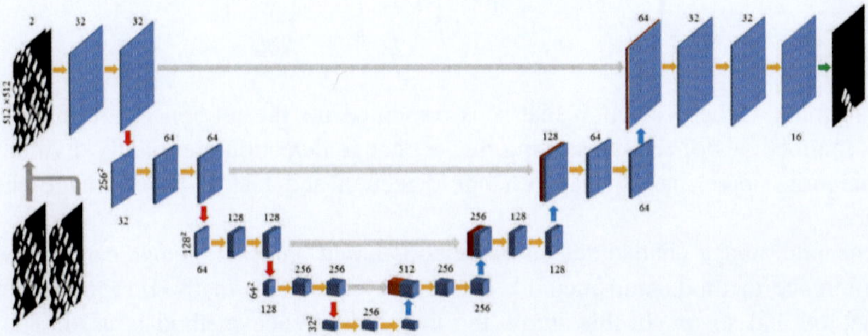

Fig. 3. GANs for visual analysis: framework and pipeline

2.3 Finalized Change Detection Framework

To make up for the shortcomings of traditional methods, object-oriented detection methods have been introduced into the research of remote sensing detection. This method mainly obtains the change information of the object by observing the change of the object unit at different times, and fully uses the information such as the spectrum, texture, and shape of the detection object to make up for the defect of the traditional object-oriented image that only sees the part and not the whole greatly improves the accuracy of observations. To improve the accuracy of remote sensing image detection, the SURF algorithm is used to eliminate the mismatch points in the process of object detection. Firstly, by comparing the images of the same object in different periods, the image spots of the target object are found, and then the stable feature points of the target image are extracted by the SURF algorithm, and the stable feature points of the target image are obtained. In formula 3, the model is described [21].

$$f(x) = \sum_k c_{j_0}(k)\phi_{j_0,k}(x) + \sum_{j=j_0}^{\infty}\sum_k d_j(k)\psi_{j,k}(x) \tag{3}$$

In the change detection, the changing area occupies a small proportion in a scene, and the constant area of the multi-temporal image is much larger than the changing area, that is, the distribution histogram of the image gray scale space obtained in the scene in different phases is consistent.

Therefore, relative radiometric correction can be achieved to some extent through the standardized processing of the multi-temporal image, and the method is simple and convenient for automatic processing.

$$\begin{pmatrix} r_0 & r_1 & r_2 & r_3 & r_4 \\ r_{-1} & r_0 & r_1 & r_2 & r_3 \\ r_{-2} & r_{-1} & r_0 & r_1 & r_2 \\ r_{-3} & r_{-2} & r_{-1} & r_0 & r_1 \\ r_{-4} & r_{-3} & r_{-2} & r_{-1} & r_0 \end{pmatrix} \begin{pmatrix} f_0 \\ f_1 \\ f_2 \\ f_3 \\ f_4 \end{pmatrix} = \begin{pmatrix} g_0 \\ g_1 \\ g_2 \\ g_3 \\ g_4 \end{pmatrix} \tag{4}$$

In formula 4, the projection matrix is presented for the references. As mentioned earlier, the methods of remote sensing image change detection are mainly divided into two categories: pixel-based image change detection and feature-based image change detection.

In general, image change detection methods based on pixel change can be divided into difference method, ratio method, correlation coefficient method, regression analysis method and so on. In this study, the image difference method is used. The difference method is to subtract the gray value of the corresponding pixel from the two registered remote sensing images, obtain a difference image, and threshold the difference image, which detects the change region. This method avoids the influence of the change information in the principal component transform by using the standard principal component transform of the principal component difference image of the normalized image, which eliminates the band inconsistency and also reduces the noise effect. Finally, the first two principal components are combined and the change information is stretched to then help determine the change threshold automatically.

3 Experiment

In this section, the experiment is conducted. By comparing with the existing algorithms, it can be seen that the noise points are effectively reduced, and the monitoring accuracy is also improved. The final experimental results show that the method proposed in this paper has improved the detection accuracy compared to the traditional pixel-based detection. Figures 4 and 5 present the results.

Fig. 4. Change detection results

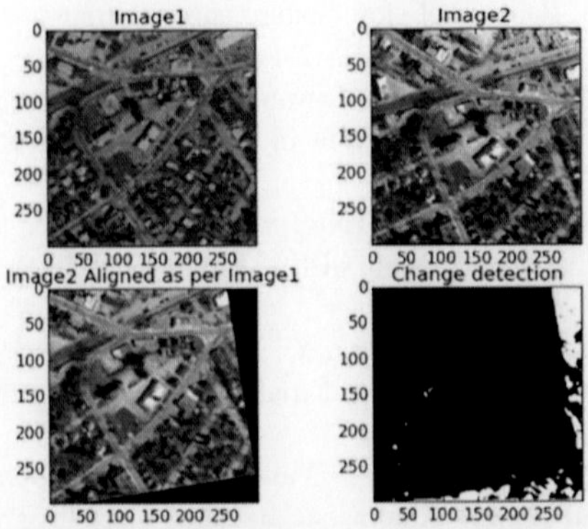

Fig. 5. Robust change detection result

4 Conclusion

Efficient GAN-based remote sensing image change detection model under noise conditions is studied in this paper. In the change detection results of the new area, the difference principal component analysis has the same problem, that is, it is impossible to determine a suitable threshold to then quickly extract the change information of different time phases. The improved principal component analysis method not only effectively detects the changes in buildings, but also correctly detects the existing roads and new roads. In this paper, the cultivated land and the water-permeable layer are not changed, and the detection problem is avoided. The proposed model is simulated and compared with the state-of-the-art approaches. The simulation results have proved the effectiveness. In our future research, the GANs will be enhanced to fix the complexity issues.

References

1. Huang, S., Cai, N., Pacheco, P.P., Narrandes, S., Wang, Y., Xu, W.: Applications of support vector machine (SVM) learning in cancer genomics. Cancer Genomics Proteomics **15**(1), 41–51 (2018)
2. Zhu, Z.: Change detection using landsat time series: a review of frequencies, preprocessing, algorithms, and applications. ISPRS J. Photogram. Remote Sens. **130**, 370–384 (2017)
3. Alcantarilla, P.F., Stent, S., Ros, G., Arroyo, R., Gherardi, R.: Street-view change detection with deconvolutional networks. Auton. Robots **42**(7), 1301–1322 (2018)
4. Lu, X., Yuan, Y., Zheng, X.: Joint dictionary learning for multispectral change detection. IEEE Trans. Cybern. **47**(4), 884–897 (2016)
5. Manogaran, G., Lopez, D.: Spatial cumulative sum algorithm with big data analytics for climate change detection. Comput. Electr. Eng. **65**, 207–221 (2018)
6. Deng, L., Wang, H.H., Li, D., Su, Q.: Two-stage visual attention model guided SAR image change detection. In: 2019 International Conference on Smart Systems and Inventive Technology (ICSSIT), pp. 1245–1249. IEEE (2019)

7. Wu, C., Zhang, L., Du, B.: Kernel slow feature analysis for scene change detection. IEEE Trans. Geosci. Remote Sens. **55**(4), 2367–2384 (2017)
8. Gong, M., Yang, H., Zhang, P.: Feature learning and change feature classification based on deep learning for ternary change detection in SAR images. ISPRS J. Photogram. Remote Sens. **129**, 212–225 (2017)
9. Wang, Q., Yuan, Z., Du, Q., Li, X.: GETNET: a general end-to-end 2-D CNN framework for hyperspectral image change detection. IEEE Trans. Geosci. Remote Sens. **57**(1), 3–13 (2018)
10. Chen, Q., Zhang, G., Yang, X., Li, S., Li, Y., Wang, H.H.: Single image shadow detection and removal based on feature fusion and multiple dictionary learning. Multimedia Tools Appl. **77**(14), 18601–18624 (2018)
11. James, M.R., Robson, S., Smith, M.W.: 3-D uncertainty-based topographic change detection with structure-from-motion photogrammetry: precision maps for ground control and directly georeferenced surveys. Earth Surf. Process. Land. **42**(12), 1769–1788 (2017)
12. Manoharan, S.: Image detection classification and recognition for leak detection in automobiles. J. Innovative Image Process. (JIIP) **1**(02), 61–70 (2019)
13. Jia, L., Li, M., Zhang, P., Wu, Y., Zhu, H.: SAR image change detection based on multiple kernel K-means clustering with local-neighborhood information. IEEE Geosci. Remote Sens. Lett. **13**(6), 856–860 (2016)
14. Yang, X., Tang, L., Stewart, K., Dong, Z., Zhang, X., Li, Q.: Automatic change detection in lane-level road networks using GPS trajectories. Int. J. Geogr. Inform. Sci. **32**(3), 601–621 (2018)
15. Huang, F., Chen, L., Yin, K., Huang, J., Gui, L.: Object-oriented change detection and damage assessment using high-resolution remote sensing images, Tangjiao landslide, three gorges reservoir. Chin. Environ. Earth. Sci. **77**(5), 183 (2018)
16. Xie, X., Huang, W., Wang, H.H., Liu, Z.: Image de-noising algorithm based on Gaussian mixture model and adaptive threshold modeling. In: 2017 International Conference on Inventive Computing and Informatics (ICICI), pp. 226–229. IEEE (2017)
17. Zhou, L., Cao, G., Li, Y., Shang, Y.: Change detection based on conditional random field with region connection constraints in high-resolution remote sensing images. IEEE J. Sel. Top. Appl. Earth Observations Remote Sens. **9**(8), 3478–3488 (2016)
18. Tan, K., Jin, X., Plaza, A., Wang, X., Xiao, L., Du, P.: Automatic change detection in high-resolution remote sensing images by using a multiple classifier system and spectral–spatial features. IEEE J. Sel. Top. Appl. Earth observations Remote Sens. **9**(8), 3439–3451 (2016)
19. Zhao, B., Huang, W., Wang, H.H., Liu, Z.: Image de-noising algorithm based on image reconstruction and compression perception. In: In 2017 International Conference on Inventive Computing and Informatics (ICICI), pp. 532–535. IEEE (2017)
20. El Amin, A.M., Liu, Q., Wang, Y.: Zoom out CNNS features for optical remote sensing change detection. In: 2017 2nd International Conference on Image, Vision and Computing (ICIVC), pp. 812–817. IEEE (2017)
21. Chen, B., Chen, Z., Deng, L., Duan, Y., Zhou, J.: Building change detection with RGB-D map generated from UAV images. Neurocomputing **208**, 350–364 (2016)

Recognition of Handwritten Digits by Image Processing Methods and Classification Models

Amelec Viloria[1(✉)], Reinaldo Rico[1], and Omar Bonerge Pineda Lezama[2]

[1] Universidad de la Costa, St. 58 #66, Barranquilla, Atlántico, Colombia
{aviloria7, rrico2}@cuc.edu.co
[2] Universidad Tecnológica Centroamericana (UNITEC), San Pedro Sula, Honduras
omarpineda@unitec.edu

Abstract. OCR (Optical Character Recognition) is a line of research within image processing for which many techniques and methodologies have been developed. Set of pixels recognized based on the digitalized image and this study presents an iterative process that consists of five phases of the OCR. For this purpose, several image processing methods are applied, as well as two variable selection methods, and several supervised automated learning methods are explored. Among the classification models, those of deep learning stand out for their novelty and enormous potential.

Keywords: Genetic algorithm · Recognition of handwritten digits · Image processing · Methods · Classification models

1 Introduction

OCR (Optical Character Recognition) consists of identifying a symbol, usually a number or a letter, from the digitization of an image [1, 2]. Currently, there are multiple processes in which character recognition is applied.

Examples of OCR include: automating the redirection of letters in the postal service, recognizing car license plates on speed cameras, or digitizing notes taken in a classroom by writing with a stylus on a tablet.

This research consists of identifying digits from a set of pixels representing handwritten numbers [3]. Its contribution is a methodology composed of 5 phases for the identification of these numbers. For this aim, the study applies dimensionality reduction, extraction of characteristics from the original image, selection of variables, and evaluation of several supervised models of automated learning.

For the selection of variables, the PCA (Principal Component Analysis) [4] and RFE (Recursive Feature Elimination) [5] techniques were incorporated for creating more efficient models for OCR. With these techniques, the size of the data set decreases and the character classification time is reduced. In the case under study, some entries always keep the same value or have very little variation in their data. These variables will be discarded.

J. I.-Z. Chen et al. (Eds.): ICIPCN 2020, AISC 1200, pp. 9–19, 2021.
https://doi.org/10.1007/978-3-030-51859-2_2

The 5-phase method (Fig. 1) incorporates image processing as well as the iterative phases corresponding to the creation of automated learning models [6–9].

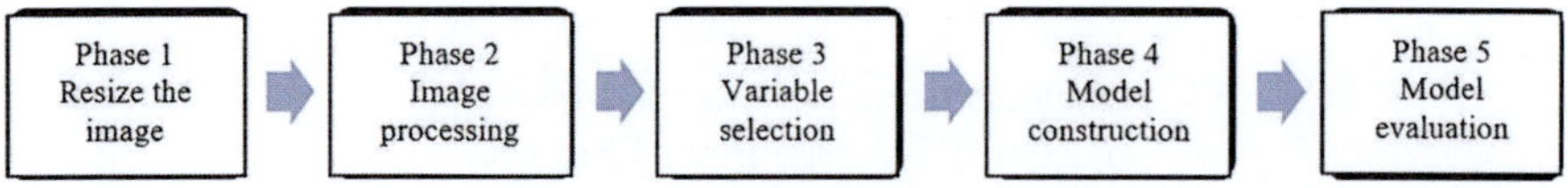

Fig. 1. Phases of the research.

For building a suitable model, the data samples must be of good quality and have a sufficient number of samples. In this sense, the study uses the level of *accuracy* or precision [10] which represents the percentage of hits in the range [0.1] to evaluate the quality of the models presented. Finally, there is a summary table with the results of the evaluation.

2 Description of the Database

Normally, the OCR process starts with the digitization of handwritten images containing the digits, which starts from a public data set in this research [11].

This data set is composed of a total of 70,000 samples [12] of which 42,000 samples were chosen at random. Each sample has 785 fields. One field is considered the output of the model and represents the digit, while the rest of the fields are the inputs. The inputs are the pixels of the image that represent the handwritten digit. Each image has a resolution of 28 × 28, making a total of 784 pixels. Each pixel is represented by a number between 0 and 255 indicating the grey level. So, the value 0 represents the white color and the value 255 represents the black color.

3 Phase 1: Resize the Image

In our data set, each image is represented by a two-dimensional array of 28 × 28 grayscale pixels.

Therefore, each image is represented by 785 attributes [13]:

- Digit in the image, its value is an integer from 0 to 9.
- 784 values in the pixels (the result of the 28 × 28 pixels), take integer values from 0 to 255, with 255 being the darkest intensity (black) and 0 being the lightest intensity (white).

Working with a data set of the size 42,000 (number of images) by 197 (number of attributes) is complicated because the larger the data dimension the more complex and slower the creation of learning models. Therefore, it is desirable to reduce the size of the data [14].

The first step for reducing the size of the data is to reduce the size of the images. That is, instead of each image is 28 × 28 pixels, the image will be of 14 × 14 pixels.

For this purpose, the values of four continuous pixels in the original image are averaged to obtain one pixel in the smaller-scale image. After this process, each image is represented by 197 attributes:

- Digit in the image, its value is an integer from 0 to 9.
- 196 values in the pixels (a result of the 14 × 14 pixels), integer values from 0 to 255.

4 Phase 2: Image Processing

Image processing is a set of techniques used to improve the quality of the image or to facilitate the search for information in them. Examples could be removing noise, locating edges, or smoothing the image.

The most used techniques that we can apply to our problem and that we will apply in phase two are [15–17]:

- Binarization: Consists of converting all the pixels of an image originally having several values to only two tones: black and white. For this, a threshold is established above which we will turn black and below which it will be white. This operation is very frequent since some algorithms work on this scale. Images take up a very small memory size since each pixel can be represented with one pixel.
- Fragmentation or segmentation: This technique consists of selecting a part of the original image. For this, algorithms are developed that are based on the detection of edges that are small changes in the intensity of colors and other properties. There are many generic fragmentation methods, although a more specific algorithm is usually applied for each type of image. For example, to detect breast cancer, a filter is first applied to the suspicious regions and then an algorithm is applied that evaluates that set of pixels.
- Component thinning: One of the most common techniques is to iteratively erase the contour points. This technique should be applied with caution so that the images do not lose their original shape and in an iterative way. This method simplifies the identification of characters and allows the extraction of characteristics such as the height of the image according to the pixels or the width of the pixels that make up the image.
- Calculate pixel average: This method of image treatment consists of taking all the samples of a certain digit and creating a new image. Where each pixel of that image is the average of all the images that represent that digit. For example, the pixel [1, 1] of the average image of eight is the average of the pixel [1, 1] of all the samples of number eight.
- The algorithm processes the images of all the digits and as a result, we obtain 10 images with the average for all the pixels. Those ten images will serve as inputs to create a new model. But its hit rate is very low so we must find a better method.
- Elimination of blank rows and columns: One of the procedures that we are going to carry out in the treatment of images is to eliminate those rows and columns that are white and almost white. In other words, we are going to remove the contours of all the images from the data set as they provide little information.

- To do this we apply a method that adds the pixels, as white is represented with zero if the sum is not greater than a certain threshold we will delete that row. The same for the columns.
- Object Characteristics Selection: At this stage, characteristics of the object present in the image are extracted after a simple binarization (pixels greater than zero were taken). The object in the image is characterized by a vector with the following components: Area, Centroid (component x, y), Length of the major axis, length of the minor axis, Eccentricity, Euler's number, Equivalent diameter, Solidity, Perimeter, Proportional centroid at the gray level (x, y components), Degree, Maximum Intensity, Minimum Intensity, and Average intensity. This vector will be the input to the classifier in the following processes.

5 Phase 3: Selection of Variables

Once the previous phases are completed, PCA and RFE are used to select the most important characteristics and discard the less relevant ones.

PCA is a statistical technique used in pattern discovery in data sets with high dimensionality. It consists of the selection of components that provide a greater contribution of information to the model [18].

- Mutual information and PCA: Entropy [19] is a measure of uncertainty in a discrete random variable, that is, it measures how uniform a variable is. In this case, if the random variable has a uniform distribution, entropy is maximum. The following function is used to calculate entropy:

$$H(X) = -\sum_{x} P(x)\ln(P(x)) \tag{1}$$

Mutual information is a measure of dependence between two discrete random variables, that is, it measures the reduction of uncertainty (entropy) of a variable, according to the knowledge of another variable. The following equations are used to calculate mutual information:

$$H(X,Y) = -\sum_{x}\sum_{y} P(x,y)\ln(P(x,y)) \tag{2}$$

$$IM(X,Y) = H(X) - H(Y) - H(X,Y) \tag{3}$$

Recursive Selection Method: The RFE (*Recursive Feature Selection) method* is a variable selection-oriented method. This method is of the *wrapper* type, that is, models are built from a combination of inputs [20].

Based on the results obtained in the evaluation of models, those variables that are less important or that generate noise are discarded. The objective is twofold: to keep a smaller number of variables and to choose the combination that provides the best results. This will allow building simpler and more precise models. The application of RFE reduced the number of variables from 196 to 148.

6 Phase 4: Construction of the Model

6.1 Methodology in Model Construction

For the construction and evaluation of models, a library called Caret will be used. Caret is an acronym for: *"Classification And REgression Training"* [21, 22]. This library allows building models in a few lines of code and in a very simple way.

Below is the code needed to build a model using *Random Forest*:

```
-1) control = trainControl (method = "cv", number
 = 10, classProbs = T) 2) grid = expand.grid (
 .mtry = c ( 5, 10, 25, 50, 75, 100 ) )
-3) modelRForest = train ( inputs, output, method
 = "rf", tuneGrid = grid ,ntree = 1000 ,trControl
 = control )
```

The first instruction, train control, is used to indicate the number of models to be executed for each configuration. The second instruction, "expand_grid", indicates the number of Random Forest type trees that will be used. The third instruction serves to create and evaluate the model. It is performed with the "train" function that has the following parameters: an array with the inputs, an array with the outputs, and the method that will be used, in this case, the random forest [RF]. The number of trees that will be used is 1000. The variables "control" and "grid" have been explained in the first two instructions. Next, the genetic algorithm is presented for the construction of the models:

Method construction algorithm

```
 -Read the Data Set
 -Select the most significant variables using
  Random forest or PCA
 -From model: 1 to N
              -Create different sets of physical
               samples using CVs
              -From configuration 1 to configuration
               N
                    -Create several models with one
                     Training Set part
                    -Test the model with the
                     remaining partition
              -End From
              -The average of the models is chosen
               and the best result
  -End From
-Sort methods according to
 best results End of
 Algorithm
```

7 Classification Methods

Deep Learning H20: The H20 Deep Learning is a package from R Studio that is based on forward-feeding multi-layer neural networks. These nets are trained with stochastic gradient descent using backpropagation.

This model downloads through the Internet a series of data used in the configuration of the nodes. These data are the result of applying the algorithm to many problems worldwide. It is possible to download, for a fee, certain libraries that significantly improve both the results and the speed in the execution of these algorithms.

Random Forest: They are a combination of predictive trees where for each tree k a random vector genera k is generated, created independently of the $\gamma 1 \ldots \gamma k{-}1$ vectors and with the same distribution used in each tree k of the set (forest). The tree is constructed using the training set y γk, resulting in a classifier h (x, γk) where x is the input vector. The most popular class for x is selected after a sufficiently large number of trees have been generated [11].

KNN (K-Nearest Neighbors): It is a classification algorithm based on the calculation of the closest distance, usually Euclidean distance, between the points of the data set to a given point, where K is the number of selected close neighbors [1].

SVN (Support Vector Machines): Is a classifier that separates the data using hyperplanes. It searches for the optimal hyperplane, which is the hyperplane that provides the maximum distance from the closest training patterns, the SVs are then the closest vectors to these patterns [3].

8 Phase 5: Evaluation of the Model and Sample Results

The evaluation consists of measuring the degree of precision of each model. This phase informs whether the chosen dimensionality is adequate for the construction of the model. The accuracy of each method is calculated for measuring the quality of the models. For calculating the accuracy, the elements of the matrix that are part of the main diagonal are added up and divided by the total of elements.

Table 1 shows the results of applying the SVM method with 10,000 training samples and 15,000 for the test.

The accuracy can be calculated with the following formula:

$$\textbf{Accuracy} = \frac{\sum_{i=1}^{k} f_{ii}}{N} \quad (4)$$

In the confusion matrix, each column represents the number of predictions of each class and each row represents the instances of the actual class. The *accuracy of* the model is 0.9752.

The results obtained in the second problem are shown in Table 2. The experiment was performed using a total of 42,000 samples with a CV of 10 partitions. The results are shown in Table 2.

Table 1. Confusion matrix with the SVM method

Predicted values

Current values

	0	1	2	3	4	5	6	7	8	9
0	1555	0	6	2	2	4	6	1	5	5
1	0	1806	3	2	3	2	0	5	5	1
2	2	7	1562	17	15	3	0	10	5	2
3	2	0	9	1524	2	30	0	7	11	8
4	1	1	3	1	1426	6	3	8	9	24
5	7	0	0	18	2	1307	7	0	10	5
6	2	3	2	2	8	10	1502	0	6	0
7	0	8	14	9	7	2	0	1556	3	16
8	6	1	11	30	2	10	3	3	1408	14
9	1	4	4	5	28	3	0	20	10	1408

Table 2. Phases in image processing

Phase 1: image reduction	Phase 2: image processing	Phase 3: selection of variables	Phase 4: model construction	Phase 5: results
Do nothing	Pixel width, Pixel height	Do not extract characteristics	KNN	0,9541
Image 14 × 14	Remove blank rows	PCA	Random forest	0,9412
Image 28 × 28	Binarize	RFE	Neuronal network	0,9758

Table 3 presents the model name, the parameters that configure it, and the average accuracy of the 10 models created.

Table 3. Comparative results of models for 28 × 28 image

Image reduction	Image treatment	Variables selection	Predictive model	Model parameters	Result
Image 28 × 28	Application of multiple methods such as density, width, size,...	RFE	Random forest	Trees: 1000 - mtry: 75	0.8521
	Pixel average	Do not select	Random forest	Trees: 5	0.2010
				Trees: 20	0.2350
				Trees: 50	0.4250

Table 4 shows the results of some traditional and Deep Learning methods. These methods are best suited to these images. Possibly because of their capacity for abstraction. Each column of the table shows the methods applied in each of the phases. As can be seen in the results, the Deep Learning methods are the most precise.

Table 4. Comparative results of models for 14 × 14 image

Image reduction	Image treatment	Selection variables	Predictive model	Model parameters	Result
Image 14 × 14	No treatment	PCA	Random forest	Trees: 1000 - mtry: 75	0.9354
		RFE	Random forest	Trees: 2000 - mtry: 75	0.9754
				Trees: 1000 - mtry: 2	0.9625
				Trees: 1000 - mtry: 75	0.9754
				Trees: 1000 - mtry: 148	0.9652
			SVM with polynomial core	Degree = 3, scale = 0.1 and C = 0.25.	0.9841
			Neural Networks with feature extraction	Size = 5 and decay = 0.1	0.7654
	Edge reduction	All	Random forest	Trees: 1000 - mtry: 75	0.9452
			Deep learning	TanhWithDropout	0.520
				Tanh	1
				RectifierWithDropout	0.9999

9 Conclusion

Having worked with R Studio and the Caret library, it is considered a very suitable environment for solving problems related to the construction and evaluation of predictive models.

The main advantages of this tool are that it allows the creation of models with different configurations, the automatic selection of the best combination, and the display of results in different metrics. Besides, it includes the *cross-validation* option, which creates different partitions to create several models for more robust calculations.

The fact that it is free software represents a great advantage because it allows students and people from the scientific community to enjoy a tool with a wide range of functions. A disadvantage of R is that they do not have any element indicating the estimation of the remaining time. This makes it difficult to plan efficiently for testing the models. It also does not give guarantees in case an error occurs as clearly indicated when initializing the shell.

The R documentation does not include all the details of the functions so, in some cases, it is difficult to find the configuration of some parameters, which has happened in the case of the RFE functions. Finally, it is possible to implement this tool with other languages such as Java.

10 Future Work

The system that we have created is part of a phase in which the digits are located and centered on an image. So we start with already cropped images that greatly facilitate recognition. But in real life the characters are not isolated, they are found within texts. One way to segment this problem is to break the texts into small images that recognize only one digit.

A next step in the investigation would be to apply our model to color images. A simple way would be to pass the images to grayscale and apply our methodology to evaluate the results.

In the same way that we have managed to create a model with great precision for number recognition, the same methodology for letter recognition could be evaluated. The study would be interesting since if we have reached a high level of precision in digit recognition, we expect good results for letter recognition.

An interesting line of research to take into account the stability of the algorithms would be to increase the noise in the images since we know that in real-life images are subject to all kinds of distortion. By this, we understand that the paper has a different background color, that the numbers may be different, etc. For this, a noise generation algorithm would be applied to the images in the data set and a model would be created to evaluate it.

Taking into account the results obtained with the deep learning methods, we find it interesting to continue delving into this type of algorithms. We think that a comparative study of deep learning methods may be of general interest. To do this, we will use several public data sets and compare both the precision of the results and the time required for the construction of models.

References

1. Boufenar, C., Kerboua, A., Batouche, M.: Investigation on deep learning for off-line handwritten Arabic character recognition. Cogn. Syst. Res. **50**, 180–195 (2018)
2. Dasgupta, J., Bhattacharya, K., Chanda, B.: A holistic approach for Off-line handwritten cursive word recognition using directional feature based on Arnold transform. Pattern Recogn. Lett. **79**, 73–79 (2016)
3. Jangid, M., Srivastava, S.: Handwritten devanagari character recognition using layer-wise training of deep convolutional neural networks and adaptive gradient methods. J. Imaging **4** (2), 41 (2018)
4. Tarawneh, A.S., Hassanat, A.B., Chetverikov, D., Lendak, I., Verma, C.: Invoice classification using deep features and machine learning techniques. In: 2019 IEEE Jordan International Joint Conference on Electrical Engineering and Information Technology (JEEIT), pp. 855–859. IEEE, April 2019

5. Niu, X.X., Suen, C.Y.: A novel hybrid CNN–SVM classifier for recognizing handwritten digits. Pattern Recogn. **45**(4), 1318–1325 (2012)
6. Wang, Z., Wang, R., Gao, J., Gao, Z., Liang, Y.: Fault recognition using an ensemble classifier based on Dempster-Shafer Theory. Pattern Recogn. **99**, 107079 (2020)
7. Zhou, B., Ghose, T., Lukowicz, P.: Expressure: detect expressions related to emotional and cognitive activities using forehead textile pressure mechanomyography. Sensors **20**(3), 730 (2020)
8. Schmidhuber, J.: Deep learning in neural networks: an overview. Neural Netw. **61**, 85–117 (2015)
9. Mohiuddin, K., Mao, J.: A comparative study of different classifiers for handprinted character recognition. Pattern Recogn. Practice **IV**, 437–448 (2014)
10. Le Cun, Y., Cortes, C.: MNIST handwritten digit database. AT&T Labs. http://yann.lecun.com/exdb/mnist. Accessed: 20 Dec 2019
11. Viloria, A., Acuña, G.C., Franco, D.J.A., Hernández-Palma, H., Fuentes, J.P., Rambal, E.P.: Integration of data mining techniques to PostgreSQL database manager system. Procedia Comput. Sci. **155**, 575–580 (2019)
12. Viloria, A., Lezama, O.B.P.: Improvements for determining the number of clusters in k-means for innovation databases in SMEs. ANT/EDI40, pp. 1201–1206 (2019)
13. Varela, N., Silva, J., Gonzalez, F.M., Palencia, P., Palma, H.H., Pineda, O.B.: Method for the recovery of images in databases of rice grains from visual content. Procedia Comput. Sci. **170**, 983–988 (2020)
14. Koresh, M.H.J.D., Deva, J.: Computer vision based traffic sign sensing for smart transport. J. Innov. Image Process. (JIIP) **1**(01), 11–19 (2019)
15. Zhang, B., Fu, M., Yan, H.: A nonlinear neural network model of mixture of local principal component analysis: application to handwritten digits recognition. Pattern Recogn. **34**(2), 203–214 (2001)
16. Ghosh, A., Pavate, A., Gholam, V., Shenoy, G., Mahadik, S.: Steady model for classification of handwritten digit recognition. In: Sharma, R., Mishra, M., Nayak, J., Naik, B., Pelusi, D. (eds.) Innovation in Electrical Power Engineering, Communication, and Computing Technology. LNEE, vol. 630, pp. 401–412. Springer, Singapore (2020). https://doi.org/10.1007/978-981-15-2305-2_32
17. Garg, A., Gupta, D., Saxena, S., Sahadev, P.P.: Validation of random dataset using an efficient CNN model trained on MNIST handwritten dataset. In: 2019 6th International Conference on Signal Processing and Integrated Networks (SPIN), pp. 602–606. IEEE, March 2019
18. El-Sawy, A., Hazem, E.B., Loey, M.: CNN for handwritten arabic digits recognition based on LeNet-5. In: International Conference on Advanced Intelligent Systems and Informatics, pp. 566–575. Springer, Cham, October 2016
19. Paul, O.: Image pre-processing on NumtaDB for Bengali handwritten digit recognition. In: 2018 International Conference on Bangla Speech and Language Processing (ICBSLP), pp. 1–6. IEEE, September 2018
20. Shamsuddin, M.R., Abdul-Rahman, S., Mohamed, A.: Exploratory analysis of MNIST handwritten digit for machine learning modelling. In: International Conference on Soft Computing in Data Science, pp. 134–145. Springer, Singapore, August 2018
21. Pujari, P., Majhi, B.: Recognition of Odia handwritten digits using gradient based feature extraction method and clonal selection algorithm. Int. J. Rough Sets Data Anal. (IJRSDA) **6**(2), 19–33 (2019)

22. Shawon, A., Rahman, M.J.U., Mahmud, F., Zaman, M.A.: Bangla handwritten digit recognition using deep CNN for large and unbiased dataset. In: 2018 International Conference on Bangla Speech and Language Processing (ICBSLP), pp. 1–6. IEEE, September 2018
23. Makkar, T., Kumar, Y., Dubey, A.K., Rocha, Á., Goyal, A.: Analogizing time complexity of KNN and CNN in recognizing handwritten digits. In: 2017 Fourth International Conference on Image Information Processing (ICIIP), pp. 1–6. IEEE, December 2017
24. Rizvi, M., Raza, H., Tahzeeb, S., Jaffry, S.: Optical character recognition based intelligent database management system for examination process control. In: 2019 16th International Bhurban Conference on Applied Sciences and Technology (IBCAST), pp. 500–507. IEEE, January 2019

Convolutional Neural Network with Multi-column Characteristics Extraction for Image Classification

Jesus Silva[1(✉)], Noel Varela[1], Janns A. Patiño-Saucedo[1], and Omar Bonerge Pineda Lezama[2]

[1] Universidad de la Costa, St. 58 #66, Barranquilla, Atlántico, Colombia
jesussilvaUPC@gmail.com, {nvarela2,jpatino8}@cuc.edu.co
[2] Universidad Tecnológica Centroamericana (UNITEC), San Pedro Sula, Honduras
omarpineda@unitec.edu

Abstract. In the last few decades, the constant growth of digital images, as the main source of information representation for scientific applications, has made image classification a challenging task. To achieve high classification yields, different pattern recognition techniques have been proposed, among which are the deep learning methods that today focus their study on image processing and computer vision. In this approach, the most popular architecture for the image classification task is the convolutional neural network (CNN), a network constructed of multiple layers and where each layer models a receptive field of the visual cortex making it much more effective in artificial vision tasks [1]. This paper proposes a convolutional network architecture with a performance-enhancing approach, a hierarchical structure that is easy to build, adaptive, and easy to train with good performance in image classification tasks.

Keywords: Convolutional neural network · Multi-column characteristics extraction · Image classification

1 Introduction

CNN combines low-level features within high-level abstract features with non-linear transformations, allowing them to learn the semantic representation of images. These networks extract generally useful characteristics from tagged and untagged data, detect and eliminate input redundancies, and preserve only essential aspects of the data in sound and discriminative representations [2]. CNN can capture the most obvious characteristics from the data [3], as a way to achieve better results in various applications. Unlike hand-created features such as SIFT [4] and HOG [1], the features extracted by CNN are generated end-to-end, which eliminates human intervention. CNN has fewer connections and parameters which makes the extraction of characteristics more efficient.

J. I.-Z. Chen et al. (Eds.): ICIPCN 2020, AISC 1200, pp. 20–30, 2021.
https://doi.org/10.1007/978-3-030-51859-2_3

Most of the convolutional network architectures such as AlexNet [5], GoogleNet [6], VGG [7], ResNet 152 [8], and many others [9–13], use the same concept to produce feature maps in the convolution layers, followed by pooling layers to reduce the dimension of the maps and, as they go deeper into the architecture, they double the number of filters to compensate for the reduction to half of the convolution of the later feature maps. The depth of the network favors the classification performance and avoids gradient fading by using class inference in consecutive convolution layers and the maximum pooling layer, or by using softmax layers that enhance gradient fading [14, 15]. Some of these architectures using new activation functions, weighting regularization methods, class inferences, previous training by layers in supervised approaches, showed very promising results [16, 17].

Increasing the number of layers in a CNN means increasing the depth and number of parameters in the network, which complicates training and significantly reduces performance, especially with small databases. On the other hand, due to the lack of unique features, the fusion of characteristics is becoming increasingly important for tasks such as classification and image retrieval. These are techniques that simply concatenate a pair of different characteristics or use methods based on canonical correlation analysis for the reduction of the joint dimensionality in the space of the characteristics [18, 19].

This paper proposes a convolutional neural network with multi-column extraction of characteristics for image classification. The network integrates the abstraction capacity of deep neural networks and the capacity to concatenate different characteristics. The network grows in both depth and width, and the image to be classified is entered through three different character extraction sections with different filters in the convolution operations. The extracted characteristics are then concatenated and entered into fully connected layers that perform the classification stage [12, 20, 21].

The results show that the network has high performance in conjunction with images such as Oliva and Torralba [13], Stanford Dogs [14], and Caltech 256 [15]. A comparison of the proposed network with existing networks is made: AlexNet [5], GoogleNet [6], and ResNet [8], whose characteristics are described in [11].

2 Proposed Architecture

The proposed architecture is composed of 18 layers, the first fifteen are for the extraction of characteristics and the last three are for the classification. The following is a description of the network layers [22–25]:

- ***Image Input Layer:*** This layer sets up a pre-processing stage of the incoming images to the network. In this layer, the images can be resized, rotated and even small random samples can be taken. This layer is designed so that the network accepts images with dimensions of 224×224 with a depth of three since the image is entered with the RGB format (see Fig. 1, left side).
- ***Characteristic extraction layers:*** the image enters three different character extraction sections, each extracting different characteristics using filters from different sizes.

Fig. 1. Example of modified images for CNN entry

- ***Classification:*** the output of each characteristic extraction section is concatenated to generate a one-dimensional vector. This way, the fully connected layers can be continued to perform the classification.
    - **Concatenation of the outputs of the 3 feature extraction sections:** The classification begins with the concatenation of the output of each extraction section, obtaining a vector with dimensions 18432 1 (image characteristics).
    - **First fully connected layer:** It has a depth of 2048 neurons and is followed by a ReLU layer and a Dropout layer.
    - **Second fully connected layer:** It consists of 1024 neurons and is followed by a ReLU layer and a Dropout layer.
    - **Third fully connected layer:** This layer is used to adjust the convolutional network to each of the training sets used as it is necessary to match the number of output neurons to the number of classes in each training set. This layer is followed by a Softmax layer, a regression function, which helps classify multiple categories.
- ***Output layer:*** This is the final layer that shows the percentage of classification success.

3 The Data Set

Three different data sets were used in this work, which is briefly described below:

- **Oliva & Torralba** [13, 26]: this data set is composed of 2,688 color scene images that belong to the same semantic category. The database has a total of 8 categories and the images were obtained from different sources: commercial databases, websites, and digital cameras.
- **Stanford Dogs** [27]: This data set consists of 20,580 color images, belonging to 120 classes or breeds of dogs worldwide. This dataset was created using Image-Net images and annotations for the fine-grained image categorization task.

- **Caltech 256** [1]: This data set consists of 30,607 color images from 256 categories plus one named "clutter", containing multiple scenes. Each category contains 80 to 827 images and most of the categories have about 100 images.

4 Training Parameters

Both the proposed network and the networks with which the comparison is made AlexNet, GoogleNet and ResNet, were trained using the ADAM (adaptive moment estimation) optimization algorithm, with a batch size of $\beta = 32$ images and a weight decay (regularization factor) of = 0.0005. The initial weights in each of the layers were initialized with a Gaussian distribution averaging 0 and a standard deviation of 0.01. The activation thresholds in each of the layers were initialized to zero. It started with a learning rate of $\mu = 0.001$ which decreased by a factor of 10 after every 50 epochs, to have more specific learning changes in 250 epochs of training [28, 29]. The networks were trained on 4 NVIDIA GTX 1080 GPUs, with 8 GB of RAM and 2560 cores, with Linux Ubuntu 16.04, Linux kernel 4.12, Python 2.7, TensorFlow 1.12, NVIDIA CUDA R 8.0, NVIDIA cuDNN v5.1.

The input image is reflected horizontally, from left to right, with a 50% probability. The brightness and contrast of the image are adjusted randomly in intervals of 63% and 0.2–1.8, respectively. Finally, the image is normalized so that the network has some independence from the image properties. In Fig. 1, some examples of images modified with data magnification are shown.

5 Experiments and Results

To assess performance, the networks were trained from scratch with each of the databases and data augmentation was used. The training sets were formed with 70% of the images from each database and the remaining 30% were used to form the test sets. The selection of the images was done randomly. For separating the training and test sets, the toolbox developed in [30] was used to create tfrecord files to avoid memory overflow when training. Table 1 shows the characteristics of each data set used to perform the experiments in this project.

Table 1. Separate images set in training and testing, using 70% of images per class for training and 30% for testing

Data set	Classes	Number of images		
		Training	Test	Total
Oliva and Torralba	10	2,421	1,037	3,458
Stanford Dogs	144	15,716	6,735	22,451
Caltech 256	258	24,821	10,637	35,458

5.1 Evaluation of the Proposed Network

Figure 2 shows the training results of the proposed network, which from now on will be called ToniNet to differentiate it from the other networks. Note that the network reaches 100% accuracy for the Oliva & Torralba training set, while for the Stanford Dogs and Caltech 256 sets it reaches only 98%.

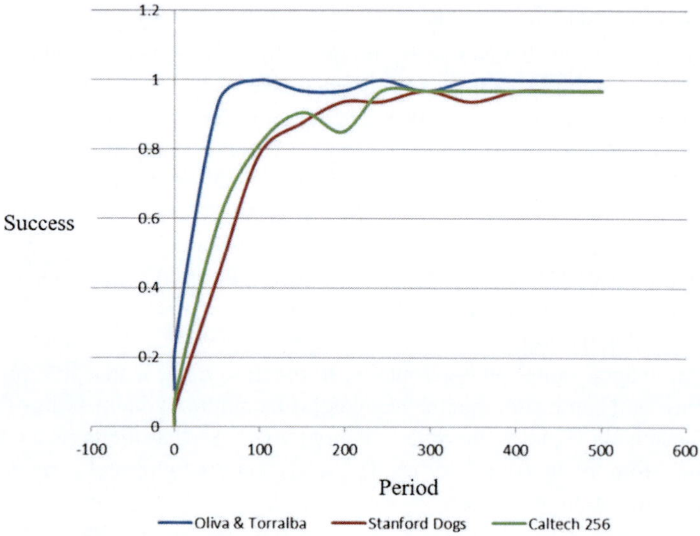

Fig. 2. Top-1 accuracy results in ToniNet network training.

5.2 Comparison

Table 2 shows a summary of the results for the test phase of the AlexNet, GoogleNet, ResNet 152, and ToniNet networks. Note that ToniNet outperforms AlexNet and GoogleNet in both top-1 and top-5 for all 3 test sets. However, ToniNet is outperformed by ResNet 152 only with the Caltech 256 test set [31]. GoogleNet outperforms AlexNet in precision because it is a network with greater depth. That is, it is a network with a greater number of extraction layers. For the same reason, ResNet 152 outperforms GoogleNet. However, although ResNet 152 also has a greater depth than ToniNet, it is only superior to ToniNet accuracy with Caltech 256; which emphasizes both the advantage and the importance of having three identical characteristic extraction sections in parallel but with different size filters.

- **Results with Oliva & Torralba:** This is the relatively simplest database, consisting of 1,879 training and 809 test images belonging to 8 classes. Analyzing the behavior of the networks in the test phase, in Fig. 3, the 4 networks learn very well and reach a good top-1 accuracy in only 150 epochs. However, it is worth mentioning that ToniNet shows a better generalization during the whole learning phase and obtains the best performance with a 94,6% of success in the 250th epoch.

Table 2. Results and comparison of networks

CNN	Oliva & Torralba			Stanford Dogs			Caltech 256		
	Minutes	Top-1	Top-5	Minutes	Top-1	Top-5	Minutes	Top-1	Top-5
AlexNet	22	91,6%	99,4%	168	45,3%	72,2%	245	58,1%	75,4%
GoogleNet	16	92,4%	99,4%	157	55,4%	82,3%	224	62,5%	80,0%
ResNet 152	88	94,2%	99,4%	621	58,5%	84,4%	841	63,3%	83,7%
ToniNet	39	96,3%	100%	301	60,3%	85,4%	406	64,8%	82,3%

With this database, ToniNet surpasses AlexNet in 4,2%, GoogleNet in 3,1% and ResNet 256 in 1,8%.

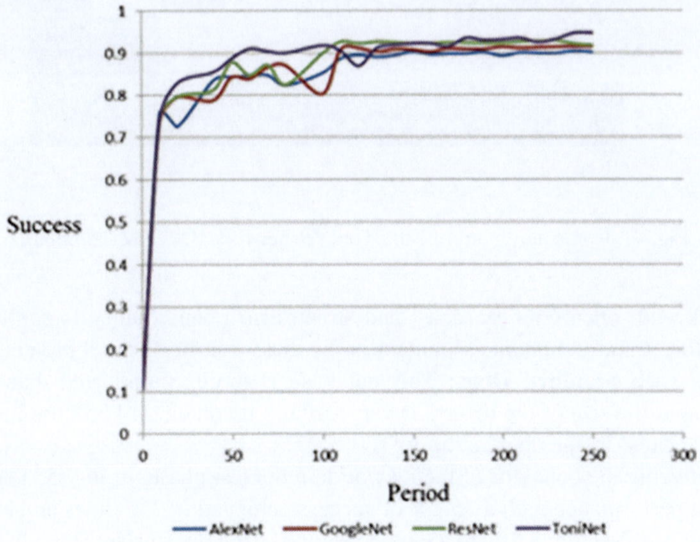

Fig. 3. Accuracy results top-1 (Oliva & Torralba).

To better visualize ToniNet's performance, Fig. 4 shows the confusion matrix with C1, C2, …, C8 that correspond to the open country, coast, forest, highway, inside city, mountain, street, and tall building respectively. In the main diagonal it is observed that the network achieves 94.56% of success with 765 correct images of 809.

The network has the best prediction with the Tall building class with 96.19% followed by Forest with 96.04%.

The class with the least prediction success is Inside city with 91.67%, confused with street and tall building. The real Highway class is the most confused by the

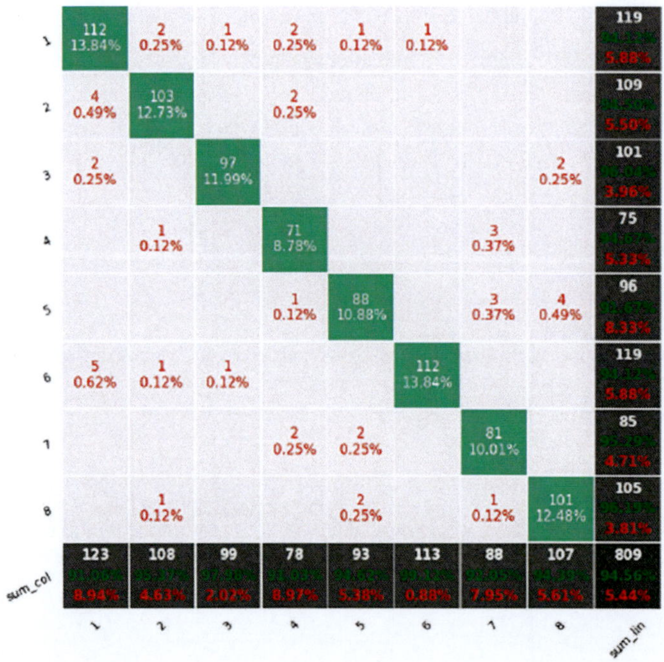

Fig. 4. Confusion matrix of the ToniNet network (Oliva & Torralba).

network with open country, coast and street, also open country is confused with coast, forest and mountain. Mountain is the class that best learns the net.

- **Results with Stanford Dogs:** This database is more complicated than Oliva & Torralba as the size of the images is very variant. It contains 14,358 training images and 6,222 test images belonging to 120 classes.

  Analyzing the behavior of the networks in the test phase, in Fig. 5, ToniNet has the best performance with a 55.2% of success achieved in 250 times and exceeds in 12% to AlexNet, in 3.3% to GoogleNet and 1.6% to ResNet 256. In this test, ToniNet trained with 500 epochs achieving a 56.2% maximum in test performance; however, the increase in performance is not significant compared to the training time.

  The network has the best prediction with the Tall-building class with 96.19% followed by Forest with 96.04%. In Fig. 6, some of the results of the test phase are shown, where the ToniNet network confuses an image of the coasting class for one of the Open country class, among others that were incorrectly classified.
- **Results with Caltech 256:** This database is more challenging than Oliva & Torralba and Stanford Dogs, as it has the largest imbalance of images by category. It contains 21,314 training and 9,293 test images from 256 classes. Analyzing the behavior of the networks in the test phase, in Fig. 6, the ResNet 152 had the best performance with 64.7% of success in 250 epochs, surpassing ToniNet by 2.2%. With this database, ToniNet is 5.0% better than AlexNet, 1.9% better than GoogleNet (Fig. 7).

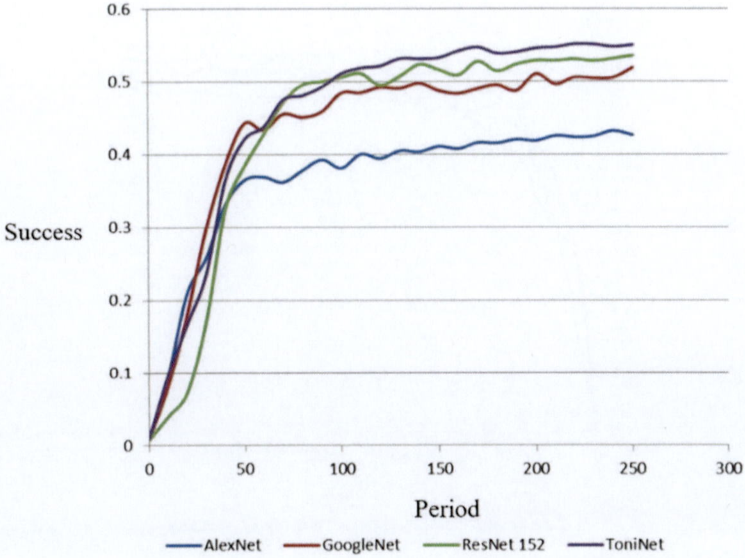

Fig. 5. Accuracy results on top-1 (Stanford Dogs).

Fig. 6. Test with Oliva & Torralba, the letter R means that it is the actual classification, the letter P means that it is the classification given by the network.

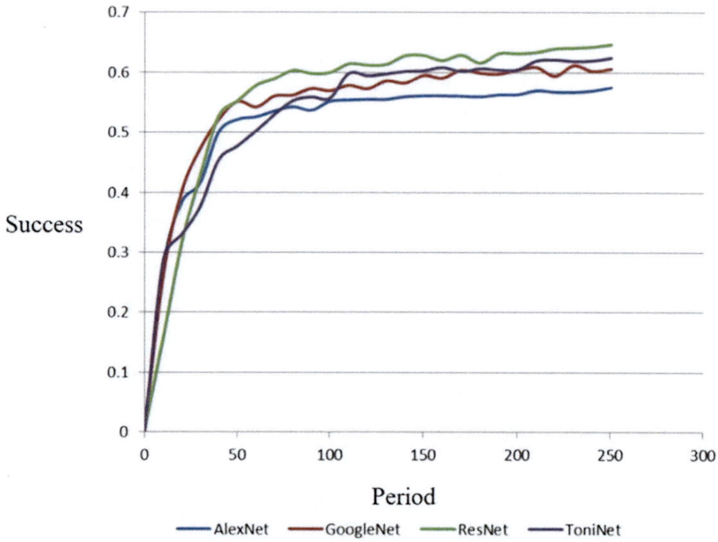

Fig. 7. Accuracy results in top-1 (Caltech 256).

6 Conclusion

The proposed network provided better results with the Oliva & Torralba and StanfordDogs image set compared to AlexNet, GoogleNet, and ResNet 152. It was surpassed by ResNet 152 with the Caltech 256 image set, but the training time of the network was less and with higher performance than AlexNet and GoogleNet. The advantage of the network, compared to the comparative ones, is that it has different characteristic extraction sections, so it needs fewer layers, like the ResNet 152 and GoogleNet. The main disadvantage is that it needs more training time than AlexNet and GoogleNet to achieve better results but with less time than ResNet.

The bases for the construction of networks are laid using different sections for the extraction of characteristics. You can add more sections depending on the hardware you have. Deep networks such as ResNet with 152 layers extract the characteristics from only one section, in the proposed network characteristics from multiple sections that can be extracted.

The network can be improved in the classification stage, mainly in the fully connected layers, also by enlarging the sections.

As future work, it is planned to use the network for object detection and image segmentation. Experimenting with more sections is also planned; to work with color, shape, texture, among others.

References

1. Wang, H., Ding, S., Wu, D., Zhang, Y., Yang, S.: Smart connected electronic gastroscope system for gastric cancer screening using multi-column convolutional neural networks. Int. J. Prod. Res. **57**(21), 6795–6806 (2019)
2. Wang, Y., Hu, S., Wang, G., Chen, C., Pan, Z.: Multi-scale dilated convolution of convolutional neural network for crowd counting. Multimedia Tools Appl. **79**(1), 1057–1073 (2019). https://doi.org/10.1007/s11042-019-08208-6
3. Li, Z., Zhou, A., Shen, Y.: An end-to-end trainable multi-column CNN for scene recognition in extremely changing environment. Sensors **20**(6), 1556 (2020)
4. Hu, Y., Lu, M., Lu, X.: Feature refinement for image-based driver action recognition via multi-scale attention convolutional neural network. Sig. Process. Image Commun. **81**, 115697 (2020)
5. LeCun, Y., Bengio, Y., Hinton, G.: Deep learning. Nature **521**, 436–444 (2015)
6. Simonyan, K., Zisserman, A.: Very deep convolutional networks for large-scale image recognition. arXiv preprint arXiv:1409.1556 (2014)
7. CireşAn, D., Meier, U., Masci, J., Schmidhuber, J.: Multi-column deep neural network for traffic sign classification. Neural Netw. **32**, 333–338 (2012)
8. Jia, Y., Shelhamer, E., Donahue, J., Karayev, S., Long, J., Girshick, R., Guadarrama, S., Darrell, T.: Caffe: convolutional architecture for fast feature embedding. In: Proceedings of the 22nd ACM International Conference on Multimedia, pp. 675–678. ACM (2014)
9. Viloria, A., Acuña, G.C., Franco, D.J.A., Hernández-Palma, H., Fuentes, J.P., Rambal, E.P.: Integration of data mining techniques to PostgreSQL database manager system. Procedia Comput. Sci. **155**, 575–580 (2019)
10. Yu, D., Wang, H., Chen, P., Wei, Z.: Mixed pooling for convolutional neural networks. In: International Conference on Rough Sets and Knowledge Technology, pp. 364–375. Springer, Cham (2014)
11. Varela, N., Silva, J., Gonzalez, F.M., Palencia, P., Palma, H.H., Pineda, O.B.: Method for the recovery of images in databases of rice grains from visual content. Procedia Comput. Sci. **170**, 983–988 (2020)
12. Ciregan, D., Meier, U., Schmidhuber, J.: Multi-column deep neural networks for image classification. In: 2012 IEEE Conference on Computer Vision and Pattern Recognition, pp. 3642–3649. IEEE (2012)
13. Oliva, T.: http://cvcl.mit.edu/database.htm, May 2016
14. Khosla, A., Nityananda, J., Yao, B., Fei-Fei, L.: Stanford Dogs Dataset, September 2017. http://vision.stanford.edu/aditya86/ImageNetDogs/
15. Caltech256: Caltech 256 Dataset, May 2016. www.vision.caltech.edu/ImageDatasets/Caltech256
16. Zhang, Y., Zhou, D., Chen, S., Gao, S., Ma, Y.: Single-image crowd counting via multi-column convolutional neural network. In: Proceedings of the IEEE Conference on Computer Vision and Pattern Recognition, pp. 589–597 (2016)
17. Cireşan, D., Meier, U.: Multi-column deep neural networks for offline handwritten Chinese character classification. In: 2015 International Joint Conference on Neural Networks (IJCNN), pp. 1–6. IEEE (2015)
18. Lu, X., Lin, Z., Shen, X., Mech, R., Wang, J.Z.: Deep multi-patch aggregation network for image style, aesthetics, and quality estimation. In: Proceedings of the IEEE International Conference on Computer Vision, pp. 990–998 (2015)

19. Ke, Q., Ming, L.D., Daxing, Z.: Image steganalysis via multi-column convolutional neural network. In: 2018 14th IEEE International Conference on signal processing (ICSP), pp. 550–553. IEEE (2018)
20. McDonnell, M.D., Vladusich, T.: Enhanced image classification with a fast-learning shallow convolutional neural network. In: 2015 International Joint Conference on Neural Networks (IJCNN), pp. 1–7. IEEE (2015)
21. Viloria, A., Lezama, O.B.P.: Improvements for determining the number of clusters in k-Means for innovation databases in SMEs. ANT/EDI40, pp. 1201–1206 (2019)
22. Yang, W., Jin, L., Xie, Z., Feng, Z.: Improved deep convolutional neural network for online handwritten Chinese character recognition using domain-specific knowledge. In: 2015 13th International Conference on Document Analysis and Recognition (ICDAR), pp. 551–555. IEEE (2015)
23. Jmour, N., Zayen, S., Abdelkrim, A.: Convolutional neural networks for image classification. In: 2018 International Conference on Advanced Systems and Electric Technologies (IC_ASET), pp. 397–402. IEEE (2018)
24. Park, T., Lee, T.: Musical instrument sound classification with deep convolutional neural network using feature fusion approach (2015). arXiv preprint arXiv:1512.07370
25. Zhong, Z., Jin, L., Xie, Z.: High performance offline handwritten Chinese character recognition using Googlenet and directional feature maps. In: 2015 13th International Conference on Document Analysis and Recognition (ICDAR), pp. 846–850. IEEE (2015)
26. Du, J., Zhai, J.F., Hu, J.S., Zhu, B., Wei, S., Dai, L.R.: Writer adaptive feature extraction based on convolutional neural networks for online handwritten Chinese character recognition. In: 2015 13th International Conference on Document Analysis and Recognition (ICDAR), pp. 841–845. IEEE (2015)
27. Bindhu, V.: Biomedical image analysis using semantic segmentation. J. Innov. Image Process. (JIIP) **1**(02), 91–101 (2019)
28. Sharma, N., Jain, V., Mishra, A.: An analysis of convolutional neural networks for image classification. Procedia Comput. Sci. **132**, 377–384 (2018)
29. Yim, J., Sohn, K.A.: Enhancing the performance of convolutional neural networks on quality degraded datasets. In: 2017 International Conference on Digital Image Computing: Techniques and Applications (DICTA), pp. 1–8. IEEE (2017)
30. Verma, A., Vig, L.: Using convolutional neural networks to discover cogntively validated features for gender classification. In: 2014 International Conference on Soft Computing and Machine Intelligence, pp. 33–37. IEEE (2014)
31. Zeng, Y., Xu, X., Fang, Y., Zhao, K.: Traffic sign recognition using extreme learning classifier with deep convolutional features. In: The 2015 International Conference on Intelligence Science and Big Data Engineering (IScIDE 2015), Suzhou, China, vol. 9242, pp. 272–280 (2015)

Face Detection Based on Image Stitching for Class Attendance Checking

Qiubo Huang(✉) and Chun Ji

School of Computer Science and Technology,
Donghua University, Shanghai, China
huangturbo@dhu.edu.cn, jonasjc@163.com

Abstract. Traditional attendance checking relies on teachers to name students in the class, which has many shortcomings, including time wastage and fake sign-ins. We use an online system to allow the students to sign in and take advantage of the face detection technology to help teachers check attendance in their classes. Because large classrooms require teachers to take multiple photos to include all students, picture stitching is required before face detection can be performed. We need to solve the problems of image registration, image fusion, adaptation of shooting angles, the proportion of overlapping parts of images, and the problem that YOLOv3 cannot detect small faces. Our research focuses on achieving a good result in stitching and improving the accuracy of face detection. The proposed system has tried to give teachers some good advice on photo shooting angles and overlapping proportions of multiple photos. Also, we make some improvements in feature point extraction and small faces detection algorithm etc. The experimental results show that the system can identify nearly 100% faces in the photos.

Keywords: Picture stitching · Face detection · Image registration · Image fusion

1 Introduction

Student attendance is an important part of classroom teaching, and most schools take student attendance as an important part of student course performance considerations. However, the high absenteeism rate of college students is a very common and easily overlooked problem, which directly affects the quality of classroom teaching and students' learning results. Studies have shown that the absenteeism rate of college students is between 10% and 20% [1], and more than two-thirds of college students have experience of skipping classes [2]. At present, there are many technical solutions to implement the attendance system. Common methods include positioning, fingerprint, and face recognition. Although positioning-based methods such as WIFI and GPS can detect the MAC address and location information of student mobile phones, the accuracy of their positioning is not high, and the inability to strongly associate students with mobile phones has always been a problem. Even if the unique MAC address of the mobile phone can be detected, the person holding the mobile phone cannot be determined to be a student in the class list. Currently, the MOOC APP and

J. I.-Z. Chen et al. (Eds.): ICIPCN 2020, AISC 1200, pp. 31–43, 2021.
https://doi.org/10.1007/978-3-030-51859-2_4

Chaoxing APP, which are used by many Chinese universities, are based on this check-in method, which cannot solve the problem of fake check-in. The fingerprint-based method requires a fingerprint recognition machine in each classroom, which is expensive. Also, it will cost a lot of class time. However, it is not uncommon to find fake check-ins through fingerprint films. With our attendance system, we overcome the shortcomings of other systems, and can accurately identify the students who are not attending.

In our system, the students use mobile apps to check-in based on positioning. If the teacher wants to confirm that the number of sign-in students matches the actual number of students, he will take one or more photos of the students, then upload them to the server. The system stitches the photos and performs face detection, and finally reports the number of students. If the reported number and the number of sign-in students are not the same, the teacher will ask the students to claim their heads. The procedure is as Fig. 1:

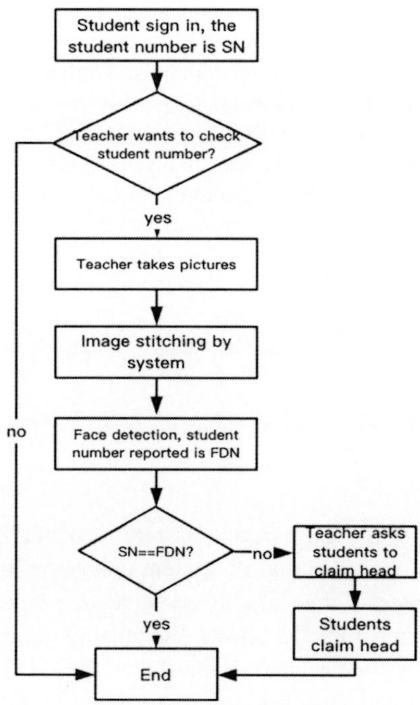

Fig. 1. General flow chart

There are four main sections followed. Section 2 is related research. Section 3 is our job. We optimized the picture stitching quality in the stitching process and improved the detection accuracy of small faces. In Sect. 4, experiments and results are showed. Section 5 gives the conclusion.

2 Related Research

To reduce the workload of teachers to check the number of students in the classroom, we use face detection to get attendance. For students in the large classrooms, to prevent some student's heads from being too small to be detected, we allow teachers to take multiple photos and automatically stitch them into one picture.

The process of image stitching is mainly divided into three parts: feature extraction, image registration, and image fusion. Among them, feature extraction is the most critical part. Now commonly used image feature extraction methods are the SIFT corner detection method [3], ORB based detection method [4], and so on.

The SIFT corner detection method uses a Gaussian Difference Pyramid (DOG) to construct a spatial scale and find extreme points in it and then extract the scale, position, and rotation invariants. This method can maintain a stable stitching effect when the picture is rotated, scaled, and changed in light, and it has a high tolerance for image noise. However, due to the use of the DOG method, the amount of calculation becomes larger and the speed becomes slower. The detection method based on ORB is mainly divided into two parts: feature point extraction and feature point description. Its feature point extraction algorithm borrows from the FAST algorithm, and the feature point description borrows from the BRIEF algorithm. The overall speed of the ORB algorithm is much faster than the SIFT algorithm, but the ORB algorithm does not perform well in terms of scale invariance. When the image resolution and direction are changed, the stitching effect is poor. We borrow the ideas of SURF [5] to improve the speed of feature point extraction and image registration while having image rotation and scaling.

There are two main types of face detection methods based on deep learning: detection frameworks based on "one-stage" and detection frameworks based on "two stages". Frameworks based on one stage mainly include SSD [6], YOLOv3 [7], and so on. Frameworks based on two stages mainly include Fast R-CNN [8] and so on. As the two-stage object detection framework is not as accurate and fast in face detection like the one stage framework and the model is too large, we use the one-stage framework for face detection. In large classrooms, the faces in the back row may be very small. SSD does not perform well for small targets, so we use YOLOv3 to improve the detection success rate of small faces.

3 Image Stitching and Small Faces Detection

We first use feature points extracted from each picture and then use the feature points between the images to register, adjust the size and shape. Finally, the configured pictures are fused to form a complete picture. After the image stitching is completed, a face detection algorithm is used to identify the faces of students in the classroom and count the number of students. Due to the effects of classroom lighting, stitching effects, face size, and student posture, there may be very few human faces that are not detected. At this time, the teacher can annotate faces manually.

Using SURF-based image stitching technology, feature extraction, image registration, and image fusion can be performed on multiple pictures with different angles and resolutions. Use YOLOv3 to detect face for stitched images. With some additional processing techniques, we can improve the accuracy of face detection to nearly 100%. In the process of image stitching, we adopt the method of scale-invariant feature transformation, which can obtain better stitching results in the case of image rotation, scaling, and light changes. In face detection, we use a convolutional neural network method to change the image at multiple scales, so that faces of different sizes can be detected.

3.1 Image Stitching

Feature Point Extraction

The method of extracting the same part of two images is as follows: feature points are extracted from the two images [10], and then the common feature points of the two images are identified. We use the integral image method to extract the feature points. This method greatly improved the speed of feature point detection compared to the method using DOG in the SIFT algorithm.

The feature points of the image are brighter or darker pixel points than their neighborhood points. We use the method of image convolution to obtain the response value of each pixel point. Due to the complexity of the Gaussian filter calculation, we use a box filter to calculate the image convolution to speed up the calculation. Since the weight of each block in the box filter is equal, the calculation speed is very fast after combining the integral graph. The response value of a pixel (i,j) is:

$$R(i,j) = \sum_{k=1,2,3\ldots} W(k)D(k) \tag{1}$$

In (1), $W(k)$ is the weight of the kth block of the filter, and $D(k)$ is the sum of pixels in the area corresponding to the kth block of the filters. After the response value of each pixel is calculated, it is compared with the response values of the surrounding pixels. If it is larger than the surrounding values, it is selected as the feature point.

The calculation of the integral image is a key step in the improvement of the algorithm. We use a two-dimensional matrix of M x N to represent the integral of an image of M x N pixels. Then the image integration of any point is:

$$S(i,j) = \sum_{\substack{0 \le x \le i \\ 0 \le y \le j}} P(x,y) \tag{2}$$

In (2), $S(i,j)$ is the grayscale integration of pixel (i,j), and $P(x,y)$ is the grayscale value at (x,y). After calculating the grayscale integral of each pixel, the sum of the pixels in any rectangular area D in the image is:

$$D(A,B,C,D) = S(A) - S(B) - S(C) + S(D) \tag{3}$$

Among them, $S(A)$, $S(B)$, $S(C)$, and $S(D)$ is the grayscale integrals of rectangular vertices A, B, C, and D.

Image Registration

After extracting feature points, we get all the feature points in each picture. We need to calculate the deformation, translation, and other changes in the picture based on these feature points. This requires calculating the homography matrix of image transformation [11]. The quality of the homography matrix and the speed of calculation have a lot to do with the quality of the feature points. To improve the accuracy of image registration [12], we will process the feature points in advance to filter out the useless feature points. Here we use the kNN algorithm to filter out useless feature points:

$$M = km(m1, m2, 2) \tag{4}$$

In (4), M represents the result of matching, $m1$, and $m2$ are the feature points of the two images, km represents the kNN algorithm, and $k = 2$ is the parameter of kNN. We choose two nearest points $p1, p2$. $p1$ is the nearest neighbor and $p2$ is the second nearest. The 2 nearest neighbor method can only get two points closest to a feature point, and we can't say that the matching feature points must be in the 2 points, so we need to delete these outlier feature points. In response to the above issues, we adopt the idea that if a match is correct, the match result will be closer to the first neighboring point. If it is a wrong match, the distance between two neighboring points to the feature point is close. We use the following method to filter out the outlier feature points:

$$CM = filter(distance1 < 0.5 \times distance2, M) \tag{5}$$

In (5), we pick out the first feature point in which distance is less than 0.5 times the distance of the second point as the result CM.

Image Fusion

Due to different shooting angles and light conditions, different shades of light may vary from image to image. After images registration, the stitching effect of images is often unsatisfactory. The main problem is that due to the different shades of light among the images, there will be a poor result at the overlapped part of the images. To solve this problem, we use the method of weighted average color values to achieve a smooth transition between images. The image pixels in the overlapped area of the image take different weighting formulas:

$$C(x, y) = \frac{\alpha1C1(x, y) + \alpha2C2(x, y)}{\alpha1 + \alpha2} \tag{6}$$

In (6), we use the lower-left corner of the overlapped part as the origin of the coordinates to establish a Cartesian coordinate system. We use weights $\alpha1, \alpha2$ for point (x, y) of the overlapped parts of the two images, respectively, to calculate the color values $\frac{\alpha1C1(x,y)}{\alpha1+\alpha2}$ *and* $\frac{\alpha2C2(x,y)}{\alpha1+\alpha2}$, then add them up to obtain $C(x, y)$. Where $C1(x, y)$ and $C2(x, y)$ are the RGB color values of the two images at (x, y).

We denote the width of the overlapped area as w. The distance of the point (x, y) from the left edge of the overlapped area is denoted as *len*. Then the color value weight $\alpha 1$ is:

$$\alpha 1 = \frac{w - len}{w} \tag{7}$$

The color value weight of $\alpha 2$ is:

$$\alpha 2 = \frac{len}{w} \tag{8}$$

The method based on the weighted average color value is accurate at the pixel level. Although it takes more time, it can ensure that the result of image fusion is excellent enough.

Far and Close Shot Photos Stitching

For students in very large classrooms, the teacher may take multiple photos. For the front and the rear students, the teacher may use different camera magnifications when shooting. He may take one photo for the front row students, and uses a 2 to 3 times magnification for the rear students to take two or more photos. This requires us to be able to stitch photos at different magnifications. We use YOLOv3 based target detection algorithm as the face detection algorithm.

3.2 Small Faces Detection

For the large classrooms, the faces of the rear students may be very small in the image, which leads to omissions in the face detection process. YOLOv3's face detection model is based on Anchor Boxes [13]. The size, proportion, and the number of Anchor Boxes will affect the detection results. At the same time, for the scene of small targets, YOLOv3 adopted a multi-scale feature fusion method to improve the success ratio of small targets. we use upsampling, downsampling, and fusion of feature maps to improve the detection of small targets. Because the large, medium, and small feature maps set in YOLOv3 detect large, medium, and small targets respectively, it is not good enough for large classroom scenes.

We use the method of automatic image enlargement to increase the size of small faces, and iteratively detect the faces in the image. So, we can ensure that all students are detected. The process is shown in Fig. 2.

As shown in Fig. 2, we will determine whether new faces are detected in one iteration. If no new face is detected, it is determined that all faces have been detected. Otherwise, there may be more faces undetected. Then the image is enlarged and the iteration continues.

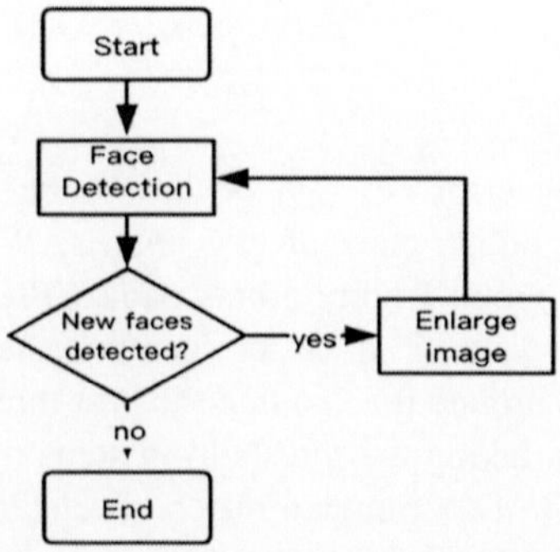

Fig. 2. Flow chart of small face detection iteration

4 Experiments and Results

4.1 Data Description

The face recognition images in student attendance mainly have the following characteristics: 1. The students' faces are dense; 2. The light is good; 3. The faces are facing straight ahead. 4. The sizes of the face in the same photo vary greatly from front to back. The front face is larger and the back face is smaller. In response to the above situation, we picked out 2,500 images as the training and test dataset from the open-source face dataset WIDER FACE [9, 14] (includes 32203 images) and Google public dataset.

4.2 Experiments and Analysis

Feature Point Extraction
First, we compared the three feature point extraction algorithms of ORB, SIFT, and SURF to decide which one to use.

From Table 1, it can be found that the improved feature point extraction algorithm based on SURF detects the largest number of descriptors and the extraction speed is also moderate. Therefore, the SURF algorithm is adopted by us for the feature point extraction process.

Table 1. Comparison of feature point extraction speed

Method	Speed (milliseconds)	Amount
ORB	3.521890	500
SIFT	27.601637	1157
SURF	11.738463	1740

When performing feature point filtering, we set a threshold in (6) to filter out those outliers after clustering. The formula is as follows:

$$E = \frac{|PN - CPN|}{CPN} \tag{9}$$

In (9), E is the matching error rate, PN is the number of pairs, and CPN is the number of common feature points pairs of two images. We use two images for our experiments, where their common feature points pairs CPN is 3600.

The results are shown in Table 2. It can be seen that the matching error rate is the lowest when the threshold is around 0.5. To find the best threshold more accurately, we use a sample of 500 images, taking the threshold in steps of 0.01 between 0.49–0.51, and calculating the error rate of each image match. The statistics are shown in Fig. 3. The x-axis represents the threshold and the y-axis represents the number of pictures with the lowest matching error rate at that threshold. The experimental results show that, for most pictures, the error rate is the lowest when the threshold is between 0.45 and 0.55, and the number of pictures with the smallest error rate when the threshold is 0.5 is the most.

Table 2. Comparison of feature point pairing

Threshold (θ)	Pairs Number (PN)	Pairing error rate
0.1	3347	0.0703
0.3	3519	0.0225
0.45	3536	0.0178
0.47	3554	0.0128
0.49	3573	0.0075
0.5	3593	0.0019
0.51	3613	0.0036
0.53	3628	0.0078
0.55	3646	0.0128
0.7	3939	0.0942

Stitching Results for Diversified Pictures

In the following, we consider the stitching effect of the pictures from three aspects: the different overlapping ratios, stitching angles, and shooting angles of two photos.

With the same shooting angle and stitching angle, we did four experiments so that the images had 1%, 5%, 10%, and 20% overlapping areas, respectively. We will find the least overlap area to make the stitching effect satisfactory.

As shown in Fig. 4, Figures (a), (b), (c), and (d) are 51%, 55%, 60%, and 70% of the left half of the original picture, and Figure (e) is 50% of the right half.

The results of the stitching are shown in Fig. 5. Where (a), (b), (c), and (d) are the stitching results of (a)-(e), (b)-(e), (c)-(e), (d)-(e) respectively. As can be seen from Fig. 5, in (a) the image quality is poor, in (b) there are black blocks in the upper left corner, and Fig. 5(c) and Fig. 5(d) have no obvious problems.

We use the method of cosine similarity [15, 16] to calculate the similarity of two pictures. When the cosine similarity between the stitched picture and the original

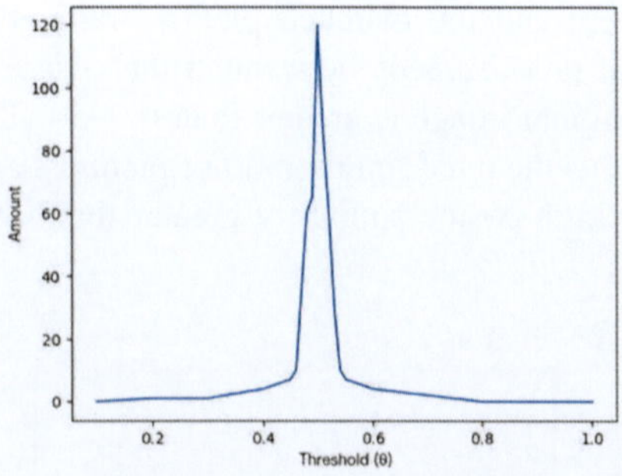

Fig. 3. Threshold experiments

Fig. 4. Pictures to be stitched

Fig. 5. Stitching results of Fig. 4

picture is greater than 0.9950, we judge it to be fine. The calculation results are shown in Table 3.

Table 3. Similarity experiments

Overlap ratio	Similarity
1%	0.8631
5%	0.8895
10%	0.9964
20%	0.9975

From the results in Table 3, we can roughly consider that the stitching effect is acceptable when the proportion of the overlapping portion is greater than or equal to 10%. To provide a reference value for the teacher, we used 500 pictures, and the overlap ratio between them and the matched pictures ranged from 5% to 15%. We observe when the overlap ratio exceeds how much the cosine similarity between the stitched image and the original image is greater than 0.9950. The results are shown in Fig. 6. The x-axis represents the overlapping ratio of pictures, and the y-axis represents the number of pictures with a cosine similarity greater than 0.9950 when the overlap ratio exceeds that value.

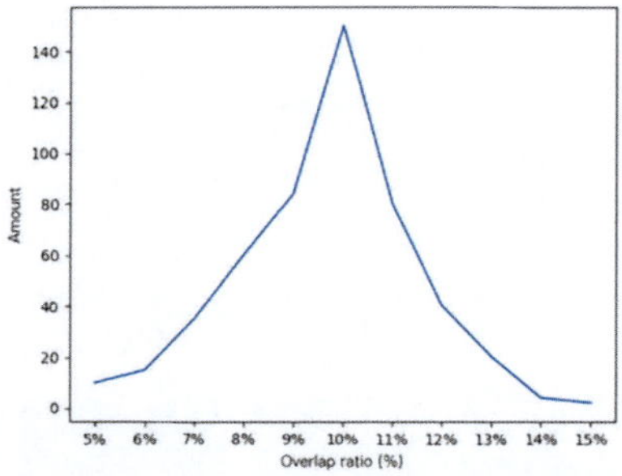

Fig. 6. Overlap ratio experiments

It can be seen from Fig. 6 that when the picture overlap ratio is greater than 15%, the stitching effect can be guaranteed If the overlapped area of uploaded photos by teachers is less than 15%, the teachers will be prompted to take photos again.

If the two pictures are the upper part and the lower part of a complete photo, the stitching result is shown in Fig. 7. Where (c) is the stitching result of (a) and (b).

Fig. 7. Picture of the upper and lower stitching experiments

As shown in Fig. 7, As long as there is enough overlapped area, the stitching result is still good.

When only the corners are overlapped, the stitching result is shown in Fig. 8. The overlapping areas of the pictures will be stitched, and the missing parts will be filled with black.

When teachers take different photos, they may be at different positions, resulting in different shooting angles. In Fig. 9, (a) and (b) are photos taken from two different angles, and (c) is the stitching result. We can see that apart from the black blocks (because of the lack of pixels), the quality of the image stitching is very good.

Fig. 8. Pictures with overlapped corner stitching experiments

Fig. 9. Stitching result of pictures with different shooting angles

For stitching when the pictures' resolutions are different, we took one picture of the front row of the classroom, two pictures of the rear row of the classroom, and stitched the three pictures. The result is shown in Fig. 10:

Fig. 10. Far and close pictures stitching

In Fig. 10, (a) and (b) are distant photos, (c) is a close-up photo, and (d) is the result of stitching. We can see that the quality of the image stitching is also good.

Small Faces Detection

For small faces detection experiments, we used YOLOv3 to perform face detection on the dataset, and the accuracy rate was 94.12%, which did not meet our requirements. The result is shown in Fig. 11.

It can be seen from Fig. 11 that persons in the back row were not detected. When we use the iterative process in Fig. 2, The detection result is shown in Fig. 12.

Two faces in Fig. 12 were not detected because they were too occluded. When teachers take pictures of students, they can ask students to fully show their faces to avoid this situation. In case this happens, the teacher can manually mark to avoid mistakes. Through the iterative detection process in Fig. 2, the face detection can reach 100% accuracy on our dataset.

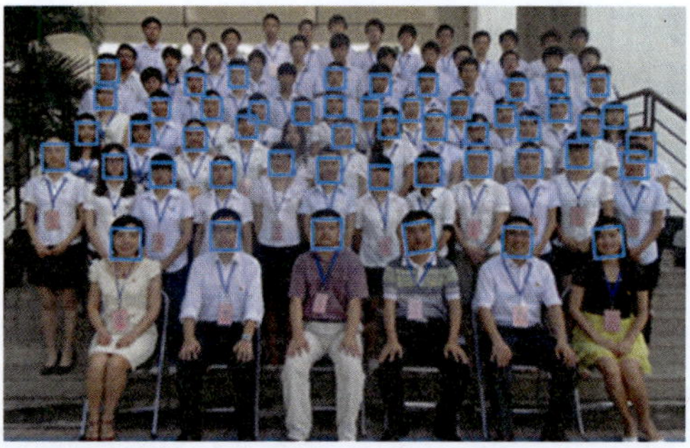

Fig. 11. Whole image detection

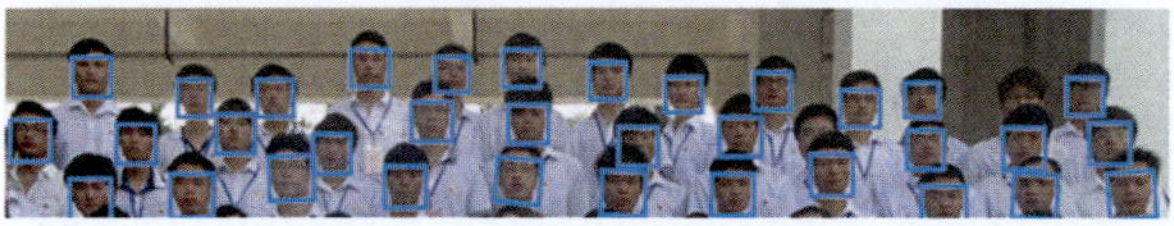

Fig. 12. Detection result after enlargement

5 Conclusion

This paper proposes a method to assist student attendance in class using picture stitching and face detection, which solved the current problems of low efficiency and high cost of class attendance. We solved the problem of stitching multiple photos in the large classrooms and detecting small faces. We used integration maps and clustering algorithms to optimize the stitching process and improved the quality of photo stitching, and ultimately improved the face detection rate to nearly 100%. The above experiments were conducted under conditions close to the ideal, such as students needed to face the camera, and have their faces fully exposed. Also, good lighting is necessary. Future works will need to take into account certain less-than-ideal conditions under which the faces in the photos will also be 100% detectable.

Acknowledgement. Informed consent to all the participants involved in this research work.

References

1. Yao, L., Zhu, L., Hu, J.: Investigation and analysis of the attendance rate of college students. Jiangsu High. Educ. **3**, 67–70 (2015). Chinese: 大学生到课率的调研与分析
2. Pan, L., Sun, Y.: Analysis of the phenomenon of college students skipping classes. Chinese Foreign Entrepreneurs **16**, 217–218 (2011). Chinese: 大学生逃课现象分析
3. Sun, J., Zhao, Y., Wang, S.: Improved SIFT feature matching algorithm based on image gradient information enhancement. J. Jilin Univ. Sci. Edn. **1**, 56–62 (2018). Chinese: 基于图像梯度信息强化的SIFT特征匹配算法改进

4. Jia, W.: Panoramic image mosaic technology based on ORB algorithm. City Survey **3**, 105–108 (2019). Chinese: 基于ORB算法的全景图像拼接技术
5. Liu, C., Shao, F., Jing, Y., et al.: Application of graph structure in feature point matching of aerial remote sensing images. CEA **1**, 54–60 (2018). Chinese: 图结构在航空遥感图像特征点匹配中的应用
6. Liu, W., Anguelov, D., Erhan, D., et al.: SSD: single shot multibox detector. In: European Conference on Computer Vision. Springer, Heidelberg (2016)
7. Redmon, J., Farhadi, A.: YOLOv3: An Incremental Improvement. Computer Science ArXiv (2018)
8. Wang, L., Zhang, H.: Application of faster R-CNN model in vehicle detection. J. Comput. Appl. **38**, 666–670 (2018). Chinese: Faster R-CNN模型在车辆检测中的应用
9. Shi, X., Shan, S., Kan, M., et al.: Real-time rotation-invariant face detection with progressive calibration networks. In: 2018 IEEE/CVF Conference on Computer Vision and Pattern Recognition (CVPR). IEEE (2018)
10. Tu, S.J., Wang, C.W., Pan, K.T., et al.: Localized thin-section CT with radiomics feature extraction and machine learning to classify early-detected pulmonary nodules from lung cancer screening. Phys. Med. Biol. **63**(6) (2018). https://doi.org/10.1088/1361-6560/aaafab
11. Hartley, R., Zisserman, A.: Multiple View Geometry in Computer Vision, pp. 32–33. Cambridge University Press, Cambridge (2003)
12. Li, Y., Chen, C., Yang, F., et al.: Deep sparse representation for robust image registration. **99**, 2151–2164 (2018)
13. Wang, C., Ji, X., Wu, W.: Node localization in wireless sensor networks based on improved particle filter algorithm. J. Nanjing Univ. Sci. Technol. **42**(3), 309–316 (2018). Chinese:基于改进粒子滤波算法的无线传感器网络节点定位
14. Chi, C., Zhang, S., Xing, J.: Selective Refinement Network for High Performance Face Detection. Association for the Advancement of Artificial Intelligence (AAAI) (2019)
15. Nguyen, H.V., Bai, L.: Cosine similarity metric learning for face verification. In: Kimmel, R., Klette, R., Sugimoto, A. (eds.) ACCV 2010. LNCS, vol. 6493, pp. 709–720. Springer, Heidelberg (2011). https://doi.org/10.1007/978-3-642-19309-5_55
16. Zhang, Y., Ji, L., Yang, F., et al.: Characterization of dual-modal infrared image fusion based on cosine similarity. Optoelectron. Eng. **10**, 46–51 (2019)

Image Processing Technique for Effective Analysis of the Cytotoxic Activity in Human Breast Cancer Cell Lines – MCF-7

K. Sujatha[1(✉)], B. Deepa lakshmi[2], B. Rajeswary Hari[3], D. Sangeetha[4], and B. Selvapriya[5]

[1] Department of Electrical and Electronics Engineering, Dr. M.G.R. Educational and Research Institute, Chennai 95, Tamil Nadu, India
drksujatha23@gmail.com

[2] Department of Electronics and Communication Engineering, Ramco Institute of Technology, Rajapalayam, Tamil Nadu, India

[3] Department of Biotechnology, Dr. M.G.R. Educational and Research Institute, Chennai 95, Tamil Nadu, India

[4] Department of EEE, Meenakshi College of Engineering, Chennai, India

[5] Dhanalakshmi College of Engineering, Chennai, India

Abstract. Breast cancer is the lethal form of cancers as it affects the adjacent organs like lungs, liver and heart very easily. The tumors in the nodules of mammary glands serve as the cause for the development of a malignant tumor. Magnetic Resonance Imaging (MRI) images are used to detect breast cancer. The objective of this work is to develop a structured scheme to analyze and evaluate the probabilities of breast cancer with the help of a typical user friendly image processing algorithms. The novelty of this work is that it has a well-developed strategy for breast cancer detection using high performance image based machine learning algorithms to extract the variations in intensity levels at the pretreatment stage followed by segmentation and feature extraction from the region of interest, present in the breast nodules. This method uses images from the open source database like The Cancer Imaging Archive (TCIA). The pre-treatment of MR images processing comprises filtering for noise removal and edge detection to extract the Region of Interest (RoI). The power spectrum of the MR images is evaluated and they play an important role to increase the sensitivity in identifying the breast tumors. These power spectrum coefficients extracted using Discrete Fourier Coefficients are used as distinct input feature set for training the Radial Basis Function Network (RBFN) to detect, identify and cross validate the activity of the drug (ethanolic extract) prepared from the traditional plant from the leaves of Excoecaria agallocha (EEEA), tested on a laboratory scale to investigate the probability of cytotoxic activity along with the anti-progressive nature to treat the apoptosis initiation and Cell cycle detention in the breast cancer MCF-7 cell lines causing malignancy in the nodules and ducts of mammary glands. The Apoptosis assessment is found to be nearly 99%.

Keywords: Image processing · Discrete Fourier Transform · Radial Basis Function Network · Breast cancer · Cytotoxicity · *Excoecaria agallocha*

J. I.-Z. Chen et al. (Eds.): ICIPCN 2020, AISC 1200, pp. 44–53, 2021.
https://doi.org/10.1007/978-3-030-51859-2_5

1 Introduction

Internationally, the top five reasons for the cause of death is breast cancer and secondly, the common type is the skin cancer reporting to approximately 10.4% in women [1]. The foremost pathological reason indicates that this disease leads to loss of cell cycle control causing uncontrolled progression and variation of cells [2]. Many breast cancers originate in the cells of the ducts present in the breast causing ductal cancers. The other type of breast cancer is that the lobular cancers while the remaining types are small in number and start in other tissues. Chemotherapy, radiotherapy and surgical extraction are used in treatment techniques. Researchers are motivated towards this study because this innovative plant based anticancer drug serves as an effective treatment with reduced or even no after effects [3]. Herbs form a large part of traditional medicine systems. The information regarding these traditional medicines serves as a vital source for the detection of anticancer agents from plants. Traditional medicine further forms a very important source of affordable and readily accessible health care system for most people in developing countries [4].

Excoecaria agallocha L. (Euphorbiaceae) is a prehistoric mangrove miniature tree found in Mangrove forests originated abundantly in Pichavaram along the regions of Indian coast [5, 6]. The common name for *Excoecaria agallocha* L. is known as "blinding tree" and it belongs to Euphorbiaceae family. This blinding tree is found to secrete a milky sap which can cause short-term blindness if it comes into contact with the eyes. In Tamil, it is famously known as Thillai, Kampetti. The sap of this tree can cause sores on the skin accompanied by irritation. Peter and Sivasothi [6]. have conducted several experiments which have stated that the plant possesses anticancer, antiviral, antibacterial and anti HIV properties. In India, women, primarily suffer from breast and cervical cancers [7, 8]. Hence, the primary objective of the present study is to inspect the antiprogressive and apoptosis activity induced by the *E. agallocha* leaves on human breast cancer cell line (MCF-7) and to analyze and cross validate the corresponding images using indigenous image processing algorithms at the test center.

Medical images have become an integral part to diagnose different diseases. Multiple imaging modalities fuelled by better imaging procedures have resulted in huge accumulation of medical data with the advent of high speed and large bandwidth communication network. The scope to share this data has increased many folds. These factors have also contributed to the risk that can be associated with the storage and transmission of medical information. The security of medical image can be perceived from different dimensions. Any medical information is sensitive which holds valuable diagnostic information. Tampering of this information even to the slightest degree can result in an unacceptable error in diagnosis. Unauthorized and unintended access to medical information may also result in a breach of privacy. Apart from holding sensitive private information with huge diagnostic value, medical data also has legal implications. Efforts to preserve the integrity of medical images are also complicated by the fact that any effort should not alter the original content of the medical data. Different morphological features reveal valuable diagnostic information and it is very essential to preserve the structural integrity of the medical image as well. This research work has addressed the concerns related to protecting the integrity of the medical image through a dual diagnosis approach [9, 10].

2 Background

The usage of imaging techniques in cancer investigation is prevalent; from the biological imaging of cells in the lab to digital pathology analysis using images of tumor section or normal images captured through body scans such as Magnetic Resonance Imaging scans and X-Rays (medical imaging). Biomarkers with imaging techniques are widely used in both cancer treatment and research. These images are often gathered as part of clinical trials. Manual diagnosis from the measurements of these imaging techniques may lead to an erroneous diagnosis which can be fatal. So, to avoid this circumstance, the entire diagnosis scheme is automated using novel image processing algorithms [11, 12].

3 Literature Survey

Due to the enormous size of medical image data repository, CBIR can be used for medical image retrieval. This chapter is envisioned to propagate the knowledge of the CBIR approach to deal with the applications of medical image management and to pull in more prominent enthusiasm from various research groups to rapidly propel research in this field [13, 14].

The most important tool in medical field is image identification and ite helps to diagnose the disease with less cost and fast noninvasive process [15, 16].

Nowadays, hard copy of images is not utilized properly due to high expense and maintenance. Digital images doesn't have any issue to store the things in storage room and that can be stored via digitally [17].

By storing more number of files in digital libraries and that lease to difficult in finding the medicinal information. To avoid this issues, text matching techniques applied to identify the exact required images in the database. And this technique called it as text base image retrieval and another name is called concept based image retrieval [17].

4 Problem Definition

The proposed work concentrates on the classification of medical images. The work aims at the following goals:

- To use the Fourier coefficients to represent medical images and develop a retrieval system that matches and retrieve images with the support of the operator.
- To cluster the cell viability using RBFN and provide new directions.

5 Materials and Methods

The objective of this work is to study its anti-progressive nature and investigate the cytotoxic probability of apoptosis initiation and Cell cycle detention in the breast cancer MCF-7 cell lines.

The second part includes Presently, mangrove plants have added significance in medication detection due to the existence of numerous phyto molecules of medicinal significance. Usually, sores, ulcers and leprosy are cured using *Excoecaria agallocha* which is extensively found in the mangrove. **Objective:** The objective of this work is to study its anti-progressive nature and investigate the cytotoxic probability of ethanolic extract from the leaves of *Excoecaria agallocha* (EEEA), apoptosis initiation and Cell cycle detention in the breast cancer MCF-7 cell lines. Inverted phase contrast Microscope (Olympus CKX41 with Optika Pro5 CCD camera) is used for obtaining the images.

The materials and methods involved in the analysis of anti-progressive nature of the EEEA extract include image processing techniques like Discrete Fourier Transform (DFT) to extract features and classified using Radial Basis function Network (RBFN) whose flowchart is shown in Fig. 1. The results obtained using image processing techniques was used to validate the results on laboratory scale identified by direct microscopic scrutiny along with MTT assay by engaging various quantities of the plant extract and the corresponding IC50 values were considered. The results yielded significant outcomes to the cytotoxic activity showed by the ethanolic extract of *Excoecaria agallocha* leaves (EEEA) was comparable with the optimal results yielded by the direct microscopic examination with an MTT assay corresponding to IC50 value of 56.5 μg/ml. For the image based analysis noticeable images of fluorescence microscopy exhibiting morphological variations by double staining method are used. It is concluded from the above results that the ethanolic leaf extract of *Excoecaria agallocha* (EEEA) induced apoptosis mediated cytotoxic and antiprogressive activity in the breast cancer MCF-7 cell lines which can be converted as a novel drug for treating breast cancer as shown in Fig. 2(a). The size of the input image in Fig. 1(a) is illustrated in Fig. 2(b).

6 Results and Discussion

In the current examination, Cytotoxicity is detected in the MCF-7 cells when ethanolic extract of *Excoecaria allogacha* (EEEA) is used for treatment. Morphological variations like rounding or reduction of cells with formation of granules and vacuoles in the cytoplasm were detected. Increase in cytotoxicity is detected in the cells when 25 mg/ml and 50 mg/ml of EEEA extracts are used for treatment. Figure 3 represents the Fourier coefficients used as inputs for training the RBFN. The Fourier coefficients in Fig. 3 denote the power spectrum values corresponding to the Classes A, B, C, D, E and F and they serve as a unique feature with appropriate information regarding the various classes for classification process by RBFN which is depicted in Fig. 4.

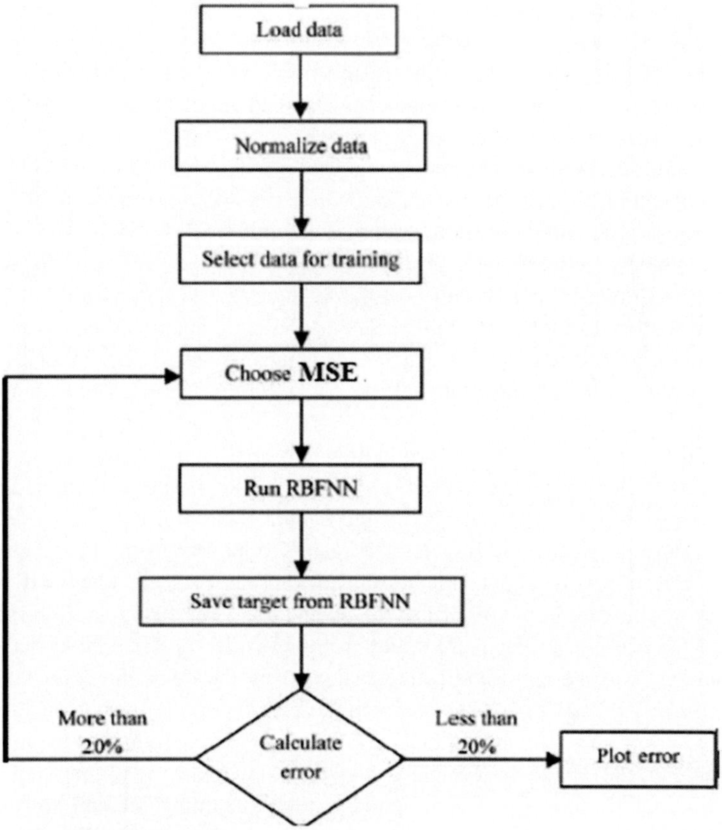

Fig. 1. Flowchart for RBFN

6.1 MTT Analysis

The MTT assay is typically accomplished to infer the toxic activity of any extract since it regulates the feasibility of cells after the process of treatment. This treatment with medicinal substance has significantly decreased the feasibility of the MCF-7 cell representing the cytotoxicity capacity of the proposed plant extract which is detected by RBFN and illustrated in Fig. 4.

6.2 Identification of Apoptosis

To differentiate between apoptosis and necrosis of the MCF-7 cells due to EEEA treatment, the DNA binding dyes acridine orange (AO) and Ethidium bromide (EB) was used to stain the nucleus. Figure 5 shows the increased cell death in the MCF-7 cells due to plant extract treatment when compared to control untreated cells. The appearance of uniform cells with a green color stained nucleus in the control group indicates the live cells with their membrane intact.

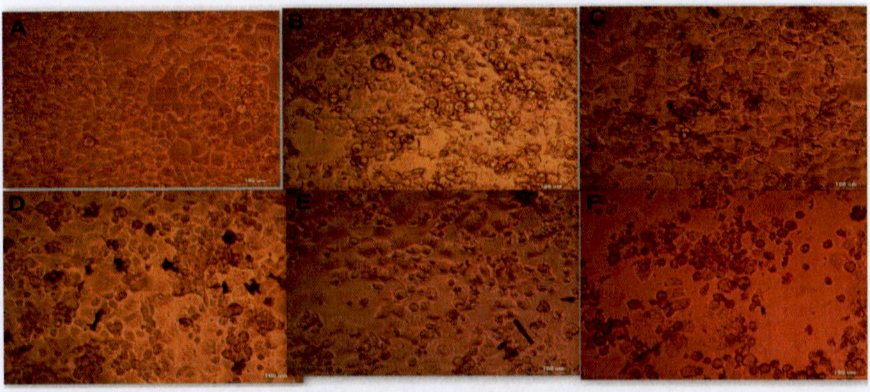

(a). Morphological changes in MCF-7 cells

Dimensions	1153 x 893
Width	1153 pixels
Height	893 pixels
Bit depth	24

(b). Resolution of images corresponding to MCF-7 cells

Fig. 2. (a) Morphological changes in MCF-7 cells. (Resolution of images corresponding to MCF-7 cells)

Treatment of cells with 56.5 μg/ml and 113 μg/ml of the EEEA chemical substance exhibited nuclei with perky green color were condensation of chromatin condensation takes place and the region of the nuclei with sunny orange color denotes the existence of early and late apoptotic cells respectively. On the other hand, the necrotic cells with an identical nucleus and undamaged membrane in sunny orange color are envisaged. On treatment with 113 μg/ml of EEEA extract exhibited an amplified increase in late apoptosis and necrosis equivalent to the positive chemical substance called doxorubicin. The above concept is evident from the classification results by RBFN depicted in Fig. 6.

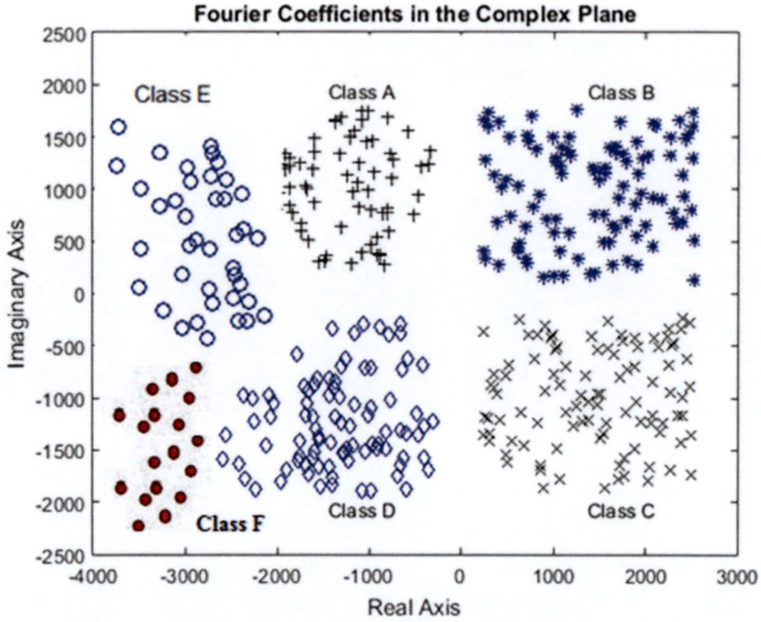

Fig. 3. Extraction Fourier coefficients

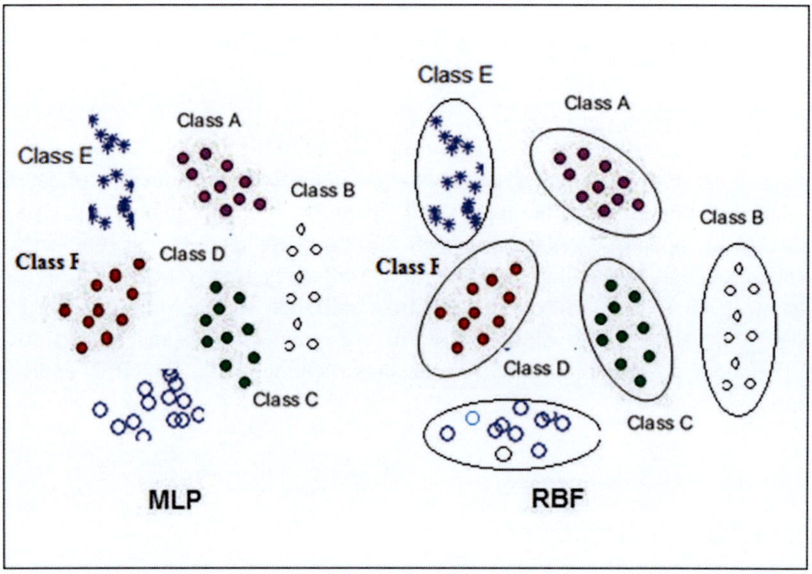

Fig. 4. Detection of cell viability of MCF-7 cell lines by MLP and RBFN

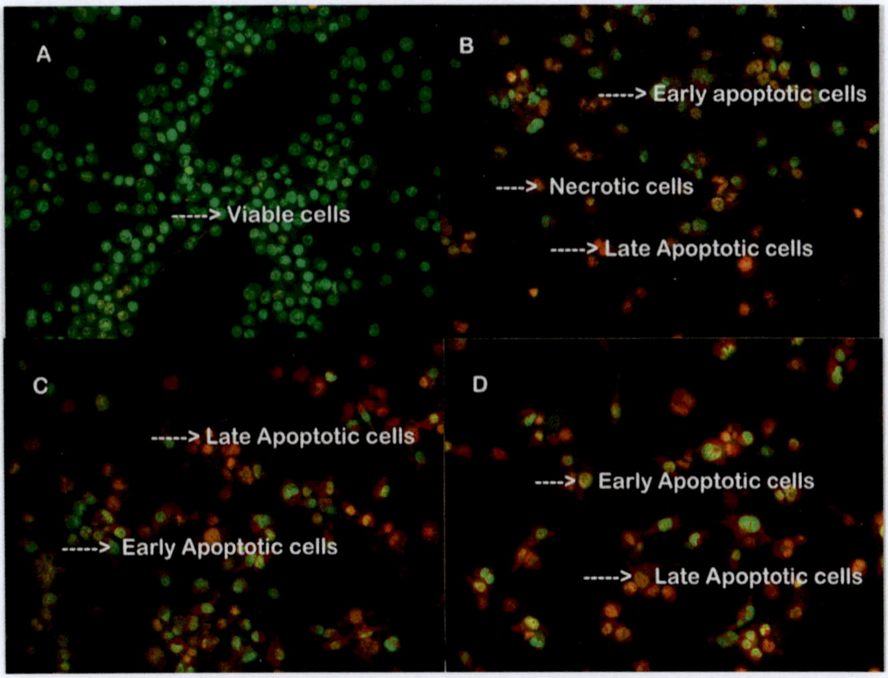

Fig. 5. Apoptosis assessment in MCF-7 cells

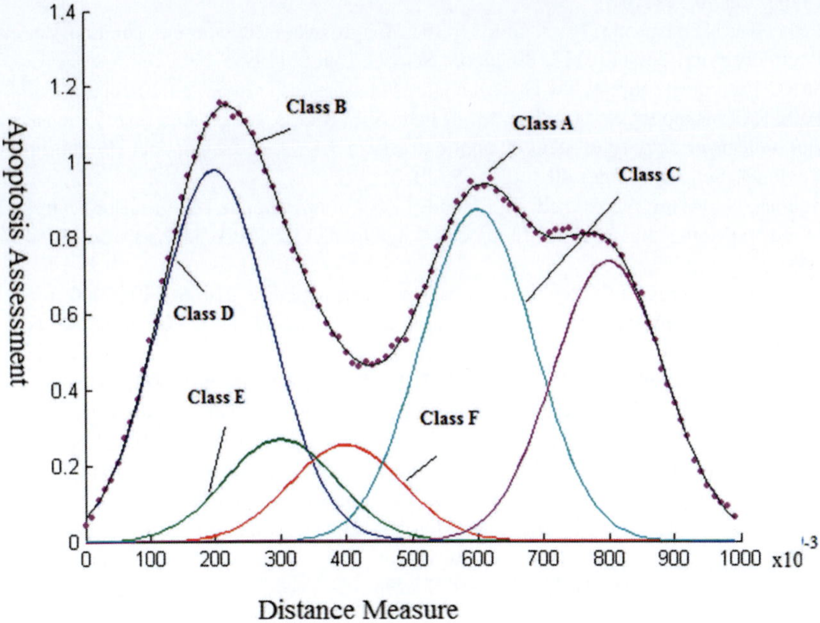

Fig. 6. Apoptosis evaluation by RBFN

7 Conclusion

Finally, it is stated that the ethanolic extract of *Excoecaria allogacha* and the existence of various phytochemicals depicted its antiprogressive property of human breast cancer cell lines by apoptosis mechanism which is detected by image analysis using Discrete Fourier Transform and Radial Basis Function Network. The results obtained by image analysis is almost comparable with that of the laboratory results. This combination of the chemical substance can be developed as a therapeutic treatment for breast cancer after, undertaking a proper clinical and preclinical assessment.

References

1. Ganesh, N., Dave, R., Sanadya, J., Sharma, P., Sharma, K.K.: Various types and management of breast cancer: an overview. J. Adv. Pharm. Technol. Res. **1**(2), 109–126 (2010)
2. Sumitra, C., Nagani, K.: In vitro and in vivo methods for anticancer activity evaluation and some Indian medicinal plants possessing anticancer properties: an overview. J. Pharmacognosy Phytochem. **2**(2), 140–152 (2013)
3. Dorai, T., Aggarwal, B.B.: Role of chemopreventive agents in cancer therapy. Cancer Lett. **215**(2), 129–140 (2004)
4. Charles, G., Jacob, G.A., Bernard, L.F., Regina, M.N.: Traditional medicine as an alternative form of health care system: a preliminary case study of Nangabo sub-county, central Uganda. Afr. J. Tradit. Complement Altern. Med. **7**(1), 11–16 (2010)
5. Satyavani, K., Gurudeeban, S.: Insight on *Excoecaria agallocha*: an overview. J. Nat. Prod. Chem. Res. **4**(2) (2016)
6. Peter, K.L.N., Sivasothi, N.: A Guide to the Mangroves of Singapore I: The Ecosystem and Plant Diversity, pp. 111–112. Singapore Science Centre (1999)
7. IARC Fact sheet. http://www.globocan.iarc.fr/factsheet.asp. Accessed 10 June 2013
8. Patil, R.C., Manohar, S., Upadhye, M., Katchi, V.I., Rao, A., et al.: Anti reverse transcriptase and anticancer activity of stem ethanol extracts of *Excoecaria agallocha* (Euphorbiaceae). Ceylon J. Sci. (Biol. Sci.) **40**, 147–155 (2012)
9. Subhan, N., Alam, A., Ahmed, F., Shahid, I.Z.: Antinociceptive and gastroprotective effect of the crude ethanolic extracts of Excoecaria agallocha Linn. Turk. J. Pharm. Sci. **5**, 143–154 (2008)
10. Lowe, S.W., Lin, A.W.: Apoptosis in cancer. Carcinogenesis **21**, 485–495 (2000)
11. Elmore, S.: Apoptosis: a review of programmed cell death. Toxicol. Pathol. **35**(4), 495–516 (2007)
12. Fadok, V.A., Voelker, D.R., Campbell, P.A., Cohen, J.J., Bratton, D.L., Henson, P.M.: Exposure of phosphatidylserine on the surface of apoptotic lymphocytes triggers specific recognition and removal by macrophages. J. Immunol. **148**(7), 2207–2216 (1992)
13. Tor, Y.S., Yazan, L.S., Foo, J.B., et al.: Induction of apoptosis through oxidative stress-related pathways in MCF-7, human breast cancer cells, by ethyl acetate extract of *Dillenia suffruticosa*. BMC Complement Alter. Med. **14** (2014). Article ID 55
14. Prakash, S., Khan, M., Khan, H., Zaman, A.: A piperidine alkaloid from *Excoecaria agallocha*. Phytochemistry **22**, 1836–1837 (1983)

15. Konishi, T., Takasaki, M., Tokuda, H., Kiyosawa, S., Konoshima, T.: Anti-tumor promoting activity of diterpenes from *Excoecaria agallocha*. Biol. Pharm. Bull. **21**, 993–996 (1998)
16. Konishi, T., Yamazoe, K., Konoshima, T., Maoka, T., Fujiwara, Y., Miyahara, K.: New bis-secolabdane diterpenoids from *Excoecaria agallocha*. J. Nat. Prod. **66**, 108–111 (2003)
17. Zou, J.H., Dai, J., Chen, X., Yuan, J.Q.: Pentacyclic triterpenoids from leaves of *Excoecaria agallocha*. Chem. Pharm. Bull. **54**, 920–921 (2006)

Development of an Algorithm for Vertebrae Identification Using Speeded up Robost Features (SURF) Technique in Scoliosis X-Ray Images

Tabitha Janumala(✉) and K. B. Ramesh

Department of Electronics and Instrumentation Engineering,
R V College of Engineering, Bengaluru-59, India
{tabithajanumala, rameshkb}@rvce.edu.in

Abstract. Vertebrae identification is one of the major parameters in extracting the features of spinal deformity. As 2–4% of the adolescents' population are suffering from spinal deformity, there is a need to detect it in the early stage itself. In this paper, an attempt is made to identify the segments of the spinal cord using speeded-up robost features (SURF) technique in X-ray images. The database is maintained with samples of X-ray images of the spine. Then with the help of support vector machine (SVM), the template matching is done to identify the vertebrae. Using this vertebrae identification many other parameters such as calculation of Cobb angle, estimation of truncal shift, and type of deformity can be classified.

Keywords: Speeded up robost features (SURF) · Support vector machine (SVM) · Scale-invariant transform (SIFT)

1 Introduction

Spinal deformity is one of the major problems faced by the adolescent population. Last Saturday of June month is celebrated as International Scoliosis Awareness Day. On this account, Dr. Krishna Kumar R from Kochi, states that world over, 2–3% of people are suffering from this condition when interviewed by Indian Express newspaper 2019.

Hence there is the need to think about the tools used to extract the features of spinal deformity for the early detection of the disease called scoliosis. The objective of this research work is to number the vertebrae of the spinal cord, to future extract the features of spinal deformity. One such parameter is the measurement of the Cobb angle which requires the identification of most upper titled vertebrae and most tilted lower vertebrae. The estimation of cobb angle using histogram technique [9] is discussed in one of the previous papers published in technical research organization India.

The paper "Development of an Image-Based Modelling Tool for the Spinal Deformity Detection Using Curve Fitting Technique", discusses the technique of tracing the spine [2]. The tracing of the spine is required to know exactly the region of deformity occurring in the spinal cord which is further used for the classification of scoliosis disease. Some of the segmentation techniques used were emphasized such as

J. I.-Z. Chen et al. (Eds.): ICIPCN 2020, AISC 1200, pp. 54–62, 2021.
https://doi.org/10.1007/978-3-030-51859-2_6

the use of using mask-based x-ray image segmentation algorithm for measurement of Cobb angle [1]. But the limitation is this technique is applicable only for C-shaped spinal curvature. Snake segmentation is also used to identify the spinal deformity, it also uses the top- hat filter in preprocessing the images [3]. The curvature estimation is done using the template matching sum of Squared Difference (SSD). This method is used to estimate the location of the vertebra by using polynomial curve fitting technique [4]. The pre-processing tools used to enhance the vertebral column specifically is Contrast Limited adjusted Histogram equalization (CLAHE) [5], which helps in highlighting the individual vertebrae. Support vector machine(SVM) techniques are used to obtain individual vertebrae by detecting the points of interest [6]. Training the SVM requires thousands of positive and negative samples, extracts their HOG (histogram of oriented gradients) features, and then use linear SVM to train it. Although it is an effective method the lack of samples causes the restrictions. The object detection techniques [7] used help in detecting the strongest feature points of an image and then detecting the desired object. Discriminative training includes the method of SVM for training in the identification of the object for various samples.

2 Methodology

The implementation of the proposed work is represented in Fig. 1. The X-ray scoliosis image is read and preprocessed. The preprocessing steps include the use of greyscale filters, median filter, Weiner filter, and Gaussian filter. Each of them is explained in detail as follows:

- **Grayscale filter:** Grayscale images are different from one-bit bi-tonal black-and-white images, which implies in terms of computer imaging that it's bi-tonal (also called *bi-level* or binary images). Grayscale images contain several types or shades of gray in between. In computing, even though the grayscale can be calculated using rational numbers, stored unsigned integers are used to quantize image pixels, to decrease the needed computation and storage. The shades are instead naturally spread out uniformly on a gamma-compressed nonlinear scale, which better approximates uniform perceptual increase for both dark and light shades, usually making these 256 shades enough to avoid obvious increments. If the image contains red, green and blue components then its linear luminance is calculated as

$$Y_{\text{linear}} = 0.2126R_{\text{linear}} + 0.7152G_{\text{linear}} + 0.0722B_{\text{linear}} \tag{1}$$

Y_{linear} = Total Linear Luminance of the image
R_{linear} = Reb colour Linear Luminance of the image
G_{linear} = Green colour Linear Luminance of the image
B_{linear} = Blue colour Linear Luminance of the image

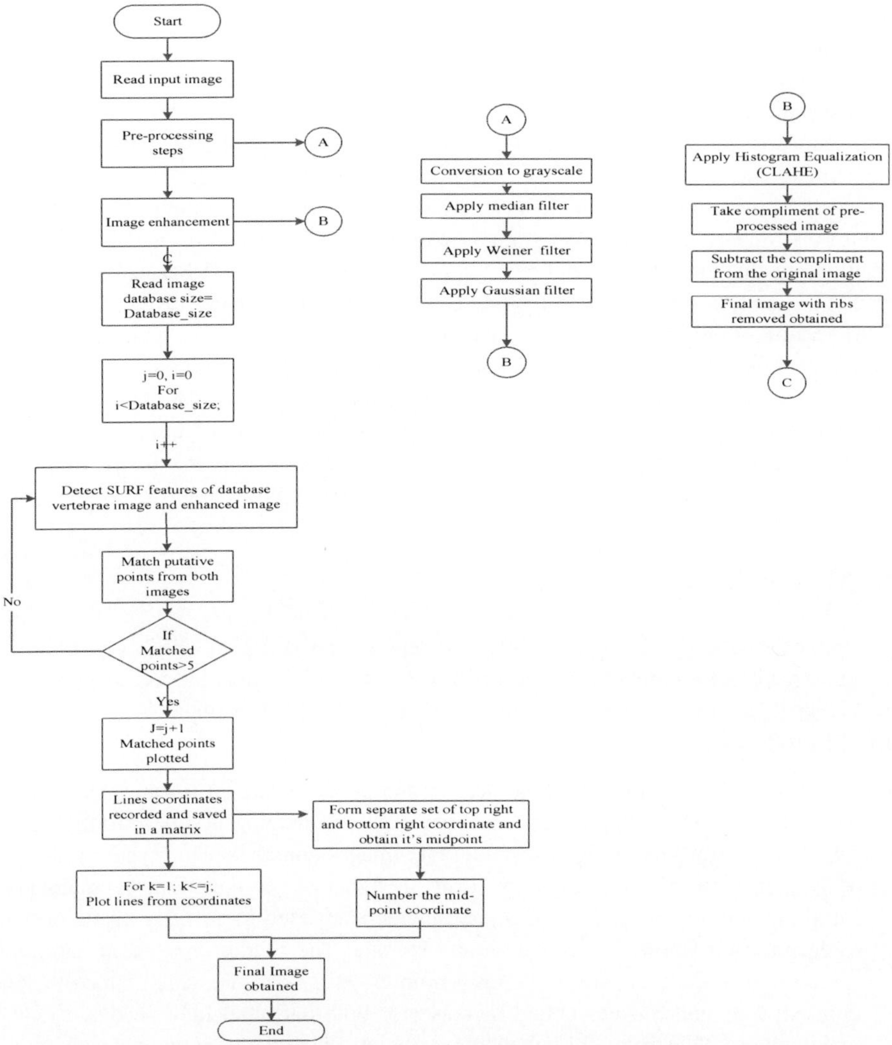

Fig. 1. Flow chart of the algorithm implemented

- **Median filter:** To remove unwanted noise from signal or image, median filter, a nonlinear digital filtering technique is used. It is a general pre-processing step for improving on later results. The main methodology is to go through each signal entry and interchange each entry with the median of neighbouring entries. Window is defined as the pattern of adjacent pixels and it slides through each entry. This is also called ad smoothing technique.
- **Wiener Filter:** The Wiener filter is a filter used to obtain an approximate of a required or target random process by linear time-invariant (LTI) filtering of a pragmatic noisy process. Littile smoothingobtained vy high variance and High smoothing obtained by less variance.

$$\mu = \frac{1}{NM} \sum\nolimits_{n1,n2 \in N} a(n1, n2) \quad (2)$$

And

$$\sigma^2 = \frac{1}{NM} \sum\nolimits_{n1,n2 \in N} a^2(n1, n2) - \mu^2 \quad (3)$$

where η is the N-by-M local neighborhood of each pixel in the image A. wiener2 then creates a pixel-wise Wiener filter using these estimates.

- **Gaussian Filter:** The convolution of the image with the Gaussian function is called the Gaussian filter. Gaussian Blur is a kind of a low pass filter. Gaussian filters have the characteristics of having zero overshoot to a step function input while decreasing the fall time and rise time. This behavior is closely attached to the fact that the Gaussian filter has the least possible group delay. It is regarded as the ideal time-domain filter. In one dimension, the Gaussian filter is applied using the function:

$$g(x) = \frac{1}{\sqrt{2\pi}.\sigma} .e^{-\frac{x^2}{2\sigma^2}} \quad (4)$$

Where sigma is the standard deviation.

When applied in two dimensions, the formula generates a Gaussian surface which has the highest at the origin, with origin as the centre, whose contours are concentric circles. A pre-processed 2-D convolution matrix is formed using the formula hence convolving with the 2-D data. Every element in the consequential matrix new value is a part of a weighted average of that element's vicinity. The heaviest weight is received by the focal element (having the highest Gaussian value) and smaller weights are received by neighboring elements as their distance to the focal point increases. In Image processing, every element in the matrix signifies a pixel quality such as colour intensity or brightness, and the total effect is known as Gaussian blur.

Image Enhancement

The major method used here is CLAHE (contrast limited adaptive histogram equalization)

- To improve contrast in images 'Adaptive histogram equalization', an image processing method is used. It works by deriving a function from the neighboring region and processing each pixel accordingly. The histogram of the square around the pixel provides the basis of transforming a pixel.
- In CLAHE, the transformation function's slope gives the contrast amplification in the region of a given pixel. It is directly proportional to the slope of the surrounding cumulative distribution function (CDF) and hence to the value of the histogram at that pixel value. Before calculating the CDF, CLAHE limits this amplification by marking the histogram at a pre-set value. Hence the slope of transformation function and the CDF are bounded. The histogram's normalization and that of the

neighboring region determines the point at which histogram has been clipped or the so-called 'clip limit'.
- 3 or 4 is the resulting amplification value.
- After applying CLAHE method on the x-ray image, a contrasted image was obtained depicting the overall features of the full image.
- The complement of the resulting image is taken.

The complemented image is then subtracted from the CLAHE image to obtain the resultant image with the vertebral features enhanced while removing the unwanted features like underlying ribs, pelvic bones etc.

Implementation of SURF Algorithm and Vertebrae Mapping
To identify each vertebra individually, the following algorithm was used involving computer learning and database management

- Vertebrae identification is implemented here by training the system using some sample vertebrae images so that it recognizes the same in lumbar x-ray images.
- SURF (speeded up robust features) can be extracted from an image by specifying the number of strongest points to be identified. SURF is an advanced version of the scale-invariant transform (SIFT) descriptor and is generally used in object identification in image processing.
- Training the system requires a database of vertebrae samples with different orientations i.e. right or left tilt etc.
- The vertebrae samples are collected from the hospital and the system is trained.
- SURF features of the enhanced image from the last block are extracted by selecting 300 strongest points.
- SURF features of the vertebrae database are extracted by selecting 10 strongest points.
- The points of vertebrae in the database and the lumbar x-ray image are matched and the vertebra is identified by drawing a box around it by detecting coordinates and going line by line.
- The geometric coordinates of the lines are stored for further purposes.

Vertebrae Numbering

- The stored line coordinates are provoked.
- The bottom right and top right line coordinates are retrieved from the stored matrix and their mid-point is obtained.
- Corresponding vertebrae is then numbered.

3 Results

An X-ray image is given as an input to the developed algorithm, the image is preprocessed and enhanced. The database of segments of the spine for scoliosis X-ray images is maintained. The system is trained with the segment samples of the spinal cord. Vertebrae identification is done by choosing 300 strongest feature points as shown in Fig. 2.

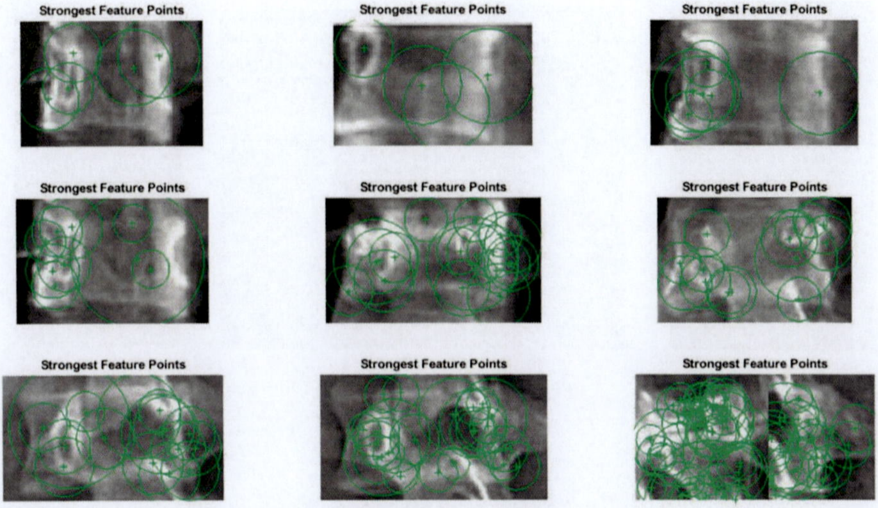

Fig. 2. SURF features database 1

The strongest points of each segment are put together in line with the spinal cord as shown in Fig. 3.

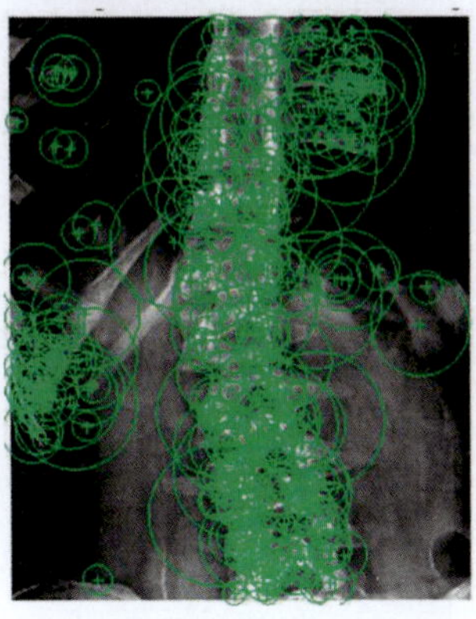

Fig. 3. SURF features sample 1

Each segment of the vertebrae from the database is matched with the input X-ray scoliosis image to identify the segments of the spinal cord as shown in Fig. 4.

After the identification of the segment of vertebrae, a rectangle is drawn and each vertebra is numbered accordingly as shown in Fig. 5.

Thus the vertebrae of the spinal cord are identified and numbered. The numbering of the vertebrae plays a vital role in identifying the exact vertebrae which is most tilted in both upper and lower regions to estimate cobb angle. Using this numbering of

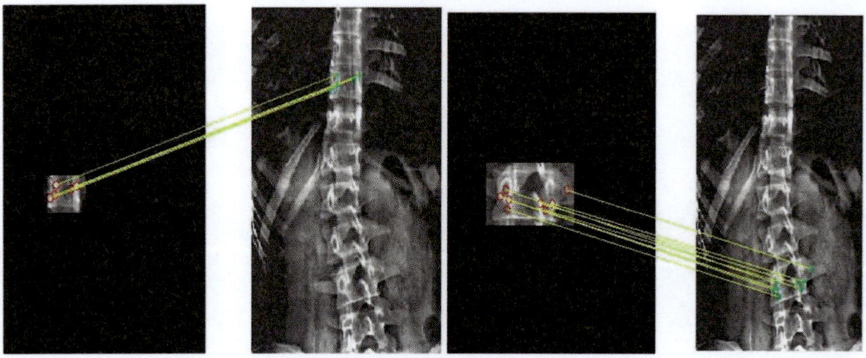

Fig. 4. Matched features 1

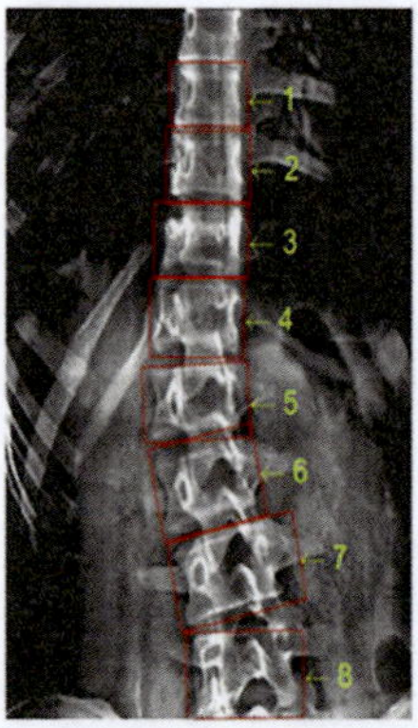

Fig. 5. Vertebra detection subject 1

vertebrae will also estimate the type of spinal deformity occurred in scoliosis X-ray images. The manual identification of Cobb angle by doctors takes at least 20 min and it also requires the second opinion to finalize the value. But the algorithm developed does not require any manual intervention in identifying the region of interest and estimating the spinal deformities with the help of vertebrae numbering.

4 Conclusion and Future Scope

An X-ray scoliosis image given as an input for this algorithm has traced the spinal cord and numbered the vertebrae as a segment. The SURF technique was used to identify the vertebrae from the database. The database of 500 samples of X-ray scoliosis images was taken from the hospital and the system was trained. Then the template matching technique was used to match each segment from the database. Considering 300 strongest points for SURF a vertebrae was identified. Then each vertebra was numbered. In the future, this can be used to extract the other features of spinal deformity.

Cobb angle can be calculated by measuring the angle between perpendicular lines drawn, selecting the most tilted upper and lower vertebrae segments. Region of interest that is vertebrae selection of most tilted ones can be automated. Another parameter of spinal deformity the truncal imbalance can also be estimated. This algorithm helps the doctors to get an insight into X-ray scoliosis images with few deformity parameters estimated.

References

1. Tariq, A.: Buket barkana department of electrical engineering, university of bridgeport, bridgeport, CT, USA: a mask based segmentation algorithm for automatic measurement of cobb angle from scoliosis X-Ray image. In: ICACC 2012.24 110. International Conference Advance Computing Communication. IEEE (2012). https://doi.org/10.1109/icacc.2012.24
2. Tabitha, J., Ramesh, K.B., Harish, A.H.: Development of an image-based modelling tool for the spinal deformity detection using curve fitting technique. Int. J. Eng. Res. Electr. Commun. Eng. (IJERECE) **5**(4), 7–11 (2018). ISSN (Online) 2394–6849
3. Madha, C.W., Tri, A.S.: Spinal curvature determination from X-ray image using GVF snake. In: International Conference Information and Communication Technology System (ICTS) (2015)
4. Bagus, A.K., Hanung, A.N., Sanu, W.: Spinal curvature determination from scoliosis X-ray image using sum of squared difference template matching. In: International Conference on Science and Technology-Computer (ICST) (2016)
5. Ili, A.M.I. (ed.) An analysis of x-ray image enhancement methods for vertebral bone segmentation. In: IEEE. 10th International Colloquium on Signal Processing Applications (2014). https://doi.org/10.1109/cspa.2014.6805749
6. Fabian, L., Mohammed, B., Saïd, M.: Fully automatic vertebra detection in X-ray images based on multi-class SVM. In: Proceedings of SPIE 8314, Medical Imaging 2012: Image Processing, p. 83142D (2012). https://doi.org/10.1117/12.911424
7. Felzenszwalb, P., Girshick, R., McAllester, D., Ramanan, D.: Object detection with discriminatively trained part-based models. IEEE Trans. Pattern Anal. Mach. Intell. **32**(9), 1627–1645 (2010). https://doi.org/10.1109/tpami.2009.167
8. Krishna Kumar, R.: Article: What you should know about scoliosis, The New Indian Express, Kochi, 01st July 2019
9. Tabitha, J., Ramesh, K.B.: Design and development of an algorithm for the detection of scoliosis using histogram technique, In: Proceeding of 16th International conference on Research in Engineering, Science Technology (AREST-2017), 2nd July 2017, Bangalore, Organized by-Technical Research Organization India (2017). ISBN 978–93-85225-94-9
10. Junhua, Z., et al.: Assessment of scoliosis using 3-D ultrasound volume projection imaging with automatic spine curvature detection. In: IEEE International Ultrasonics Symposium (2015). https://doi.org/10.1109/ultsym.2015.0485
11. Dong, X., Zheng, G.: Automated vertebra identification from XRay images. In: Image Anal. Recogn. LNCS, pp. 6112–6119 (2010)
12. Liming, D., Yong, H., Jason, P., Yin, C.: A data-driven decision support system for scoliosis prognosis. IEEE, June 2017
13. Bogdan, A.P., Bogdan, D., Elena, L.: Scoliosis risk assessment in children with pelvic asymmetry - a simple screening test based on gluteal area asymmetry. In: The 6th IEEE International Conference E-Health Bioengineering, June 2017

14. Alomari, R.S., et al.: Lumbar spine disc herniation diagnosis with a joint shape model. In: Computational Methods and Clinical Applications for Spine Imaging, pp. 87–98. Int. Publishing (2014)
15. Ghosh, S., et al.: A supervised approach towards segmentation of clinical MRI for automatic lumbar diagnosis. In: Computational Methods and Clinical Applications for Spine Imaging, pp. 185–195. Int. Publish (2014)
16. Hong, L.: IEEE department of research texas scottish rite hospital, Dallas, TX, USA: review of advances in Cobb angle calculation and image-based modelling techniques for spinal deformities. ISPRS Annals of the Photogrammetry, Remote Sensing and Spatial Information Sciences, vol. III-5, 2016 XXIII ISPRS Congress, July, Prague, Czech Republic (2016)

Comparison of Machine Learning Algorithms for Smart License Number Plate Detection System

Anjali Suresan(✉), Divyaa Mahalakshmi G, Meenakshi Venkatraman, Shruthi Suresh, and Supriya P

Department of Electrical and Electronics Engineering, Amrita School of Engineering, Amrita Vishwa Vidyapeetham, Coimbatore, India
anjali3498@gmail.com, divyaamaha@gmail.com

Abstract. Identification of License plates of vehicles is significant in various monitoring and security applications. This paper involves two major processes: image processing of the License plate and classification of the individual characters. Preprocessing of the number plate image is done for noise removal by converting it to a grayscale image followed by conversion to a binary image. The Bounding Box method is implemented for the segmentation of individual characters that are to be classified. Two Supervised machine learning algorithms namely Support Vector Machine (SVM) and Decision Tree is used for the classification. The performance indices of the two algorithms are analysed to determine the more accurate method. It is proposed to implement the algorithm on a stand-alone model to employ it for real-time applications after improving the accuracy with more training and test data.

Keywords: License plate · Preprocessing · Segmentation · Bounding box · Classification · Supervised machine learning · Decision tree · Support Vector Machine

1 Introduction

With increasing technological advancements in the field of transportation and population growth, the use of vehicles is increasing exponentially. License plate recognition of these vehicles is essential for various applications like traffic management and identification of unregistered vehicles.

An efficient recognition system for number plates will particularly be useful in cases where certain computation is involved. For example, in Delhi, India, the odd-even rule is applicable to reduce the number of vehicles on the road thus reducing pollution. The number plate recognition system will make it easier to detect the vehicles that violate the rule to penalise them accordingly. In places where high security is required, such as those where highly important officials reside and where military bases are located, there is a necessity for tracking and identifying the vehicles that move around in such areas. In these cases, the number plate recognition system can

J. I.-Z. Chen et al. (Eds.): ICIPCN 2020, AISC 1200, pp. 63–75, 2021.
https://doi.org/10.1007/978-3-030-51859-2_7

be used to track unknown vehicles and thus prevent the possibility of any threat to security.

In the present study, the license plate images of cars from various Indian states were classified using the Classification Learner in MATLAB R2017b. Linear Support Vector Machine and Decision Tree classifier are used for the classification. The training time and accuracy are used as the parameters to compare the results obtained. The objective is to determine the faster and more accurate classifier for the identification of characters in the number plate of vehicles that will increase the efficiency of a surveillance or traffic management system where it is employed.

Image processing techniques are crucial for noise removal and distinct segmentation of individual characters in the number plate image. The identification of these characters is performed using the two supervised machine learning algorithms for an improved success rate and minimal computation time. Section Two reviews the previous work implemented in the area. The methodology of the work is reviewed in Section Three. Section Four brings out the simulation results of the work. The conclusions and future work are deliberated in Section Five. The overall proposal of the work is provided in Fig. 1.

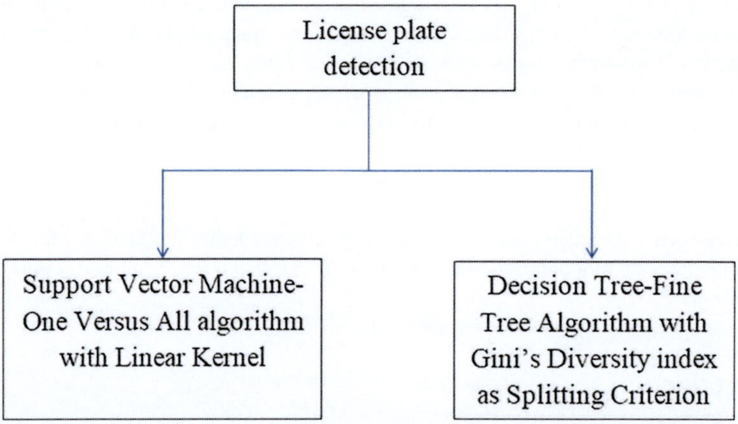

Fig. 1. Detection of license plate characters using two supervised machine learning methods

2 Related Work

Various literature is available on machine learning and image processing algorithms for license plate detection. The background study is discussed in detail in this work. Kajale et al. [1], proposed the use of supervised machine learning algorithms for character identification to combat the challenge of dealing with dynamic data and also to increase accuracy in the results obtained. The method is found to give fairly accurate results for large datasets as well. A dataset of characters was generated using a series of pre-processing techniques including noise removal and character segmentation. The dataset of alpha numerals and special characters was trained and a decision tree classifier was developed. After the testing phase, the results were found to be almost 95% accurate. The classification accuracy can be improved by the use of Support Vector Machines (SVM) for character recognition as is found in [5].

Bostik et al. [2], shows a comparison of the conventional machine learning algorithms for Optical Character Detection of CAPTCHA codes. Experimentation is carried out in the MATLAB computing environment to compare the accuracy of K-Nearest Neighbors, Pattern Recognition Artificial Neural Networks, Feed-forward Artificial Neural Networks, Decision Trees and Support Vector Machines. The preprocessing of the image is done in three steps. The individual characters in the image are separated using predetermined vertical lines. Colour segmentation of the separated character is achieved using colour thresholding followed by linearization of the binary character image. The input to the machine learning models is in terms of a vector that is the result of the feature extraction stage. The algorithms were compared based on the success rate and time consumption or computational complexity. Support Vector machines were found to have a success rate of 98.80%, whereas the decision tree algorithm gives a better result in terms of classification time.

Mahesh Babu et al. [3], has used the following steps which include Image Preprocessing, extraction of the license plate, obtaining individual characters and their identification. The captured image has been brightened and the noise has been removed using filters and converted into a grayscale image. Extraction is done by finding the edges of the license plate and it is cropped into a rectangular frame. The Bounding box method is used for the segmentation of characters.

Yang [4] has proposed a machine learning method where he has used SVM for character recognition. To enhance the character recognition accuracy and speed, a Least Squares SVM based wavelet kernel is used. Instead of the general kernel function like RBF kernel or Gauss kernel, Mexico hat Wavelet has been used to have a better generalizing capacity. The method used produced a better result by having a recognition rate of 98.3% with a testing time of 0.13 s when experimented over 500 vehicle images.

As mentioned by Ambika et al. [6], Support Vector Machine classification method is a supervised learning method suitable for plant classification wherein the datasets are separated by an optimal hyperplane which is used for their classification. The boundary of the training data is maximized by the separating hyperplane. The work includes preprocessing the image captured and converting it into its binary form. The features that are extracted are then used for identifying the species of the leaf.

Vignesh Kumar et al. [7] have compared SVM classifier and Decision Tree for single and multiple fault detection in a gearbox. SVM is found to be a better classifier for multi-class problems. The efficiency and accuracy of the two classifiers are compared by creating confusion matrices. In the Motor Electrical Signature Analysis, the SVM model with Gaussian kernel was found to be more efficient than Decision Tree. SVM is also preferred as a classification method as it involves the inclusion of the complete data. Decision tree uses if/then rule and the decision-making process at each node to increase consistency.

Bhardwaj et al. [8] have implemented a six-step process for the Automated Number Plate System which includes Image Acquisition, Preprocessing, Edge Detection, sectionalisation to obtain individual characters, Extraction of image features and distinguishing them. The data identification has been done with Kstar Machine learning algorithm and the efficiency is evaluated using the Precision score, Recall score and F-measure.

Qiao et al. [9] suggested a method for classifying species of birds using Multiscale decision tree and SVM framework. A new feature based on the physical appearance of the bird species is defined and integrated into the machine learning approach to give better recognition results.

Bagade et al. [10] has developed a system where the input image is obtained using an infra-red camera. Preprocessing of the image is done using a MATLAB function to eliminate the noise. The interesting region, which is the number plate location, is captured using a contour tracing algorithm. The edges are sharpened using Canny Edge detection algorithm and then this output is filtered using Gaussian Kernel. Each character is extracted with Segmentation algorithm and then the classification of these characters is done using artificial neural network. The final output of the system with the entry and exit time of the vehicle is noted to calculate the total cost for the toll.

Manoharan [11], has used input images that were captured with the help of a camera. The image is pre-processed to convert the RGB image to gray scale, enhancement of the image and the noise is removed using a filter. Edge detection is done using the Sobel operator and irrelevant parts of the image are eliminated using shrinking operation. Grey level threshold process is used to distinguish between background and text. Segmentation of each character is done using Bounding Box analysis and character recognition is done using Machine Learning Algorithms. Hence the number plate is detected and only the authenticated vehicle will be allowed entry and the rest(the un-authenticated vehicle that is not entered in the database) will be denied entry with a buzzing of alarm.

Ghadage and Khedkar [12], introduces a system where the conversion of text to digital form is done with a camera and the processing is done using LattePanda Alpha system. The extraction of the information is done using Optical Character Recognition which uses the grayscale version of the images to distinguish between the background and text. The OCR performs its operation on one word at a time and character segmentation is done. The recognition of characters is done using feature detection and pattern recognition techniques. Each character is then converted to its ASCII value and the final content is converted to audio using a speech synthesizer.

Automatic Number Plate Recognition can be carried out using several approaches. As described by Kumari et al. [13], the number plate recognition is carried out using the calculation of HOG features of the images present in the dataset. With the help of cropping and histogram equalization, the number plate portion can be filtered from the provided image. The various characters are classified using Support Vector Machines and in this way, the Number Plate characters are recognized.

In the methodology described by Meghana et al. [14], the Optical Character Recognition of the image after pre-processing is carried out using Template Matching wherein the characters present in the number plate are filtered out and compared using pixels with the template images of the various characters provided in the dataset. This technique is also mentioned to implemented more easily as compared to the others present. It helps tackle challenges faced in recognition due to weather conditions.

Sannakki et al. [15] have used image processing and SVM to detect diseases in pomegranate leaf accurately. K- means clustering is done to draw out the interesting region from the image. Two sets of features have been compared in the classification of the image. The classification set with saturation and entropy features has higher accuracy than the feature set with hue and energy.

3 Methodology

The number plate recognition is carried out as mentioned in Fig. 2. The first block represents the database of number plate images employed for the classification purpose. The second block highlights the preprocessing the image undergoes, namely, converting it to a grayscale image and subsequently to a binary image. The third block represents the character extraction using Bounding box where individual characters on the number plate are obtained for classification. The final block represents the classification of the obtained characters using supervised Machine Learning algorithms; Decision Tree and SVM.

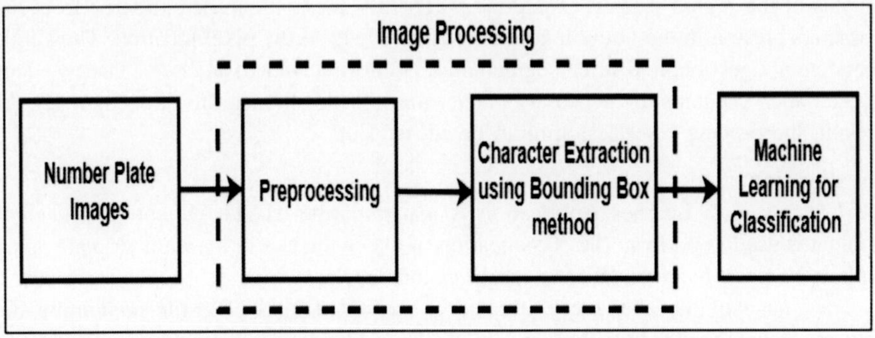

Fig. 2. Process flow for license plate recognition

The number plate recognition is done in four main steps: the preprocessing of the license plate image, bounding box method for segmentation, extraction of individual numbers or letters and their classification using SVM and Decision tree algorithms.

Collection of Number Plate Data:
The dataset used is the EMNIST Dataset consisting of alphabets and digits obtained from the NIST Special Database 19. The characters are converted to a 28*28 pixel image format. There are 36 samples, i.e. 10 digits (0-9) and 26 capital letters(A-Z). Each character is represented by a set of 1016 variations out of which 100 have been taken.

Preprocessing:
Preprocessing is carried out to reduce the noise present in the input images. The images are first taken from the file containing the number plate images and the following steps

are carried out for preprocessing using the image type conversion functions of the Image Processing toolbox in Matlab2017b:

i) The RGB image is converted to its grayscale form.
ii) The grayscale image is converted to a binary image.

The threshold for conversion to a binary image is determined by Otsu's method [16].

Bounding Box:
Connected component labeling takes an image and categories the pixels based on similar pixel intensity values. In this paper, the connected component labeling is used to label the different characters that are located in different areas of the image. To create the bounding boxes, the region properties of the image some of which include the area, the height and the width are analyzed and created.

Text Extraction:
A single character image is extracted from the pre-processed license plate image using bounding box segmentation method. A bounding box is created for every individual character present in the image with the help of connected component labeling and the analysis of the region properties is done. After this, the text can be extracted from the characters present in the bounding boxes with the help of the pixel location. The single character image obtained after segmentation is normalized to a 28 * 28 array. This reduced array contains the necessary pixel values of the character to be recognized. As a result, the process takes less time to be carried out.

Classification:
The Classification Learner Application is a part of the MATLAB computing environment is used to perform the classification of the characters in the number plate using Support Vector Machines (SVM) and Decision Tree.

The image of the character extracted is converted to a CSV file containing the features of the character image.

In the Classification learner, the generated dataset containing the alphabets and numerals with the response in categorical format is used to train the supervised machine learning algorithms. The trained model is exported as a structure to test the algorithm with the number plate dataset.

In the decision tree algorithm, at each step of the prediction, the feature of the input image is compared with a predictor variable giving a true or false response. The decision- making process in the tree starts from the root and extends until the leaf node, giving the required response. The splitting criterion used in the decision tree is Gini's diversity index.

For multi-class classification, the Gini impurity is calculated using the formula:

$$G(p) = \sum_{i=1}^{J} p_i \sum_{k \neq i}^{J} p_k = 1 - \sum_{i=1}^{J} (p_i)^2$$

Where:

G(P) = Gini impurity

J = Number of classes

P_i = Probability of an item in the ith class

P_k = Probability of wrongly categorizing an item in that class;

$$\sum_{k \neq i} Pk = 1 - Pi$$

The value of the Gini impurity varies from zero to one, where zero denotes perfectly classified. It is suitable in the partition of large data like the dataset of characters used for the recognition.

The Support Vector Machine algorithm is a commonly used supervised learning method applicable to classification problems. An optimal hyperplane to separate and categorize the dataset is found, at the same time making the separating interval to be maximized.

In MATLAB Classification Learner, SVM with a Linear Kernel function is found to give efficiency of 90.8%. A multiclass method of One-vs-All is implemented.

The Linear Kernel function is chosen for the character classification due to a large number of features in the dataset used. Text categorization largely involves linearly separable data. Training of SVM with Linear Kernel function is faster than that with any other kernel as it involves only the optimization of the C regularization parameter. The C parameter is chosen such that a higher margin is set and the misclassification rate on the testing data is minimized. For training of other kernels, the gamma parameter also has to be optimized. Non-linear hyperplanes use the gamma parameter. The training data that is fit increases with the increase in gamma value, thereby making the computation more complex and increasing the processing time.

In this paper, One vs All classification recognizes the various characters extracted from the license plate. In this method of classification, different variations of a single character are taken as one class and this is similarly followed for all other characters thereby obtaining multiple classes. This method makes use of multiple binary classifiers. For each classifier, the class under consideration is taken as positive class and the remaining are together taken as negative class. When an input is given in this method the probability that it belongs to a particular class is calculated and it is set to belong to the class with the highest probability. One vs All which uses N classifiers is preferred over One vs One which uses N * (N − 1)/2 classifiers since it requires lesser number of computations.

4 Simulation Results

Based on the performance indices, the supervised machine learning algorithm for classification that gives more accurate results is identified.

The work is implemented on MATLAB R2017b on a computer with an i7 CORE Processor.

Figure 3 shows the computing environment. The image of a particular license plate that has been considered for the processing and recognition is shown in Fig. 4. The image is converted to grayscale and binarized for noise removal. A threshold value is given. Area opening is done to remove objects which have pixel lesser than 30 to reduce the noise in the image. The resulting image is shown in Fig. 5.

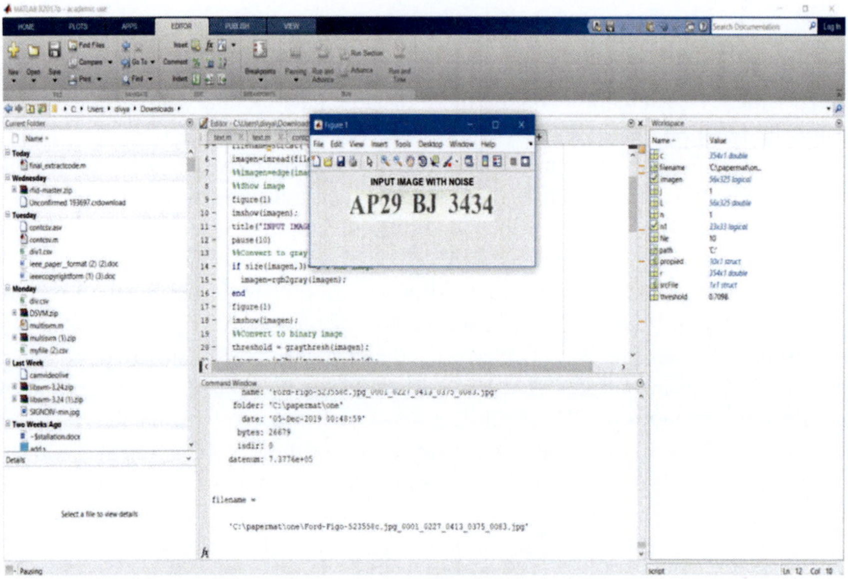

Fig. 3. A snapshot of the working environment is shown

Fig. 4. License plate image before pre-processing

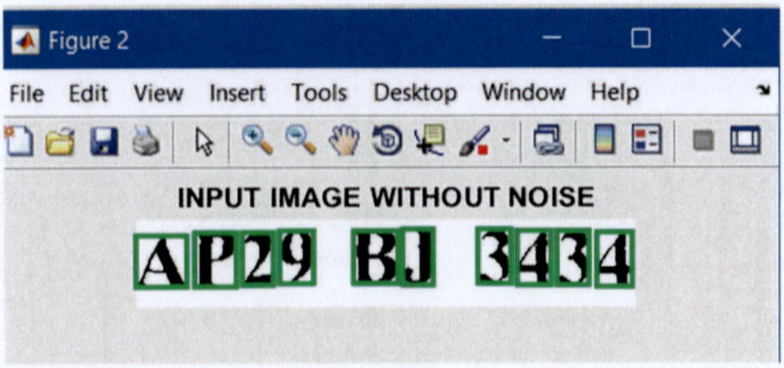

Fig. 5. License plate image with segmented characters after the pre-processing that has been done for noise removal

The processed image has been segmented using the Bounding Box method. In this method, the objects in the processed image are found using Connected Components Labelling and the pixels that are connected and share similar pixel intensity values are grouped and labelled. Since the image is binarized, the character pixels with an intensity of 1 are grouped with 8-connectivity. In 8-connectivity all neighbouring sides of the pixel are checked for similar pixel value, whereas in 4-connectivity the diagonal pixels in the neighbouring area are not checked. Hence 8-connectivity helps in separating each character as an object. The region property gives the corner pixel location, height and width of each separated object. With the parameters that are obtained, the bounding box is drawn around each character and the character is extracted. The individual characters in the license plate are extracted as shown in Fig. 6. The parameters of the trained SVM model and Decision tree model shown in Fig. 7.

Fig. 6. The characters that are individually extracted from the number plate using the Bounding Box method

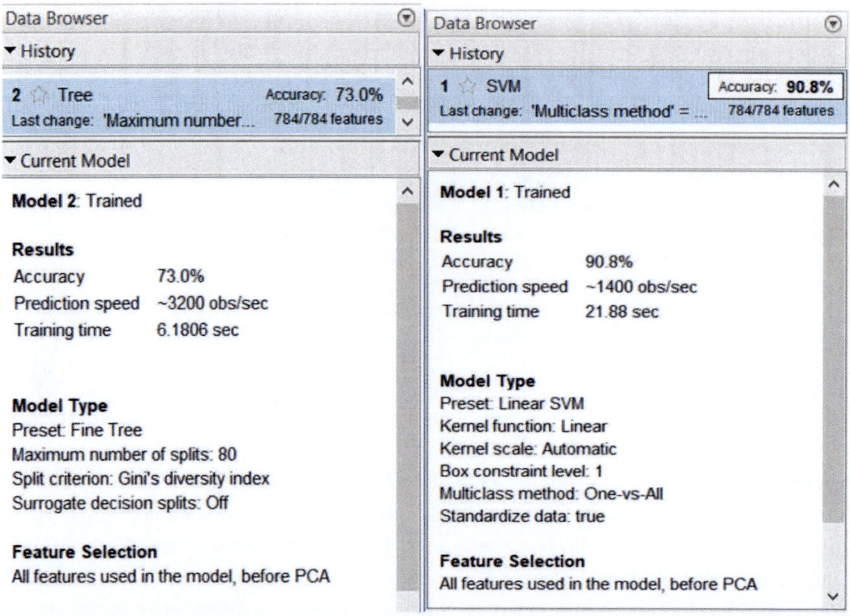

Fig. 7. Results and parameters of the training of the Decision Tree Model and the Linear SVM Model

The predictions for the extracted license plate characters are done using the two models. The sample results of the individual characters are shown in Fig. 8.

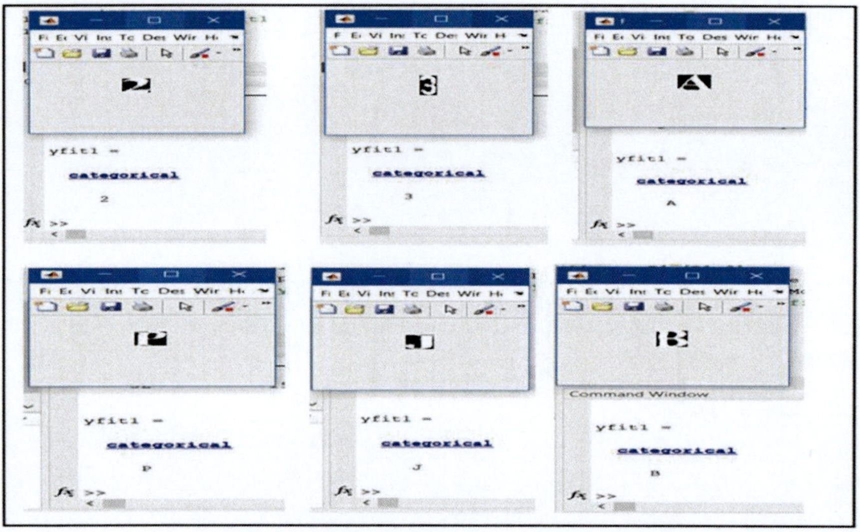

Fig. 8. The outputs predicted by the Linear SVM Model

Confusion matrix for a classification problem gives a summary of the predicted results. It evaluates the classifier performance based on the number of correct and incorrect predictions made by the classification model. The confusion matrix for the Decision tree and SVM model is plotted in Matlab. The true (Target) class is represented by the rows and the predicted (Output) class is represented by the columns. The diagonal cells show the correctly classified observations and the off-diagonal cells correspond to incorrect classifications.

The confusion matrices of the two models (Linear SVM and Decision Tree) are shown in Fig. 9 and Fig. 10:

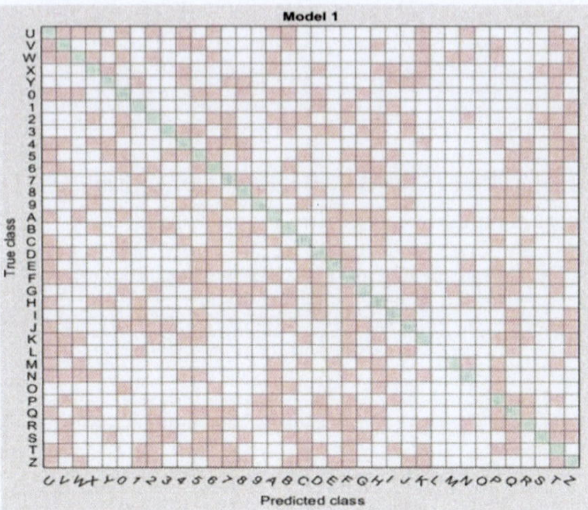

Fig. 9. Confusion matrix for the Decision Tree Model

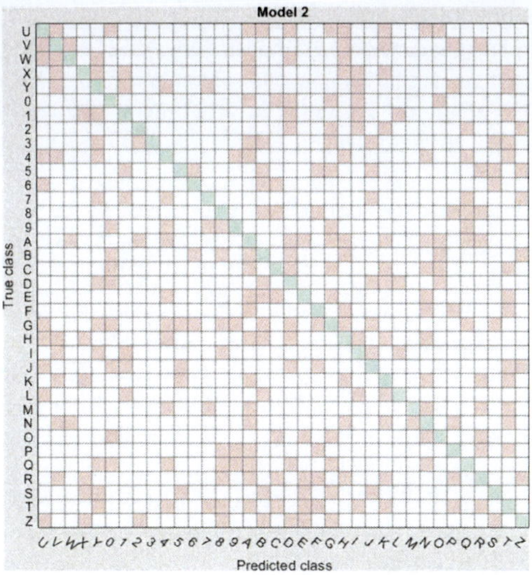

Fig. 10. Confusion matrix for the Linear SVM Model

5 Conclusions and Future Scope

The pre-processing of the license plate image and recognition of the characters has been performed in MATLAB. The accuracy obtained for the trained model of Linear SVM and Decision Tree is 90.8% and 73.8% respectively. Although, the accuracy for Linear SVM is high the training time for Decision tree is less. The comparison of the classification accuracy and training time for the two models is shown in Table 1.

Table 1. Comparison of accuracy and training time

	Decision Tree	SVM (Linear Kernel)
Accuracy	73.4%	90.8%
Training time	6.18 s	21.88 s

This work can be further extended by importing it to RaspberryPi to create a standalone system for implementation in real-time. The results can be made more accurate with a larger dataset.

References

1. Kajale, R., Das, S., Medhekar, P.: Supervised machine learning in intelligent character recognition of handwritten and printed nameplate. In: 2017 International Conference on Advances in Computing, Communication and Control (ICAC3). IEEE (2017)
2. Bostik, O., Klecka, J.: Recognition of CAPTCHA characters by supervised machine learning algorithm. IFAC-PapersOnLine **51**(6), 208–213 (2018)
3. Mahesh Babu, K., Raghunadh, M.V.: Vehicle number plate detection and recognition using bounding box method. In: 2016 International Conference on Advanced Communication Control and Computing Technologies. IEEE (2016)
4. Yang, G.: License plate character recognition based on wavelet kernel LS-SVM. In: 2011 3rd International Conference on Computer Research and Development. IEEE (2011)
5. Catunda, J.P.K., da Silva, A.T., Berton, L.: Car plate character recognition via semi-supervised learning. In: 2019 8th Brazilian Conference on Intelligent Systems (BRACIS). IEEE (2019)
6. Ambika, N.K., Supriya, P.: Detection of Vanilla species by employing image processing approach. In: 8th International Conference on Advances in Computing and Communication (ICACC 2018) (2018)
7. Vigneshkumar, S., Shankar, V.K., Krishna, P.N., Supriya, P.: Fault detection in gearbox using motor electrical signature analysis. In: 2018 9th International Conference on Computing, Communication and Networking Technologies (ICCCNT). IEEE (2018)
8. Bhardwaj, D., Kaur, H.: Comparison of ML algorithms for identification of automated number plate recognition. In: Proceedings of 3rd International Conference on Reliability, Infocom Technologies and Optimization (2014)
9. Qiao, B., Zhou, Z., Yang, H., Cao, J.: Bird species recognition based on SVM classifier and decision tree. In: 2017 First International Conference on Electronics Instrumentation & Information Systems (EIIS). IEEE (2017)

10. Bagade, J.V., Kamble, M., Pradeshi, K., Punjabi, B., Singh, R.: Automatic number plate recognition system: machine learning approach. In: Second International Conference on Emerging Trends in Engineering (SICETE), pp. 34–39 (2013)
11. Manoharn, S.: A smart image processing algorithm for text recognition, information extraction and vocalisation for the visually challenged. J. Innovative Image. Process. (JIIP) **1** (1), 31–38 (2019)
12. Ghadage, S.S., Khedkar, S.K.: A review paper on automatic number plate recognition system using machine learning algorithms. Int. J. Eng. Res. Technol. (IJERT), **8**(12) (2019)
13. Kumari, R., Sharma, S.P.: A machine learning algorithm for automatic number plate recognition. Int. J. Comput. Appl. **174**(1), 6–9 (2017). (0975–8887)
14. Meghana, P., Sagar Imambi, S., Sivateja, P., Sairam, K.: Image recognition for automatic number plate surveillance. Int. J. Innovative Technol. Explor. Eng. (IJITEE) **8**(4) (2019)
15. Sannakki, S.S., Rajpurohit, V.S., Nargund, V.B.: SVM-DSD: SVM based diagnostic system for the detection of pomegranate leaf diseases. In: International Conference on Advances in Computing. Advances in Intelligent Systems and Computing, vol. 174. Springer New Delhi (2013)
16. Otsu, N.: A threshold selection method from gray-level histograms. IEEE. Trans. Syst. Man. Cybernet. **9**(1), 62–66 (1979)

Sign Language Identification Using Image Processing Techniques

Amelec Viloria[1(✉)], Evelyn Sanchez[1], and Omar Bonerge Pineda Lezama[2]

[1] Universidad de la Costa, St. 58 #66, Barranquilla, Atlántico, Colombia
aviloria7@cuc.edu.co
[2] Universidad Tecnológica Centroamericana (UNITEC), San Pedro Sula, Honduras
omarpineda@unitec.edu

Abstract. In everyday life, computers handle a large amount of data from different sources and formats such as sensors, databases, social networks, texts, etc. In addition to this process, people need to use different communication devices that enrich and facilitate human-computer interaction (HCI). As a result, there is a need to develop computational techniques that allow the search for patterns or characteristic data in images, audio waves, or electrical pulses, among others, to carry out tasks that only humans can do better so far. In this way, to improve both the User Experience (UE) and the ease of interaction with computers, various approaches to natural interaction have been proposed, including digital image processing and acquisition from various data sources such as a sensor like Kinect. In this study, the processing of images obtained from a digital camera is approached to characterize them by using basic computer vision techniques. The paper presents the development of a prototype for supporting people who speak sign language to know if the sign they are doing is correct.

Keywords: Genetic algorithm · Sign language identification · Image processing techniques

1 Introduction

Sign language, or sign language, is a natural language of speech and gestural-spatial arrangement and visual perception by which deaf people can create a means of communication with their social world, whether established by other deaf persons or by someone who knows the sign language used. Although contact with the oral language is formed in a vocal-auditory system, sign language does so in a gesture-visual-spatial system [1].

The set of symbols that make up the language includes static signs and signs with movement, as well as signs for the alphabet. In this work, the images that represent the signs of the alphabet are used, and specifically those that do not have movement for their representation, as the first stage of the project. In total there are 23 images; Fig. 1 shows the set of images used for training the neural network.

J. I.-Z. Chen et al. (Eds.): ICIPCN 2020, AISC 1200, pp. 76–86, 2021.
https://doi.org/10.1007/978-3-030-51859-2_8

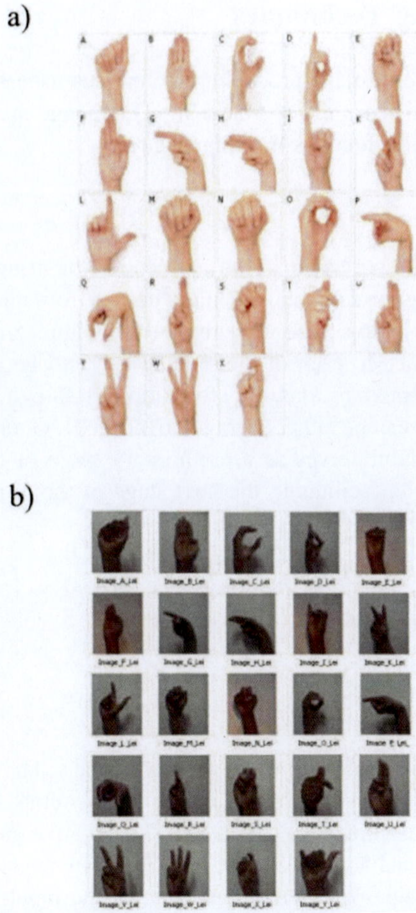

Fig. 1. a) Trained (ideal) alphabet symbols b) Recognized alphabet symbols (captured in low light) [2, 3].

This paper proposes the identification of signs derived from Spanish Sign Language by using basic algorithms for image processing such as spatial transformations (grayscale transformation, binarization, the use of thresholds, etc.) [4]. The objective proposed after using different techniques for image processing is to develop a prototype through which a person with or without Sign Language (SL) training can validate whether he or she carries out the sign correctly. According to [2, 5], this process is called "signing". On the other hand, SL is defined as that visuospatial language in which communication is generated using the hands and their positions concerning the body and the intentional gaze [6].

Thus, the structure of this paper considers an introduction, description of the image processing techniques, the design of a prototype, identification of the sign, conclusions, and references.

2 Image Processing Techniques

In this section, the different image processing techniques used for the development of the prototype that recognizes a set of Static Sign Language are presented and briefly explained. Each of these techniques is explained below.

2.1 Greyscale

The first technique to be explained is the application of a greyscale transformation filtering. To transform a greyscale image while it comes from the RGB model, it can be done in different ways. In this case, the input RGB values are averaged and equal averaging value is assigned to each of the pixels of the image.

The function represented by this transformation is shown in Eq. (1), where the transformed RGB values can be either averaged over the R, G, or B values. In this case, the grayscale transformation serves as a preliminary stage of the binarization, as all pixels are in the grayscale, facilitating the next stage of the process [7]:

$$\left(r', g', b'\right) = \begin{cases} (r, r, r), & 0 \\ (g, g, g), & 0 \\ (b, b, b). \end{cases} \tag{1}$$

2.2 Negative

Another basic filter is the transformation of negative values. The negative filter consists of subtracting the maximum value of the image (255) minus the RGB value of the pixel, which is done for each pixel of the image. The negative transformation is defined from a range of intensity values between [0, $L - 1$] and by applying Eq. (2). The result is the inversion of the values given the maximum of the input range $L-1$ minus the input pixel r [8]:

$$s = L - 1 - r. \tag{2}$$

2.3 Threshold

The threshold filter, or better known as a binary filter, allows transforming an image to binary or black and white. This filter requires the user to enter a value greater than 0 and less than 255, which allows the filter to catalog the colors to know whether to change it to white or black. It is used in this case to highlight the hand with the black color and the background in white [9]:

$$s = \begin{cases} 0 \\ L - 1 \end{cases} \tag{3}$$

$$\text{Yes}\, r \leq c, \text{Otherwise}.$$

2.4 Binary Trimming

A function called binary trimming was defined to remove the hand, which is in black. In this way, the binary clipping goes through the image looking for the black points, establishing the minimum and maximum points that make up the image to discard the rest of the background [1].

2.5 Zoom

One of the most important filters of the application is the application of digital magnification, as well as its corresponding digital diminution. The digital magnification consists in increasing the size of the image depending on a value introduced by the user. From there, an equation must be made to separate the pixels and fill them with values from a neighbor [10]. The digital decrease is similar to the increase but in its inverse function. It is used to provide height to the image that the user enters, in this case, the image size is handled at 150 pixels [11].

2.6 Rotation

The rotation function is used as a tolerance test mechanism when identifying an image. In this case, tolerance of $\pm 5°$ was considered. The rotation function is as described in Eq. (4), where is defined as in (5) [12, 13]:

$$x' = Rx + t, \tag{4}$$

$$R = \begin{bmatrix} \cos\theta & -\sin\theta \\ \sin\theta & \cos\theta \end{bmatrix}. \tag{5}$$

2.7 Closing and Opening

The opening filter allows to enlarge small spaces in the images, and it is achieved by applying the erosion filter and then dilation. The closing filter, unlike the opening one, focuses on closing the small holes that can be considered as noise in an image. This is achieved by applying the dilation filter and then the erosion filter. Thus, the application of these filters allows the removal of noise from the image, as well as the noise generated by applying various filters for treatment [14].

3 The Prototype

The development of the prototype was carried out in Java language, with a graphic interface that allowed us to see the functionality of the system and that, employing its use, the user could know if when making a certain sign it was made correctly, see Fig. 2.

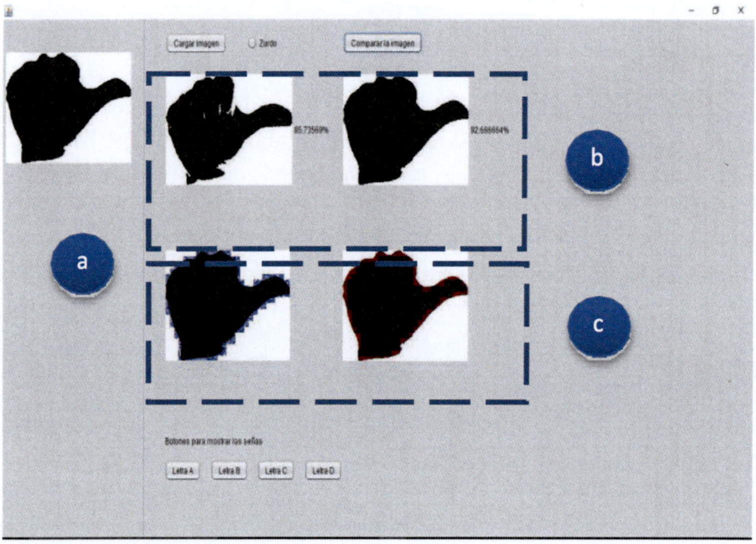

Fig. 2. Prototype screen. a) user image, b) images identified as correct, c) comparison errors.

For this version of the prototype, the images to be verified can be taken from a cell phone camera or a webcam and then saved to a.jpg or .png file that must be entered into the system for validation. The application only consists of a single window that is divided into four functionalities described in detail in the following section.

3.1 Use Cases

The functionalities that can be performed by the user in the system are limited to four activities that are denoted in the use case diagram shown in Fig. 3. Each use case is briefly explained below [15–19].

Show Signs. The purpose of this function is to show the letters that correspond to a certain password [20]. In this case, for testing purposes, the set of signs is reduced and limited to four letters of the dactylological alphabet. Thus, when the user wants to know how the letter "a" is signed, the system will show the letter "a" in area b of Fig. 4.

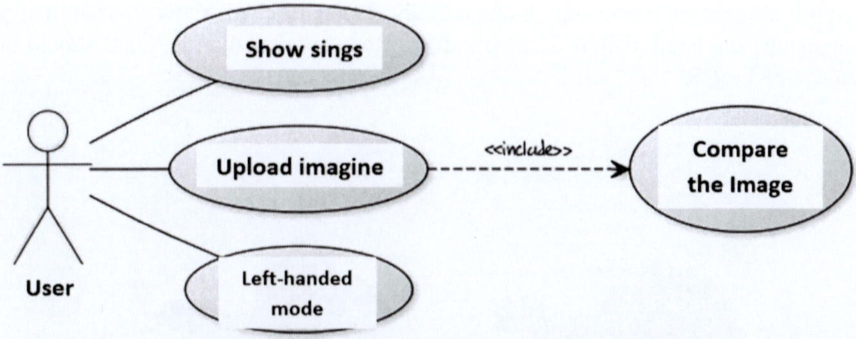

Fig. 3. Use case diagram for the developed prototype.

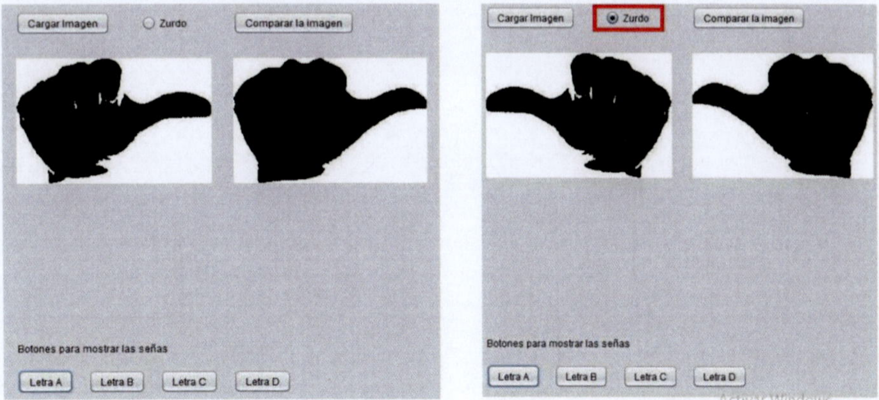

(a) (b)

Fig. 4. Application window (a) showing the letter A, (b) showing the letter A in left-handed.

Load Image. The upload image functionality, as its name indicates, allows the user to upload an image of his hand to the system by making a password. For proper operation, it is expected that the hand is visible and the background is dark. When the uploading process is performed, the different steps of the image processing described in the previous section will also be called. In particular, when loading, the image is binarized: greyscale, negative, and threshold.

Left-Handed Mode. The left-handed mode, as its name indicates, is a function that when activated applies a horizontal reflection to the images. The purpose of activating the southpaw mode is to recognize those images that have been recorded by left-handed people [21].

Compare the Image. Finally, the image comparison function performs a calculation that consists of determining the percentage of similarity between a set of signs previously registered in the system concerning the image loaded by the user [14].

In this case, when comparing the images, the application will show a red outline for the areas that the hand should occupy and blue for the areas that the hand should not occupy. See Fig. 5.

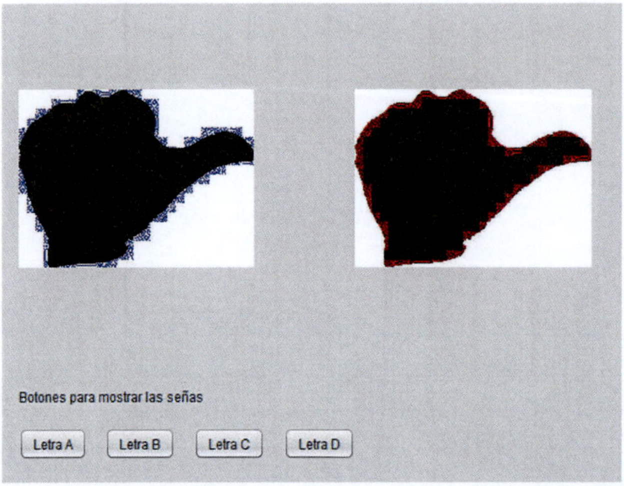

Fig. 5. Area of the application that shows the results of the image comparison

4 Identification of the Sign

The process of identifying the password is based on the comparison of two images, the first one is acquired by the user through the camera and the second one is previously stored in the system. From the two images, the system goes through them pixel by pixel comparing the tone value of these. In the case that the binary values coincide, the system will increase a counter, otherwise, the system will identify which was the expected result based on the system's image [22–24]. Thus, depending on the expected value, the system keeps a record of the excess or lack of the match. In this way, it will indicate with a red color those pixels that represent the area it should have occupied or in blue the area it should not occupy, that is to say, both are a type of natural error since it is very complex that the hand always occupies the same position even when the same person does it [25]. Once the system has finished scanning the pixels, it averages the number of pixels scanned to determine the percentage of accuracy. The explained process must be repeated to compare the user's image with all the images that the system has, and from the highest average, the user's signal is deduced. In this way, the algorithm will compare all variations to tolerate certain imprecision derived from the generation of the sign by the user.

5 Tests

To confirm that the algorithm proposed in Fig. 6 was functional for the identification of the sign, two types of tests were carried out.

```
Compare function (image A, image B)

double percentage=0 integer

counter=0 integer

numberPixel=0

From integer I=0, to I == width(A)

   From integer J=0, to J == high(A)

integer shadeA = shade(I,J) of A integer shadeB = shade(I,J) of B If
shadeA == shadeB

Counter = counter + 1 numberPixel= numberPixel + 1

      End from

End from

Percentage = counter/numberPixel Percentage = percentage*100
returns percentage

End function
```

Fig. 6. Pseudo-code of the function compares image.

In the first test, which worked as a control experiment, the previously identified image was entered into the system hoping that when it was compared with its corresponding one, it would have 100% accuracy. However, it was found that the algorithm only provided accuracies above 90% without reaching 100%. After an analysis, it was observed that the function implemented in Java generated a small noise, but this was treated by applying a threshold filter so that these errors disappeared producing a 100% certainty [26, 27].

For the second type of test, different images were used considering a normal execution of the system. In other words, new signals were captured and entered into the system to identify them against those registered. For the verification, several images were captured by applying rotation, which was considered as it was approximately between 10 and 40° [28]. In this way, groups of five images were entered for each letter, which was made by two people, for a total of twenty test images as a point of comparison and recognition [29].

As a result of the identification of the signs, it was found that the system correctly recognized sixteen of the twenty images. Among these images, it did not identify the letter a and the letter d which corresponded to the recording of the sign made by a person not registered in the system, while for the same person, even when rotations and natural variations were applied, it generated 100% accuracy. It is worth making some clarifications, the variation in the rotation is because the person who entered the image had little knowledge of Sign Language and therefore the emission of the sign presented error, while the person with whom it had better accuracy was a user who knows the SL and therefore produced fewer variations at the time of entering the sign.

6 Conclusion

In this research, image processing was addressed by applying essential techniques such as those described in Sect. 2, and a procedure was implemented to identify the signal made using the processed image as the source. In the study, techniques derived from computer vision were used to demonstrate their potential in the recognition of signs from Spanish Sign Language, particularly static signs, i.e. those that do not require continuous movement for their execution. According to the results of the test, it was found that the recognition percentage was higher than 75%. Thus, it is considered that the prototype fulfills 1) the function of processing images for digital processing, and 2) identify the images through a simple algorithm as described in Fig. 6, at the end of the paper.

Finally, as future research, it is considered to expand the image base used for testing, i.e. firstly to provide the whole set of letters of the alphabet and then to test various static signs of the sign language. In this way, a means is provided for people to practice the use of sign language on the one hand, and on the other hand, the inclusion of hearing-impaired people is promoted by generating an awareness attitude in the hearing people through the learning of sign language.

References

1. Deshpande, A.M., Kalbhor, S.R.: Video-based marathi sign language recognition and text conversion using convolutional neural network. In: Emerging Trends in Electrical, Communications, and Information Technologies, pp. 761–773. Springer, Singapore (2020)
2. Paudyal, P., Lee, J., Kamzin, A., Soudki, M., Banerjee, A., Gupta, S.K.: Learn2Sign: explainable AI for sign language learning. In: IUI Workshops (2019)
3. Naren, J., Venkatesan, R., Rajendran, P., Vasudha, G.S.: Indian sign language spelling finger recognition system. In: Smart Systems and IoT: Innovations in Computing, pp. 845–855. Springer, Singapore (2020)
4. Ibrahim, N.B., Zayed, H.H., Selim, M.M.: Advances, challenges and opportunities in continuous sign language recognition. J. Eng. Appl. Sci. **15**(5), 1205–1227 (2020)
5. Oz, C., Leu, M.C.: American Sign Language word recognition with a sensory glove using artificial neural networks. Eng. Appl. Artif. Intell. **24**(7), 1204–1213 (2011)

6. Ahuja, R., Jain, D., Sachdeva, D., Garg, A., Rajput, C.: Convolutional neural network based american sign language static hand gesture recognition. Int. J. Ambient Comput. Intell. (IJACI) **10**(3), 60–73 (2019)
7. Sombandith, V., Walairacht, A., Walairacht, S.: Hand gesture recognition for Lao alphabet sign language using HOG and correlation. In: 2017 14th International Conference on Electrical Engineering/Electronics, Computer, Telecommunications and Information Technology (ECTI-CON), pp. 649–651. IEEE (2017)
8. Pansare, J.R., Ingle, M.: Vision-based approach for American sign language recognition using edge orientation histogram. In: 2016 International Conference on Image, Vision and Computing (ICIVC), pp. 86–90. IEEE (2016)
9. Varela, N., Silva, J., Gonzalez, F.M., Palencia, P., Palma, H.H., Pineda, O.B.: Method for the recovery of images in databases of rice grains from visual content. Procedia Comput. Sci. **170**, 983–988 (2020)
10. Zafrulla, Z., Brashear, H., Starner, T., Hamilton, H., Presti, P.: American sign language recognition with the kinect. In: Proceedings of the 13th International Conference on Multimodal Interfaces, pp. 279–286 (2011)
11. Cheok, M.J., Omar, Z., Jaward, M.H.: A review of hand gesture and sign language recognition techniques. Int. J. Mach. Learn. Cybern. **10**(1), 131–153 (2019)
12. Viloria, A., Acuña, G.C., Franco, D.J.A., Hernández-Palma, H., Fuentes, J.P., Rambal, E.P.: Integration of data mining techniques to PostgreSQL database manager system. Procedia Comput. Sci. **155**, 575–580 (2019)
13. Bhavsar, H., Trivedi, D.: Indian sign language alphabets recognition from static images using correlation-coefficient algorithm with neuro-fuzzy approach, 18 May 2019
14. Starner, T., Weaver, J., Pentland, A.: Real-time american sign language recognition using desk and wearable computer-based video. IEEE Trans. Pattern Anal. Mach. Intell. **20**(12), 1371–1375 (1998)
15. Hassan, S.T., Abolarinwa, J.A., Alenoghena, C.O., Bala, S.A., David, M., Enenche, P.: Intelligent sign language recognition using image processing techniques: a case of Hausa Sign language. ATBU J. Sci. Technol. Educ. **6**(2), 127–134 (2018)
16. Dreuw, P., Stein, D., Deselaers, T., Rybach, D., Zahedi, M., Bungeroth, J., Ney, H.: Spoken language processing techniques for sign language recognition and translation. Technol. Disabil. **20**(2), 121–133 (2008)
17. Rajam, P.S., Balakrishnan, G.: Real time Indian sign language recognition system to aid deaf-dumb people. In: 2011 IEEE 13th International Conference on Communication Technology, pp. 737–742. IEEE (2011)
18. Jiang, S., Gao, Q., Liu, H., Shull, P.B.: A novel, co-located EMG-FMG-sensing wearable armband for hand gesture recognition. Sens. Actuators A: Phys. **301**, 111738 (2020)
19. Shah, T.J., Banday, M.T.: Empirical performance analysis of wavelet transform coding-based image compression techniques. In: Examining Fractal Image Processing and Analysis, pp. 57–99. IGI Global (2020)
20. Viloria, A., Lezama, O.B.P.: Improvements for determining the number of clusters in k-Means for innovation databases in SMEs. ANT/EDI40, pp. 1201–1206 (2019)
21. Munib, Q., Habeeb, M., Takruri, B., Al-Malik, H.A.: American sign language (ASL) recognition based on Hough transform and neural networks. Expert Syst. Appl. **32**(1), 24–37 (2007)
22. Jin, C.M., Omar, Z., Jaward, M.H.: A mobile application of American sign language translation via image processing algorithms. In: 2016 IEEE Region 10 Symposium (TENSYMP), pp. 104–109. IEEE (2016)
23. Jo, J., Jadidi, Z.: A high precision crack classification system using multi-layered image processing and deep belief learning. Struct. Infrastructure Eng. **16**(2), 297–305 (2020)

24. Li, K.F., Lothrop, K., Gill, E., Lau, S.: A web-based sign language translator using 3d video processing. In: 2011 14th International Conference on Network-Based Information Systems, pp. 356–361. IEEE (2011)
25. Nikam, A.S., Ambekar, A.G.: Sign language recognition using image based hand gesture recognition techniques. In: 2016 Online International Conference on Green Engineering and Technologies (IC-GET), pp. 1–5. IEEE (2016)
26. Al-Jarrah, O., Halawani, A.: Recognition of gestures in Arabic sign language using neuro-fuzzy systems. Artif. Intell. **133**(1–2), 117–138 (2001)
27. Kartheek, M.N., Prasad, M.V., Bhukya, R.: Local optimal oriented pattern for person independent facial expression recognition. In: Twelfth International Conference on Machine Vision (ICMV 2019), vol. 11433, p. 114330R. International Society for Optics and Photonics (2020)
28. Gupta, R., Rana, S., Gupta, S., Pandey, K., Dabas, C.: Comparative analysis of various classifiers for gesture recognition. In: Intelligent Computing Techniques for Smart Energy Systems, pp. 85–94. Springer, Singapore (2020)
29. Sokhib, T., Whangbo, T.K.: A combined method of skin-and depth-based hand gesture recognition. Int. Arab J. Inf. Technol. **17**(1), 137–145 (2020)

Hybrid Speckle Reduction Filter for Corneal OCT Images

H. James Deva Koresh[1(✉)] and Shanty Chacko[2]

[1] Department of Electronics and Communication Engineering, Karunya Institute of Technology and Sciences, Coimbatore, India
jamesdevakoresh@gmail.com

[2] Department of Electrical and Electronics Engineering, Karunya Institute of Technology and Sciences, Coimbatore, India
shanty@karunya.edu

Abstract. Optical Coherence Tomography (OCT) is a non-invasive imaging device helpful in capturing the layers of cornea and retina. The OCT images are supportive in understanding the health of an eye during the surgical assessment. Due to the movement of scattering waves from the OCT device, the captured images are usually disturbed with speckle noise. The paper suggests an efficient hybrid filtering technique by combining anisotropic filter, Gaussian filter and bilateral filter to reduce the speckle noise present in the corneal OCT images. The proposed technique is compared with the existing filtering techniques in terms of coefficient correlation and PSNR ratio for performance evaluation.

1 Introduction

The layers of cornea and retina are made up of several biological tissues. The corneal layer consists of five different tissues on its structure. Similarly, there are eleven various tissues are exists in the retinal layer [1–7]. The cornea is the outmost transparent layer supportive in transmitting the light waves to enter and focus over the surface of the retina. The retinal nerves further transmit the focused light waves to the brain in terms of electric signals. The abnormal thickness in the corneal layer may lead to a false focus of light waves falling towards the retina. This will cause a blurry vision and it results in refractive error (short-sightedness or long-sightedness). To correct up the blurry vision caused by refractive error, the thickness of the corneal layer is needed to be changed with the help of laser treatment. For measuring the existing corneal layer thickness and for the amount of corneal layer to be changed in laser operation is estimated from the OCT corneal images [8, 9].

The OCT device can capture the cross-section view of both retina and corneal portions by changing the capturing wavelength produced by the device [9–12]. The cross-section view of the corneal image helps identify the thickness variations, corneal edema and ocular hypertension. The OCT device is similar to an ultrasound scan but instead of sound waves, a light source is transferred into an eye. The time-delay of such light wave to travel inside an eye is calculated by the device to mark out the cross-sectional view of the retina as well as cornea. The general wavelength of the light source in the OCT device is 1310 nm for capturing the surface of the retina and the

J. I.-Z. Chen et al. (Eds.): ICIPCN 2020, AISC 1200, pp. 87–99, 2021.
https://doi.org/10.1007/978-3-030-51859-2_9

wavelength is adjusted to 820 nm for capturing the surface of the cornea [13]. There are two different kinds of OCT devices are available namely Time-Domain OCT (TDOCT) and Spectral-Domain OCT (SDOCT). The TDOCT was introduced first and now the SDOCT is replaced everywhere due to improvement in imaging speed. The thickness measurement of the cornea is not yet fully automated because of poor reliability in the existing segmentation techniques [1]. To make a reliable segmentation technique, it is must to clear out the corneal OCT image from noises.

The speckle noise present in the OCT images is because of continuous infrared wave transmission between the device and surface to be captured. The device transmits the refractive index and its corresponding waves to the surface. This will lead to an uncommon delay in the waves travels back from the surface to the device. These multiple forward and backward waves generate the speckle noise all over the image captured by the OCT device [14]. The mathematical expression for the speckle noise is:

$$f(x, y) = g(x, y) \cdot h(x, y) \tag{1}$$

Where, f(x, y) is real noise level, g(x, y) is ideal clear signal and h(x, y) is the actual speckle signal [15].

The speckle noises are needed to be cleared using some filters for further image processing purposes like image enhancement and segmentation. But removing speckle noise may lead to poor image quality and loss of information [16]. As the speckle noise is not equal in space all over the image, there are chances for a filter to remove the required information also. This will lead to a false boundary estimation on segmentation. So developing a speckle filtering technique for an OCT image is a challenging one. But still, the researches are going on to remove the speckle noises in a better way.

2 Related Work

The corneal OCT images and retinal OCT images are having some kind of speckle noises. There are some more images like Synthetic Aperture Radar (SAR), ultrasound images and several medical images are also affected by speckle noises. The literature has been carried to sort out the existing filtering techniques implemented over the speckle noises. Chen et al. (2020) proposed a generative adversarial network with the help of low pass Gaussian filters and a multi-scale average pooling function to train up their network for speckle noise reduction. The results of the proposed method give Peak Signal to Noise Ratio (PSNR) of 25.4566 (dB) [16]. Menon et al. (2020) developed a deep neural network-based speckle reduction algorithm called Gated Convolution Deconvolution Structure (GCDS) for reducing the speckle noises on retinal OCT images. The PSNR of the GCDS technique is 26.060 db on certain noise level images [17]. Pan et al. (2020) implemented a multi-scale vessel enhancement filter based on the Hessian matrix to remove the speckle noises from the retinal image before the segmentation process [18]. Migliori et al. (2020) utilized intensity thresholding and morphological operations for the de-speckling process along with that Sobel filter was taken to identify the clear view on the edges of OCT images [19].

George et al. (2019) developed an algorithm for the corneal layer segmentation from the OCT image. In that, to remove the speckle noise rotating kernel transformation filter was employed due to its efficiency and easiness [1]. Karthik et al. (2019) introduced a hybrid filter methodology for speckle noise removal. The hybrid methodology consists of a Gaussian filter along with low pass filter for noise removal and Wiener filter and Wavelet filter for reconstructing the blurred edges from the retinal OCT images [20]. Elsawy et al. (2019) developed an algorithm to segment the corneal layers with the help of optical coherence tomography images. A median filter along with average filter of different weights was implemented here to remove the speckle noise from the image. The average filter was utilized here to smoothen the edges for the segmentation process [21].

Kumar et al. (2018) geometric mean filter for removing the speckle noise from the retinal OCT image. Before applying the grayscale morphological method for segmentation, the filter was applied [22]. Sahu et al. (2018) proposed a statistical despeckling model for speckle noise reduction by combining the features of wavelet transform along with statistical modeling, Gaussianization and Bayesian estimators in retinal OCT images [23]. Choi et al. (2018) utilized an anisotropic diffusion filter for removing the noises in SAR images. The remaining speckle noises in the image are converted into addictive noise with the help of log transformation. A discrete wavelet transform is added to the image for reducing the addictive noise and guided filter was added in the final stage to preserve the edge information [24].

Rong et al. (2018) proposed a convolution neural network-based retinal image classification. At the time of training, the sparse representation method was applied here to remove the speckle noises [25]. Sui et al. (2018) employed the wavelet thresholding for speckle noise reduction in retinal OCT images [26]. Lee et al. (2018) removed the speckle noise by the median filter for segmenting the blood vessels from the human skin image captured by an OCT device [27]. Xiang et al. (2018) engaged an anisotropic filter for de-speckling the noises on the retinal OCT image. The anisotropic filter reduced the speckle noise without changing the edge line information on the OCT image. So that, the proposed technique segmented the eleven boundaries on retinal layer precisely [28].

Fang et al. (2017) introduced a deep learning-based retinal layer segmentation using a graph search algorithm to segment nine retinal boundaries. The algorithm employed median filter to clear up the speckle noise presence [29]. Ben-Cohen et al. (2017) segmented the retinal layers with convolution networks in OCT images. The network is equipped with a bilateral filter for noise reduction and Sobel edge detector along with graph search for the segmentation process. The network was able to find four retinal boundaries [30]. Roychowdhury et al. (2017) de-speckled the OCT image of diabetic macular edema with high pass filter for segmenting the seven sub-layers of the retina. Several gradient search methods were used for segmenting the layers [31].

3 Proposed Method

In the proposed scheme, there are two filters are combined for reduction of speckle noise from OCT images. The challenge in filtering the OCT image is not only removing the speckle noises. A good filtering algorithm must remove the speckles without losing the information of the edges in the image. So there are two filtering methods are combined in this work for making the edge to preserve for other segmentation processes. To preserve the edges, the bilateral filter was employed along with anisotropic diffusion filter. The anisotropic filter blocks out the speckle noises present in the OCT image. In the process of speckle noise movement, the edges of the OCT images also become to unsmoothed condition. These unsmoothed edges may result in uneven edge prediction while the image is taken for segmentation. So, the bilateral filter is taken together with the diffusion filter for leveling the edges of the cornea in the OCT image.

The proposed hybrid scheme of noise removal technique is compared with the traditional median filter, wiener filter and Gaussian filter for showing the betterment in the proposed methodology. The OCT images which are used for the proposed system are the same used for validating with the existing methodologies. The OCT images are the same images which were taken out for the segmentation work [32] carried out by Rabbani et al. (2016). The images are taken through SD-OCT device [38] and the images are kept under 15 different folders. Each of these folders is having corneal images taken from different patients. From those folders, a better image is taken for the proposed work to verify the algorithm. The following section describes the filters which were used to compare with the proposed work.

3.1 Median Filter

The median filter is the modified version of the mean filter which was used to remove the noises from an image without losing the edge information. The images like OCT and several medical images are always used to employ for the segmentation process. To give a better segmentation tool the edges in the images must be visible without the noises on its edges area. Because the noises covered over the edges may result in identifying the false edges during the segmentation. The median filter is one such a filter can able to clear out the noises from an image by preserving the edge information. The major difference of the median filter from the mean filter is, the median filter carries the information of the neighbor pixels. This will smoothen the edges of the OCT images. Figure 1a indicates the original raw OCT image with zoom in portion at where the pixel values are going to inspect. Similarly, Fig. 1b explores the values of the pixels in the inspected portion which was mentioned the zoom-in images as a blue box.

(a)

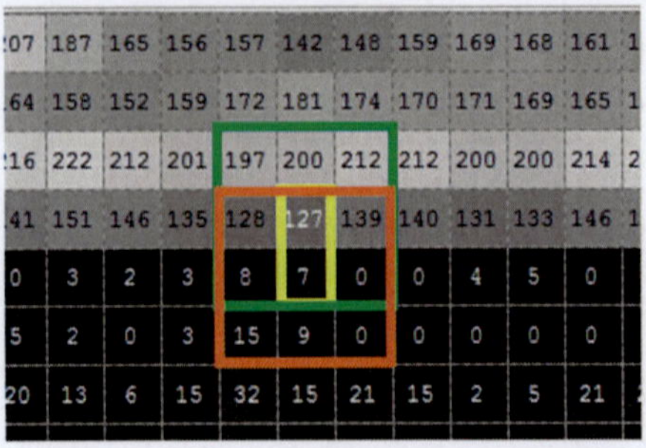

(b)

Fig. 1. (a) Raw OCT image with noise. (b) Inspected pixel values before the filter.

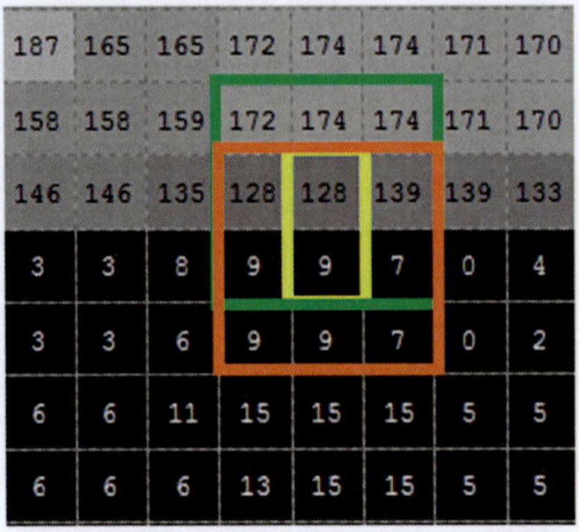

Fig. 2. Inspected pixel values after the filter.

The raw OCT image was applied with 3 × 3 modified median filter [33] and the pixel values are inspected as shown in Fig. 2. The values of the pixels are varied for the neighbor pixel $N(m, n)$. Each pixels $E(m, n)$ in the image has its information $On(m, n)$ along with noise values of $No(m, n)$ and it is represented as in the Eq. 1. The modified median filter will take the values of the neighbor pixels concerning 45° of change in all of its direction $Ne_1(m, n)$ is taken as 0-degree pixel values and it goes till $Ne_4(m, n)$ for the 135° diagonal pixels. The function taken care to the neighbor pixel is represented as in Eq. 3 to 6. The filter window size is $(2K+1) \times (2K+1)$.

$$E(m,n) = On(m,n) + No(m,n) \tag{2}$$

$$Ne_1(m,n) = [No(m,n+k), -K \leq k \leq K] \tag{3}$$

$$Ne_2(m,n) = [No(m+k,n+k), -K \leq k \leq K] \tag{4}$$

$$Ne_3(m,n) = [No(m+k,n), -K \leq k \leq K] \tag{5}$$

$$Ne_4(m,n) = [No(m+k,n+k), -K \leq k \leq K] \tag{6}$$

3.2 Wiener Filter

The Wiener filter with Stein's unbiased risk estimate (SURE) strategy [34] is implemented to the raw OCT image to see the comparison performance with the proposed technique. By reading the details of the neighbor pixels coefficients the image is de-speckled to the Wiener Shrinkage Eq. 7.

$$E_s(m,n) = \omega(m,n) \cdot E(m,n) \tag{7}$$

Where,

$$\omega(m,n) = \frac{\max\left(F(m,n) - \lambda\sigma_n^2, 0\right)}{\max(F(m,n) - \lambda\sigma_n^2, 0) + \lambda\sigma_n^2} \tag{8}$$

$\lambda\sigma_n^2$ is the variance of the noise value in the pixel and similarly λ is the adjusting parameter added to the pixel for tuning the pixel value. The $F(m,n)$ is the squared value averaging of the $E(m,n)$ as the middle pixel. The intrapixel values are expressed by integrating the thresholding function as in Eq. 9.

$$\partial(E(m,n)) = \alpha E(m,n) + \beta E_s(m,n) \tag{9}$$

Where α and β are values assumed during the testing process. The raw coefficient of the $E(m,n)$ is carried with the coefficient of $No(m,n)$ for performing the wavelet discriminating function to the high signal to noise ratio. For this, Eq. 9 is added to form the Eq. 10 as de-speckling function.

$$E^w(m,n) = \sum\nolimits_{K=1}^{4} \partial(E_k(m,n)) \cdot E_s(m,n) \tag{10}$$

3.3 Gaussian Filter

Gaussian filter is another existing technique generally used for filtering the noise from all kind of images from normal 2D image to satellite, medical and other sources of images. An efficient expanded operand decomposition [35] is combined with Gaussian filter for minimizing the level of noise in the OCT images. The one-dimensional operator for Gaussian filter expressed in Eq. 11 and for two-dimensional operator is expressed in Eq. 12.

$$G_a(x) = \frac{1}{\sqrt{2\pi\alpha}} e^{-\left(\frac{x^2}{2\alpha^2}\right)} \tag{11}$$

$$G_a(x, y) = \frac{1}{\sqrt{2\pi\alpha^2}} e^{-\left(\frac{x^2+y^2}{2\alpha^2}\right)} \tag{12}$$

The standard deviation of the Gaussian function is represented as α. The smoothness of an image is varied to the value of α. The expanded operand decomposition has binary numbers of two n bits called x and y as expressed in Eq. 13 and 14.

$$x = x_{k-1}x_{k-2}x_{k-3}\ldots x_2x_1x_0. \tag{13}$$

$$y = y_{k-1}y_{k-2}y_{k-3}\ldots y_2y_1y_0. \tag{14}$$

The expanded operands of x and y value is further compounded as three unknown values like a, b and c. The following equations are used for calculating the individual bits for the compounded parameters.

$$a_k = x_k + y_k \tag{15}$$

$$b_k = \bar{x}_k + y_k \tag{16}$$

$$c_k = x_k + \bar{y}_k \tag{17}$$

Where, + and – indicates the 'AND' and 'NOT' logic gate operations. The expanded operand performs the noise reduction process based on the three logic operations. The logarithmic function is added to the Gaussian filter for improving the betterment of the traditional Gaussian filter.

3.4 Proposed Hybrid Speckle Reduction Filter

The proposed Hybrid Speckle Reduction (HSR) filter is multiple combinations of various filters added together to perform well in reducing the noises on OCT images. An anisotropic diffusion filter is occupying the major portion in the proposed technique. Before applying the anisotropic diffusion filter to an image, it is applied with a Gaussian filter for minimizing the noise gradients over the image pixels. To the local gradients of the image, a threshold function based diffusion [36] is performed to the image after the Gaussian filter. The raw OCT image $E(m, n)$ pixel is further mentioned as $E_g(m, n)$ after the implementation of the Gaussian filter.

Then $E_g(m,n)$ is applied with a manual modification value to the pixels affected with noise and without noise. The modified value $E_{mod}(m,n)$ of a pixel can be determined by the following expression 18.

$$E_{mod}(m,n) = \left(E(m,n) - E_g(m,n)\right) \tag{18}$$

Threshold value $E_{th}(m,n)$ is applied to identify the pixels in an image which are not affected by noise. The threshold range must be as expressed in Eq. 19.

$$E_{th}(m,n) = 10\%\left(E_g(m,n)\right) \tag{19}$$

If the pixel values of $E_{mod}(m,n)$ is higher than the $E_{th}(m,n)$ then the pixel is assumed as the noise infected pixels. $E_g(m,n)$ is applied over such noise infected pixels. The algorithm for such operation is extracted below. Where, $E_o(m,n)$ are the pixels reduced the noise level using $E_g(m,n)$.

if

$$\left(E(m,n) - E_g(m,n)\right) > E_g(m,n)$$

$$E_o(m,n) = E_g(m,n);$$

else if

$$\left(E(m,n) - E_g(m,n)\right) \leq E_g(m,n)$$

$$E_o(m,n) = E(m,n);$$

When $E_o(m,n)$ is the pixel value carried to the anisotropic diffusion filter [37] as the input pixel value with the intensity of $E_i(m,n)$. The input pixel coordinates are conceded with the t iteration for estimating the variance on each pixel from its 3 × 3 neighbor pixels. The variance is expressed as in Eq. 20.

$$\rho_t^2(m,n) = \frac{1}{9}\sum\nolimits_{i=-1}^{1}\sum\nolimits_{i=1}^{1}\left[E_t(m+i,n+j) - \bar{E}_t(m,n)\right]^2 \tag{20}$$

The gray level mean is mentioned as $\bar{E}_t(m,n)$, for the change in variance is higher to the gray level in the gradient. So that the normalized gray level variance is

$$\rho_t^2 N(m,n) = 1 + \frac{\rho_t^2(m,n) - Min\ \rho_t^2}{Max\rho_t^2 - Min\ \rho_t^2} \cdot 254 \tag{21}$$

The iterated minimum and maximum variance level is given by min ρ_t^2 and max ρ_t^2. The diffusion gradient along with gray level variance feature makes the diffusion function as

$$E_{t+1}(m,n) = E_t(m,n) + \frac{1}{4}\sum\nolimits_{i=1}^{4} d(\nabla E_t^i(m,n), \rho_t^2 N(m,n) \cdot \nabla E_t^i(m,n)) \tag{22}$$

The bilateral filter is added in following the anisotropic diffusion filter for preserving the edges of the corneal OCT image. The bilateral filter acts as a weighted average filter by taking the neighbor pixel values to their sum to the distance of intensity and spatial values. The results of the bilateral filter at the pixel location of (m,n) is mathematically represented as

$$E_{bi}(m,n) = \frac{1}{K}\sum\nolimits_{n \in N(x)} e^{\frac{-\|n-m\|^2}{2\rho_d^2}} e^{\frac{-\|E(n)-E(m)\|^2}{2\rho_t^2}} \tag{23}$$

4 Results and Discussion

The performance of the proposed HSR filter is implemented to the corneal OCT images which are affected with speckle noises. The performance analysis is taken through by measuring the 2D correlation coefficient (r_2) between $E_{bi}(m,n)$ and $E(m,n)$. The 2D correlation coefficient is computed as by the following equation.

$$r_2 = \frac{\sum_m \sum_n (E_{bi}(m,n) - \overline{E_{bi}(m,n)})\left(E(m,n) - \overline{E(m,n)}\right)}{\sqrt{\left(\sum_m \sum_n \left(E_{bi}(m,n) - \overline{E_{bi}(m,n)^2}\right)\right)\left(\sum_m \sum_n (E(m,n) - \overline{E(m,n)^2})\right)}} \tag{24}$$

Similarly, Peak Signal to Noise Ratio (PSNR) is also taken for doing the performance analysis of the proposed algorithm with the existing techniques. The PSNR is tabulated for the image pixels of $E_{bi}(m,n)$ to the pixel values of $E(m,n)$ as the reference value as to give in the equation below.

$$PSNR = 10log_{10}\left[\frac{P_V^2}{MSE}\right] \tag{25}$$

Where,
P_v is the peak pixel value of an image.

$$MSE = \sum\left[\frac{[2(E_{bi}(m,n))] - [2(E(m,n))]}{\sum[(m)(n)]}\right] \tag{26}$$

The proposed algorithm and the existing algorithms are performed and verified with MATLAB 2018a software version. Figure 3 indicates the performance of the existing algorithm along with the proposed HSR algorithm.

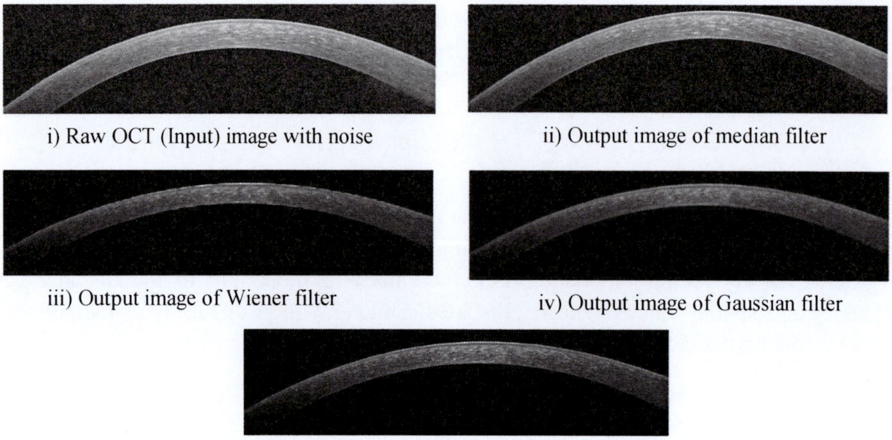

i) Raw OCT (Input) image with noise

ii) Output image of median filter

iii) Output image of Wiener filter

iv) Output image of Gaussian filter

v) Output image of the proposed HSR filter

Fig. 3. Comparison of existing filter output with the proposed HSR filter.

It shows the output of the median filter technique minimizes the speckle noise to a certain level only. In comparing to the median filter, the rest of the existing Wiener and Gaussian filters are performing better in removing the speckle noises. In terms of edge preservation, the Gaussian filter performance is notable over than the Wiener filter. But the edges of Wiener filtered OCT image is not smooth enough for undergoing the segmentation process. Similarly, the edges of the Gaussian filter is over smoothed and it results in loss of information in the edges. The proposed HSR filter overcomes such drawbacks and limitations of the existing techniques by preserving the edge information along with remarkable noise level reduction.

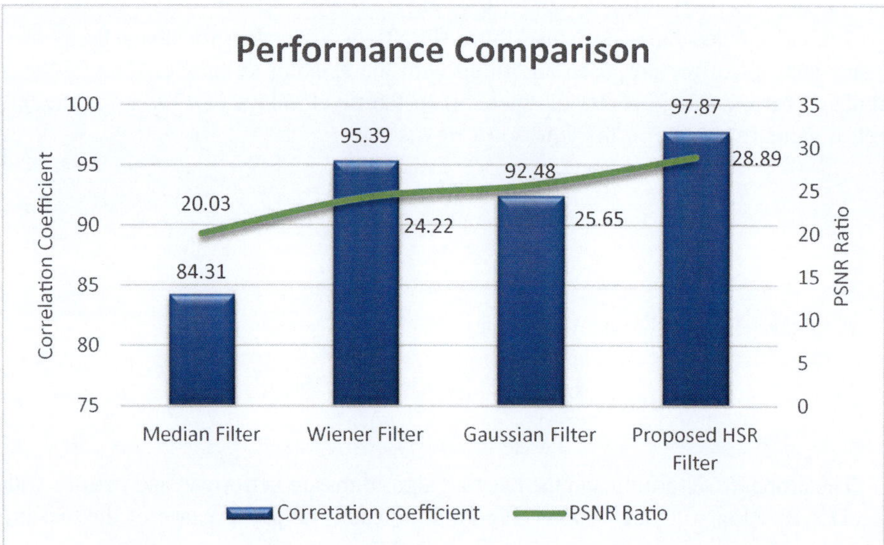

Fig. 4. Performance chart of existing filters and the proposed HSR filter

Figure 4 labels the performance chart of the 2D correlation coefficient and PSNR ratio. In both performance metrics, the proposed HSR filter performs better than the existing filters. The performance of the Wiener filter and Gaussian filter is inverted in terms of correlation coefficient and PSNR ratio.

5 Conclusion

Removal of speckle noise from corneal OCT image is a needed one for edge estimation and segmentation process for surgical design and post-surgical health estimation. Minimizing the noise without losing the edge information is a challenging one while designing an efficient filter. The proposed HSR filter blocks the speckle noises and efficiently preserves the edge information for further segmentation process. The proposed HSR filter designates a better performance over the existing filters in terms of coefficient correlation and PSNR ratio.

References

1. George, N., Jiji, C.V.: Two stage contour evolution for automatic segmentation of choroid and cornea in OCT images. Biocybern. Biomed. Eng. **39**(3), 686–696 (2019)
2. Vargas, V., Arnalich-Montiel, F., del Barrio, J.L.A.: Corneal healing. In: Corneal Regeneration, pp. 13–22. Springer, Cham (2019)
3. Pircher, N., Beer, F., Holzer, S., Gschließer, A., Donner, R., Pircher, M., Hitzenberger, C.K., Schmidinger, G., Lammer, J.: Large field of view corneal epithelium and Bowman's layer thickness maps in keratoconic and healthy eyes. Am. J. Ophthalmol. **209**, 168–177 (2020)
4. Yousaf, S., Keshel, S.H., Farzi, G.A., Momeni-Moghadam, M., Ahmadi, E.D., Asencio, I. O., Mozafari, M., Sefat, F.: Scaffolds for corneal tissue engineering. In: Handbook of Tissue Engineering Scaffolds: Volume Two, pp. 649–672. Woodhead Publishing (2019)
5. Bekalo, L., Niu, S., He, X., Li, P., Okuwobi, I.P., Yu, C., Fan, W., Yuan, S., Chen, Q.: Automated 3-D retinal layer segmentation from SD-OCT images with neurosensory retinal detachment. IEEE Access **7**, 14894–14907 (2019)
6. Nunes, A., Silva, G., Alves, C., Batista, S., Sousa, L., Castelo-Branco, M., Bernardes, R.: Textural information from the retinal nerve fibre layer in multiple sclerosis. In: 2019 IEEE 6th Portuguese Meeting on Bioengineering (ENBENG), pp. 1–4. IEEE (2019)
7. Gao, E., Shi, F., Chen, H., Chen, X.: Quantitative analysis of retinal layers' optical intensities based on optical coherence tomography. In: Retinal Optical Coherence Tomography Image Analysis, pp. 169–191. Springer, Singapore (2019)
8. Wang, D., Li, Y., Sun, M., Guo, N., Zhang, F.: Lenticule thickness accuracy and influence in predictability and stability for different refractive errors after SMILE in Chinese myopic eyes. Curr. Eye Res. **44**(1), 96–101 (2019)
9. Sella, R., Zangwill, L.M., Weinreb, R.N., Afshari, N.A.: Repeatability and reproducibility of corneal epithelial thickness mapping with spectral-domain optical coherence tomography in normal and diseased cornea eyes. Am. J. Ophthalmol. **197**, 88–97 (2019)
10. Shirazi, M.F., Wijesinghe, R.E., Ravichandran, N.K., Kim, P., Jeon, M., Kim, J.: Dual-path handheld system for cornea and retina imaging using optical coherence tomography. Opt. Rev. **24**(2), 219–225 (2017)

11. Ouanezar, S., Sandali, O., Atia, R., Temstet, C., Georgeon, C., Laroche, L., Borderie, V., Bouheraoua, N.: Contribution of Fourier-domain optical coherence tomography to the diagnosis of keratoconus progression. J. Cataract Refract. Surg. **45**(2), 159–166 (2019)
12. James Deva Koresh, H., Chacko, S.: Identification of optimum image capturing technique for corneal segmentation—a survey. In: International Conference on ISMAC in Computational Vision and Bio-Engineering, pp. 715–726. Springer, Cham (2018)
13. Acquistapace, A., Cereda, M.G., Cigada, M., Staurenghi, G., Bottoni, F.: Imaging of tangential traction types in lamellar macular holes. Graefe's Arch. Clin. Exp. Ophthalmol. **255**(12), 2331–2336 (2017)
14. Amini, Z., Kafieh, R., Rabbani, H.: Speckle noise reduction and enhancement for OCT images. In: Retinal Optical Coherence Tomography Image Analysis, pp. 39–72. Springer, Singapore (2019)
15. Song, D., Liu, Y., Lin, X., Liu, J., Tan, J.: Research and comparison of OCT image speckle denoising algorithm. In: 2019 IEEE 8th Joint International Information Technology and Artificial Intelligence Conference (ITAIC), pp. 1554–1558. IEEE (2019)
16. Chen, Z., Zeng, Z., Shen, H., Zheng, X., Dai, P., Ouyang, P.: DN-GAN: denoising generative adversarial networks for speckle noise reduction in optical coherence tomography images. Biomed. Sig. Process. Control **55**, 101632 (2020)
17. Menon, S.N., Reddy, V.B.V., Yeshwanth, A., Anoop, B.N., Rajan, J.: A novel deep learning approach for the removal of speckle noise from optical coherence tomography images using gated convolution–deconvolution structure. In: Proceedings of 3rd International Conference on Computer Vision and Image Processing, pp. 115–126. Springer, Singapore (2020)
18. Pan, L., Shi, F., Xiang, D., Yu, K., Duan, L., Zheng, J., Chen, X.: OCTRexpert: a feature-based 3D registration method for retinal OCT images. IEEE Trans. Image Process. **29**, 3885–3897 (2020)
19. Migliori, S., Chiastra, C., Bologna, M., Montin, E., Dubini, G., Genuardi, L., Aurigemma, C., Mainardi, L., Burzotta, F., Migliavacca, F.: Application of an OCT-based 3D reconstruction framework to the hemodynamic assessment of an ulcerated coronary artery plaque. Med. Eng. Phys. **78**, 74–81 (2020)
20. Karthik, K., Mahadevappa, M., Patel, K.: Removal of speckle noise in images through a hybrid methodology with application in optical coherence tomography (OCT) images. In: TENCON 2019-2019 IEEE Region 10 Conference (TENCON), pp. 1746–1751. IEEE (2019)
21. Elsawy, A., Abdel-Mottaleb, M., Sayed, I.-O., Wen, D., Roongpoovapatr, V., Eleiwa, T., Sayed, A.M., Raheem, M., Gameiro, G., Abou Shousha, M.: Automatic segmentation of corneal microlayers on optical coherence tomography images. Transl. Vis. Sci. Technol. **8** (3), 39 (2019)
22. Senthil Kumar, T., Helen Prabha, K.: Geometric mean filter with grayscale morphological method to enhance the RNFL thickness in the SD-OCT images. Multimed. Tools Appl. **77** (8), 10285–10301 (2018)
23. Sahu, S., Singh, H.V., Kumar, B., Singh, A.K.: Statistical modeling and Gaussianization procedure based de-speckling algorithm for retinal OCT images. J. Ambient Intell. Humanized Comput. **9**(3), 1–14 (2018)
24. Choi, H., Jeong, J.: Despeckling images using a preprocessing filter and discrete wavelet transform-based noise reduction techniques. IEEE Sens. J. **18**(8), 3131–3139 (2018)
25. Rong, Y., Xiang, D., Zhu, W., Yu, K., Shi, F., Fan, Z., Chen, X.: Surrogate-assisted retinal OCT image classification based on convolutional neural networks. IEEE J. Biomed. Health Inform. **23**(1), 253–263 (2018)

26. Sui, X., Ishikawa, H., Selesnick, I.W., Wollstein, G., Schuman, J.S.: Speckle noise reduction in OCT and projection images using hybrid wavelet thresholding. In: 2018 IEEE Signal Processing in Medicine and Biology Symposium (SPMB), pp. 1–6. IEEE, December 2018
27. Lee, P.-H., Chan, C.-C., Huang, S.-L., Chen, A., Chen, H.H.: Extracting blood vessels from full-field OCT data of human skin by short-time RPCA. IEEE Trans. Med. Imaging **37**(8), 1899–1909 (2018)
28. Xiang, D., Chen, G., Shi, F., Zhu, W., Liu, Q., Yuan, S., Chen, X.: Automatic retinal layer segmentation of OCT images with central serous retinopathy. IEEE J. Biomed. Health Inform. **23**(1), 283–295 (2018)
29. Fang, L., Cunefare, D., Wang, C., Guymer, R.H., Li, S., Farsiu, S.: Automatic segmentation of nine retinal layer boundaries in OCT images of non-exudative AMD patients using deep learning and graph search. Biomed. Opt. Express **8**(5), 2732–2744 (2017)
30. Ben-Cohen, A., Mark, D., Kovler, I., Zur, D., Barak, A., Iglicki, M., Soferman, R.: Retinal layers segmentation using fully convolutional network in OCT images. RSIP Vision (2017)
31. Roychowdhury, S., Koozekanani, D.D., Reinsbach, M., Parhi, K.K.: Automated OCT Segmentation for Images with DME. In: Medical Image Analysis and Informatics, pp. 85–101. CRC Press (2017)
32. Rabbani, H., Kafieh, R., Kazemian Jahromi, M., Jorjandi, S., Mehri Dehnavi, A., Hajizadeh, F., Peyman, A.: Obtaining thickness maps of corneal layers using the optimal algorithm for intracorneal layer segmentation. Int. J. Biomed. Imaging **2016**, 1–11 (2016)
33. Tang, J., Wang, Y., Cao, W., Yang, J.: Improved adaptive median filtering for structured light image denoising. In: 2019 7th International Conference on Information, Communication and Networks (ICICN), pp. 146–149. IEEE (2019)
34. Zhang, X.: An effective SURE-based wiener filter for image denoising. In: International Conference in Communications, Signal Processing, and Systems, pp. 889–895. Springer, Singapore (2016)
35. Nandan, D., Kanungo, J., Mahajan, A.: An error-efficient Gaussian filter for image processing by using the expanded operand decomposition logarithm multiplication. J. Ambient Intell. Humanized Comput. **9**(4), 1–8 (2018)
36. Xu, J., Jia, Y., Shi, Z., Pang, K.: An improved anisotropic diffusion filter with semi-adaptive threshold for edge preservation. Sig. Process. **119**, 80–91 (2016)
37. Chao, S.-M., Tsai, D.-M.: An improved anisotropic diffusion model for detail-and edge-preserving smoothing. Pattern Recogn. Lett. **31**(13), 2012–2023 (2010)
38. James Deva Koresh, H., Deva, J.: Computer vision based traffic sign sensing for smart transport. J. Innov. Image Process. (JIIP) **1**(01), 11–19 (2019)

Digital Image Restoration Using Modified Richardson-Lucy Deconvolution Algorithm

J. Jency Rubia(✉) and R. Babitha Lincy

Department of Electronics and Communication Engineering,
Sri Venkateswara College of Engineering, Sriperumbudur 602 117, India
jencyrubia@gmail.com, rblincy@gmail.com

Abstract. Since the Surveillance camera came into the market, the advanced digital image processing is getting more research importance. One of the major issues of digital photographing is motion and image blur. This can be occurring when the camera is in motion, the shape of the blurring is constant, but only the spatial extent is varying. Here, a novel algorithm is proposed for deconvoluting the image samples in the filtering process. Moreover, the proposed algorithm is successfully useful for compressed domain operations such as downsampling, translation, and filtering, etc. The recommended deconvolution algorithm produces the approximated output images that are perceptually closer to the actual image with reduced computational complexity. In this paper, the proposed multiplier less convolution algorithm is employed in the Richardson-Lucy filter for the Deblurring analysis. Also, this proposed method was compared with the existing techniques and stated that the novel multiplier less deconvolution algorithm provides better performance in terms of PSNR.

Keywords: Approximation algorithm · Linear convolution algorithm · Mean square error · PSNR · Richardson-Lucy filtering

1 Introduction

The advancement of compressed domain processing of the digital images and video stream leads to a very critical problem. But the compressed domain processing is an inevitable one since an enormous study devoted to many concrete operations with some specific applications such as filtering, downsampling, resampling, motion estimation, feature extraction, image segmentation, inverse motion compensation, volume rendering, transcoding and other [1, 2]. These applications of image processing could be realized through the convolution operation of the digital images. The convolution process can be done by two methods; they are exact algorithm and approximate algorithm [9, 10]. Most of the researches so far has been directed towards the exact algorithm, since it provides the precisely desired images [11, 12]. But on the other hand, with little attention, the approximate algorithm has been used with reduced computational hardware and without compromising the quality [4–9]. In literature, the work done by Bhaskaran et al. [3] falls under the context of the approximate algorithm for linear filtering operation which will be used for image sharpening by designing various gain factors in the Discrete Cosine Transform (DCT) domain. The core idea of

J. I.-Z. Chen et al. (Eds.): ICIPCN 2020, AISC 1200, pp. 100–112, 2021.
https://doi.org/10.1007/978-3-030-51859-2_10

[3] is, multiply each coefficient of the DCT with the selective constant to produce fewer operations of the linear filtering and hence it is computationally costless. Subsequently, this idea has been further consolidated and extended in [13], where the gain factors of the DCT will be designed optimally by the 2D filter that will be mainly used in image processing can be approximated in this work [13]. Another compressed-domain approximate algorithm was proposed by Natarajan and Bhaskaran in [14], which will be used for downsampling in the spatial domain process. In that, quantize each element of the matrices to the closing value of the power of two in the linear transformation. The multiplication of the matrices done with the help of simple shift and addition operation and this technique is implemented in Precision Architecture Reduced Instruction Set (PA-RISC) processor efficiently with a shift-and-add single operation, which is combined both shift and add operation.

In this paper, we consolidate the idea of [3, 13] and [14] and further extend their scope to general linear operations in the compressed-domain. The modified multiplier less approximation algorithm is applied in the deconvolution algorithm called Richardson-Lucy for the filtering process. The restored image is obtained with improved accuracy by using the proposed deconvolution algorithm. This paper has been organized as follows: Sect. 2 summarizes the relative works of the Multiplierless approximation algorithm and the Richardson-Lucy deconvolution. Section 3 describes the design methodology of the proposed work. Section 4 has shown the simulation results and discussed the comparative analysis of the proposed work with the existing work. Finally, Sect. 5 concludes the present research work and provides future research directions.

2 Related Works

Before introducing the proposed design, the background on the modified multiplier less approximation algorithm and the traditional Richardson-Lucy deconvolution process are given.

2.1 Multiplierless Approximation Algorithm

The digital image is processed in the form of the matrix, and various row-column operations expressed the images by using the weighted sum of the input blocks. Consider input is digital image is processed as two $N \times N$ spatial domain blocks (matrix) $x_1, x_2, \ldots$ and the respective outputs are $y_1, y_2, \ldots$,

$$y_t = \sum_k a_k x_{f(t,k)} b_k^T \tag{1}$$

Where a_k and b_k represents the fixed $N \times N$ matrices, $f(t, k)$ denotes the integer function that is based on t and k, and the superscript T signifies the matrix transposition. In the process of downsampling, a_k and b_k blocks consist of antialiasing filtering and avoid every second sample. The antialiasing filter can be performed where the rows of a_k and the columns of b_k shifted towards the convolution kernel [14, 16]. The compressed domain operation relates the DCT function, and by using the distributive property the matrix multiplication of the DCT operator is given as,

$$Y_t = \sum\nolimits_k A_k X_{f(t,k)} B_k^t \tag{2}$$

Where A_k, B_k, X_t and Y_t are the two-dimensional DCT of a_k, b_k, x_t and y_t respectively.

To facilitate the depiction of the proposed approximation algorithm, we consider the one-dimensional case [15]. Thereby we achieved two dimensions processing by merely employing the 1-D operation for every row and column [19, 21]. Hence the notation $x_1, x_2, \ldots.$ and $y_1, y_2, \ldots.$ represents the 1-D input and output signals. Similarly, the symbolization $X_1, X_2, \ldots$ and $Y_1, Y_2, \ldots$ denotes the 2-D input and output signals. So the 1-D linear relation is provided by,

$$y_t = \sum\nolimits_k a_k x_{f(t,k)} \tag{3}$$

And hence

$$Y_t = \sum\nolimits_k A_k X_{f(t,k)} \tag{4}$$

Where A_k is the 2-D DCT of the $N \times N$ matrix of a_k. The approximation is between each A_k with the product of three $N \times N$ matrices [20].

$$\widehat{A_k} = D_2 \widehat{A_k} D_1 \tag{5}$$

Where D_l and D_2 are the two diagonal matrices that do not depend on 'k' and the elements of $\widehat{A_k}$ are included in the set $S \triangleq \{0, \pm 1, \pm \frac{1}{2}, \pm \frac{1}{4}, \pm \frac{1}{8}\}$. The (i, j)th element of $\widehat{A_k}$ is represented by,

$$\widehat{A_k}(i,j) = D_2(i,j) \widehat{A_k}(i,j) D_l(i,j) \tag{6}$$

So the optimal choice of $\widehat{A_k}$ is given by quantizing the blocks of A_k to the defined set S. This rationale choice is because of the two reasons. First, it guarantees the optimization concerning both diagonal matrices produces the results that are better than any other existing results [14]. Secondly, we are likely to avoid the unity which is occurred when processing the diagonal matrices, because it will affect the quantization tables and so the compression ratio degradation [24].

The next step will be selecting the elements of the diagonal matrix D_l and D_2, this is possible when the block $\widehat{A_k}$ would be close as possible to A_k. Let us assume that,

$$R_{kl} = E X_{f(t,k)} X_{f(t,l)}^T \tag{7}$$

This relation is independent of 't' and the approximation criteria as follows,

$$
\begin{aligned}
\in^2 &= E\|Y - Y\|^2 \\
&= \operatorname{tr} E(Y - Y)(Y - Y)^T \\
&= \operatorname{tr} E\left[\sum\nolimits_k \left(A_k - \widehat{A_k}\right) X_{f(t,k)}\right]\left[\sum\nolimits_l \left(A_l - \widehat{A_k}\right) X_{f(t,l)}\right]^T \\
&= \sum\nolimits_k \sum\nolimits_l \operatorname{tr}\left(A_k - D_2 \widehat{A_k} D_1\right) R_{kl} D_2 \left(A_l - D_2 \widehat{A_k} D_1\right)^T
\end{aligned}
\tag{8}
$$

To reduce the size of the expression (8) utilizing simplicity, this modified algorithm assumes an autoregressive model in the time domain. This indicates the autocorrelation function between i th and j th samples of the 1-D input signal is given by,

$$r(i,j) = \sigma^2 \rho^{|i-j|} \tag{9}$$

In this Eq. (9), where $\sigma^2 > 0$ is a constant and $|\rho| \le 1$.

2.2 Richardson-Lucy Deconvolution Algorithm

This algorithm helps to reimpose the source image and also which belongs to an iterative method. Initially, the deconvolution process has been performed using the blurred image and the Point Spread Function (PSF). The point spread function can be estimated while the camera moves during the exposure [26]. Then the clear image with input blur is given to the deconvolution operation. The restoration of the desired image can be realized by doing several times of the deconvolution operation. The blurred image is expressed as [23],

$$I = L \otimes k + n \tag{10}$$

Where I is distorted image, L is Latent unblurred image, k denotes the point spread function, n represents some noise and $\otimes$ is known as a convolution operator. The Richardson-Lucy algorithm for the blurred image is expressed by [22],

$$L^{t+1} = \left(k * \frac{I}{L^t \otimes k}\right) L^t \tag{11}$$

Where $*$ is the correlation coefficient operator and $\otimes$ is the convolution operator, 't' denotes the number of iteration, L^t is the present iteration of the deblurred image, L^{t+1} is the next iteration of the deblurred image. By this method, the clearer image will be getting from the blurred image, but ringing artifacts present around the edge. This ringing artifact is the tiresome effect in image and video, which looks like a rippling artifact near the keen boundary of the image. This effect is led by deformation or loss of high-frequency data in the image [17, 18].

The ripping artifact effect can be reduced by the deconvolution algorithm treating the soft part and edge part differently since the ringing always appears near the smooth part of the edge. In [25], recommend the smooth part of the blurred image is the same as the deblurred image. This process is called a local prior method, and this is defined as,

$$p_l(L) = \prod_{i\in\Omega} N\left(\partial_x L_i - \partial_x I_i | 0, \sigma\right) N\left(\partial_y L_i - \partial_y I_i | 0, \sigma\right) \tag{12}$$

Where i is the position of pixels in the image, L is the latent unblurred image, I is the blurred image, Ω is the smooth region in the image.

2.3 Features and Issues of the Related Work

The conventional exact algorithm for the deconvolution process provides an enhancement in the computational complexity. But the approximation algorithm gives a considerable reduction in the computational hardware such as adder, multipliers, and delay element. Since the convolution and deconvolution process is the main source of hardware complexity, many researchers tried to reduce the algorithm procedure without affecting the quality. As a result, an improved multiplier less algorithm has emerged. By utilizing this new algorithm, many multipliers have been replaced with simple shift-and-add operators. Thereby, it reduced power consumption and area. Also, this approximation algorithm provides fast computation due to the reduction of the algorithm steps. Though it has a good capability, it had a challenging factor also. That is, it allows a negligible amount of mean square error which is a little troublesome.

Richardson-Lucy deconvolution algorithm is a popular method and simple to implement. This technique is considered effective for compressed-domain operations. But the ringing artifact effect is the only issue of using this method. To eliminate the artifact from the deconvoluted image, many modified Richardson-Lucy algorithms have been found in the literature [27–29].

3 Proposed Work

The proposed work presents a novel deconvolution algorithm is developed by combining multiplier less approximation in the Richardson-Lucy algorithm. Moreover, this proposed algorithm is implemented in the filtering process and restored the original image with improved accuracy.

3.1 Design Methodology

The design methodology of this proposed work has been illustrated in Fig. 1. This block diagram represents the proposed workflow of the research. This research is mainly used for Deblurring the image while capturing the image or video in motion. Hence this work aims to rectify the distortion and to get the desired image. Initially, the input to the filter is given as the original image, which is superimposed with the Gaussian blur to become a distorted image. Then the digital image is processed via a linear filter. In filter, the compressed domain operation has been performed such as downsampling, convolution, translation, deconvolution, etc. The deconvolution process is a very important operation in digital image processing because this computation is responsible for getting the restored original image signal. This usually requires the features of the convolution (i.e., the impulse or frequency response) to be known. The standard of the deconvolution process can be measured with the terms PSF and PSNR.

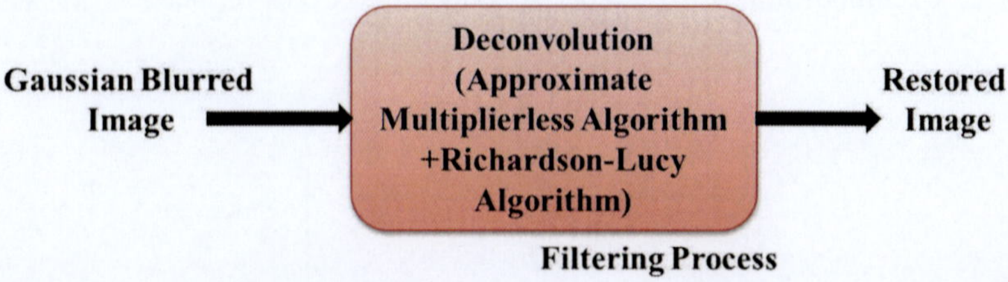

Fig. 1. Block diagram of the proposed work

The compressed domain functions in the filter have been computed by our proposed algorithm. The proposed method is applied to a multiplier less approximation technique in the popular Richardson-Lucy algorithm. The benefits of using the multiplier less approximation technique are to reduce the computational complexity without sacrificing the quality and also Richardson-Lucy algorithm is very familiar for the deconvolution process. This Richardson-Lucy algorithm used to restore the original image and also this technique belongs to an iterative method. Initially, the deconvolution process has been performed using the blurred image and the point spread function. The PSF can be estimated while the camera moves during the exposure. Then the clear image with input blur is given to the deconvolution operation. In the deconvolution process, the proposed algorithm is employed to obtain the restored image. Thus, the final output image can be restored as the input image.

3.2 Work Flow

Figure 2 represents the design flow of the proposed work. In the real-time environment, the input image is captured by the digital camera which is a blurred one due to the shaking of the camera or some other reasons. However, in the experimental environment, we purposefully add some blur manually to make the input with distortion. The sample image can be blurred using the Gaussian function which is mainly utilized for reducing the image noise. Since the Fourier transform of the Gaussian function has eliminates the high-frequency image, the Gaussian blur can be used for low pass filters. During the MATLAB simulation, the 'Gaussian blur' from the Math work has been superimposed to the input image. Our proposed system aims to restore the image as the original image with less computational complexity and improved accuracy. The restoration of the digital image from the noisy image is possible with the concept of Deconvolution. The deconvolution process is a very important operation in digital image processing because this computation is responsible for getting the restored original image signal. Deconvolution is the process of filtering a signal to compensate for an undesired convolution. The compressed domain operations utilize more hardware complexity and also it has to enhance the accuracy of the desired output data.

These limitations are tackled by using the proposed modified deconvolution algorithm. This proposed deconvolution technique comprises of multiplier less approximation algorithm and implements the algorithm in the Richardson-Lucy method. If the selection criterion does not meet, the iteration once again reaches the Extrema-based Multiscale Decomposition (EMD) block to smooth the edges further.

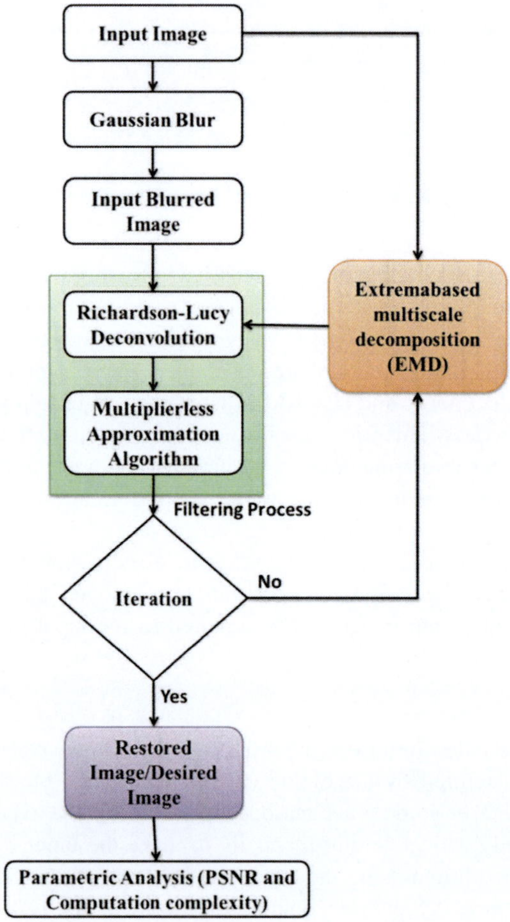

Fig. 2. Flow chart of the proposed design

The EMD considered the edge mask contains the structure of the image and not the ringing artifacts. At first, the obtained edge mask and the blurred image are given as the first input to the deconvolution process. Then the processed image analyzed by the improved Richardson-Lucy deconvolution algorithm to produce the deblurred image and it is considered as the restored image. Finally, the results have been analyzed by the parameters of computation complexity and Peak SNR.

3.3 Deconvolution

The restored image is obtained with improved accuracy by using the proposed deconvolution algorithm. Downsampling in image processing is a way of reducing the spatial resolution but does not affect the original 2-D representation of the image. This phenomenon is very convenient to reduce the storage capacity and transmission demands of images. First, the downsampling has been performed for 1-D case by the factor of two, in which every consecutive two input vectors x_{2t-1} and x_{2t} are mapped into one output vector namely $y_t = a_1 x_{2t-1} + a_2 x_{2t}$. The antialiasing filter can do the distribution uniformly the weighted averaging between each successive sample. Now, consider the filter implementation equation,

$$U = \widehat{A_1} V + \widehat{A_2} W \tag{13}$$

Where U, V and W denotes the 8-D column vectors whose i th coordinates are represented by u_i, v_i and w_i, $i = 1, 2, \ldots 8$, respectively. The elements of the $\widehat{A_1}$ and $\widehat{A_2}$ could be differ only by the sign from each other. Moreover, the implement Eq. (13) has been effectively changed by the following steps, where $z_1, \ldots\ldots z_7$ are defines to store the intermediate results temporarily.

$$u_1 = (v_1 + w_1)/2 \tag{14}$$

$$z_1 = (v_1 - w_1)/2 \tag{15}$$

$$z_2 = (v_2 + w_2)/2 \tag{16}$$

$$u_2 = z_1 + z_2/2 \tag{17}$$

$$u_3 = \left[v_2 - w_2 + \frac{w_8 - v_8}{4}\right]/2 \tag{18}$$

$$z_3 = (v_3 - w_3)/4 \tag{19}$$

$$z_4 = (v_7 - w_7)/8 \tag{20}$$

$$u_4 = z_2 + z_3 - z_4 - \left[z_1 + \frac{v_8 + w_8}{2}\right]/4 \tag{21}$$

$$u_5 = \left[v_3 + w_3 - \frac{v_7 + w_7}{2}\right]/2 \tag{22}$$

$$z_5 = (v_4 + w_4)/4 \tag{23}$$

$$z_6 = (v_6 + w_6)/4 \tag{24}$$

$$z_7 = -z_4 + (z_1 - z_2)/4 \tag{25}$$

$$u_6 = z_3 + z_5 + z_7 - z_6/2 \tag{26}$$

$$u_7 = \left[v_4 - w_4 + \frac{w_6 - v_6}{2}\right]/2 \tag{27}$$

$$u_8 = z_5 - z_6 - z_7 - z_3/2 \tag{28}$$

These equations require a total of 18 additions and 7 shift-and-adds and 6 shifts. However, in the case of 2-D, each four 8 × 8 input matrix provides one output matrix. This can be realized by repeating the implementation of (13) for each column and rows for 24 times. Thus, for the 2-D case, the consolidated computational complexity will be 500 additions, 210 shift-and-adds, and 187 shifts. That is a total of 1023 basic operations per output block.

4 Results and Discussion

The simulation result has been taken from the MATLAB Simulink software. Figure 3 represents the output waveform of the Richardson-Lucy filter response with iteration. The waveform is taken between brightness intensity and the number of pixels. This is because both the parameters are essential to viewing the image. The brightness denotes the entire lightness and the darkness of the image. Besides, every image contains an array of pixels and each pixel represents a small block of the matrix. A pixel provided, the amount of grayscale of the particular portion of the image which is represented by the integers range from 0 to 255. There are 256 combinations of grayscale of the image.

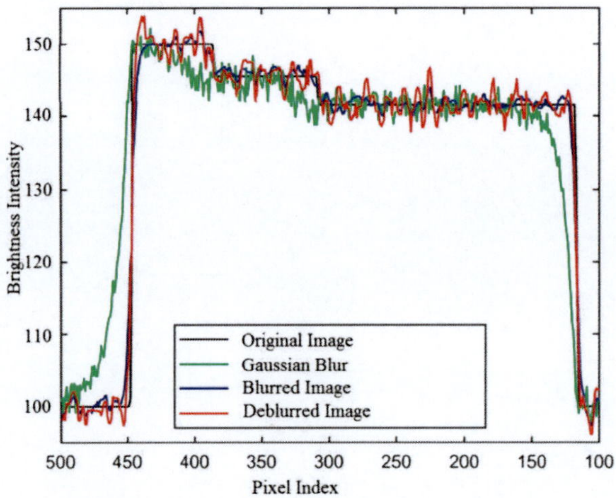

Fig. 3. Flow chart of the proposed design

The graph is consists of the original image profile, Gaussian blur noise, blurred image profile, and the deblurred image profile. Also, this graphical analysis shows that the performance of the different stage profile of the proposed algorithm. The other notable design parameter is known as PSNR since it determines the compressed or reconstructed image quality. PSNR defines the ratio of maximum signal power to the corrupted signal power, and its unit is decibels (dB).

Figure 4 consists of four different images of the proposed design, which represents the various kinds of images while computed through the proposed filtering process. An image has been given as the source image to the digital image processor that will be taken from the digital camera and this image is known as our original or source image in Fig. 4(a). For verification of the proposed logic, the Gaussian noise will be forcefully superimposed to the original image that will be shown in Fig. 4(b).

(a) (b) (c) (d)

Fig. 4. Modified Richardson-Lucy deconvolution result ımages (a) Original image (b) Gaussian blur (c) Blurred image (d) Restored image

The superimposed input signal changed as the blurred image and given to the filtering process for the compressed domain operations as depicted in Fig. 4(c). After that, the blurred image has been processed by our proposed modified deconvolution algorithm called the modified Richardson-Lucy algorithm and the obtained restored image can be shown in Fig. 4(d). Further, the mean square error of the resultant image will be 1.123, which is very less compared to other methods. The MSE is defined as the amount of squarer error between the original image and the compressed/restored image. As well, the value of PSNR will be 37 dB, and the desired value of the PSNR will be 35–45 dB. From the definition, the performance measure PSNR is high when the image looks significantly cleaner and smooth. Hence, the quality of the restored image can be evaluated by two factors, namely MSE and PSNR, and both the results are stated that the proposed filter can restore the image with good quality.

The comparative analysis of the computational complexity of the compressed domain operations has been shown in Table 1. Our proposed work has been compared with the exact deconvolution algorithm works [22] and the approximate deconvolution works [14, 21]. The goal of this research work is to reduce the computational complexity by minimizing the number of arithmetic operations. In the conventional exact deconvolution method, instead of shift-and-add, the multiplier circuit was used which is, area and power consumption. However, in the approximate algorithm, the multipliers were replaced by shift-and-add operations which is to reduce the hardware components effectively. From the comparative table, we can conclude that of all methods our proposed system provides better performance in terms of computational complexity. This comparison has been performed during the two-dimensional case of the deconvolution computation. This is realized by combining the optimization of the diagonal elements of the image matrix using the quantized operator matrices. This will enhance the quality without affecting the computational complexity.

Table 1. The comparative analysis of the computational complexity

Types of algorithms	Number of additions	Number of shift-and-adds	Number of shifts	Total no of basic operations
Exact algorithm [14]	1024	525	236	1524
Approximate algorithm [21]	992	448	192	1632
Approximate Algorithm [22]	696	320	236	1524
Proposed work	500	210	187	1023

5 Conclusion and Future Work

In this paper, the digital image has been successfully restored by using the modified Richardson-Lucy deconvolution algorithm. The modified Richardson-Lucy method has been implemented by the multiplier less approximation algorithm. This proposed method will be used for compressed domain operations such as filtering, downsampling, convolution, and deconvolution, etc. This research work has been investigated through parametric and comparative analysis. The simulation of the proposed design can be done using MATLAB Simulink software. The quality of the restored image can be evaluated by two factors namely MSE and PSNR. The MSE and PSNR of the proposed design are 1.123 and 37 dB. From the comparative analysis, we can conclude that of all existing methods, our proposed system provides better performance in terms of computational complexity.

Shortly, this work will be extended for real-time applications such as the highway vehicle counting method. Moreover, it will be interesting if the proposed work is combined with any advanced multiplication schemes for the convolution process and investigates the computational complexity and the accuracy rate. Besides that, the proposed work can be used for the autonomous optical navigation system in space organization. İn Defense, the proposed model will be used to the unmanned vehicle for moving target detection and tracking, and vehicle navigation.

References

1. Bosch, M., Zhu, F., Delp, E.J.: Spatial texture models for video compression. In: IEEE International Conference on Image Processing, TX, San Antonio, pp. I:93–I:96 (2007)
2. Assuncao, P.A.A., Ghanbari, M.: Transcoding of MPEG-2 video in the frequency domain. In: IEEE International Conference on Acoustics, Speech, and Signal Processing, Munich, vol. 4, pp. 2633–2636 (1997)
3. Bhasharan, V., Konstantinides, K., Beretta, G.: Text and image sharpening of scanned images in the JPEG domain. In: Proceedings of International Conference on Image Processing, Santa Barbara, CA, vol. 2, pp. 326–329 (1997)
4. Chang, S.F., Messerschmitt, D.G.: A new approach to decoding and compositing motion-compensated DCT-based images. In: IEEE International Conference on Acoustics, Speech, and Signal Processing, Minneapolis, MN, USA, vol. 5, pp. 421–424 (1993)
5. Chang, S.-F., Messerschmitt, D.G.: Manipulation and compositing of MC-DCT compressed video. IEEE J. Sel. Areas Commun. **13**(1), 1–11 (1995)
6. Kresch, R., Merhav, N.: Fast DCT domain filtering using the DCT and the DST. IEEE Trans. Image Process. **8**(6), 821–833 (1999)
7. Sanchez, V., Garcia, P., Peinado, A.M., Segura, J.C., Rubio, A.J.: Diagonalizing properties of the discrete cosine transforms. IEEE Trans. Sig. Process. **43**(11), 2631–2641 (1995)
8. Lee, J.B., Lee, B.G.: Transform domain filtering based on pipelining structure. IEEE Trans. Sig. Process. **40**(8), 2061–2064 (1992)
9. Cai, J., Xu, W., Yang, Y.: Large scale 2D spectral compressed sensing in continuous domain. In: IEEE International Conference on Acoustics, Speech and Signal Processing (ICASSP), New Orleans, LA, pp. 5905–5909 (2017)
10. Raghavan, V., Prabhu, K.M.M., Sommen, P.C.W.: An analysis of real-Fourier domain-based adaptive algorithms implemented with the Hartley transform using cosine-sine symmetries. IEEE Trans. Sig. Process. **53**(2), 622–629 (2005)
11. Julio, R.O., Soares, L.B., Costa, E.A.C., Bampi, S.: Energy-efficient Gaussian filter for image processing using approximate adder circuits. In: IEEE International Conference on Electronics, Circuits, and Systems (ICECS), Cairo, pp. 450–453 (2015)
12. Smith, B.C., Rowe, L.A.: Algorithms for manipulating compressed images. IEEE Comput. Graph. Appl. **13**(5), 34–42 (1993)
13. Merhav, N., Kresch, R.: Approximate convolution using DCT coefficient multipliers. IEEE Trans. Circ. Syst. Video Technol. **8**(4), 378–385 (1998)
14. Natarajan, B.K., Vasudev, B.: A fast approximate algorithm for scaling down digital images in the DCT domain. In: Proceedings of International Conference on Image Processing, Washington, DC, USA, vol. 2, pp. 241–243 (1995)
15. Rubia, Jency, Kumar, Sathish: Design and analysis of low power fixed-width multiplier using reduced precision redundancy block. IOSR J. VLSI Sig. Process. **9**(6), 01–08 (2019)
16. Jency Rubia, J., Sathish Kumar, G.A.: A survey paper on modern technologies in fixed-width multiplier. In: 4th IEEE International Conference on Signal Processing, Communications and Networking, pp. 1–6 (2017)
17. Merhav, N.: Multiplication-free approximate algorithms for compressed-domain linear operations on images. IEEE Trans. Image Process. **8**(2), 247–254 (1999)
18. Merhav, N., Bhaskaran, V.: Fast algorithms for DCT-domain image downsampling and for inverse motion compensation. IEEE Trans. Circ. Syst. Video Technol. **7**(3), 468–476 (1997)
19. Jency Rubia, J., Sathish Kumar, G.A.: FIR filter design using floating point radix 4 algorithm. Int. J. Electr. Eng. 1–13 (2019). https://doi.org/10.1177/0020720919891064

20. Jency Rubia, J., Sathish Kumar, G.A.: FIR filter design using floating point column bypassing technique. Int. J. Recent Technol. Eng. (IJRTE) **8**(2S4), 409–413 (2019)
21. Wang, Y., Feng, H., Xu, Z., Li, Q., Dai, C.: An improved Richardson–Lucy algorithm based on local prior. Opt. Laser Technol. **42**(5), 845–849 (2010)
22. Wu, J., Chang, C., Chen, C.: An improved Richardson-Lucy algorithm for single image deblurring using local extrema filtering. In: International Symposium on Intelligent Signal Processing and Communications Systems, Taipei, pp. 27–32 (2012)
23. Richardson, W.H.: Bayesian-based iterative method of image restoration. J. Opt. Soc. Am. **62**(1), 55–59 (1972)
24. Kou, W., Fjallbrant, T.: A direct computation of DCT coefficients for a signal block taken from two adjacent blocks. IEEE Trans. Sig. Process. **39**(7), 1692–1695 (1991)
25. Liu, D., Chen, X., Liu, X.: An improved Richardson-Lucy algorithm for star image deblurring. In: IEEE International Instrumentation and Measurement Technology Conference (I2MTC), Auckland, New Zealand, pp. 1–5 (2019)
26. Jency Rubia, J., Sathish Kumar, G.A.: A high speed fixed-width floating point multiplier using RLNS algorithm. Int. J. Electr. Eng. Educ. 1–9 (2018). https://doi.org/10.1177/0020720918813836
27. Yamauchi, H., Somha, W.: A filter design for blind deconvolution to decouple unknown RDF/RTN factors from complexly coupled SRAM margin variations. In: IEEE 7th Latin American Symposium on Circuits & Systems (LASCAS), Florianopolis, pp. 247–250 (2016)
28. Yamauchi, H., Somha, W.: A technique to solve issue of Richardson-Lucy deconvolution for analyzing RTN effects on SRAM margin variation. In: IEEE 5th Latin American Symposium on Circuits and Systems, Santiago, pp. 1–4 (2014)
29. Zha, Y., Huang, Y., Yang, J., Wu, J., Zhang, Y., Yang, H.: An improved Richardson-Lucy algorithm for radar angular super-resolution. In: IEEE Radar Conference, Cincinnati, pp. 0406–0410 (2014)

A Study of Electronic Health Record to Unfold Its Significance for Medical Reforms

Rugved V. Deolekar[1(✉)] and Sunil B. Wankhade[2]

[1] Vidyalankar Institute of Technology, Wadala, Mumbai, India
rugved.deolekar@vit.edu.in
[2] Rajiv Gandhi Institute of Technology, Andheri, Mumbai, India
sunil.wankhade@mctrgit.ac.in

Abstract. In today's digitized era, data is one of the most important entities. Every system generates a lot of data and even it has to deal with bulk data. The efficiency of the system depends on how the data is available and in which form the data is stored. With the advent of newer techniques of storing and processing the data, the medical domain has leaped the form of the electronic patient record (EPR) or electronic health record (EHR). This paper is based on a study of EHR which aims at highlighting its significance and emphasize its importance for medical reforms. In this paper, our sincere efforts were to introduce the concept of EHR with the help of literature that was already in place. Further, we have focussed on the proposed structure of EHR along with the different categories. Even though the EHR system has added a lot of benefits, there were many challenges in its adoption and implementation. We had tried enlisting the challenges for the EHR system and sincerely tried suggesting some measures to mitigate the same. Keeping aside the obstacles in a way of EHRs, numerous existing applications have been discussed further. Finally, we have given some novel and evolving usages of the EHR system which has increased its productivity to a greater extent. Overall our paper tries to give a complete overview of the EHR system which shall be useful in earning a basic understanding of the concept. It shall prove to be good literature for the novices working on the EHR related research.

Keywords: Adverse drug events · Clinicians · Drug malfunctioning · EHR · EPR · Healthcare · Medico-legal · Medical non-adherence · Medical records · Opioids abuse · Pandemic · Tailored medicines

1 Introduction

These days the world is facing problems due to the use of medicines or possibly we should say due to the misuse of medicines. People are using the prescribed drugs for recreational purposes or even taking an overdose of such medicines. This leads to serious health issues or life-risking problems. In the efforts of handling such problems, it was needed to keep complete track of treatment given to patients. This includes symptoms experienced by patients, diagnosis by the doctor, the treatment suggested, and finally, the medication prescribed. The concept of medical records exists from the inception of healthcare. But the earlier method was to record only the disease and the

J. I.-Z. Chen et al. (Eds.): ICIPCN 2020, AISC 1200, pp. 113–123, 2021.
https://doi.org/10.1007/978-3-030-51859-2_11

cause of the disease. Even such information was stored on a paper chart. When such records were started getting stored online, it has led the concept of electronic patient record (EPR) or electronic health record (EHR). Hence electronic health records can be defined as electronically stored patient information in an organized manner. From the inception of EHR, it took a long time for the world to understand its significance. It was the mid of the twentieth century where the governments of various developed nations started taking the medical and healthcare industry seriously [1]. This was the time where many third-party manufacturers and service providers entered the market. Governments started making various laws regarding the medical industry. Hence the era came when the process of organizing the health record became essential and the first electronic health record had taken shape [2].

The pace with which electronic health record system got adapted was very slow. There were a variety of reasons for the same including the absence of standardized structure, extensive cost of the system, fulfilling the government requirements, and the need of taking the precautions for keeping the privacy of patients [3]. Most of the developed countries realized that they have cutting edge technology for medical practice but the old traditional paperwork system for maintaining health records. So, it was clear that electronic health records can prove to be the greatest finding of the era in the health care domain. To boost the spread and use of the EHR concept, the corresponding governments had offered incentives to all healthcare providers. The incentives were given only after fulfilling the minimum set of requirement criteria instilled by the governments. It includes digitizing the order entry done by the physicians, enhancing the quality of reporting the data, and performing the orders of the medicines electronically. It must also be possible for health care providers to make access of EHR available to other health care providers [1].

2 Evolution of EHR Structure

To begin with, the need was to organize patient-related data into a digitized form in such a way that the retrieval of that data should become easy. At that point, there was no one thinking about the standards to be used to achieve uniformity. EHR was simply a collection of information electronically stored in the format specified by the provider of information. The type of data that was getting stored use to depend on the method of capturing the data employed during the record-keeping process at that medical establishment. Even though the majority of the data was uniform, but there were many data items which were location-specific, person-specific, and time-specific. The storing of this voluminous data was the next challenge. As there would be a need for retrieving this data in the future, it was supposed to be stored either in a distributed manner or at a centralized location [4].

2.1 Traditional Structure of EHR

The traditional structure of EHR had surely contained a lot of features. But there was a lot of scope for improvements. The following enlist the general observations about the traditional structure of EHR.

a) There were no efforts incurred for the design, hence it could just be visualized as the accumulation of data.
b) In case any information is required to be retrieved, then the probability of that information is present on this basic EHR is very low.
c) Even if the information were present on the basic EHR, retrieving the same used to be very challenging. As it may be on any random page which would need extensive search.
d) After finding the given information, there was a chance that some more important information may be available in the latter part of the EHR. There was a probability that such information might not be mapped with the first instance. This might result in obtaining partial or incorrect information for the given query.

This much elaboration was sufficient enough to understand that the basic structure of the EHR discussed above was insignificant. Hence, there was a need for organizing the data in a structured manner to make it usable and reach a level of quality. Along with improving the structuring of data, it was dreadfully needed to generate appropriate views according to the usages [5]. It was also required to understand the expectations of a usable EHR.

2.2 Features of EHR

The traditional EHR needed a complete reform. Figure 1 shows the block diagram of the EHR system which shows a glimpse of the various sources to provide the data into the system. It also shows some of the significant usages of the stored data of EHR. This should surely help in understanding the features of EHR.

i) EHR should be reconfigurable
With the invention of new medicines and new treatments, additions in EHR shall be inevitable. But the EHR system should be created in such a way that it becomes adaptive to frequent changes done with its structure. Hence, EHR structure modification and addition of its views should not require to completely change the current structure of the EHR. It means the system should be able to adapt to new changes or additions with minimum efforts.
ii) User interaction must be intuitive
In the preliminary phase, the patients should be able to interact with the EHR console to create his/her instance of EHR. To make this intuitive, a simple form-like application can be designed to acquire all the basic information along with the necessary critical details. EHR should be able to select the appropriate views which are needed to store the information of the corresponding patient [6].
iii) Upgradation of EHR data must be incremental
After initiating the entry in EHR during the first visit, the patient will need to upgrade the EHR during the subsequent visits. This new information of the patient must be merged with the previously filled information of that patient. The system might need to create supplementary views as a result of the addition of new treatments taken by the patients [7]. Even a separate application can be thought of for assisting patients in upgrading their EHR most appropriately.

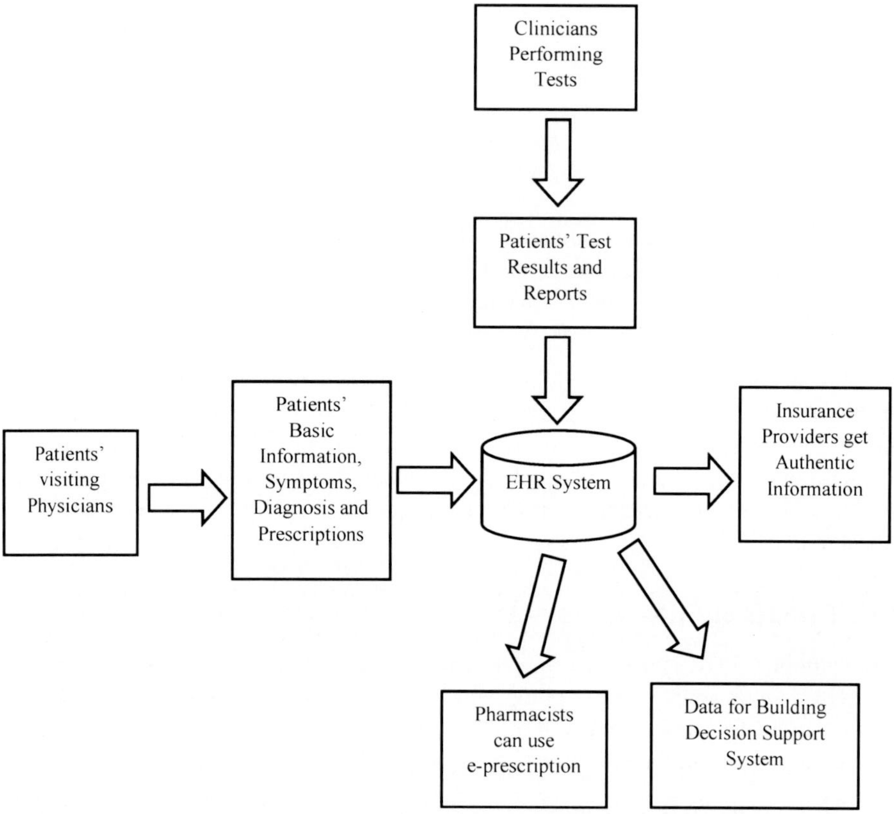

Fig. 1. Block diagram of EHR system

2.3 EHR Methodologies

The efficient EHR system can be designed by adopting an appropriate methodology. There can be two approaches to create an EHR system. One of them could be defining the EHR from the root by identifying the data which will be relevant to the system and then designing the corresponding views. The other approach could be to use a default EHR and then make it adaptable to our environment by performing necessary structural modifications. Let us discuss some of the methodologies that can be used for designing EHR.

The basic characteristic required in a general EHR methodology is portability. The methodology used must be independent of the geographical location. It must be recognized and adopted on a larger scale. The methodology must also be supporting the development of generic EHR which can be applied to different areas of health care. It must provide the strength of getting reapplied at any needed time, as the new inventions will need the addition of new data and views. The methodology must be easily understandable by an average health practitioner which will ensure the participation of general users in the development process.

EHR content can be defined by identifying User type i.e. patients, emergency patients, general physicians, emergency physicians, surgeons, cardiologists, etc. Then use types can be defined concerning every User type like the emergency patient care, patient's first visit, patient's follow-up visit, accidental cases, medico-legal cases, etc. Then each of the user types can be mapped to one or more use types. Similarly, each of use type can also be mapped to one or more User types [4]. This results in the creation of a many-to-many relational model for the EHR definition. This can be enhanced further by giving weights to each User type. High weights can indicate higher priority which may be less frequent or bearing more risk. Low weights can indicate general treatments that may be occurring more frequently and hence bearing lesser risks. Further, the weights for the various data elements may vary depending on the User and Use type mapping. Hence one of the data which might have less weight with one User type may need to have higher weight with other user types. There are other methodologies also proposed for the designing of EHR which have defined different models for data mapping. It is also being observed that the application of multiple models in a combined manner results in a better methodology. This is because it derives benefits from each of the models and hence proves to be better than applying any single model.

Now we will look at the different categories of EHR. The main criteria for categorizing EHR should be based on creator, storage, maintenance, and access [8].

2.4 Categories of EHR

i) Local EHR
Local EHR is also called as institutional EHR. Such EHR is maintained by the health-care facilities and the health-care providers. This can be stored either manually or electronically or by using the combination of both. The main advantage of such type of EHR is that it generally stores in-depth information about the patient which is collected during the patient's consecutive visits. It contains very useful information like medical diagnosis, recommendations by the doctors, etc. Such EHR can be accessed by local health officials related to the corresponding health-care providers only. This proves to be the limitation of such type of EHR.

ii) Shared EHR
Shared EHR is the concept used for the intention of helping and caring the patients over a large community. It must be able to support the exchange of extracts between two communicating facilities. Even shared EHR can also be restricted within a geographic region but comparing to local EHR the region should necessarily be very large. It can be spanning to a country or even to a global extent. This type of system may not contain detailed information about the patient's medical history. But it can contain information about the general practitioners, popular care clinics, medical institutes, the attached health professionals, and even the alternative options of doctors. This type of EHR can act supplementary to the other types of EHR. Hence this type is considered to be one of the very important types in the medical domain.

iii) Personal Health Record
In some countries, it was thought of giving partial access to EHR to the patient also. In such a case, only the information related to the patient can be accessed by the patient. This type of EHR can be considered as Personal Health Record (PHR). Some of the information on the PHR is partially entered by the corresponding patient only. The argument at old-time was whether PHR should be a subset of EHR or they should be completely separate entities. The strong thought was to have EHR and PHR designed with the same structure.

iv) Public Health EHR
Public Health EHR can be considered as just an aggregate collection of information combined from different sources like other EHRs or digitally stored information. The main objective of such public health EHRs is to monitor the health of masses in case of a pandemic. It can also be useful for research purposes as it can provide a good dataset. It proves to be very helpful for authorities to maintain health statistics which in turn helps in the development of various health-related policies.
The most suitable type of all those discussed above shall be 'Shared EHR'. Because that has got a larger spread and greater usability. But designing and maintaining shared EHRs are equally challenging. It must be longitudinal i.e. it must span lengthwise as well as it must be granular to cover significant depth in detailing [9]. The information stored must complete in all aspects and must also be persistent. A good amount of effort is needed in designing the policies regarding the privacy and security of the data stored. The far most important issue would be about providing access to EHR to the different types of users. These users must be categorized based on the level of access that should be granted to them.

3 Challenges in EHR

With a plethora of benefits of EHR in the healthcare domain, there are equally important challenges that must be identified and resolved [1]. Some of the challenges are enlisted with a brief discussion.

i) Huge Cost of EHR
The initial cost of implementation of the EHR system should be huge. Similarly, appreciable costs would also be involved in maintaining the same. After spending on setting up the EHR system, the cost would also be incurred in giving appropriate training in medical staff. This would be of utmost importance to attain a good level of usability of the EHR system.

ii) Overhead in Data Maintenance
With the implementation of the EHR system, the traditional methods of keeping records must be replaced by advanced techniques. Many a time, it increases the overall process of record-keeping that burdens the clinical staff. This can gradually lead to creating a confrontation with the EHR system and the urge to revert to the traditional system. The incapability of implementing a newer process of documentation can lead to a significant drop in the overall efficiency of the EHR system.

iii) Declining Speed of the System
To handle the load of the EHR system, the underlying network must be strong. It should be able to provide the necessary speed for the smooth functioning of EHR based communication. If the network could not handle the overhead caused then it may lead to a system crash. This will hamper the ongoing process of updating EHR also. Hence, sufficient arrangements of backup systems and replication must be implemented.

iv) Security of EHR System
The electronically stored information is considered to be easy targets for illegal access. In addition to that, EHRs contain in-depth personal information of the patients which can be easily plagued by hackers. The other medical information of the patients can also be easily misused by the intruders. This is a horrible challenge that must be addressed with the appropriate intervention by the authorities. This is important to safeguard the trust of the masses in the EHR system.

v) Alignment with Government Policies
As discussed above, to nurture the use of EHRs and to uphold the security measures, government authorities must dictate a lot of rules and policies. One bigger challenge would be to follow all these policies and adapt them by doing necessary changes in the EHR system.

3.1 Suggestions for Mitigating Challenges

For developing an efficient EHR system, the above-mentioned challenges must be addressed and mitigated. The various funding agencies can help in raising funds for the development of EHRs. The medical staff must be convinced about the overall benefits of EHRs which should make them bear the extra burden of change in processes. This must be handled with proper brainstorming sessions by the experts.

As discussed, the implementation of EHR would burden the current network. So, it is expected that a strong foundation of the network must be devoted to the implementation of EHR. Considering the possible scaling of the data and communication between the users, the expected bandwidth can be calculated. The arrangements for the same can be made as a part of the initial plan.

As far as security is concerned, there has to be a trade-off between security and access. Keeping a high level of security can make the EHR inaccessible to most of its stakeholders. But an appropriate level of security standards must be employed for successful implementation. Even the extraction of time expression from the patients' test records will significantly improve the accuracy of the stored data. It will also help in arranging the records chronologically improving their usability [10].

Government authorities make strict laws and policies to restrict malpractices. But authorities must consider the development of the EHR system as a revolution in the medical domain and must act supportive to create a conducive environment. All these shall aid in the successful implementation and spread of the EHR system.

4 Applications of EHR

In the traditional system, when EHR was not invented; the medical staff still used to complete all their routine processes. But with the inception of the EHR system, the same processes had become efficient with the consumption of less human efforts. Hence it is important to brief about different areas getting benefitted by the use of EHR.

1) For Administrative Purpose
   The EHR contains patient registration which includes detailed information about the patient. It contains the name of the patient, age, gender, contact information, address, and other allied information. Along with this, it also contains the patient's symptoms and complaints which should also be useful information from an administration point of view. The insurance-related information available with EHR should smoothen up the claim process for both the hospital staff and the patient.
2) Clinical Applications
   One of the important utilities which prove as an essential requirement for the EHR system is Computerized Physician Order Entry (CPOE). This is the application which physicians use to prescribe medicines online, even suggest tests to the patients online without using paper. It can also be used to order radiology, laboratory, and other allied services. It highly improves the accuracy of the process as the health care staff knows what all tests to be performed.
   Any existing systems for laboratory records and radiology records can also be integrated into the EHR system. This should help in storing the tests prescription, test results, expert comments, generated images along with their analysis directly into the EHR system.
   Other clinical documentation can also be the part of the EHR system which includes the various remarks added by doctors, nurses, RMOs, and other medical practitioners. This contains a lot of vital components like doctors' assessments, signs showed by the patients, medication followed, tests performed, and discharge summary.
3) Digitized Medicinal Records
   The medical shops or pharmaceutical establishments can also get benefitted by connecting to the EHR system. This should enable tracking of medication applied to a patient, medical allergies, drug malfunction related problems. Digital medicine prescription should also eradicate the errors that used to cause due to the untidy handwriting of the medical practitioner. EHR system should make the digital prescription given to the patient directly available to the corresponding medical shop.

5 Use of EHR in Medical Reforms

Along with these general applications of EHRs discussed so far, many new evolving applications are acting as solutions to some of the existing problems. Slowly and steadily, the medical fraternity has started understanding that this plethora of data can be put to more significant use. To aid this cause, various data processing mechanism had also evolved over the years. These mechanisms namely machine learning and deep learning have shown how to extract information from the raw data and use the same for decision support.

1) Medico-Legal Evidence
There are certain medical cases of injuries or illnesses where investigation by the legal authorities is essential. This is required may be to fix up criminal responsibility as per the law and regulations of the country. EHR can greatly help in such cases. Through the EHR system, the authorities can not only see the detailed information about the current illness or injuries but also can have full access to the complete history of the patient. This will surely aid in bringing more clarity towards the case.
2) Personalized Medicines
It is the concept of providing tailored medicines to the patients by analyzing their medical history, genetics, illness patterns, inheritance, allergies, and the overall lifestyle. Most of these will be easily possible by the use of the EHR system. Data processing tools like machine learning and deep learning can be used to analyze the bunch of information available through EHR and the same can be used for meaningful interpretations [11].
3) Prediction of Medical Non-adherence
The extent to which each patient religiously follows the treatment prescribed by the doctor may not be the same. It is very much intuitive to understand that the available data can easily be used to study patients with medical non-adherence. The various patterns of such patients can be identified which will help in understanding the various factors of non-adherence. Some of the factors are patient's age, the complexity of the treatment, low income, lack of education, patient's mental health, etc. This will surely help in the future for tracking such non-adherence by the patients in advance [12].
4) Adverse Drug Event Prediction
Generally, patients tend to dutifully follow the medication prescribed by the physician. But still, there are some cases where the prescribed drug has caused a lot of problems for the patient. One of the simple reasons might be the patient having allergies of such kind drugs. There might be some problems with that drug that are identified post-approval. Such problems might have caused an adverse drug event (ADE). EHR system shall greatly help in making another system for analyzing ADE. The huge data sets of patient diagnosis and medication will prove to be very useful for the prediction of ADE [13].
5) Tracking Opioid Abuse
Opioids can be termed as general medicines used as painkillers or pain-reliever. But many patients can experience adverse effects due to these medicines. These effects range from simple nausea to life-threatening diseases. The continued use of opioids may result in long-lasting back pain, severe pain, and osteoarthritis. EHRs provide a wealth of patient information that could aid in the identification of long-term opioid use [14]. Some of the patients use opioids for recreational purposes which also can result in life-risking scenarios. Such patients do symptoms faking at the emergency clinic to get such opioids prescribed by the doctors. EHR can help in providing a decision support system for the identification of opioid abuse. The patient's EHR can be used to analyze the patient's medical records. This should result in

determining the set of symptoms the patient is likely to have. These symptoms can be compared with the symptoms given by the patients at the emergency clinic. The level of similarity shall aid the doctor in deciding whether the patient is faking the symptoms or not. The variety of deep learning neural networks can be analyzed to decide the one that can be best suited for the purpose [15].

6 Conclusion

After shedding light on the concept and structure of the EHR system, we have seen the various types of EHRs that can be designed along with their significance. We have even seen the major challenges to the EHR system and our proposed suggestion to mitigate the same. Towards the end, we have elaborated on the existing applications of the EHR system. To conclude, we have even listed the various evolving applications of the EHR system which will help for medical reforms. Each one of these is still getting refined with a lot of research and new findings. The advanced techniques like machine learning are proving to be very efficient for processing of data. Further advancement in the form of deep learning is aiding the analysis of patients' test results which are in the form of images.

EHR system is the future of the medical domain. The fact cannot be denied that the concept of an EHR system has not percolated yet in all parts of the world. Some of the developed countries had adopted this system and shown a significant improvement in the efficiency of the overall processes. The benefits enjoyed by those countries should prove to be the inspiration for the other countries. The challenges like unifying the format of the EHR system worldwide is going to be the future challenge once the use of EHR becomes prevalent.

Acknowledgments. We are thankful to all those who helped us make this paper, for the valuable information provided by them in their respective fields. We are grateful for their co-operation during the writing of the paper.

References

1. Seymour, T., Frantsvog, D., Graeber, T.: Electronic Health Record (EHR). Am. J. Health Sci. **3**(3), 201–210 (2012)
2. Sun, H., Depraetere, K., Roo, J.D., Mels, G., Vloed, B.D., Twagirumukiza, M., Colaert, D.: Semantic processing of EHR data for clinical research. J. Biomed. Inform. **58**, 247–259 (2015)
3. Gazzarata, G., Gazzarata, R., Giacomini, M.: A standardised SOA based solution to guarantee the secure access to EHR Science Direct. Procedia Comput. Sci. **64**, 1124–1129 (2015)
4. Covvey, H.D., Zitner, D., Berry, D.M., Cowan, D.D., Shepherd, M.: Formal structure for specifying the content and quality of the electronic health record. In: 11th IEEE International Requirements Engineering Conference, pp. 162–168. IEEE (2003)

5. Jillian, B.D.N., Carranza, K.A.L.R., Lin, L.M.S., Ponce, A.R., Reyes, W.R.O., Bugtai, N.T., Baldovino, R.G.: Design of a web-based and electronic health record management system for medical teleconsultation. In: IEEE 10th International Conference on Humanoid, Nanotechnology, Information Technology, Communication and Control, Environment and Management (HNICEM), pp. 1–5. IEEE (2018)
6. Zhang, J., Walji, M.F.: TURF: toward a unified framework of EHR usability. Sci. Direct, J. Biomed. Inform. **44**, 1056–1067 (2011)
7. Konnoju, P.K., Sridhar K.V., Prasad, K.S.R.: A new paradigm of electronic health record for efficient implementation of healthcare delivery. In: IEEE EMBS Conference on Biomedical Engineering and Sciences (IECBES), pp. 352–354. IEEE (2010)
8. Kwak, Y.S.: International standards for building electronic health record (EHR). In: 7th International Workshop on Enterprise Networking and Computing in Healthcare Industry, HEALTHCOM, pp. 18–23. IEEE (2005)
9. Zhang, J., Kowsari, K., Harrison, J.H., Lobo, J.M., Barnes, L.E.: Patient2Vec: a personalized interpretable deep representation of the longitudinal electronic health record. IEEE Access **6**, 65333–65346 (2018)
10. Najafabadipour, M., Zanin, M., Rodríguez-González, A., Gonzalo-Martín, C., García, B.N., Calvo, V., Bermudez, J.L.C., Provencio, M., Menasalvas, E.: Recognition of time expressions in Spanish electronic health records. In: IEEE 32nd International Symposium on Computer-Based Medical Systems (CBMS), pp. 69–74. IEEE (2019)
11. Zhang, S., Bamakan, S.M.H., Qu, Q., Li, S.: Learning for personalised medicine: a comprehensive review from deep learning perspective. IEEE Rev. Biomed. Eng. **12**, 194–208 (2019)
12. Janssoone, T., Bic, C., Kanoun, D., Rinder, P., Hornus, P.: Machine learning on electronic health records: models and features usages to predict medication non-adherence. https://arxiv.org/abs/1811.12234v1 [cs.LG] (2018)
13. Rebane, J., Karlsson, I., Papapetrou, P.: An investigation of interpretable deep learning for adverse drug event prediction. In: IEEE 32nd International Symposium on Computer-Based Medical Systems (CBMS), pp. 337–342. IEEE (2019)
14. Che, Z., St. Sauver, J., Liu, H., Liu, Y.: Deep learning solutions for classifying patients on opioid use. In: AMIA Annual Symposium Proceedings, pp. 525–534 (2017)
15. Bashar, A.: Survey on evolving deep learning neural network architectures. J. Artif. Intell. Capsule Netw. **1**(02), 73–82 (2019)

Texture Analysis in Skull Magnetic Resonance Imaging

Amelec Viloria[1(✉)], Ethel de la Hoz[1], and Omar Bonerge Pineda Lezama[2]

[1] Universidad de la Costa, St. 58 #66, Barranquilla, Atlántico, Colombia
{aviloria7, edelahoz3}@cuc.edu.co
[2] Universidad Tecnológica Centroamericana (UNITEC), San Pedro Sula, Honduras
omarpineda@unitec.edu

Abstract. Traumatic brain injury (TBI) represents a serious public health problem worldwide. It is the most common cause of death and disability in the young population (aged 15–45 years), with major family, social and economic implications [1]. In medical terms, the human body can be studied as an object. The reconstruction of bone structures after physical damage generated by such an unfortunate event as disease or trauma can range from the implementation of prostheses to the engineering of artificial bone implants [2]. To make a virtual or physical model of any human anatomy, it must first be captured in three dimensions in a way that can be used by computational processes. Most hospital scanners capture data from the entire body both internally and externally. These machines are typically medical imaging devices capable of scanning the entire human body, among which, the most common is the magnetic resonance imaging (MRI) equipment [3]. The goal of the research is to analyze the texture in magnetic resonance imaging and its relationship to bone mineral content (BMC) using simple linear regression.

Keywords: Genetic algorithm · Texture analysis · Skull magnetic resonance imaging

1 Introduction

Traumatic brain injury (TBI) represents a serious public health problem worldwide. It is the most common cause of death and disability in young people (aged 15–45), with major family, social and economic implications [4]. In medical terms, the human body can be studied as an object. The reconstruction of bone structures after physical damage caused by an unfortunate event such as disease or trauma can range from the implementation of prostheses to the engineering of artificial bone implants [5].

To make a virtual or physical model of any human anatomy, the image must first be captured in three dimensions in a way that can be used by computational processes. Most hospital scanners capture data from the entire body both internally and externally. These machines are typically medical imaging machines capable of scanning the entire human body, and the most common equipment is magnetic resonance imaging (MRI) [6].

J. I.-Z. Chen et al. (Eds.): ICIPCN 2020, AISC 1200, pp. 124–134, 2021.
https://doi.org/10.1007/978-3-030-51859-2_12

MRI is a unique and powerful tool for medical diagnosis, as it is a non-invasive technique that allows the visualization of soft tissues. Diagnostic information is often included in the image texture, so higher-order image statistics must be taken into account. Quantification of texture, i.e. its description by precisely defined parameters (characteristics), is necessary to extract information on tissue properties. The numerical values of the texture parameters can be used for the classification of different regions in the image, e.g. representing tissues of different origin or normal and abnormal tissues of a given type. Changes of properly selected texture parameters over time can quantitatively reflect changes in the physical structure of the tissue [7].

Bone mineral density (BMD), on the other hand, determines the amount of calcium and other types of minerals present in an area of the bone, which can detect osteoporosis and predict the risk of bone fractures [8]. The calculation of density is carried out by double energy radiological densitometry (DXA) and is determined through a mathematical process that begins with the differentiation of the bone tissue concerning the soft tissues – differential of the low and high energy beam capture – determination of the explored area (cm^2), determination of the mineral content (CMO, g of HA) and with the quotient of both. The density per unit area (BMD, g of HA/cm^2) is obtained in each subsector of the explored bone region. Once the BMD is obtained in a given patient, it must be considered concerning the values of its control population, either concerning the peak bone mass of the young healthy population (T score) or concerning its age group and sex (Z score). In both cases, the BMD value is transformed into standard deviations from the population mean value (Fig. 1) [9].

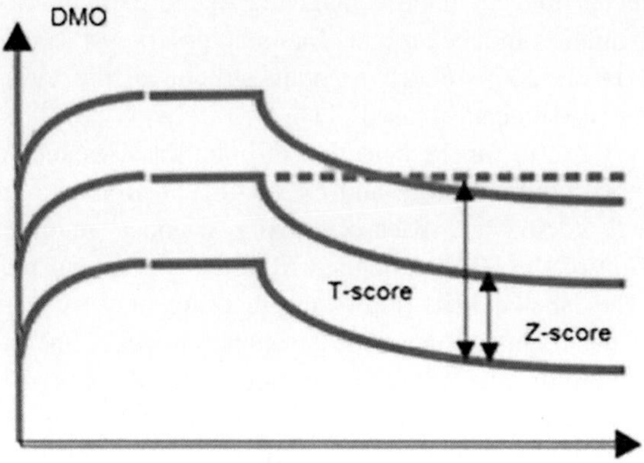

Fig. 1. Graphical representation of the calculation of T and Z scores depending on the age of the patient.

The fundamentals of the co-occurrence matrix are presented in Fig. 2. Figures 2a and 2b show three levels of gray, numerical gray levels-transition from one to two levels of gray and assigned in the corresponding position in the co-occurrence matrix, respectively.

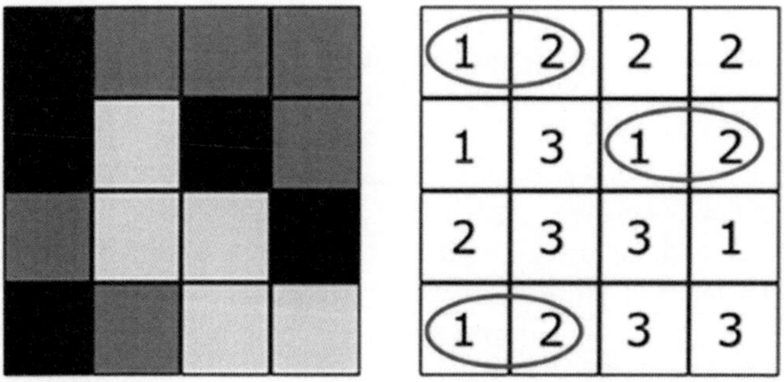

Fig. 2. Calculation of a co-occurrence matrix for a given 4 × 4 pixel image.

2 Background

The imaging technique called magnetic resonance imaging (MRI) is a very important medical imaging modality, from a clinical point of view, due to the exceptional contrast exhibited by the images generated by it. In contrast, this imaging technique has as main disadvantages of its high cost and the inability to access the patient during the acquisition process. MRI exploits the existence of an induced magnetic field in the patient's body. Materials that have an odd number of protons exhibit a nuclear magnetic moment which, despite being very weak, is observable [10].

Normally, this type of moment is randomly oriented, however, when the referred materials are subjected to very intense magnetic fields, their moments tend to align according to an orientation induced by the dominant polarity of said fields. When MRI is used to extract information related to brain structures, the term Brain Magnetic Resonance Imaging (BMR) can be used [11].

The BMR is very useful for the detection of brain diseases such as hydrocephalus (communicating, non-communicating and ex vacuo), aneurysms, cysts, parenchymal lesions, perilesional edema in space-occupying lesions, among others. Another important application of the BMR is tagged MRI and functional magnetic resonance imaging (fMRI). The labeled MRI allows the marking of various planes containing specific regions of the brain, on magnetic resonance images. For this, to disturb the magnetization of the protons present in one or more areas of the brain tissue, a selective radiofrequency pulse is applied before the acquisition process [10]. A resonance image is obtained, in a plane that is orthogonal to a particular labeled plane, before all protons regain magnetization. After a time T, the disturbed protons they recover the original magnetization and generate a radiofrequency pulse that is detected by the acquisition team, which produces a labeled image of the analyzed structure. The displacement that occurs between the resonance image and the labeled image is reflected as a set of stripes or areas of different intensities, in the image provided by this technique. The number, position, thickness, and intensity of these zones depend both on the angle of the radio frequency pulse and on the intensity and direction of the applied magnetic field [11, 12].

Thus, a current MRI is capable of producing brain images made up of a variable number and, generally, a large number of sections (layers) with a spatial resolution, usually 256 × 256 pixels and with a depth, relative to the levels gray, 16-bit. By applying methods of reconstruction of the transaxial planes, obtained during the acquisition process, the system can generate, for example, a volumetric representation of the brain structures under study [13, 14].

However, the generation of a large number (of high spatial resolution images) results in clinical experts working with brain structures having to develop manual segmentation processes that become cumbersome tasks and, additionally, have the disadvantage of being dependent operator, so, in general, such segmentations should be restricted to those situations in which they can be developed in a way simple and robust [15].

The development of automatic segmentation methods has the potential to substantially reduce the time taken by some medical procedures, which could thus be carried out with greater effectiveness and less risk, for example, the planning of surgeries or complex treatments such as radiotherapy, among others [16]. For this reason, to overcome the problems involved in manual segmentation processes, and with the purpose to detect certain pathologies, some methodologies have been proposed for the segmentation of the most important brain structures. Thus, [19], they propose a technique based on normalization processes, gray level profiles, filtering (based on anisotropic diffusion and sigmoidal functions), co-occurrence matrices, and textures for the two-dimensional and automatic segmentation of the cerebral ventricles in computed tomography (CT) images.

They consider 30 layers of data from healthy subjects reporting excellent values for the metrics they considered, including the comparison of volumes occupied by manual segmentations, generated by an expert neurologist, and the automatic segmentations generated by the method they present. Also, [17] consider magnetic resonance imaging to automatically segment the cerebral ventricles using a methodology based on thresholding + growth of regions. These authors present the Dice coefficient as a metric, reporting values higher than 0.90 [18].

Additionally, ****Bhanu et al. 9, detect intra and extra ventricular effusions in skull images acquired through the CT modality using a technique based on regularized distance level set methods. These researchers report a correspondence of automatic and manual volumes greater than 0.9. All these antecedents are directed towards the detection of various types of diseases that directly or indirectly affect the functioning of the human brain [19].

Furthermore, this work is an extension of [20]. The main contributions are a) Use of a grouping technique, based on the growth of regions, for the efficient segmentation of brain structures. b) Automatic segmentation of certain structures linked to the cerebral ventricles system, without the use of insulating planes.

3 Methods

Images of 5 anonymous patients between 44 and 69 years of age were used, who was provided by the Medellin mayor's office in Colombia. The Brivo MR355 1.5T Inspire digital magnetic resonance imaging equipment was used to perform the study. Five magnetic resonance imaging sequences were obtained from the skull of five patients with known bone mineral density (BMD) and T-score, four men (H1, H2, H3, and H4) and one woman (M1); in standard digital format in Digital Imaging and Communications in Medicine (DICOM), which is the international standard for transmitting, storing, retrieving, printing, processing information and viewing medical images. Table 1 shows the information on each patient.

Table 1. Gender, age, and bone mineral density of each patient.

Patient key	Gender	Age	DMO (g of HA/cm^2)	T-score
H1	Male	45	1.033	0.032
H2	Male	46	1.214	0.758
H3	Male	47	0.901	−0.769
H4	Male	67	0.914	−0.555
M1	Female	50	0.902	−0.914

Once the MRI scans were recovered in DICOM format, the transformation of the sequences to a MaZda compatible format (bitmap 256×256) was performed through the ImageJ software, which is a Java image processing program in the public domain.

Subsequently, the images corresponding to the central axial section of the skull of each of the patients were displayed and selected, and each of the histograms was obtained to compare them and subsequently analyze the properties corresponding to the segmented image of the bone. Once the information from the selected images was validated, the regions of interest (ROIs) of the bone structures were chosen using the MaZda package. The name of the program is an acronym derived from 'Macierz Zdarzen' which is the Polish counterpart of the English term 'co-occurrence matrix' (one of the most popular techniques for extracting texture characteristics). This software was developed during the European project named COST B11 "Quantitative Analysis of Magnetic Resonance Imaging" [21].

By selecting the region of interest (ROI) or segmentation, which is the separation in an image of the different structures or tissues, the quantitative analysis of the texture parameters of the skull regions of each sequence and through a linear regression analysis using the Minitab statistical package was performed.

4 Results

The intensity of the grey levels of the axial magnetic resonance images corresponding to the central slices of the skull of each of the patients was measured to compare each of the histograms (see Fig. 3).

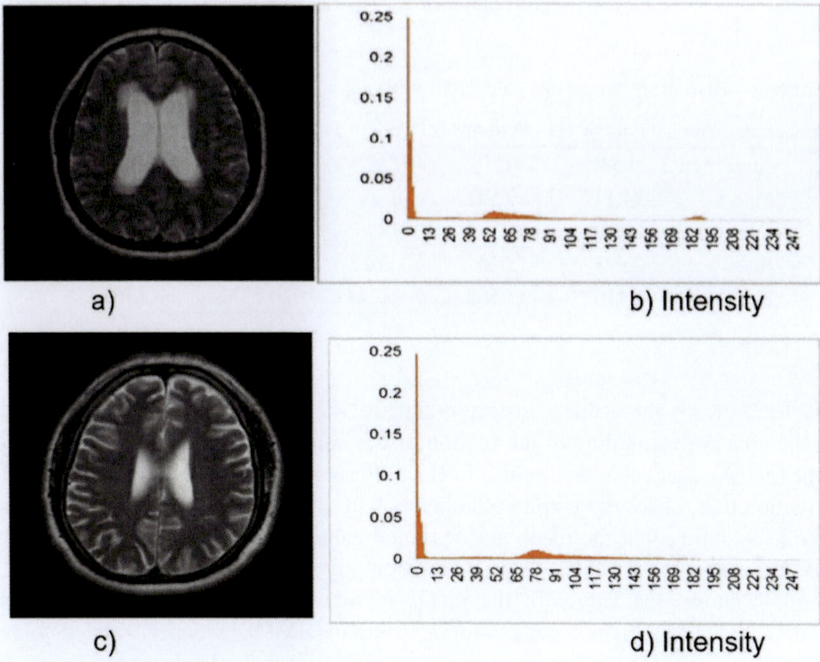

Fig. 3. Images of two patients and their respective histograms.

Figure 3 shows the images of two patients and their respective histograms, where the distribution of the intensity of the grey levels and the similarity between both distributions can be seen. Figure 3a shows the axial slice corresponding to the slice 14 enhanced in T2 of patient H1, with repetition time and spin-echo time RT/ET 6000 ms/85.6 ms. Figure 3b shows the histogram of grey level intensities for patient H1. Figure 3c corresponds to the image of the axial section of the 14th slice enhanced in T2 of the skull of patient H2, with RT: repetition time and ET: spin-echo time RT/ET 6000 ms/85.7 ms. Figure 3d corresponds to the histogram of grey level intensities of the image of patient H2.

The regions of interest (ROI) corresponding to the bone in each of the selected images were then selected and the texture parameters of each patient's skull were measured using the MaZda package. Figure 4 shows the segmentation of each patient's bone. Table 2 shows the texture parameters corresponding to the histogram of the standardized gray levels $\pm 3\sigma$.

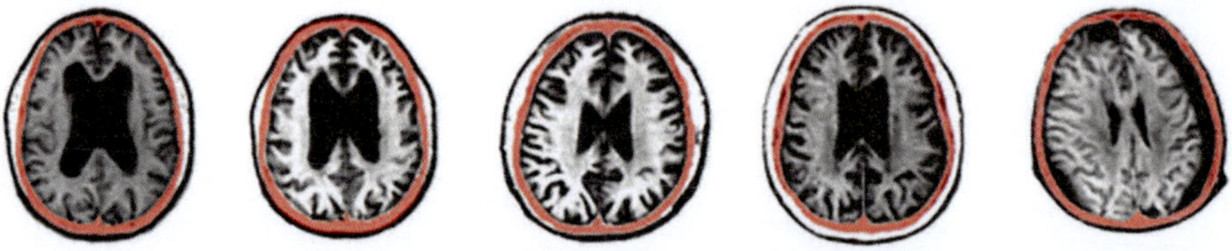

Fig. 4. Segmentation of each patient's flat bone.

Table 2. Histogram properties calculated with a standardized histogram vector $\pm 3\sigma$.

Parameter	Patient H1	Patient H2	Patient H3	Patient H4	Patient M1
Area (mm^2)	3954	5021	5123	3852	4002
Media	24.147	39.541	44.251	54.214	51.20
Variance	288.2	947.24	718.52	611.25	520.36
Bias	0.9987	1.520	0.3021	0.4852	0.4023
Kurtosis	1.4102	1.9562	−0.74120	−0.3352	−0.13652

Table 3 shows the values corresponding to the co-occurrence matrix (GLCM). Once the area corresponding to the section of the bone was obtained, the bone mineral content (g HA) was calculated and a linear regression analysis was performed to find the equation that relates the texture parameters with the bone quality for each section of the skull, obtaining that the mean and variance values are significant. The results and ANOVA are shown in Tables 4 and 5, respectively [22–24].

Table 4 shows the values of the image texture parameters (mean and variance) together with each patient's bone mineral content (BMC), which is related to bone mineral density (BMD) and fracture toughness (T-value). In Table 5, the values corresponding to the ANOVA can be found and the P values are lower than 0.05, which indicates that the factors that are included in the model are statistically significant [25, 26]. The equation presents a square R of 99.22%, indicating that the texture parameters of the skull images explain 99.22% of the variation in the response of the bone mineral content (BMC). Likewise, the adjusted square R of 98.47% indicates that the correct factors were included in the model, in this case, mean and variance. Equation 1 relates the texture parameters (mean and variance) to the bone mineral content.

$$CMO\,(g\,HA) = 42.8478 - 0.55214\,Media + 0.0400236\,Varianza \tag{1}$$

Where:
S = 1.2035
R-Sq = 99.22%
R-Sq(adj) = 98.47%
PRESS = 50.896
R-Sq (pred) = 86.33%

Table 3. Co-occurrence matrix parameters calculated with a standardized histogram vector ±3σ for S (1.0)*.

Parameter	Patient H1	Patient H2	Patient H3	Patient H4	Patient M1
Area S(1,0)	7412	9452	8501	6852	7123
S(1.0)AngScMom	0.003214	0.004412	0.0044123	0.00201	0.00145
S(1,0)Contrast	65.35	21.365	18.46	65.85	63.99
S(1,0)Correlation	0.7100	0.9236	0.7952	0.8125	0.74123
S(1,0)SumOfSqs	104.23	107.58	44.258	118.25	116.25
S(1,0)InvDfMom	0.2236	0.3025	0.3352	0.2485	0.2035
S(1,0)SumAverg	64.25	64.369	21.054	66.258	69.25
S(1,0)SumVarnc	365.25	410.25	146.25	446.25	410.25
S(1,0)SumEntrp	1.921	1.7952	1.6895	1.8985	1.9123
S(1,0)Entropy	2.88123	2.85332	2.698	3.214	3.025
S(1,0)DifVarnc	32.25	10.25	8.251	22.369	30.658
S(1,0)DifEntrp	1.2458	0.99369	0.9752	1.1685	1.2412

*Parameters are computed up to 20 times for (d, 0), (0, d), (d, d), (d, −d) where distance d can take values of 1, 2, 3, 4 and 5 [6, 27, 28].

Table 4. Bone mineral content corresponding to the image area of the segmented flat bone.

Patient	Area (mm^2)	Media	Variance	CMO (g HA)
Patient H1	3950	24.25	300.25	40.25
Patient H2	5002	35.25	985.25	58.36
Patient H3	4426	44.35	710.32	41.36
Patient H4	3658	54.58	620.35	38.25
Patient M1	4036	50.36	541.25	35.01

Table 5. Analysis of Variance (ANOVA) corresponding to the linear regression model.

Source of variation	DF	Seq SS	Adj SS	Adj Ms	F	P
Regression	4	334.25	334.69	170.25	120.5	0.0098
Media	2	57.25	134.25	135.269	94.23	0.0114
Variance	2	290.25	288.25	288.25	201.35	0.0055
Error	4	3.01	3.21	1.52		
Total	6	340.25				

5 Discussion

The texture analysis of magnetic resonance images of 5 patients with known bone densities was performed and the surface corresponding to the section of the bone was characterized to subsequently design 3D models. The patients' bone density and

T-values are within the normal range, indicating that the patients have a healthy bone structure with fracture resistance. Histogram data and co-occurrence matrix values allowed for uniform segmentation of the flat bone for each of the different patient images. Through the segmentation of magnetic resonance images, non-destructive evaluation of the bone quality (BMD and T-value) can be performed. The proposed model of linear regression can be used to estimate the bone mineral content (BMC), the equation obtained relates the texture of the segmented images of the skull bone (mean and variance) to parameters such as the composition and mineralization of the bone tissue (g HA) and the strength of the bone (T-value).

References

1. Galm, Brandon P., Buckless, C., Swearingen, B., Torriani, M., Klibanski, A., Bredella, Miriam A., Tritos, Nicholas A.: MRI texture analysis in acromegaly and its role in predicting response to somatostatin receptor ligands. Pituitary **23**(3), 212–222 (2020)
2. Mitra, A., Tripathi, P. C., Bag, S.: Identification of astrocytoma grade using intensity, texture, and shape based features. In: Soft Computing for Problem Solving, pp. 455–465. Springer, Singapore (2020)
3. Sabino, M.: Techniques for the manufacture of polymeric scaffolding with applications in fabric engineering. Rev. Latin Am. Metal, 120–146 (2017)
4. Felsenberg, D.: Structure and function of bone: supporting structure of collagen and hydroxyapatite. Pharm. Our Time **30**(6), 488–494 (2001)
5. Bahadure, N.B., Ray, A.K., Thethi, H.P.: Image analysis for MRI based brain tumor detection and feature extraction using biologically inspired BWT and SVM. Int. J. Biomed. Imag. (2017)
6. Latha, M., Kavitha, G.: Segmentation and texture analysis of structural biomarkers using neighborhood-clustering-based level set in MRI of the schizophrenic brain. Magn. Reson. Mater. Phys. Biol. Med. **31**(4), 483–499 (2018)
7. Avizenna, M.H., Soesanti, I., Ardiyanto, I.: Classification of brain magnetic resonance images based on statistical texture. In: 2018 1st International Conference on Bioinformatics, Biotechnology, and Biomedical Engineering-Bioinformatics and Biomedical Engineering, vol. 1, pp. 1–5. IEEE (2018)
8. Ali, A.H., Al-hadi, S.A., Naeemah, M.R., Mazher, A.N.: Classification of brain lesion using K-nearest neighbor technique and texture analysis. In: Journal Physics: Conference Series, vol. 1178, no. (1), p. 012018. IOP Publishing (2019)
9. Somwanshi, D.K., Yadav, A.K., Roy, R.: Medical images texture analysis: a review. In: 2017 International Conference on Computer, Communications and Electronics (Comptelix), pp. 436–441. IEEE (2017)
10. Paul, T.U., Ghosh, A., Bandhyopadhyay, S.K.: Brain tumor texture analysis-using wavelets and fractals. Int. J. Med. Imag. **4**(4), 23 (2016)
11. Kawashima, Y., Fujita, A., Buch, K., Li, B., Qureshi, M.M., Chapman, M.N., Sakai, O.: Using texture analysis of head CT images to differentiate osteoporosis from normal bone density. Eur. J. Radiol. **116**, 212–218 (2019)

12. Ta, D., Khan, M., Ishaque, A., Seres, P., Eurich, D., Yang, Y.H., Kalra, S.: Reliability of 3D texture analysis: a multicenter MRI study of the brain. J. Magn. Reson. Imag. **51**(4), 1200–1209 (2019)
13. Varela, N., Silva, J., Gonzalez, F.M., Palencia, P., Palma, H.H., Pineda, O.B.: Method for the recovery of images in databases of rice grains from visual content. Procedia Comput. Sci. **170**, 983–988 (2020)
14. Chondro, P., Hu, H.C., Hung, H.Y., Chang, S.Y., Li, L.P.H., Ruan, S.J.: An effective occipitomental view enhancement based on adaptive morphological texture analysis. IEEE J. Biomed. Health Inform. **21**(4), 1105–1113 (2016)
15. Nair, J.K.R., Vallières, M., Mascarella, M.A., El Sabbagh, N., Duchatellier, C.F., Zeitouni, A., Shenouda, G., Chankowsky, J.: Magnetic resonance imaging texture analysis predicts recurrence in patients with nasopharyngeal carcinoma. Can. Assoc. Radiol. J. **70**(4), 394–402 (2019)
16. Li, Y., Li, M.M., Zhang, Y., Cheng, J.L., Shang, Z.G., Bu, C.X.: Utility of texture analysis of magnetic resonance imaging in differential diagnosis of common pediatric cerebellar tumors in children. Zhonghua yi xue za zhi **96**(23), 1853–1855 (2016)
17. Lee, K.M., Kim, H.G., Lee, Y.H., Kim, E.J.: mDixon-based texture analysis of an intraosseous lipoma: a case report and current review for the dental clinician. Oral Surg. Oral Med. Oral Pathol. Oral Radiol. **125**(3), e67–e71 (2018)
18. Viloria, A., Lezama, O.B.P.: Improvements for determining the number of clusters in k-means for innovation databases in SMEs. Procedia Comput. Sci. **151**, 1201–1206 (2019)
19. Chang, H.H., Hsieh, C.C.: Brain segmentation in MR images using a texture-based classifier associated with mathematical morphology. In 2017 39th Annual International Conference of the IEEE Engineering in Medicine and Biology Society (EMBC), pp. 3421–3424. IEEE 2017
20. Lee, S., Lee, H., Kim, K.W.: Magnetic resonance imaging texture predicts progression to dementia due to Alzheimer disease earlier than hippocampal volume. J. Psychiatry Neurosci.: JPN **45**(1), 7 (2020)
21. Ke, C., Chen, H., Lv, X., Li, H., Zhang, Y., Chen, M., Hu, D., Ruan, G., Zhang, Y., Zhang, Y., Liu, L.: Differentiation between benign and nonbenign meningiomas by using texture analysis from multiparametric MRI. J. Magn. Reson. Imag. **51**(6), 1810–1820 (2019)
22. Iqbal, S., Khan, M.U.G., Saba, T., Rehman, A.: Computer-assisted brain tumor type discrimination using magnetic resonance imaging features. Biomed. Eng. Lett. **8**(1), 5–28 (2018)
23. Kim, D., Wang, N.C., Ravikumar, V., Raghuram, D.R., Li, J., Patel, A., Wendt, R.E., Rao, G., Rao, A.: Prediction of 1p/19q codeletion in diffuse glioma patients using preoperative multiparametric magnetic resonance imaging. Front. Comput. Neurosci. **13**, 52 (2019)
24. Basheera, S., Ram, M.S.S.: Convolution neural network-based Alzheimer's disease classification using hybrid enhanced independent component analysis based segmented gray matter of T2 weighted magnetic resonance imaging with clinical valuation. Alzheimer's & Dementia: Transl. Res. Clin. Interv. **5**, 974–986 (2019)
25. GGrist, J.T., Withey, S., MacPherson, L., Oates, A., Powell, S., Novak, J., Abernethy, L., Pizer, B., Grundy, R., Bailey, S., Mitra, D., Arvanitis, T.N., Auer, D.P., Avula, S., Peet, A. C.: Distinguishing between paediatric brain tumour types using multi-parametric magnetic resonance imaging and machine learning: a multi-site study. NeuroImage: Clin. **25**, 102172 (2020)

26. Baiocco, S., Sah, B.R., Mallia, A., Kelly-Morland, C., Neji, R., Stirling, J.J., Jeljeli, S., Bevilacqua, A., Cook, G.J., Goh, V.: Exploratory radiomic features from integrated 18 F-fluorodeoxyglucose positron emission tomography/magnetic resonance imaging are associated with contemporaneous metastases in oesophageal/gastroesophageal cancer. Eur. J. Nucl. Med. Mol. Imag. **46**(7), 1478–1484 (2019)
27. Viloria, A., Acuña, G.C., Franco, D.J.A., Hernández-Palma, H., Fuentes, J.P., Rambal, E.P.: Integration of data mining techniques to postgreSQL database manager system. Procedia Comput. Sci. **155**, 575–580 (2019)
28. Su, C.Q., Zhang, X., Pan, T., Chen, X.T., Chen, W., Duan, S.F., Ji, J., Hu, W.X., Lu, S.S., Hong, X.N.: Texture analysis of high b-value diffusion-weighted imaging for evaluating consistency of pituitary macroadenomas. J. Magn. Reson. Imag. **51**(5), 1507–1513 (2019)

Security Analysis for Machine Learning and Image Processing Related Information Systems

Ruchi Sharma and Kiran Davuluri(✉)

Department of Information Technology, NMIMS University-MPSTME, Mumbai, India
dkiran238@gmail.com

Abstract. With the exponential increase in data available online, the need for its safeguarding is manifold. Privacy models have been built to give sophistication in the expense of data in every aspect. The information has become the new asset. Machine Learning and Image Processing techniques replace traditional modes of data privacy. There is a need to involve autonomous systems in privacy mechanisms. From the storage of data to its analysis, there is always a threat of data theft or malpractice from intruders. A novel model based on data fragmentation and increased associates in the authentication system which ensures enhanced security of information is proposed. The model brings out layered protection of data using autonomous systems.

This paper analyzes key Machine Learning and Image Processing based privacy models and contrasts them. Challenges related to data privacy and security are also discussed.

Keywords: Data · Privacy · Security · Image processing · Information systems

1 Introduction

The privacy of information and preservation of data has become one of the most important aspects of conglomerates in recent times. The organization of information resources has become a crucial challenge in the day to day affairs of multinational institutions [16].

There is a need to discuss the current trends that are being used to secure data across systems. Data fragmentation along with secured data access is necessary. With the arrival of virtual storage spaces like computers, information is accessible only by authorized users.

1.1 Definitions

Artificial Intelligence – Authorize computer systems to perform tasks that require human intelligence, such as decision making, visual design, language processing, speech recognition, etc. [3].
Image Processing – Operation on image to extract data and its analysis.

J. I.-Z. Chen et al. (Eds.): ICIPCN 2020, AISC 1200, pp. 135–147, 2021.
https://doi.org/10.1007/978-3-030-51859-2_13

Bandwidth – Maximum capacity of data transfer across a given path. Bandwidth may be network bandwidth, data bandwidth, or digital bandwidth.
Big Data – Data sets on a large scale, which can be used for analytical or computational purposes.
Encryption – Encoding of sensitive information.
Information System – A large hierarchy of servers and controls using information technology to help manage business processes.
Privacy of Data – Internal and external safeguard of data repositories.

1.2 Concept Explanation

Privacy Governance – Privacy of data requires the following of some standard procedures. Rules and regulations are adhered to offer the highest quality of information security
Data protection – Defending significant data from inadvertent threats is called data protection. Varied mechanisms are employed for data protection.
Information Management – The method undertaken to furnish information in the most facile way, for business processes and decision making.

1.3 Need for Data Privacy

The need for standards in information provisioning and its access has seen a great leap. The developments and industry practices have changed the way users perceive information. Privacy and security of the virtual resources need to follow practical application to produce outcomes that come from the amalgamation of theories [1].

Data protection techniques that apply Machine learning and Image Processing are required. To address the potential threats and breach of trust in the security of information systems, there have been many academicians as well as industry experts who have, time to time issued regulations to be kept in practices when dealing with computerized data-keeping and its access [2].

Flexible and secure retrieval of data is significant in images resources, as well as ensuring that organizations collect, analyze, and diffusion of images without exposing sensitive information to third parties or mistrusted sources [3].

2 Role of Machine Learning and Image Processing in Data Privacy: Related Works

2.1 Machine Learning Techniques

Some of the prominent encryption techniques in machine learning for data privacy will be discussed here. The description of techniques with examples is given below.

2.1.1 Homomorphic Encryption

An encrypted result is presented for a query instead of the original result [12].

Hence there is no need to decrypt the data.

For example:

Original Encrypted Result
$8 + 2 = 10$ $16 + 4 = 20$
To know the output divide the encrypted result by 2.
Hence this does not reveal the input set i.e. {8, 2}.

2.1.2 Secret Sharing

Also known as a threshold scheme, different players are required to co-operate to share parts of the key for the maintenance of private information.

Example: A key K is divided into various parts K1, K2, etc., say K4 to the stakeholders in the authentication system. A threshold of key K2 out of K4 i.e. 50% must be validated for the disclosure of protected data.

2.1.3 Differential Privacy

The identity of the user is not revealed. But the result of the query is got by adding noise to the data or partial computations [7].

Let us take a case and understand it better:

Table 1 shows the information of certain users with their names and output, calculated concerning a certain algorithm.

Table 1. Sample user information with output.

S no.	Name	Output (O)
1	Alex	1
2	Bella	0
3	Chang	1

The query Q is to find the status of Bella. The differential computation will not directly reveal information about Bella, instead, it will perform Output1 minus Ouput2. This results in deciphering the output without identifying her (Bella) information.

2.1.4 Federated Learning

This model allows data to be present in the user's device and send only a summary of the operational data. An increase in data privacy is seen. The algorithm trains itself from the summary sent to the server. Server administrators are mostly reluctant to allow their data for transactional or analytical purposes. Hence this is an efficient technique for such purposes [4].

Autonomous Vehicles: Many people do not want their travel data to be saved on centralized servers. Federated Learning can help store data in users' devices and only sharing the summary. This increases privacy.

2.1.5 Personalized Privacy

Data breach can only be to some extent. This is completely on the owner's rights over the information. This model gives the owner the discretion to decide over the partial sharing of data [6].

For Example:

A certain information set has a record owner. Assuming there is a sensitive attribute called DISEASE. The record owner may be willing to disclose that he is ILL, but protect which type or the type of ill. This is implemented through tree taxonomy over nodes. The node CANCER value belongs to the subtree of TERMINAL DISEASE. The guardian owner of node TERMINAL DISEASE is under the subtree of ILL. The protection of the value of node 'type of disease' is at the discretion of the owner. The nodes contain sensitive information, which flows through the owner's right to open them [6].

2.2 Image Processing Techniques

2.2.1 Pixel Ciphering and Spatial Watermarking

It uses image processing and pictorial analysis for the allocation and retrieval of classified images.

It is an authentication system to provide security and privacy to sensitive images of medical patients, like x-ray, scanned images stored in medical information systems. The system gives access by validating the clients' accounts in the system. Patients' unique information is used in the encryption and watermarking of the medical images. Figure 1 displays the modules involved in image ciphering [12].

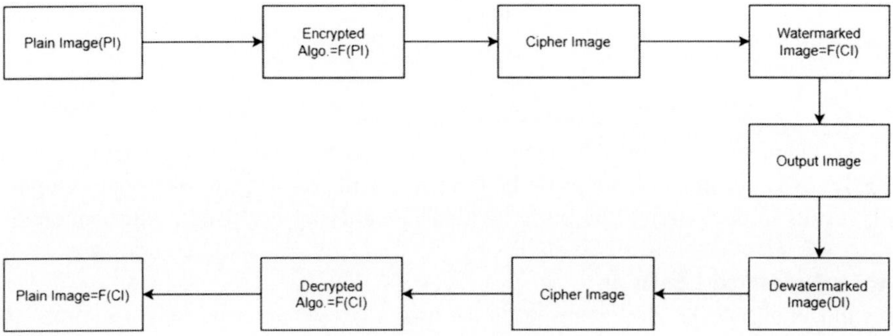

Fig. 1. Image ciphering process.

The Methodology of Pixel Ciphering

Encryption Process

Access grant to the user is carried out by using symmetric encryption process. Unique information of patients is used in the encryption and watermarking.

(a) Data from the image is imported and the image graphic object is created by interpreting each element from an image to a matrix.
(b) Calculate size as [c, p], entropy, and mean of the image.

(c) Extract red, green, and blue components as 'r', 'g', 'b' respectively.
(d) Let r, g, b be Transpose of 'r', 'g', 'b'.
(e) Reshape r, g, b with (r, c, p), (g, c, p) and (b, c, p).
(f) Concatenate the arrays r, g, b into the same dimension of 'r', 'g', 'b' of the original image.
(g) Data is converted into an image format to get an encrypted image [12].

A secret key is obtained by calculating He, the entropy value, and arithmetic mean for all pixels in the image.

Decryption Process
Decryption takes place by calculating the inverse of the algorithm. Plain image is obtained back from Encrypted image.

Watermarking Process
Spatial watermarking along with one channel, channel R is employed to watermark the Cipher Image.

(a) Let d = Aij be the data to be embedded.
(b) Let size(d) = [cl, pl].
(c) Embedding data into CI

```
for i = 1:c1
        for j = 1:p1
        r(i,j) = (Aij + r(i,j)) mod 256;
        end
end
```

2.2.2 Secure Steganography Creation Algorithm

The work utilizes the dual 'RSA' [16, 17] to encrypt the plain and original images. Stego/cover images [18] are used to protect data and maintain their data coherence.

The algorithm is described as follows:

At the transmitter's side
(a) The plain image(PI) is input
(b) The plain image is encrypted with Dual RSA.
(c) Apply steganography to produce Stego Image [19].
At the receiver's side
(d) Stego information is received and the encrypted information is extracted. The encrypted code is subjected to RSA decryption.

3 Privacy in Image Processing Related Information Systems

Images and imaging devices are on the rise. With the rise in technology and social media, people are becoming insecure about the storage of images in information systems. The need for securing the images with adequate algorithms and standards is seen to go up. Software-based interventions and methods for neutralization of images are currently in the markets. Secure programming and processing of image resources are also essential [13].

3.1 Opportunities and Threats in Image Processing Privacy

Opportunities

1. Large scope for image retraction and methods for storage, allocation, and analysis of the image.
2. Methods like image stitching and image stenography are being used for the retrieval of images after electronic transference [14].
3. The abstracting capacity of important data is optimum with the usage of data implantation in the spatial domain.
4. Multi-tasking methods have been displayed in image processing and capsule networks that enhance image identification by feature extraction training[15] and object fitting [20, 21].

Threats

1. Risk of losing the source of resources.
2. Increase in certificates related to cipher information.
3. Limited usage of the standard practice of securing data in images or the images themselves.
4. Figure and persona analysis require a large amount of time for privacy-related operations.
5. Ciphered spatial information is more unguarded and easy to decode than normal data.
6. Image authentication requires expensive resources to be successfully carried out.

4 Comparative Analysis of Security Techniques

Table 2 discusses various security techniques and comparative analysis.

Table 2. Security techniques and comparative analysis.

Technique	Features	Advantages	Disadvantages	Applications
Technique 1: Homomorphic Encryption [11]	An encrypted result is generated when a set of data is queried. This does not reveal the identity of native data Used for privacy-preserving data, redistributing, and computational purposes [9]	There is no need to decrypt data	Slow and computationally expensive	Educational Data [11] Healthcare Systems – clinical workflows Predictive Modelling
Technique 2: Secret Sharing [10]	Secret i.e. key is shared between multiple users. The keys up to a threshold are required to combine the secret and reveal the data Often the secret is constructed at a server	The secrets can be reused A highly secure scheme as the interceptor has to find the threshold secrets	Minimum 't' players must be active in the model Large resource allocation A dedicated and trusted admin must control the system to avoid	Distributed RSA Algorithm
Technique 3: Differential privacy [7]	This algorithm does not allow a particular user's information to be known even if it used in statistical computation The identification of the individual is kept closed whose information is present in a database	It doesn't compromise on the speed of the system The original dataset remains the same. There is no modification of the original values	More noise is required based on the requirement of privacy in the model	RAPPORT – reports usage statistics for Google Chrome [8]
Technique 4: Federated Learning [5]	It is suitable when data in remote nodes (devices) are significant and privacy-sensitive Algorithms are trained directly on the user's device and only the	A more privacy led solution No need to collect user's data No data charges: Since only some information is sent to the server, battery, or data	Bias/Eligible devices: There is a certain bias over the devices that can be deployed e.g. memory of 2 GB and recent android phones	Google Gboard [5] Self Driving Cars

(*continued*)

Table 2. (*continued*)

Technique	Features	Advantages	Disadvantages	Applications
	necessary information is sent to the servers in the form of a summary This process takes place numerous times and the algorithm is updated	charges of the device will be reduced	Convergence Time: Increase in parallelism can be achieved i.e. only 100 s of devices are in parallel but many more are available Bandwidth: Compression techniques can be applied to reduce the bandwidth requirement. The models can be shown in quantized representation to save bandwidth	
Technique 5: Personalized Privacy [6]	The owner has the discretion of reveling or sharing the data. The privacy of the owner concerning his/her data is maintained There are some types of data that the owner does not want to reveal	The owner has full rights on the amount of outsourcing of data The privacy measure can be defined by the owner [10] Highly efficient for sensitive attributes	Difficult to implement	Maintaining illness records in medical centers
Technique 6: Pixel ciphering and spatial watermarking [12]	Image pixels ciphered by encrypting them with cipher image containing the patient's unique information	Helpful in the storage and authentication of highly confidential images	Large storage space required Excessive amount of Certificated to be stored	Medical images Intelligence and anti-terrorist organizations
Technique 7: Secure Steganography Algorithm[19]	Dual RSA is applied to get stego images for transmission	Secure Transmission of information Fewer memory options and faster encryption process [19]	Huge Amount of data and file size. So someone may suspect it	Storage of sensitive data in images for secure transmission

4.1 Motivation for a New Privacy Model

Table 2 discusses the prominent security models that implement Machine Learning and Image Processing in the privacy and security of information.

The issues that have been seen in the above models are:

- Resource requirement is there in terms of bandwidth, memory, etc.
- Privacy mechanisms often have a certain level of complexity.
- The computations and operations generally affect the device's health, namely data, memory, and battery.

Solutions that may be taken into account to withdraw the challenges:

- A simple yet efficient mechanism is asked for.
- Minimize the number of actors and increase the activity of autonomous systems.
- A data privacy concept for information systems must be in sync and incremental with basic factors like time, keys, etc.

5 Proposed Model

The proposed work concentrates on the dissemination of data into fragments based on the vulnerability of the privacy of sensitive information in information systems.

Factors that need to be considered while designing a data privacy model:

- Identity of the data users should not be disclosed – on the discretionary of stakeholders of data.
- Data protection
- Simple mechanism
- Seamless transfer of data on demand.

Features of the proposed model, as shown in Fig. 2:

- Data encryption – standard encryption
- Access hierarchy- data owner, associates, third party users
- Data protection from threats – Fragments
- Real time key granting system.

Actors:

Figure 3 depicts the stakeholders that have access to the data.

Data owner/proprietor – data is owned by this person.

Associates-Confidants of the data owner. They may be co-data owners

Third party users – They are the data requestors. Data access requests will be made by the users.

Key Generation System (KGS):-

Real time key generation is performed by the KGS, on-demand by the data owner or associates.

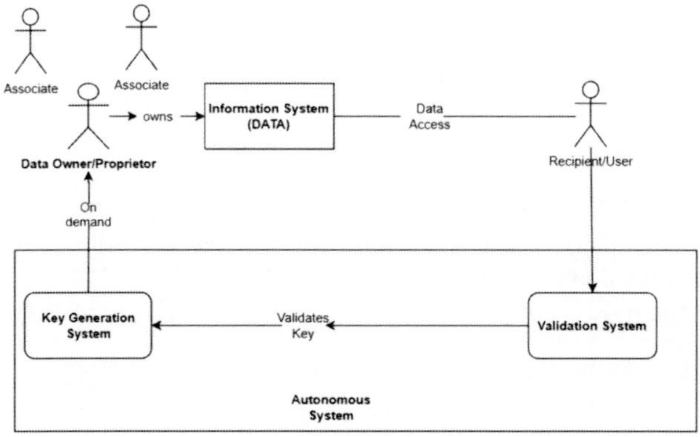

Fig. 2. Proposed model for data fragment based privacy model.

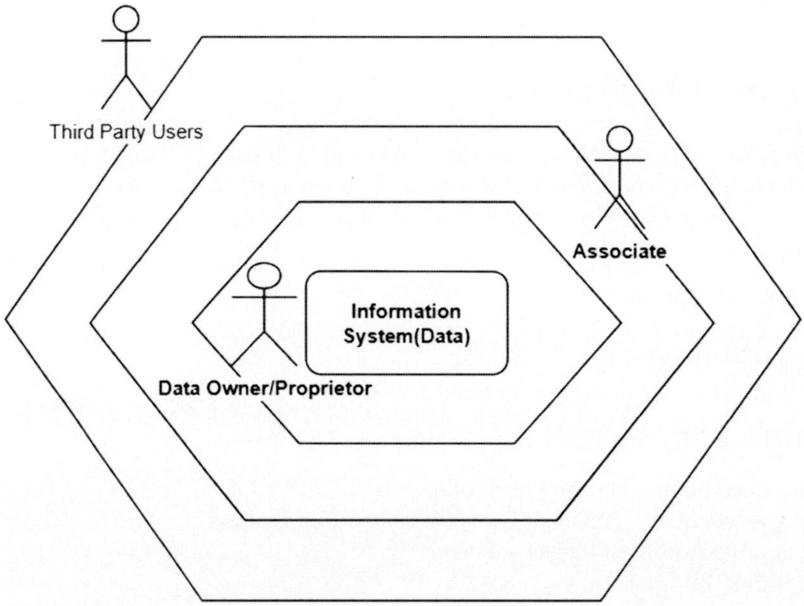

Fig. 3. Actors in the proposed model with different access levels.

Privileges:-

Different privileges to the data users will be set according to the data server.

Validation System (VS):-

The keys provided here are validated internally through the KGS.

Figure 4 discusses the fragment division of a sensitive file/data based on its increasing vulnerability.

Data access by Associates:-

A standard privacy model for the associates of data owner will be implemented with encryption of the data which may include the following processes:-

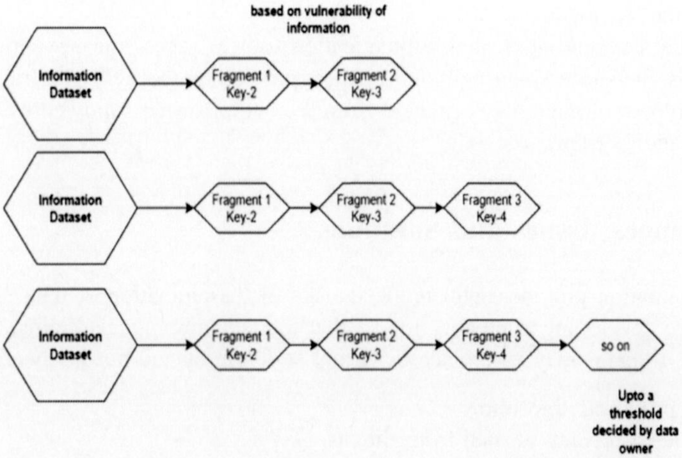

Fig. 4. Fragment division based on the vulnerability of data/file.

- Validation of data requestor i.e. third party user.
- Encryption of data by a standard algorithm.

Data access for the data owners/associates:-

Associates can log into the information system using their credentials and real-time authentication (biometrics, etc.).

If the user is an associate, he/she may access the entire information set by gaining the decryption key from the proprietor(owner) and validating it at the VS. Different privileges to the data users will be set according to the data server.

Data access by third party users:-

Third-party users must request a master key (MK) which is combinations of keys from the proprietor as well as each of the associates.

$$\text{MK} = \text{Key}(\text{proprietor}) + \text{Key}(\text{Associate } 1) + \text{Key}(\text{Associate } 2)\ldots$$

Encrypted Data will be available only when the master key (MK) is granted by the proprietor and associates. This must be validated at the Validation System.

The data will be decrypted using a key that will be granted by the owner followed by validation at the validation system.

Data Protection from Threats:-

Based on the vulnerability of a file/dataset, data will be stored in different fragments with key for the next fragment.

Consider a scenario: If the privacy of data has been violated for the initial phase, it will be divided into two divisions, with the above-mentioned specifics. If the vulnerability increases, more will the no. of divisions be. Once a threshold (decided by the proprietor) is reached, the file will cease to be available for third party recipients. Figure 4 shows the description of data fragments based on their vulnerability.

Key in the fragment:-

It will be a system-generated authentication tool to access the next fragment. In simple terms, the successive fragment of data is protectively locked. The fragment data will be decrypted through a key granted from the author/owner followed by validation at the validation system.

6 Challenges: Issues and Solutions

Data fragmentation provides with reduced risks of dissemination of data. The issues that arise are developing a network model that trains autonomous systems. Storage of keys and certificates may take place in virtual storehouses and not hardware systems.

Pros of the proposed algorithm

- Guaranteed privacy of data from threats.
- Secure information systems with incremental data.
- Use of autonomous systems to generate keys.

Some of the challenges in the proposed model

- The complexity of implementation.
- An excessive amount of keys and certificates.

Capsule networks and deep neural networks must be implemented with the training of autonomous systems for allowing automatic extraction of hierarchical keys and storage of files in the cloud. Efficient analysis of the threshold can be performed sorting and ordering mechanisms in an unstructured way [21].

7 Conclusion

Data privacy is a fundamental aspect of any information system to fulfill its needs. Artificial computing is the way forward to secure our data assets. Data security and privacy that involve Machine Learning and Image Processing models are being implemented widely given their advanced algorithms.

This paper highlighted the need for privacy in the information system. It also compared and contrasts the privacy models that are applied in the present scenarios. The challenges and issues faced in safeguarding sensitive data are seen. Finally, a privacy model is built keeping in mind all the aspects of security. The future scope may include refining the model with the above mentioned technologies like capsule networks and neural networks.

References

1. Plato.stanford.edu.: Privacy and information technology (stanford encyclopedia of philosophy) (2020). https://plato.stanford.edu/entries/it-privacy/. Accessed 28 Feb 2020
2. Bobgellman.com (2020). https://bobgellman.com/rg-docs/rg-FIPshistory.pdf. Accessed 28 Feb 2020

3. Cate, F.H., Dockery, R.: Artificial intelligence and data protection: observations on a growing conflict. J. Law Econ. Regul. 1–30 (2018). Seoul National University. https://ostromworkshop.indiana.edu/pdf/seriespapers/2019spr-colloq/cate-paper.pdf
4. Keith, B., et al.: Towards federated learning at scale: system design. arXiv.org, February 2019
5. Yang, T., Andrew, G., Eichner, H., Sun, H., Li, W., Kong, N., Ramage, D., and Beaufays, F.: Applied federated learning: improving Google keyboard query suggestions. arXiv preprint arXiv:1812.02903 (2018)
6. Ricardo, M., Joao, V.: Privacy-preserving data mining: methods, metrics, and applications. IEEE Access (2017)
7. Begum, S.H., Nausheen, F.: A comparative analysis of differential privacy vs other privacy mechanisms for big data. In: 2018 2nd International Conference on Inventive Systems and Control (ICISC), Coimbatore, pp. 512–516 (2018)
8. Xiong, P., Zhu, T.-Q., Wang, X.-F.: A survey on differential privacy and applications. Jisuanji Xuebao/Chin. J. Comput. **37**, 101–122 (2014). https://doi.org/10.3724/SP.J.1016.2014.00101
9. Xu, K., Zhang, W., Yan, Z.: A privacy-preserving mobile application recommender system based on trust evaluation. J. Comput. Sci. **26**, 87–107 (2018)
10. Fung, B.C.M., Wang, K., Chen, R., Yu, P.S.: Privacy-preserving data, publishing: a survey of recent developments. ACM Comput. Surv. **42**(4), 14:1–14:53 (2010)
11. Archer, D., et al.: Homomorphicencryption.org (2020). http://homomorphicencryption.org/white_papers/applications_homomorphic_encryption_white_paper.pdf
12. Kester, Q., Nana, L., Pascu, A.C., Gire, S., Eghan, J.M., Quaynor, N.N.: A security technique for authentication and security of medical images in health information systems. In: 15th International Conference on Computational Science and Its Applications, Banff, AB, pp. 8–13 (2015)
13. Padilla-López, J.R., Chaaraoui, A.A., Flórez-Revuelta, F.: Visual privacy protection methods. Expert Syst. Appl. **42**(9), 4177–4195 (2015). https://doi.org/10.1016/j.eswa.2015.01.041
14. Qin, J., Luo, Y., Xiang, X., Tan, Y., Huang, H.: Coverless image steganography: a survey. IEEE Access **7**, 171372–171394 (2019)
15. Saxena, S., Kapoor, B.: An efficient parallel algorithm for secured data communications using RSA public key cryptography method. In: 2014 IEEE International Advance Computing Conference (IACC), pp. 850–854. IEEE (2014)
16. Sharma, R., Srinath, P.: Business intelligence using machine learning and data mining techniques-an analysis. In: Second International Conference on Electronics, Communication and Aerospace Technology (ICECA) (2018)
17. Sun, H.-M., Wu, M.-E., Ting, W.-C., Jason Hinek, M.: Dual RSA and its security analysis. J. Innov. Image Process. (JIIP) **01**(01), 20–30 (2019). https://www.irojournals.com/iroiip/. https://doi.org/10.36548/jiip.2019.1.003. 30 ISSN 2582-4252, IEEE Trans. Inf. Theory **53** (8), 2922–2933 (2007)
18. Ansari, A.S., Mohammadi, M.S., Parvez, M.T.: A comparative study of recent steganography techniques for multiple image formats. Int. J. Comput. Netw. Inf. Secur. **11**(1), 11 (2019)
19. Vinothkhanna, R.: A secure steganography creation algorithm for multiple file formats. J. Innov. Image Process. (JIIP) **01**(01), 20–30 (2019)
20. He, K., Zhang, X., Ren, S., Sun, J.: Deep residual learning for image recognition. In: Proceedings of the IEEE Conference on Computer Vision and Pattern Recognition, pp. 770–778 (2016)
21. Bashar, A.: Survey on evolving deep learning neural network architectures. J. Artif. Intell. **1** (02), 73–82 (2019)

The High Performance Image Encryption Method Based on HAD-L

P. Mithun and J. Indumathi(✉)

Department of Information Science and Technology, Anna University, CEG, Chennai, Tamilnadu, India
mithunpalanimuthu@gmail.com, indumathi@annauniv.edu, indumathi.j@gmail.com

Abstract. With the vast development of the communication medium, the use of digital data is increased rapidly. The lots of development increased and also side by side illegal disruptions are also be developed for misuse of digital data. For that case specifically for digital images, developed the HAD-L (Harmonic Level, systematic Block Level, and Additional Data Level) of encryption system. The first level with customized encryption with specified harmonic encryption properties. The second level of the encryption process with the higher block level vector scheme. Then that additional data also combined with the encryption scheme to deliver the third level of the encryption process. The overall effect of the scheme will deliver a robust encryption scheme among the communication channels. The experimental results show that the good randomness effect of the scheme and also deliver the pseudorandom properties for the key evaluations. Furtherly results from the security analysis and capacity analysis show much more performance among transmission channel.

Keywords: Encryption · HAD-L

1 Introduction

The major concern for digital media is data privacy. For enhancing the stability of privacy in an open-air space medium, it is very difficult to achieve. Many types of intruders are awaking to hack the original data during transmission over the network. For that case, some more technologies are always introduced to bar the attack while transmission. The individual security of personal data is considered for protection. Using that, some cryptographic, stenographic and watermarking technologies are improvised and used for the protection of data. The case of security is involved at the time of transmission and also uploading certain types of data in the network. There are the intruders are always ready to hack the data and also include some interrupts to the communication medium and also storage medium. The privacy of certain types of data is always crucial for all the stages. The major protection involves the cryptographic stages to protect personal data in the process of communication. In the modern world, each stage of privacy is enhanced and provides an efficient way of protection. The support of the mechanism for protection is that the case not only gives protection and also gives easy access to the environment which is already kept. The conservation of the medium which accompanies the value of data and also the medium. In this case of

J. I.-Z. Chen et al. (Eds.): ICIPCN 2020, AISC 1200, pp. 148–159, 2021.
https://doi.org/10.1007/978-3-030-51859-2_14

protection, we involve the major security aspects of the data that carry over is mostly of images related data. The data protection specifically for images is taken out in the scenario and also evaluating the practical measures over the previously developed mechanisms. There is much more enhancement are developed and used in the protection of data, specifically for images. The secret mechanism is evolved to protect personal data from the outside world. The meaning full measures are provided using stability in all the current aspects of security and also improvises for future attacks. The major portion for attacks is in the image data to acquire the pixel data and use of illegal manners. The combined methods are always arising and ensure privacy protection. But the day to day development in the communication medium, side by side illegal entry and acquisition of data is also be increased. The development is well reached the major portions of the interactions. The development in the security channels is improvising every day. For the special case, we developed the mechanism to provide the security of personal image data. The personal image data are sensitive data that evolves and usages in the modern world are more and more. For each social media, every data we provided in this is more acquiring level data for the hackers. For the level of the image, usage is increased rapidly. The image is the collection of pixel data, which is located easily using acquiring a single part of an image or block of image pixel data. The corresponding person would not realize the actual image is hacked or not. The illegal practices using images in social media is increased in the day to day life. Personal care for the usage of images is also very important. The images are the sensitive handler for every person because of illegal and wrong consumption, modification in the communication medium. The proposed system would capable of ensuring a better encryption effect and also high secrecy over the communication medium. The proposed work has three-level (HAD-L) of encryption to protect personal data. Especially image data are handled significantly and illustrate the work in the proposed system section. Hence, the remaining section of the work is organized as, related works in section two, the proposed system in section three, the results and performance evaluation in section four, and the conclusion is described in section five.

2 Related Works

The digital medium can have access to all the details from the person and also give effective solutions whatever we need. For that case every good thing there is some bad thing to handle in every situation. The digital medium is also having the same type of situation. For that case there some specific methodologies are taken out and point out the cases which are not solved by it. Deng et al. [1] proposed the image encryption system which is capable of chaotic mapping technologies. There some logical confusion in the segmentation of images during transmission is also be solved by the methodologies. The took care of the chosen plaintext attack over the communication also be considered. The enhanced cryptographic methodologies are used by the parties and efficient protection for the images is handled. The logical connections in the pixel separation are also be calculated and then apply the chaotic mapping to ensure the encryption effect. Alireza Arab et al. [2] proposed the image encryption system which is capable of using the AES mechanism and also Arnold sequence for the separation of pixel values and key generation mechanism is also be handled. The differential attacks are also highly valued taken by the proposed system. The enhanced AES mechanism for the development practices and round also be taken care of the systematic approach. The digital image encryption system can resist the brute force attacks during the collection of each data. The greater keyspace also discussed in this digital method. Tejas Duseja et al. [3] proposed the image encryption scheme with the Chinese remainder theorem for the distribution of the keys in the corresponding image pixel values. The corresponding co-prime value is calculated in the scheme to retain the randomness for the grayscale images. The image encryption scheme is specifically for the grayscale images, into two-pixel blocks only calculates and also used the hap map knowledge. There are 255 hash maps are used to store the pixel values with the shared secret key. The randomness of the encryption system is greatly discussed and for specifically grayscale images. In the actual scenario, the digital medium is highly oriented with RBG based colour images. So that there is a low impact in the current scenario. H.Y. Wang et al. [4] proposed the reversible data encryption scheme with the block truncation method. In the proposed method there is the sequence of reversible blocks are arranged and the image encryption scheme with a block-level sequence is discussed. The number of bit planes is used to truncate the original bit planes and to embed the original data. The sequence of variable bit plane is highly efficient in hiding perspective secret data scheme. The original bits are secretly stored with the embedded plane and to restore the necessary values by re-embedded the collective values. This causes the necessary time complexity to arise and values reconstruction in the high-level significance. The overall perspective of the scheme to ensure the original bit values to the corresponding truncated level planes with zero significant points. S. Ma et al. [5] proposed the plaintext information that is valued by the capacity to ensure secrecy with a block-level encryption system. The combined methodologies for the image encryption scheme with the chaotic key generation with the distribution of chines remainder theorem-based sharing among the pixel block. The individually valued subsequence in the pixel blocks is combined and the diffusion process is also carried out for the original block-level image encryption systems. For every stage in the sequence blocks, much more improved iterations are carried out in the specific combinational area for the correct assumption strategy. The speeding chaotic generator is the more capacity for that the scheme is resisted from the data attacks. The differential attacks are also be compromised in the specific digital chaos-based image encryption methodologies. The measurement for the high accuracy in the entropy calculation far good as the secrecy maintaining over the digital enhancements. Y. Liao et al. [6] studied the mechanism of the identity system and provide the fined grained approach. In this test results shows that the identity schemes are not much more efficient and also pubic encryption system combination with this identity schemes are having a poor illustration. The rapid inventions of the digital era, H.T. Wu et al. [7] proposed that the reversible data hiding schemes for the images. In this, all other schemes acquisition of the same bit plane is carried to accompanying the original bit-plane values. Instead of this the whole portioned value, the least significant bit is carried out. The LSB bits with reverse embedding capacity are carried and to ensure privacy. The hiding original is recovered only by extracting the hiding bits into it. The measured value from the higher embedded bits is also be discussed in this case. The digital empowerment holds a greater discussion for the images and shows the effective calculation in the experimental area. For that case, the constraints of the values are effectively in the low medium. The pre-processed cipher image is getting for the addition of more bit plane to the original bits to increase strength for the encryption system. For the development of social media, the usage level also increased. In that case, the privacy among the data which was handled by the authority of the medium must satisfy and also legible conditions to the user. For the protection over the communication medium, Chuman et al. [8] proposed the block level encryption system with the scrambling of jpeg images. The corresponding original three colour condition-based images and two-colour condition-based images are matching efficiently and a collapse scenario is taken place to avoid illegal consumption. The encryption system with a description of the upload and download scenario is taken place. The secrecy maintains between two parties are handled efficiently by the proposed system. The security against the solver attacks is also be taken the major constraint in the proposed methodologies. Encryption and then compression are the major aspects of the scheme, from that less acquisition of original matters are retrieved and also try to retrieve the actual content necessarily. This type of work also major practices in the modern era. K. H. Jung et al. [9] proposed the image encryption with data hiding. The overall methodology used in the scheme is to ensure the secrecy of images using differencing and significant bit enhancement. The quality of the original image is discussed with the convolution study. The process flow in the scheme, which would initially partition the region and firstly applied the embedded data into it. The quality loss is less to ensure the strengthening of the proposed scheme. For over, not only the social media uses image-based data, instead of the medical sector are also used images. The effective proposal from Liu et al. [10]. The much more constraint of the scheme to reconstruct the original contents without any loss because it never uses normal images. It uses electrocardiogram images of medical patients. In this method, image encryption followed by compression methodologies is enabled. The quality control throughout the data is also be checked to enduring the capacity of secrecy. The overall methodologies are conventional schemes to provide security and also enabling the version style computation. Y. Wang et al. [11] proposed the encryption system with data hiding methodologies to ensure the protection of privacy data into it. There are some major illustrations are carried out to describing the system which is capable of a highly randomized system. Initially, the classification of blocked and non-blocked bit values is separated. For each case, individual separation of mechanisms is followed to ensure the highness of the scheme. The original image is encrypted and then further embedded with another pixel image. The valued result is further constrained under the specific modulation to prove the randomness of the method. The major illustration is arising and enhancing every invention. For better provision is the necessary needed for each individuality. Ma S et al. [12] proposed the method of image encryption scheme with the residue number system and the permutation polynomials to generated the sequence of the key for each pixel value in the consortium. The correct actual iterations are calculated and provide a systematic approach to the legal system to enduring the protection. There is some lookup table accompanying the correct keys to the specific pixel value. In this computational complexity are also be increased. Not only the randomness key generation but also having better stability during the attacks. J. Wu et al. [13] proposed a scheme is having that type of illustrations and also the good cryptographic performances. In this three-dimensional matrix permutational model is used to provide secrecy for the images. This additional cat map is used to map the original pixel values and the keys in the appropriate manner. The collection of the sequence pixel value in the randomized cat map linear segregation, the block of every pixel value is highly enriched. The speed test also is illustrated in the proposed method. Not only the logistic scheme but also the cat map is also be used for key mapping is also be studied efficiently. The shortcoming of the early three-dimensional mechanisms is to be studied and effective measures are evaluated in the use of cat maps. There is some specific block-level encryption has entitled the secrecy of the data. For these conditions, Vimalraj et al. [14] proposed the three-level of encryption systems. In this, the subset hill algorithm and the one-time pad are efficiently used, and also three levels of encryption for each scenario are delivered by the intermediate results. The conditions for each level of encryption is handled efficiently and also acquired highness in the communication environment. But the major backlogs for the constraint-based system is that providence for the whole system is only for grayscale images. Those further actions are calculated the timing sequence to ensure the practice of the encryption system efficiently. Finally, the possible improvements are also be considered in the proposed scheme for the logistic scheme to be considered. W. Feng et al. [15] proposed the encryption methodologies for the images in the manner of providing a scheme with advanced logistic mas to use of it. The existing problems are identified and clarified in the advanced logistic scheme. The cryptanalysts take the attack algorithm and also employed in the scheme. Furthermore, evaluations are studied.

3 Proposed System

The overall process of the proposed encryption scheme is discussed in Fig. 1. Initially, the image encryption is carried out followed by data hiding is ensure to protection. The corresponding pixel values are taken out and then apply for image encryption. The image encryption is carried to acquire the pixel values appropriately.

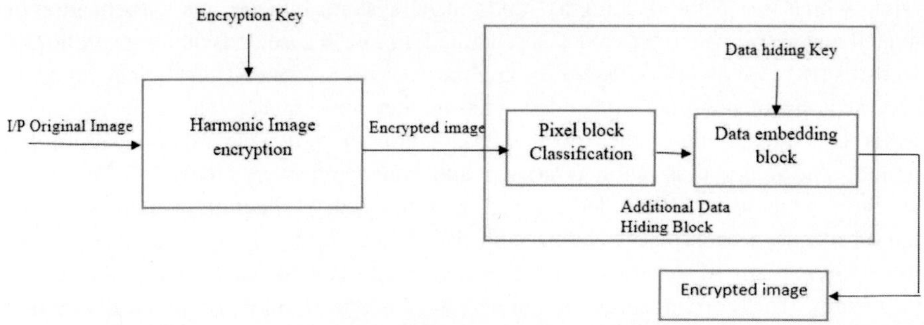

Fig. 1. System design for the encryption process

3.1 Image Encryption

The special scenario images are taken part then apply the harmonic image encryption. There is an advanced and enhanced procedure that is followed to encrypt the image initially. For that case sequence of steps is raised by the information security procedure. The major part of every encryption system is the key generation for every equal set of the image block. Then the classification is the major constraint in handling the pixel values. The image encryption is carried out by the initial separation of pixel values into m and n block. The sequence of blocks is called as row and column blocks. For each separation of blocks, the specific keys are used to accompany the pixel values. The key generation is carried by pseudorandom standard mapping for special cases to enhances the stability of keys in the communication area. The certain random numbers and their inverse calculation are made to use it as the keys. The keys which are capable of holding every pixel value in the specific block itself. For the generation of every key, the corresponding inverse key is calculated for the decryption process and it will be stored in the file format. The RBG values are raised and the corresponding matrix distribution is speaking for the continuous separation and performance of the encryption system. In the image encryption system, initially, the pixel values are composed of eight-bit integers. The corresponding integers will be calculated with every positive integer and key matrix and then modulated with the big prime number. The corresponding result will be the first level of the encrypted image. The sequence of steps to elaborate the mechanism.

- Encryption (E), Decryption (D).
- Initially separate m and n blocks and convert the bits into eight-bit integers.
- Then apply $E = (mk + i) \bmod p$, and $(nk + i) \bmod p$ here, E = encryption, m = m odd block, n = n even block, k = corresponding random matrix key, i = input matrix and p = total no of positive integer. This condition followed by a sequence of operations and to get an encrypted image.
- For decryption, $D = (Ek - i) \bmod q$, D = decryption process, E = encrypted image, after the segregation of m and n blocks, the k = inverse key is used and i = input matrix can be differentiated with continuous modulation of a total number of positive integers.

3.2 Block Scrambled

The continuous flow after the first level of encryption, the block scrambled process is taken out. The sequence of steps to elaborate on the process.

- Convert the RGB pixel values of M*N blocks into 3 vectors.
- The generating simultaneous vector for the RGB values.
- Compare the individual pixel values form block.
- Apply corresponding pixel value and get the new pixel value.

For the generation of pixel values from the index vector, there are the formulae to apply for the creation of the index vector. $V = s \bmod l$, here v is the vector created and s is the pixel value, and l is the corresponding length of the pixel value. Finally, the second level of the encrypted image is getting.

3.3 Embedded the Image and Data Hiding Block

For the secrecy among the communication era, the withstand of the encryption system without any back loss must be considered. For the acquisition of the original methodologies for the pixel segregation and separation is mandatory. In this circumstance, we proposed the encryption scheme with high-end performance in the illegal partitions from outside communication is a major aspect of adding this concept. Adding the additional image into the second level of encryption using specific methodologies. The sequence of steps illustrates and elaborates on the process of adding an image.

- The corresponding sized image is added to the second level of the encrypted image.
- The new image and the encrypted image is segmented into a sequence of m and n blocks.
- By choosing the edge LSB bits the corresponding new image of LSB bits is changed. Otherwise, the most significant bits of the new image are equally changed.
- For each segregation, the new image is adapted to them individually in the encrypted image.
- The process going for n number of iterations and finally get the third level of the encrypted image.

3.4 Separation of Blocks

The separation of the block is a very crucial one to be implemented. Because in the 2*2 square image, the separations of blocks in four aspects. The first one separation from edges, the second one is the separation from the centre, third is the separation from the vertex and finally separation of blocks without the vertex. In this, every pixel block is segregated as odd and even blocks from choosing any one of the four aspects and then converting as 2*2 or 3*3 matrix blocks. The separate the pixel values from the images of each block. Analyze the block sequence and separate individual blocks. This separation and segregation of pixel and blocks are sequentially used in the encryption process.

3.5 Decryption Process

The decryption process is shown in Fig. 2. The important thing we want to analyze is the decryption process. The reversibility of every value is a must and we want to provide each encryption system with good decryption of effect. Every practice is undergoing the conditions that will meet with less amount of quality loss. The sequence of steps is followed to decryption.

- Each decryption must need the decryption key.
- Two types of keys are used in the method, i.e. decryption key and data hiding key.
- If having a decryption key, the block separation is the first step. Then apply the inverse key and formulae to decrypt. After that block-level separation is decrypted by the inverse vector process.
- If having a data hiding key, the additional data is must remove for getting the original pixel values. The separation of the pixel block is carried out and then

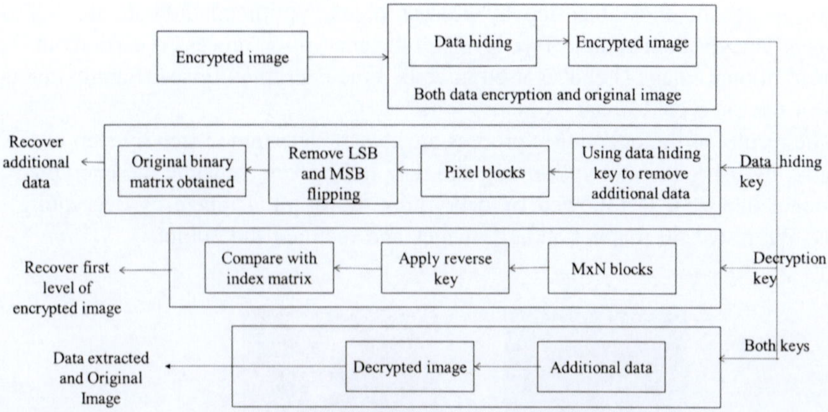

Fig. 2. System design for decryption process

remove the changing of LSB and MSB connections. After that, the hiding image is recovered appropriately.

- Finally, the original image and additional image is obtained.

4 Experimental Results

The overall encrypted process is shown in Fig. 3.

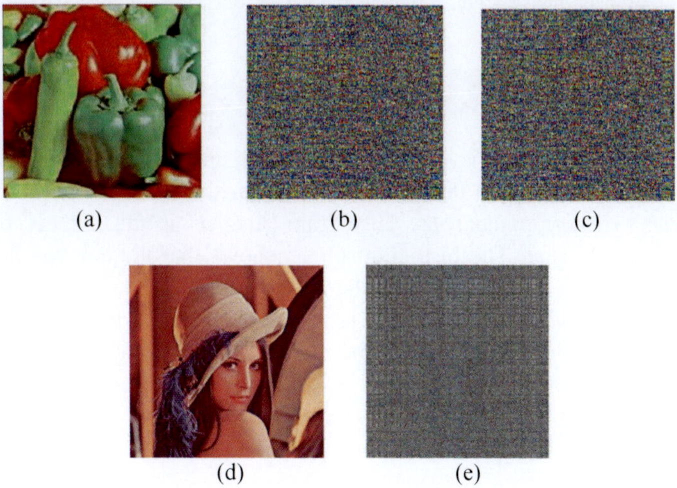

Fig. 3. Results of encryption process: (a) Original peppers color image, (b) First level encryption, (c) Second level encryption, (d) Additional image, (e) Final encrypted image.

As the results show the step by process of encryption method. In this Original 'Peppers' image taken and the Two levels of the encryption process is carried out. Then lastly additional image 'Lena' is used to embed the encrypted image. Finally, the third level of the encrypted image is getting out.

The results of the decryption process are clearly determined step by step in Fig. 4. Initially, by using the encryption key and data hiding key, separate the encrypted and additional image. It is followed by decrypting the original image by decrypting two levels. We tested 50 major standard images and rectified the output.

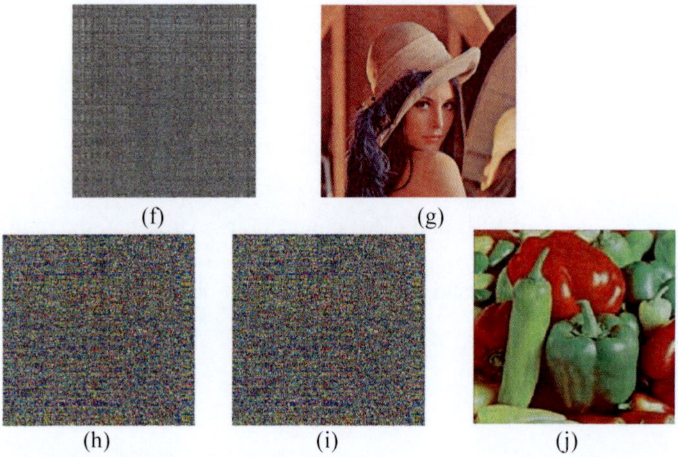

Fig. 4. Results of decryption process: (f) encrypted image, (g) recovering additional image, (h) First level of decrypted image, (i) second level of decrypted image, (j) Original image retrieved.

4.1 Performance Evaluation

The performance evaluation is used to understand the capacity and withstand the mechanism during attacks. For this, we analyze the method using some of the analysis methodologies. Histogram analysis: Histogram analysis is the content partitioning among the data set value. The histogram Fig. 5 shows that the encryption effect has good randomness. The peak value for the encrypted image is the same for all the images but the original image histogram is variant in the level and it causes collapse among the data set during the transmission. For that, the encryption shows good randomness.

4.2 Embedding Capacity

The embedding capacity of every data hiding scheme illustrates the strengthening of the proposed methodologies. The list of values shows that the proposed system has a higher embedding capacity. The other proposed schemes have less capacity value than

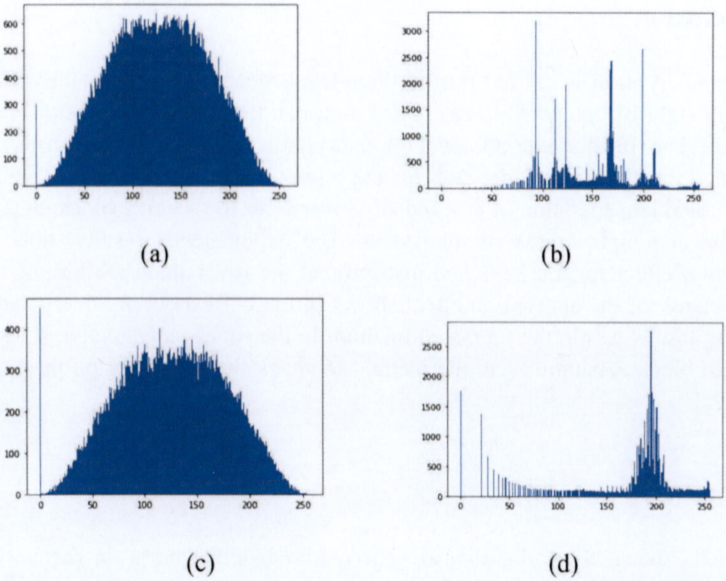

Fig. 5. Histogram analysis: (a, c- encrypted image, b, d- original image)

the proposed scheme. It not only increases the embedding capacity but also strengthens the encryption system during the outside attacks.

The embedding capacity of the system is shown in Table 1. From the analysis, the proposed method has a higher sequence rate in the distribution of values in the embedding stage. The results are compared with proposed methods [16–18], and shows that the three-level HAD-L has higher embedding capacity.

Table 1. Embedding capacity with various methods

Image	Liao	Zhang	Bhardwaj	Proposed
Peppers	1296	1296	1875	32485
Boat	1296	1024	1875	32194
Lena	1296	1764	1875	32684
Barbara	196	225	225	32334

4.3 Security Analysis

The major aspects of the security analysis the security performance. The usage of different types of keys in each module increases the stability of mechanisms in the communication medium. The decrypting key has a higher priority for the reversible data collection. For that case of adding additional images, enhancements lead to higher priority against the attacks. The different keys enhance the distortion levels and withstand the attacks.

5 Conclusion

The image encryption is carried out by three-level mechanisms. The individual pixel values with eight-bit integers are calculated and then the block level vector is used for scrambling. The further second level of encryption with additional protection, we implemented the systematic approach for the hiding of additional images to encrypted images. Individual separation of key and key inverse for the specific block enhances the randomness and high degree of robustness. The experimental results show that the good random effect for the keys and also protect the reversible data hiding scenario. The correctness of the encryption effect shows in the corresponding decrypted image. The quality loss is less in the proposed method. In the future, an effective study for the randomized block separation and the method to check the streaming purpose data will be proposed.

References

1. Deng, Z., Zhong, S.: A digital image encryption algorithm based on chaotic mapping. J. Algorithms Comput. Technol. **13**, 1–11 (2019)
2. Arab, A., Rostami, M.J., Ghavami, B.: An image encryption method based on chaos system and AES algorithm. J. Supercomput. **75**, 6663–6682 (2019)
3. Duseja, T., Deshmukh, M.: Image compression and encryption using Chinese remainder theorem. Multimed. Tools Appl. **78**, 16727–16753 (2019)
4. Wang, H.-Y., Lin, H.-J., Gao, X.-Y., Cheng, W.-H., Chen, Y.-Y.: Reversible AMBTC-based data hiding with security improvement by chaotic encryption, special section on security, privacy, and trust management in smart cities. IEEE Access **7**, 38337–38347 (2019)
5. Ma, S., Zhang, Y., Yang, Z., Hu, J., Lei, X.: A new plaintext-related image encryption scheme based on chaotic sequence. IEEE Access **7**, 30344–30360 (2019)
6. Liao, Y., Fan, Y., Liang, Y., Liu, Y., Mohammed, R.: Cryptanalysis of an identity-based encryption scheme with equality test and improvement. IEEE Access **7**, 75067–75072 (2019)
7. Wu, H.-T., Yang, Z., Cheung, Y.-M., Xu, L., Tang, S.: High-capacity reversible data hiding in encrypted images by bit plane partition and MSB prediction. IEEE Access **7**, 62361–62371 (2019)
8. Chuman, T., Sirichotedumrong, W., Kiya, H.: Encryption-then-compression systems using grayscale-based image encryption for JPEG images. IEEE Trans. Inf. Forensics Secur. **14**(6), 1515–1525 (2019)
9. Jung, K.-H.: Data hiding scheme improving embedding capacity using mixed PVD and LSB on bit plane. J. Real-Time Image Process. **14**(1), 127–136 (2017)
10. Liu, T.Y., Lin, K.J., Wu, H.C.: ECG data encryption then compression using singular value decomposition. IEEE J. Biomed. Health Informatics **22**(3), 707–713 (2018)
11. Wang, Y., Cai, Z., He, W.: A new high capacity separable reversible data hiding in encrypted images based on block selection and block-level encryption. IEEE Access **7**, 175671–175680 (2019)
12. Ma, S., Liu, J., Yang, Z., Zhang, Y., Hu, J.: A pseudo-random sequence generation scheme based on RNS and permutation polynomials. Sci. China Inform. Sci. **61**, 0823041–08230412 (2018)

13. Wu, J., Liao, X., Yang, B.: Cryptanalysis and enhancements of image encryption based on three-dimensional bit matrix permutation. Sig. Process. **142**, 292–300 (2018)
14. Vimalraj, T.J., Mithun, P., Mathiyarasu, K.: Cryptosystem using chaotic shuffling hill pad encryption scheme for secure data transmission. J. Comput. Sci. Eng. **6**(4), 174–178 (2018)
15. Feng, W., He, Y., Li, H., Li, C.: Cryptanalysis and improvement of the image encryption scheme based on 2D logistic-adjusted-sine Map. IEEE **7**, 12584–12597 (2019)
16. Bhardwaj, R., Aggarwal, A.: An improved block based joint reversible data hiding in encrypted images by symmetric cryptosystem. Pattern Recogn. Lett. (2018, to be published)
17. Liao, X., Shu, C.: Reversible data hiding in encrypted images based on absolute mean difference of multiple neighbouring pixels. J. Vis. Commun. Image Represent. **28**, 21–27 (2015)
18. Zhang, X.: Reversible data hiding in encrypted image. IEEE Sig. Process. Lett. **18**(4), 255–258 (2011)

Iclique Cloak Approach for Protecting Privacy of Mobile Location with Image Processing Techniques

A. Ayyasamy[1], L. Sai Ramesh[2(✉)], and V. Sathiyavathi[3]

[1] Department of Computer Engineering, Government Polytechnic College, Nagercoil, Tamilnadu, India
samy7771@yahoo.co.in

[2] Department of Information Science and Technology, CEG, Anna University, Chennai, Tamilnadu, India
sairamesh.ist@gmail.com

[3] Department of Computer Science, JBAS College for Women, Chennai, Tamilnadu, India
sathya.mca2006@gmail.com

Abstract. The usage of handheld devices with network connectivity has become the usual one in everyone's daily life. In this routine usage, almost everyone has an experience of misplacing or losing his or her cell phone. Recently, mobile devices such as cell phones have been equipped with Global Positioning System (GPS) receptors that allow us to get the Location Based Services (LBS)-based device coordinates that are considered a key feature of many future mobile applications with image processing techniques. In LBS technology, the IClique Cloak Algorithm helps to prevent from the attackers to locate the mobile clients. It has been used to incrementally maintain the maximal clique that has been needed for location cloaking in an undirected graph by travesring the path the device and find the similarity between the images to find the traversed path of the device. In this way, the system can update the location of the mobile device continuously and sends the location whenever it requested by the authenticated user.

Keywords: Location based services · Global positioning system · IClique cloak · Image processing · Service provider

1 Introduction

The mobile devices are the essential one in the current internet scenario where it becomes the needed communication all over the world. And this become more advanced in many perspectives which is also linked with some vulnerable attacks. One of the advances in mobile device is Global Positioning System (GPS) tracking. In this routine usage, almost everyone has an experience of misplacing or losing the cell phone. Hence, to solve this problem the proposed has been developed the application in the smart phone. The lost mobile has been retrieved by IMEI number using Telephony Manager Method based on the LBS Technology.

J. I.-Z. Chen et al. (Eds.): ICIPCN 2020, AISC 1200, pp. 160–168, 2021.
https://doi.org/10.1007/978-3-030-51859-2_15

On the client side, sending a location based query request has the major drawback of location privacy. In existing, a straightforward method has to be replaced the identity with a identity value before sending the query to the service provider [1]. Most of the existing location privacy metrics are location k-anonymity and cloaking granularity.

The proposed system find the mobile device using Telephony Manager which access the phone and network properties directly. Then client side adopting the privacy metrics needed for maintains the cloaking granularity and fix the range in both high and low to locate the device when request is sent by the user. The location of the user is taken as the snapshot and stored in the server side database to retrieve the user device when its needed. The rest of the paper is organized as follows: Sect. 2 describes the literature survey done for the location privacy LBS. Section 3 describes the system design of the proposed work. Section 4 describes the system implementation details and results of each module. Finally, the conclusion of the paper is presented in Sect. 5.

2 Related Work

The previous researches focused on mobile tracking and privacy protection in Location Based Service (LBS). These existing work focus on application providing in mobile, prevent the user location and protect the location disclosure.

Chandra et al. [2] proposed the GPS Locator system to locate the devices with its current location by to execute the system commands easily. This system also allows the user to locate the nearby another location through these devices. Al-Suwaidi and Zemerly [3] proposed the user friendly system to locate the family and friends mobile devices by using LBS technology. This system enables with the alert system which alert the famile members and friends who are nearby your location and it helps to protect your device which is steal any other unknown person. Chandra et al. [4] discussed about one more kind of location sharing system using GPS with the maintenance of web server which maintains the user data. Manav Singhal and Anupam Shukla [5] proposed a location based services in the handheld android devices using Global positioning system and the web services. This work uses the web services to located the devices with the information provided to them to identify in a secure manner. This work also perceives less time to locate the devices when compared to previous existing systems.

Gruteser and Grunwald [6] provides the model to prevent the location details of the mobile user from an unauthorized user. This system uses both spatial and temporal location of the user when it locate them. It also helps to locate the path traversed by the device with the help of sptio-temporal information of the device. The basic idea is to extend an exact user location to a cloaked region that covers at least k users each region is its association with protection technique that preserve the true of the data. M.F. Mokbel et al. [7] has been proposed the cloaking granularity methodology which is used for prevent the location disclosure.

Raghuraman et al. [8] shows the image retrieval technique for using relative location based on region of interest of the user specified regions. This will help in this system to store only the needed images for the specific users. Ghinita et al. [9] proposed the system to identify the behavior of the requested user by the way it can distinguish the normal user and attackers. This work using the postpone request wantedly to check whether the requested user already known about their location to prevent themselves from the unauthorized user. Xu et al. [10] provide the query based location sharing between the users.

The image retrieval concept is mentioned in [11] helps to clustered the images which are relevant. When you are using image processing techniques, then segmentation is the important for efficient processing and the work proposed in [12] explains about segmentation with K- means clustering approach. Another work in [13] explains about to secure the image files using steganography. And also efficient retrieval process helpful to fast retrieval of image as mentioned by Raghuraman et al. [14]. And also image has to be compressed for better transmission as discussed in [15]. From the survey of existing systems, it is clearly noted that the location prevention is the important security needed for the user to avoid unknown user to known about the behavior of themselves. So, the proposed system in this paper uses the sensor devices to locate the system through the IoT connected devices with secure transmission of sharing the information between the users.

3 Proposed Work

System architecture depicts the components and its relationship in a system. Figure 1 shows the overall system architecture for protecting location privacy through IClique Cloak in LBS for mobile. In this architecture, anonymity proxy act as middleware between mobile clients and service provider. Within the anonymity proxy, cloaking engine, cloaked repository are maintained along with IClique Cloak approach.

The service provider has directly connected with the telephony manager to update the status of the mobile through the internet providers. Using that internet provider, the GPS location of the device is updated in periodic manner and also when there is any change in locations of mobile. IClique Cloak approach maintains the maximal clique and updated that in other modules of anonymity proxy. Service provider gets the query location of the mobile devices through the anonymity proxy whenever it receives the query.

And also, the clients doesn't know about the connectivity between the GPS satellite and service provider through the telephony manager maintained in the server. Once the clients send the requests service provider to know about the location of the another clients mobile, the service has to send the request to the server and through that they create the connection with GPS satellite to retrieve the accurate location of the requested client id. The working conditions of each modules are explained in the following sections.

3.1 Access IMEI Number

The Access IMEI Number is used for accessing the device unique Id using Telephony Manager. In this module IP address, password as the input fields. And Mobile information stored in Database will be the output field.

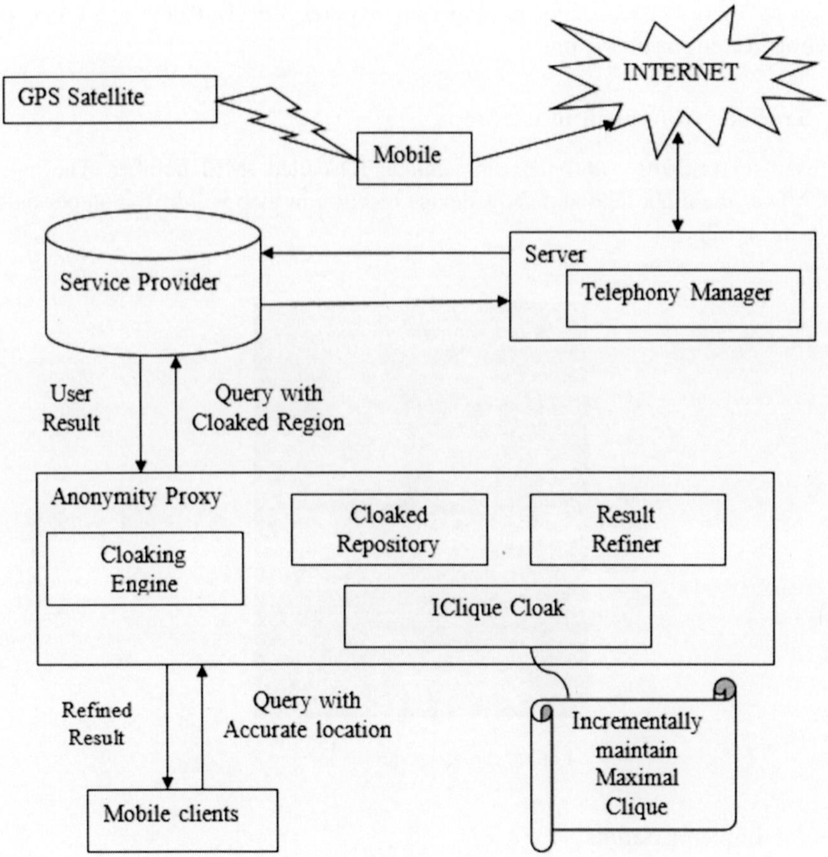

Fig. 1. System architecture for a tracking mobile using GPS

Steps for Implementation

In this module reading the phone, network, GPS states and using intent to share with other application. The detailed steps for implementation is expressed as follows:

Step 1: In Location-based services using Location Manager and Location Providers technologies to find a device's current location.
Step 2: LBS is handled by the Location Manager system Service. To access Location Manager, request an instance of the LOCATION_SERVICE using the getSystemService method.
Step 3: Before use the Location Manager need to get uses_permission from manifest xml. The GPS provider requires fine permission, while the Network provider require only coarse.
Step 4: Using getSystemService method to access the telephony APIs is managed be Telephony Manager which provides direct access the IMEI number of the device and network state details.

Step 5: To read each of those properties requires READ_PHONE_STATE uses permission in manifest xml.

3.2 Tracking Movement in the Map

It is used for tracking a mobile using unique ID called IMEI number. The mobile unique ID as the input field and show device location in map will be the output field is shown in Fig. 2.

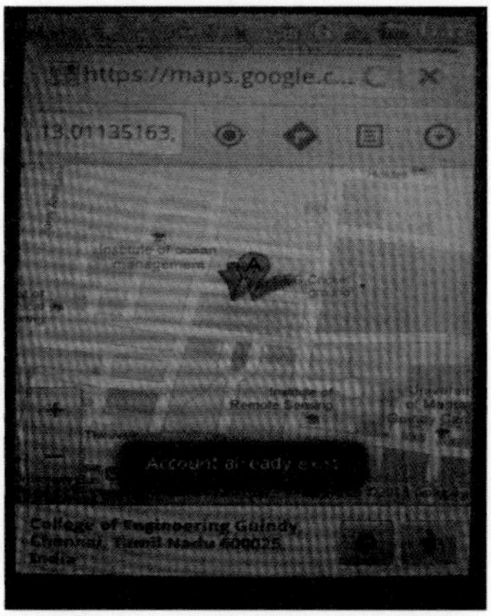

Fig. 2. Tracking device in map

Steps for Implementation

In this module updates the current location changes using Location Listener. The detailed steps for implementation is expressed as follows:

Step 1: Use the request Location Updates method to get updates whenever the current location changes, using a Location Listener. It accepts either a specific Location Provider name or a set of Criteria to determine the provider to use.
Step 2: Interact with database using url JDBC connector to update into row.
Step 3: The Map View provides an ideal user interface option for presenting geographical data and it stores the image snapshot of current location.
Step 4: Get permission from application manifest for internet access to download the Google Maps to show the mobile location and it group the users in similar locations.

3.3 Location Dependent Attack

It is used for considered cloaking granularity and K-Anonymity as the privacy metrics in client side. Location dependent attack as uses Query with accurate location as input and Detection of user identification and location linking attack as the output fields.

Steps for implementation

In this module hacker may easily access the location data and sensitive information due to LBS. The detailed steps for implementation is expressed as follows:

Step 1: Take a snapshot of user location using location k-anonymity and cloaking granularity.

Step 2: K-anonymity is associated with protection technique that preserve the truthfulness of the data and hide user information and send only query to service provider.

Step 3: User locations range are extended to cover the k users to obtain location k-anonymity

Step 4: Cloaking granularity doesn't prevent the system from attack but it will prevent the location privacy.

Step 5: Considered cloaking of the current user location with respect to the last cloaked region and gives the information only to the authenticated user or service provider.

3.4 IClique Cloak Cluster Process

It used for identify the neighbour location using Max Clique Set and create cluster process. This process uses max-clique set MCSet, a new query request as input and Update max-clique set, a set cloaked set as the output fields. It is to find a candidate cloaking set first, and then extend some sides of the cloaked region until every user in the cloaking set is free of location-dependent attacks.

4 Results and Analysis

Table 1 shows the route numbers and route names along with the connecting and disconnecting range distances in metres.

Table 1. Summary of the result

Route Number	Route Name	Distance (in meters)	
		Connecting Range	Disconnecting Range
1	Guindy	2000	2800
2	Saidapet	2500	3000
3	Mylapore	3000	3500
4	Adyar	2800	3400

Table 2. Comparison of time taken for locating mobile devices

Route number	Route name	Connecting range (metres)	Time taken for connection (secs) without IClique cloak	Time taken for connection (secs) with IClique cloak
1	Guindy	2000	10	8
2	Saidapet	2500	13	9
3	Adyar	2800	15	12
4	Mylapore	3000	20	15

The result shows that, if route number 1 has less than 2000 m then mobile can track the latitude and longitude of the other mobile. If the range is above 2800 m then the other mobile can't be tracked. Similarly for Route 2, the minimum range is 2500 m to find the other mobile and 3000 m will be the non trackable range. Similarly for route 3 and 4 the trackable range is 3000 m and 2800 m, Where as 3500 m and 3400 m will be the out of range to track the other mobile. Table 2 shows the time taken taken for locating handheld devices and compared the results with and without the usage of sensor devices in devices. In this, the devices connected with IoT system will located faster than the devices which is not using the sensor devices with them. Figure 3 shows the pictorial representation of the time taken comparison.

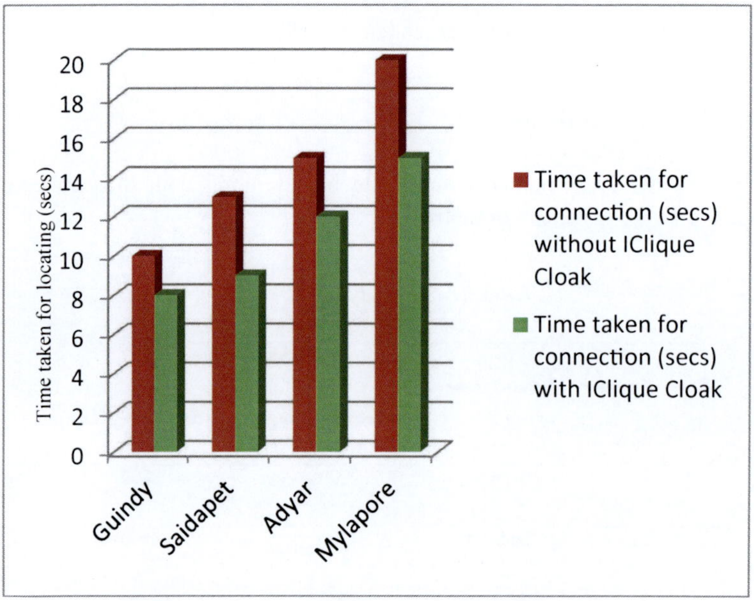

Fig. 3. Time taken comparison for locating devices

5 Conclusion

This proposed system can track the mobile exact location coordinates by the image processing techniques which extract location similarity between the mobile device with the centralized system and send to the database as a service provider. The next issue that is considered in Client side application some privacy protection is there because of using LBS. This proposed system receives the continuous location for the mobile device but can't located by unauthorized users. Only, the authorized user's can get the location of the device when they send the request through the IcliqueCloak approach. It validate the request and send the current location to the requested user. This will maintain the privacy of the mobile device where it is located and also user details. This work may further extended by implementing intelligent agents with fuzzy rules for effectively locating the handheld device.

References

1. Pan, X., Xu, J.: Protecting location privacy against location-dependent attacks in mobile services. IEEE Trans. Knowl. Data Eng. **24**(8), 123–130 (2012)
2. Chandra, A., Jain, S., Qadeer, M.A.: Design and implementation of location awareness and sharing system using GPS and 3G/GPRS. Int. J. Multimedia Ubiquitous Eng. **7**(4), 167–172 (2012)
3. Al-Suwaidi, G.B., Zemerly, M.J.: Locating friends and family using mobile phones with global positioning system (GPS). In: 2009 IEEE/ACS International Conference on Computer Systems and Applications, pp. 145–152. IEEE (2009)
4. Chandra, A., Jain, S., Qadeer, M.A.: GPS locator: an application for location tracking and sharing using GPS for JAVA enabled handhelds. In: 2011 International Conference on Computational Intelligence and Communication Networks, pp. 234–240. IEEE (2011)
5. Singhal, M., Shukla, A.: Implementation of location based Services in Android using GPS and Web services. Int. J. Comput. Sci. Issues **9**(1), 78–83 (2012)
6. Gruteser, M., Grunwald, D.: Anonymous usage of location-based services through spatial and temporal cloaking. In: Proceedings of First International Conference Mobile Systems, Applications and Services, (MobiSys 2003), pp. 163–168 (2003)
7. Mokbel, M.F., Chow, C.Y., Aref, W.G.: The new casper: query processing for location services without compromising privacy. In: Proceedings of 32nd International Conference Very Large Data Bases (VLDB 2006), pp. 95–100 (2006)
8. Raghuraman, G., Sabena, S., Sairamesh, L.: Image retrieval using relative location of multiple ROIS. Asian J. Inform. Technol. **15**(4), 772–775 (2016)
9. Ghinita, G., Damiani, M.L., Silvestri, C.: Preventing velocity-based linkage attacks in location-aware applications. In: Procceedings of 17th ACM SIGSPATIAL International Conference Advances in Geographic Information Systems (GIS 2009, pp. 56–61 (2009)
10. Xu, J., Tang, X., Hu, H., Du, J.: Privacy-conscious location-based queries in mobile environments. IEEE Trans. Parallel Distrib. Syst. **21**(3), 313–326 (2010)
11. Sabena, S., Yogesh, P., SaiRamesh, L.: Image retrieval using canopy and improved K mean clustering. In: International Conference on Emerging Technology Trends (ICETT), pp. 15–19 (2011)

12. De, S., Rakshit, S., Biswas, A., Saha, S., Datta, S.: Implementation of real-time skin segmentation based on K-Means clustering method. In: Smys, S., Tavares, J.M.R.S., Balas, V.E., Iliyasu, A.M. (eds.) ICCVBIC 2019. AISC, vol. 1108, pp. 964–973. Springer, Cham (2020)
13. Vinothkanna, R.: A secure steganography creation algorithm for multiple file formats. J. Innovative Image Process. (JIIP) **1**(01), 20–30 (2019)
14. Raghuraman, G., Ananth, J.P., Shunmuganathan, K.L., Sairamesh, L.: Krawtchouk moment based interactive image retrieval algorithm. J. Comput. Theor. Nanosci. **12**(12), 5562–5565 (2015)
15. Sharieff, A.H., Sabena, S., Sathiyavathi, V., SaiRamesh, L.: Intelligent framework for joint data hiding and compression using SMVQ and fast local image in-painting. Int. J. Sci Technol. Res. **9**(2), 2267–2271 (2020)

Deep Learning of Robust Representations for Multi-instance and Multi-label Image Classification

Jesus Silva[1(✉)], Noel Varela[1], Fabio E. Mendoza-Palechor[1], and Omar Bonerge Pineda Lezama[2]

[1] Universidad de la Costa, St. 58 #66, Barranquilla, Atlántico, Colombia
jesussilvaUPC@gmail.com, {nvarela2, fmendozal}@cuc.edu.co
[2] Universidad Tecnológica Centroamericana (UNITEC), San Pedro Sula, Honduras
omarpineda@unitec.edu

Abstract. In multi-instance problems (MIL), an arbitrary number of instances is associated with a class label. Therefore, the labeling of training data becomes simpler (since it is done together, instead of individually) with the disadvantage that a weakly supervised database is produced [9]. In the PCRY, each restaurant is represented by a set of images that share the attribute label(s) of that establishment. This paper explores the use of previously learned attribute extractors, trained in 3 different databases that are similar and complementary to the PCRY database.

Keywords: Deep learning · Multi-instance · Multi-label · Image classification

1 Introduction

In recent years, relying on increased computational capabilities such as GPUs, machine vision tasks have been dominated by Deep Learning (DL) models [1] and, particularly, by convolutional neural networks (CNNs). ImageNet Large Scale Visual Recognition Challenge (ILSVRC) [2] has stimulated competition to propose convolutional models such as VGG [3], GoogleNet [4], and Residual Networks [5]. These consist of two sections: a convolutional section, for the extraction of features at hierarchical levels of complexity, and the classification section, where the interrelationship between features is learned.

One advantage of DL is that models are automatically trained from start to finish, avoiding the use of human-designed attribute extractors. This is possible because low-level attribute extractors are used in higher layers to build high-level attributes that are specific to the database under study. Additionally, the weights learned by these models can be used in other problems such as object detection and segmentation [6] or other classification problems. One technique that allows the use of previously trained weights on a database other than the task at hand is enhanced transfer-learning with

J. I.-Z. Chen et al. (Eds.): ICIPCN 2020, AISC 1200, pp. 169–178, 2021.
https://doi.org/10.1007/978-3-030-51859-2_16

fine-tuning [7]. There are still unanswered questions about learning representations, such as how good the optimized representation is for a problem, given different training databases.

Most large-scale CNN models are focused on the task of image classification. The Yelp Restaurant Classification Challenge (YRCC)1, defined on a database of labeled restaurants, is a multi-instance (the number of images per restaurant is not fixed) and multi-label image classification problem (there are 9 possible attributes to be selected). This allows exploring ideas in representation learning, in particular, how to induce robust characteristics for restaurants through pre-trained CNNs on different training sets.

Following the reports in [8], this question is addressed by aggregating high-level characteristics obtained from a VGG network to generate a characteristic vector per restaurant. This is done with the feature extractors pre-trained in 3 different databases. Once the characteristics vectors per restaurant are calculated, the fine-tuning procedure is applied in each of the classification modules of the VGG networks to predict, using a system of binary classifiers, if an attribute applies or not for a restaurant.

To improve system performance, swarm learning is applied using a 4-fold structure, resulting in a total of 36 trained grading modules. This method offers the advantage of maximizing learning that is transferred using 3 complementary databases, inducing a robust representation per restaurant. Another advantage is the inclusion of a final classifier that integrates the predictive capabilities of each representation per label.

2 Background

In multi-instance problems (MILs), an arbitrary number of instances is associated with a class tag. Therefore, labeling of training data becomes easier (as it is done together, rather than individually) with the disadvantage that a weakly supervised database is produced [9]. In the PCRY each restaurant is represented by a set of images that share the attribute label (s) of the said establishment; some examples of training sets are shown in Fig. 1. A function is required that relates the labels per instance to the labels per set. Under the multi-instance standard assumption (SMIA) a set belongs to the 'positive' class if at least one of its instances is 'positive' [10]. Following SMIA, a MIL problem can be seen as a single instance classification problem, by assigning the labels of the set to the instances associated with it. However, this is not the only possible assumption for MIL problems [11], and our problem, it is not advantageous. An alternative was proposed in [6], where the representation of attributes of a set is obtained by adding the characteristics extracted from the instances associated with it, and said aggregated representation becomes a prototype vector for the set. The advantage is that the network can learn which of these characteristics are useful for discriminating between classes. In this work, we follow this last scheme.

Fig. 1. Examples of training images for 3 random restaurants (in rows) [5].

In addition to being a multi-instance problem, PCRY is also a multi-label problem. One approach to addressing multi-label problems is to train as many classifiers as the cardinality of the power set of the labels [12]. Thus, a system could learn the details of each possible configuration. However, this approach could lead to a severe lack of representatives of each configuration, due to its exponential nature: for our particular problem, it would imply training 29 = 512 classifiers using only 2000 attribute vectors where not all the configurations are present and therefore they cannot be learned. A simpler approach is to convert the multi-label problem to a single-label problem [13]. This can be accomplished through a system of 'binary classifiers' apply'/'not apply', one for each available tag. This last approach is the one we follow in this work.

Transfer of high-level attribute extractors, learned from large-scale databases such as ImageNet, can be useful when addressing a similar visual recognition problem that presents a limited amount of training data [6]. One of the first and most exhaustive works on the transfer of parameters learned from a CNN is [14], where the authors found that performance degradation between the original task and the new task is dictated by the dissimilarity between them. Additionally, they concluded that low-level extractors (first convolutional layers) are generic in machine vision tasks, while high-level extractors (last convolutional layers and fully connected layers) are database-specific of training. In this work, we explore the use of previously learned attribute extractors, trained in 3 different databases that we consider to be similar and complementary to the PCRY database.

3 Proposal

The method consists of the following steps:

1. Take advantage of the representation capabilities of the previously trained CNN architectures, by using the transfer-learning technique on their weights. For this task it was decided to use the VGG network because it is the standard CNN architecture with a competitive accuracy in large-scale visual recognition tasks [15].
2. Once the pre-trained weights are loaded, the high-level attributes for each image in the FC6 layer are extracted. The FC6 layer (more specific attributes) is chosen instead of the last convolutional layer (more generic attributes) as a way of regularization, considering that a large proportion of the VGG-16 network parameters connect the last convolutional layer with the FC6 layer (100 million parameters). It is assumed that the learning of so many parameters using a slightly supervised database would lead to an over-adjustment [16].
3. The representation extracted for the *j*-th instance (imagen) of the *i*-th set (restaurant) is denoted as *hij*. According to this notation, the aggregate representation of a set is given by $\hat{hi} = (hi1, hi2, \ldots .hin)$. The function *f* encodes the mapping of the attributes at the instance level to the attributes at the set level. In this paper, the average function $\hat{hi} = (hij)$ is used to obtain the prototype attribute vector for the *i*-th restaurant. This choice follows the assumption that similar triggers will be present in restaurants that share a class label [17].
4. Once the attribute vectors by the restaurant are obtained, the researcher begins to apply fine-tuning over the FC7 layer and the prediction layer. The weights in the FC7 layer are initialized with the optimal values learned for the original task. The prediction layer is modified to contain a single sigmoid unit to produce a binary classifier by attribute [18].
5. Finally, since not every representation is equally useful in predicting a label, a final classifier is implemented with 9 output sigmoid neurons and 27 input units (9 for each representation) [19]. In this way, the final module is added to an automatic end-to-end system that avoids the use of human heuristics to select the appropriate weights of each representation for a given label.

4 Experimental Design

The trained weights from the ImageNet and MIT Places 2 databases were downloaded from Caffe Model Zoo2. For Food-101, there is no such weight, so the ImageNet weights were fine-tuned using Food-101 until the loss of validation converged [20, 21].

The validation folders were used for the adjustment of hyper-parameters during the training of the binary classifiers. A learning rate (*lr*) was selected with a high reduction of validation loss during the initial times that considerably decreases as training progresses. Given the power of representation of the network and the potential for over-adjustment, the classifiers were regularized with a $dropou\_rate = 0.5$ for the input attributes and FC7 activations. The hyper-parameters used are: lr = 10–6, epochs = 100. These parameters were kept fixed for all trained classifiers. The pseudocode for the training of the binary classifiers is shown in Fig. 2 and that of the final classifier is shown in Fig. 3.

```
# Training of Representation-Dependent Binary Classifiers
for dataset ∈ {ImageNet, Places, Food-101} do:
    DBFC ← aggregated bag features, per restaurant
    shuffle the businesses index BI
    Instances n ← number of restaurants / number of folds
    for fold ∈ {0, 1, 2, 3} do:
        folds[fold] = BI[n×fold to n×(fold + 1)]
    end for
    for attribute j ∈ {0, 1, ..., 8} do:
        for validation fold k ∈ {0, 1, 2, 3} do:
            for fold ∈ {0, 1, 2, 3} do:
                if fold is not k do:
                    append folds[fold] to train_idx
                end if
            val_idx ← folds[k]
            end for
            Train classifier_j,k on DBFC[train_idx]        // DBFC: Dataset Business FC
            ValProbs_j,k ← predict probabilities using val_idx
            ValClasses_j,k ← predict classes using val_idx
        end for
        datasetValProbs_j ← concatenate the ValProb_j,k on k axis
        datasetValClasses_j ← concatenate the ValClases_j,k on k axis
    end for
    datasetValProbs ← concatenate datasetValProbs_j on j axis
    datasetValClasses ← concatenate datasetValClasses_j on j axis
end for
Calculate validation accuracy with datasetValClasses
valProbs ← concatenate datasetValProbs on dataset
```

Fig. 2. Pseudocode of the binary classifier for each representation

```
# Final classifier
valTargets ← get the restaurant target attributes
for validation fold k ∈ {0, 1, 2, 3} do:
    for fold ∈ {0, 1, 2, 3} do:
        if fold is not k do:
            append folds[fold] to train_idx
        end if
    end for
    val_idx ← folds[k]
    Train FinalClassifier_k on ValProbs[train_idx]
    Evaluate FinalClassifier_k on ValProbs[val_idx]
end for
Get the validation accuracy for the final classifier
```

Fig. 3. Final classifier pseudo code

The adjustment process on the weights in the FC layers is as follows: train the weights in the prediction layer while keeping the FC7 weights fixed. Once the loss of validation is stabilized, the FC7 weights are defrosted [22].

The objective is to obtain a better representation through the FC6 input attributes in the FC7 layer. Adam was chosen as the optimizer because it favorably compares to other optimizers [22]. The official metrics for the challenge is the F1 score, defined for the *C* possible restaurant attributes based on the metrics of *Precision* and *Recall* which are calculated as shown below [23, 24]:

$$Precision = \frac{\sum_{k \in C} TP_k}{\sum_{k \in C} TP_k + FP_k}, \tag{1}$$

$$Recall = \frac{\sum_{k \in C} TP_k}{\sum_{k \in C} TP_k + FN_k}, \tag{2}$$

where *TP*: True Positives, *FP*: False Positives and *FN*: False Negatives. The *F*1 *Score* is the harmonic mean between *Precision* and *Recall* [24]:

$$F1\,Score = \frac{2 \times Precision \times Recall}{Precision \times Recall}. \tag{3}$$

5 Results

The precision results in the validation set obtained for each representation is first presented individually: ImageNet, MIT Places 2, Food-101, and for the final classifier, the results on the evaluation set, obtained from the competition's website.

Figures 4 and 5 (ImageNet, Food101) show the validation accuracy, averaged for all 4 folds, during the training phase. The solid line indicates frozen weights in the FC7 layer, where the dotted line indicates that all network parameters are free to be updated. Also, note that the label box is arranged in descending order for validation accuracy: top-down, left-right.

One of the assumptions when using lower-level input features FC6 (instead of FC7) is that they could lead to better representation in the hidden layer FC7 by using the fine-tuning technique; however, the use of the extracted features in the layer FC7 (represented by the solid line) does not affect the validation accuracy, meaning that this is because the training set is not large enough to train so many parameters and the optimizer does not induce a better representation.

Because fine-tuning is applied to the ImageNet weights for the food recognition task, the attribute extractors became more robust, including knowledge of the original ImageNet 1000 classes and the target task containing 101 food varieties.

Another trend that persists is the lower performance of classifiers trained with the MIT Places 2 representation. The hypothesis is that this is due to the discrepancy between the tasks: in the YRCC, the image sets consist mainly of food images and rarely of natural or human scenes. Although it is assumed that the label 'outdoor seating' could make use of attributes found in MIT Places 2, it was noted that this label is consistently the most difficult to predict because it contains little, and even ambiguous, visual information about whether the label applies or not. The prediction of the 'has table service' attribute is considerably better in all representations, particularly with the ImageNet/Food-101 representations. This makes sense for two reasons: firstly, there is a consistent correlation between the label and the presence of menus which makes it an easy class to predict, and secondly, the menu class is part of the original 1000 classes in ImageNet. Relevant to this same idea is that even with the fine-tuning over the original representation of ImageNet, its representation capabilities in Food-101 were not lost, which leads to conclude that the learning capability of VGG-16 exceeds the Food-101 database and therefore most of the attribute extractors kept their original weights (Fig. 6).

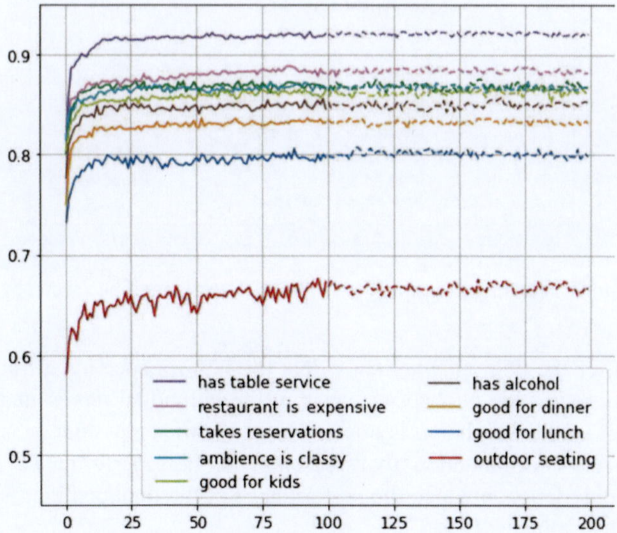

Fig. 4. Results (validation) accuracy for ImageNet representation (accuracy vs. time)

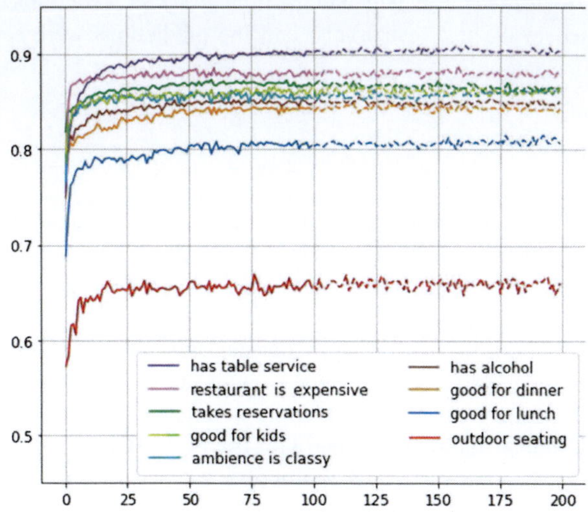

Fig. 5. Results (validation) accuracy for the MIT places 2 representation.

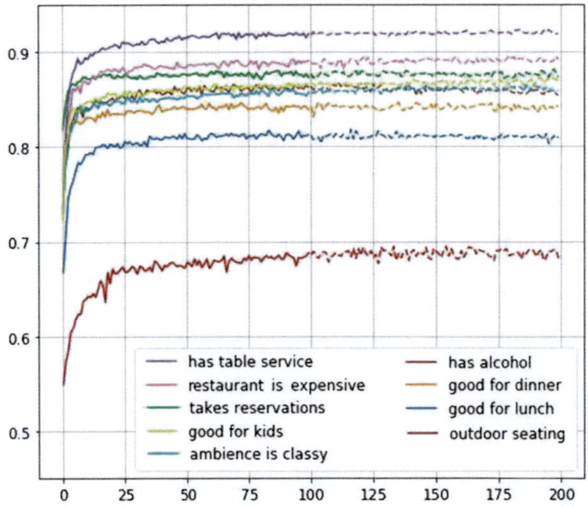

Fig. 6. Results (validation) accuracy for Food-101 representation (accuracy vs. time)

The results for the final qualifier show that the performance by attribute, in almost all cases, is as good as the best performance of the individual representations. From a detailed analysis of the weights, It is observed that the final classifier is assigning more importance to the representation that presents the best performance in validation accuracy. The only label on which the performance of the final qualifier is significantly worse is ‘good for dinner’ [25–27].

This result is due to the strong dependence of the attribute on images of food, and specifically on images of expensive dishes, such as porcelain plates (a class that is not represented in ImageNet/MIT Places 2 but appears in Food-101 as a useful attribute). Once the swarm of attribute classifiers and the final classifier were trained, the complete model was applied in the test restaurants, and the predictions were sent to the competition page [10, 28].

6 Conclusion

The representation of characteristics by restaurants, obtained by aggregating high-level characteristics from their image set, is useful when addressing a multi-instance problem due to the conformity with the assumption that certain characteristics with similar magnitude appear for restaurants with the same class label. The high-level representation directly influences the performance of each attribute classifier. This is exemplified by the ‘has table service’ label, where weights learned from the original database extract attributes from objects that are strongly correlated (i.e., menus) with the label [29, 30]. The additional fine-tuning stage with the Food-101 database, using weights pre-trained with ImageNet, was useful to improve the performance of the representation, due to its ability to discriminate classes in both databases.

As more pre-trained CNN weights become available it is expected to achieve better performance using end-to-end models that extract optimal performance by representation.

Although more evidence and argumentation are needed, the present method could be extended and applied in other classification tasks. Future research contemplates the inclusion of a data augmentation scheme to circumvent one of the main factors that limited the performance of this model.

References

1. Chen, Z., Chi, Z., Fu, H., Feng, D.: Multi-instance multi-label image classification: a neural approach. Neurocomputing **99**, 298–306 (2013)
2. Zhang, Y., Wang, Y., Liu, X.Y., Mi, S., Zhang, M.L.: Large-scale multi-label classification using unknown streaming images. Pattern Recogn. **99**, 107100 (2020)
3. Li, P., Chen, P., Xie, Y., Zhang, D.: Bi-modal learning with channel-wise attention for multi-label image classification. IEEE Access **8**, 9965–9977 (2020)
4. Yu, W.J., Chen, Z.D., Luo, X., Liu, W., Xu, X.S.: DELTA: a deep dual-stream network for multi-label image classification. Pattern Recogn. **91**, 322–331 (2019)
5. Wang, S., Zhu, Y., Yu, L., Chen, H., Lin, H., Wan, X., Fan, X., Heng, P.A.: RMDL: recalibrated multi-instance deep learning for whole slide gastric image classification. Med. Image Anal. **58**, 101549 (2019)
6. Loukas, C., Sgouros, N.P.: Multi-instance multi-label learning for surgical image annotation. Int. J. Med. Robot. Comput. Assist. Surg. **16**, e2058 (2019)
7. Zhang, M., Li, C., Wang, X.: Multi-view metric learning for multi-label image classification. In: 2019 IEEE International Conference on Image Processing (ICIP), pp. 2134–2138. IEEE, September 2019
8. Varela, N., Silva, J., Gonzalez, F.M., Palencia, P., Palma, H.H., Pineda, O.B.: Method for the recovery of images in databases of rice grains from visual content. Procedia Comput. Sci. **170**, 983–988 (2020)
9. Song, L., Liu, J., Qian, B., Sun, M., Yang, K., Sun, M., Abbas, S.: A deep multi-modal CNN for multi-instance multi-label image classification. IEEE Trans. Image Process. **27**(12), 6025–6038 (2018)
10. Yang, Y., Fu, Z.Y., Zhan, D.C., Liu, Z.B., Jiang, Y.: Semi-supervised multi-modal multi-instance multi-label deep network with optimal transport. IEEE Trans. Knowl. Data Eng. (2019)
11. Viloria, A., Acuña, G.C., Franco, D.J.A., Hernández-Palma, H., Fuentes, J.P., Rambal, E.P.: Integration of data mining techniques to PostgreSQL database manager system. Procedia Comput. Sci. **155**, 575–580 (2019)
12. Tsoumakas, G., Katakis, I., Vlahavas, I.: Data Mining and Knowledge Discovery Handbook (2009)
13. Yosinski, J., Clune, J., Bengio, Y., Lipson, H.: How transferable are features in deep neural networks? In: Ghahramani, Z., Welling, M., et al. (eds.) Advances in Neural Information Processing Systems, pp. 3320–3328 (2014)
14. Hu, H., Cui, Z., Wu, J., Wang, K.: Metric learning-based multi-instance multi-label classification with label correlation. IEEE Access **7**, 109899–109909 (2019)
15. Bossard, L., Guillaumin, M., Van Gool, L.: Food-101: mining discriminative components with random forests. In: European Conference on Computer Vision (2014)

16. Zeng, T., Ji, S.: Deep convolutional neural networks for multi-instance multi-task learning. In: 2015 IEEE International Conference on Data Mining, pp. 579–588. IEEE, November 2015
17. Li, J., Liu, J., Yongkang, W., Nishimura, S., Kankanhalli, M.: Weakly-supervised multi-person action recognition in 360° videos. In: The IEEE Winter Conference on Applications of Computer Vision, pp. 508–516 (2020)
18. Zhu, M., Li, Y., Pan, Z., Yang, J.: Automatic modulation recognition of compound signals using a deep multi-label classifier: a case study with radar jamming signals. Sig. Process. **169**, 107393 (2020)
19. Yang, H., Tianyi Zhou, J., Cai, J., Soon Ong, Y.: MIML-FCN + : multi-instance multi-label learning via fully convolutional networks with privileged information. In: Proceedings of the IEEE Conference on Computer Vision and Pattern Recognition, pp. 1577–1585 (2017)
20. Viloria, A., Lezama, O.B.P.: Improvements for determining the number of clusters in k-means for innovation databases in SMEs. ANT/EDI40, pp. 1201–1206 (2019)
21. Li, D., Wang, J., Zhao, X., Liu, Y., Wang, D.: Multiple kernel-based multi-instance learning algorithm for image classification. J. Vis. Commun. Image Represent. **25**(5), 1112–1117 (2014)
22. Feng, S., Xiong, W., Li, B., Lang, C., Huang, X.: Hierarchical sparse representation based multi-instance semi-supervised learning with application to image categorization. Sig. Process. **94**, 595–607 (2014)
23. Zhu, F., Li, H., Ouyang, W., Yu, N., Wang, X.: Learning spatial regularization with image-level supervisions for multi-label image classification. In: Proceedings of the IEEE Conference on Computer Vision and Pattern Recognition, pp. 5513–5522 (2017)
24. Shang, J., Hong, S., Zhou, Y., Wu, M., Li, H.: Knowledge guided multi-instance multi-label learning via neural networks in medicines prediction. In: Asian Conference on Machine Learning, pp. 831–846, November 2018
25. Wu, J.S., Huang, S.J., Zhou, Z.H.: Genome-wide protein function prediction through multi-instance multi-label learning. IEEE/ACM Trans. Comput. Biol. Bioinform. **11**(5), 891–902 (2014)
26. Ding, X., Li, B., Xiong, W., Guo, W., Hu, W., Wang, B.: Multi-instance multi-label learning combining hierarchical context and its application to image annotation. IEEE Trans. Multimed. **18**(8), 1616–1627 (2016)
27. Cheplygina, V., de Bruijne, M., Pluim, J.P.: Not-so-supervised: a survey of semi-supervised, multi-instance, and transfer learning in medical image analysis. Med. Image Anal. **54**, 280–296 (2019)
28. Shakya, S.: Machine learning based nonlinearity determination for optical fiber communication-review. J. Ubiquit. Comput. Commun. Technol. (UCCT) **1**(02), 121–127 (2019)
29. Laib, L., Allili, M.S., Ait-Aoudia, S.: A probabilistic topic model for event-based image classification and multi-label annotation. Sig. Process. Image Commun. **76**, 283–294 (2019)
30. García-Domínguez, M., Domínguez, C., Heras, J., Mata, E., Pascual, V.: FrImCla: a framework for image classification using traditional and transfer learning techniques. IEEE Access **8**, 53443–53455 (2020)

Classification of Mitochondrial Network Images Associated with the Study of Breast Cancer

Jesus Silva[1(✉)], Noel Varela[1], Esperanza Diaz Arroyo[1], and Omar Bonerge Pineda Lezama[2]

[1] Universidad de La Costa, St. 58 #66, Barranquilla, Atlántico, Colombia
jesussilvaUPC@gmail.com, {nvarela2,ediaz5}@cuc.edu.co
[2] Universidad Tecnológica Centroamericana (UNITEC), San Pedro Sula, Honduras
omarpineda@unitec.edu

Abstract. Within various cellular processes, an increase in fission (a division of a single organelle into two or more independent structures) causes mitochondrial fragmentation and an increase in fusion (the opposite reaction of fission) produces a network of mitochondria that counteracts metabolic processes [1]. A balance between fission and fusion defines a mitochondrial morphology whose purpose is to meet metabolic demands and ensure removal of damaged organelles. These events have been associated with proliferation and redistribution of mitochondria, allowing the study of different breast cancer subtypes [2, 3]. This study presents a classification method for images of mitochondrial networks extracted from different cellular lines (MCF10A, BT549, MDAMB23, and CMF) belonging to different breast cancer subtypes.

Keywords: Classification of mitochondrial · Network images · Breast cancer

1 Introduction

This study aims to classify images of mitochondrial networks (more than one mitochondrion), based on the four types of morphology shown in Fig. 1: (i) completely tubular, (ii) tubular with fragments, (iii) Fragmented with tubules, (iv) completely fragmented, which were defined for the study of breast cancer [4]. This classifier allows experts in the field to speed up the classification process normally applied in a visual way and also aims to reduce the error that could be derived from factors related to

J. I.-Z. Chen et al. (Eds.): ICIPCN 2020, AISC 1200, pp. 179–189, 2021.
https://doi.org/10.1007/978-3-030-51859-2_17

perception parameters or human conditions. The set of images were extracted from the following cell lines (MCF10A, BT549, MDAMB23, and CMF), which are the result of applied studies [5].

2 Method

Figure 2 represents the block diagram of the processing stages. In the first stage, two algorithms are applied (segmentation and thinning) to maintain the morphology of the mitochondrial network in the image and treat this morphology as a graph [6]. In the second stage, measurements of complex networks are obtained from the morphology obtained in the previous stage. Finally, the unsupervised algorithm is defined, which uses the measurements obtained to use the K-means method [7].

2.1 Stage 1

The images used are represented by the RGB channels (red, green, blue). The first step is to obtain an image in grayscale, for which the average of the value of each channel for a unit of the pixel is calculated, the Eq. 1 determines the function f (x, y) that is obtained when calculating the average value of the intensity of the channels r(x, y), g(x, y) and b(x, y) [8]:

$$f(x.y) = \frac{r(x,y) + g(x,y) + b(x,y)}{3} \tag{1}$$

2.1.1 Segmentation

In this section, the aim is to apply an algorithm to obtain the region of interest of the image, i.e. the mitochondrial network. This procedure is applied because the images contain elements (pixels) that correspond to the nucleus of the cell, and also represent part of the chemical reaction that dyes the mitochondrial network during the image capture [9]. This segmentation is processed on the f (x, y) function, for which the Otsu segmentation method is used [10].

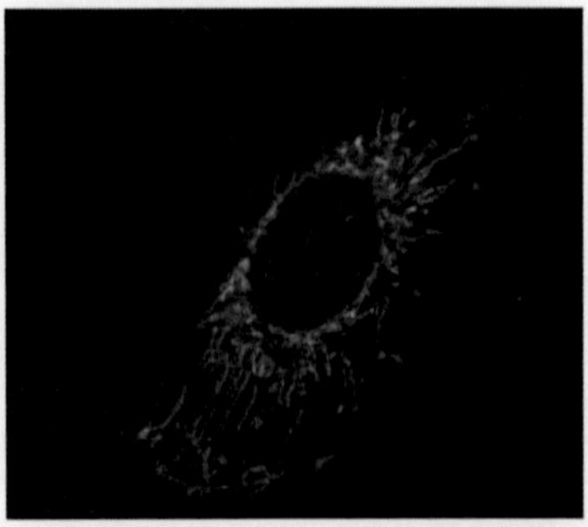

(a) Completely tubular

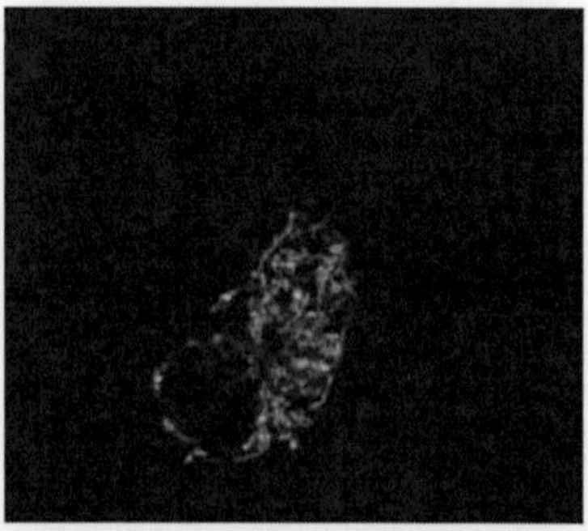

(b) Tubular wit fragments

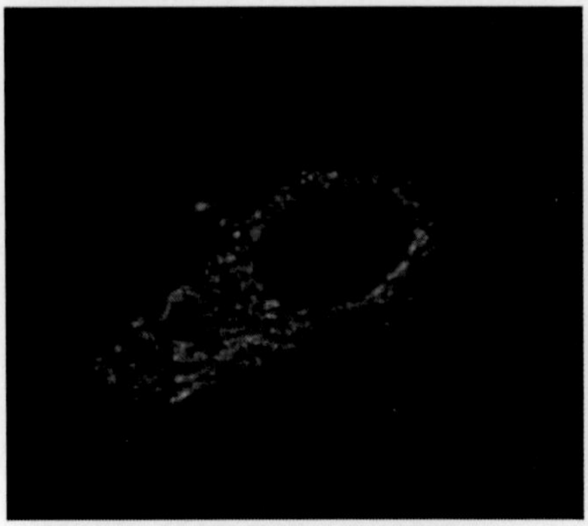

(c)Fragmented with tubules

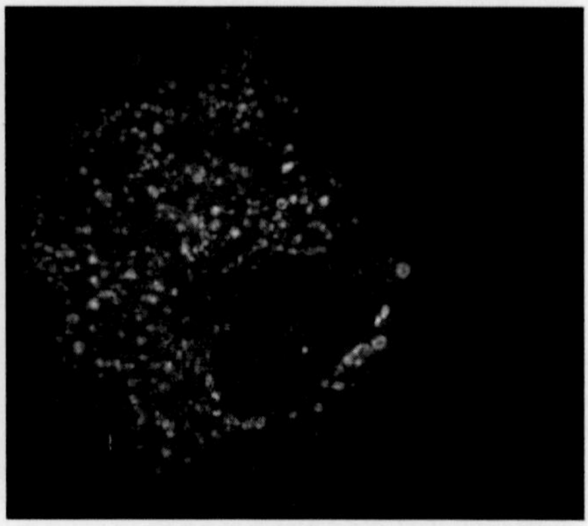

(d) Completely fragmented

Fig. 1. Classes defined for the classification process.

The Otsu algorithm consists in obtaining segmentation thresholds within a range of intensities 0–255, being the number of thresholds the number of sections between which the greyscale range is to be divided. For this study, two thresholds, t1 and t2, are defined, which allow to extract the geometry of the mitochondrial network. For finding the thresholds t1 and t2, a histogram is obtained with the probabilities of intensity of each level in the image, which are separated obtaining the probability of three classes with borders established by t1 and t2 with the objective of finding the values t1 and t2 for which the variance of each of the three classes is the maximum. These thresholds define the segmentation of the intensities in the range 0–255. Equation 2 shows the function obtained after applying the thresholds found using the Otsu method. This new image g(x, y) contains only the geometry of the network on a binary scale [11, 12].

Figure 3 shows the image obtained after applying the Otsu segmentation process, and the region of interest is defined in the image with the white intensity level [13]:

$$g(x,y) = \begin{cases} 0 & f(x,y) < t1 \\ 0 & t1 \leq f(x,y) \leq t2 \\ 1 & f(x,y) > t2 \end{cases} \quad (2)$$

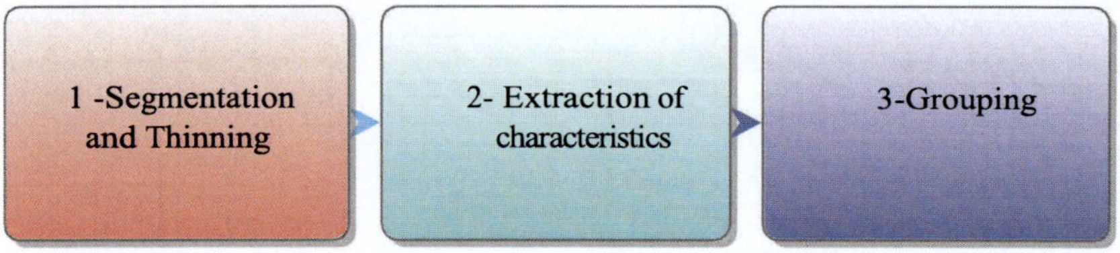

Fig. 2. Stages of the grading method.

Figure 3 shows the image f(x, y), and the binary image g(x, y) obtained in the segmentation process, which is of interest since it contains the morphology of the mitochondrial network [14].

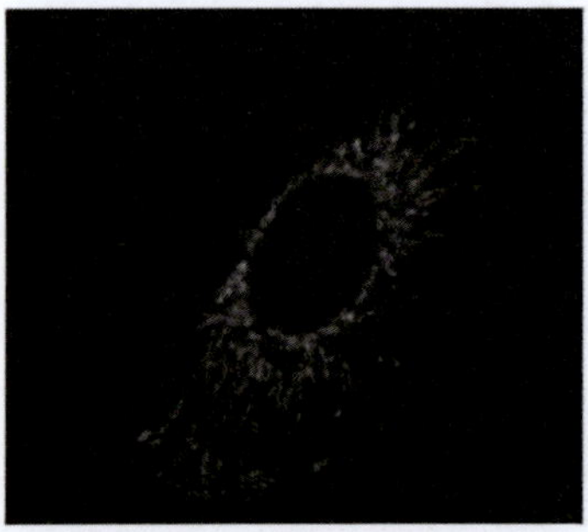

(a) Grey scale image f(x,y).

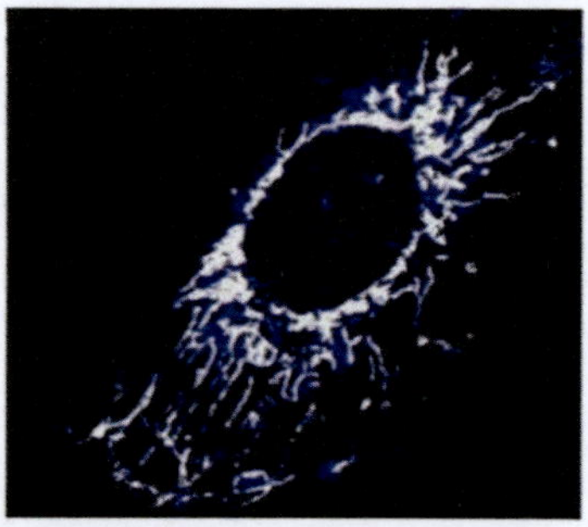

(b) Image g(x,y) obtained using Otsu

Fig. 3. Segmentation process

2.1.2 Thinning

Before the extraction of features, a thinning algorithm of the image morphology is used. The objective of this process is to transform the morphology of the mitochondrial network into a graph, keeping the parts connected. The thinning method used is based on the mathematical morphology applied to images, this type of morphology consists of defining an image as a set of elements (pixels) and applying certain operations using established sets [5], these established sets are called structural elements.

The algorithm consists of performing an operation called success-failure g(x, y) b using eight structural elements. This success-failure operation is applied to the image using each of the eight structural elements, and applying the operation to the result of operating the element 1 on the image, as defined in equation [15]:

$$\boldsymbol{h}(\boldsymbol{x},\boldsymbol{y}) = (..((\boldsymbol{g}(\boldsymbol{x},\boldsymbol{y}) \otimes \boldsymbol{b}1) \otimes \boldsymbol{b}2 \ldots \otimes \boldsymbol{b}8), \tag{3}$$

where g(x, y) is the image obtained in stage 1, and b1…b8 are the sets or structural elements operating on the image, allowing to obtain a thinning of the geometry without affecting its related properties. Figure 4 shows the image h(x, y) obtained from g(x, y), geometry that determines the graph of the mitochondrial network in the image [16].

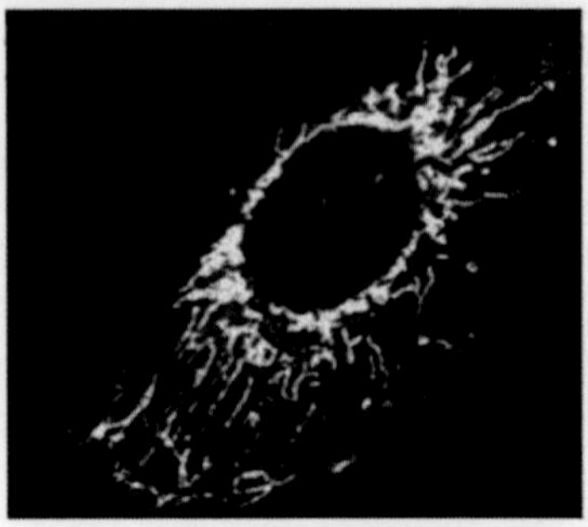

(a) Image g(x,y)

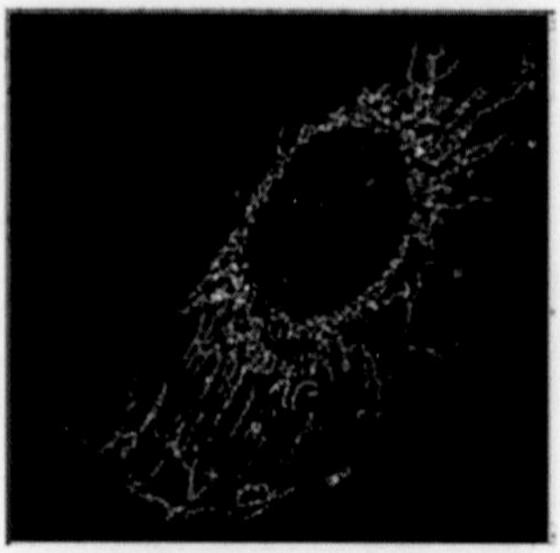

(b) Image h(x,y)

Fig. 4. Thinning process [7]

2.2 Stage 2

In this stage, there is a set of characteristics or properties that define the morphology of the mitochondrial network and are usually applied to complex networks that are topologically graphical. These characteristics define a quantitative difference in each group of images. The proposed measures are obtained by establishing a node v as the minimum unit within the image (pixel), on the other hand, the adjacencies of v with its neighboring pixels are defined as the edges e; a node v within the image is taken into account if the value of the pixel is equal to the level of white intensity in binary scale. The characteristics are described below [17, 18]:

2.2.1 Path Length

A w is defined as the average number of steps through the distance (shortest path) of all possible node pairs, where v is the total number of nodes, and d_{vivj} is the distance from node v_i to node v_j:

$$w = \frac{1}{N(N-1)} \sum_{i \neq j} d_{vi\,vj}. \tag{4}$$

2.2.2 Centrality

d_{ij} is the shortest possible path between the v_i and v_j nodes (also known as geodesic distance). The x centrality measures the distance of a node from the nearest node [19]:

$$x = \frac{N - 1}{\sum_{i \neq j} dij} \tag{5}$$

2.2.3 Eccentricity

Eccentricity y is defined as the maximum geodetic distance between nodes v_i and v_j, where $i = j$. In other words, it is the maximum distance of all node pairs [20]:

$$y = max^{N}_{vivj}\{dvivj\}. \tag{6}$$

2.2.4 Density

Z can be defined as the radius of edges in network tt concerning the maximum number of possible edges. Where N is the total number of nodes in the network and E the total number of edges [21]:

$$z = \frac{2E}{N(N - 1)} \tag{7}$$

2.3 Stage 3

In this process, the scalar value obtained in each characteristic w, x, and z is assigned to a four-dimensional vector (w, x, y, z) for each image. Then, four groups are defined corresponding to the types of morphologies defined [22]. Later, the K-means algorithm is applied to establish the grouping of the four classes [23]. Equation 8 shows the terminology established for the algorithm.

$$\begin{gathered} I(w, x, y, z)\,Vector\ image \\ mk\ centroid\ of\ cluster\ k \\ rnk\ Element\ rn\ assigned\ to\ cluster\ k \\ Xn\ Vector\ X\ of\ image\ n \\ J\ Distance\ of\ the\ vectors\ of\ group\ k\ from\ their\ centroid \end{gathered} \tag{8}$$

Once the vector (w, x, y, z) is established, the algorithm seeks to minimize the error J determined by the sum of the Euclidean distances of the vectors r concerning the centroid mk, belonging to class k, obtaining each of the centroids for each iteration, until the total error is less than an established stop condition. In this way, the four groups of images corresponding to the four types of morphology (completely tubular, tubular with fragments, fragmented with tubules, completely fragmented) are obtained [24–26].

Algorithm 1. Classification of processes

while J >error
for 1 to N

$$r_n k = \begin{cases} 1 & k = min||x_n - m_k||^2 \\ 0 & otro \end{cases}$$

$m_k = \frac{\sum_N r_{nk} x_n}{\sum_N r_{nk}}$
$J = \sum_{n=1}^{N} \sum_{k=1}^{K} ||x_n - m_k||^2$
end for
end while

3 Analysis of Results

A set of 30 test images corresponding to the four types of the morphology of the mitochondrial networks were used. This set was classified by the expert in the area after presenting the results obtained by the proposed algorithm. For testing the efficiency of the classifier and for obtaining some quality measures (accuracy, error, sensitivity, specificity) the algorithm was implemented and the results were compared against the results assigned by the expert: 5 completely tubular, 8 tubular with fragments [27], 10 fragmented with tubules and 7 fragmented. With these data, the confusion matrix shown in Table 1 was calculated. The rows of the table represent the class to which the image belongs, and the columns represent the class to which the image was assigned by the classifier. In this way, the success and failure percentage of the test set can be obtained [28, 29].

Table 1. Confusion matrix

.	Tubular	Tubular with fragments	Fragmented with tubules	Fragmented
Tubular	5	2	1	1
Tubular with fragments	2	2	1	0
Fragmented with tubules	1	4	6	0
Fragmented	0	2	0	3

Figure 5 shows the confusion matrix graph, detailing the number of successes and failures accumulated. For example, in class 1 (Completely tubular) and class 4

(Fragmented), the best classification results are obtained, showing the highest number of success. On the other hand, it is observed that the highest percentage of failures in the classification is derived from classes 2 (Tubular with fragments) and 3 (Fragmented with tubules).

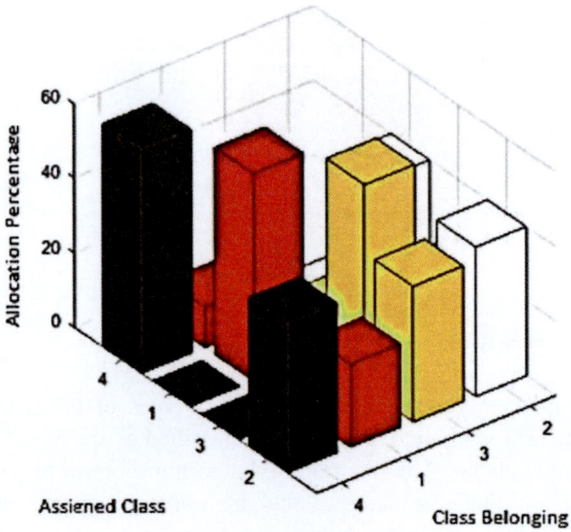

Fig. 5. Confusion matrix graph

As can be noted in the graph in Fig. 5, the highest percentage of success corresponds to classes 1 and 4. This result is because some of the proposed measures applied in the morphology graph show a notable difference from the proposed measures. However, measures such as eccentricity, centrality, and average length are vulnerable to scalability in the image, which causes a decrease or increase in the obtained magnitude, a factor that influences the percentage of success.

$$\text{Error} : 38.24\%. \tag{9}$$

The highest percentage of error is obtained in the classification of intermediate groups (2 and 3), because they present similarities in the proposed measures, in these circumstances, the grouping algorithm plays an important factor since when applying the K-means algorithm clusters are not obtained and the error does not decrease.

$$\text{Sensitivity} : 60.25\%. \tag{10}$$

The number of correct answers for each class was determined from the comparison of the results and the classification made by the expert for this test set, the sensitivity obtained is mostly defined by class 1 and 4, which is where the greatest number of real positives was obtained (items classified correctly).

$$\text{Specificity} : 85.85\%. \tag{11}$$

This percentage was obtained due to the incorrect classification in the intermediate classes (2 and 3), the magnitudes in the scalars obtained for the characteristics in these classes do not present a notable difference, therefore, an element can be poorly classified with a higher probability.

4 Conclusion

The measures proposed for the characterization of the mitochondrial networks are of great support to define the morphologies established by the researchers in the area. However, the error obtained (36.66%) should be reduced to produce better results in the studies [1]. The properties used to obtain characteristics are vulnerable to factors such as image scalability, processing time, and rotation. An example of this effect is the distance between pairs of nodes, the magnitude of which can be affected by image scalability or rotation. Due to these factors, the procedure proposes to implement supervised learning, in which the described affectation factors do not play an important role in the classification process. Also, it is possible to train the system with a determined set to increase the percentage of accuracy and to diminish the error.

References

1. Fang, Y., Zhao, J., Hu, L., Ying, X., Pan, Y., Wang, X.: Image classification toward breast cancer using deeply-learned quality features. J. Vis. Commun. Image Represent. **64**, 102609 (2019)
2. Zhang, Z., Chen, L., Humphries, B., Brien, R., Wicha, M.S., Luker, K.E., Chen, Y.-C., Yoon, E.: Morphology-based prediction of cancer cell migration using an artificial neural network and a random decision forest. Integr. Biol. **10**(12), 758–767 (2018)
3. Terao, M., Goracci, L., Celestini, V., Kurosaki, M., Bolis, M., Di Veroli, A., Vallerga, A., Fratelli, M., Lupi, M., Corbelli, A., Fiordaliso, F.: Role of mitochondria and cardiolipins in growth inhibition of breast cancer cells by retinoic acid. J. Exp. Clin. Cancer Res. **38**(1), 1–20 (2019)
4. Caino, M.C., Seo, J.H., Aguinaldo, A., Wait, E., Bryant, K.G., Kossenkov, A.V., Hayden, J. E., Vaira, V., Morotti, A., Ferrero, S., Bosari, S.: A neuronal network of mitochondrial dynamics regulates metastasis. Nat. Commun. **7**(1), 1–11 (2016)
5. Gonzalez, C.R., Woods, R.: Digital Image Processing, pp. 78–135. Prentice Hall, Upper Saddle River (2007)
6. Bishop, C.M.: Pattern Recognition and Machine Learning, pp. 423–428. Springer, Heidelberg (2006)
7. Iqbal, M.S., El-Ashram, S., Hussain, S., Khan, T., Huang, S., Mehmood, R., Luo, B.: Efficient cell classification of mitochondrial images by using deep learning. J. Opt. **48**(1), 113–122 (2019)
8. Aggarwal, S., Gabrovsek, L., Langeberg, L.K., Golkowski, M., Ong, S.E., Smith, F.D., Scott, J.D.: Depletion of dAKAP1–protein kinase A signaling islands from the outer mitochondrial membrane alters breast cancer cell metabolism and motility. J. Biol. Chem. **294**(9), 3152–3168 (2019)

9. Bindhu, V.: Biomedical image analysis using semantic segmentation. J. Innov. Image Process. (JIIP) **1**(02), 91–101 (2019)
10. Escala-Garcia, M., Abraham, J., Andrulis, I.L., Anton-Culver, H., Arndt, V., Ashworth, A., Auer, P.L., Auvinen, P., Beckmann, M.W., Behrens, S.: A network analysis to identify mediators of germline-driven differences in breast cancer prognosis. Nat. Commun. **11**(1), 1–14 (2020)
11. Reis, Y., Bernardo-Faura, M., Richter, D., Wolf, T., Brors, B., Hamacher-Brady, A., Eils, R., Brady, N.R.: Multi-parametric analysis and modeling of relationships between mitochondrial morphology and apoptosis. PLoS ONE **7**(1), e28694 (2012)
12. Hamacher-Brady, A., Stein, H.A., Turschner, S., Toegel, I., Mora, R., Jennewein, N., Efferth, T., Eils, R., Brady, N.R.: Artesunate activates mitochondrial apoptosis in breast cancer cells via iron-catalyzed lysosomal reactive oxygen species production. J. Biol. Chem. **286**(8), 6587–6601 (2011)
13. Wang, L., Ward, J., Bouyea, M., Barroso, M.: Heterogeneity of mitochondria morphology in breast cancer cells. In: Multiscale Imaging and Spectroscopy, vol. 11216, p. 112160P. International Society for Optics and Photonics (2020)
14. Darvishi, K., Sharma, S., Bhat, A.K., Rai, E., Bamezai, R.N.K.: Mitochondrial DNA G10398A polymorphism imparts maternal Haplogroup N a risk for breast and esophageal cancer. Cancer Lett. **249**(2), 249–255 (2007)
15. Crudele, F., Bianchi, N., Reali, E., Galasso, M., Agnoletto, C., Volinia, S.: The network of non-coding RNAs and their molecular targets in breast cancer. Mol. Cancer **19**(1), 1–18 (2020)
16. Jin, J., Lu, J.Q., Wen, Y., Tian, P., Hu, X.H.: Deep learning of diffraction image patterns for accurate classification of five cell types. J. Biophotonics **13**(3), e201900242 (2020)
17. Vernier, M., Dufour, C.R., McGuirk, S., Scholtes, C., Li, X., Bourmeau, G., Giguère, V.: Estrogen-related receptors are targetable ROS sensors. Genes Dev. **34**, 544–559 (2020)
18. Yang, W.S., Moon, H.G., Kim, H.S., Choi, E.J., Yu, M.H., Noh, D.Y., Lee, C.: Proteomic approach reveals FKBP4 and S100A9 as potential prediction markers of therapeutic response to neoadjuvant chemotherapy in patients with breast cancer. J. Proteome Res. **11** (2), 1078–1088 (2012)
19. Ezzati, M., Yousefi, B., Velaei, K., Safa, A.: A review on anti-cancer properties of Quercetin in breast cancer. Life Sci. 117463 (2020)
20. Viloria, A., Acuña, G.C., Franco, D.J.A., Hernández-Palma, H., Fuentes, J.P., Rambal, E.P.: Integration of data mining techniques to PostgreSQL database manager system. Procedia Comput. Sci. **155**, 575–580 (2019)
21. Smolková, K., Bellance, N., Scandurra, F., Génot, E., Gnaiger, E., Plecitá-Hlavatá, L., Rossignol, R.: Mitochondrial bioenergetic adaptations of breast cancer cells to aglycemia and hypoxia. J. Bioenerg. Biomembr. **42**(1), 55–67 (2010)
22. Smolková, K., Bellance, N., Scandurra, F., Génot, E., Gnaiger, E., Plecitá-Hlavatá, L., Ježek, P., Rossignol, R.: Mitochondrial bioenergetic adaptations of breast cancer cells to aglycemia and hypoxia. J. Bioener. Biomembr. **42**(1), 55–67 (2010)
23. Viloria, A., Lezama, O.B.P.: Improvements for determining the number of clusters in k-means for innovation databases in SMEs. ANT/EDI40, 1201–1206 (2019)
24. Hannemann, J., Velds, A., Halfwerk, J.B., Kreike, B., Peterse, J.L., van de Vijver, M.J.: Classification of ductal carcinoma in situ by gene expression profiling. Breast Cancer Res. **8** (5), R61 (2006)
25. Dekker, T.J., Balluff, B.D., Jones, E.A., Schöne, C.D., Schmitt, M., Aubele, M., Kroep, J.R., Smit, V.T., Tollenaar, R.A., Mesker, W.E., Walch, A.: Multicenter matrix-assisted laser desorption/ionization mass spectrometry imaging (MALDI MSI) identifies proteomic differences in breast-cancer-associated stroma. J. Proteome Res. **13**(11), 4730–4738 (2014)

26. Giedt, R.J., Feruglio, P.F., Pathania, D., Yang, K.S., Kilcoyne, A., Vinegoni, C., Mitchison, T.J., Weissleder, R.: Computational imaging reveals mitochondrial morphology as a biomarker of cancer phenotype and drug response. Sci. Rep. **6**, 32985 (2016)
27. Shermis, R.B., Wilson, K.D., Doyle, M.T., Martin, T.S., Merryman, D., Kudrolli, H., Brenner, R.J.: Supplemental breast cancer screening with molecular breast imaging for women with dense breast tissue. Am. J. Roentgenol. **207**(2), 450–457 (2016)
28. Varela, N., Silva, J., Gonzalez, F.M., Palencia, P., Palma, H.H., Pineda, O.B.: Method for the recovery of images in databases of rice grains from visual content. Procedia Comput. Sci. **170**, 983–988 (2020)
29. Jones, M.M., Manwaring, N., Wang, J.J., Rochtchina, E., Mitchell, P., Sue, C.M.: Mitochondrial DNA haplogroups and age-related maculopathy. Arch. Ophthalmol. **125**(9), 1235–1240 (2007)

Review on Brain Tumor Segmentation: Hard and Soft Computing Approaches

Prashant Mishra, Adhesh Garg, Diwanshi Gupta, Mayur Tuteja, Prajawal Sinha, and Sanjay Saxena(✉)

Department of Computer Science and Engineering, IIIT Bhubaneswar, Bhubaneswar, Odisha, India
sanjay@iiit-bh.ac.in

Abstract. Segmenting the brain of the human is essential to obtain an exact analysis of tumors from the brain medical image dataset. A lot of research works have conceded in novice algorithms and approaches for exact segmentation of tumor, however, as of now, no solitary standard techniques have been proposed. This article aims to emphasis over the computing methods that are soft and hard for the segregation of medical image data set on tumor cause din brain. The taxonomy of hard-computing approaches of segmentation includes edge-based segmentation, normalized cut method, mean shift method, and so on. Soft computing methods include approaches based on fuzzy logic, ANN, and genetic algorithms. Nowadays, deep learning plays a very vital for brain tumor segmentation significantly. This provides an efficient structure of the review of different pieces of literature based on soft and hard computing methods for brain tumor segmentation. After going through this paper, researchers can easily analyze the methods based on data used, observation, research results, advantages, and disadvantages.

Keywords: Brain tumor segmentation · Hard computing · Soft computing · Medical imaging

1 Introduction

Abnormal, uncontrolled, unnatural growth of cells with the potential to spread to other body parts is often associated with cancer. A malignant tumor is cancerous. A tumor or neoplasm is an anomalous lump of tissue that can be solid or fluid-filled. All tumors are not cancers but some of them can grow to be one. Brain tumors are growth of abnormal cells. Primary brain tumors are the tumors originating from the brain. The most common among them are gliomas. Others include meningiomas, pituitary adenomas, etc. Their name comes from the segment of the brain or the brain cell type where they have originated. Cancerous cells form a tumor elsewhere in the body that involves the risk of metastatic brain tumors and the process of transmission of such cells from that tumor to the brain is called metastasis. Medical image processing techniques have significantly contributed to the improvement of preliminary diagnosis and prognosis of gliomas. Some of them are "Magnetic Resonance Imaging (MRI), Single-Photon Emission Computed Tomography (SPECT), Computed Tomography (CT), Positron

J. I.-Z. Chen et al. (Eds.): ICIPCN 2020, AISC 1200, pp. 190–200, 2021.
https://doi.org/10.1007/978-3-030-51859-2_18

Emission Tomography (PET), Magnetic Resonance Spectroscopy (MRS)". The aforementioned methods give crucial data such as position, extent, metabolism, and shape of brain tumors serving well in its detection. MRI utilizes radio frequency (RF) signals to excite the targeted tissues producing its inside images in the presence of a very powerful magnetic field. Different MR sequences pictures are produced during image acquisition by varying recurrence and excitation times. Four of the standard MRI modal are used for diagnosis of gliomas and these are "Fluid Attenuated Inversion Recovery (FLAIR), 'T1'weighted MRI, 'T2'weighted MRI and 'T1' weighted MRI with gadolinium contrast enhancement (T1-Gd)". Among these "T1-weighted and T2-weighted" scans are common "MRI" sequences. T1-weighted scanned images are generated by consuming Repetition Time (TR) and short Time to Echo (TE). The intensity and disparity of the "MRI" are mainly determined by tissues' T1 properties. Contrarily, T2-weighted scanned images are generated by using longer TE and TR times. The intensity and disparity in these images are mainly determined by the tissues' T2 properties. Normal Cerebrospinal Fluid (CSF) is manifested as shady on 'T1' weighted imaging and vivid on 'T2' weighted imaging. Thus, T1 and T2-weighted scanned images can be easily distinguished by going after the signal intensity of the CSF. To protect healthy tissues from the risk of getting cancerous and eliminating cells of tumor during the treatment. Brain tumor segmentation involves analyzing, outlining, and segregating tumor tissues from regular ones.

2 Hard Computing and Soft Computing

Computing is nothing but is a mapping function mapping from set of input to output. Now, this mapping is also alternatively called as formal method also it is called an algorithm, so basically algorithm to solve a problem. The base of hard computing lies on crisp systems, solid binary logic and numerical analysis. Soft computing is based on Neural Networks, Fuzzy Logic, Heuristic Programming, Evolutionary, Genetic Computing and Probabilistic Reasoning. Unlike this, Hard Computing requires a precisely state analytic model. Categoricity and precision is maintained by Hard Computing whereas dispositional and approximation is maintained by Soft Computing. Soft Computing has the ability to upgrade it's programs on its own whereas programs need to be written in Hard Computing. Soft Computing incorporates stochasticity and deals with ambiguous, noisy data whereas Hard Computing has a deterministic nature and has exact input data requirements. Soft Computing allows parallel computations whereas Hard Computing is strictly sequential. Traditional control systems fail to manage the ever-growing complexity of modern machinery due to a lack of accurate models that describe their behavior. A shift to soft computing methodologies is proving effective in tackling these problems.

3 Analytical Review of the Papers

3.1. "A Novel Data Mining Approach for Brain Tumor Detection" [1] - Technique: After pre-processing, segmentation, and feature extraction of both the testing and training set, the extracted features are compared with an ID3 classifier.

Observation: A Java-based simulator is used to log the experiment results and validate them with the existing approach. Result: It gave positive results and the application can be effectively integrated with software-based healthcare products.

Input: Both normal images and classified brain tumor images are collected from the Internet. Pre-processing: Median filtering techniques were used considering the brightness, darkness, slight darkness, and original image. Advantages: This prototype can be used for the efficient segmentation process and increasing QoS by the healthcare industry.

3.2. "Classification of MRI Brain Image and Tumor Detection using SVM and Wavelet Transform" [2] -Technique: Morphological operators and Ostu's thresholding is used to extract tumor region and goes through anisotropic diffusion filters. Observation: The accuracy of the proposed algorithm is measured.

Result: The results were quite satisfied with an accuracy of 86%. The tumor (if detected) was successfully classified as critical, medium, or normal. Input: The dataset used are T2, FLAIR, and T1 MRI images. Pre-processing: Tumor boundary is detected using a morphological operator and Otsu's thresholding. The noise is removed by anisotropic diffusion filter. Advantages: Brain tumor was classified using SVM which provided better results than the previously proposed methods like fuzzy c-means algorithm.

3.3. "Extract tumor section from Brain using MR Images Soft-Computing assisted tool [3] -Technique: Social Group Optimization assisted Kapur's multi-thresholding is used to enhance the tumor of the MRI", followed by application of morphological operations for grouping pixels. Observation: The ground truth and the segmented tumor region are compared based on jaccard, sensitivity, dice, accuracy, and specificity. Result: The proposed approach gives the following accuracy- Dice(90.74%), specificity (91.55%), sensitivity (97.81%), Jaccard (83.31%), and accuracy (96.16%) for the considered dataset. Input: BRAINIX, CEREBRIX, and MICCAI BRATS challenge dataset is used. Pre-processing: An optimal threshold vector is found by integrating Kapur's function and Social Group Optimization (SGO) by increasing the overall entropy of the image. Advantages: The above method extracts the tumor section very efficiently and gives improved data's of statistical measures, GLCM and image resemblance.

3.4. Novel Approach with Non-Negative Matrix Factorization for Brain Tumor Segmentation [4] -Technique: Fuzzy clustering and Non-Negative Matrix Factorization (NMF) is used for Segmentation. Observation: Segmentation accuracy with ground truth is found using Specificity, Dice coefficient, Sensitivity and Jaccard coefficient. Result: On a ($3 \times 3 \times 3$) window, it secures mean Dice Similarity Coefficient, mean Sensitivity, mean Specificity coefficient, mean Jaccard coefficient as 0.77, 0.85, 0.96 and 5.35 respectively for tumor and 0.81, 0.87, 0.97 and 4.7 respectively for edema. The proposed algorithm outperforms for ($3 \times 3 \times 3$) window in comparison to

(5 × 5 × 3), (7 × 7 × 3), (9 × 9 × 3) window sizes. Input: Performance is assessed on training dataset of BRATS 2012 consisting of fully annoted 20 real HGG, 10 real LGG, 25 HGG (synthetic) and 25 LGG (synthetic) images. Each image has "ground truth marked as necrotic, edema and whole tumor". Pre-processing: Skull stripping, de-noising and intensity normalization is performed. De-noising using anisotropic diffusion filter is used on T2 and FLAIR images. ITK N3 is used for Bias field normalization.

Advantages: This technique outdoes existing methods.

3.5. "Fully connected CRF with Data-driven prior for multi-class brain tumor" segmentation [5] -Technique: Tumorous and non-tumorous tissues are separated using FC-CRF model in the first layer followed by distinction of tumor core and edema present in the tumorous region by FC-CRF model. Observation: Sensitivity, Dice Coefficient and Positive Predictive Value are calculated for the tumorous region. Results: The proposed approach has the Dice coefficient as 0.86, 0.73, and 0.74 for Complete, Enhancing and Core respectively for HGG images and 0.88 for Complete and 0.70 for Core for LGG images. Input: The proposed method is evaluated on co-registered and skull-stripped images from BRATS 2013 clinical dataset with 4 modalities – FLAIR, T1, T1c and T1. Pre-processing: The intensity values are clipped to quantile range [0.01, 0.99] followed by further rescaling to [0,1]. Advantages: FC-CRF proves to be effective in significant improvement of the precision of the segmentation of tumor boundaries. Disadvantages: Performance is not improved by prior on small dataset.

3.6. Using Threshold Techniques and Hidden Markov Random Fields for Detecting Brain Tumor in Magnetic Resonance Images [6] -Technique: It proposes to improve the detection of brain tumor using a hybrid method which is a combination between Observation: Comparative study with other existing techniques are done. Results: A higher accuracy is achieved through this method. Input: Dataset of T1 MRI images of patients of different gender and ages is taken. Pre-processing: K-means clustering is executed on the grey-level intensities before using HMRF-EM. Advantages: Shape of the tumor is observed by the segmentation which can be used for tumor size calculation.

3.7. Improved Fuzzy c-means Clustering and Watershed algorithm for Brain Tumor Segmentation of MR Brain Images [7] -Technique: Histogram is used for improving the FCM's initial centroid selection method. Observation: Tanimoto and Dice coefficient are used to compare the result of the technique with the ITK Snap tool's created ground truth. Results: The Tanimoto and Dice coefficients of the improved FCM are 81.56 and 88.91 respectively while for the same for the Watershed algorithm is 86.64 and 93.13 respectively. Input: For the experiment, 90 T1 axial glioma tumored MR brain images were collected with 512 × 512 dimensions. Pre-processing: Skull stripping and contrast enhancement based on mathematical morphological operation and noise removal is performed. Advantages: The proposed method of using histogram for initial centroid selection for FCM gives better results.

3.8. Convolution neural network for Brain tumor image segmentation [8] -Technique: A new architecture model for deep CNN is used which combines multi-modality images and learns useful features to produce segmented tumor maps as results. Observation: A better Dice value is obtained by taking classification results of 8 adjacent pixels for use by a de-noise mechanism assisting brain tumor voxels' recognition. Results: Larger patch size and deeper architecture gave better results, especially for 28 * 28 patch size. Input: FLAIR, T2 and T1 datasets of 30 patients and 50 synthetic ones having ground truth from BRATS 2013 were taken. Pre-processing: Tri-planar 2D CNNs are made from the x, y, z axil slices that centroid for each voxel. Advantages: This method has better accuracy and can be reliably used for clinical purposes.

3.9. Mathematical Morphological Reconstruction for Brain Tumor Extraction From MRI Images [9] -Technique: Tumor is extracted from impulse noise affected MRI images by applying global thresholding method followed by hole-filling. Observation: To validate, there is an addition of 0.015 density of salt and pepper noise as well as a threshold value of 0.6. Results: The PSNR value for uniform intensity images are averaged as 28.835 and for non-uniform image is 29.27. Input: Dataset contains labeled brain MRI images with impulse noise. Pre-processing: Grayscale is converted to binary using global thresholding. Advantages: This technique is efficient if the tumor width is larger than the radius of SE.

3.10. Modified FCM and Level Set Algorithms for Brain Tumor Detection [10] - Technique: It combines FCM and Level Set Algorithm. FCM, a modified form of FCM, containing spatial information for an improved estimation of the cluster centers is used where a 3x3 window gives statistical information of the neighboring pixels to the K-means algorithm. Observation: SA was calculated to validate the results for the sections - sagittal, axial and coronal. Results: For FCM_S1 + K-means the segmentation accuracy is 87.42 for axial, 91.84 for coronal, 92.19 for sagittal. For FCM_S2 + K-means the segmentation accuracy is 89.36 for axial, 93.75 for coronal, 94.02 for sagittal. Input: The dataset contains 256 $\times$ 256 pixel T1-MRI images representing the brain sections - axial, coronal, and sagittal. Advantages: capable of handling noise.

3.11. "PNN technique for Brain Tumor Classification and Segmentation in MRI Images" [11] -Technique: In this approach, GLCM is used to for feature extraction. PCA is applied for image recognition and compression. Observation: The proposed method is analyzed using- Sensitivity, Specificity, Precision, Negative predictive value, Accuracy and F-measures. Results: The performance parametric gave an accuracy of 80%. The classification accuracy of PNN is good and fastest.

Input: All MRI images are used. Being in JPEG formats, they are converted to gray level images.. Advantages: instantly learns data.

3.12. Using "Classification of Super-Voxels for Automatic Segmentation of Multimodal Brain Tumor Images" [12] -Technique: After preprocessing, 3D super-voxels are used for segmentation of the enhanced and normalized images. Intensity based clustering is done by the super-voxel algorithm by utilizing the 3D correlation of MR images. Observation: 3-fold cross validation is used while evaluating performance using standard deviation and mean of Dice Score for both HGG and LGG. Results: The Dice score for mean and standard deviation for Low Grade tumor is 68 and 12 respectively and for High Grade Tumor is 59 and 25 respectively. LGG has better average dice score while HGG has comparable results. Input: T2, FLAIR and T1c from BRATS 2012 was used as dataset. Pre-processing: Image enhancement is done based on unsharp-masking and then every single channel is divided by its maximum intensity for intensity normalization. Advantages: It outperformed comparable method in terms of Dice score.

3.13. Automatic brain tumor tissue detection in T-1 weighted MRI [13] - Technique: After de-noising and skull stripping, pixel intensity is used for segmentation based on k-means clustering following which a hierarchical centroid shape descriptor is applied to remove misclassified healthy tissues during clustering. Observation: Performance metrics like False Negative, False Positive, True Positive, True Negative and are used to compare the result with ground truth images. Results: This technique has an accuracy of 95% and is successful in detection even in low pixel intensity as well as of irregular tumor shapes. Input: The experiment was performed on T1-weighted MR images. Pre-processing: Brain is extracted using thresholding (based on the image histogram) and morphological operation (open and dilation). Advantages: It outperforms ANN, SVM in terms of accuracy since the verification used improves it.

3.14. Automatic Brain Tumor Extraction and Segmentation in MR Images [14] - Technique: The proposed method segments the tumor using FCM clustering which also labels it as per three intensities as 3, 2 and 1 for the segmented regions, the rest of the brain image and the background respectively. Observation: Dice coefficient was measured for comparing the results with the Ground truth. Results: It gave 0.79 as average dice coefficient value. Input: The dataset used is acquired from BRATS through MICCAI 2013 Grand Segmentation Challenges. Here, T1-contrast MRI modality is used. Pre-processing: Bias field is removed using N3 algorithm. Advantages: As per the results, the proposed method is efficient. Disadvantages: For tumors have linear shape, the circularity criterion doesn't work well.

3.15. "Exploration of Histogram Characteristics of the Cancerous Hemisphere for Automated Brain Tumor Segmentation from MRI Data" [15] -Technique: The intensity histogram enables detection of the hemisphere containing tumor. Determination and application of threshold intensity is done followed by median filtering and second threshold (if necessary). Observation: The segmented results are weighed up against the actual values to figure out the dice value, sensitivity and accuracy. Results: The average dice coefficient value obtained was 0.8056, for sensitivity was 0.8649 and for specificity was 0.9317. Input: The dataset from BRATS "consists of 3D-MRI data of 20 HGG patients consisting of 4 types of 3D volumetric data: T1, T1c, FLAIR and T2. Pre-processing: The 3D data present in BRATS is disintegrated into 2D slices of image. The 2D slices are taken along the XY plane. Advantages: The proposed method guarantees the recognition of tumor, with an 80.56% of average dice similarity coefficient, and also reduces the computation time extremely".

3.16. Learning population and patient-specific feature sets for 3D brain tumor segmentation in multimodal MR images [16] -Technique: In this method, a graph is constructed by learning the population and patient-specific features in multimodal MRI images and graph-cut is utilized to achieve the concluding segmentation. Observation: For quantitative comparison of the performance, additional error metrics like DSC, Jaccard similarity, accuracy and sensitivity are calculated. Results: The DSC value is 84.5 ± 9.4, Jaccard is 74.10 ± 14.5, Sensitivity is 87.22 ± 7.9 and Specificity is 83.11 ± 14.8. Input: It contains 45 HGG and 35 LGG images with the modality of – FLAIR, T1, T1c, T2 images of MRI. Pre-processing: The intensity is normalized for all the input images to [0,1] through min-max method. Advantages: Adaboost technique is helpful in distinguishing lesions, high-dimension features and normal tissues.

3.17. “A graph cut distribution matching approach: 3D multimodal MRI brain glioma tumor and edema segmentation” [17] -Technique: The segmentation is an optimization of cost functions having a prior matching distribution, and smoothness to remove of unnecessary regions from the result. Observation: Dice Metric is used for quantitative evaluation by computing harmony between real segmentations provided in BRATS 2012 dataset and the results obtained. Results: the evaluation of the “Dice Metric measure” of the entire tumor was done to obtain HGG and LGG respectively”. The proposed algorithm produces i advantageous with a very reasonable performance. Disadvantages: In some LGG, images having low contrast regions of interest can affect the segmentation results.

3.18. “A low cost approach for brain tumor segmentation based on intensity modeling and 3D Random Walker” [18] -Technique: It exploits MR modalities to identify and segment brain tumor. The MR sequences are combined with ML algorithms this technique is applied to three different datasets, which include “T1, T2, FLAIR and T1c MR Images from the MICCAI 2013”. Pre-processing: The images of first dataset were pre-processed using noise reduction, rigid intra-subject registration and bias-field correction. Advantages: The results depict that the proposed method achieves a lesser cost of computation and higher degree of automation. Disadvantages: The method is not very good for segmenting non-enhancing tumors and necrosis.

3.19. “A deep learning model integrating FCNNs and CRFs for brain tumor segmentation “ [19] -Technique: This brain tumor segmentation technique is modeled by integrating FC-NNs and Conditional Random Fields (CRFs) in a united framework to get results in consistency. Observation: Evaluation is done by Dice, Positive Value and Sensitivity. Results: The results demonstrate that the tumor segmentation model achieved a competitive performance. Input: The image dataset used is taken from the BRATS 2013, 2015 and 2016. The BRATS 2013 provided 65 glioma patients, consists of LGG and HGG images Advantages: The united framework of FCNNs and CRF-RNN improves the segmentation process with robust parameters included.

3.20. “Brain tumor segmentation from multimodal magnetic resonance images via sparse representation” [20] -Technique: This approach involves marking each voxel as the maximum posterior probability (MAP) calculated by calculating sparse probability representation and prior probability of Markov random field (MRF. Input: The dataset is obtained from BRATS 2013 consisting of HGG and LGG images having the various modalities Advantages: The proposed algorithm results and ranks good in comparison to the algorithms in MICCAI BRATS-2013 challenge. Disadvantages: Analogy of HG data with LG data shows a slight decrease in accuracy for real LG data as it is harder to recognize LGG.

3.21. Brain tumor “detection and segmentation in MR images: Gabor wavelet vs. statistical features” [21] -Technique: The two texture-based feature extraction techniques, namely statistical and Gabor wavelet feature extraction technique, such as GLCM, GLRLM, HOG, and LBP are used for segmentation and detection. Input: The data used is taken from NCI-MICCAI 2013 Challenge having modality as T1-2 and Flair MR images with HGG. Pre-processing: MRI intensity is normalized using-contrast stretch, histogram, Gaussian kernel normalization and intensity scaling, histogram stretching, and histogram equalization. Advantages: The use of single-spectral MRI and independence from atlas registration, advance anatomical understanding, and bias correction is a benefit that highlights this method.

3.22. "Concurrent tumor segmentation and registration with uncertainty-based sparse non-uniform graphs" [22] -Technique: "The segmentation and registration issues are modeled using a unified pairwise discrete MRF on a sparse grid superimposed to the image domain". Observation: The results were assessed by analysing AS to MS and Dice Score, FPR, TPR and MAD. Results: The results achieved a mean Dice score of 70% to 72% and median of 72% to 73%, with and without uncertainties respectively. Input: The dataset consists of 110 3D flair MRI volumes in varieties featuring LGG before any treatment. "Advantages: This approach achieved state-of-the-art results and showed continuous performance and sharply reduced complexity in the model. Disadvantages: A restriction of this approach is its reliance on the enhancing classification output".

3.23. Accurate brain lesion segmentation, Efficient multi-scale 3D CNN with fully connected CRF [23] -Technique: Convolutional Neural Network 11-layers deep a dual pathway, three-dimensional is used. Observation: Each fold contains both HG and LG images a 5-fold cross validation. determined is, the mean Dice similarity coefficient, Precision, Sensitivity. Results: Average performance of the system in term of DSC is 89.8 for whole, 75.0 for core and 72.1 for enhanced, in terms of Precision is 91.5 for whole, 84.4 for core and 75.9 for enhanced and in terms of Sensitivity is 89.1 for whole, 72.1 for core and 72.5 for enhanced.

3.24. Mathematical Morphological Reconstruction based image Segmentation for Early Stage Brain Tumor Detection [24] -Technique: To remove noise Image is artefacts and pre-processed and then to discover areas of interest with probable tumor segmented. Then a normalized threshold is found and MRI is changed to binary from grayscale followed by reconstruction by opening. Observation: From a high-peak-signal-to-noise-ratio (PSNR) the accuracy of segmentation can be realized. Results: The Proposed technique identifies the cancer with accuracy of 92% which is better than existing technique and classifier has accuracy of 86.6%. Input: The segmentation is applied in a set of images of four categories of tumor- Glioma, Sarcoma and 19 brain images, Metastatic Adenocarcinoma and Meningioma. Pre-processing: Median filter is used to remove labels, noise and artifacts. By skull stripping method Skull tissues are stripped. Advantages: Experimental results show that while substantially reducing the computation time the segmented images have a high accuracy.

3.25 Image segmentation and Brain tumor detection in a CRF (conditional random fields) framework with super pixel level features and pixel-pairwise affinity [25] - Technique: Using algorithms Multimodal MR images are segmented into superpixels. Next, with the help of multi-level Gabor wavelet filters, features were extracted from the super pixels. Observation: For real data sets: all the other images are trained except just one image used for testing. Similarly, For the synthetic images, five images are left for validation and others are trained. Advantages: reasonable performance.

3.26. Brain Tumor Segmentation Using Random Forest Trained on Iteratively Selected Patients [26] - Technique: In training the RDF an iterative approach is used. Instead of selecting dataset randomly, some patients are added in every iteration in the training data by some euristics approach. Advantages: can be performed on variety of cases as in [27, 28].

4 Conclusion

Fully automatic system for brain tumor segmentation for cancer identification, is a thought-provoking task. At a recent time, easy accessibility of datasets publically provided a mutual platform for the scientist and research field workers to progress and quantitatively appraise their respective implementation techniques. In this article, we provided a review of the most recent methods based on different hard and soft computing techniques, and a brief indication of outmoded procedures for brain tumor segmentation. We have concluded now a days, deep learning methods can be well-thought-out as the present best technique for tumor segmentation. Further, In this paper, numerous mechanized segmentation techniques for brain tumor CT, MRI, etc., viewed. The approaches, rewards, their boundaries, pre-processing, data sets and further challenges are deliberated to deliver insight into numerous methods. Brain tumor segmentation on MR images methods are employed comparatively more for brain tumor segmentation owed to the upright soft tissue divergence and noninvasive of MR scans. Future developments and alterations in several architectures and accumulation of corresponding information from additional imaging modalities may progress the present existing approaches, ultimately foremost to the progress of clinically adequate involuntary tumor segmentation techniques for healthier diagnosis and treatment planning.

References

1. Kiranmayee, B.V., Rajinikanth, T.V., Nagini, S.: A novel data mining approach for brain tumor detection. In: 2016 2nd International Conference on Contemporary Computing and Informatics (IC3I), pp. 46–50. IEEE (2016)
2. Mathew, A.R., Anto, P.B.: Tumor detection and classification of MRI brain image using wavelet transform and SVM. In: 2017 International Conference on Signal Processing and Communication (ICSPC), pp. 75–78. IEEE (2017)
3. Priyedarsni, P., Nandhini, B., Catherine, A.R., Sahana, K., Sundaravadivu, K.: Soft-computing assisted tool to extract tumor section from brain MR images. In: 2017 IEEE International Conference on Power, Control, Signals and Instrumentation Engineering (ICPCSI), pp. 2776–2780. IEEE (2017)
4. Baid, U., Talbar, S., Talbar, S.N.: Novel approach for brain tumor segmentation with non negative matrix factorization. In: 2017 International Conference on Innovations in Electronics, Signal Processing and Communication (IESC), pp. 101–105. IEEE (2017)
5. Shen, H., Zhang, J.: Fully connected CRF with data-driven prior for multi-class brain tumor segmentation. In: 2017 IEEE International Conference on Image Processing (ICIP), pp. 1727–1731. IEEE (2017)

6. Abdulbaqi, H.S., Mat, M.Z., Omar, A.F., Mustafa, I.S.B., Abood, L.K.: Detecting brain tumor in magnetic resonance images using hidden markov random fields and threshold techniques. In: 2014 IEEE Student Conference on Research and Development (SCOReD), pp. 1–5. IEEE (2014)
7. Benson, C.C., Deepa, V., Lajish, V.L., Rajamani, K.: Brain tumor segmentation from MR brain images using improved fuzzy c-means clustering and watershed algorithm. In: 2016 International Conference on Advances in Computing, Communications and Informatics (ICACCI), pp. 187–192. IEEE (2016)
8. Lang, R., Zhao, L., Jia, K.: Brain tumor image segmentation based on convolution neural network. In: International Congress on Image and Signal Processing, BioMedical Engineering and Informatics (CISP-BMEI), pp. 1402–1406. IEEE (2016)
9. Sharma, Y., Meghrajani, Y.K.: Brain tumor extraction from MRI image using mathematical morphological reconstruction. In: 2nd International Conference on Emerging Technology Trends In Electronics, Communication and Networking (2014)
10. Beddad, B., Hachemi, K.: Brain tumor detection by using a modified FCM and level set algorithms. In: 2016 4th International Conference on Control Engineering & Information Technology (CEIT), pp. 1–5. IEEE (2016)
11. Lavanyadevi, R., Machakowsalya, M., Nivethitha, J., Kumar, A.N.: Brain tumor classification and segmentation in MRI images using PNN. In: 2017 IEEE International Conference on Electrical, Instrumentation and Communication Engineering (ICEICE), pp. 1–6. IEEE (2017)
12. Kadkhodaei, M., Samavi, S., Karimi, N., Mohaghegh, H., Soroushmehr, S.M.R., Ward, K., Najarian, K.: Automatic segmentation of multimodal brain tumor images based on classification of super-voxels. In: 2016 IEEE 38th Annual International Conference of the Engineering in Medicine and Biology Society (EMBC), pp. 5945–5948. IEEE (2016)
13. Raj, C.S., Shreeja, R.: Automatic brain tumor tissue detection in T-1 weighted MRI. In: 2017 International Conference on Innovations in Information, Embedded and Communication Systems (ICIIECS), pp. 1–4. IEEE (2017)
14. Sehgal, A., Goel, S., Mangipudi, P., Mehra, A., Tyagi, D.: Automatic brain tumor segmentation and extraction in MR images. In: IEEE Conference on Advances in Signal Processing (CASP), pp. 104–107. IEEE (2017)
15. Akter, M.K., Khan, S.M., Azad, S., Fattah, S.A.: Automated brain tumor segmentation from MRI data based on exploration of histogram characteristics of the cancerous hemisphere. In: 2017 IEEE Region on Humanitarian Technology Conference (R10-HTC), vol. 10, pp. 815–818. IEEE (2017)
16. Jiang, J., Wu, Y., Huang, M., Yang, W., Chen, W., Feng, Q.: 3D brain tumor segmentation in multimodal MR images based on learning population-and patient-specific feature sets. Comput. Med. Imaging Graph. **37**(7–8), 512–521 (2013)
17. Sathesh, A.: Performance analysis of granular computing model in soft computing paradigm for monitoring of fetal echocardiography. J. Soft Comput. Paradig. (JSCP) **1**(01), 14–23 (2019)
18. Njeh, I., Sallemi, L., Ayed, I.B., Chtourou, K., Lehericy, S., Galanaud, D., Hamida, A.B.: 3D multimodal MRI brain glioma tumor and edema segmentation: a graph cut distribution matching approach. Comput. Med. Imaging Graph. **40**, 108–119 (2015)
19. Kanas, V.G., Zacharaki, E.I., Davatzikos, C., Sgarbas, K.N., Megalooikonomou, V.: A low cost approach for brain tumor segmentation based on intensity modeling and 3D random walker. Biomed. Sig. Process. Control **22**, 19–30 (2015)
20. Zhao, X., Wu, Y., Song, G., Li, Z., Zhang, Y., Fan, Y.: A deep learning model integrating FCNNS and CRFS for brain tumor segmentation. Med. Image Anal. **43**, 98–111 (2018)

21. Li, Y., Jia, F., Qin, J.: Brain tumor segmentation from multimodal magnetic resonance images via sparse representation. Artif. Intell. Med. **73**, 1–13 (2016)
22. Nabizadeh, N., Kubat, M.: Brain tumors detection and segmentation in MR Images: gabor wavelet vs. statistical features. Comput. Electr. Eng. **45**, 286–301 (2015)
23. Parisot, S., Wells III, W., Chemouny, S., Duffau, H., Paragios, N.: Concurrent tumor segmentation and registration with uncertainty-based sparse non-uniform graphs. Med. Image Anal. **18**(4), 647–659 (2014)
24. Ain, Q., Jaffar, M.A., Choi, T.S.: Fuzzy anisotropic diffusion based segmentation and texture based ensemble classification of brain tumor. Appl. Soft Comput. **21**, 330–340 (2014)
25. Devkota, B., Alsadoon, A., Prasad, P.W.C., Singh, A.K., Elchouemi, A.: Image segmentation for early stage brain tumor detection using mathematical morphological reconstruction. Procedia Comput. Sci. **125**, 115–123 (2018)
26. Wu, W., Chen, A.Y., Zhao, L., Corso, J.J.: Brain tumor detection and segmentation in a CRF (Conditional Random Fields) framework with pixel-pairwise affinity and superpixel-level features. Int. J. Comput. Assist. Radiol. Surg. **9**(2), 241–253 (2014)
27. Ellwaa, A., Hussein, A., Alnaggar, E., Zidan, M., Zaki, M., Ismail, M.A., Ghanem, N.M.: Brain tumor segmentation using random forest trained on iteratively selected patients. In: International Workshop on Brainlesion: Glioma, Multiple Sclerosis, Stroke And Traumatic Brain Injuries, pp. 129–137. Springer, Cham (2016)
28. Bindhu, V.: Biomedical image analysis using semantic segmentation. J. Innov. Image Process. (JIIP) **1**(02), 91–101 (2019)

The Appraised Structure for Improving Quality in the Compressed Image Using EQI-AC Algorithm

M. Durairaj and J. Hirudhaya Mary Asha(✉)

Department of Computer Science, Bharathidasan University, Tiruchirapalli, Tamilnadu, India
durairaj.bdu@gmail.com, hirudhaya_ashaa20@yahoo.co.in

Abstract. The electronic form of data is compressed practically using techniques to acquire information which are not redundant is called image compression. After compression, the image gains high visual quality with reduced bit rate compared to the raw image. In the Internet of Things (IoT) environment, efficient transmission and storage of images are required for security purposes. Block splitting (BS) and discrete cosine transform (DCT) is applied initially on the raw images to reduce the actual bit rate which minimizes the mean square value (MSE) and the noise in the near-constant region. In this paper, EQI-AC (Enhanced Quality Image After Compression) algorithm is used to enhance contrast, intensity and brightness of the compressed image, which is balanced, for visualization. Peak signal noise ratio (PSNR) values are calculated and found that 1/3rd of the psnr value is decreased comparing to the original image as displayed in Table 1. Structure similarity index (SSIM) values are calculated and found that there is slight degrade from 0.01 to 0.1 range in the quality of the compressed image compared to the original image.

Keywords: Raw image · Compressed image · JPEG · Enhanced image · Balanced for visualization

1 Introduction

In order to make our living environment productive and smart, the evolution of smart homes takes place. Every people spend their day-to-day life maximum at their school, colleges and their workplace. So for their convenience, apartment homes had been bought to the use, where there is a collaboration of people in a single colony. Hence in order to preserve the feeling of safety, technologies that include several sensors along with the machine learning techniques could be adapted to identify, detect and respond to the potential threats. But this system is said to be stand-alone security-based used for many years and they could not be capable of obtaining or working through the plenty of information, which is provided by the sensors combined all over the building that is not capable of providing the right information. Hence smart doors could be designed that does the face recognition serving as the video surveillance system. Since the storage of the device increases and it will not suffice if we did not compress the video file although hard disk of capacity 2 Terabytes are used. The aim of this paper is to

J. I.-Z. Chen et al. (Eds.): ICIPCN 2020, AISC 1200, pp. 201–217, 2021.
https://doi.org/10.1007/978-3-030-51859-2_19

compress the image captured from the video using EQI-AC algorithm and enhancing brightness, contrast and intensity. Every image should be represented in the appropriate value of brightness and contrast for human eye visualization. Brightness refers to the limited darkness needed to visualize the image. If the brightness value of the image is high as usual, the white color ratio of pixels is dripped, if the brightness value of the image is low, the black color ratio of the pixels is dripped and hence the actual details of the image will be destroyed. Contrast means the percentage of color or grayscale representation exists in images. The level of contrast helps to differentiate the object. If the contrast value of the image is set high as compared to average, the black color in the pixels become too black and the white color in the pixels become too white. If the contrast value is set as low, the pixels of the image fades. So the contrast and brightness are the parameters used to measure the intensity of the image. The proposed system of this paper focuses more on contrast and brightness level of the compressed image compared to the raw image. JPEG (Joint Photographic Experts Group) is an international standard for continuous-tone still image compression both the color and the grayscale and helps with the large varieties of application. This standard is now in progress with the JPEG XS that forms a solution for the interoperable solution for visually lossless compression, low-latency and light-weight.

2 Related Work

The smaller digital camera with maximization of picture elements tends to rise a critical problem. A novel method had been implemented in [1] for the detection and elimination of aberration in the images. The refractive effects of the lens tend to affect the optical systems by causing aberration. Chromatic aberration is of two types: lateral aberration - the occurrence of geometry shift or the displacement occurs among each color components that are well focused and longitudinal aberration - blurring of images occurs due to the foci of several color components. The color variation occurs on every transition region and the variation of the RGB is given as (226, 210, 200) to (71, 30, 40). Z(p) is the transition region retrieved from the pixel set l(p) and r(p) (the color varies from the start point l(p) and ends at r(p)). The chromatic aberration is not seen at the normal edges, where the chromatic correlation exists. Then the gradient operation is applied in detecting the color intensity variation with the special operators. The proposed algorithms in [1] will separate the data for processing in both the vertical and in the horizontal direction. The transition region is determined with the search procedure that involves 3 steps:

1) The initial point p is obtained by incrementing the pixel location j, which is kept initially at 1. This is done till the green gradient magnitude becomes greater than or equal to T. This is given as,

$$
\begin{aligned}
j &\leftarrow j+1, \text{ while } |EG(j)| < T, \\
\text{Set}\, p &= j, \text{ if } |EG(j)| > T.
\end{aligned} \tag{1}
$$

   The location of the pixel p should be within the transition area.

2) Next, the transition region Z(p) is calculated by the initial point p to the point m and n, which is the left and the right place of the transition width. At the initial point p, the largest gradient value between R, G and B with the similar edge G is given by,

$$H(x|p) = \text{Max}\{s(p)\,ER(x),\ s(p)\,EG(x),\ s(p)\,EB(x)\} \quad (2)$$

3) Then the next transition value search begins at r(p) + 1 with the gradient value.

The chromatic aberration is detected and eliminated for the individual signals by conducting the pixel search within the transition region. In this proposed method [1], the color fringes are eliminated directly from the given image. This algorithm also reduces the computation of parameter estimation process complexity. In [2], the influence of chroma sub-sampling on the apparent video quality had been analyzed subjectively. The video codecs used in this paper belongs to H.264/AVC and H.265/HEVC (High-Efficiency Video Coding) that is commonly used. The process of image encoding with the implementation of lower resolution for chroma information is termed as chroma subsampling (the concentration is provided more to chroma information rather than luma). The visual systems of the human have a minimum sensitivity to the color details. The evaluation of the video quality depends on two important factors. The perceived quality of an image obtained from the observers or people, who rate the quality of the video by watching them is termed as subjective quality assessment. This is the most fundamental way in determining the quality of the video that is termed as Quality of Experience (QoE). The duration is the drawback of this method but the accuracy quality could be enhanced. This method is further divided as single and double stimulus methods. The objective quality assessment is not concentrated when compared to the subjective type. Then the coding procedure is carried out with the FFmpeg tool, where the chroma sub-sampling is done at the test video sequence with the change in the resolution carried out from UHD to FHD. This procedure is carried out at all the test sequences. Then, the encoding process takes place and all the sequences are encoded to H.264/AVC and H.265/HEVC standard for the compression. "Bund Nightscape" and the "Construction Field" are denoted as the sequences with the low SI and TI values that have low MOS score and bit rates [2].

The modification of the luminance quantization table with the enhanced JPEG compression algorithm with two most important components are considered in the quantization table namely chrominance and luminance. As stated in the previous work, luminance table changes could not be detected by the human visual system easily. Different image color had been chosen that utilizes Baboon, Pepper, Brick-house etc. The modified performance JPEG is assessed by peak signal and noise ratio (PSNR), compression ratio (CR) and mean square error (MSE). The proposed JPEG algorithm is shown in Fig. 1. Due to the modification of quantization table, the PSNR, CR and MSE values had been improved by the proposed JPEG algorithm [3].

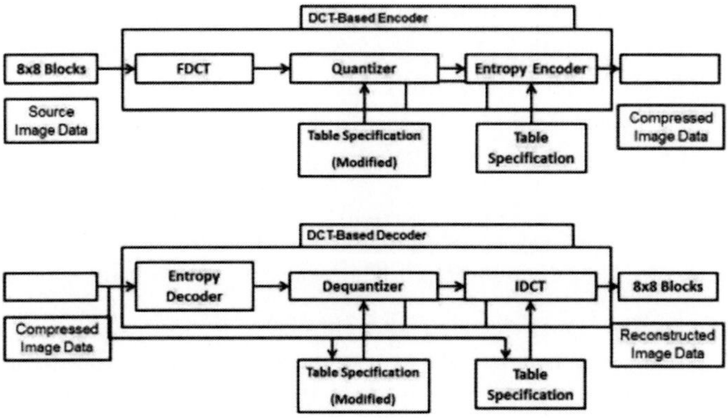

Fig. 1. DCT based encoder and decoder processing stages in JPEG image compression algorithm [3]

The most important process of capturing, storing, sharing, rendering and utilizing the images belongs to the standards that include JPEG, JPEG 2000, JPEG XT and JPSearch. In [4], the archiving and digitization application with these standards had been presented with certain efforts of ongoing standard. The author had also stated the advancement of each of these standards. Among these, JPEG 2000 has good compression efficiency with the versatility and good rate-distortion control. This standard could be adapted in the field of satellite imaging, medical imaging, digital cinemas and many more due to their wavelet-based compression technique. By piggybacking with the other standards, JPEG XT finds its application into the widen consumer markets with the advanced functionalities. These standards play a role in the portability of image, searching and accessing the image along with the metadata handling. The images are stored in the personal devices, social media and online repositories. Due to this wide growth of image data, interoperability should be supported among the systems. This could be possible with the JPSearch that helps in search and retrieval of still images. Certain ongoing activities include JPEG PLENO and JPEG LS that should support low-latency along with the new image modalities.

The reduction of data correlation is the basic idea behind the development of data compression techniques. Discrete Cosine Transform (DCT) is used that transfers the data from the time domain to the frequency domain. But the image at higher frequency is not clearly visible to the human vision due to the low sensitivity. In case of video captured with the 3-D space, reduction of time correlation along with the spatial correlation is necessary. Motion Estimation (ME) method had been proposed in [5] in which the past and the new evolving frames are compared to get the similar parts. Then the time correlation factor is reduced by replacing the image with the Motion Vector (MV). The compression standard in this paper includes JPEG and MPEG respectively.

The detection of Contrast Enhancement (CE) with the laundering attacks is said to be a very difficult task. The CE detectors had been misleading by the compression standard JPEG that is found to be a detrimental laundering attack. In paper [6], compression standard JPEG had been utilized along with the adaptive histogram (AH) to determine the CE by means of training a JPEG-aware SVM detector with the color SPAM feature. Results have also been estimated based on this SVM trained detector and this works well with the imposed Quality factor (QF) along the compressed images under tests. If this QF factor could not be determined from the image then idempotency properties of JPEG compression for extracting QF could be adapted. The enhancement in the contrast is done by Contrast-Limited implementation of AHE (CLAHE). Then the SVM classifier was given with 1372 dimensional features that are determined from the color images. The parameters of the Gaussian kernel were obtained by 5-fold cross-validation. This approach could also be adapted with the Convolutional Neural Networks (CNN) that could provide higher significant results in many applications.

To detect the presence of histogram equalization, a novel machine learning concept had been proposed in this paper [7] and the statistical parameters of DC Discrete Cosine Transform (DCT) coefficients are considered. The features used for the classification of original and histogram equalized image is done by the Gaussian Mixture Model (GMM) statistical parameters. The accuracy is obtained with the SVM classifier that classifies the original and histogram equalized images.

Enhancing the contrast of the image could be possible with the Contrast-Limited implementation of AHE (CLAHE). In Adaptive Histogram Equalization (AHE), an image will be sub-divided into several tiles and the enhancement will be carried out by applying HE to every individual tiles that concentrates on every small details by improving the image local contrast. Sometimes, AHE could increase the amplification noise factor that could be prevented by CLAHE in the relative homogenous region. Prior to the cumulative distribution function (CDF) computation, the above procedure will be carried out that could clip the histogram at the particular value. The function of slope transformation will be limited that obtains the contrast amplification. This slope transformation function will be estimated by the CDF. Clip limit is said to be the clipped value of the histogram and this will rely on the histogram normalization value as well as the default neighborhood region size 8×8 [10, 11]. The RGB image will be converted to HSV color space in spite of applying the CLAHE directly for each channel that color result in the improper color balance and also produces unbalanced image presentation. By converting the image to HSV color space, the above cons could be avoided and the CLAHE could be applied focusing on the luminance channel. After this process, the image will be converted again back to the original RGB domain image forgeries are highly possible due to the contrast enhancement inconsistency.

A novice method had been introduced in [9] to analyze the contrast enhancement for the particular image. According to this method, mapping is done among the pixel values and also a distinct characteristic is introduced to the histogram of the image pixel. The method is said to be highly robust that considers the additive noise perturbations that affects the contrast enhancement.

3 Novel Structure Addressing Image Compression

The clarity and certain features of the images are not guaranteed during the image compression technique done by JPEG standard. Hence the below technique should be carried out to enhance the quality and control over the image that could be better than the source image.

3.1 Color Space Conversion from RGB to YCbCr

The RGB color representation of the image is converted in the form of Luminance (Y) and Chrominance (Cb & Cr) in order to achieve the delicacy and for interpreting the spatial content better for an image since human visual system is more sensitive to luminosity [12]. This methodology is done by the JPEG compression standard to limit the amount of information stored in the image. Usually, the color of the image could be represented by RGB and the transformation to Y, the luminosity component will enhance the intensity and the brightness of the pixels in the image. Cb and Cr represent the chrominance of the color blue and green respectively. Through this conversion, high compression could be achieved and this is mathematically represented as,

$$[Y\,Cb\,Cr] = [R\,G\,B] = \begin{bmatrix} 0.299 & 0.169 & 0.499 \\ 0.5870 & 0.3310 & 0.418 \\ 0.114 & 0.500 & 0.081 \end{bmatrix}$$

3.2 Chrominance Components - Sub-sampling Process

Our human visual system is composed of rods and cons. Rods indicates the brightness and cons detect the color sensitivity [13]. The density of the rods is very high when compared to the cons that make them go well with the brightness rather than the color. This concept had been utilized by the JPEG standard in reducing the spatial resolution of the chrominance components. The subsampling process takes place on the Cb and Cr chrominance components formed due to the color transformation. The advantage of this method is that humans do not have effective diligence to find chrominance changes [14]. Every horizontal and the vertical components will be removed from the image with the JPEG standard that results in the subtraction of 3/4 of the information from each of these components [15]. In simple words, certain pixels could be merged that have the same chrominance values done at a factor of 2 in both the vertical and horizontal directions.

3.3 Block Splitting

The entire matrix will be sub-divided into the smaller individual matrix of 8X8. These sub-matrices is said to be known as the blocks. After this splitting process, it is necessary to fill the empty block by pixel edge data.

3.4 Discrete Cosine Transformation

The redundancy in the image data is removed by applying DCT in every pixel block. For lossy image compression, DCT is the best-adapted method due to their robust energy compaction [16]. DCT will store a large amount of data with the low-frequency component and the rest of the smaller data will be stored at a low bit rate level. DCT transformation formula is given as:

$$F_{u,v} = \frac{c(u)c(v)}{4} \sum_{x=0}^{7} \sum_{y=0}^{7} f(x,y) cos \frac{(2x1)u\pi}{16} cos \frac{(2x1)v\pi}{16}$$

Where,

$$C(w) = \begin{cases} \frac{1}{\sqrt{2}} for\ w = 0 \\ 1 : for\ w = 1, 2, \ldots 7 \end{cases} \quad (3)$$

u, v – spatial frequencies in both horizontal and vertical direction and the integers should be in the range of $0 \leq u < 8$

$$F\,u, \text{ – DCT coefficient at the coordinates (u, v)} \quad (4)$$

f(x, y) – the value of the pixel at the coordinates (x, y).

The information regarding the image file should be obtained before performing the DCT transformation [17]. The information file consists of the pixel value in terms of integers ranging from 0–255 that is divided in 8X8 matrix block and then DCT is applied in every block of data. Before transforming the image to the frequency domain, JPEG should subtract 128 to all the pixel value of the image in order to minimize the DC coefficient size. After DCT transformation it is found that 90% of the data will fall below the lower frequency component. In order to enhance the compressed image, CLAHE metric is applied to the images.

3.5 CLAHE Transformation on Images

In order to reduce noise in the near-constant region Contrast Limited AHE (CLAHE) is adapted that works better with adaptive histogram equalization by limiting the contrast amplification factor thereby reducing the noise factor. The contrast amplification in CLAHE is given by the transformation function slope for the sufficient pixel value. This will be proportional to the slope of neighborhood cumulative distribution function (CDF) and hence to the histogram of that particular pixel value. Prior to the CDF computation, the histogram amplification will be clipped at a predefined value thereby limiting the transformation function and CDF slope. This clip limit value depends on the size of the neighborhood region. The clip limit amplified value lies at a range of 3 and 4. This does not eliminate the histogram value rather it will distribute the values to the histogram bins (Fig. 2).

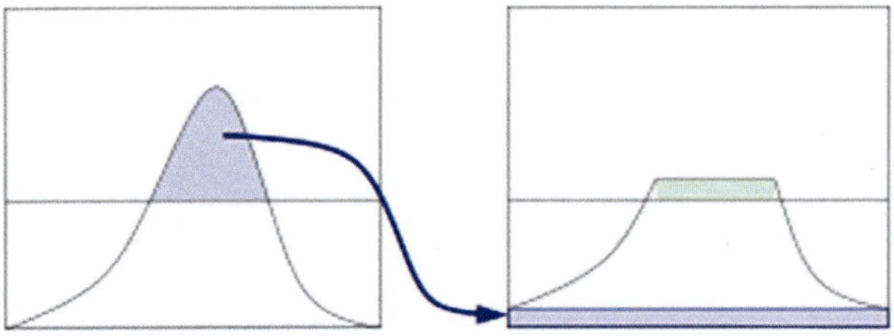

Fig. 2. Clip limit of the histogram value

In the above figure, it is seen that range above the clip limit will be pushed to the bin through the redistribution process (shaded in green color) that results in the effective clip limit of an exact value. This redistribution process could be repeated until the prescribed effective value is obtained. Hence the only one difference that separated CLAHE with AHE is that an extra procedure should be carried out to clip the histogram prior to the mapping function of CDF computation [18]. An optimal parameter should be considered with CLAHE that specifies the level to be clipped for the histogram. At the initial stage, clipping operation is not done. The range of the clip value validity will be 1 to 1/bins. Clipping is not a straight forward method since the excess portion after clipping should be redistributed among the other bins. This could increase the histogram clipping level [18]. Therefore, this process should be carried out at the lower range of the specific clipping level. This tends to maintain the clip level equal to the histogram at the maximum level during the redistribution process. To determine the clipping point, the binary search method should be adapted as mentioned in [18]. From Eq. (1), (2), (3), (4), the image is transformed and compressed using EQI-AC algorithm.

4 Methodology

Based on the novel structure addressing image compressing, EQI-AC algorithm is found for compressing and enhancing the raw image. Every horizontal and the vertical components will be removed from the image with the JPEG standard that results in the subtraction of 3/4 of the information from each of these components. Certain pixels could be merged that have the same chrominance values done at a factor of 2 in both the vertical and horizontal directions. The entire matrix will be sub-divided into the smaller individual matrix of 8×8. These sub-matrices is said to be known as the blocks. After this splitting process, it is necessary to fill the empty block by pixel edge data. The redundancy in the image data is removed by applying DCT in every pixel block. After the DCT transformation, CLAHE metrics are applied. The following EQI-AC algorithm is used to improve the color, brightness and intensity of the compressed image. The raw images of varying size are selected automatically and stored in the temporary folder and considered for compression. For compressing the raw images, the folder which contains the raw images are imported as an input to the source codes. For example.png files are compressed and saved with.jpg extension. The dataset of nearly 98 images is given as an input to the source codes for compression. From the Table 1, it is easy to visualize the extension of the input file and its actual file size and what is the extension of the saved output file and its compressed file size.

4.1 Pseudocode for EQI-AC Algorithm for Image Compression with Enhanced Quality

Input: Raw images with .png Extention

Output: Compressed images with enhanced colour, brightness and intensity

---

Step1: Take an image and divide it up into 8-pixel by 8-pixel blocks. If the image cannot be divided into 8-by-8 blocks, then you can add in empty pixels around the edges, essentially zero-padding the image.

Step 2: For each 8-by-8 block, get image data such that you have values to represent the color at each pixel.

Step 3: Take the Discrete Cosine Transform (DCT) of each 8-by-8 block.

Step 4: After taking the DCT of each block, multiply the block by a mask called matrix value that will zero out certain values from the DCT matrix.

Step 5: Finally, to get the data for the compressed image, take the inverse DCT of each block. All these blocks are combined back into an image of the same size as the original.

Step 6: Now we have the compressed Images.

Step 7: Calculate a grid size based on the maximum dimension of the image. The minimum grid size is 32 pixels square.

Step 8: If a window size is not specified chose the grid size as the default window size.

Step 9: Identify grid points on the image, starting from top-left corner. Each grid point is separated by grid size pixels.

Step 10: For each grid point calculate the cdf of the region around it, having area equal to window size and centered at the grid point.

Step 11: After calculating the mappings for each grid point, repeat steps 6 to 8 for each pixel in the input image.

Step 12: For each pixel find the four closest neighboring grid points that surround that pixel.

Step 13: Using the intensity value of the pixel as an index, find its mapping at the four grid points based on their cdfs.

Step 14: Interpolate among these values to get the mapping at the current pixel location. Map this intensity to the range [min:max) and put it in the output image.

Step 15. Thus, output images formed are of high quality when compared to the JPEG Compressed Images.

Step 16: Stop the process

4.2 Flowchart for EQI-AC Algorithm Methodology

See Fig. 3

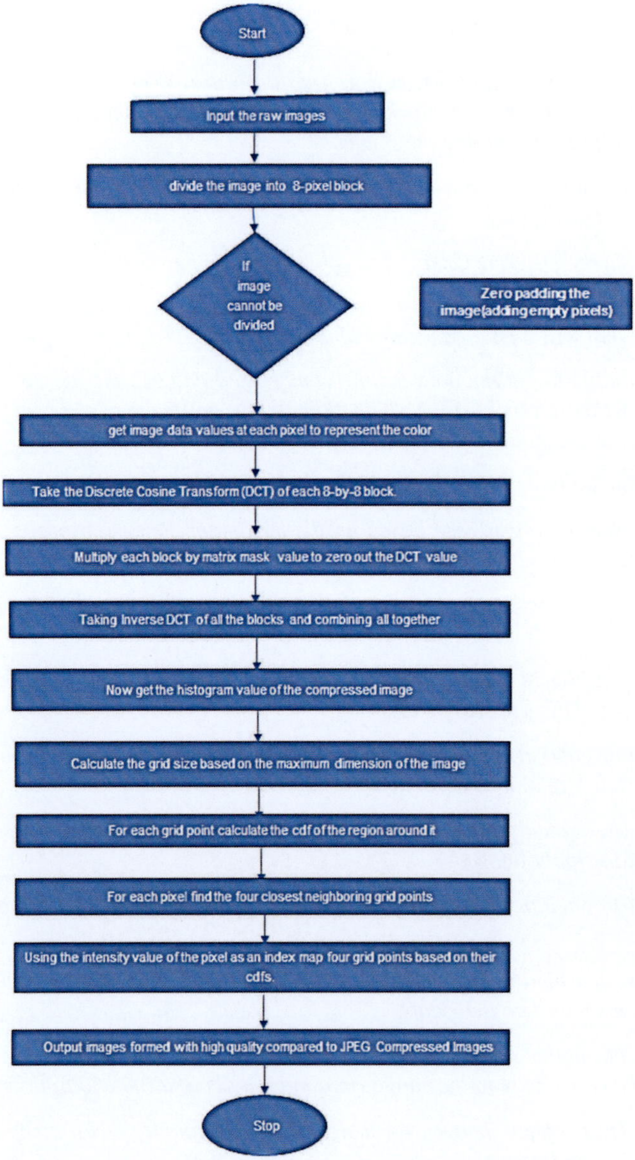

Fig. 3. Flowchart of EQI-AC algorithm

5 Results and Discussion

The contrast, brightness and intensity of the raw image is changed for human visualization and represented in the chart. The size of the raw image and the size of the compressed images are compared and tabulated below (Table 1).

Table 1. Compression table shows the name of the raw image, its actual size, the compressed image with its shrunk file size.

S. no	Raw image name	Raw image size	Compressed image name	Compressed image size
1	1.06mag-06firstwords-t_CA0-articleLarge-v2.png	182.9 KB	1.06mag-06firstwords-t_CA0-articleLarge-v2.jpg	41.3 KB
2	1.forest4_612251.png	435.0 KB	1.forest4_612251.jpg	155.0 KB
3	10.Untitled-design-9-796x417.png	189.9 KB	10.Untitled-design-9-796x417.jpg	136.1 KB
4	10.forestsmain.png	2.7 MB	10.forestsmain.jpg	742.9 KB
5	11.csm_Finland_forest_by_Ninara_flickr_42b14a06aa.png	669.9 KB	11.csm_Finland_forest_by_Ninara_flickr_42b14a06aa.jpg	184.3 KB
6	11.when-will-little-people-big-world-return-1546445203.png	320.4 KB	11.when-will-little-people-big-world-return-1546445203.jpg	86.7 KB
7	12.730_ThinnedForest_UnthinnedForest_background.png	826.8 KB	12.730_ThinnedForest_UnthinnedForest_background.jpg	275.2 KB
8	12.bianca-devins.png	2.0 MB	12.bianca-devins.jpg	765.6 KB
9	14.inner_slider.png	2.1 MB	14.inner_slider.jpg	637.0 KB
10	15.FOD.png	1.1 MB	15.FOD.jpg	351.5 KB
11	15.denis5-1-400x400.png	180.7 KB	15.denis5-1-400x400.jpg	37.1 KB
12	16.theni%20forest.png	767.0 KB	16.theni%20forest.jpg	202.9 KB
13	17.ron-golden-1.png	275.8 KB	17.ron-golden-1.jpg	208.1 KB
14	18.river.png	429.8 KB	18.river.jpg	102.1 KB
15	19.aboutus-our-people-tab-careers.png	132.4 KB	19.aboutus-our-people-tab-careers.jpg	127.0 KB

The following chart shows the size chart of how the raw image is compressed (Fig. 4).

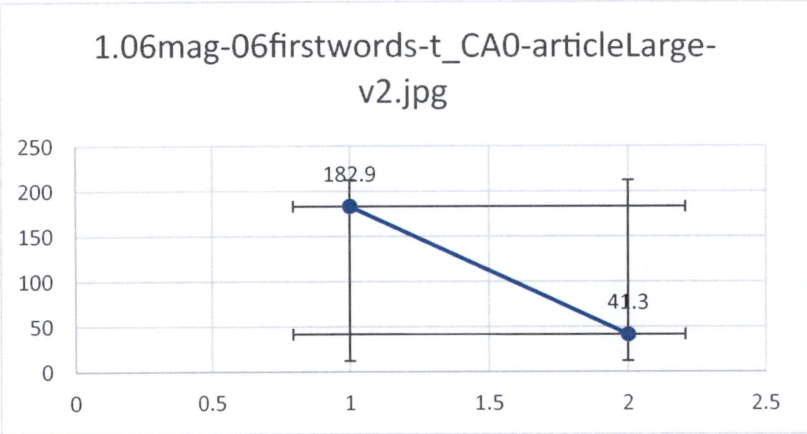

Fig. 4. shows how the image is compressed from 182.9 kb size to 41.3 kb size.

5.1 PSNR

For measuring the quality of the compressed image, MSE (Mean Square Error) and PSNR values are calculated. Peak Signal Noise Ratio (PSNR) is the rational value calculated between the signal or original data and the noise or error which corrupts or affects the quality of the image after compression. The Quality comparison table is drawn to display how the compressed image is enhanced and its PSNR values are quoted in Table 2.

Table 2. Quality comparison table with PSNR values

S. no	Compressed Image Name	Compressed image size	PSNR for compressed image	PSNR for image enhanced using EQI-AC
1	1.06mag-06firstwords-t_CA0-articleLarge-v2.jpg	41.3 KB	42.01572425	33.5922751
2	1.forest4_612251.jpg	155.0 KB	31.03455893	28.37294619
3	10.Untitled-design-9-796x417.jpg	136.1 KB	37.53878891	28.67509408
4	10.forestsmain.jpg	742.9 KB	39.03112413	28.15049873
5	11. csm_Finland_forest_by_Ninara_flickr_42b14a06aa.jpg	184.3 KB	32.27701443	28.21003005
6	11.when-will-little-people-big-world-return-1546445203.jpg	86.7 KB	36.29536386	29.16766387
7	12.730_ThinnedForest_UnthinnedForest_background.jpg	275.2 KB	32.2505882	28.55379357
8	12.bianca-devins.jpg	765.6 KB	49.48272976	28.37872612
9	14.inner_slider.jpg	637.0 KB	34.09973061	28.35485278
10	15.FOD.jpg	351.5 KB	31.60122669	28.16294106
11	15.denis5-1-400x400.jpg	37.1 KB	40.55023649	30.59407807
12	16.theni%20forest.jpg	202.9 KB	38.95896213	28.21869
13	17.ron-golden-1.jpg	208.1 KB	33.83155675	28.44046449
14	18.river.jpg	102.1 KB	42.87173892	28.49909941
15	19.aboutus-our-people-tab-careers.jpg	127.0 KB	37.69856963	30.45363582

The below bar chart is the graphical representation for the PSNR value of the compressed image and the PSNR value of the enhanced image by applying EIQ-AC algorithm. From the output images it is visibly seen that the color, contrast and intensity of the given images are enhanced (Fig. 5).

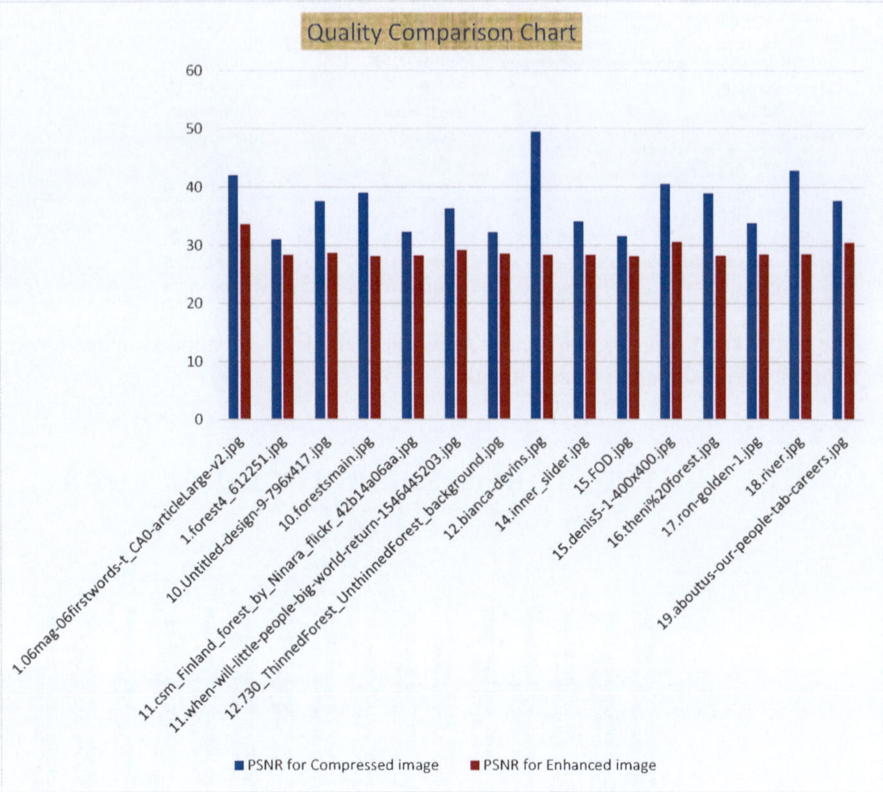

Fig. 5. The quality comparison chart for the raw image and compressed image based on its PSNR value.

5.2 SSIM

Structure Similarity Index is the quality measure used to find the similarity quality between two same images, the raw image and the compressed image. The following table shows the structural similarity between the 15 raw images taken as a sample input for compression (Table 3).

Table 3. The structure similarity index of the compressed image and enhanced images

S. no	Compressed image	SSIM value of compressed image	SSIM value of EQI-AC applied image
0	1.06mag-06firstwords-t_CA0-articleLarge-v2.jpg	0.993276488	0.967669013
1	1.forest4_612251.jpg	0.980494321	0.831007905
2	10.Untitled-design-9-796x417.jpg	0.976760044	0.818730254
3	10.forestsmain.jpg	0.988178904	0.723306708
4	11.csm_Finland_forest_by_Ninara_flickr_42b14a06aa.jpg	0.97671113	0.808361255
5	11.when-will-little-people-big-world-return-1546445203.jpg	0.97784198	0.863056503
6	12.730_ThinnedForest_UnthinnedForest_background.jpg	0.970369237	0.794251695
7	12.bianca-devins.jpg	0.99559195	0.86142213
8	14.inner_slider.jpg	0.992146256	0.85905047
9	15.FOD.jpg	0.964105035	0.762102643
10	15.denis5-1-400x400.jpg	0.991242482	0.899955901
11	16.theni%20forest.jpg	0.992998901	0.842987241
12	17.ron-golden-1.jpg	0.944776986	0.776488431
13	18.river.jpg	0.990599087	0.828948821
14	19.aboutus-our-people-tab-careers.jpg	0.970074594	0.881192121
15	2.forestblog3.jpg	0.971007348	0.773670275

The below chart (Fig. 6) graphically shows how the EQI-AC applied image having the similarity with the compressed image.

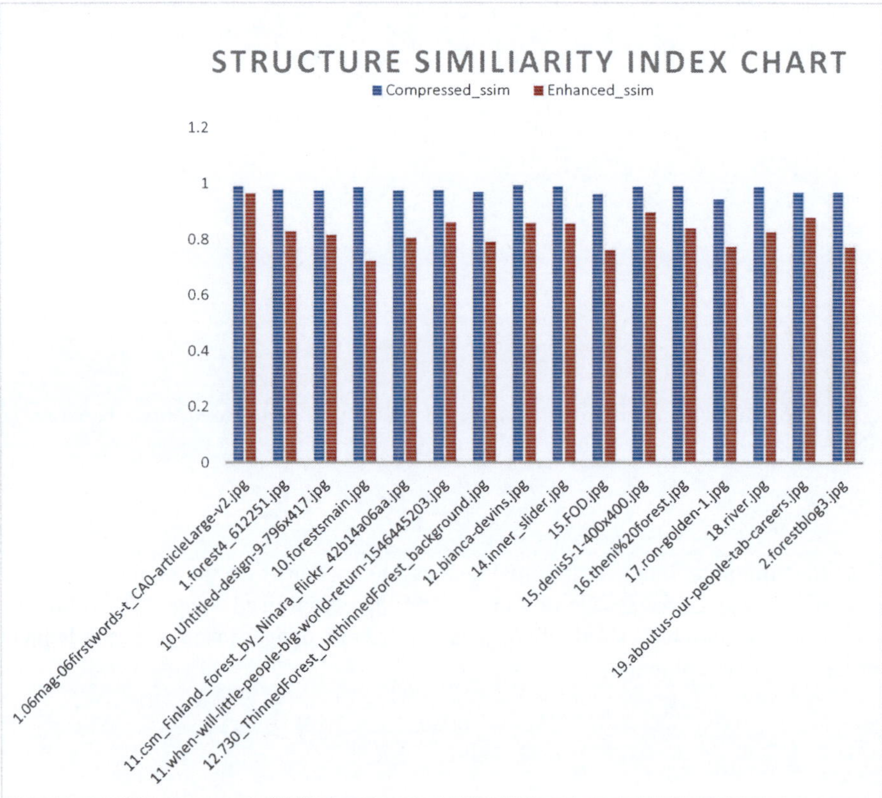

Fig. 6. The structure similarity index chart for the above table

6 Output

The following images shows how the raw images and the enhanced images are differed in contrast, brightness and intensity before and after applying EQI-AC algorithm for compression. Here the enhanced output of the 6th image, 12th image and 98th image of the dataset are displayed for the reference (Fig. 7).

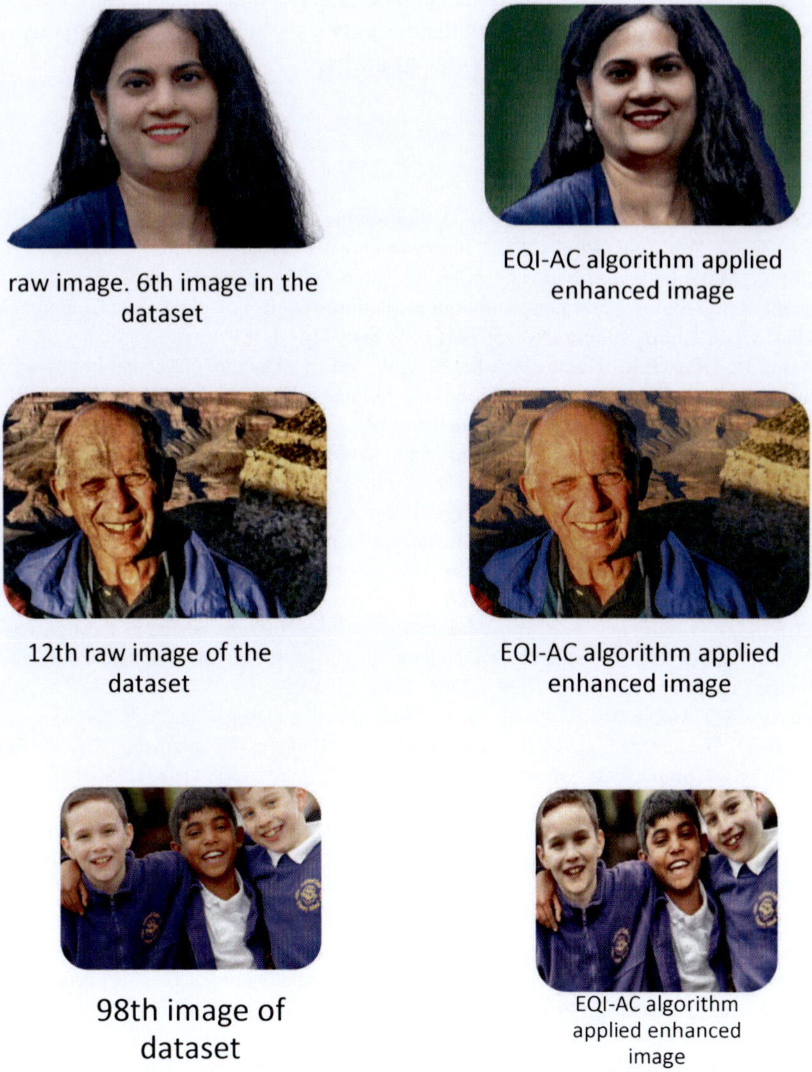

Fig. 7. Compressed image with JPEG standard and CLAHE

7 Conclusion

While attempting image compression, the size of the image decreases been a fact, Using EQI-AC algorithm, the file size is reduced and the quality of the image is also improved. By calculating the PSNR & SSIM of the original images, it is acceptable that the image color, contrast and intensity of the image is improved even after compression. In this paper, EQI-AC (Enhanced Quality Image After Compression) algorithm is used to enhance contrast, intensity and color of the compressed image, which is balanced, for visualization. Peak signal noise ratio (PSNR) and Structure similarity index (SSIM) is also calculated to prove that there is only a slight degrade in the quality of the compressed image compared to the original image.

References

1. Chung, S.-W., Kim, B.-K., Song, W.-J.: Removing chromatic aberration by digital image processing. Opt. Eng. **49**(6) (2010). https://doi.org/10.1117/1.3455506
2. Uhrina, M., Bienik, J., Mizdos, T.: Chroma subsampling influence on the perceived video quality for compressed sequences in high resolutions. Digit. Image Process. Comput. Graph. **15**(4) (2017). https://doi.org/10.15598/aeee.v15i4.2414
3. Alam, L., Dhar, P.K., Hasan, M.A.R., Bhuyan, M.G.S., Daiyan, G.M.: An improved JPEG image compression algorithm by modifying luminance quantization table (2019)
4. Temmermans, F., Bruylants, T., Schelkens, P., Ebrahimi, T.: An introduction to JPEG standards for digitization and archiving applications (2009)
5. Barni, M., Nowroozi, E., Tondi, B.: Detection of adaptive histogram equalization robust against JPEG compression (2018). https://doi.org/10.1109/iwbf.2018.8401564
6. Chen, Z., Zhao, Y., Ni, R.: Detection for operation chain: histogram equalization and dither-like operation. KSII Trans. Internet Inf. Syst. 9(9), 3751–3770 (2015). ISSN 1976-7277. https://doi.org/10.3837/tiis.2015.09.026
7. Cozzolino, D., Gragnaniello, D., Verdoliva, L.: Image forgery detection through residual-based local descriptors and block-matching. In: 2014 IEEE International Conference on Image Processing (ICIP), pp. 5297–5301. IEEE (2014)
8. Pizer, S.M., Amburn, E.P., Austin, J.D., Cromartie, R., Geselowitz, A., Greer, T., ter Haar Romeny, B., Zimmerman, J.B., Zuiderveld, K.: Adaptive histogram equalization and its variations. Comput. Vis. Graph. Image Process. **39**(3), 355–368 (1987). ISSN 0734-189X. https://doi.org/10.1016/S0734-189X(87)80186-X
9. Cao, G., Zhao, Y., Ni, R.: Forensic estimation of gamma correction in digital images. In: 2010 17th IEEE International Conference on Image Processing (ICIP), pp. 2097–2100. IEEE (2010)
10. Cao, G., Zhao, Y., Ni, R., Li, X.: Contrast enhancement-based forensics in digital images. IEEE Trans. Inf. Forensics Secur. **9**(3), 515–525 (2014)
11. Barni, M., Fontani, M., Tondi, B.: A universal technique to hide traces of histogram-based image manipulations. In: Proceedings of the on Multimedia and Security, pp. 97–104. ACM (2012)
12. De Rosa, A., Fontani, M., Massai, M., Piva, A., Barni, M.: Second-order statistics analysis to cope with contrast enhancement counter-forensics. IEEE Sig. Process. Lett. **22**(8), 1132–1136 (2015)

13. Pan, X., Zhang, X., Lyu, S.: Exposing image forgery with blind noise estimation. In: Proceedings of the Thirteenth ACM Multimedia Workshop on Multimedia and Security, pp. 15–20. ACM (2011)
14. Li, H., Luo, W., Qiu, X., Huang, J.: Identification of various image operations using residual-based features. IEEE Trans. Circuits Syst. Video Technol. **28**, 31–45 (2016)
15. Singh, N., Gupta, A.: Analysis of contrast enhancement forensics in compressed and uncompressed images. In: 2016 International Conference on Signal Processing and Communication (ICSC), pp. 303–307. IEEE (2016)
16. Barni, M., Nowroozi, E., Tondi, B.: Higher-order, adversary-aware, double jpeg-detection via selected training on attacked samples. In: 2017 25th European Signal Processing Conference (EUSIPCO), pp. 281–285, August 2017
17. Bestagini, P., Allam, A., Milani, S., Tagliasacchi, M., Tubaro, S.: Video codec identification. In: 2012 IEEE International Conference on Acoustics, Speech and Signal Processing (ICASSP), pp. 2257–2260. IEEE (2012)
18. Fridrich, J., Kodovsky, J.: Rich models for steganalysis of digitalimages. IEEE Trans. Inf. Forensics Secur. **7**(3), 868–882 (2012)
19. Verdoliva, L., Cozzolino, D., Poggi, G.: A feature-based approach for image tampering detection and localization. In: 2014 IEEE International Workshop on Information Forensics and Security (WIFS), pp. 149–154. IEEE (2014)
20. Goljan, M., Fridrich, J., Cogranne, R.: Rich model for steganalysis of color images. In: 2014 IEEE International Workshop on Information Forensics and Security (WIFS), pp. 185–190. IEEE (2014)
21. Kodovsky, J., Fridrich, J., Holub, V.: Ensemble classifiers for steganalysis of digital media. IEEE Trans. Inf. Forensics Secur. **7**(2), 432–444 (2012)

ExclusiveOR-Discrete Cosine Transform-A Chaotic Algorithm for Image Encryption and Decryption

M. Durairaj and J. Hirudhaya Mary Asha(✉)

Department of Computer Science, Bharathidasan University, Tiruchirappalli, Tamilnadu, India
durairaj.bdu@gmail.com, hirudhaya_ashaa20@yahoo.co.in

Abstract. Image encryption is an operative method to shield images or videos from disrupting, while transporting them in a shared link for security reasons. The security levels are increased by applying various algorithms. In this literature, image encryption is done with chaotic methods. Chaotic representation like XOR representation is used to scramble the image data in a pixel by pixel limit. The input image is scrambled using P Fibonacci-XOR transform with masking coefficients and Discrete Cosine Transform (DCT). Later, the scrambled images got unscrambled by applying the inverse of the applied function. As a result, Chaotic concepts avoids iteration of the image data by the unintended persons and the performance of the ExclusiveOR-Discrete Cosine Transform XORDCT algorithm is analyzed by PSNR, NPCR, UACI and Correlation coefficient factors.

Keywords: Compressed image · Discrete Cosine Transform (DCT) · XOR transform · Masking coefficients · Inverse operation · Image scrambling and unscrambling

1 Introduction

Secret communication is expected between the users to ensure privacy. The contents may be hide from exposing to the public and delivered via authentication acknowledgement. Unintended persons may intercept our messages in order to aware of it or destroy it. Many encryption algorithms are found and suggested in the past. Thus, hiding messages are serious problem to consider. XOR Cipher encrypts image data using mathematical parameters for making the encryption stronger [1]. XOR cipher combines matrix data of an image with a random key to make the process highly efficient and unbreakable. Chaos based scrambling gives more number of permutation to re-align the pixels of the image which can be non-recognizable and dis-order. Chaos-based algorithms are simple and highly secured to adopt. The chaotic or mathematical formulae are widely applied in image diffusion and image confusion [2]. The P-Fibonacci XOR operation is applied for scrambling the image. These mathematical techniques are called scrambling techniques follows randomization using numerical data of the image but unchanging the actual representation of the image data.

J. I.-Z. Chen et al. (Eds.): ICIPCN 2020, AISC 1200, pp. 218–232, 2021.
https://doi.org/10.1007/978-3-030-51859-2_20

P-Fibonacci XOR transformation is the technique. This encryption process is formulated by giving the hash value of the raw image and compressed image as input. Transformation in the image data is done using DCT (discrete cosine transform) [2]. Encryption process encodes the original image using cipher key and the decryption process reverses the encryption to get the original image back. There are two methods Confusion and Diffusion to perform encryption. The masking key and image is confused and rearranged in the distribution process [3]. High redundant pixels in the image have strong correlations in the encrypted image also. Chaos based iterations are applied [5–13]. Chaotic system have two types of integer order, the proper fractional-order and improper-fractional order. Parameters of the matrix image data are arranged in fractional orders. While transferring secret image or text, the density of the image pixel and security of the data are interrelated [17, 18].

2 Related Works

Due to the in-determinant rise in digital communication, manifold technologies are used to transfer private particulars like images and videos. Every individual, companies and officials use encryption methods. Encrypting or transforming the data into an altered format is a constructive method to protect images or videos. The authorized users can decode encrypted objects [1]. Many applications required security as a basement for shielding privacy to hide confidential commercial matters. Authorization is allocated to others via fingerprint matches, iris matches etc., are such type of security applications. Depends upon the expecting security, the data can be either partially or fully altered by following different Technologies [2]. Fresnel Transform (Fr-T) were used to transform the input image [3, 4]. Fractional and Fourier Transform (Fr-FT) are followed to scramble the image input [5–7]. Spatial domain calculus is used to protect data with high security and high quality. Naïve-based algorithms assume the data stream to encrypt the data [8]. Des (Data Encryption Standard) encrypts data part of the sequence [9] Without scrambling the input, AES (Advanced Encryption Standard) method performs expandable output [10–12]. Chaos or mathematical systems use numeric data and provide results with real numbers which guide the algorithm for scrambling [13–18]. Recursive routines like Fibonacci sequences encrypt images with less security [19, 20]. The efficiency and security of the encryption process are improved using P-Fibonacci transform sequence and P-recursive function [21, 22]. Permutation changes the pixel values of image blocks vulnerable to plain-text attacks [23, 24]. Exclusive-OR functions followed for bit plane encryption [25]. In handheld devices, transforms are done using Asynchronous standards for encryption (AES). Images are arranged with matrix of pixels of Red Green Blue color. Each color value starts from 0 to 255 for each bit. Base picture based method is similar to the random matrix based methodology. Here, instead of generating the random matrix, we take a base picture and extract the pixel values from it and use it to XOR each value in the other picture matrix. If the size of the base picture is smaller than the size of the original picture, we repeat the same values from start as in a loop [26].

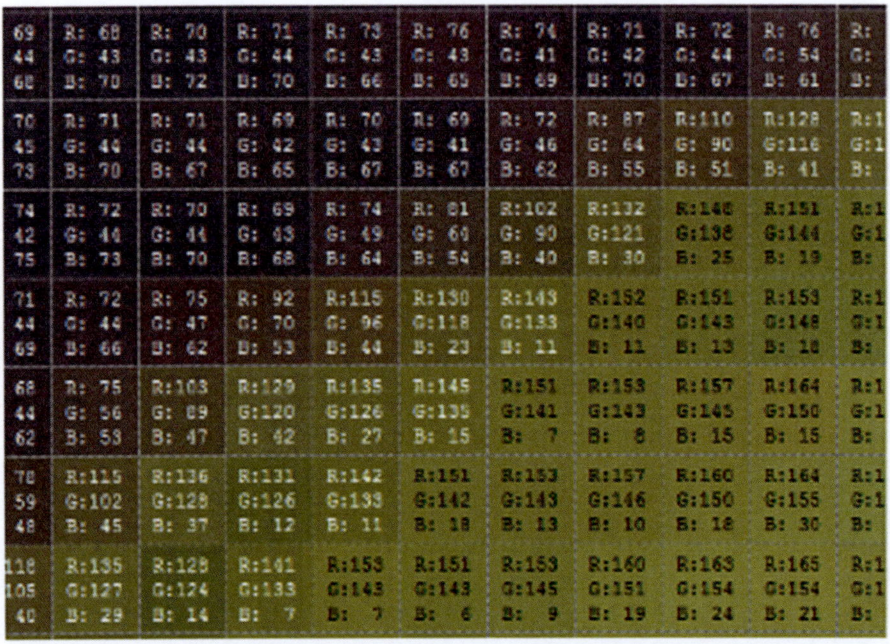

Fig. 1. Shows how the images are represented in matrix form

From the above Fig. 1, can aware of how the images are represented in matrix form. The matrix form of an image is transformed using DCT(discrete cosine transform) and the output is XORed with the masking coefficients to get the scrambled image.

2.1 P-Fibonacci Transform Function

P-Fibonacci transforms method is used to select an initial sequence from the inputted image matrix. Let, consider some of the properties of the P-Fibonacci transform properties for selecting the input sequence. To encrypt the image, the pixel location of the images is scrambled initially using P-Transform series.

Definition 1.1: The Fibonacci sequence is a recursive function

$$f(p) = \begin{cases} 0 & p < 0 \\ 1 & 1 \\ f(p-1) \wedge (p-2)^2 & p > 0 \end{cases} \quad (1)$$

Where, p is an indexing parameter of the Fibonacci XOR transform sequence. This parameter is used to mark the distance vector. From the Eq. (1), the Fibonacci sequence changes with the difference in the values of p and calculated by representing the image data in binary form, the result will be true or 1 if either bit of data is 1 otherwise false [19, 20].

Definition 1.2: Fibonacci sequence generator is formed by

$$Fn = Fn - r + \ldots . + Fn - s \ (\mathrm{mod}\, q) \text{ where } r > s > 0 \tag{2}$$

Initially, two elements are needed to find the next element in the sequence. From the previous expression, the value of r and s must be set greater than 0. Adding all the terms, the nth term of the function is calculated. The interest of function Fn is calculated by the power of two that is,

Definition 1.3: Golden spiral-ratio is continued based on Eq. (2). to select the next term of the fibonacci series. The Fibonacci sequence varies depending on the p and q value. Fibonacci series is an infinite series of integer numbers, where each term is the sum of the two previous term. The total pixel ratio of an image is found by the summing of the pixel values of the whole image are called the Fibonacci-golden ratio. It is represented mathematically with p and q where $p > q > 0$,

$$q = 2^q \tag{3}$$

$$p/q = (p + q)/p = 1.6180339887498948420\ldots \tag{4}$$

If $p = 0$, the Fibonacci-0 coding sequence is a power sequence of 2 that 1, 2, 4, 8, 16,
If $p = 1$, the Fibonacci-1 coding sequence is the classical Fibonacci sequence 1, 1, 2, 3, 5, 8, 13, 21,...
If $p > 1$, the Fibonacci-2 coding sequence is a recursive function [21–23] (Figs. 2 and 3).

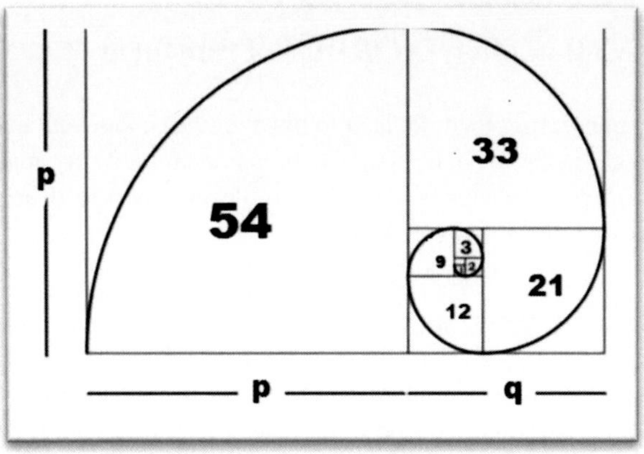

Fig. 2. The golden spiral representation of an equation

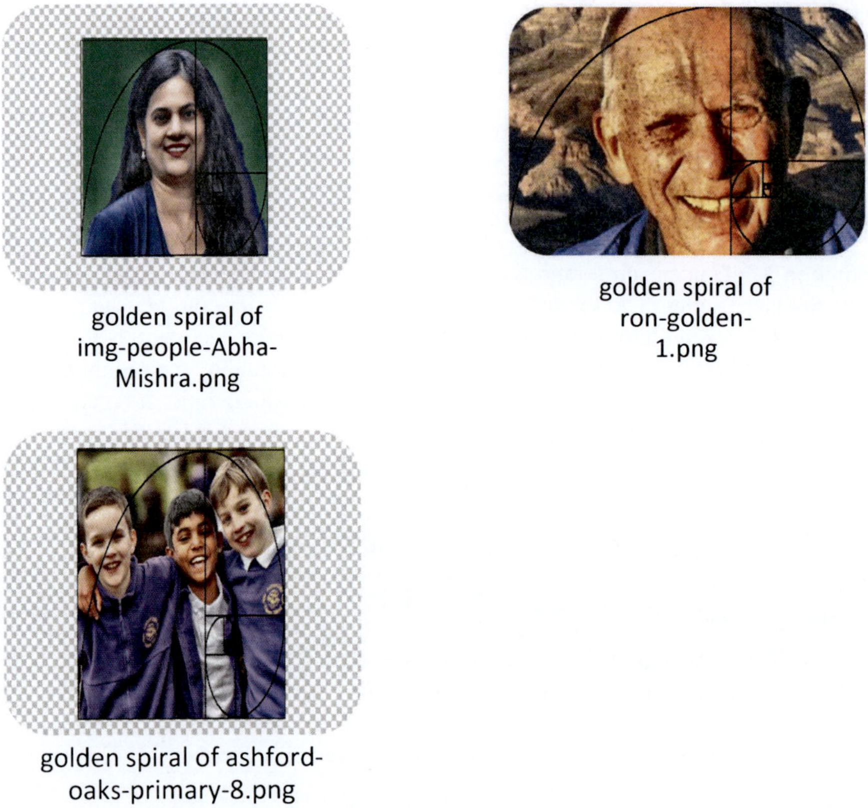

Fig. 3. Golden ratio spiral representation of the taken inputs

3 Proposed Methodology-XORDCT Transform

XOR cipher is an encrypting method used to make cipher image and hard to find by the intruders. Every character of the image can be scrambled by applying the XOR operation using a given cipher key. To descramble the encrypted image, reapply the XOR function along with the cipher key.

Proof

Let the secret image is represented as I. Based on the Fibonacci series concept, an image is splitted into sequence of images such as I1, I2, I3,….In. The XOR operator in the source side transforms the inputted single secret image into set of images before transmitting it to the destination side. The masking coefficients are discovered by performing discrete cosine transform (DCT) on the input image. The XOR-ed result of the input images are,

$$R <= \Sigma\{I1 \oplus I2 \oplus I3,\ldots \oplus In\} \tag{4}$$

Where R and $\sum\{.\}$ denotes the coefficients for masking the input images. The masked image is represented as Mi and

$$Mi = M1, M2, M3, \ldots Mn \tag{5}$$

The masked image Mi is XOR - ed using the Ii and R. Likewise $Mi = Ii \oplus R$ (6)

Where i = 1, 2, 3,…n. Then the masked images Mi are formed and transmitted to the destination side. After receiving, the destination part joins all the masked images M1, M2…Mn and perform the XOR to recover the original input image.

$$\overline{Ii} = Mi \oplus \bar{R} \tag{7}$$

Theorem 1
The XOR sequence is reversible if and only if $R = \bar{R}$.

Proof: The P-fibonacci XOR sequence is reversible if it satisfies $Ii = \overline{Ii}$, where i = 1, 2, …n. This condition is solved when $Ii = \overline{Ii}$ then

$$\overline{Ii} = Mi \oplus \bar{R} \tag{8}$$

From the above equation know that, $Mi = Ii \oplus R$, where i = 1, 2,…n then

$$\overline{Ii} = Ii \oplus R \oplus \bar{R} \quad \text{(substitute Eq.(9) here)} \tag{9}$$

whereas $Ii \oplus 0 = \overline{Ii}$.

By using the XOR property $\boxed{R \oplus \bar{\bar{R}} = 0}$ multiply both side by $\bar{R}$

$$\begin{gathered} R \oplus \bar{R} \oplus \bar{R} = 0 \oplus \bar{R} \text{ and} \\ 0 \oplus \bar{R} = \bar{R}, \end{gathered} \tag{10}$$

The above equation is simplified as $R = \bar{R}$

$$\text{Now the value of } \bar{R} = \sum\{I1 \oplus I2 \oplus \ldots In \oplus \underbrace{R \oplus R \oplus R}_{\text{n is odd}}\} \tag{11}$$

Finally $R \oplus \bar{R} = 0$ equals $R = \bar{R}$. If the masking coefficients are identical for both encryption and decryption, the process is reversible.

Theorem 2
The process is reversible, if n is even number.

Proof: If n is even number, the masking coefficient is

$$\bar{R} = \sum\{M1 \oplus \mathrm{M2} \oplus \ldots \oplus \mathrm{Mn}\} \text{ By fact} \tag{12}$$

Mi = Ii ⊕ R (Mi- masking coefficient) Where R can be written as

$$\bar{R} = \sum\{I1 \oplus \mathrm{I2} \oplus \ldots. \oplus In \oplus \mathrm{Rn}\} \text{ By regrouping of R} \tag{13}$$

$$\bar{R} = \sum\{I1 \oplus \mathrm{I2} \oplus \ldots. \oplus In \oplus \underbrace{R \oplus R \oplus \ldots\ldots \oplus R}_{\text{n is even}}\} \tag{14}$$

$$\bar{R} = \sum\{I1 \oplus \mathrm{I2} \oplus \ldots \oplus \mathrm{In} \oplus 0\}, \text{Where } \mathrm{R} \oplus \mathrm{R} \oplus \ldots \oplus \mathrm{R} = \mathrm{R} \oplus \mathrm{R} = 0\} \text{ it can be rewritten as } \sum\{I1 \oplus \mathrm{I2} \oplus \ldots \oplus \mathrm{In}\} \tag{15}$$

The value of $\bar{R}$ is equal to the value of R. (i.e.) $\bar{R} = R$.
If n is odd number, the recovered masking coefficient $\bar{R}$ is calculated as

$$\bar{R} = \sum\{M1 \oplus \mathrm{M2} \oplus \ldots \oplus \mathrm{Mn}\}$$

Initially M1 = Ii ⊕ R then $\bar{R}$ can be rewritten as

$$\bar{R} = \sum\{I1 \oplus \mathrm{R} \oplus \mathrm{I2} \oplus \mathrm{R} \oplus \ldots \oplus \mathrm{In} \oplus \mathrm{R}\} \tag{16}$$

By using XOR property $\underbrace{R \oplus R \oplus \ldots \oplus R}_{\text{n is odd}} = R \oplus R \oplus R = 0 \oplus R = R$

The value of $\bar{R}$ can be formed by $\bar{R} = \sum\{I1 \oplus \mathrm{I2} \oplus \ldots \oplus \mathrm{In} \oplus \mathrm{R} \oplus \mathrm{R} \oplus \ldots \oplus \mathrm{R}\}$

$$\bar{R} = \sum\{I1 \oplus \mathrm{I2} \oplus \ldots \oplus \mathrm{In} \oplus \mathrm{R}\} \tag{17}$$

The $\bar{R}$ is not identical to original R. From comparing Eq. (11) and Eq. (17) it is clear that $\bar{R}$ is identical to R when the inputted images are scrambled into even number of images. Thus scrambling the image is an alternative choice for enhancing the security of the image. The images are altered in unreadable mode. The original image of the jpeg format is transformed using a Discrete Cosine Transform (DCT).

The coefficients of the image are formed by the 2D P-Fibonacci XOR Transform. The parameters of P-Fibonacci XOR transform are p and i. Lossless image scrambling method is build using P-Fibonacci XOR transform.

Encryption Using XORDCT Algorithm

Step 1: Based on Fibonacci series, The image Ii is converted to three color components: R, G, B such as 'n' number of shared images I1, I2, I3,…In, before sending to the destination.
Step 2: Apply DCT to each color component and combining together all components. (Ii = I1 ⊕ I2 ⊕ I3 ⊕ …⊕In)
Step 3: XOR-ing the matrix component of an image for calculating the masking coefficient R (R <=XOR (Ii))
Step 4: The shared images Mi are scrambled by XOR-ing the Ii and R. (Mi <= Ii ⊕ R)
Step 5: If the scrambled images have 'n' odd shared images then the original image is ir-reversible otherwise reversible.
Step 6: The scrambled images are got.

Decryption Using Inverse- XORDCT Algorithm

Step 1: The destination module XOR-ed the scrambled image with the masking coefficient Mi.
Step 2: Calculate the XOR inverse function for the masking coefficient matrices of an image $\bar{R}$. ($\bar{R} <= (M1 \oplus M2 \oplus M3 \oplus \ldots \oplus Mn)$)
Step 3: After that apply inverse DCT to each color component and combining together all components. ($\overline{Ii} = Mi \oplus \bar{R}$)
Step 4: Recovering the color components together to get the original R:G:B image.
Step 5: The actual image is reconstructed.

3.1 Block Diagram of XORDCT Algorithm

The image compressed using EQIAC algorithm is taken as an input for XORDCT encryption algorithm (Fig. 4).

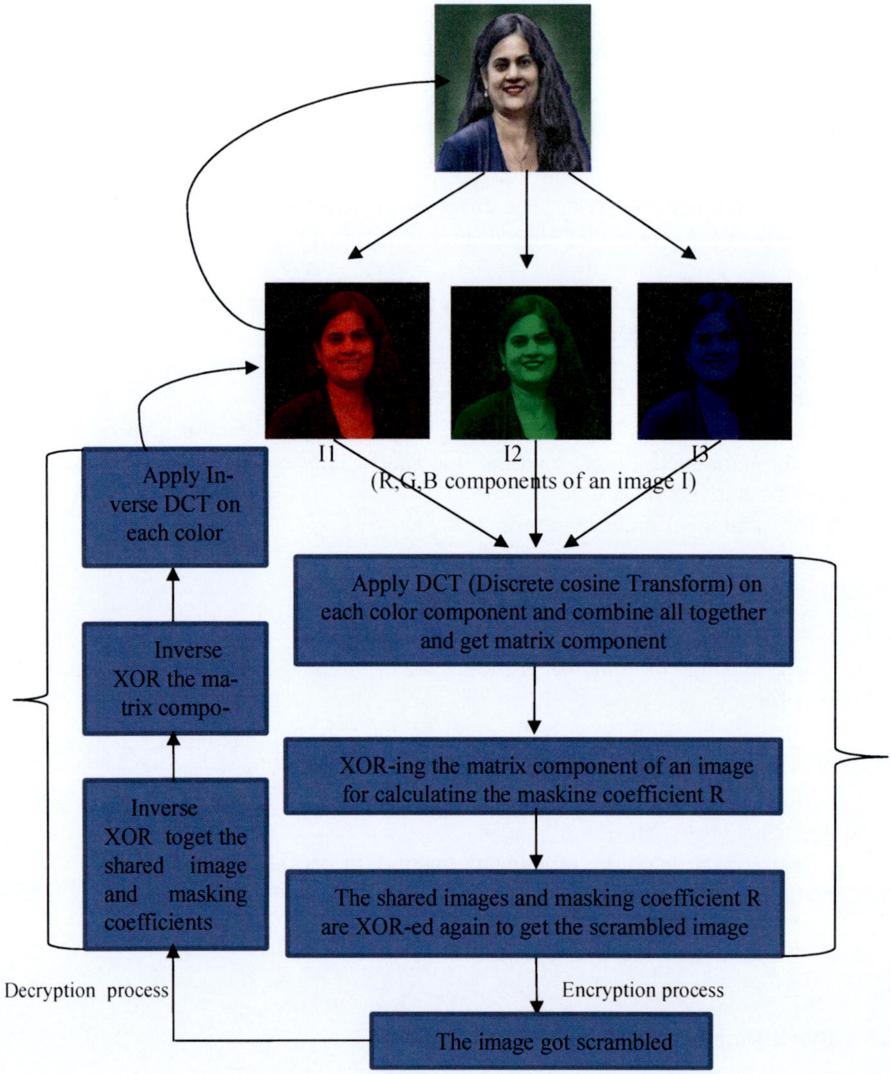

Fig. 4. Block diagram of XORDCT algorithm

4 Experimental Results

4.1 Statistical Result Analysis

Here more than 95 images are considered. First of all, consider an image which we want to encrypt. Initially convert the image into a binary data which want to encrypt. The image with .png format is taken for compression. The iteration made in the images are seen visually. In order to quantify the encryption process two metrics are used.

1. Net Pixel Change Ratio (NPCR)

Change in the rate of the pixel number comparing the original and the encrypted image.

$$\mathrm{NPCR} = \frac{1}{MxN}\sum\nolimits_{i=1}^{M}\sum\nolimits_{j=1}^{N} \mathrm{D(i,j)}\, x\, 100\% \tag{18}$$

where D(i, j) = 0 when C1(i, j) = C2(i, j)
D(i, j) = 1 when C1(i, j) ≠ C2(i, j)

The parameters C1 be the original image and C2 be the modified image, D be the array used for storing the image size before and after encryption M and N are the height and width of the images. NPCR is used to analyze the image and find the bit change by comparing pixel by pixel.

2. Unified Average Changing intensity (UACI)

This metric is used to measure the average of the intensity differences in pixels of the raw image and the original images.

$$\mathrm{UACI} = \sum\nolimits_{i=1}^{M}\sum\nolimits_{j=1}^{N} \frac{|(C1(i,j) - C2(i,j)|}{255}\, x\, \frac{100\%}{M\, x\, N} \tag{19}$$

3. Correlation coefficient analysis

The pixels of the encrypted image is intercorrelated with others. The correlation exists in the adjacent pixels can be used by the intruder for creating the sequence of the image. The range of the correlation coefficient lies within −1 and +1.

$$\mathrm{Cov}(\alpha, \beta) = \frac{1}{N}\sum\nolimits_{i=1}^{N} (\alpha \mathrm{i} - \mathrm{E}(\alpha))(\beta \mathrm{i} - \mathrm{E}(\beta)) \tag{20}$$

$$\mathrm{r}\alpha\beta = \frac{\mathrm{Cov}\ (\alpha, \beta)}{\sqrt{\sqrt{D(\alpha)}\sqrt{D(\beta)}}}$$

$$\mathrm{D}(\alpha) = \frac{1}{N}\sum\nolimits_{i=1}^{N} (\alpha \mathrm{i} - \mathrm{E}(\alpha))^2$$

$$\mathrm{E}(\alpha) = \frac{1}{N}\sum\nolimits_{i=1}^{N} \alpha \mathrm{i}$$

The correlation covariance of the original image and the encrypted image is calculated using Cov(α,β) and D(α) is the variance range value of the image and E(α) is assumed as the mean deviation of the image pixel values (Tables 1 and 2).

Table 1. UACI and NPCR value of the taken input image

Image	Image Name	NPCR (in %)	UACI (in %)
	Image A-img-people-Abha-Mishra.jpg	99.585	28.621
	Image B-ron-golden-1.jpg	99.6078	50.1931
	Image C-ashford-oaks-primary-8.jpg	99.6094	27.4092

Table 2. Correlation coefficient of original image and cipher image.

(img-people-Abha-Mishra.jpg)				
	Horizontal	Vertical	Diagonal	Counterdiagonal
Original image	0.972953	0.970462	0.916925	0.938441
Encrypted image	-4.097226×10^{-5}	1.158832×10^{-4}	4.620716×10^{-5}	4.539076×10^{-4}
(ron-golden-1.jpg)				
	Horizontal	Vertical	Diagonal	Counterdiagonal
Original image	0.915352	0.922622291	0.868221	0.857255
Encrypted image	-3.811868×10^{-6}	$- 7.131676 \times 10^{-4}$	-9.372164×10^{-4}	-3.496642×10^{-4}
(ashford-oaks-primary-8.jpg)				
	Horizontal	Vertical	Diagonal	Counterdiagonal
Original image	0.94514	0.950725	0.919702	0.937272
Encrypted image	−0.001678	3.514572×10^{-4}	-9.329940×10^{-4}	-4.626904×10^{-5}

4. PSNR and MSE analysis

It is the ratio between the original image (signal) and the encrypted image (noise). PSNR scale is used to represent the signal. It is calculated by

$$\text{PSNR} = 20 \text{ x } \log_{10} \frac{(255)}{\sqrt{MSE}} \text{dB}$$

Where *MSE* is the average of the errors $\text{MSE} = \sum_{i=1}^{M} \sum_{j=1}^{N} \frac{[R(i,j) - S(i,j)]}{MxN}$

Where R(i, j) is the pixel value of the taken input image, S(i, j) is the pixel value of the encrypted image (Table 3).

Table 3. Shows the MSE and PSNR of the input image, encrypted image and decrypted image

Image	Image name	MSE for encrypted image	PSNR for decrypted image
	Image A-img-people-Abha-Mishra.jpg	0.2698	28.621
	Image B-ron-golden-1.jpg	0.2711	50.1931
	Image C-ashford-oaks-primary-8.jpg	0.2013	27.4092

d. Graphical result

a) Image A (original) b) Image A (encrypted) c) Image A (Decrypted)

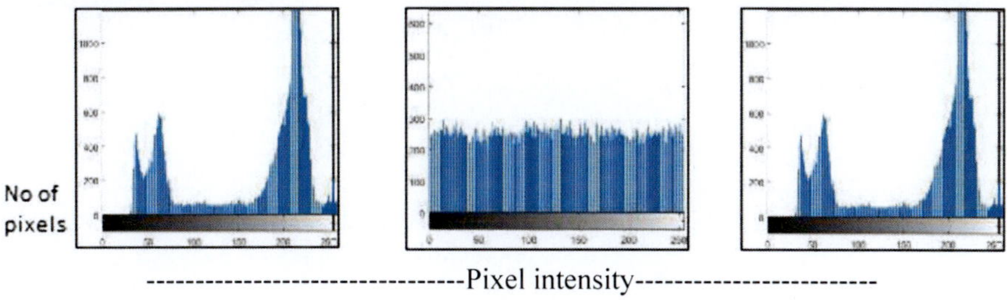

d) Histogram of original imageA e) Histogram of encrypted imageA f) Histogram of decrypted imageA

d. Histogram analysis

An image can be represented graphically using the intensity values of the pixels as their axis values. Histogram analysis is a type of representation of image format.

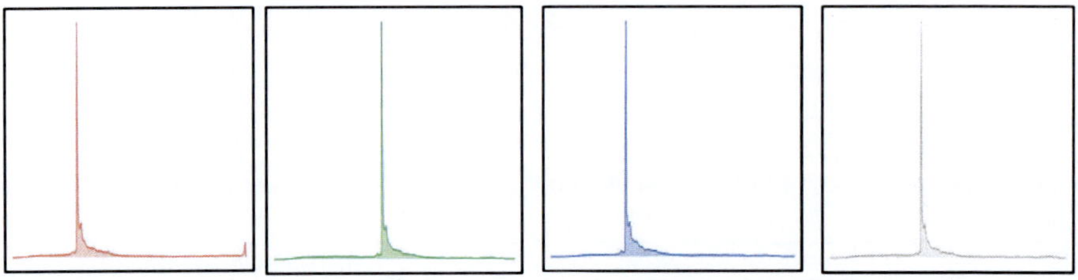

a. b. c. d.

a) Red histogram of image A b) green histogram of image A

c) Blue histogram of image A d) luminosity of image A

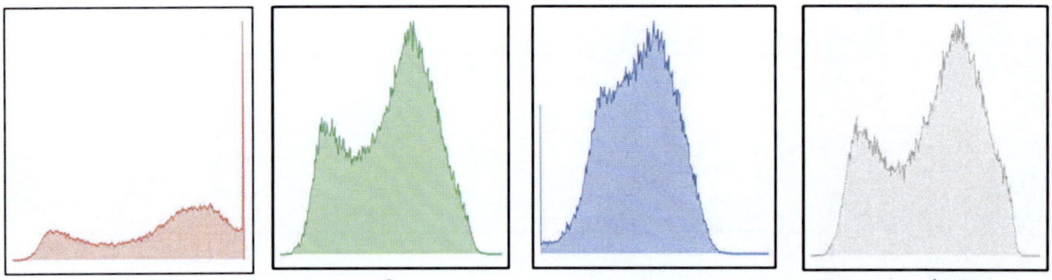

e. f. g. h.

e) Red histogram of image B f) green histogram of image B

g) Blue histogram of image B h) luminosity of image B

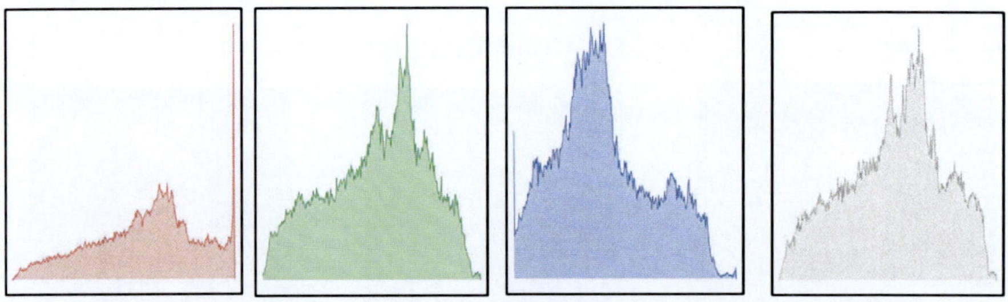

i. j. k. l.

i) Red histogram of image C j) green histogram of image C

k) Blue histogram of image C l) luminosity of image C

5 Conclusion

An image is scrambled using Fibonacci series, XOR operation, masking coefficients and DCT (discrete cosine transform). XORing the image matrix component makes changes in the input also brings drastic variation in the encrypted image. The masking coefficients used in this experiment is also XORed to bind more with the input data. The working feature of the encryption and decryption algorithm is experimented and described in this literature. he Experimental results of all the images (image A, image B, image C) are tabulated in this literature. Depends on the ratio of the color accommodated in the image, the PSNR, UACI, NPCR and correlation coefficients varied. From the experimental results, UACI of the taken image is calculated as 30% and NPCR of the taken image is 90%. It is clear that XORDCT algorithm is suitable for encryption and avoids attacks which violates security brute-force attacks.

6 Future Enhancement

In future, this algorithm can be used in real-time devices used in smart homes, smart environments, which uses audios and videos as a primary key for security. The presented algorithms enables high security for secret and confidential data transfer. It can be applicable for text message transferring also. It can be adapted with face recognition algorithms to store images in large databases like cloud for future references.

References

1. Kadhim, Q.K.: Image compression using discrete cosine transform method. Int. J. Comput. Sci. Mobile Comput. **5**(9), 186–192 (2016)
2. Anupriya, E., Soni, S., Agnihotri, A., Babelay, S.: Encryption using XOR based extended key for information security – a novel approach. Int. J. Comput. Sci. Eng. **3**(1), 146–154 (2011). ISSN: 0975-3397
3. Schneier, B.: Applied Cryptography: Protocols, Algorithms, and Source Code in C, 2nd edn. Wiley, New York (1996)
4. Daemen, J., Rijmen, V.: The Design of Rijndael: AES-The Advanced Encryption Standard. Springer, Germany (2002)
5. Draksharam, S., Katravulapalli, D., Rohith Krishna, K., Thanikaiselvan, V.: Analysis of hybrid chaotic image encryption, vol. 18, pp. 697–703. IEEE (2018)
6. Kumar, M., Aggarwal, A., Garg, A.: A review on various digital image encryption techniques and security criteria. Int. J. Comput. Appl. (0975 – 8887), **96**(13) (2014)
7. El Abbadi, N.K., Abaas, S.T., Alaziz, A.A.: New image encryption algorithm based on Diffie-Hellman and singular value decomposition. Int. J. Comput. Appl. (0975 – 8887), **96** (13), 19–26 (2014)
8. Zhou, Y., Panetta, K., Agaian, S.: Image encryption using binary key-images. In: Proceedings of the 2009 IEEE International Conference on Systems, Man, and Cybernetics, vol. 09, pp. 4569–4574 (2009)
9. Modak, P.M., Pawar, V.: A comprehensive survey on image scrambling techniques. Int. J. Sci. Res. **4**(12), 814–818 (2015)

10. Kocarev, L., Jakimoski, G., Stojanovski, T., Parlitz, U.: From chaoticmaps to encryption schemes. In: Proceedings of the IEEE International Symposium on Circuits and Systems (ISCAS 1998), pp. 514–517. IEEE, Monterey (1998)
11. Shannon, C.E.: Communication theory of secrecy systems. Bell Syst. Tech. J. **28**(4), 656–715 (1949)
12. Pareek, N.K., Patidar, V., Sud, K.K.: Discrete chaotic cryptography using external key. Phys. Lett. A **309**(1–2), 75–82 (2003)
13. Wang, X., Teng, L.: An image blocks encryption algorithm based on spatiotemporal chaos. Nonlinear Dyn. Int. J. Nonlinear Dyn. Chaos Eng. Syst. **67**(1), 365–371 (2012)
14. Xiao, D., Liao, X., Wei, P.: Analysis and improvement of a chaos-based image encryption algorithm. Chaos, Solitons Fractals **40**(5), 2191–2199 (2009)
15. Wang, X.-Y., Bao, X.-M.: A novel block cryptosystem based on the coupled chaotic map lattice. Nonlinear Dyn. **72**(4), 707–715 (2013)
16. Jakimoski, G., Kocarev, L.C.: Analysis of some recently proposed chaos-based encryption algorithms. Phys. Lett. A **291**(6), 381–384 (2001)
17. Behnia, S., Akhshani, A., Mahmodi, H., Akhavan, A.: A novel algorithm for image encryption based on mixture of chaotic maps. Chaos, Solitons Fractals **35**(2), 408–419 (2008)
18. Mirzaei, O., Yaghoobi, M., Irani, H.: A new image encryption method: parallel sub-image encryption with hyper chaos. Nonlinear Dyn. **67**(1), 557–566 (2012)
19. Wang, Y., Wong, K.-W., Liao, X.F., Chen, G.R.: A new chaos based fast image encryption algorithm. Appl. Soft Comput. **11**(1), 514–522 (2011)
20. Ismail, A.I., Mohammed, A., Hossam, D.: A digital image encryption algorithm based a composition of two chaotic wavelet function map. Int. J. Netw. Secur. **11**(1), 1–10 (2010)
21. Rhouma, R., Belghith, S.: Cryptanalysis of a new image encryption algorithm based on hyper-chaos. Phys. Lett. Sect. A: Gen. Atomic Solid State Phys. **372**(38), 5973–5978 (2008)
22. Li, X.-F., Chlouverakis, K.E., Xu, D.-L.: Nonlinear dynamics and circuit realization of a new chaotic flow: a variant of Lorenz, Chen and Lü. Nonlinear Anal. Real World Appl. Int. Multidiscip. J. **10**(4), 2357–2368 (2009)
23. Zhao, J.F., Wang, S.Y., Chang, Y.X., Li, X.F.: A novel image encryption scheme based on an improper fractional-order chaotic system. Nonlinear Dyn. **80**(4), 1721–1729 (2015)
24. Wu, X., Wang, H., Lu, H.: Modified generalized projective synchronization of a new fractional-order hyperchaotic system and its application to secure communication. Nonlinear Anal. RealWorld Appl. **13**(3), 1441–1450 (2012)
25. Mannai, O., Bechikh, R., Hermassi, H., Rhouma, R., Belghith, S.: A new image encryption scheme based on a simple firstorder time-delay system with appropriate nonlinearity. Nonlinear Dyn. **82**(1–2), 107–117 (2015)
26. Liu, Q., Li, P.-Y., Zhang, M.-C., Sui, Y.-X., Yang, H.-J.: A novel image encryption algorithm based on chaos maps with Markov properties. Commun. Nonlinear Sci. Numer. Simul. **20**(2), 506–515 (2015)

Privacy Preservation for Continuous Decremental Data Publishing

Surapon Riyana[1(✉)], Nattapon Harnsamut[2], Uratcha Sadjapong[2], Srikul Nanthachumphu[1], and Noppamas Riyana[1]

[1] Maejo University, Chiangmai, Thailand
{surapon_r, srikul, noppamas}@mju.ac.th
[2] University of Phayao, Phayao, Thailand
{nattapon.ha, uratcha.sa}@up.ac.th

Abstract. The privacy violation is a serious issue that must be considered when datasets are published to the outside scope of the data-collecting organization for gaining more data utility. To address this issue, a well-known privacy preservation model, (α, k)-Anonymity, is proposed. Unfortunately, this privacy preservation model is generally proposed to address privacy violation issues in datasets that are focused on performing one-time data publishing. Therefore, if published datasets are dynamic, i.e., the data of them can be always changed and performing multiple time data publishing such as continuous decremental datasets, (α, k)-Anonymity could be insufficient. For this reason, the vulnerabilities of (α, k)-Anonymity in continuous decremental datasets are identified in this work. Moreover, an exhaustive decremental privacy preservation algorithm for addressing privacy violation issues in continuous decremental datasets is also proposed. The aim of the proposed algorithm is that aside from privacy preservation constraints, the data utility of datasets is also maintained as much as possible. Furthermore, the experimental results are shown which indicate that the proposed algorithm is highly effective.

Keywords: Privacy preservation · Continuous decremental datasets · Anonymization models

1 Introduction

A serious issue in datasets which collects the person-specific data, the privacy data, to be published for utilizing in the business collaborators, it is the privacy violation [1–7, 12–14]. To address this issue, the common privacy principle suggests that before datasets are published, all explicit identifiers of users, e.g., names, addresses, and phone numbers, are removed or encrypted. However, in [2] and [7], the authors demonstrate that this privacy preservation idea still has privacy violation concerns that must be addressed [2]. That is because the adversary can still violate the privacy data of users by using identity linkage attacks and attribute linkage attacks.

k-Anonymity [1–6] is an effective privacy preservation model which is proposed to address privacy violation issues in published datasets. It prevents identity linkage attacks by requiring every equivalence class of published datasets must contain at least

J. I.-Z. Chen et al. (Eds.): ICIPCN 2020, AISC 1200, pp. 233–243, 2021.
https://doi.org/10.1007/978-3-030-51859-2_21

k indistinguishable quasi-identifier tuples. However, this privacy preservation idea cannot handle attribute linkage attacks. To rid this vulnerability of k-Anonymity, in [7], an extended privacy preservation model of k-Anonymity is proposed, so-called as l-Diversity. To address attribute linkage attacks, the authors introduce that every equivalence class of published datasets must include the sensitive values to be at least l different values. Although this privacy preservation idea can handle identity linkage attacks and attribute linkage attacks. It is quite difficult for the data holder to set the appropriate value of l, and it cannot preserve the privacy data in published datasets from using probability inference linkage attacks.

With probability inference linkage attacks, they can appear is when the sensitive data in equivalence classes to occur frequently. To limit probability inference linkage attacks in equivalence classes, (α, k)-Anonymity [8] is proposed. With this privacy preservation model, it uses α to limit the confidence of inferring the protected sensitive values in equivalence classes. That is, every protected sensitive value in equivalence classes have the inferred confidence to be at most α.

Unfortunately, to the best our knowledge about k-Anonymity, l-Diversity, and (α, k)-Anonymity, they are generally proposed to address privacy violation issues in datasets that are focused on performing one-time publishing. For this reason, they could be insufficient to preserve the privacy data in datasets which always allow decreasing their data-collection. An example of the real-life scenarios of continuous decremental data publishing can be found in hospital at California, i.e., it publishes all discharged patient data to researchers for every six months [9].

Example 1 (Privacy violation issues in continuous decremental datasets): Let Table 1 be the raw dataset at the timestamp 1, and its anonymized dataset version is shown in Table 2 such that it satisfies the (0.5, 2)-Anonymity privacy preservation constraint by using generalization technique, DGH or VGH [11]. Let Table 3 be an anonymized dataset of Table 1 such that it is published after the tuple t_6, Alice's profile tuple was deleted, it denotes the anonymized dataset version of Table 1 to be published at the timestamp 2. Also, Table 3 satisfies the (0.5, 2)-Anonymity privacy preservation constraint by using DGH or VGH and assume that the adversary received Table 2 and 3. Given Alice is the target user of the adversary, and Alice's diagnosis is the target data. Moreover, it is supposed that the adversary knows that Alice is a female person who is 52 years old. Furthermore, the adversary highly believes Alice's profile tuple is collected in Table 2, and he/she further knows that it does not available in Table 3. For this situation, the adversary can highly ensure that Alice's diagnosis is collected in Table 2 to be HIV.

To address privacy violation issues in continuous data publishing, in [10], m-Invariance is proposed. It can address privacy violation issues in published datasets which allow changing the data of them by using the insertion and deletion operations. The privacy preservation idea of m-Invariance is that if an arbitrary tuple is collected in published datasets does not satisfy the given privacy preservation requirement, the counterfeit tuples are added to published datasets until it satisfies the given privacy preservation requirement. However, m-Invariance privacy preservation constraints do not consider the set of protected sensitive values. For this reason, if the sensitive values are collected in an arbitrary equivalence class of published datasets to only be

Table 1. An example of raw datasets (the patient datasets).

#	Age	Sex	Diagnosis
t_1	56	Male	Flu
t_2	57	Male	Fever
t_3	58	Male	HIV
t_4	59	Male	Cancer
t_5	51	Female	Fever
t_6	52	Female	HIV
t_7	53	Female	Flu

Table 2. An example published table version of Table 1.

#	Age	Sex	Diagnosis
t_1	[56–59]	Male	Flu
t_2	[56–59]	Male	Fever
t_3	[56–59]	Male	HIV
t_4	[56–59]	Male	Cancer
t_5	[51–53]	Female	Fever
t_6	[51–53]	Female	HIV
t_7	[51–53]	Female	Flu

Table 3. A published table version of Table 1 after t_6 was deleted.

#	Age	Sex	Diagnosis
t_1	[51–59]	*	Flu
t_2	[51–59]	*	Fever
t_3	[51–59]	*	HIV
t_4	[51–59]	*	Cancer
t_5	[51–59]	*	Fever
t_7	[51–59]	*	Flu

disgusting, the adversary can infer that the data owner of theses sensitives to be disgusting. For this situation, it may even have a detrimental effect on the sensitive data owner. To rid this vulnerability of m-Invariance, a new decremental privacy preservation algorithm is proposed in this work. With the proposed algorithm, aside from privacy preservation constraints, the data utility is also maintained as much as possible. Moreover, published datasets of the proposed algorithm can guarantee the "confidence bounding" of re-identification of every protected sensitive value.

The organization of this work is as follows. Section 2 presents the problem definition of this work. Subsequently, a privacy preservation algorithm is based on the problem definition to be proposed in Sect. 3. Then, Sect. 4 is devoted to evaluating the effectiveness of the proposed algorithm. Finally, the conclusion will be discussed in Sect. 5.

2 Problem Definition

(α, k)-Anonymity [8] is a well-known privacy preservation model which is extended from k-Anonymity [2, 11]. For privacy preservation, the tuples of published datasets are firstly partitioned by α and k. Finally, the unique quasi-identifier values of each partition are generalized by their less specific values which are represented by DGH or VGH [11], to be indistinguishable. For this reason, all possible re-identification conditions to published datasets that are satisfied by (α, k)-Anonymity constraints, they always have at least k tuples to be satisfied. Moreover, they can guarantee that the confident re-identification of every protected sensitive value is at most α. Unfortunately, (α, k)-Anonymity is proposed to address privacy violation issues in datasets that are focused on performing one-time publishing. Therefore, such an arbitrary published dataset is satisfied by (α, k)-Anonymity constraints, it still has privacy violation issues that must be addressed when the data of them are changed by deletion and released by performing multiple times. That is because the privacy data is collected in (α, k)-Anonymity published datasets can be violated by using such an appropriate comparison data attacking. To address this vulnerability of (α, k)-Anonymity, a new privacy preservation algorithm for continuous decremental datasets is proposed in this work. Before the proposed algorithm will be presented, its basic definitions are first defined.

Definition 1 (Raw datasets): Let $U = \{u_1, u_2, \ldots, u_i\}$ be the set of users. Let $QI = \{qi_1, qi_2, \ldots, qi_m\}$ be the set of quasi-identifier attributes. Let S be the sensitive attribute. Let $D_{qi_x} = \{d_1, d_2, \ldots, d_q\}$ be the data domain of $qi_x \in QI$, where $1 \leq x \leq m$. Also, let $D_S = \{s_1, s_2, \ldots, s_p\}$ be the data domain of S. Let $T = \{t_1, t_2, \ldots, t_i\}$ be the raw dataset such that it is represented by multiset notations. Every tuple $t_y \in T$ represents the profile tuple of the user $u_y \in U$, such that $t_y = (v_{qi_1}, v_{qi_2}, \ldots, v_{qi_m}, v_s)$, where $v_{qi_1} \in D_{qi_1}, v_{qi_2} \in D_{qi_2}, \ldots, v_{qi_m} \in D_{qi_m}$, and $v_s \in D_S$. Let T[S] be projecting the sensitive values in S of T.

For example, let $U = \{u_1, u_2, u_3, u_4, u_5, u_6, u_7\}$ be the set of users. Let $QI = \{Age, Sex\}$ be the set of quasi-identifier attributes such that $D_{Age} = [50, 60]$ and $D_{Sex} = \{Male, Female\}$ is the data domain of Age and Sex respectively. Let Diagnosis be the sensitive attribute such that its data domain is $D_{Diagnosis} = \{$Flu, HIV, Cancer, Fever$\}$. With these given instances, Table 1 is an example of raw dataset version T which is satisfied by Definition 1.

Definition 2 ((α, k)-Anonymity [8]): Let $PS \subseteq D_S$ be the set of sensitive values that must be protected. Let α be a real number in the range between 0.0 and 1.0. Let k be a positive integer, where $k \geq 2$. Let $f_{GEN1}(\alpha, k, PS, T) : T \rightarrow_{\alpha,k,PS} T'$ be a generalized function for transforming T to become T'. That is, all unique quasi-identifier values of T' are generalized by their DGH or their VGH, to be at least k indistinguishable tuples. Moreover, the re-identified proportion of every protected sensitive value $s \in PS$ which is available in T' must be at most α. Also, each group of tuples is satisfied by the given value of α and k to call an equivalence class of T'. Therefore, T' can also be represented by the set of its equivalence classes $EC = \{ec_1, ec_2, \ldots, ec_g\}$. Without loss of generality, $ec_1 \cap ec_2 \cap \ldots \cap ec_g = \emptyset$.

For example, let Table 1 be the raw dataset T. Let α be set at 0.5, and k be set at 2. Let $PS = \{HIV, Cancer\}$ be set of the protected sensitive values. Therefore, a published dataset version of Table 1 is constructed from the given instances and satisfies Definition 2, it is shown in Table 2.

Definition 3 (Decremental privacy preservation based on (α, k)-anonymity): Let T_{j-1} be the raw dataset T at the timestamp $j - 1$. Let T_j be the raw dataset T at the timestamp j. Let T'_{j-1} be the published dataset of T_{j-1} such that it satisfies Definition 2. Let $f_{GEN2}\left(\alpha, k, PS, T_{j-1}, T_j, T'_{j-1}\right) : T_j \rightarrow_{\alpha,k,PS,T_{j-1},T'_{j-1}} T'_j$ be the generalized function for transforming T_j to become T'_j such that T'_j also satisfies Definition 2. Moreover, T'_j must be satisfied by both limitations as follows.

- Let T_{Delete} be the set of tuples which occur in T_{j-1} but is not available in T_j. Let SEC_{j-1} be the set of the specific equivalence classes of T'_{j-1} such that each $ec_{ee} \in SEC_{j-1}$ must collect the tuple(s) of $t_{Delete} \subset T_{Delete}$. Let ec_e be an arbitrary equivalence class of T'_j such that it relates to $ec_{ee} \in SEC_{j-1}$. For privacy preservation, $ec_{ee} - ec_e$ must also be satisfied by α and k, and
- Let D_{Loss} cost be the penalty cost of T'_j; it must be minimized. In addition, D_{Loss} will be explained in Definition 4.

Besides, it is assumed that all ordered pairs of published dataset versions of T such that they are published at the timestamp $j - 1$ and j respectively, they are always satisfied by Definition 3. For this reason, only T'_{j-1} is considered in conjunction with α and k to construct T'_j. For example, let α be set at 0.5, and k be set at 2. Let Table 1 be the raw dataset at the timestamp $j - 1$, and its published dataset version is shown in Table 2. Let $PS = \{HIV, Cancer\}$ be a set of protected sensitive values. Furthermore, after Table 2 was published and supposed that, the tuple t_6 and t_7 are removed from Table 1. For this situation, a published dataset version of Table 1 at the timestamp j such that it does not have privacy violation concerns from using comparison data attacking, i.e., it satisfies Definition 3, to be shown in Table 4.

Table 4. A published table version of Table 1 after t_6 and t_7 were deleted.

#	Age	Sex	Diagnosis
t_1	[51–59]	*	Flu
t_2	[51–59]	*	Fever
t_3	[51–59]	*	HIV
t_4	[51–59]	*	Cancer
t_5	[51–59]	*	Fever

Definition 4 (Data utility metric): The general impact on the data quality in $T_j^{'}$ can be defined by D_{Loss}. More cost of D_{Loss} means more penalty or less data utility in $T_j^{'}$. The penalty cost of D_{Loss} is in the range between 0.0 and 1.0.

$$D_{Loss} = \frac{\sum_{x=1}^{m} \frac{h(qi_x)}{H(qi_x)} + \sum_{\forall t_{Delete} \in T_{Delete}} m}{|T_j| \cdot m}$$

Where $h(qi_x)$ is the height of the generalized value in the quasi-identifier attribute $qi_x \in QI$ and $H(qi_x)$ is the height of DGH of qi_x.

2.1 Problem Statement

Let α and k be the privacy preservation constraints. Let T_{j-1} and T_j be both raw dataset versions of T at the timestamp $j - 1$ and j respectively. Moreover, let $T_{j-1}^{'}$ be the generalized dataset version of T_{j-1}. Let PS be the set of the protected sensitive values. Let DGH_{qi_x} or VGH_{qi_x} be the domain generalization hierarchy or the a value generalization hierarchy for each $qi_x \in QI$, where $1 \leq x \leq m$. For privacy preservation, T_j is transformed to become $T_j^{'}$ such the quasi-identifier values of $T_j^{'}$ are generalized by DGH_{qi_x} or VGH_{qi_x} to be at least k indistinguishable tuples. Moreover, the re-identified proportion of every protected sensitive value $s \in PS$ in $T^{'}$ must be at most α. And the penalty cost D_{Loss} of $T^{'}$ must be minimized. Furthermore, all compared results between the related equivalence classes of $T_{j-1}^{'}$ and $T_j^{'}$ must also be satisfied by α and k.

3 Proposed Algorithm

Algorithm 1 shows the pseudo-code of the exhaustive decremental privacy preservation algorithm. For privacy preservation, the algorithm firstly investigates all deleted tuples such that "Can they satisfy the given value of α?", as shown in the lines between 2 and 9. If the deleted sensitive tuples cannot satisfy the given value of α, they are deleted, i.e., they are not considered to publish at the timestamp j. If not the next step of the algorithm is processed. That is, all possible generalization dataset versions of T_j are constructed, so denoted as T_j^*, such that every generalization dataset version of T_j^* must satisfy Definition 2. Then, the desired generalized dataset versions $t_w^* \in T_j^*$, $T_j^{'}$ of T_j, is determined by using D_{Loss} metric, i.e., an arbitrary generalized dataset version $t_w^* \in T_j^*$ has the penalty cost of D_{Loss} to be minimized. Moreover, all compared results between the related equivalence classes in $T_{j-1}^{'}$ and $T_j^{'}$ must also be satisfied by α and k, Finally, the algorithm returns $T_j^{'}$.

```
Algorithm 1: Decremental privacy preservation algorithm
Input: T'_{j-1}, T_{j-1}, T_j, α, and k.
Output: T' satisfies Definition 3.
1.Initialize MIN_Loss as the maximized value;
2.Let IsPassed be Boolean;
3.IF T_{j-1} is not NULL THEN
4.   T_Delete = T_{j-1} - T_j;
5.   Let T_Delete[S] is projecting the sensitive values in T_Delete;
6.   IF T_Delete[S]does not satisfy α THEN
7.      T_j = T_{j-1};
8.   END
9.END
10. FOR all generalized datasets T*_j of T_j DO
11.     IF T*_w ∈ T*_j satisfies α and k THEN
12.        Let LS be the defined D_LOSS of T*_w;
13.        IsPassed = TRUE;
14.        FOR all equivalence classes EC_{j-1} of T'_{j-1} THEN
15.           FOR all equivalence classes EC_w of T*_w THEN
16.              IF ec_{j-1} - ec_w, where ec_{j-1} ∈ EC_{j-1} and ec_w ∈
                    EC_w does not satisfy α and k THEN
17.                    IsPassed = FALSE;
18.                    BREAK;
19.              END
20.           END
21.           IF IsPassed = FALSE THEN
22.              BREAK;
23.           END
24.        END
25.        IF LS < MIN_Loss and IsPassed = TRUE THEN
26.           T'_j = T*_w;
27.         MIN_Loss = LS;
28.        END
29.     END
30. END
31. Return T'_j;
```

4 Experiment

In this section, the experimental results are based on the effectiveness of the proposed algorithm to be discussed by the comparison with (α, k)-Anonymity [8].

4.1 Experimental Setup

All experiments of this work are conducted on an Intel(R) Xeon(R) Silver 4110 @2.10 GHz CPU with 16 GB memory and 1 TB HDD running Microsoft Windows 10 64-bit professional edition. All the implementation is built and executed on Microsoft Visual Studio 2017 Community Edition and SQL Server 2017 RTM.

Moreover, all experimental results are discussing in this section; they are conducted on Adult dataset available at the UCI Machine Learning Repository [15]. This dataset is constructed from 32,561 user profile tuples such that each user profile tuple consists of 14 attributes such that Age, Work class, Fnlwgt, Education, Education-num, Marital-status, Occupation, Relationship, Race, Sex, Capital-gain, Capital-loss, Hours-per-week, and Native-country. For conducting the experiments effectively, all user profile tuples include the missing values, "?"; they are removed. Thus, the user profile tuples only remain in the experimental dataset to be 30162 tuples. Moreover, given the attributes as Age, Sex, Race, Native-country are the quasi-identifiers, and given the attribute as Occupation is the sensitive attributes of the experimental dataset. The generality of users' age in the experimental dataset is in the range between 17 and 90 years old, but 54.88% of the users' age is in the range between 25 and 45. In the attribute Sex of the experimental dataset, it collects Male to be more than Female, i.e., Male is available in the experimental dataset to be 67.55%. The users mostly are collected in the experimental dataset to be white person, i.e., they are 85.97% of data which is available in its attribute Native-country. The proportion of users' occupation is available in the experimental dataset, is as follows. Prof-specialty is 13.39%. Craft-repair is 13.35%. Exec-managerial is 13.23%. Adm-clerical is 12.33%. Sales are 11.88%. And other occupations are 35.82%. Moreover, Let PS = {Prof-specialty, Exec-managerial} be the set of protected sensitive values. All experimental results are discussing in this section; it is constructed from the n-randomized-tuples of the experimental dataset and based on D_{Loss} which is explained in Definition 4 of Sect. 2.

4.2 Discussion

In the first experiment, the effect of the proposed model which based on the dataset size is evaluated by comparison with (α, k)-Anonymity. With this experiment, the value of k and α is fixed at 10 and 0.5 respectively. Moreover, the tuples in experimental datasets are randomly chosen by varying between 100 tuples and 1600 tuples. That is, firstly random to chosen 100 tuples for constructing the first experimental dataset of this experiment. Then, the experimental datasets are constructed by randomly to select doubling the number of tuples of its previous experimental dataset version, and they are further combined.

From the experimental results, as shown in Fig. 1, observed that all experimental results indicate that (α, k)-Anonymity is slightly more effective than the proposed model. The cause of (α, k)-Anonymity is more effective is that it does not consider comparison privacy violation issues in its published datasets. For this reason, it influences to be the small size of equivalence classes and the low level of generalized values than the proposed model. However, it is observed that the proposed model is more effective when the number of tuples is increased, i.e., the D_{Loss} costs of the proposed model in the experimental datasets which are constructed by more than 400 tuples to be the very slight difference from (α, k)-Anonymity. Therefore, it is concluded that although the proposed model is the less data utility in published datasets than (α, k)-Anonymity, it can be more effective in term of privacy preservation than (α, k)-Anonymity, i.e., it can address privacy violation issues in published datasets from using continuous decremental attacking.

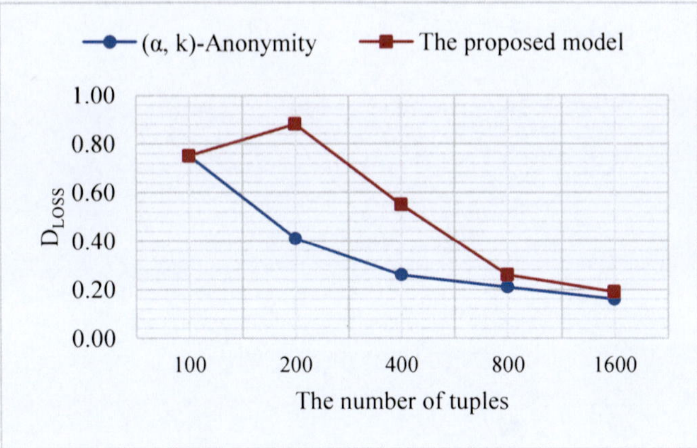

Fig. 1. The effect of dataset size.

In the second experiment, the effect of k is evaluated. For the experiment, the value of α is fixed at 0.5. The experimental dataset is constructed by randomly selecting 800 tuples from the prepared Adult dataset. Moreover, the value of k is varying between 2 and 10. From the experimental results, as shown in Fig. 2, conclude that the more value of k leads to be less data utility in published datasets. That is because the more value of k often leads to the larger size of equivalence classes and the high level of generalized values in published datasets.

In the final experiment, it is proposed to evaluate the effect of α. For the experiment, the value of k is fixed at 10. Also, the experimental dataset is constructed by randomly selecting 800 tuples from the prepared Adult dataset. Moreover, the value of α is varying between 0.20 and 1.00. From the experimental results, as shown in Fig. 3, observed that α is very little impact on the data utility in published datasets, especially, α is when increased. The cause of very little impact of α is that after the published datasets are satisfied by the given values of k, they often satisfy the given value of α.

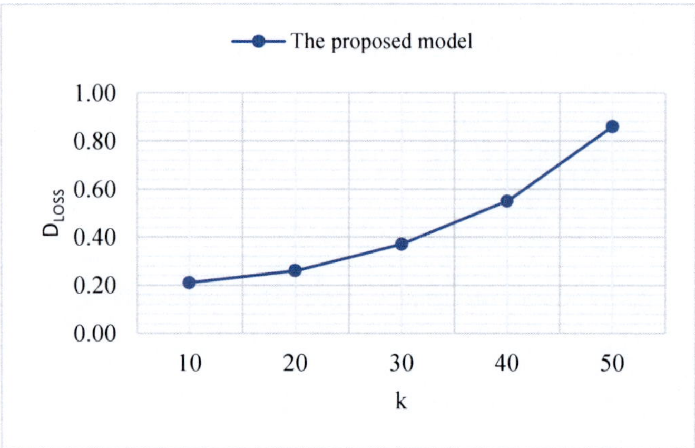

Fig. 2. The effect of k.

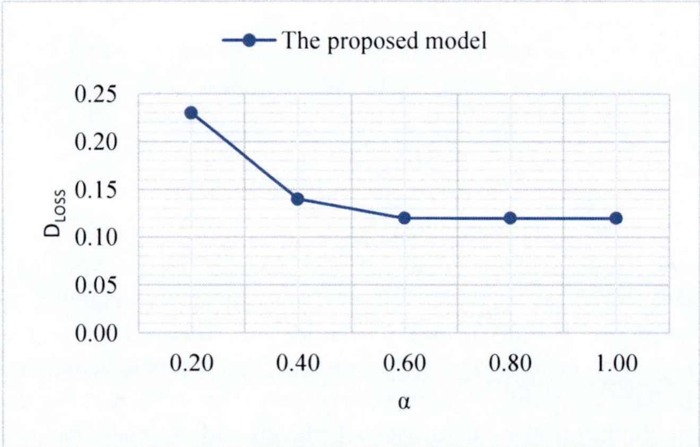

Fig. 3. The effect of α.

5 Conclusion

This research work foucses the (α, k)-anonymization issue and it is addressed in different circumstance that the transformed dataset is used to ensure the continuous decremental publishing of the data. The main contributions are twofold. Firstly, the observation is generalized for the continuous decremental data release problem. Secondly, an exhaustive decremental privacy preservation algorithm is proposed and it helps to detect the optimal solution to preserve privacy and sustain the data quality of continuous decremental data publishing. Also, from our proposed algorithm, shown that the proposed algorithm is not only preserved privacy but also maintain data quality.

References

1. Samarati, P.: Protecting respondents' identities in microdata release. IEEE Trans. Knowl. Data Eng. **13**(6), 1010–1027 (2001)
2. Sweeney, L.: K-anonymity: a model for protecting privacy. Int. J. Uncertain. Fuzziness Knowl.-Based Syst. **10**(5), 557–570 (2002)
3. Fung, B.C.M., Wang, K., Yu, P.S.: Top-down specialization for information and privacy preservation. In: Proceedings of the 21st International Conference on Data Engineering, ICDE 2005, pp. 205–216. IEEE Computer Society, USA (2005)
4. Harnsamut, N., Natwichai, J.: A novel heuristic algorithm for privacy preserving of associative classification. In: Proceedings of the 10th Pacific Rim International Conference on Artificial Intelligence: Trends in Artificial Intelligence, PRICAI 2008, pp. 273–283. Springer, Heidelberg (2008)
5. Harnsamut, N., Natwichai, J., Sun, X., Li, X.: Data quality in privacy preservation for associative classification. In: Proceedings of the 4th international conference on Advanced Data Mining and Applications, ADMA 2008, pp. 111–122. Springer, Heidelberg (2008)
6. Riyana, S., Natwichai, J.: Privacy preservation for recommendation databases. Serv. Oriented Comput. Appl. **12**(3–4), 259–273 (2018)
7. Machanavajjhala, A., Kifer, D., Gehrke, J., Venkitasubramaniam, M.: L-diversity: privacy beyond k-anonymity. ACM Trans. Knowl. Discov. Data **1**(1), 3-es (2007)
8. Wong, R., Li, J., Fu, A., Wang, K.: (α, k)-anonymous data publishing. J. Intell. Inf. Syst. **33** (2) 209–234 (2009)
9. Fung, B.C.M., Wang, K., Fu, A.W., Pei, J.: Anonymity for continuous data publishing. In: Proceedings of the 11th International Conference on Extending Database Technology: Advances in Database Technology, EDBT 2008, pp. 264–275. Association for Computing Machinery, New York (2008)
10. Xiao, X., Tao, Y.: M-invariance: towards privacy preserving re-publication of dynamic datasets. In: Proceedings of the 2007 ACM SIGMOD International Conference on Management of Data, SIGMOD 2007, pp. 689–700. Association for Computing Machinery, New York (2007)
11. Sweeney, L.: Achieving k-anonymity privacy protection using generalization and suppression. Int. J. Uncertain. Fuzziness Knowl.-Based Syst. **10**(5), 571–588 (2002)
12. Riyana, S., Harnsamut, N., Soontornphand, T., Natwichai, J.: (k, e)-anonymous for ordinal data. In: Proceedings of the 2015 18th International Conference on Network-Based Information Systems, NBIS 2015, pp. 489–493. IEEE Computer Society, USA (2015)
13. Riyana, S., Riyana, N., Nanthachumphu, S.: Enhanced (k, e)-anonymous for categorical data. In: Proceedings of the 6th International Conference on Software and Computer Applications, ICSCA 2017, pp. 62–67. Association for Computing Machinery, New York (2017)
14. Sitti, S., Riyana, S., Riyana, N.: Scenario of privacy violation within the recommendation databases. In: 2017 International Conference on Digital Arts, Media and Technology (ICDAMT), Chiang Mai, pp. 383–388 (2017)
15. UCI Machine Learning Repository. https://archive.ics.uci.edu/ml/datasets/Adult

An Effective and Efficient Heuristic Privacy Preservation Algorithm for Decremental Anonymization Datasets

Surapon Riyana(✉), Noppamas Riyana, and Srikul Nanthachumphu

Maejo University, Chiangmai, Thailand
{surapon_r, noppamas, srikul}@mju.ac.th

Abstract. (α, k)-Anonymity is a well-known anonymization model that is extended from k-Anonymity. It is proposed to address privacy violation issues in published datasets from using identity linkage attacks, attribute linkage attacks, and probability inference linkage attacks. Unfortunately, (α, k)-Anonymity is generally sufficient to preserve the privacy data in datasets that are focused on performing one-time data publishing. Thus, if published datasets, e.g., decremental datasets, are dynamic, i.e., the data of them is always changed by using deletion methods and multiple time data publishing, then the privacy data of users is collected in these published datasets could be violated by using such an appropriate comparison data attacking. To rid this vulnerability of (α, k)-Anonymity, an effective decremental privacy preservation algorithm is available in existence. Although this algorithm can address privacy violation issues in published decremental datasets, it has a vital vulnerability that must be improved, i.e., it is highly complex in terms of transforming the data which is available datasets to satisfy the specific privacy preservation constraints. For this reason, a heuristic privacy preservation algorithm for publishing decremental datasets based on clustering techniques to be proposed in this work. With the proposed algorithm, aside from privacy preservation, the data utility and execution time are also maintained as much as possible. Furthermore, we show the experimental results which indicate that the proposed algorithm is highly effective and efficient.

Keywords: Heuristic algorithm · Clustering algorithm · Anonymity model · Decremental datasets · Privacy preservation

1 Introduction

In recent decade years, there are several well-known privacy preservation models to be proposed such as k-Anonymity [1, 2], l-Diversity [3], t-Closeness [4], (α, k)-Anonymity [5], and (k, e)-Anonymous [10, 11]. Unfortunately, these privacy preservation models generally propose to address privacy violation issues in datasets that are focused on performing one-time data publishing. For this reason, they could be insufficient to address privacy violation issues in datasets which always change the data of them by using deletion methods and published multiple times. An example of the real-life scenarios of the data in published datasets are changed by using deletion

J. I.-Z. Chen et al. (Eds.): ICIPCN 2020, AISC 1200, pp. 244–257, 2021.
https://doi.org/10.1007/978-3-030-51859-2_22

methods, this event occurred in hospital at California, i.e., it publishes all discharged patient data to researchers for every six months [6]. To address privacy violation issues, in [7], the authors suggest that before datasets are published, all explicit identifier values are collected in datasets to be removed. Moreover, the unique quasi-identifier values are generalized or suppressed to satisfy the given value of α and k. Furthermore, all compared results between the specific published dataset and its previously related published dataset versions must also be satisfied by the given value of α and k. Although this privacy preservation algorithm is more effective in term of privacy preservation, it still has a vital vulnerability must be improved, i.e., it is highly complex in term of transforming the data which is available datasets to satisfy the specific privacy preservation constraints. For this reason, an effective and efficient heuristic privacy preservation algorithm for publishing decremental datasets is proposed in this work, it will be presented in Sect. 4.

The organization of this work is as follows. Section 2 presents the traditional (α, k)-Anonymity. Subsequently, the privacy violation issues in published decremental datasets are explained in Sect. 3. In Sect. 4, an effective and efficient heuristic privacy preservation algorithm for publishing decremental datasets is proposed. Then, Sect. 5, it is devoted to evaluating the effectiveness and efficiency of the proposed algorithm. Finally, the conclusion of this work will be discussed in Sect. 6.

2 (α, k)-Anonymity [5]

Let $U = \{u_1, u_2, \ldots, u_i\}$ be the set of users. Let $D = \{d_1, d_2, \ldots, d_i\}$ be the raw dataset such that it is the multiset. Each $d_x \in D$ is proposed to represent the profile tuple of the user $u_x \in U$ such that it consists of the quasi-identifier attributes, $QI = \{qi_1, qi_2, \ldots, qi_m\}$, and a sensitive attribute S. Let α and k be the privacy preservation constraints such that α is the real number in the range between 0.0 and 1.0, and k is the positive integer to be equal or greater than 2. Moreover, let PS be the set of protected sensitive values. For privacy preservation, before the raw dataset D is published, D is transformed to become D' by the limitations as follows.

- The tuples of D are partitioned by the given value of α and k. That is, every partition must include at least k tuples. Moreover, every protected sensitive value $ps \in PS$ in each partition has the re-identified proportion to be at most α. All partitions of D are denoted as $PAR = \left\{par_1, par_2, \ldots, \text{and } par_p\right\}$, where $par_1 \cup par_2 \cup \ldots \cup par_p = D$ and $par_1 \cap par_2 \cap \ldots \cap par_p = \emptyset$, and
- Moreover, the unique quasi-identifier values in $qi_y \in QI$, where $1 \le y \le m$, of each partition $par_g \in PAR$, where $1 \le g \le p$, is distorted by an appropriate less specific value which is represented by DGH [2, 8], or NDGH [2] to be indistinguishable. Each indistinguishable quasi-identifier tuples calls an equivalence class. For this reason, D' can also be represented by the set of its equivalence classes $EC = \{ec_1, ec_2, \ldots, ec_e\}$. Besides, only data distortion based on NDGH to be focused on this work because it can be more effective and simpler [2].

For example, let Table 1 be the raw data D. Let PS = {Cancer, HIV} be the set of protected sensitive values. Let the value of α and k be set at 0.70 and 2 respectively. For privacy preservation, the tuples of Table 1 are firstly partitioned by the given value of α and k. Thus, an output table version of this step is shown in Table 2. With this table, the tuples are partitioned to be three partitions such that each partition includes at least 2 tuples, and the re-identifiable proportion of every protected sensitive value ps $\in$ PS is at most 0.70. Finally, the quasi-identifier values in each attribute of each partition are distorted by their related set, the less specific value of them, to be indistinguishable, as shown in Table 3.

Table 1. An example of raw datasets.

#	Age	Sex	Diagnosis
t_1	40	Male	Flu
t_2	40	Male	Cancer
t_3	42	Male	HIV
t_4	44	Female	Cancer
t_5	44	Female	Fever
t_6	45	Female	HIV
t_7	45	Female	Flu

Table 2. A partitioned data version of Table 1 is satisfied by α = 0.70 and k = 2.

#	Age	Sex	Diagnosis	PAR
t_1	40	Male	Flu	par_1
t_2	40	Male	Cancer	
t_3	42	Male	HIV	
t_4	44	Female	Cancer	par_2
t_5	44	Female	Fever	
t_6	45	Female	HIV	par_3
t_7	45	Female	Flu	

Table 3. A published data version of Table 1 is satisfied by the (0.70, 2)-Anonymity constraint.

#	Age	Sex	Diagnosis	EC
t_1	{40, 42}	Male	Flu	ec_1
t_2	{40, 42}	Male	Cancer	
t_3	{40, 42}	Male	HIV	
t_4	44	Female	Cancer	ec_2
t_5	44	Female	Fever	
t_6	45	Female	HIV	ec_3
t_7	45	Female	Flu	

With Table 3, all possible re-identified conditions over on its quasi-identifier attributes always have at least two tuples to be satisfied, moreover, every specific protected sensitive value ps ∈ PS has the re-identified proportion to be at most 0.70. For this situation, privacy violation issues in Table 3 (or published dataset versions are satisfied by (α, k)-Anonymity constraints) seem to be impossible. Unfortunately, in the next section of this work, we will demonstrate that Table 3 still has privacy violation issues that must be addressed.

3 Privacy Violation Issues in Re-published Decremental Datasets

Let D'_{j-1}, and D'_j be both related published dataset versions of the raw dataset D such that they are published at the timestamp j − 1 and j respectively. We assume that the adversary has the background knowledge about the target user to be enough. Moreover, the adversary highly believes that the profile tuple of the target user is only collected in D'_{j-1}, i.e., the adversary knows that after D'_{j-1} was published, the profile tuple of the target user is deleted from the raw dataset D. With this situation if the adversary can use his/her background knowledge to specify an arbitrary equivalence class $ec_g \in EC$ in D'_{j-1} and its related equivalence classes $ec_{g_1}, \ldots, ec_{g_a} \in EC$ which are available in D'_j such that the compared result between the sensitive values which are available in ec_g and $ec_{g_1}, \ldots, ec_{g_a}$ does not satisfy the given value of α and k, then the privacy data of the target user is collected in D'_{j-1} could be violated by the adversary.

For example, let Table 1 be the raw dataset D at the timestamp j − 1, and its published dataset version D'_{j-1} is shown in Table 3. We suppose that after Table 3 is published, the tuple t_2 is deleted from Table 1 by such an appropriate business reason. After that, at the timestamp j, the raw dataset D is generalized to satisfy the (0.70, 2)-Anonymity and published to the data analyst again, as shown in Table 4. Also, we can see that Table 4 seems not to have privacy violation concerns because it can guarantee that all possible re-identification conditions determine over on its quasi-identifier attributes always have at least two satisfied tuples. Moreover, every protected sensitive value which is available in Table 4 has the re-identified proportion to be at most 0.70.

Table 4. A published data version of Table 1 is satisfied by the (0.70, 2)-Anonymity constraint after the tuple t_2 was deleted.

#	Age	Sex	Diagnosis	EC
t_1	{40, 42}	Male	Flu	ec_1
t_3	{40, 42}	Male	HIV	
t_4	44	Female	Cancer	ec_2
t_5	44	Female	Fever	
t_6	45	Female	HIV	ec_3
t_7	45	Female	Flu	

Unfortunately, in this work, we show that these published datasets, Table 3 and 4, still have privacy violation issues must be addressed. We suppose that the adversary received Tables 3 and 4. Let Bob be the target user of the adversary such that he/she needs to reveal Bob's diagnosis from Table 3. We assume that the adversary knows that Bob is a **male who is 40 years old**. Moreover, the adversary knows that Bob's profile tuple is deleted after Table 3 to be published. For this situation, Bob's diagnosis is collected in Table 3 can be revealed by the steps as follows.

The adversary firstly verifies Bob's age and Bob's sex in Table 3 by matching his/her background knowledge. With this step, the adversary can ensure that such a diagnosis, i.e., **Flu, Cancer, or HIV**, is available in ec_1 must be Bob's diagnosis because only this equivalence class is accorded with the adversary's background knowledge.

Subsequently, the adversary tries to specify the related equivalence classes of ec_1 such that it is available in Table 4, i.e., the arbitrary equivalence class(es) of Table 4 is fully covered by ec_1 of Table 3. With this step, the adversary can see that only ec_1 of Table 4 relates to ec_1 of Table 3. With this equivalence class of Table 4, it relates to **Flu and HIV**.

Finally, the adversary compares the list of diagnoses which are collected in ec_1 of Table 3 and the list of their related diagnoses which is available in equivalence classes ec_1 of Table 4, i.e., **Flu, Cancer, and HIV** and **Flu and HIV**. For this situation, the adversary can be highly confident that Bob's diagnosis collects in Table 3 to be **Cancer**.

From the example, it is clear. Published datasets are satisfied by (α, k)-Anonymity constraints; they still have privacy violation issues that must be addressed. To address these issues, in [7], a decremental privacy preservation algorithm to be proposed. For privacy preservation with this algorithm, aside from distorting the unique quasi-identifier values, all compared results between their previously related published dataset versions must also be satisfied by (α, k)-Anonymity constraints. Although this privacy preservation algorithm can address privacy violation issues in published datasets from using compared data attacking and more effective, it is not efficiency, i.e., it is highly complex in term of transforming the data which is available datasets to satisfy the specific privacy preservation constraints because this algorithm is based on brute force notations. For this reason, an effective and efficient heuristic privacy preservation algorithm for publishing decremental datasets is proposed in this work, it will be explained in Sect. 4.

4 Proposed Model

In this section, it is devoted to proposing an effective and efficient privacy preservation idea for addressing privacy violation issues in published decremental datasets from using compared data attacking.

4.1 Notation

Let PS be the set of sensitive values that must be protected. Let α and k be the privacy preservation constraints such that α is the real number in the range between 0.0 and 1.0, and k is the positive integer to be equal or greater than 2. Let D_{j-1} and D_j be the raw dataset at the timestamp $j - 1$ and j respectively. Let D'_{j-1} be the published dataset version of D_{j-1}. Let $f_{Decremental}\left(\alpha, k, PS, D_{j-1}, D'_{j-1}, D_j\right) : D_j \rightarrow_{\alpha,k,PS,D_{j-1},D'_{j-1}} D'_j$ be a generalized and suppressed function for transforming D_j to become D'_j. That is, all unique quasi-identifier values of D'_j are generalized by their less specific values which are presented by NDGH or suppressed to be at least k indistinguishable tuples. Moreover, every protected sensitive value $ps \in PS$ relates to each group of the indistinguishable quasi-identifier tuples has the re-identified proportion to be at most α. Furthermore, all compared results between the specific equivalence class of D'_{j-1} and its related equivalence classes which are available in D'_j must also be satisfied by the given value of α and k. Also, every group of tuples is satisfied by the given value of α and k to call an equivalence class of D'_j. Therefore, D'_j can also be represented by the set of its equivalence classes $EC = \{ec_1, ec_2, \ldots, ec_g\}$. Without loss of generality, $ec_1 \cap ec_2 \cap \ldots \cap ec_g = \emptyset$. Moreover, the penalty cost, it will be presented in Sect. 4.2, of D'_j must be minimized.

4.2 Data Utility Metric

Although (α, k)-Anonymity can address privacy violation issues in D'_j from using compared data attacking, it generally leads to losing some data utility of D'_j. Thus, the metric for defining the penalty cost of data generalizations, Eq. 1, and data suppressions, Eq. 2, in published datasets is necessary. With Eq. 1, it is proposed to define the penalty cost of each equivalence class ec_g of D'_j such that the equivalence classes are larger can lead to more penalty costs than the smaller equivalence classes. Another penalty cost in D'_j is the suppressed penalty cost, as shown in Eq. 2. This penalty cost is based on the size of quasi-identifier attributes and the number of suppressed tuples of D'_j, i.e., the number of suppressed tuples * the number of quasi-identifier attributes of D'_j. Also, more suppressed tuples lead to be more penalty cost in D'_j. Therefore, the penalty cost of D'_j can be defined by Eq. 3 such that more cost of D_{Loss} lead to being more penalty or less data utility in D'_j. The penalty cost of D_{Loss} is in the range between 0.0 and 1.0.

$$P_{LOSS}(D_j, ec_g) = \sum_{x=1}^{|ec_g|} \sum_{y=1}^{|m|} \frac{|ec_g[d_x[qi_y]]|}{|D_j[qi_y]|} \quad (1)$$

$$T_{LOSS}(D_{j-1}, D_j) = \sum_{\forall d \in (D_{j-1} - D_j)} m \quad (2)$$

$$D_{LOSS}\left(D_{j-1}, D_j, D_j'\right) = \frac{\sum_{g=1}^{|e|} P_{LOSS}(D_j, ec_g) + T_{LOSS}(D_{j-1}, D_j)}{|D_j| \cdot m}, \text{ where } ec_g \in D_j' \quad (3)$$

Where, $ec_g[d_x[qi_y]]$ is selecting the values which are a violation in the quasi-identifier attribute qi_y of the tuple d_x in the equivalence class ec_g. $D_j[qi_y]$ is the set of all possible quasi-identifier values of the attribute qi_y in D_j.

4.3 Proposed Decremental Privacy Preservation Algorithm

In this section, the pseudo-code of an effective and efficient heuristic privacy preservation algorithm for publishing decremental datasets is proposed. With the proposed algorithm, aside from privacy preservation, the data utility and execution time are also maintained as much as possible. To achieve the proposed privacy preservation idea, clustering techniques [9] are applied. That is, all tuples are available in the specific equivalence class; they must be more similar than other tuples are available in D_j'. The similarity score of tuples in each equivalence class can be defined by Eq. 1.

Algorithm 1 is the pseudo-code the proposed effective and efficient heuristic privacy preservation algorithm for decremental datasets. With the proposed algorithm, the inputs are D_{j-1}, D_{j-1}', D_j, α, and k. That is, D_{j-1} is the raw dataset D at timestamp $j - 1$ and its published dataset is D_{j-1}'. D_j is the raw dataset D at timestamp j. α and k are privacy preservation constraints.

For privacy preservation, the algorithm firstly investigates D_j. If D_j can transform to satisfy α and k, the algorithm ignores its pseudo-code between line 2 and line 8. If not, the algorithm returns D_{j-1}' if D_{j-1}' is not null, otherwise, the algorithm returns failure.

In the second step, all tuples are collected in D_j to be iterated such that each iteration the appropriate tuple group of $d \in D_j$ is that it satisfies the given value of α and k to be constructed by the GenBestPar() function, as shown in Function 1, until D_j does not satisfy the given value of α and k.

```
Algorithm 1: The effective and efficient heuristic privacy
preservation algorithm for decremental anonymization da-
tasets.
Input: Dj-1, D'j-1, Dj,,  , and k.
Output: D'do not have privacy violation concerns from using
data comparison techniques.

Let Λ be the set of partitions and Ψ be temporary;
IF D'j-1 is not NULL THEN
    IF Dj-1[S] - Dj[S] do not satisfy α and k THEN
        Return D'j-1;
    END IF
ELSE IF Dj do not satisfy α and k THEN
    Return failure;
END IF
FOR all tuples of Dj DO
    Ψ = GenBestPar(D'j-1, Dj - d, d,  , k), where d ∈ Dj;
    Λ = Λ ∪ Ψ;
    Dj = Dj - Ψ;
    IF Dj does not satisfy α and k THEN
        BREAK;
    END IF
END FOR
IF Dj is not NULL THEN
    FOR all tuples of Dj DO
        Ψ = FindBestPar(D'j-1, Dj, Λ, d,  , k), where d ∈
    Dj;
        λ = Ψ ∪ d;
        Λ = (Λ - Ψ) ∪ λ;
    END FOR
END IF
GENERILAZATION(NDGHqi1,...,NDGHqim,Λ), where qi1,..., qim ∈ QI;
Return D'j;
```

In the third step, this procedure of the algorithm is optional. It is enabled when D_j remains the tuples that cannot assign to the appropriate partition by using the second step. The appropriate partition of every remained tuple is determined by the FindBestPar() function, as shown in Function 2.

Finally, the quasi-identifier values of each attribute of every partition are generalized by their related value set and returned.

Function 1: GenBestPar(D'$_{j}$-1, D$_j$, d, , k)
Input: D'$_{j-1}$, D$_j$, d, , and k.
Output: Return the best partition of D$_j$ is constructed from the nucleus d to satisfy and k such that its penalty cost P$_{LOSS}$ is minimized.

```
Let Υ be the data loss score, let Ψ1 and Ψ2 be temporary;
Ψ1 = d;
Initialize Υ as the maximized value;
WHILE all tuples of Dj DO
   IF Ψ1 do not satisfy α and k THEN
      Ψ1 = Ψ1 ∪ dd, where dd ∈ Dj;
   END IF
   IF Ψ1 satisfies k THEN
    FOR all equivalence classes EC of D'j-1 DO
         IF ec[S] - Ψ1[S] satisfies α, where ec ∈ EC THEN
            IF Υ > PLoss(Ψ1) THEN
               Ψ2 = Ψ1;
               Υ = PLoss(Ψ1);
               Ψ1= d;
            END IF
         END IF
      END FOR
   END IF
END WHILE
Return Ψ2;
```

Function 2: FindBestPar(D'$_{j}$-1, D$_j$, Λ, d, , k)
Input: D'$_{j-1}$, D$_j$, Λ, d, , and k.
Output: Return the best partition of d that it is available in Λ

```
Let Υ be the data loss score;
Let Ψ be temporary;
Initialize Υ as the maximized value;
FOR all partitions of Λ DO
   FOR all equivalence classes EC of D'j-1 DO
      IF ec[S] - λ ∪d[S] satisfies α, where ec ∈ EC THEN
         IF Υ > PLoss(λ ∪d) THEN
            Ψ = λ;
            Υ = PLoss(λ ∪d);
         END IF
      END IF
   END FOR
END FOR
Return Ψ;
```

5 Experiment

In this section, the experimental results are based on the effectiveness and efficiency of the proposed algorithm to be discussed by the comparing with the re-published decremental privacy preservation algorithm, the optimal algorithm is proposed by [7].

5.1 Experimental Setup

All experiments of this work are conducted on an Intel(R) Xeon(R) Silver 4110 @2.10 GHz CPU with 16 GB memory and 1 TB HDD running Microsoft Windows 10 64-bit professional edition. All the implementation is built and executed on Microsoft Visual Studio 2017 Community Edition and SQL Server 2017 RTM.

Moreover, all experimental results are discussing in this section; they are conducted on Adult dataset available at UCI Machine Learning Repository [12]. This dataset is constructed from 32561 user profile tuples such that each user profile tuple consists of 14 attributes such that age, work class, education, education-num, marital status, occupation, relationship, race, sex, capital-gain, capital-loss, hours-per-week, and native-country. For conducting the experiments effectively, all user profile tuples include the missing values, "?"; they are removed. Thus, the user profile tuples only remain in the experimental dataset to be 30162 tuples. Moreover, given the attributes as Age, Sex, Race, Native-country are the quasi-identifiers, and given the attribute as Occupation is the sensitive attributes of the experimental dataset.

5.2 Discussion

In this section, the experimental results for evaluating the effectiveness and efficiency of the proposed algorithm are discussed.

With Figs. 1, 2, and 3, they are proposed to evaluate the effectiveness of the proposed algorithm. From all experimental results, we can observe that they indicate that the compared algorithm is more effective than the proposed algorithm, however, they are only a little different. The cause of the proposed algorithm is less effective than the compared algorithm is that the proposed algorithm based on the greedy notations. Therefore, the proposed algorithm only defines the best equivalence class in each iteration cycle to construct the desired released datasets. For this reason, the size of equivalence classes which are constructed from the proposed algorithm to be larger than the compared algorithm.

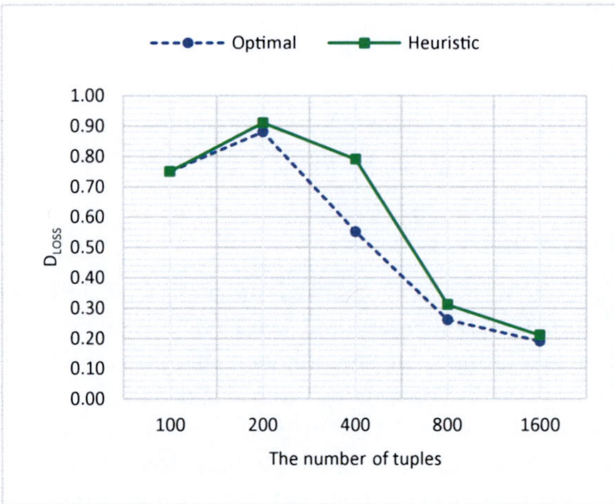

Fig. 1. The data utility based on dataset sizes.

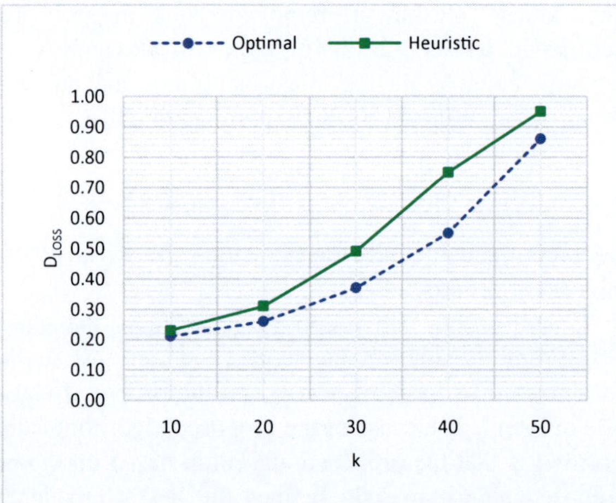

Fig. 2. The data utility based on k

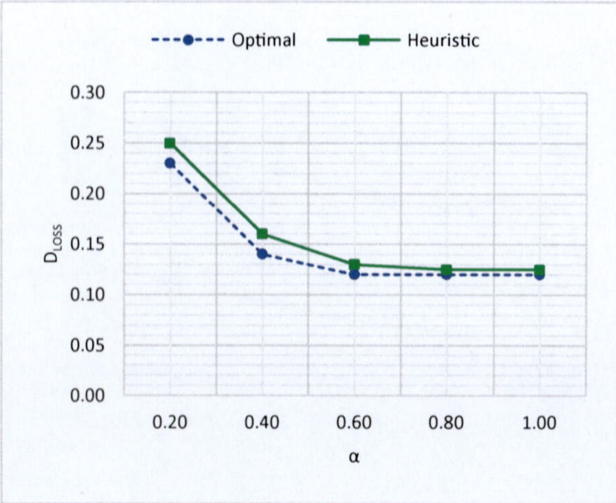

Fig. 3. The data utility based on α

However, the experimental results are shown in Figs. 4, 5, and 6 to indicate that the proposed algorithm is more efficient than the compared algorithm. Especially, the number of experimental tuples is increased. The cause of more efficiency of the proposed algorithm is that the space of the data processing, the number of equivalence classes, is so much less than the compared algorithm.

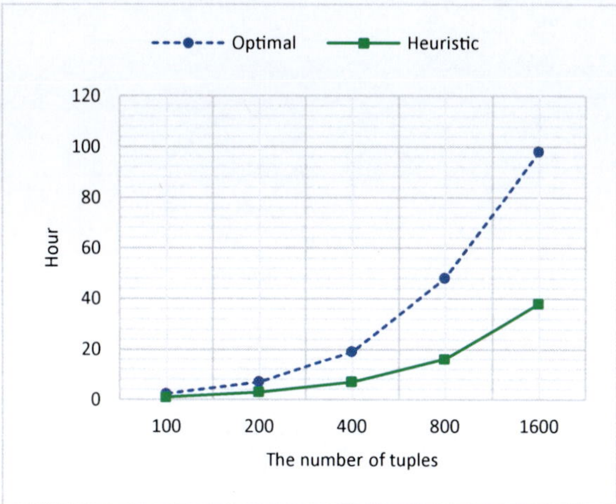

Fig. 4. The execution time based on the dataset sizes

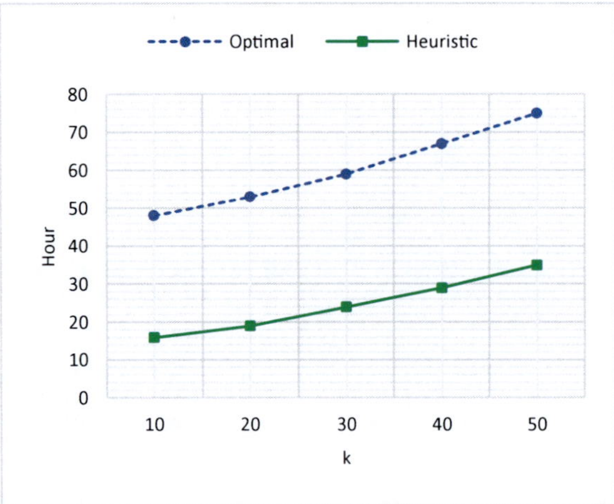

Fig. 5. The execution time based on k

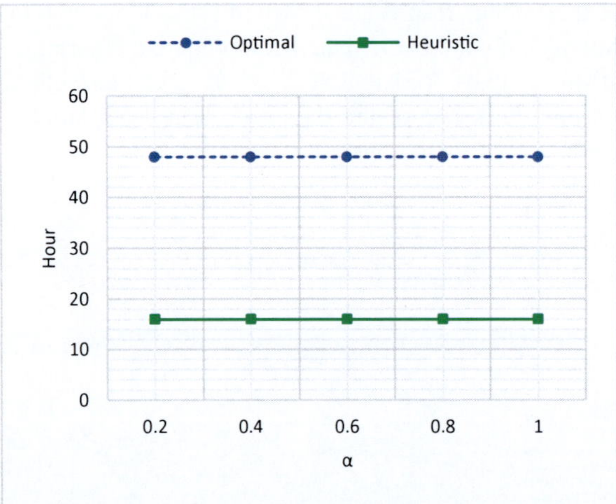

Fig. 6. The execution time based on α

6 Conclusion

In this work, a heuristic privacy preservation algorithm for publishing decremental datasets based on (α, k)-Anonymity constraints in conjunction with NDGH to be proposed. The aim of the proposed algorithm is that aside from privacy preservation, the data utility and execution time are also be maintained as much as possible. To achieve these constraints in published datasets, clustering techniques are applied in the proposed algorithm, i.e., all tuples are available in each equivalence class must be more similar than others. From the experimental results, they indicate that the proposed algorithm is more effective and efficient.

References

1. Sweeney, L.: K-anonymity: a model for protecting privacy. Int. J. Uncertain. Fuzziness Knowl.-Based Syst. **10**(5), 557–570 (2002)
2. Riyana, S., Natwichai, J.: Privacy preservation for recommendation databases. Serv. Oriented Comput. Appl. **12**(3–4), 259–273 (2018)
3. Machanavajjhala, A., Kifer, D., Gehrke, J., Venkitasubramaniam, M.: L-diversity: privacy beyond k-anonymity. ACM Trans. Knowl. Discov. Data **1**(1), 3-es (2007)
4. Li, N., Li, T., Venkatasubramanian, S.: t-closeness: privacy beyond k-anonymity and l-diversity. In: 2007 IEEE 23rd International Conference on Data Engineering, Istanbul, pp. 106–115 (2007)
5. Wong, R., Li, J., Fu, A., Wang, K.: (α, k)-anonymous data publishing. J. Intell. Inf. Syst. **33** (2), 209–234 (2009)
6. Fung, B.C.M., Wang, K., Fu, A.W.-C., Pei, J.: Anonymity for continuous data publishing. In: Proceedings of the 11th International Conference on Extending Database Technology: Advances in Database Technology, EDBT 2008, pp. 264–275. Association for Computing Machinery, New York (2008)
7. Riyana, S., Harnsamut, N., Sadjapong, U., Nanthachumphu, S., Riyana, N.: Privacy preservation for continuous decremental data publishing. In: International Conference on Image Processing and Capsule Networks, ICIPCN 2020. Springer (2020)
8. Sweeney, L.: Achieving k-anonymity privacy protection using generalization and suppression. Int. J. Uncertain. Fuzziness Knowl.-Based Syst. **10**(5), 571–588 (2002)
9. Berkhin, P.: A Survey of clustering data mining techniques. In: Kogan, J., Nicholas, C., Teboulle, M. (eds.) Grouping Multidimensional Data. Springer, Heidelberg (2006)
10. Riyana, S., Harnsamut, N., Soontornphand, T., Natwichai, J.: (k, e)-anonymous for ordinal data. In: Proceedings of the 2015 18th International Conference on Network-Based Information Systems, NBIS 2015, pp. 489–493. IEEE Computer Society, USA (2015)
11. Riyana, S., Riyana, N., Nanthachumphu, S.: Enhanced (k, e)-anonymous for categorical data. In: Proceedings of the 6th International Conference on Software and Computer Applications, ICSCA 2017, pp. 62–67. Association for Computing Machinery, New York (2017)
12. UCI Machine Learning Repository. https://archive.ics.uci.edu/ml/datasets/Adult

A Brain Computer Interface Based Patient Observation and Indoor Locating System with Capsule Network Algorithm

D. A. Janeera(✉) and S. Sasipriya

Department of ECE, Sri Krishna College of Engineering and Technology, Coimbatore, India
{janeera, sasipriya}@skcet.ac.in

Abstract. Brain Computer Interface (BCI) offers direct communication and controlling channel between the human brain and various physical devices. This paper introduces a mind controlled device that enables monitoring of the patient and indoor location using Capsule Network (CapsNet) Algorithm. This system offers huge benefits to patients who are restrained to wheelchairs and ventilators but have their cognitive ability intact. The electroencephalogram (EEG) signals are extracted and processed using Arduino Uno Microcontroller and classified with deep learning techniques. Comparison of the EEG signals is done using capsule network algorithm for feature extraction and to predict the brain attention and meditation state. The proposed architecture based on CapsNet outperforms the existing ANN approaches.

Keywords: Brain Computer Interface (BCI) · Capsule Network · Microcontroller · EEG · Embedded system · ANN

1 Introduction

The capacity to perform daily activities by oneself is the key to independence in mobility. Nevertheless, for people with severe motor-disabilities who cannot comfortably use conventional interfaces, it is challenging to make use of powered wheelchairs. A promising solution is offered for such people using a non–invasive brain-computer interface (BCI) that assists them in overcoming the problem of interacting with the devices [1]. This paper provides an efficient solution for operating wheelchairs, smart hospital beds and other smart environment devices with precision by an intelligent shared control architecture that works on the commands received based on the thoughts of the user.

Rehabilitation regimens and adaptive training techniques are customized based on the assessment of brain activity in a real-time environment thereby improving their effectiveness. BCI technology aids in improving focus by adjusting the brain state according to the environment [2]. Neural activities of the brain are assessed for identification of brain based disorders and diseases even before the appearance of behavioral symptoms. These technologies also enable improved communication by assessment of the neural activity of the brain. In addition to these applications in

J. I.-Z. Chen et al. (Eds.): ICIPCN 2020, AISC 1200, pp. 258–268, 2021.
https://doi.org/10.1007/978-3-030-51859-2_23

medical assistive devices, BCI technology also has diverse applications in the gaming sector. The influence of BCI in human computer interaction, interaction with the devices in the environment and with other humans is discussed briefly in this paper.

The EEG headband used for this purpose is the Neurosky Technologies Mindwave Headset from which the EEG signals are gathered based on the electrical activity of the brain. Further, these signals are transferred to the microcontroller with the help of a Bluetooth device. A battery is used for powering the mind wave headset. The components of the headset include an ear clip for ground connection, a flexible ear arm, a sensor tip and a power switch. Since the headset uses non-invasive sensors, the user headset enables pain free usage. An inbuilt LED indicator in the mindwave headset provides information regarding the active, in use, off state, the Bluetooth connectivity and the battery status of the device [3].

The EEG signals are classified using CapsNet classifier. This technique uses dynamic routing between capsules and activity vectors. In CapsNet, the vector outputs are used to replace the Convolution Neural Network (CNN) based the scalar-output feature detectors. Also, routing-by-agreement is used for replacing the technique of max-pooling subsampling thereby enabling replication of the obtained information across space [4]. The convolution layer groups are present in the primary capsule layer. The higher level layers cover the extensive regions of the image.

2 Existing Literature

Smart devices and computers can be controlled by promising BCI technology without the requirement of human interaction. A BCI system monitors the neural activity of the user and records the electrical signals generated by the brain [5]. These signals undergo pre-processing, feature extraction and classification through image processing algorithm and further applied for detecting the intent of the user in controlling a machine. Electroencephalography (EEG) is the most efficient, non-invasive and low cost technique used to capture the neural activity of the brain and provide input signals to the BCI systems [6]. Brain waves such as gamma (>32 Hz), beta (16–31 Hz), alpha waves (8–15 Hz) and sensorimotor rhythm (SMR), steady state evoked potential (SSEP) and event-related potential (ERP) are the major EEG signal types. Steady-state sensory evoked potential (SSSEP) and steady-state visual evoked potential (SSVEP) are the major classifications of SSEP. The SMR or mu rhythm is used for the execution of motor related tasks like the movement of tongue, feet and hands. This feature makes it the most commonly used signal in BCI applications. event-related synchronization and (ERS) and event-related desynchronization (ERD) occur in the beta band (16–31 Hz) and the mu band (8–12 Hz) of SMR [7]. In the BCI motor imagery applications, the ERS and ERD concepts can be used for attenuation and restorations of motor images.

The steps performed by BCI systems based on Motor Imagery (MI) are as follows: (i) EEG signal recording and pre-processing, (ii) Extraction of features from the signals measured, (iii) training the models for classification and (iv) using the trained models for testing purpose. To improve the performance of BCI systems based on MI, feature classification and extraction techniques are used [8]. Feature extraction is performed by the common spatial pattern (CSP) algorithm. The motor imagery EEG signals of two classes are used for extraction of discriminatory features effectively and to maximize the variance by the spatial filters designed with the CSP algorithm. Based on the individual characteristics, manual selection of frequency band is to be done in CSP which is the major drawback of the algorithm. This drawback is overcome by a filter bank common spatial pattern (FBCSP) algorithm [9]. The performance of the classification of MI based EEG is improved by applying various frequency bands in parallel to each other to overcome the problems of the frequency band using FBCSP algorithm. Common machine learning techniques such as Naïve Bayes (NB) algorithm, linear discriminant analysis (LDA) and support vector machines (SVMs) are commonly used for classification purpose [10].

3 System Design

The architecture of the proposed BCI system for patient observation and the indoor location is as represented in Fig. 1. Sensor nodes for monitoring the heart rate and temperature are used along with the PIR and IR sensors. The Bluetooth device is used for transmitting the EEG signals from the headband to the microcontroller. An attainable and smaller intermediate step is to opt while working towards attaining large futuristic goals. In this regard, the Arduino board is used to process the information from the headset and the data received from the sensors as well as to control the smart system. This system insists on using the "divide and conquer" paradigm approach to solve the smaller issues and simplifying the system thereby avoiding future complications. Optical centering of the system on the ground with a painted line as the line of focus, it is possible to simplify the problems related to movement. Attention is the most important factor in the interpretation of brain signals. It is essential to block dangerous responses and misinterpretation of brain signals.

3.1 Data Processing Unit

The Bluetooth module receives the data transmitted by the EEG headband and transfers it to the microcontroller. The Level Analysis platform analyses the received information obtained at the specific port-pin that is connected to the Bluetooth device.

3.2 EEG Signals

Electrodes are used for the collection of EEG signals from the surface of the scalp. Silver/silver chloride (Ag/AgCl) electrodes are most commonly used due to their sensible stability, low contact electric resistance and low price. Successful clinical diagnosis and research can be done with the modern commercial acquisition systems that consist of Ag/AgCl electrodes based graphical recording cap that is aligned with electronic equipment. These electrodes require a filling gel as an interface between the scalp and electrodes. Such electrodes are termed as wet electrodes. Operations using wet electrodes that require longer time duration may make the users uncomfortable [11]. The limitations of these electrodes are overcome by dry electrodes which do not require the application of the gel. Constant electric resistance is the major drawback of

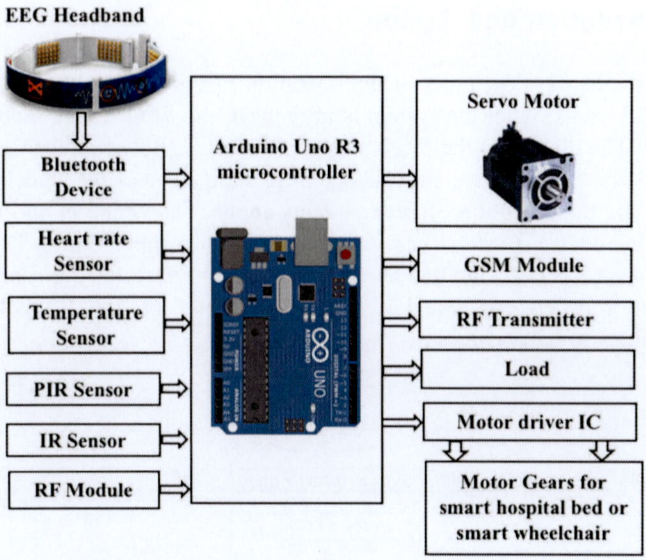

Fig. 1. Block diagram of the proposed BCI system

dry electrodes. Signals in the specific area are recorded graphically using these electrodes. The dry electrode based acquisition systems are used for commercial purposes. The Neurosky Mindwave headsets are used in telephone receiver applications. However, there are concerns regarding the acquired physiological signals by these systems. Due to these drawbacks, wet electrodes are preferred in operations concerning brain-controlled systems to gather the neural data.

3.3 Patient Monitoring System

Tracking of triage status, storage of medical records, location tracking and monitoring of vital signs are the four key functionalities of wearable sensors. The non-invasive vital signs sensors namely blood pressure sensor and pulse oximeters are integrated. The body temperature and heart rate of a patient are measured by the pulse oximeter which can be attached to the finger of the patient. The systolic and diastolic blood pressure can be measured by the cuff pressure sensor which is fixed to the upper arm of the patient.

3.4 Indoor Positioning System (IPS)

An indoor location sensing system is integrated for detection of the location where GPS signals are not accessible. This helps in determining the current position based on the initial position of the system by tracking the position advances. Determination of position advances from the previous position is identified accurately using gyroscopes and accelerometers. This can be used in hospital or home environment for tracking the current position of the patient and enable automation of devices in the specific room or location.

4 Implementation and Testing

Figure 2 represents the flow chart of the operation of the system based on the inputs from the EEG signals. After the system is initialized, the EEG signal is obtained from the headband. The quality of the received signal is tested. Based on the signal quality, the attention level and distance are calculated. If the quality of the signal is poor, the system waits for the reception of better quality input. The attention and distance are assessed and the device is made to move forward based on these values. The values are also displayed on the LCD module as an additional interface for the caretaker. If the value of attention is between 50 and 65, the device moves forward. If it is less than 49, the device comes to halt position.

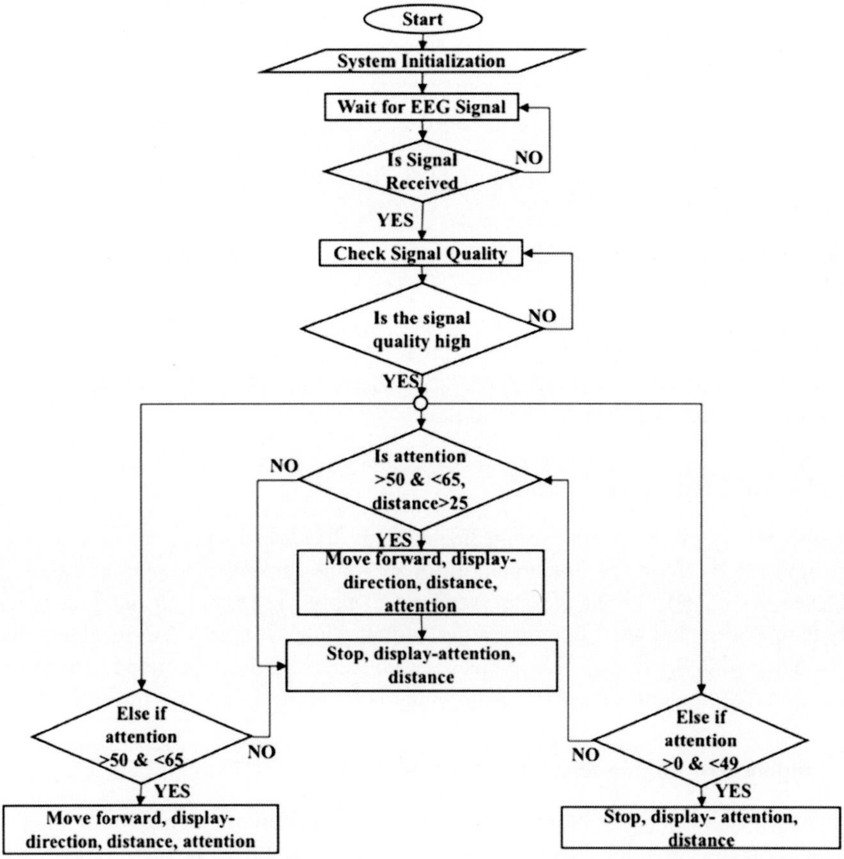

Fig. 2. Flow chart of the proposed system operation

4.1 Brain Sense Module

Based on the brain state, the typical frequency values of brain wave and its range varies. The high frequency beta waves associated with concentration ranges between 12 to 30 Hz and alpha waves or low frequency waves that decides the clam state of mind ranges between 8 to 12 Hz. The mental images and unconscious state or non-rem sleep are associated with the low frequency waves ranging between 0.1 to 7 Hz. The frequencies between 0.5 to 50 Hz can be sensed by the NeuroSky MindSets and transmitted the data every second using a Bluetooth connection using a proprietary protocol. Electromyogram (EMG) waves are the unique wave patterns generated by muscle contractions. These waves enable identification of eye blinking and often shadow other brainwaves [12]. They are commonly used in eSense algorithms and meters for obtaining the status of the brain based on attention and meditation values (Fig. 3)..

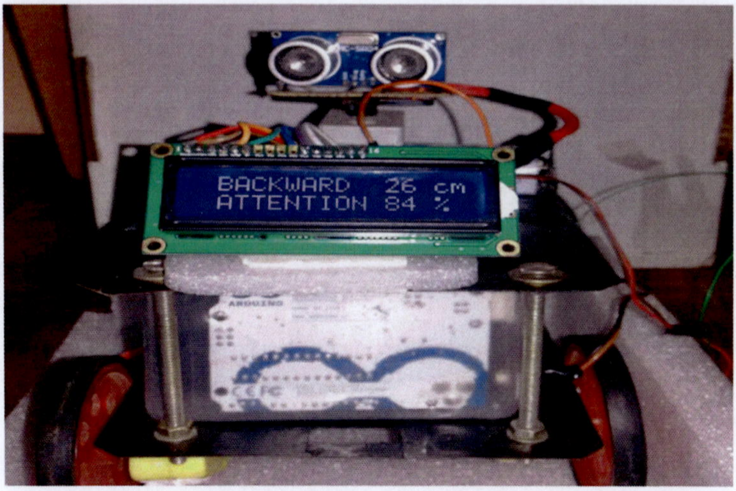

Fig. 3. Prototype of the BCI module based system

Attention Level. The attention or focus level of the user and its intensity is indicated by the eSense Attention meter. This provides information regarding the stable mental activity and intensity of concentration. The value of attention ranges between 0 and 100. Anxiety, lack of focus, wandering thoughts and distraction reduces the level of attention in the meter.

Meditation Level. The level of relaxation or calmness of the user's mental state is indicated by the eSense Meditation meter. The value of meditation ranges between 0 and 100. Relaxation of body muscles or the physical levels does not result in improved meditation level. The mental state of the person is the prime factor that decides the meditation level. Nevertheless, relaxation of body aids in the relaxation of the mind in most circumstances for many people. Reduction in the active mental processes and activity of the brain leads to the meditation state. The processing of images from the eyes can be stopped by closing one's eyes and thereby reducing the mental activities [13]. Hence, closing the eyes is an effective solution to increase the level of meditation meter. Anxiety, lack of focus, wandering thoughts and distraction reduces the level of meditation in the meter. The values of meditation and attention are categorized into five classes namely 1 to 20 as agitation or distraction level, 20 to 40 as reduced value level, 40 to 60 as average value level, 60 to 80 as slightly elevated level and 80 to 100 as elevated level. Wider order of ranges is chosen to enable learning by slow adaptive algorithms without providing much significance to fluctuations.

Despite different people and varying environmental conditions, this system offers enough tolerance maintaining the efficiency of output. Figure 4 represents the output of processed brain signals based on the eSense meters. The focus level or attention level of the subject is measured by the eSense Attention meter. Anxiety, distraction or lack of focus may lower the attention level [14]. The mental calmness level is measured by the eSense Meditation meter. Training these mental states is essential for improving the efficiency in the performance of the system and avoid noises like sensory stimuli, agitation, anxiety, wandering thoughts and distractions.

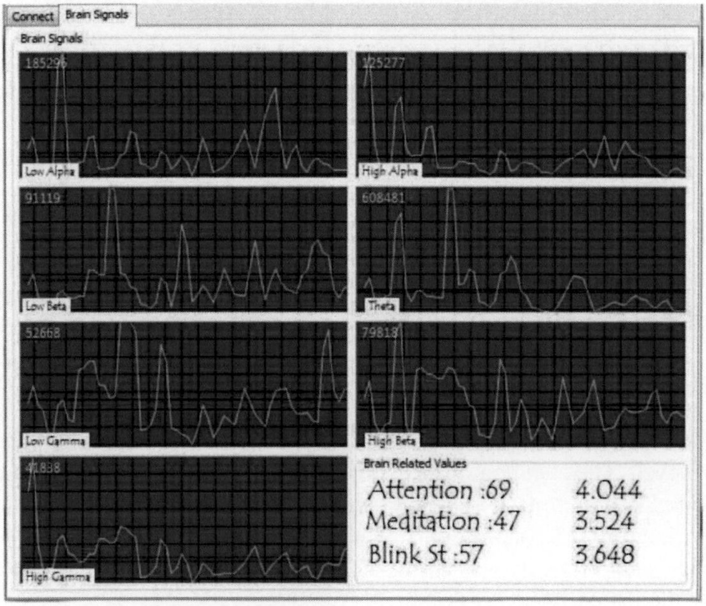

Fig. 4. Output of processed brain signals

4.2 Arduino Module

The Wiring library can be used for programming and Arduino microcontroller using C or C++ languages. The benefits of Arduino include multiplatform programming, open source software and hardware, USB programming, user friendly development environment and Processing Programming IDE interface. The Arduino controlled system receives input data from the NeuroSky MindSet through a Bluetooth connection as ThinkGear packets [15]. This data is transmitted asynchronously as a byte stream with a length of 4 to 173 bytes and a checksum of 1 byte. The prototype developed for this purpose makes use of a traditional rover robot simulated by the neural signals.

In the front portion of the cart, two light sensors are attached to enable simplification and to avoid steering. For simulation and simplicity, the prototype operates as a simple line follower that trails on the white line painted on the ground. The two sensors are used to calculate the distance and keep track of the white line based on the differences in the reflected light. These values are used for monitoring the path and speed of the cart and to reposition or slow down the cart if required. Four DC motors in continuous rotation are used for moving the four independent wheels of the cart as they are highly compatible with the Arduino hardware.

The motors are adjusted to set identical responses using the external potentiometers adjustment screws. This enables the deterministic movement of the cart. Experimental results show that fatigue of concentration also impacts the outcomes and results of the model and causes a reduction in the performance of the system. The performance can be improved with steady training rather than forcing a relaxed state. Figure 5 represents the attention level, clearance in centimeter and direction of motion of the prototype on an Arduino console.

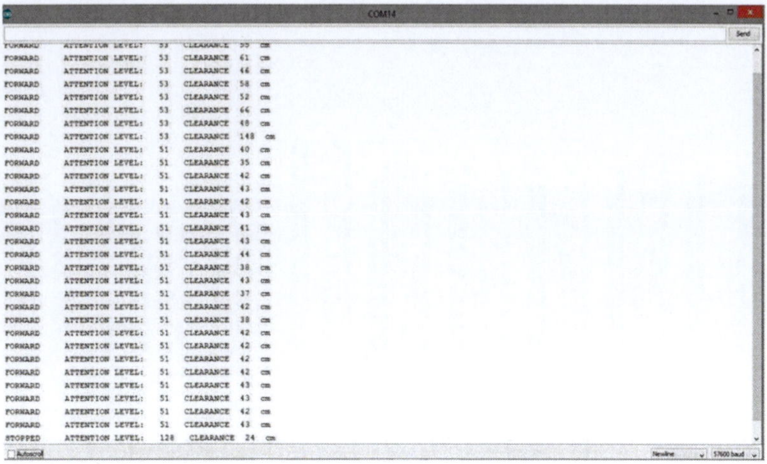

Fig. 5. Attention level and clearance of the prototype on Arduino console

4.3 Capsule Network

Capsule Networks outperforms the deep learning architectures with their strong characteristics in learning from the input data and identifying the hierarchical relationships [16]. Group of neurons termed as capsules are used for capturing information that is important and of higher level based on dynamic routing. In this paper, a capsule network is used for capturing the attention mechanism. Feature representation and training of samples are performed by the layers of the capsule network.

The training and testing accuracy of the proposed system is compared to other existing neural network algorithms for ten subjects. Table 1 provides the comparison of accuracy in the classification of various approaches including the proposed CapsNet, CNN, EEGNet, DeepNet, ShallowNet and Support Vector Machine (SVM) techniques. It is observed that the proposed technique outperforms the existing systems (Fig. 6).

Table 1. Comparison of accuracy in classification of various approaches

Subject	Caps Net	CNN	EEGNet	DeepNet	ShallowNet	SVM
1	87.18	80.91	80	79.25	82.18	74.2
2	91.25	87.56	89.68	87.75	85.93	71.9
3	75.62	69.75	76.87	77.03	76.56	76.1
4	83.43	78.03	77.5	74.46	76.87	75.1
5	83.12	85.91	86.87	84.59	85	79
6	95.93	95.31	95.31	94.5	95.93	94
7	55	52.88	55.62	54.87	53.12	53.8
8	55.71	54.75	58.21	56.10	53.57	56.4
9	78.75	69.78	67.18	67.25	71.56	70.1
10	83.24	86.02	86.91	84.67	85.12	79.03
Avg	78.92	76.09	77.42	76.05	76.58	72.96

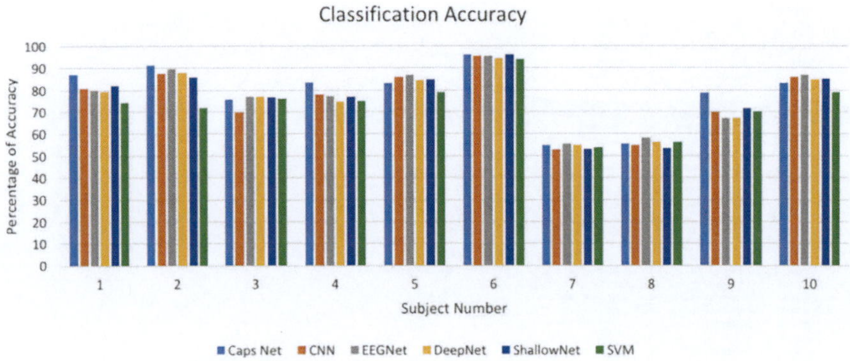

Fig. 6. Comparison of classification accuracy

5 Conclusion and Future Scope

The quality of life of people with contractile organ or limb disorders can be improved by the study and development of the BCI controlled environment. This paper provides problem analysis, key procedures and the complete system review for brain computer interface based device control. The proposed system enables patient observation and indoor location based on the EEG signals gathered by the Neurosky Mindwave headband, eSense and Arduino microcontroller. This system has promising applications in wheelchairs, ventilator assistance and automation systems. The inherent drawbacks of brain signal processing are overcome by the CapsNet algorithm providing acceptable information bit rate and significant accuracy. The accuracy is improved with multiple iterations and reduced training time, thereby making this system more. BCI has widespread applications such as neuroprostheses, environmental control, accessing email, browsers or word processing and other simple daily activities of disabled people. In the case of driving based on neural signals only, it is essential to address issues regarding circumstances where the driver feels sleepy, gets agitated or loses concentration based on experimentation. The future scope is in the direction of enhancing the algorithms in BCI based systems, to reduce the delay in processing the signals and to further reduce the training time.

References

1. Krishna, D.H., Sengupta, R., Pasha, I.A., Savithri, T.S.: Brain computer interface for mobility-a case study with ALS patient
2. Norani, N.A.Md., Mansor, W., Khuan, L.Y.: A review of signal processing in brain computer interface system. In: 2010 IEEE EMBS Conference on Biomedical Engineering and Sciences (IECBES), pp. 443–449. IEEE (2010)
3. Ramakuri, S.K., Peddi, A., Rao, K.S.N., Gupta, B., Ghosh, S.: Performance and analysis of human attention using single-channel wireless EEG sensor for medical application. In: First International Conference on Artificial Intelligence and Cognitive Computing, pp. 355–362. Springer, Singapore (2019)
4. Acı, Çİ., Kaya, M., Mishchenko, Y.: Distinguishing mental attention states of humans via an EEG-based passive BCI using machine learning methods. Expert Syst. App. **134**, 153–166 (2019)
5. Yuan, Y., Su, W., Li, Z., Shi, G.: Brain–computer interface-based stochastic navigation and control of a semiautonomous mobile robot in indoor environments. IEEE Trans. Cogn. Dev. Syst. **11**(1), 129–141 (2018)
6. Wang, K., Wang, X., Liang, B., Li, G., Liang, R.: Review of brain computer interface components and corresponding research topics. In: International Conference on Education, Management, Computer and Society. Atlantis Press (2016)
7. de Arruda Leite, H.M., de Carvalho, S.N., da Silva Costa, T.B., Attux, R., Hornung, H.H., Arantes, D.S.: Analysis of user interaction with a brain-computer interface based on steady-state visually evoked potentials: case study of a game. Comput. Intell. Neurosci. **2018**, 10 (2018)
8. Pinheiro, P.G., Cardozo, E., Pinheiro, C.G.: Anticipative shared control for robotic wheelchairs used by people with disabilities. In: 2015 IEEE International Conference on Autonomous Robot Systems and Competitions, pp. 91–96. IEEE (2015)
9. Anh, N.T.H., Hoang, T.H., Thang, V.T., TT, Q.B.: An artificial neural network approach for electroencephalographic signal classification towards brain-computer interface implementation. In: 2016 IEEE RIVF International Conference on Computing & Communication Technologies, Research, Innovation, and Vision for the Future (RIVF), pp. 205–210. IEEE (2016)

10. Chae, Y., Jeong, J., Jo, S.: Toward brain-actuated humanoid robots: asynchronous direct control using an EEG-based BCI. IEEE Trans. Rob. **28**(5), 1131–1144 (2012)
11. Shantala, C.P., Rashmi, C.R.: Mind controlled wireless robotic arm using brain computer interface. In: 2017 IEEE International Conference on Computational Intelligence and Computing Research (ICCIC), pp. 1–8. IEEE (2017)
12. Mishchenko, Y., Kaya, M., Ozbay, E., Yanar, H.: Developing a 3-to 6-state EEG-based brain-computer interface for a robotic manipulator control. bioRxiv (2017). 171025
13. Paiva, H.A.D.: Human-machine interface modules to support indoor navigation of a robotic wheelchair. Master's thesis (2015)
14. Reis, L.P., Faria, B.M., Vasconcelos, S., Lau, N.: Multimodal interface for an intelligent wheelchair. In: Informatics in Control, Automation and Robotics, pp. 1–34. Springer, Cham (2015)
15. Bhalchandra, A.: A mobile interface for real-time EEG monitoring. Ph.D. diss., University of Cincinnati (2017)
16. Solari, F.: Advances in human-computer interactions: methods, algorithms, and applications

Approach for the Classification of Polliniferous Vegetation Using Multispectral Imaging and Neural Networks

Jesus Silva[1(✉)], Noel Varela[1], Jorge L. Díaz-Martinez[1], Javier Jiménez-Cabas[1], and Omar Bonerge Pineda Lezama[2]

[1] Universidad de la Costa, St. 58 #66, Barranquilla, Atlántico, Colombia
jesussilvaUPC@gmail.com,
{nvarela2,jdiaz5,jjimenez41}@cuc.edu.co
[2] Universidad Tecnológica Centroamericana (UNITEC), San Pedro Sula, Honduras
omarpineda@unitec.edu

Abstract. Beekeeping has suffered a serious deterioration in the regions of the world. The possibility of nectar-polliniferous resources has decreased and, therefore, the feeding of bees, with the consequent decrease in production. There is, therefore, a need to improve marketing and diversification systems and to update production techniques and the management of the production process by beekeepers to obtain the quality of honey required by the market [1]. This work proposes the use of spectral information to identify the different pollen-producing plants using remote vision, image processing, and artificial neural networks.

Keywords: Classification of polliniferous vegetation · Multispectral imaging · Neural networks

1 Introduction

One way to improve the production and quality of honey is to characterize the vegetation around a colony of bees. The bees' diet is very specific (Table 1), so it is expected that 5 km around an apiary there will be abundant vegetation suitable to avoid the problems of honey production [2].

Sorting plants into field studies is a complex and inefficient process. Commonly, satellite images are used for the evaluation and classification of the vegetation because of their high information content [4]. However, the distance of the sensors avoids obtaining good resolution images, because in a multispectral satellite image one pixel can represent an area of 30 m^2, and obtaining aerial images with a drone enhances the spatial resolution [5]. The Sequoia Parrot camera allows images to be obtained with a ratio of 3.7 and 18.6 cm per pixel for flights of 30 and 150 m in height respectively [6].

With the information on the indices of vegetation, the average values can be calculated between spectral values indices that are used as the vector of characteristics of the neuronal network [7]. In this paper, the description of the problem with the

J. I.-Z. Chen et al. (Eds.): ICIPCN 2020, AISC 1200, pp. 269–279, 2021.
https://doi.org/10.1007/978-3-030-51859-2_24

Table 1. Most important plant species for honey production [3].

Common name	Scientific name
Ts'its'ilche'	Gymnopodium floribundum
Taj	Viguiera dentata
Ja'abin	Piscidia piscipula
Box katsim	Acacia gaumeri
Sac katsim	Mimosa bahamensis
Tsalam	Lysiloma latisiliquum
X-tabentun	Corymbose Turbine
Kitimche	Caesalpinia gaumeri
Vines	Several species of the family convolvulaceae

polliniferous flora, and the importance and benefits from the classification of this flora have been presented in the introduction.

2 Works in the Spectral Image Classification Area

In the consulted literature we can find similar works, in which machine learning algorithms are used for surface recognition using satellite images. The works focused on solving coverage classification problems using deep learning methods are mentioned below; These works are characterized by making use of hyperspectral images. Hyperspectral images are composed of contiguous images in the entire electromagnetic spectrum and can become a set of more than 100 images.

The most recent work on the general classification of the land cover uses deep learning techniques and is found in the investigations of [8], who propose to use autoencoders for the classification of hyperspectral images [9]. In [10] the same experiment was performed, changing the autoencoder method for a Deep Belief Network (DBN) [11]. The results were quite good, the network was able to identify the same regions that the autoencoder technique detected, however, the architecture of the autoencoder slightly exceeds DBN. On average the autoencoder had an effectiveness of 96% and DBN of 93%.

It is also possible to find works that use Coevolutionary Neural Networks (CNN) techniques, as they are effective methods for image processing and feature extraction. In the tests of [12] and [13] results of the effectiveness of 99% are obtained.

The previous works describe the methodology to classify the coverage using deep learning methods using hyperspectral images, which turned out to be highly efficient. However, in the literature some methodologies use multispectral images, this type of image consists of a set between 4 and 20 spectral images, depending on the satellite or sensor that is used. For the processing of multispectral images in vegetation, 4 bands of the light spectrum are usually used: green (530–570 nm), red (640–680 nm), red border (730–740 nm) and near-infrared (770–810 nm). Some of the works cited using the spectral bands mentioned for the identification of the different covers are mentioned below; Besides, it is worth mentioning that an Artificial Neural Network is used as a method of classification and satellite sensors to obtain spectral information [14].

The use of artificial neural networks in the classification of data from remote sensors has grown in recent years [15]. In the cited literature they have been used to identify different land covers, such as water, asphalt, bare soil, vegetation.

Among the most important applications is multitemporal analysis [16] that consists of the classification of coverage but analyzing spectral variations over time to determine changes in the biophysical characteristics of the regions, determine the progress of the restoration of areas ecological [17] or the monitoring of national parks [18]. Similarly, the technique of combining multispectral images with neural networks has led to the analysis of crops, for example, that of sugar cane [19] or the analysis of soil conditions, such as the level of salinity in the soil.

Because various land covers can present similar spectral responses, variants have been made that complement the image information with cartographic information [20] such as digital elevation models (DEM), thematic maps of the region, or atmospheric precipitation data [1], improving the effectiveness by approximately 10%.

In the works presented above, the advantages of using multispectral images and neural networks for the classification of different ground covers are described, however, one of the limitations of using satellite images is the cost of image acquisition and processing, because an image can have a dimension greater than 27000 pixels2 with an 11-bit resolution (QuickBird) for each of the bands it operates [8].

The sensor on the satellite. Another limitation that is common to observe is the resolution of the pixel, in the QuickBird satellite you have a resolution of 61 cm of the resolution, this means that a pixel represents an area of 0.36 m^2; This resolution can be useful when it comes to classifying large areas, however, to detect specific plants it requires a more precise resolution [20].

Additionally, in the works presented, the general process for the classification of coverage using spectral images is identified: acquisition of the images, a pre-processing in the image, such as the calculation of the vegetation indexes or the corrections and calibration of the satellite images, finally. The use of machine learning algorithm techniques are made for classification.

3 Materials and Methods

Unlike the literature [8, 12, 14], this paper proposes to use a drone and the Sequoia Parrot multispectral camera to sample areas of interest for the collection of beekeeping flora samples, since this camera improves the spatial resolution of the images and avoids the influence of gases in the atmosphere and clouds. Subsequently, it is proposed to calculate vegetation indices from the sampled images. These indices are the normalized difference vegetation index, the normalized green vegetation index, and the chlorophyll index. From the calculated vegetation index, the vector of characteristics will be constructed from statistic data: average, mode, minimum, and maximum of each of the indices [14]. In total, a vector of 12 characteristics for each of the images. Then, a database labeled with the calculated vectors of characteristics will be generated. A fully connected neural network is then designed, programmed, and trained for the classification of the vegetation using the tagged database. Once the network is trained, the classifier will be tested and finally, the results will be analyzed. The following subsections describe the materials needed to carry out the proposed methodology.

3.1 Determination of the Spectral Data

The images to be used correspond to the Sequoia Parrot multispectral sensor (see Fig. 1) [20]. This sensor is composed of two modules: the body and the light sensor. The body is composed of four 1.2 Mpx spectral cameras, a 16 Mpx RGB camera, WiFi, IMU, and magnetometer. The light sensor has 4 spectral sensors with the same filters as the body, GPS, IMU, and magnetometer. The information of the different frequencies of the light that can be obtained is the following bands: green (530–570 nm), red (640–680 nm), red edge (730–740 nm) and near-infrared (770–810 nm), in addition to taking color photos (RGB images). The images of the 4 spectral bands are in 10-bit RAW format in a TIFF file. The RGB image is saved in JPG format. In total, there are 5 images for each picture.

Fig. 1. Sensor body (Left). Light sensor (Right).

With the information from the multispectral images, it is possible to calculate spectra indices that constitute a useful tool to evaluate large areas [16]. Recent advances in deep learning have made it possible to use these indices to solve problems related to the identification of elements in an image, with coherent, precise, and reliable methods [9].

3.2 Data Analysis

There are combinations of the spectral images obtained to obtain relevant information about the vegetation, this is known as vegetation indices. These indices allow estimating biophysical variables of the vegetation from the light that the plants emit or reflect, minimizing the influence of disturbances such as those due to the soil and atmospheric conditions [10]. The selection of multispectral bands depends on the index to be calculated [21]; however, it is frequent to use the spectral bands of red (from the visible field) and near-infrared, since plants strongly reflect the near-infrared light band due to a thick layer on the lower surface of the leaf (Fig. 2). Below is a description of the vegetation indices to be used for the extraction of characteristics.

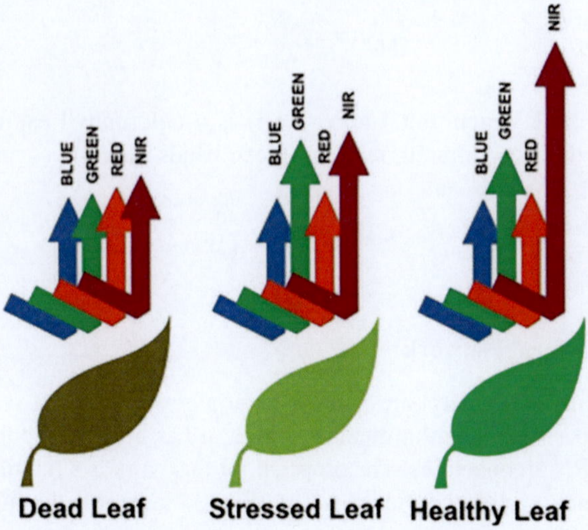

Fig. 2. The plant's leaf pigment, chlorophyll, strongly absorbs visible light for use in photosynthesis. The cell structure of the leaves, on the other hand, intensely reflects near-infrared light [2].

The **Normalized Difference Vegetation Index (NDVI)** was proposed by Rouse in 1974 and is one of the most widely used to date [2]. An interesting aspect of this index compared to the simple ratio is that it takes values between 1 and 1, which facilitates its interpretation [17]. The NDVI is the normalized difference of the infrared (NIR) and red (R) bands (Eq. 1):

$$NDVI = \frac{NIR - R}{NIR + R} \tag{1}$$

This index is one of the most used because it is strongly related to plant health. A healthy plant has cell walls in the leaves that are full of water; therefore, they strongly reflect infrared light. On the other hand, a plant that suffers from pest stress or droughts, reflects the infrared band weakly, because its leaves do not have enough water.

The **Green Normalized Difference Vegetation Index (GNVI)** is an index of photosynthetic activity and is one of the most commonly used indices of vegetation to determine the absorption of water and nitrogen in the foliage of the crop [22]. It is similar to the NDVI; however, the red band of the visible spectrum is replaced by the green band of the visible spectrum (V) (Eq. 2) [23]:

$$GNVI = \frac{NIR - V}{NIR + V} \tag{2}$$

The **Chlorophyll Vegetation Index (CVI)** is a chlorophyll estimator [24]. It is calculated using the near-infrared, red, and green bands (Eq. 3):

$$CVI = NIR \cdot \frac{R}{G^2} \tag{3}$$

3.3 Artificial Neural Network

Artificial neural networks (RN) are mathematical algorithms based on how the brain works. It is a supervised learning method, that is, it has a training phase. There are a large number of RN architectures; The proposal for this work is a multilayer perceptron with two completely connected hidden layers (Fig. 3) since it is the most widely used in remote sensing [8] for being more efficient than other methods for classifying land cover [25].

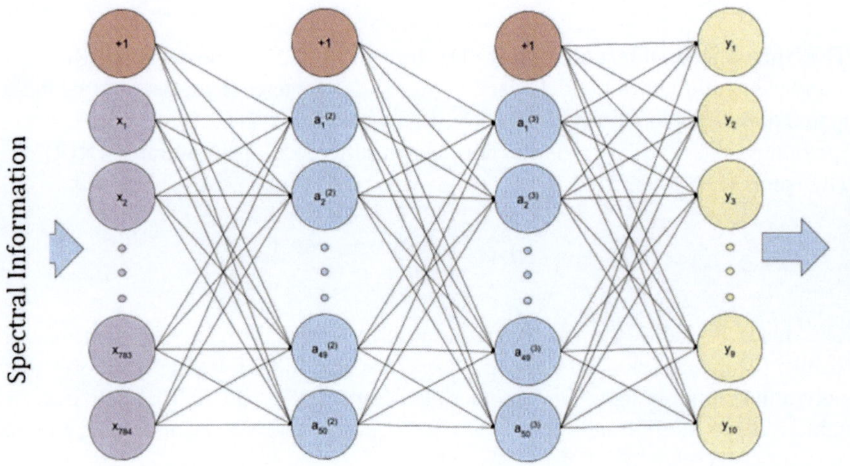

Fig. 3. Structure of the neural network.

For the training of the network, it is proposed to extract statistical information from the vegetation indices to generate the inputs to the neural network. These inputs are called the characteristics vector. The proposed statistical parameters are the mean, mode, maximum and minimum values of each index used. In total, the feature vector is composed of 12 parameters: mean, mode, maximum and minimum of NDVI, GNDVI, and CIR.

4 Spectral Analysis

As mentioned before, the literature cited uses satellite images for the classification of different textures, however, within the texture "vegetation", the classification of specific plants has not been reported, possibly due to the spatial resolution of the sensors on the satellites. Section 2 explains the reasons for using multispectral aerial images in the drones to obtain a better spatial resolution, with the spectral differences of each plant serving as a unique indicator of information for each type of plant, which is known as the spectral signature. A comparison of different spectral data is made between 2 citric plants, lemon, and orange (Fig. 4), to mark the spectral difference that exists in the diversity of the flora, which will avoid biased data in the training of the neural network and therefore, offering better efficiency in the classifier [26].

Fig. 4. In the upper left of the image, there is an orange tree, the lower one is a lemon tree.

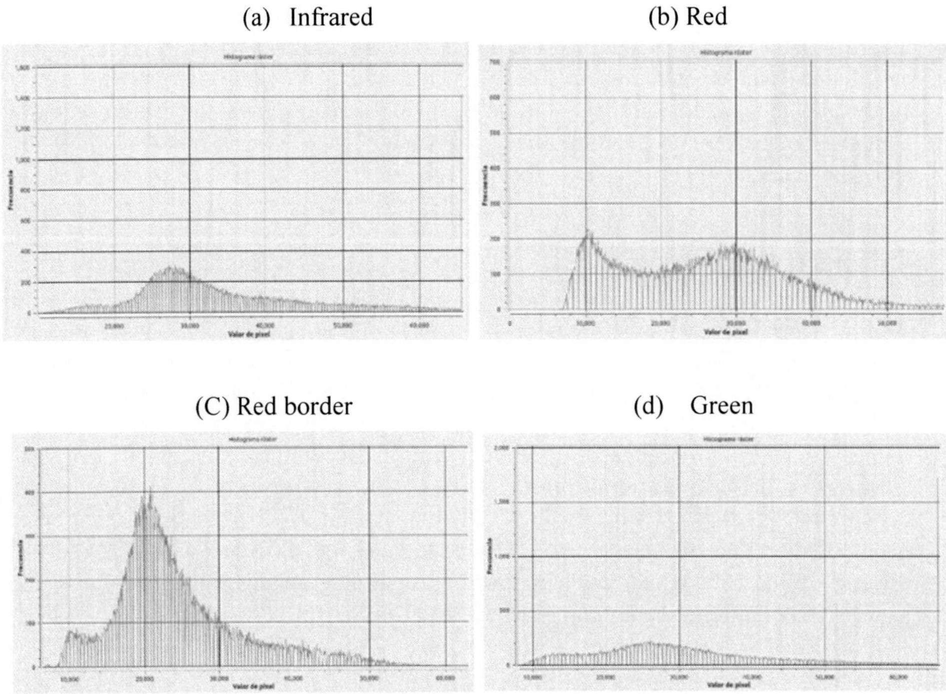

Fig. 5. Histograms for the different spectral bands.

The histograms of the different spectral bands were calculated (Fig. 5), showing that there are two mounds in the frequency of some histograms, being the most notorious in the red band. However, it is also possible to observe in the red border band but with a greater difference in the number of the frequency. In the infrared and green bands, only the frequencies that are related to the lemon tree are noticeable. Then the vegetation indices were calculated: NDVI, GNVI, and CVI to analyze the biophysical parameters of the trees (Fig. 6). From the images of the calculated indices, it can be noted that the lemon tree has higher biomass density and a higher concentration of nitrogen, water, and chlorophyll in its leaves than the orange tree. Therefore, both have unique information that will allow building an efficient classifier.

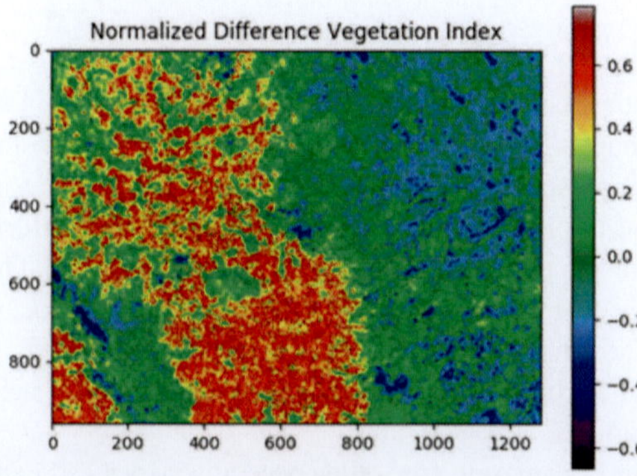

(A) NDVI

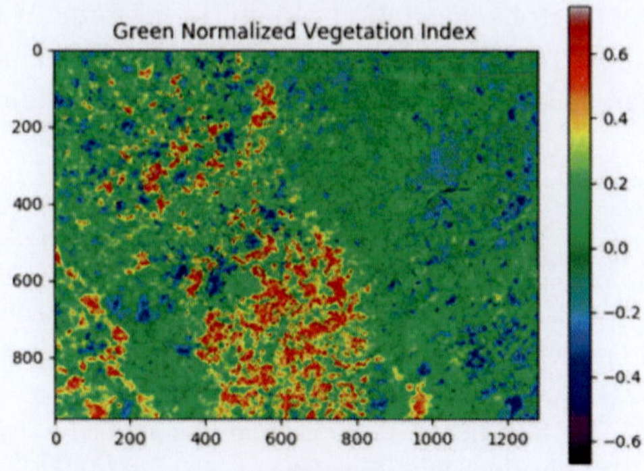

(B) GNVI

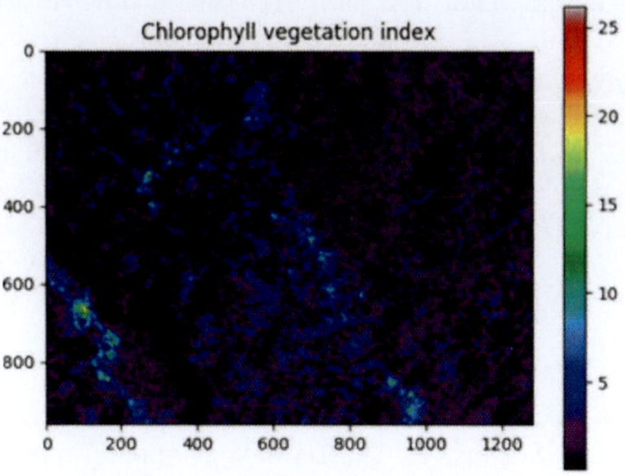

(C) CVI

Fig. 6. Pictures of the different vegetation indices

5 Conclusion

The study proposes the use of a multispectral camera placed in a drone for image capture. By using images within the Earth's atmosphere, the interference of gases, atmosphere, or clouds on spectral images can be eliminated. The short distance allows for better resolution of the images and therefore more detailed information on the land cover. Using a vector of descriptors minimizes training time. The use of machine learning techniques, in this case, neural networks, and image processing improves the analysis of spectral information. This work contributes to process innovation in beekeeping activities, enhancing the organization of producers for achieving an optimal distribution of the colonies.

In future work, the information from the spectral sensor used can be complemented with spectral information from the cameras in the satellites; and improve or update the information of the polliniferous vegetation bloom. In the same way, an analysis could be made to observe how pesticides affect the behavior of plants and a study of the reduction of aquatic areas in search of improving the distribution of apiaries.

References

1. Li, H.C., Wang, W.Y., Pan, L., Li, W., Du, Q., Tao, R.: Robust capsule network based on maximum correntropy criterion for hyperspectral image classification. IEEE J. Sel. Top. Appl. Earth Obs. Remote Sens. **13**, 738–751 (2020)
2. Grigoroiu, A., Yoon, J., Bohndiek, S.E.: Deep learning applied to hyperspectral endoscopy for online spectral classification. Sci. Rep. **10**(1), 1–10 (2020)
3. He, Z., Shi, Q., Liu, K., Cao, J., Zhan, W., Cao, B.: Object-oriented mangrove species classification using hyperspectral data and 3-D siamese residual network. IEEE Geosci. Remote Sens. Lett. (2020)
4. Chang, J., Clay, D.E., Dalsted, K., Clay, S., O'Neill, M.: Corn (L.) yield prediction using multispectral and multidate reflectance. Agron. J. **95**(6), 1447–1453 (2003)
5. Jia, P., Zhang, M., Yu, W., Shen, F., Shen, Y.: Convolutional neural network-based classification for hyperspectral data. In: 2016 IEEE International Geoscience and Remote Sensing Symposium (IGARSS), pp. 5075–5078. IEEE, July 2016
6. Chen, Y., Zhao, X., Jia, X.: Spectral–spatial classification of hyperspectral data based on deep belief network. IEEE J. Sel. Top. Appl. Earth Obs. Remote Sens. **8**(6), 2381–2392 (2015)
7. Viloria, A., Lezama, O.B.P.: Improvements for determining the number of clusters in k-means for innovation databases in SMEs. ANT/EDI40, pp. 1201–1206 (2019)
8. Modvala, S., Eremia, N., Scripnic, E.: The Technology of pastoral beekeeping. Simpozionul ştiinţific internaţional "Modern animal husband-Food safety and durable development". În: Lucrări Ştiinţifice, Universitatea de Ştiinţe Agricole şi Medicină Veterinară "Ion Ionescu de la Brad", Iaşi, Seria Zootehnie, pp. 50–55 (2016)
9. Adgaba, N., Al-Ghamdi, A., Shenkute, A.G., Ismaiel, S., Al-Kahtani, S., Tadess, Y., Abdulaziz, M.Q.A.: Socio-economic analysis of beekeeping and determinants of box hive technology adoption in the Kingdom of Saudi Arabia. J. Anim. Plant Sci. **24**(6), 1876–1884 (2014)
10. ElMasry, G., Wang, N., Vigneault, C.: Detecting chilling injury in red delicious apple using hyperspectral imaging and neural networks. Postharvest Biol. Technol. **52**(1), 1–8 (2009)

11. Jolivot, R., Vabres, P., Marzani, F.: Reconstruction of hyperspectral cutaneous data from an artificial neural network-based multispectral imaging system. Comput. Med. Imaging Graph. **35**(2), 85–88 (2011)
12. Menad, H., Ben-Naoum, F., Amine, A.: Deep convolutional neural network for pollen grains classification. In: JERI (2019)
13. Viloria, A., Acuña, G.C., Franco, D.J.A., Hernández-Palma, H., Fuentes, J.P., Rambal, E.P.: Integration of data mining techniques to Postgresql database manager system. Proc. Comput. Sci. **155**, 575–580 (2019)
14. Lin, Z., Chen, Y., Zhao, X., Wang, G.: Spectral-spatial classification of the hyperspectral image using autoencoders. In: 2013 9th International Conference on Information, Communications, and Signal Processing (ICICS), pp. 1–5. IEEE (2013)
15. Madani, H., McIsaac, K.: Spectral perturbation method for deep learning-based classification of remote sensing hyperspectral images. In: Image and Signal Processing for Remote Sensing XXV, vol. 11155, p. 111550 V. International Society for Optics and Photonics (2019)
16. Maggiori, E., Tarabalka, Y., Charpiat, G., Alliez, P.: Convolutional neural networks for large-scale remote-sensing image classification. IEEE Trans. Geosci. Remote Sens. **55**(2), 645–657 (2017)
17. Safari, K., Prasad, S., Labate, D.: A multiscale deep learning approach for high-resolution hyperspectral image classification. IEEE Geosci. Remote Sens. Lett. (2020)
18. Varela, N., Silva, J., Gonzalez, F.M., Palencia, P., Palma, H.H., Pineda, O.B.: Method for the recovery of images in databases of rice grains from visual content. Proc. Comput. Sci. **170**, 983–988 (2020)
19. Ševo, I., Avramović, A.: Multispectral scene recognition based on dual convolutional neural networks. In: Proceedings of the 10th International Symposium on Image and Signal Processing and Analysis, pp. 126–130. IEEE (2017)
20. Cao, X., Li, R., Ge, Y., Wu, B., Jiao, L.: Densely connected deep random forest for hyperspectral imagery classification. Int. J. Remote Sens. **40**(9), 3606–3622 (2019)
21. Paola, J.D., Schowengerdt, R.A.: A review and analysis of back propagation neural networks for classification of remotely-sensed multi-spectral imagery. Int. J. Remote Sens. **16**(16), 3033–3058 (1995)
22. Noh, H., Zhang, Q., Shin, B., Han, S., Feng, L.J.B.E.: A neural network model of maize crop nitrogen stress assessment for a multi-spectral imaging sensor. Biosys. Eng. **94**(4), 477–485 (2006)
23. Wu, D., Yang, H., Chen, X., He, Y., Li, X.: Application of image texture for the sorting of tea categories using multi-spectral imaging technique and support vector machine. J. Food Eng. **88**(4), 474–483 (2008)
24. Vijayakumar, T.: Neural network analysis for tumor investigation and cancer prediction. J. Electron. **1**(02), 89–98 (2019)
25. Hollaus, F., Brenner, S., Sablatnig, R.: CNN based binarization of multispectral document images. In: 2019 International Conference on Document Analysis and Recognition (ICDAR), pp. 533–538. IEEE, September 2019
26. Rana, A., Yauney, G., Lowe, A., Shah, P.: Computational histological staining and destaining of prostate core biopsy RGB images with generative adversarial neural networks. In: 2018 17th IEEE International Conference on Machine Learning and Applications (ICMLA), pp. 828–834. IEEE, December 2018

Adaptive Ternary Pattern Based on Supervised Learning Approach for Ground-Based Cloud Type Classification

Vinh Truong Hoang(✉)

Ho Chi Minh City Open University, Ho Chi Minh City, Vietnam
vinh.th@ou.edu.vn

Abstract. Ground-based cloud images recognition is a difficult task because of extreme varieties in cloud appearance under various atmospheric conditions. Local Binary Pattern and its variants are widely applied due to its efficiency for texture classification. Among them, Local Ternary Pattern (LTP) descriptor is the most common to characterize cloud images. In this approach, an additional parameter is used to define a tolerance for measuring the similarity between different gray intensities. This paper proposes a new approach to determine a suitable value of the parameter for LTP descriptor by the supervised learning approach. The proposed method is evaluated on the SWIMCAT database.

Keywords: Ground-based cloud images · LTP · Adaptive ternary pattern · Cloud images recognition

1 Introduction

Over the last decades, clouds have been extensively investigated within the research community and have a various application. Thus, the ground-based cloud image recognition is a major approach for climate research including weather prediction, solar energy generation, meteorology research, and solar energy collection. In the past, cloud image weather observation relies primarily on human vision and technical expert with limited accuracy and efficiency. The advances development of computer technology and the progress of image processing have brought our daily lives several benefits.

To address this issue, texture classification methods have recently been introduced. Local Binary Pattern is among the efficient methods for representing texture image with many application [1–3]. Various works apply this operator to characterize cloud images for a classification task. For example, Liu et al. [4] introduce the salient local binary pattern which captures the most descriptive information patterns. Wang et al. [5] propose a method to select and rank frequencies of all rotation invariant patterns defined in the LBPs of cloud images. Wang et al. [6] use Kullback–Leibler divergence to measure the different resolutions between the LBP histogram and cloud textured images. Zhang et al. [7] introduced a new extracted approach from convolutional neural network and it will adjust with illuminations, occlusions, resolutions and location phenomena.

J. I.-Z. Chen et al. (Eds.): ICIPCN 2020, AISC 1200, pp. 280–286, 2021.
https://doi.org/10.1007/978-3-030-51859-2_25

In existing approaches, various color description schemes were introduced for image discrimination and from that one of most important scheme is local binary pattern scheme designed but Ojala et al. and he has implemented in various statistical approaches and calculated the efficacy and illumination changes [8]. LBP algorithm is more sensitive to noise and it is analyzed based on the thresholding operation. Many techniques are involved to analyze the power discrimination, robustness noise gain and quantization level such as Neighborhood intensity LBP [9], Improved LBP [10], Local Median LBP [11] and Threshold Modified LBP [12]. Among these approaches, the Local Ternary Patterns (LTP), proposed by Tan et al., have achieved a great success [13]. In this approach, an additional parameter is used to define a tolerance based on the different grey level. This ternary is splitted into two different parts such as positive and negative sides, based on these two sides histograms are computed. The most difficult thing in the local binary pattern algorithm is setting an appropriate values for LBP. Yang et al. [14] proposed an adaptive Local Ternary Pattern (ALTP) and it selects the threshold values automatically. Sandid and Douik [15] discussed the local combination adaptive ternary pattern (LCATP) descriptor and it encodes the texture images based on color and information.

This paper proposes a new method to choose a suitable parameter for LTP descriptor based on information in the supervised learning context. The rest of this paper is organized and structured as follows. Section 2 introduces the feature extracting methods based on LTP descriptors and the proposed approach. Section 3 presents the experimental results on SWIMCAT dataset. Finally, the conclusion is discussed in Sect. 6.

2 Proposed Approach

This section briefly reviews the Local Ternary Patterns and the proposed approach for finding its parameter.

2.1 Local Ternary Pattern

Ojala et al. [16] propose to compute LBP via the local neighborhood structure by characterizing the texture around each pixel of the image from a square neighborhood of 3×3 pixels. LBP features capture microscopic local image texture. The gray level values of each neighboring pixel are first compared with the value of the central pixel, that allows to extract a binary vector of 8 bits. This binary vector is called "local binary pattern".

The definition of the original LBP operator has then been generalized to explore intensity values of points on a circular neighborhood. The circular neighborhood is defined by considering the values of radius R and P neighbors around the central pixel.

The $LBP_{P,R}(x_c, y_c)$ code of each pixel (x_c, y_c) is computed by comparing the gray value g_c of the central pixel with the gray values $\{g_i\}_{i=0}^{p-1}$ of its P neighbors, as follows:

$$LBP_{P,R}(x_c, y_c) = \sum_{i=0}^{P-1} \Phi(g_i - g_c) \times 2^i \tag{1}$$

where Φ is the threshold function which is defined as:

$$\Phi(g_i - g_c) = \begin{cases} 1 & \text{if } (g_i - g_c) \geq 0, \\ 0 & \text{otherwise} \end{cases} \tag{2}$$

LBP is resistant to lighting effects in the sense that they are invariant to monotonic gray-level transformations, and they have been shown to have high discriminative power for texture classification. However, because they threshold at exactly the value of the central pixel g_c they tend to be sensitive to noise, especially in near-uniform image regions. Many authors have proposed several LBP variants by changing the thresholding scheme or the number of quantization level to gain noise robustness and discrimination power. In order to limit this drawback, Tan and Triggs propose LTP to encode the central pixel by three values $\{-1, 0, 1\}$ as follows [13]:

$$\Phi'(g_i - g_c) = \begin{cases} 1 & \text{if } (g_i - g_c) \geq t, \\ 0 & \text{if } |g_i - g_c| < t \\ -1 & \text{if } (g_i - g_c) \leq t \end{cases} \tag{3}$$

where t is a given threshold defined by user which can make LTP code more robust against noise.

2.2 Proposed Approach

Several works proposed to use a fixed value of t for LTP descriptor in their experiment. In this approach, supervision information is applied to find an appropriate value of this parameter. The proposed approach requires to divide the initial image dataset into three subsets such as validation, training and testing set by cross-validation decomposition. Figure 1 explains in detailed this process. During the learning stage, LTP descriptor is applied for validation and training images to extract two sets of features. Each value of $t \in \{1, 2,.., 255\}$ is investigated to perform feature extraction step. The SVM classifier is applied for each value of t to predict the label of the validation set. Then, the optimum value of t is determined by the maximum accuracy obtained. During the classification stage, this value can be used for extracting LTP features from the testing set. It is worth to note that all images of training and validation set are used for building the final training model which is used to predict the images of the testing set. This method can be applied for solving the real-world problem since this value is mainly depended on the characteristic of images.

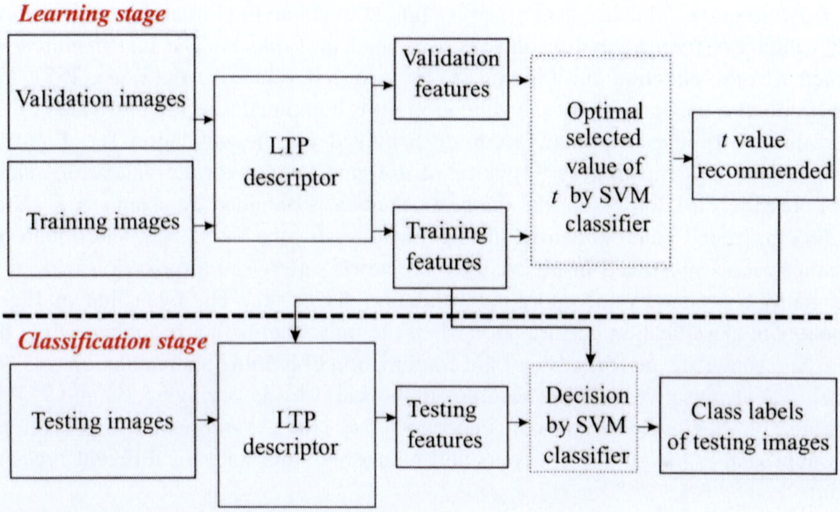

Fig. 1. Illustration of the proposed approach.

3 Experimental Results

Although the ground-based cloud images classification problem has been studied for many years, few publicly available clouds image datasets exist. SWIMCAT [17] stands for Singapore Whole sky IMaging Categories Database, containing 784 images categorized in 5 distinct classes: clear sky, patterned clouds, thick dark clouds, thick white clouds, and veil clouds (see Fig. 2).

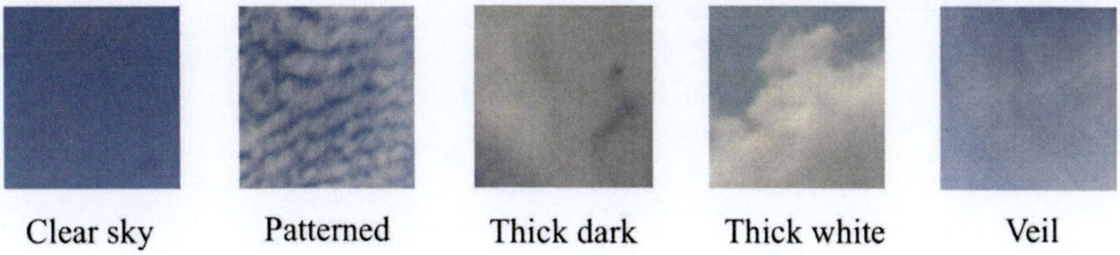

Fig. 2. Example images of SWIMCAT dataset.

Table 1. Number of images divided by 5-fold cross validation.

Class	Validation	Training	Testing	Total images
Clear sky	45	45	134	224
Pattern	18	18	53	89
Thick dark	51	51	149	251
Thick white	28	28	79	135
Veil	18	18	49	85

A 5-fold cross validation technique is applied to create the initial dataset into three sets: validation, training and testing (as presented in Table 1). The LTP features are extracted from validation and training set by using all value of $t \in \{1, 2.., 255\}$. For each value of t, the accuracy of the validation set is computed, and the best value of t is determined with the maximum accuracy achieved on the validation set. Figure 3 illustrates the classification performance of different value t on the validation set by SVM classifier. It is observed that the best accuracy is obtained by using $t = 1$. Using this recommended value to extract the LTP features for the testing set, and obtain the classification as illustrated in Fig. 4. Two scenarios supervised information is used to feed SVM classifier by incorporating validation set or not. The cyan line of Fig. 4 presents the classification performance when the only training set is applied while the green line shows the performance of the combination of training and validation set. The value $t = 1$ always gives the best accuracy in two cases by achieving 91.8% and 95.4%, respectively. Previous works usually choose $t = 5$ in [13, 18] without any explanations. This approach is the simplest way to find a suitable value of t for different types of texture.

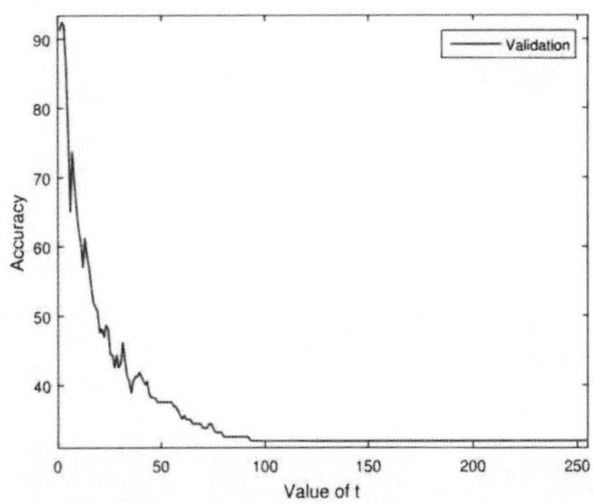

Fig. 3. Classification results on the validation set.

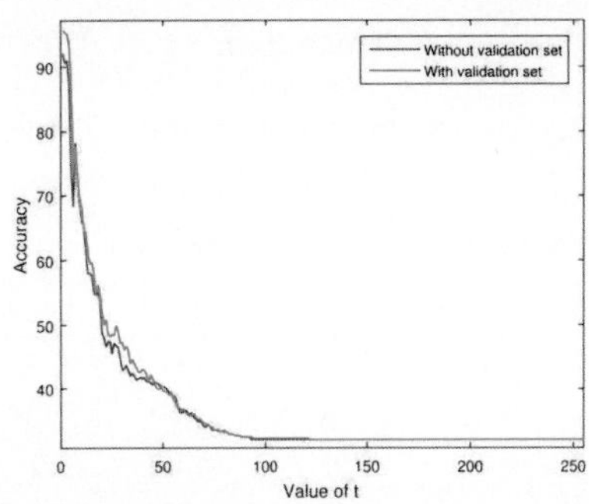

Fig. 4. Classification results on the testing set.

4 Conclusion

This paper presents a new method for finding an appropriate parameter of LTP descriptor by using SVM classifier to extract LTP features on a validation set. The experimental results show the efficiency of the proposed approach for ground-based cloud images classification. This work is now extended to select the relevant LTP histogram to achieve a compact representation for cloud images.

References

1. Van, T.N., Hoang, V.T.: Early and late features fusion for kinship verification based on constraint selection. In: 2019 25th Asia-Pacific Conference on Communications (APCC), pp. 116–121 (2019)
2. Nhat, H.T.M., Hoang, V.T.: Feature fusion by using LBP, HOG, GIST descriptors and Canonical Correlation Analysis for face recognition. In: 2019 26th International Conference on Telecommunications (ICT), pp. 371–375 (2019)
3. Van, T.N., Hoang, V.T.: Kinship verification based on local binary pattern features coding in different color space. In: 2019 26th International Conference on Telecommunications (ICT), pp. 376–380 (2019)
4. Liu, S., Wang, C., Xiao, B., Zhang, Z., Shao, Y.: Salient local binary pattern for ground-based cloud classification. Acta Meteorol. Sinica **27**(2), 211–220 (2013)
5. Wang, Y., Shi, C., Wang, C., Xiao, B.: Ground-based cloud classification by learning stable local binary patterns. Atmos. Res. **207**, 74–89 (2018)
6. Wang, Y., Cunzhao, S., Chunheng, W., Baihua, X.: Measure for the difference between LBP features extracted from original and resized cloud images with varying resolutions. IEEE Geosci. Remote Sens. Lett. **14**(7), 1106–1110 (2017)
7. Zhang, Z., Li, D., Liu, S., Xiao, B., Cao, X.: Multi-view ground-based cloud recognition by transferring deep visual information. Appl. Sci. **8**(5), 748 (2018)
8. Ojala, T., Pietikäinen, M., Mäenpää, T.: Multiresolution gray-scale and rotation invariant texture classification with local binary patterns. IEEE Trans. Pattern Anal. Mach. Intell. **24** (7), 971–987 (2002)
9. Liu, L., Zhao, L., Long, Y., Kuang, G., Fieguth, P.: Extended local binary patterns for texture classification. Image Vis. Comput. **30**(2), 86–99 (2012)
10. Jin, H., Liu, Q., Lu, H., Tong, X.: Face detection using improved LBP under Bayesian framework. In: Proceedings of the Third International Conference on Image and Graphics, December 2004, pp. 306–309 (2004)
11. Hafiane, A., Seetharaman, G., Zavidovique, B.: Median binary pattern for textures classification. In: Proceedings of the 4th International Conference on Image Analysis and Recognition, pp. 387–398. Springer, Berlin (2007)
12. Heikkila, M., Pietikainen, M.: A texture-based method for modeling the background and detecting moving objects. IEEE Trans. Pattern Anal. Mach. Intell. **28**(4), 657–662 (2006)
13. Tan, X., Triggs, B.: Enhanced local texture feature sets for face recognition under difficult lighting conditions. In: International Workshop on Analysis and Modeling of Faces and Gestures, pp. 168–182. Springer (2007)
14. Yang, W., Wang, Z., Zhang, B.: Face recognition using adaptive local ternary patterns method. Neurocomputing **213**, 183–190 (2016)
15. Douik, A., Sandid, F.: Texture descriptor based on local combination adaptive ternary pattern. IET Image Proc. **9**(8), 634–642 (2015)

16. Ojala, T., Pietikäinen, M., Harwood, D.: A comparative study of texture measures with classification based on featured distributions. Pattern Recogn. **29**(1), 51–59 (1996)
17. Dev, S., Lee, Y.H., Winkler, S.: Categorization of cloud image patches using an improved texton-based approach. In: 2015 IEEE International Conference on Image Processing (ICIP), September 2015, Quebec City, Canada. IEEE (2015)
18. Tan, X., Triggs, B.: Enhanced local texture feature sets for face recognition under difficult lighting conditions. IEEE Trans. Image Proc. **19**(6), 1635–1650 (2010)

Analyzing User Behavior and Sentimental in Computer Mediated Communication

Abdulrahman Alrumaih[1,6], Ruaa Alsabah[2], Hiba J. Aleqabie[2], Ahmed Yaseen Mjhool[3], Ali Al-Sabbagh[4,5](✉), and James Baldwin[6]

[1] Law College, Al-Nahrain University, Baghdad, Iraq
[2] College of Science, Computer Science Department, University of Kerbala, Kerbala, Iraq
[3] Information Technology Research Development Center, University of Kufa, Najaf, Iraq
[4] Computer Engineering Department, Al-Taff College University, Kerbala, Iraq
aalsabbagh2014@my.fit.edu
[5] Ministry of Communication, ITPC, Babylon, Iraq
[6] Sheffield Hallam University, Sheffield, UK

Abstract. Social media platforms are witnessing significant growth in both size and purpose. One specific aspect of social media platforms is sentiment analysis, by which insights into the emotions and feelings of a person can be inferred from their posted text. Twitter is a widely known social media platform in which users from different cultural and linguistic backgrounds. It is also valuable in studying a large amount of data and turning it into valuable information that can be useful in decision making. This study aims at investigating the sentiment orientation of the textual features as well as the emojis-based features of Arabic comments posted on social media, Twitter, during the World Cup 18 event. It also measures whether different results will be obtained when the text is analyzed in isolation from emojis. This paper presents the approach that involves extracting thousands of tweets, implementing the sentiment analysis, on the texts, and the emojis separately. This analysis has been conducted once via a using an artifact and once again manually, then a comparison between the results is made. This comparison shows that emojis support the sentiment orientation in the text. However, it can be relied solely neither on the text nor on emojis as they complement each other.

Keywords: Visual content analysis · Complex networks · Sentiment analysis · Social media · Tweets

1 Introduction

Communication networks have a significant advantage among which are: firstly, being a source of news, entertainment, and education. Secondly, they are an essential tool that could lead to a dramatic change in cultures and societies. Thirdly, they can guide the political elections in some countries to a specific destination [1]. Furthermore, social media provide users with a virtual social environment that focuses heavily on common

J. I.-Z. Chen et al. (Eds.): ICIPCN 2020, AISC 1200, pp. 287–298, 2021.
https://doi.org/10.1007/978-3-030-51859-2_26

interests. It is also flexible and accessible through different devices, making it easier to share resources among individuals, communities, and organizations [2].

Besides, the goal of social media is to harness random big data to obtain important information regarding public opinion, below stands for the diverse communication methods people can rely on in their daily life. Upon communicating using these platforms, there is a strong tendency to use abbreviated expressions that helps in expressing feelings and emotions. Recently there has been an increasing interest in measuring these emotions using diverse methods of analysis, among these methods is sentiment analysis (SA), as shown in Fig. 1, that gave a general understanding of SA.

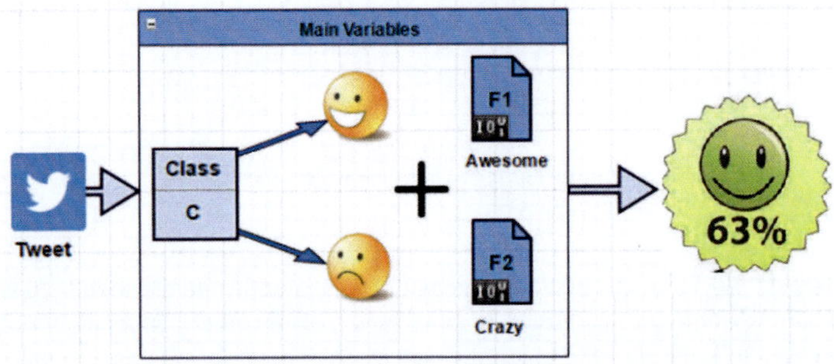

Fig. 1. Sentiment analysis in social media [9].

Also, [3] defines sentimental analysis as "the field of study that analyses people's opinions, sentiments, evaluations, appraisals, attitudes, and emotions towards entities such as products, services, organizations, individuals, issues, events, topics, and their attributes". The research demonstrates that there is a strong adherence to social media. People also tend to incorporate emojis and abbreviated expressions in their computer communicative interaction [4]. Heart codes, smiley faces, or crying faces are reflecting the sentiment orientation of the speaker which, on many occasions, substitute the text or enhance the emotional intention. For that reason, it is necessary to take into consideration the role of emojis in any sentiment query carried out on computer-mediated interaction.

Furthermore, there has been little research on analyzing emotions using emojis. To the best of the researcher's knowledge, no recent study has been devoted to addressing the sentiment queries of Arabic comments alongside the inserted emojis posted on twitter during world cup competition. From this perspective, this research is an attempt to fill this gap in the literature. It also aims at answering the following research questions: to what extent can Sentiment Analysis measure the difference in the emotions obtained from textual data and that of emojis in the Arabic comments posted in social media [34]?

Many studies have also intensified the importance of considering the influence of emojis when conducting sentiment analysis. As in [5] remarks, emojis can transmit relevant information that compensates the deficiency of short text in social media platforms by conveying and reflecting sentiments. The same text can be interpreted differently when negative emojis are inserted from when positive ones are inserted [6]. Likewise, relying on emojis solely is not sufficient in arriving at the intended meaning of the writer [7]. Furthermore, the order of emojis within the text is influential in directing the meaning due to the difference in the relation of words to each other and the emojis incorporated [8]. Based on these views, emojis and text complement each other in addition to the influence of their location within the text. This paper aims to investigate the sentiment orientation of the textual features as well as the emojis-based features of Arabic comments posted on social media, Twitter, during the World Cup event. It also measures whether different results will be obtained when the text is analyzed in isolation from emojis.

2 Background and Literature Review

With the recent heavy interest in emojis, several inquiries come into the mind, some of which are related to the source of these images, the identity of the person who created them, whether he/she bears in mind the variations between societies, cultures, and civilizations, and others if the use of emoji differs in culturally and linguistically differentiated societies. Then, emojis, their definition, evolution, their impact on social media, and the breadth of its use in the expression of feelings is investigated. This paper ends with an eye on sentiment analysis, focusing on the contemporary related studies that targeted Arabic speakers.

Many researchers argued that social and cultural norms have an integral role in the manners of using the internet or social media [9]. It has been proved that individuals and their methods of using social media are variant under the cultural and the social level on the basis that the basic cultural norms vary from community to society ref [10–14, and 15]. Concerning the variation between cultures in the heavy use of social networking sites, there has been a significant indication of Arabic interest, especially among youths [16]. It is worth mentioning that the Arab youth generation is under two types of influence: the traditional Islamic culture and the modern culture [17]. For social and religious reasons, the former leads to very limited space in expressing views and/or engagement in activities, while the latter calls for liberty and statement of personal perspectives. These effects are often contradictory, especially concerning the expansion of social networks, which may lead to more sources of identity [18].

An example of this contradiction is the Saudi preacher Ali al-Maliki, who is the most vocal critic of Facebook and social networks, saying: "the network is corrupting the youth of the nation: 'Facebook is a door to lust, and young women and men are spending more on their mobile phones and the Internet than they are spending on food'" (Facebook, 2009) One of the most popular social networking platforms in Arab societies is Facebook, which is used and traded among different ages [19]. The literature agrees that the time spent on Facebook by Arab youth has increased after 2009, i.e. after the launch of the Facebook interface in the Arabic world [20]. Throughout the results of much research, it is found that the time spent by the youth on Facebook is more than three hours a day [21], exceeding Twitter in this perspective [22].

In comparing the intensity of using the communication sites as a means of free communication. In [22] (2015) reports a study conducted on university students in

Saudi Arabia. It was found that Facebook was the head with 76.5%. Then, Twitter was 54.4%, followed by WhatsApp, which was 37.6%. As far as the number of users is concerned, statistics published on [23] illustrate that Facebook contains 1.310.000000 active Arab participants. Up till 01/05/2014, Twitter was used by 645.750000 participants, adding 135000 new subscribers daily, whilst WhatsApp contains 600.000000 effective users.

After considering the emergence and development of social media, it is necessary to explore a substantial element of communication in this platform, i.e. emojis. It is taken for granted that the written language descends from drawing images inside caves, developed into hieroglyphs, which depend on symbols to record novels and epics. Currently, despite the development and complexity of the written language and its multiplicity, millions of people employ symbols in their messages to manifest their emotions. The current strong interest in computers and mobile phones has assisted in the widespread of these symbols.

Besides, these symbols enhance people's interaction and conversation whatever is their linguistic and cultural background. In 2006, Google began converting Japan's emoji into Unicode-specific codes. In 2007, the first-ever transferable Emoji by characters became apparent, leading to the global interest in emoji and the focus of many studies [24].

There is often confusion between the use of the term emoji and posters in what is related to the methods of composition and purposes of use. Posters are a series of text characters (punctuation or symbols) that can be used to show some facial animations, which preceded emojis in expressing emotions [24]. Stickers were modified to include Unicode characters. This modification is not used only with ASCII characters but includes (‿ (‿), ︵ (︵), ◉ (◉) and ʘ̆ (ʘ̆)) through which facial impressions can be configured as follows: (^‿^ ◉︵◉ ʘ̆_ʘ̆). These stickers were the pre-emojis, being the custom images in many messenger platforms such as Facebook, Messenger, Snapcash, and Font. However, it is not a standard language that can be used on different platforms. They are also not adaptable as they are used only in specific applications they are created by. Also, in [25] discusses a modern and innovative method of adding fun to the language of communication and building social relationships through the digital world by giving the recipient a picture of the writer's psychological situation. This promotion of communication is motivated by economic and political motives. Through emojis, expressions can support psycho-social interactions, such as joy and sadness. They also reflect the phenomenon of "emotional Contagion". That is to say, people's intentions, ethnic differences, or sexual orientation can be represented via emojis [6]. However, the wrong choice of emoji would not improve the expression, instead, it deceives the transference of the required meaning.

An important research area that emerged in natural language processing and text mining is known as Sentiment analysis (SA). It is the latest type of text analysis that aims at determining the reviewers' opinions and subjectivity [25]. Sentiment Analysis (SA) and Opinion Mining (OM) are two terms that can be used interchangeably to understand any side of tendencies, feelings, or the inclined opinion regarding a particular topic [26]. The two terms are used in representing the process of extracting the polarity of an opinion on a specific object or sentiment orientation [26]. Such an object can either be a person, product, service, or event. This analysis can be conducted by a certain person, a service, an event, and so forth. Both Sentiment Analysis and Opinion Mining determine the positivity or negativity of a sentence message, comment, or document. These views may be stated via variant manners including articles, discussions in web forums, reviews, short comments on Facebook and Twitter, or by using Emojis that incorporate special meanings.

It has been observed by [6] that a plain text on digital communication if does not encompass emojis, could lead to misinterpretation and/or sometimes to misunderstanding. It has a significant impact on avoiding ambiguity and elucidating the meaning. Therefore, the main purpose behind adopting sentiment analysis is to exploit the modern technologies in collecting data in a wide range of fields. Researchers have resorted to the advanced techniques after noticing that the traditional methods of querying data are exhausting and not sufficient to extract useful information from various types of large databases. Besides, the human ability to store this large amount of data and extract suitable knowledge without using techniques that were used for this purpose was not sufficient either [27]. As far as time is concerned, SA assists in stating the sentiment polarities of thousands of comments on a particular product or service within a relatively short duration. After the evaluation of these sentiments, companies will have the opportunity of improving their products and/or services [28]. Further, since SA can be a quick and more accurate alternative for public polls, it might be adopted by the governments in evaluating the public's opinion regarding controversial issues. By analyzing people's internet posts about a particular issue, SA can automatically and accurately estimate the public's opinion on that issue. Among this limited number and to fill the gap in the literature that concerns Sentiment Analysis (SA) in the Arabic text, [29] have adopted a lexicon-based approach for Arabic Sentiment analysis using two main components this approach: the lexicon and the SA tool.

After developing a dictionary measuring words polarity (negative-positive) of 2000 tweets in Arabic, [29] found that there are several algorithms through which emotion analysis can be analyzed. They also figured out that the accuracy level of lexicon-based SA was low and have proposed to expand the lexicon to be dialect-specific and domain-specific, adding more details concerning the strength of the positive/negative sentiment, developing the combination/aggregation techniques of individual words polarities to calculate the overall polarity of a text, studying neutral words more explicitly, examining more words using the corpora, finally improving the way sarcasm and other cases handled in a better way. It must be remarked here that this study was helpful in the steps carried out in the data processing.

3 Research Methodology

With the expanding dominating role internet plays in people's lives, addressed in the previous sections, technology now enables them to express opinions, share knowledge and experience, and state comments on diverse topics. This chapter presents the steps adopted in pursuing this research.

It starts by describing the reason why social media is widely used as a source of data, particularly twitter. Then, different types of sampling and methods of data collection will be presents, naming the process of data collection adopted in this work.

After presenting the merits and drawbacks of harvesting data on social media platforms, this section is related to the types of data and the possible strategies in accessing samples. In [30] lists three types of data on social media, namely: content data, behavior data, and network structure data. Content data can be users' comments on forums, social network sites, news articles, photos, and so forth. This data can either be structured or unstructured. The structured ones are in the form of tables that has rows and columns. The majority, though, is unstructured wherein researchers are obliged to performing text mining stages ahead of processing this data. Due to the nature of this study, content data has been adopted in which users' posts and comments will be analyzed after the text mining process. In behavioral data, users' behaviors can be recorded on social media with precise timestamps, such as when the users read online news, comment on blogs write reviews for the product, watch the video, etc. Network structure data, on the other hand, refers to visual or hidden hyperlinks among users and/or content.

After determining the type of data, it is necessary to mention sampling strategies. According to [30], "Sampling is the initial stage of the sample under study." They also remark that there are two techniques for selecting sampling seeds: convenience-based seeds and randomly selected seeds. As its name indicate, the former select few users or content pages that are handy to locate such as targeted websites or based on search results. After locating the page, additional webpages will be followed via the existing hyperlinks among them. This strategy is known in social media as 'snowballing' and in information retrieval, it is called 'breadth-first search'.

As to Random selection of seeds, it requires a sampling frame, an operational version of the given social media understudy. In this project, a mixture of the two approaches has been adopted whereby samples have been selected according to the content of the comments, i.e. world cup event, regardless of the users' identity or nationality and provided they are written in the Arabic language and contain emojis.

A positive point in the convenience samples approach is that it is fast to finish and easy to implement, thus it has been widely recruited in harvesting data on social media. Nevertheless, a significantly mentioned drawback is the possibility of involving selection biases particularly when the convenience-based seeds might lead to users that are more active, visible, and connected. As such, despite its difficulty to carry out, the random-selection approach is recommended by communication scholars because it is subject to fewer selection biases (Ibid).

Before proceeding into the description of data collection, it is relevant to illustrate the diverse methods adopted in obtaining data on the social media platform. The current section starts with the levels of sentiments analysis, then the different approaches will be presented, ending with the one adopted in this project. As in [25, 31] remark, there are three levels of opinion or sentiment analysis: document level, sentence level, or word/feature level classification. A classification that relies on the structure of the text. On the document level, it is aimed to assign either a positive or negative polarity for the whole document or review by the information and opinions stated in it. On the sentence level, the work is carried on two steps: firstly, to decide whether the sentence carries opinions and, secondly, to determine the positive and the negative polarity in the investigated sentence. The word/feature analysis is the major method that is applied to the level of adjectives or features. The sentence will be analyzed into its elements, then every element will be assigned a polarity, either positive or a negative one.

In the latest method, [31] notes that text can be analyzed differently:

- Words are categorized according to their importance. To distinguish between the importance of words. A word is assigned number one if it appears in the text and if it appears twice, it will be assigned number two and so on. The sentence or the text is categorized into its elements, i.e. verbs, nouns, adjectives, prepositions, adverbs…etc. Adjectives are mostly chosen because they demonstrate opinions.
- Words that state opinion. There are plenty of words that reflect the speaker's or the writer's feeling such as hate, like, good, bad. These words play a substantial role in the speed and accuracy of the sentimental analysis.
  Negation. It is one of the essential elements of sentimental analysis as when a negative tool appears in the text it oppositely changes its direction. For example, when the text carries a positive word, if a negative element appeared ahead of it, it is changed into an opposite implication. However, these words must be treated carefully as negative words don't need to change the direction of the opinion.

Sentiment analysis can be classified according to the way it is implemented: the technique used, the view of the text, the level of detail of text analysis, the rating level, … etc. The emphasis in this section is on the technique adopted in implementing the sentiment analysis. There are three approaches identified: machine learning, lexicon-based, statistical, and rule-based approaches. In [25] list these methods: The machine learning method, the lexicon-based approach, and the rule-based approach.

The machine statistic method is also called supervised or corpus-based. It is conducted without human interference and uses machine learning classifiers trained on a labeled corpus. After the construction of the model, the inputted text is classified into one of the predefined classes. The lexicon-based method, on the other hand, is unsupervised and relies on human effort in which a small group of words is formulated, leading to an expansion of the related words and synonyms. It involves a list of words that are associated with their polarities, either positive or negative (+1 or −1). The rule-based approach seeks opinion words in a text. Then, this text is classified based on the number of positive and negative words, considering different rules for classification such as dictionary polarity, negation words, booster words, idioms, emoticons, mixed opinions, … etc.

According to [29], the lexicon-based approach "calculates the total polarity of the inputted text from the individual polarities of the words/phrases comprising the inputted text." Despite the proven accuracy of the first approach, there are significant drawbacks: firstly, the requirement of building a huge corpus (dataset) and labeling it manually by human experts. Manual annotation might be a challenging process even for native speakers of the studied language due to the pragmatic use of language including sarcasm and cultural references. Secondly, this approach can be of high cost and time-consuming [32].

Thirdly, and most importantly, the supervised or corpus-based model could be a domain-biased. That is to say, it might provide a lower accuracy if it is applied in a domain that is different from the one where it was learned [33, 34]. By contrast, the lexicon-based approach requires less enormous, manually annotated corpus. The challenge in this method lies in lexicon construction and labeling processes. The lexicon can be constructed either manually or automatically [26]. The following Fig. 2 illustrates the methodology of analysis explained so far.

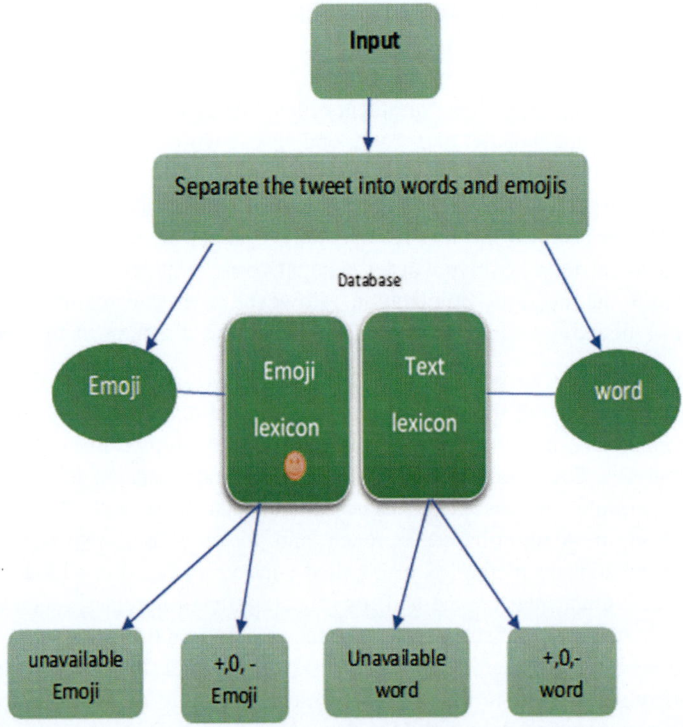

Fig. 2. Data processing.

As a summary, this section presented the methodological steps in this study, the core area in this work. It starts by highlighting the advantage of harvesting data using social media followed by describing the types of sampling, the methods of data collection, and its analysis. It ends with an illustration of the tool design and its implementation. In the next section, there will be a display for the results arrived at throughout this study. As Fig. 2 shows, data analysis involved three stages.

4 Results and Discussion

After presenting the steps of data processing, below is a description of the results arrived from the case mentioned below, then polarities have been carried out. In the first place, results arrived at using python artifact will present. The next table and the subsequent ones display the results in 10 hashtags. Each arrow shows the polarity of a hashtag in which 1000 tweets are analyzed.

The first column is related to the more positive polarity, the second to the positive ones, the third shows the neutral, the forth displays the negative words, and finally, the fifth column counts the more negative ones. After processing text separately from emojis the following results were obtained in Table 1 (left side). The below table one and Fig. 3 (left) illustrate the total polarity is positive orientation. The tenth hashtag involves the highest positive polarity, meanwhile carries the least negative value. In the fifth hashtag, however, positive and negative values are equal.

As to emoji analysis, it is mentioned earlier that the same steps have been followed in their analysis. The highest positive polarity obtained from the fourth hashtag, followed by the third and fifth hashtag. However, the more positive polarity exceeds the positive ones.

The highest rate was in the seventh and tenth hashtags, followed by the ninth and tenth hashtags. Below are the obtained polarity calculations for ten hashtags as shown in Table 1 (right side) and Fig. 3 (right side).

Table 1. Text polarity calculations (left) vs Emoji polarity (right).

Positive+	Positive	Neutral	Negative	Negative−	Positive+	Positive	Neutral	Negative	Negative+
20	62	0	16	2	40	40	6	10	2
10	56	10	22	2	38	30	6	14	12
6	60	14	16	4	20	48	4	22	6
12	48	22	18	0	30	58	6	6	0
4	34	24	34	4	34	42	10	8	6
4	56	6	26	8	40	28	8	6	18
4	54	10	24	8	52	28	2	8	10
4	46	18	32	0	30	34	16	8	12
8	48	10	28	6	44	26	6	14	10
6	68	8	14	4	52	26	4	10	8

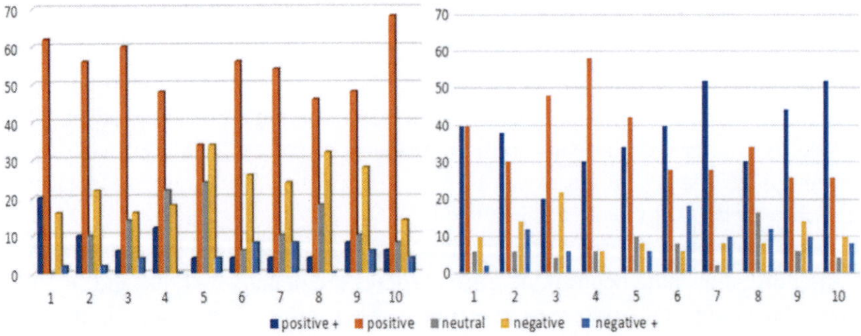

Fig. 3. Results text calculations (left) vs results of emoji calculations (right).

5 Conclusion

It is taken for granted that twitter is a widely known social media platform in which users from different cultural and linguistic backgrounds post comments on whether for their followers or to the public. Throughout observing the large number of comments posted by Arab users, in particular those from Saudi Arabia, it is found that they use emojis instead of text. Western users, in contrast, mostly support their text with an emoji. Therefore, upon analyzing comments in a particular platform, emojis of that platform must be analyzed. Further, this artifact can be applied in any context at any event, not only the World Cup, used in this study, in which all the tweets were related to sports.

Future research can rely on more than one text or emoji lexicon in analyzing data to avoid having a long list of neutral words. Further, sarcasm and the level of politeness are also valuable areas of investigation. Emojis can be a relevant indicator that assists in detecting the sentiment of the in tweets, an area that has not touched upon. By studying the polarity of (im)politeness may result in interesting analysis.

References

1. Cook, C.: Mobile marketing and political activities. Int. J. Mob. Mark. **5**(1) (2010)
2. Leung, L.: Predicting internet risks: a longitudinal panel study of gratifications-sought, internet addiction symptoms, and social media use among children and adolescents. Health Psychol. Behav. Med. J. **2**(1), 424–439 (2014). https://doi.org/10.1080/21642850.2014.902316
3. Liu, B.: Sentiment analysis and opinion mining. Synth. Lect. Hum. Lang. Technol. **5**(1), 1–167 (2012)
4. Phillips, C.: Millennials Tech-Dependent, But Not Necessarily Tech-Savvy. Millennial Marketing Insights. http://millennialmarketing.com/2010/04/millennials-tech-dependent-but-not-necessarily-tech-savvy/
5. Han, J., Pei, J., Kamber, M.: Data Mining: Concepts and Techniques. Elsevier, Amsterdam (2011)

6. Hakami, S. A. A.: The importance of understanding Emoji: an investigative study. In: Research Topics in HCI, pp. 1–20 (2017)
7. Miller, H., Thebault-Spieker, J., Chang, S., Johnson, I., Terveen, L., Hecht, B.: Blissfully "happy" or "ready to fight": varying interpretations of Emoji. In: Proceedings of ICWSM (2016)
8. Seyednezhad, S.M., Menezes, R.: Understanding subject-based emoji usage using network science. In: Workshop on Complex Networks CompleNet, pp. 151–159. Springer, Cham, March 2017
9. Mjhool, A.Y.: Evaluation of user perceived QoE in mobile systems using social media analytics. Doctoral dissertation (2016)
10. Bumgarner, B. A.: You have been poked: exploring the uses and gratifications of Facebook among emerging adults. First Monday **12**(11) (2007). https://doi.org/10.5210/fm.v12i11.2026
11. Sheldon, P.: The relationship between unwillingness-to-communicate and students' Facebook use. J. Media Psychol.: Theories Methods Appl. **20**(2), 67 (2008). https://doi.org/10.1027/1864-1105.20.2.67
12. Urista, M.A., Dong, Q., Day, K.D.: Explaining why young adults use MySpace and Facebook through uses and gratifications theory. Hum. Commun. **12**(2), 215–229 (2009)
13. Hew, K.F., Cheung, W.S.: Use of Facebook: a case study of Singapore students' experience. Asia Pac. J. Educ. **32**(2), 181–196 (2012). https://doi.org/10.1080/02188791.2012.685560
14. Pai, P., Arnott, D.C.: User adoption of social networking sites: eliciting uses and gratifications through a means-end approach. Comput. Hum. Behav. **29**(3), 1039–1053 (2013). https://doi.org/10.1016/j.chb.2012.06.025
15. Alhabash, S., Chiang, Y.H., Huang, K.: MAM & U&G in Taiwan: differences in the uses and gratifications of Facebook as a function of motivational reactivity. Comput. Hum. Behav. **35**, 423–430 (2014). https://doi.org/10.1016/j.chb.2014.03.033
16. Salem, F., Mourtada, R.: Social media in the Arab world: influencing societal and cultural change. Arab Soc. Media Rep. **2**(1), 1–29 (2012)
17. Solberg, C.A.: Culture and industrial buyer behavior: the Arab experience. In: 18th IMP Conference (2002). http://impgroup.org/uploads/papers/522.pdf
18. Miladi, N.: New media and the Arab revolution: citizen reporters and social activism. J. Arab Muslim Media Res. **4**(2/3), 3–5 (2011). http://www.intellectbooks.co.uk/journals/view-jour
19. Aljasir, S., Bajnaid, A., Elyas, T., Alnawasrah, M.: University students usage of Facebook: the case of obtained gratifications and typology of its users. J. Manag. Strategy **8**(5), 30 (2017). https://doi.org/10.5430/jms.v8n5p30
20. Arab Social Media Report: A Report on Facebook Usage: Factors and Analysis, vol. 1, no. 1 (2011). http://unpan1.un.org/intradoc/groups/public/documents/dsg/unpan044212.pdf. Accessed 17 Sept 2016
21. Ahmari, Ali bin Said, the purpose of the use of university students to social networking sites: a field study on students University of Imam Muhammad bin Saud Islamic University of Imam Muhammad bin Saud Islamic University, Unpublished MA Thesis, 1432 AH. https://doi.org/10.13189/ujp.2015.030302
22. Alsanie, S.I.: Social media (Facebook, Twitter, WhatsApp) used, and it's relationship with the university students contact with their families in Saudi Arabia. Universal Journal of Psychology **3**(3), 69–72 (2015). https://doi.org/10.13189/ujp.2015.030302
23. https://www.shamra.sy/academia/download_document/document_cf5588e58ae3cc4deb18d1d0c87130ab.pdf. http://www.statisticbrain.com/twitter-statistics.5.1.2014.s
24. Davis, M., Edberg, P.: Unicode Emoji, November 2016. http://unicode.org/reports/tr51/

25. Stark, L., Crawford, K.: The conservatism of Emoji: work, affect, and communication. Soc. Media + Soc. **1**(2), 2056305115604853 (2015). https://doi.org/10.1177/2056305115604853
26. Taboada, M., Brooke, J., Tofiloski, M., Voll, K., Stede, M.: Lexicon-based methods for sentiment analysis. Comput. Linguist. **37**(2), 267–307 (2011). https://doi.org/10.1162/COLI_a_00049
27. Preisendorfer, M., Sengupta, S., White, J., Tekeoglu, A.: Social media Emoji analysis, correlations and trust modeling. Doctoral dissertation (2018). http://hdl.handle.net/1951/69574
28. Alsanie, S.I.: Social media (Facebook, Twitter, WhatsApp) used, and it's relationship with the university students contact with their families in Saudi Arabia. Univ. J. Psychol. **3**(3), 69–72 (2015). https://doi.org/10.13189/ujp.2015.030302
29. Abdulla, N.A., Ahmed, N.A., Shehab, M.A., Al-Ayyoub, M., Al-Kabi, M.N., Al-rifai, S.: Towards improving the lexicon-based approach for arabic sentiment analysis. Int. J. Inf. Technol. Web Eng. (IJITWE) **9**(3), 55–71 (2014). https://doi.org/10.4018/ijitwe.2014070104
30. Alshenqeeti, H.: Are emojis creating a new or old visual language for new generations? A socio-semiotic study. Adv. Lang. Literary Stud. **7**(6), 56–69 (2016). http://journals.aiac.org.au/index.php/alls/article/view/2823
31. Hajjous, M.: Opinions analysis in Tweeter. J. Alba'ath Univ. **38**(41), 31–63 (2016)
32. He, Y., Zhou, D.: Self-training from labelled features for sentiment analysis. Inf. Process. Manag. **47**(4), 606–616 (2011). https://doi.org/10.1016/j.ipm.2010.11.003
33. Read, J., Carroll, J.: Weakly supervised techniques for domain-independent sentiment classification. In: The 1st International CIKM Workshop on Topic-Sentiment Analysis for Mass Opinion, pp. 45–52. ACM (2009). https://doi.org/10.1145/1651461.1651470
34. Aleqabie, H.J., Safoq, M.S., Raoof, I., Alsabah, R., Abd, E.H.: Sentiment analysis for movie reviews using deep learning with semantic orientation. In: 8th International Conference on Applied Science and Technology, ICAST 2020, April 2020

Diabetes Analysis and Risk Calculation – Auto Rebuild Model by Using Flask API

Akkem Yaganteeswarudu(✉) and Prabhakar Dasari

Pyramid Softsol Pvt Limited, Hyderabad, India
yaganteeswaritexpert@gmail.com,
prabhakar.dasari@infosharesystems.com

Abstract. The existing diabetes analysis techniques are considering few parameters like age, sex, BMI, insulin, glucose, blood pressure, diabetes pedigree function, pregnancies. But in this paper, we considered in addition to the basic parameters, we also included serum creatinine, potassium, GlasgowComaScale, heart rate/pulse rate, respiration rate, body temperature, low-density lipoprotein (LDL), high-density lipoprotein (HDL), TG (Triglycerides). Our paper includes analysis of Pima Indian diabetes datasets which are available in UC Irvine (UCI) machine learning repository, the data set which was acquired from a hospital in Frankfurt, Germany, and also visited some local hospitals to get Data sets for a diabetes analysis. Since the analysis includes all parameters which cause diabetes, it may also help to detect diseases like heart disease, neuropathy, retinopathy, hearing loss, and dementia. The main aim of this research work is to analyze the datasets by using different machine learning algorithms along with parameter tuning. The proposed research work also analyzes diabetes risk factors and based on the risk factor, it provides suggestions to the patients. The importance of this paper is continuously monitoring the model if model accuracy reduces, it automatically rebuilds the model and suggests the best model. We prepared the Flask API (Application Programming Interface) to consume this model, which is primarily integrated into the front-end. So that if a user calls this API it will display the diabetes status of the patients. We have collected live samples of datasets from apple phone readings and reached many hospitals for live data sets and in each time if the accuracy reduces in the current model it will automatically reload and considers the model that produces better accuracy.

Keywords: Flask · API · Dataset · Lipoprotein · Triglycerides · GlasgowComaScale

1 Introduction

In the Diabetic analysis process we used python language because python providing a lot of libraries like NumPy, pandas, Scikit learns, Matplotlib, and so on. Import the required libraries like pandas, NumPy, seaborn, matplotlib. From sklearn import train_test_split, GradientBoostingClassifier, accuracy_score, roc_auc_score, roc_curve, AUC (Area under the Curve), SelectKBest, chi2, RandomForestClassifier and so on. Not only machine learning algorithms, but to be more confident in the results produced

J. I.-Z. Chen et al. (Eds.): ICIPCN 2020, AISC 1200, pp. 299–308, 2021.
https://doi.org/10.1007/978-3-030-51859-2_27

deep learning techniques also implemented. For deep learning, we used Tensorflow Library from python. Once users got comma-separated value (CSV) file format or excel file format pandas read_csv or read_excel method is used to read CSV or excel file. Sample code: diabetes_df=pd.read_csv ("diabetesIndia.csv") #This code will read diabetes [1] CSV file and creates a DataFrame.

Initially, we used the Pima Indian Diabetes [PID] data set for analysis, and later we collected data set which consists of all parameters which will find the type of diabetes and also which will help us to predict future diseases which will cause from diabetes. In addition to the Pima Indian Diabetes [PID] data set, we used the data set acquired from a hospital in Frankfurt, Germany. In the existing system, few parameters are used which are mainly used to analyze whether the patient has diabetes are not. But in this paper, we considered maximum parameters which will cause diabetes-like age, sex, BMI, insulin, glucose, blood pressure, diabetes pedigree function, pregnancies, serum creatinine, potassium [2], GlasgowComaScale, heart rate/pulse rate, respiration rate, body temperature, low-density lipoprotein (LDL), high-density lipoprotein (HDL), TG (Triglycerides). Due to including all parameters which cause diabetes, it is possible to identify other diseases like heart disease, neuropathy, retinopathy, hearing loss, and dementia.

When analyzed previous analysis of diabetes there is no auto rebuild of the model. But whenever we have new data set there may be chances that the existing model may produce less performance than other models. In this case, whenever we will get a new data set, it is better to reconstruct the model and select the best model among the existing models. In this paper, we used the python method to reload the python script whenever a new data set arrives. This article aims to produce better results for the patients so that they can prevent the diseases which are occurring due to diabetes.

2 Dataset Pre-processing Include Test Set and Train Set Preparation

Initially, Pima Indian Diabetes [PID] data set is used in model development. Features in the dataset are pregnancies (number of times pregnant), glucose level [3], blood pressure (diastolic blood pressure (mm Hg)), Skin thickness, insulin, BMI (Body Mass Index), diabetes pedigree function, age, outcome (class variable (0 or 1)). Later the model is tested with practice fusion diabetes data set with 9946 patient's details, data set which was acquired from a hospital in Frankfurt, Germany, and also live data collected from various hospitals. Live data sets collected from Hospitals includes a lot of additional parameters like serum creatinine, potassium, Glasgow coma scale (GCS), heart rate/pulse rate, respiration rate, body temperature, low-density lipoprotein (LDL), high-density lipoprotein (HDL), TG (Triglycerides). GCS score [16] calculated for 15 by using different parameter values like best eye response (1 to 4), best verbal response (1 to 5), best motor response (1 to 6). GCS score 3 will be the worst and 15 will be best. Like this LDL, HDL, respiration rate [4], potassium, creatinine, and so on measured based on standards specified by the corresponding doctors. To collect the standards visited a lot of hospitals [15] and consulted doctors and considered normal and abnormal readings for each feature.

From the given dataset prepared Trainset and Test set as per industry standards. From the sklearn library used the train_test_split method to divide the data into 70% train and 30% test data or based on the required data can be split [14]. Separated features (Inputs) and Labels (outputs). Sample code is given below.

Sample code: diabetes_features = diabetes_df.iloc [: 0:8]; diabetes_label = diabetes_df.iloc [: 8]; diabetes_feature_train, diabetes_feature_test, diabetes_label_train, diabetes_label _test = train_test_split (diabetes_features, diabetes_label, test_size=0.3, random_state=0).

3 Dataflow Diagram for the Suggested Model

3.1 Existing Model

When the existing diabetes analysis models analyzed and observed that once the model building is done, there is no provision to auto rebuild. Henceforth, if accuracy reduces, manually the developer has to check and rebuild the model again. All existing systems are mainly considering only a few parameters like age, BMI, bp, insulin, diabetes pedigree function due to that it is not possible to detect any other diseases which are caused due to diabetes.

3.2 Proposed Model

Figure 1 represents the proposed model.

The importance of the suggested model is to build a flask API to consume the model and auto rebuild model whenever model performance decreases, then the model will be auto rebuild to select the best model for analysis [5]. The model used to calculate risk by using logistic regression. Once risk is calculated patient is advised with different doctor suggestions based on risk factors [13].

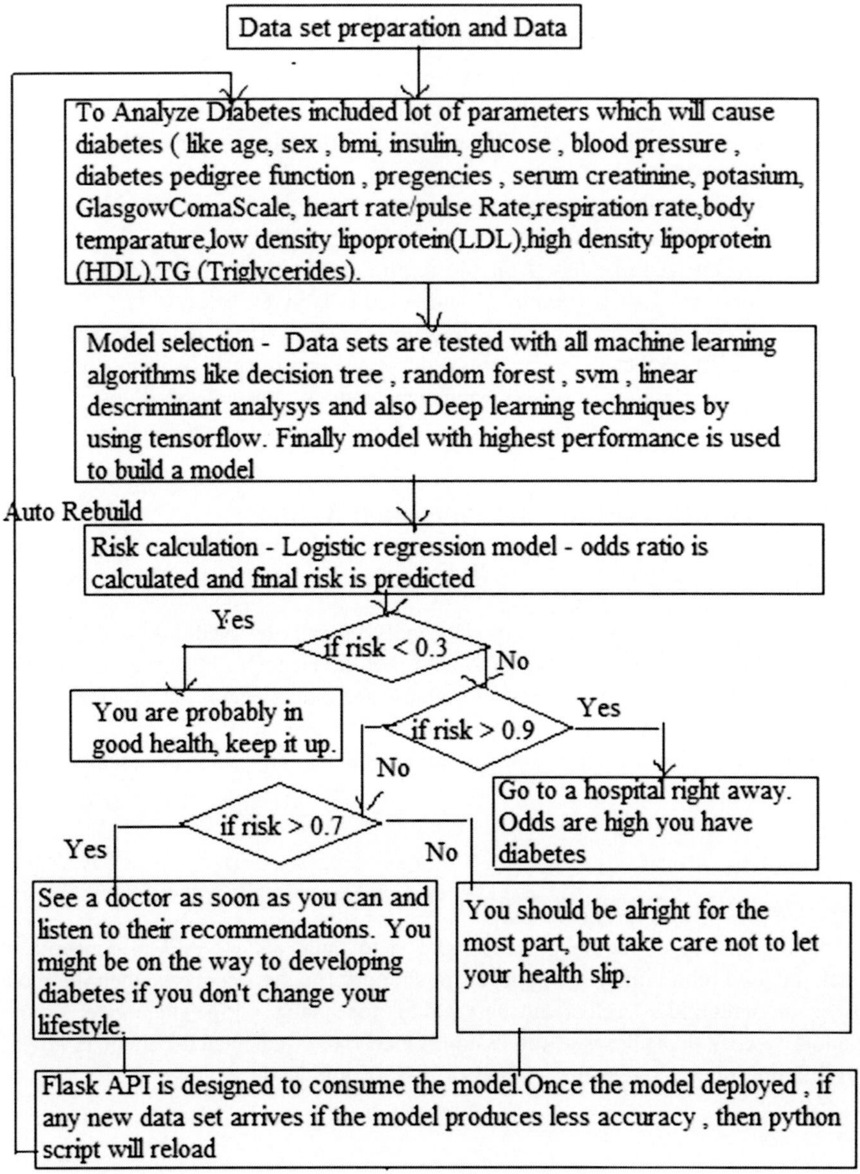

Fig. 1. Data flow in the proposed model

4 Model Selection and Construction

4.1 Machine Learning

Model selection or algorithm selection phase is the most exciting and the heart of machine learning [4]. Various algorithms are performed and accuracy is verified in each algorithm. Initially, model selection algorithms are applied in Pima Indian diabetes and later all these techniques [6] are applied to practice fusion data set to select the appropriate model.

RandomForest: Create a random forest classifier instance. Sample code: rmfr = RandomForestClassifier (). Fit the features and labels by using the fit method. Sample code: rmfr.fit (diabetes_feature_train, diabetes_label_train). Predict future values by using label_pred = rmfr.predict (diabetes_features_test). And after the execution of the above python code, and achieved accuracy is 0.75229. Different Parameters Tuning in RFC: n_estimators [7] which will tune the number of trees in the forest. When n_estimators are more, usually the model will give more accuracy. But adding more trees also affect model performance, which will slow down the accuracy Consider sample n_estimators as [1, 2, 4, 8, 16, 32, 64, 100, and 200]. Random forest classifier repeated for the above number of estimators. AUC curve is drawn for the different number of trees and finally, the tree that gives more accuracy will be selected. Another parameter max_depth tuning in random forest which will define max depth of which will capture more information Sample code: max_depth = np.linspace (1, 32, 32, endpoint=True). min_samples_split (minimum number of samples to split an internal node), max_features in the random forest are tuned and the AUC curve drawn for train and test set and tree with maximum accuracy is selected.

DecisionTreeClassifier: From sklearn import DecisionTreeClassifier and create an instance for DecisionTreeClassifier like diabetes_df = DecisionTreeClassifier (). Train the model by using a fit method. Sample code: diabetes_df. Fit (diabetes_features_train, diabetes_labels_train). After training accuracy achieved is 0.6605504587155964. Parameters tuned in the decision tree are, the criterion by default it is gini, we can change it to entropy, max_depth, minimum_samples_split. AUC curve is drawn for trainset and test set and observed the performance after tuning.

XGBClassifier: Create an object for XGBClassifier like xgb = XGBClassifier (). Train the model with the train data [1]. The following accuracy is achieved 0.76147. Perform logistic Regression. Create an object for LogisticRegression like logmodel = LogisticRegression (). After training the model and achieved the accuracy is 0.786. Evaluate the training and test set by using KNeighborsClassifier. First Import KNeighborsClassifier and create an object like knn = KNeighborsClassifier (n_neighbors=1). After Training accuracy is 0.72727.

Create an object for SVM (Support Vector Machine). Sample code: sv = svm.SVC (kernel='linear') and train the model accuracy achieved is 0.87013. And tune Parameters in svm: Change Kernel to "rbf" Example: sv = svm.SVC (kernel='rbf') and the accuracy achieved is 0.85714. Change Kernel to "poly" like sv = svm.SVC (kernel='poly') and Train the model following accuracy is achieved is 0.80519. And also in SVM gamma and parameter C is tuned and constructed AUC curve and observed graph. GradientBoostingClassifier: Import gradient boosting classifier and create an object for GradientBoostingClassifier and LinearDiscriminant Analysis are performed and accuracy is measured.

4.2 Deep Learning

A neural network with hidden layers constructed with the data set collected from the live data set. Measured accuracy resulted in 0.7214. By using a Tensorflow convolution neural network [12] with convolution layers and pooling layers are constructed. For Example in max-pooling patients' values are compared and the patient with abnormal pools are constructed, filters are applied. Final accuracy achieved is 0.7345.

5 Risk Factor Calculation

5.1 Risk Calculation Process

Here a list of steps followed are described

- Written a diabetes risk function which will take parameters like glucose, BP, SkinThickness, insulin, BMI, and age
- Scikit learn logistic regression model is used for risk calculation
- Logistic regression, predict_proba method will return the probability for diabetes or no diabetes
- In theoretical, 1 means diabetes and 0 means no diabetes
- So predict_proba method will return probability for diabetes by considering all the parameters
- Based on the probability of risk value (that is having diabetes) the patient is warned

  If risk is less < 3 patient is having good health
  If risk is > 0.9 suggested that go to the hospital right away
  If risk > 0.7 suggested that take the recommendations from the doctor.

5.2 Sample Code for Risk Calculation

```
def diabetes_risk_prediction (glucose, BP, skin thickness, insulin, BMI, age):

indicator_list = [glucose, BP, skin thickness, insulin, BMI, age]
predictions = log_model.predict_proba (np.array (indicator_list).reshape (1, −1))
# logistic regression method is used
Risk = predictions [0,1]
if risk < 0.3: the patient is having good health
if risk > 0.9: Go to a hospital right away
if risk > 0.7: See a doctor as soon as you can and listen to their recommendations.
    you might be on the way to developing diabetes if you don't change your lifestyle.
```

6 Final Model Development

After Applying machine learning algorithms and deep learning techniques [8] it is observed that SVM achieves maximum accuracy. Construct a Diabetes model (Diabetesmodel.py). Import necessary libraries and read CSV file and divide the train_test_split [1] and since SVM giving more accuracy so create an object for SVM and train the model. Python pickle [8] is used to save the model behavior in a file Sample code: sv = svm. SVC (kernel='linear'); sv.fit (x_train, y_train); Pred = sv. predict (x_test); p.dump (sv, open ('final_diabetes.pickle', 'wb')).

Construct Flask API (diabetesapi.py): To consume the model which already built, construct Flask API [9], Import required flask methods to write Flask API and read the behavior of model from pickle file. Develop diabetes main page: Develop HTML page for required controls to enter input values for insulin, BMI, age, etc.

Sample code:

```
@app.route ("/diabetesprediction", methods= ['POST']); def predictdiabetes ():
    Prediction=np.array2string (model.predict (data));@app.route ('/')
def index():
    return render_template("diabetesmainpage.html")
if __name__=='__main__':    Load pickle file
```

7 Automatic Reloading of Model

The main aspect of this paper is to reload a model when accuracy reduced. Initially, the model constructed with Pima Indian Diabetes dataset [2]. SVM achieved better accuracy when compared with other algorithms [10]. Whenever we collected live datasets re-built the model with the new data and again the model accuracy tested against a new data set. Python provides an automatic reload of the module. The model which built is constructed as a module and reload function is applied if the model produces accuracy less than the specified value. The automatic reload of the model tested with various inputs of data [11]. The first model tested with Pima Indian diabetes data set and later tested with practice fusion data set and also many live samples of data sets collected from many hospitals. Reload module is working fine whenever accuracy got changed in the current model deployed.

8 Experiments and Results

Initially, a model built with Pima Indian Diabetes data set and data set acquired from a hospital in Frankfurt, Germany. The Model tested for train data and test data. All the results are documented and tested for accuracy.

Initially, data sets are pre-processed and all classification machine learning algorithms are applied and Deep learning techniques are applied to know the performance of the models. Table 1 represents the performance of the different algorithm's execution on the data set.

Table 1. Results of machine learning algorithm's accuracy

Algorithm	Accuracy
GradientBoostingClassifier	0.85
RandomForestClassifier	0.77
XGBClassifier	0.79
LogisticRegression	0.80
KNeighborsClassifier	0.72
KNeighborsClassifier Tuned with 16 neighbors	0.84
SVM	0.92
LinearDiscriminantAnalysis	0.88
Tensorflow	0.79

From the above Table 1, SVM is having the highest accuracy when compared with other algorithms. So the model is designed with SVM to predict future data.

Figure 2 shows the input fields to be entered by patients.

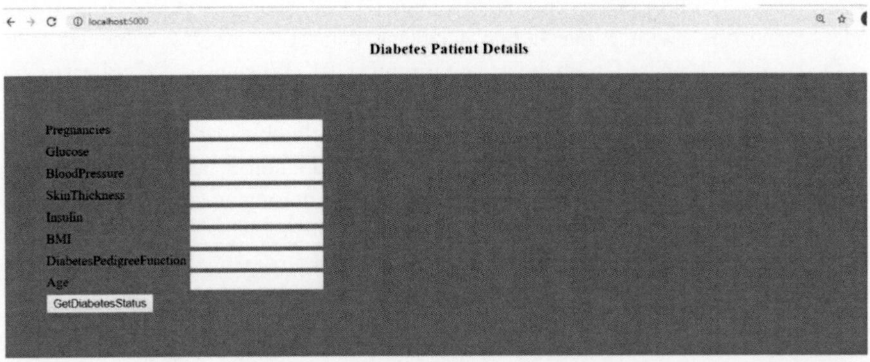

Fig. 2. Diabetic patient details input screen

Figure 3 shows the sample input entered by the user. Once the user enters all the details user can click on get diabetic status to know the status of the patient along with risk value.

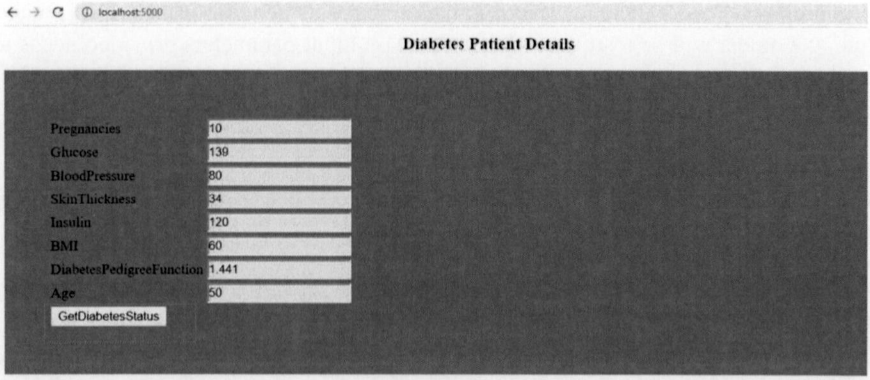

Fig. 3. User inputs

Figure 4 shows once the user enters the details of the patient, it will display diabetic status, whether the patient is affected with diabetes or not. And also risk analysis. If the patient with the highest risk factor will be suggested to go to the hospital right away and also it will display the percentage of risk.

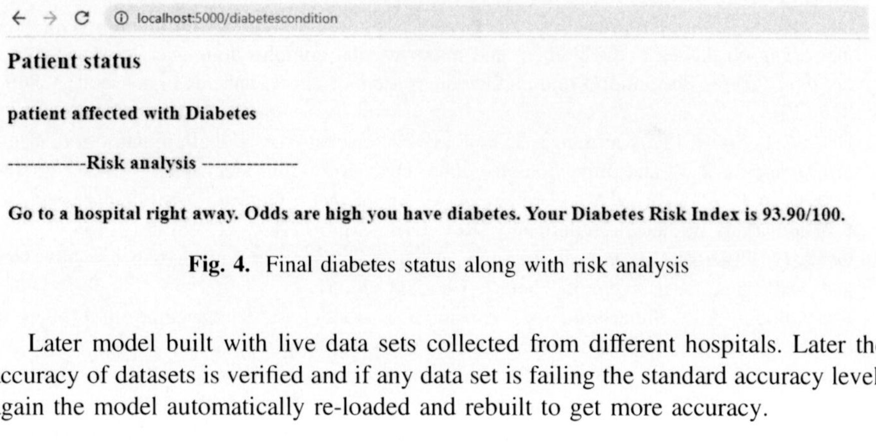

Fig. 4. Final diabetes status along with risk analysis

Later model built with live data sets collected from different hospitals. Later the accuracy of datasets is verified and if any data set is failing the standard accuracy level, again the model automatically re-loaded and rebuilt to get more accuracy.

9 Conclusion

The importance of the proposed system is to consider all the parameters which will cause diabetes. Due to considering all the parameters which cause diabetes, it is possible to detect other diseases that are caused due to diabetes. One more important feature of the proposed model is to auto rebuild the model so that find the best model which gives more accuracy whenever data set changes. The current analysis limited to calculate the risk by using logistic regression and in future risk can be calculated in other ways like by consulting doctors based on parameter standard values if any abnormality observed then the risk is calculated and we can warn the patient. Currently in the proposed model includes risk calculation and auto rebuild of the model, but in future, the model will be restructured that based on input parameters provided needs to identify any chance of occurring the disease like heart disease, neuropathy, retinopathy, hearing loss, and dementia due to diabetes so that we can reduce mortality ratio.

Acknowledgment. Thanks to Prabhakar Dasari, Vice president and Manager, Info share systems Pvt Ltd, for support and guidance. Throughout my research, Prabhakar sir helped me in various ways like collecting data sets, model selection, and model building.

References

1. Glechner, A., Harreiter, J., Gartlehner, G., et al.: Sex-specific differences in diabetes prevention: a systematic review and meta-analysis. Diabetologia **58**(2), 242–254 (2015)
2. Li, H., Li, W., Gu, Y., et al.: Comparison of continual insulin or secretagogue treatment in type 2 diabetic patients with alternate insulin-secretagogue administration. Diab. Res. Clin. Pract. **84**(2), 158–162 (2009)
3. American Diabetes Association: Microvascular complications and foot care. Diab. Care **42** (Suppl 1), S124–S138 (2019)
4. UK Prospective Diabetes Study (UKPDS) Group: Effect of intensive blood-glucose control with metformin on complications in overweight patients with type 2 diabetes (UKPDS 34). Lancet **352**, 854–865 (1998)
5. Diabetes Prevention Program Research Group: Long-term effects of lifestyle intervention or metformin on diabetes development and microvascular complications over 15-year follow-up: the Diabetes Prevention Program Outcomes Study. Lancet Diabetes Endocrinol. **3**, 866–875 (2015)
6. Cho, N.H., Shaw, J.E., Karuranga, S., et al.: IDF Diabetes Atlas: global estimates of diabetes prevalence for 2017 and projections for 2045. Diab. Res. Clin. Pract. **138**, 271–281 (2018)
7. Jenkins, A.J., Joglekar, M.V., Hardikar, A.A., Keech, A.C., O'Neal, D.N., Januszewski, A. S.: Biomarkers in diabetic retinopathy. Rev. Diab. Stud. **12**(1–2), 159–195 (2015)
8. Geiss, L., Engelgau, M., Pogach, L., et al.: A national progress report on diabetes: successes and challenges. Diab. Technol. Ther. **7**, 198–203 (2005)
9. Rosenbloom, A.L., Silverstein, J.H., Amemiya, S., Zeitler, P., Klingensmith, G.J.: Type 2 diabetes in children and adolescents. Pediatr. Diab. **10**(Suppl 12), 17–32 (2009)
10. Cakan, N., Kizilbash, S., Kamat, D.: Changing spectrum of diabetes mellitus in children: challenges with initial classification. Clin. Pediatr. (Phila) **51**, 939–944 (2012)
11. Wilkin, T.J.: The accelerator hypothesis: a review of the evidence for insulin resistance as the basis for type I as well as type II diabetes. Int. J. Obes. (Lond.) **33**, 716–726 (2009)
12. Pradhan, A.D., Rifai, N., Buring, J.E., Ridker, P.M.: Hemoglobin A1c predicts diabetes but not cardiovascular disease in nondiabetic women. Am. J. Med. **120**, 720–727 (2007)
13. Edelman, D., Olsen, M.K., Dudley, T.K., Harris, A.C., Oddone, E.Z.: Utility of hemoglobin A1c in predicting diabetes risk. J. Gen. Intern. Med. **19**, 1175–1180 (2004)
14. Sola, D., Rossi, L., Schianca, G.P., et al.: Sulfonylureas and their use in clinical practice. Arch. Med. Sci. **11**, 840–848 (2015)
15. Orchard, T.J.: The effect of rosiglitazone on overweight subjects with type 1 diabetes. Diabetes Care **29**, 746–747 (2006)
16. Henry, R.R., Thakkar, P., Tong, C., Polidori, D., Alba, M.: Efficacy and safety of Canagliflozin, a sodium-glucose cotransporter 2 inhibitor, as an add-on to insulin in patients with type 1 diabetes. Diab. Care **38**, 2258–2265 (2015)

Impact of Class Imbalance on Convolutional Neural Network Training in Multi-class Problems

Ahmad Ilham[1], Jesus Silva[2(✉)], Nohora Mercado-Caruso[3], Donato Tapias-Ruiz[3], and Omar Bonerge Pineda Lezama[4]

[1] Department of Informatics, Universitas Muhammadiyah Semarang, Semarang 50254, Indonesia
ahmadilham@unimus.ac.id

[2] Universidad Peruana de Ciencias Aplicadas, Lima, Peru
jesussilvaUPC@gmail.com

[3] Universidad de la Costa, St. 58 #66, Barranquilla, Atlántico, Colombia
{nmercadol, dtapias}@cuc.edu.co

[4] Universidad Tecnológica Centroamericana (UNITEC), San Pedro Sula, Honduras
omarpineda@unitec.edu

Abstract. Image classification is the process of assigning an image one or multiple tags that describe its content. To perform the classification, a model must be designed for learning the labels to be assigned to a given image. The assignment is made through a learning process that uses a set of previously labeled training images, which must be large enough to guarantee efficient training. Many approaches have been researched to find optimal solutions to classification problems, however, databases with large amounts of images and the increased processing power of GPUs have made convolutional neural networks (CNNs) the best choice, as they outperform traditional algorithms. This paper presents a systematic analysis aimed at understanding how the issue of class inequality affects the efficiency of a convolutionary neural network trained for a task of image classification, and presents a technique for correcting the overtraining and that the network generalization.

Keywords: Impact of class imbalance · Convolutional neural network · Multi-class problems

1 Introduction

Currently, there is a variety of publicly available tagged image databases that are used for developing and comparing classification models with CNN. Some of these reference sets contain the same number of images per class and are known as balanced data sets [1]. However, most of these datasets contain an unequal number of images and are known as imbalanced data sets. Dealing with class imbalance is a great challenge because, in these sets, examples from some classes significantly outnumber examples from other classes. Furthermore, minority classes are of greatest interest (from a

J. I.-Z. Chen et al. (Eds.): ICIPCN 2020, AISC 1200, pp. 309–318, 2021.
https://doi.org/10.1007/978-3-030-51859-2_28

learning point of view) because they represent rare cases or because acquiring data from these examples is costly. In [2] and [3] the authors state that class imbalance can have a significant detrimental effect on the training of CNN-based classifiers. To alleviate this effect, several proposals have emerged in the last three years for adapting traditional techniques to combat data imbalance in the context of deep learning. These techniques are divided into three groups: (a) data level sampling, where the training set is modified to decrease the majority class examples [4–6] or increase the minority class examples [7]; (b) algorithmic level modification, which aims to modify the deep learning algorithm at the target function level [4, 8, 9]; and (c) cost-sensitive, which aims to further penalize errors made in the minority classes [10, 11]. The simplest approach in data-level sampling solutions focused on the majority class is to randomly eliminate examples from the majority class, but some techniques select examples to be eliminated in a way that minimizes information loss [8]. In the case of sampling centered on the minority class, the simplest approach is to randomly duplicate examples, however, there are more intelligent solutions that generate new synthetic examples by interpolating between pre-existing positive examples that are very close together (SMOTE) [12] and solutions where examples are synthesized in an attempt not to alter the distribution of classes [13]. Although random sampling techniques are easy to implement and considerably reduce the level of imbalance, they can lead to an over-adjustment of the classification model. In the case of intelligent techniques for generating or eliminating examples, these are generally complex and computationally expensive [14].

In this paper, a systematic study is presented to analyze the impact of class imbalance on the classification performance of a CNN. For the analysis, databases with different imbalance ratios are generated using a sub-sampling method that eliminates random samples from the minority classes. Also, the presented methodology consists of the application of a series of state-of-the-art techniques and algorithms in a proposed order to correct the over-training caused by random sampling and increase the generalization of the network.

2 Methods

This section provides the details of the method implemented for class imbalance, describes the learning algorithm selected for this study and its training parameters, describes the over-adjustment phenomenon and the regularization process to avoid it, finally, the section describes the experiments performed.

2.1 Class Imbalance

Class imbalance can take many forms, particularly in the context of multi-class classification, typical of CNN. In [4], the imbalance is defined in steps, in which the number of examples is equal within the majority and minority classes, but differs between the majority and minority classes. This type of imbalance is characterized by two parameters: the fraction of the minority classes defined in Eq. 1 and the ratio of the imbalance between the number of examples in the majority classes and the number of examples in the minority classes defined in Eq. 2 [15–17].

$$\varphi = \frac{\{iI\{1, \ldots N\} : C_1 \, es \, minoritario\}}{N} \tag{1}$$

$$\rho = \frac{max_i\{|C_i|\}}{min_i\{|C_i|\}} \tag{2}$$

Where Ci is a set of examples in class I, and N is the total number of classes. In this study, the Caltec101 data set [9] is used, which consists of 9,144 high-resolution color images belonging to 103 categories. Each category contains 31 to 800 images, and most categories have about 50 images. The size of each image is approximately 300 × 200 pixels. The 5 smallest classes (minority classes) and the 5 largest classes (majority classes) were selected to form a subset with 1973 images. The subset was divided into 4 sets with class imbalance, using the parameter μ = 0.5, and different relations of ρ (1:1, 3:1, 7:1, 15:1) to implement the imbalance by steps [18]. For the 1:1 ratio, the number of images per class is standardized to the class with the least number of images to form a set of 320 images. This is the base set for the formation of the different imbalance ratios. For the 3:1 ratio, 16 images are removed from each of the minority classes and 16 images are added to each of the majority classes. For the 7:1 ratio, 8 of the remaining 16 images are removed and added to the minority and majority classes respectively; for the 15:1 ratio, the same procedure was followed (Table 1). Standardization, removal, and addition of images were performed randomly.

Table 1. Summary of the sets with image imbalance.

Clase	Num. De imágenes x clase	Relación de desequilibrio			
		1:1	3:1	7:1	15:1
Aviones	900	33	51	59	62
Caras	521	33	51	59	62
Reloj	301	33	51	59	62
Leopardos	285	33	51	59	62
Bonsai	132	33	51	59	62
Lirio de agua	41	33	18	10	6
Fresa	36	33	18	10	6
Gato montés	33	33	18	10	6
Binocular	31	33	18	10	6
Metrónomo	31	33	18	10	6

2.2 Learning Algorithm

A CNN was used for the classification which has a depth of 6 layers, 4 convolution layers, and 2 fully connected layers. The convolution layers handle 32 to 512 filters ranging in size from 7 × 7 to 3 × 3. A ReLu unit is applied to the output of each convolution layer and fully connected layer. A maximum pooling layer is applied to the output of each layer of ReLu in each convolution layer. The output of the second fully connected layer is fed to a Sofmax that produces a distribution on the labels of the *NN* classes (see Table 2) [19–21].

Table 2. CNN proposal.

Capa	Dimensión			Kernel	Paso	Relleno
	Ancho	Alto	Prof			
0	230	230	4	–	–	–
	115	115	33	8	1	–
1	115	115	33	–	–	–
	61	61	33	5	2	–
	61	61	68	5	2	2
2	61	61	68	–	–	–
	30	30	68	4	2	–
	30	30	132	4	1	3
3	30	30	132	–	–	–
	15	15	132	4	2	–
	15	15	132	4	1	1
4	15	15	132	–	–	–
	8	8	132	–	–	–
5	2	2	516	–	–	–
	1	1	516	–	–	–
	2	2	NC	–	–	–
6	1	1	NC			
	2	2	NC	–	–	–

CNN was trained using stochastic gradient descent with a batch size of 10, a moment of 0.9, and a decrease in weights (regularization factor) of 0.0005. The initial weights in each of the layers were initialized with a Gaussian distribution with a mean of 0 and a standard deviation of 0.01. The activation thresholds in each of the layers were initialized to zero. The number of training epochs was set at 50. It started with a learning rate of 0.001 which decreased by a factor of 10 after every 10 epochs. CNN was trained on an NVIDIA GeForce GTX 980 M GPU with 1536 cores and 4.3 GB of memory, using the MatLab 2018a Neural Network Toolbox [22–24].

2.3 Over-Adjustment

When training a CNN, there is a compromise between optimization (accuracy) and generalization of the network. Optimization refers to the process of adjusting the network to get the best possible performance from the training data, while generalization refers to how well the trained network is performing with data it has never seen. In machine learning, over-adjustment is defined as the effect of over-training a learning algorithm with the data used for training. During over-training the optimization increases while the generalization of the network worsens indicating that the network has been adjusted to very specific characteristics of the training data [25].

The process to fight CNN overfitting is known as regularization and includes techniques such as:

- To find the appropriate network size for the data, different architectures must be evaluated, starting with few layers and parameters, and increasing the size of the layers or adding new layers to the network until finding diminishing returns concerning the loss of validation [26].
- Indeed, this regularization encourages small and scattered weight distributions over large values concentrated in only a few neurons.
- Dropout: It is a technique in which the outputs of some neurons, randomly selected, are set to zero so that they do not contribute to the forward step and do not participate in the backward propagation of the training algorithm; which reduces complex co-adaptations of neurons, generated in the standard learning of back-propagation in neural networks. The selected neurons together with their incoming and outgoing connections are temporarily marginalized, which is equivalent to sampling a "reduced" network from the original network; which is formed with the n neurons that survived the marginalized or isolation. The network with n units can be seen as a collection of 2 *nn* possible reduced neural networks, which share weights so that the total number of parameters is less. In this way, each time a training example is presented, the neural network samples and trains a different architecture.
- Increased data: Training a CNN with a large amount of data is the key to increasing the generalization of the network. However, in practice, the amount of data available is limited, which may impede the proper development of the model. To avoid this problem, it is necessary to create false data and add it to the training set, a technique known as data augmentation. The simplest approaches to increasing the data include adding noise and applying random transformations to existing data. In image classification, Gaussian noise can be added and simple (random) transformations can be applied to images, including flipping (horizontally and vertically), rotation, scaling (image size), translation (along the X or Y direction), etc. Increasing data can increase the size of the training set 10 times or more, making the model more robust (and avoiding overfitting) [27].

2.4 Experiments

For evaluating CNN's performance, the network was trained from scratch on each of the imbalance ratios. For the tests, each database was divided into training and test sets. The training set was formed with 70% of the database images and the remaining 30% was used to form the test set. Three tests were performed to test the correction of the CNN overlay. In the first one, CNN was used without any regularization to observe the over-adjustment. In the second one, weight regularization, dropout, and data augmentation were added to CNN to see their effect in reducing or eliminating the over-adjustment. Finally, to the network implemented in test two, normalization was added in different layers to increase the generalization of the network. In these tests, the parameters for data enhancement were: rotation range of −20 to 20°, random reflection in the left-right direction, horizontal and vertical translation range of −3, and 3 pixels respectively. The weight regularization rate was set at 0.0005 [28, 29].

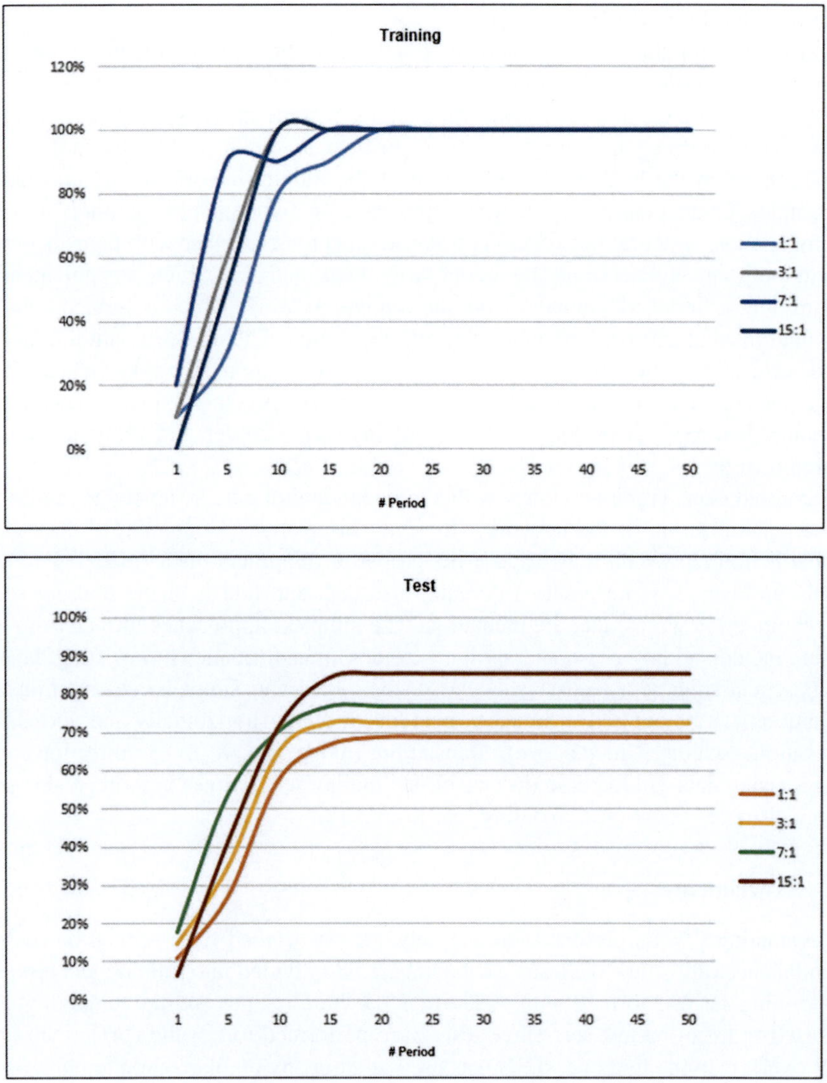

Fig. 1. Accuracy graphs of CNN trained 50 times.

3 Results

Class Imbalance. Figure 1 shows the results of the stepwise imbalance process implemented with the proposed CNN. The accuracy graphs for the training and test phases are shown. In training, CNN achieves 100% accuracy, in all relationships, in approximately 20 time periods; however, in the test phase, it obtains 69% to 85% accuracy, evidencing the lack of generalization of the network. Table 3 presents the recovery yields at the class level (accuracy and recall), to obtain separate information for the minority (in blue) and majority classes. Accuracy is defined as the number of samples correctly recognized as class i (true positives, PAs) divided by the number of samples recognized as class i (PAs plus type 1 errors), while recall is defined as the number of PAs divided by the number of samples that should have been recognized as class i (PAs plus type 2 errors).

Table 3. Class performance comparison (CNN).

	1:1		3:1		7:1		15:1	
Class	Exac	Rec	Exac	Rec	Exac	Rec	Exac	Rec
Aircraft	0.854	0.741	0.875	0.930	0.920	0.951	1.000	0.954
Faces	1.000	0.735	0.696	0.795	0.910	0.725	0.869	1.000
Clock	0.562	1.000	0.903	0.620	0.810	0.736	0.875	0.735
Leopards	0.534	0.945	0.821	0.950	0.810	0.895	0.825	0.765
Bonsai	1.000	0.400	0.865	0.895	0.752	0.954	0.775	0.895
Water lilies	0.620	0.525	0.535	0.632	0.000	0.000	0.541	1.000
Strawberry	0.914	0.936	0.500	0.632	0.385	0.495	0.000	0.000
Wild cat	0.520	0.645	0.320	0.424	0.000	0.000	0.000	0.000
Binoculars	1.000	0.669	0.702	0.465	0.000	0.000	0.490	1.000
Metronome	0.950	0.802	1.000	0.445	0.620	0.459	1.000	1.000

3.1 Over-Adjustment

Figure 2 shows the results of the tests carried out to correct the network over-adjustment, the accuracy and error graphs for the test phase. It can be noted that the accuracy increases and the error decrease when the simple CNN is regularized with weight regularization, data enhancement and dropout (RP + AD + Droupout) and accuracy is further enhanced when standardization layers are added to CNN (RP + AD + Droupout + Normalization).

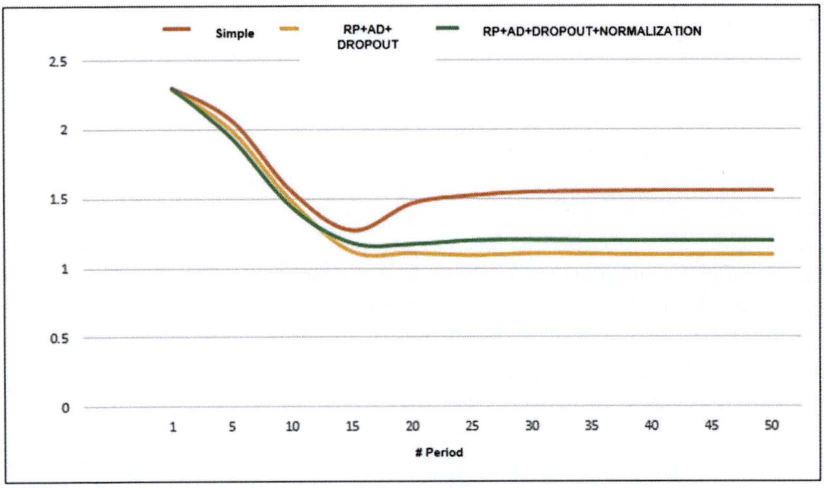

Fig. 2. CNN over-adjustment correction results for the error.

4 Discussion

The class imbalance has a significant detrimental effect on the training of an NPC, affecting both convergences during the training phase and the generalization of the network into the test set. In this sense, the use of the standard accuracy measure used to guide the CNN learning process, maybe providing an advantage to the classification of the majority classes; furthermore, the minority classes, as the imbalance ratio increases, may be identified as noise and, therefore, may be erroneously discarded by the classifier.

Using a static unbalanced relationship in the proposed form is not equivalent to transforming the multi-class classification problem into multiple binary classification problems, rather, it accentuates the cases of multiple minorities and majority classes. As discussed in [3], these cases negatively affect the overall performance of the classifier over the minority classes. The reason for using a static imbalance ratio is to work in the worst-case scenario that can occur in multi-class imbalance problems. The above indicates that the metrics must be properly adjusted to measure the performance of the classifier to the pursued objectives. As an example of this, the 15:1 ratio of the experiments where the majority classes with 60 images per class and the minority classes with only 4 images. The aircraft class is 18.75% of the 320 images and the wildcat class is 1.25% of the set, but the study is interested in identifying cats and not aircraft. In this case, the network reached an accuracy of 94.4% predicting the aircraft class and an accuracy of 0% predicting the wildcat class, all the time, which generates a useless classifier for the intended purpose. In contrast, a properly adjusted metrics could achieve lower accuracy predicting aircraft but would have a substantially higher recovery rate, which is the measure for which it should have been optimized.

5 Conclusion

Since many of the CNNs used for image classification problems use imbalanced training sets, this can generate sub-optimal classification models; that is, good coverage of majority class examples, while minority class examples are often misclassified. Therefore, those algorithms that perform well in the CNN classification framework do not necessarily achieve the best performance for the imbalanced data sets. Class imbalance results in an over-adjustment of CNN to the training data, which affects its ability to generalize. The techniques of weight regularization, data enhancement, and dropout work well to reduce CNN overtraining but do not help increase test performance. The normalization layers built into the CNN increase test accuracy by 10%, i.e. they improve the CNN generalization.

References

1. Khan, S.H., Hayat, M., Bennamoun, M., Sohel, F.A., Togneri, R.: Cost-sensitive learning of deep feature representations from imbalanced data. IEEE Trans. Neural Netw. Learn. Syst. **29**(8), 3573–3587 (2017)
2. Kalinin, A.A., Iglovikov, V.I., Rakhlin, A., Shvets, A. A.: Medical image segmentation using deep neural networks with pre-trained encoders. In: Deep Learning Applications, pp. 39–52. Springer, Singapore (2020)
3. Buda, M., Maki, A., Mazurowski, M.A.: A systematic study of the class imbalance problem in convolutional neural networks. Neural Netw. **106**, 249–259 (2018)
4. Dong, Q., Zhu, X., Gong, S.: Single-label multi-class image classification by deep logistic regression. In: Proceedings of the AAAI Conference on Artificial Intelligence, vol. 33, pp. 3486–3493, July 2019
5. Talo, M., Yildirim, O., Baloglu, U.B., Aydin, G., Acharya, U.R.: Convolutional neural networks for multi-class brain disease detection using MRI images. Comput. Med. Imaging Graph. **78**, 101673 (2019)
6. Nguyen, T.D., Kasmarik, K.E., Abbass, H.A.: An exact transformation from deep neural networks to multi-class multivariate decision trees. arXiv preprint arXiv:2003.04675 (2020)
7. Varela, N., Silva, J., Gonzalez, F.M., Palencia, P., Palma, H.H., Pineda, O.B.: Method for the recovery of images in databases of rice grains from visual content. Procedia Comput. Sci. **170**, 983–988 (2020)
8. Raghu, S., Sriraam, N., Temel, Y., Rao, S.V., Kubben, P.L.: EEG based multi-class seizure type classification using convolutional neural network and transfer learning. Neural Netw. **124**, 202–212 (2020)
9. Fidon, L., Li, W., Garcia-Peraza-Herrera, L. C., Ekanayake, J., Kitchen, N., Ourselin, S., Vercauteren, T.: Generalised wasserstein dice score for imbalanced multi-class segmentation using holistic convolutional networks. In: International MICCAI Brainlesion Workshop, pp. 64–76. Springer, Cham, September 2017
10. Benegui, C., Ionescu, R.T.: Convolutional neural networks for user identification based on motion sensors represented as images. IEEE Access **8**(6), 61255–61266 (2020)
11. Talo, M.: Convolutional neural networks for multi-class histopathology image classification. arXiv preprint arXiv:1903.10035 (2019)
12. Shahtalebi, S., Asif, A., Mohammadi, A.: Siamese Neural networks for EEG-based Brain-computer Interfaces. arXiv preprint arXiv:2002.00904 (2020)

13. Zhang, S., Wen, L., Bian, X., Lei, Z., Li, S.Z.: Single-shot refinement neural network for object detection. In: Proceedings of the IEEE Conference on Computer Vision and Pattern Recognition, pp. 4203–4212 (2018)
14. Jmour, N., Zayen, S., Abdelkrim, A.: Convolutional neural networks for image classification. In: 2018 International Conference on Advanced Systems and Electric Technologies (IC_ASET), pp. 397–402. IEEE, March 2018
15. Park, T., Lee, T.: Musical instrument sound classification with deep convolutional neural network using feature fusion approach. arXiv preprint arXiv:1512.07370 (2015)
16. Zhong, Z., Jin, L., Xie, Z.: High performance offline handwritten chinese character recognition using googlenet and directional feature maps. In: 2015 13th International Conference on Document Analysis and Recognition (ICDAR), pp. 846–850. IEEE, August 2015
17. Viloria, A., Acuña, G.C., Franco, D.J.A., Hernández-Palma, H., Fuentes, J.P., Rambal, E.P.: Integration of data mining techniques to PostgreSQL database manager system. Procedia Comput. Sci. **155**, 575–580 (2019)
18. Zhang, Y., Zhou, D., Chen, S., Gao, S., Ma, Y.: Single-image crowd counting via multi-column convolutional neural network. In: Proceedings of the IEEE Conference on Computer Vision and Pattern Recognition, pp. 589–597 (2016)
19. Jacob, I.J.: Capsule network based biometric recognition system. J. Artif. Intell. **1**(02) 83–94 (2019)
20. McDonnell, M.D., Vladusich, T.: Enhanced image classification with a fast-learning shallow convolutional neural network. In: 2015 International Joint Conference on Neural Networks (IJCNN), pp. 1–7. IEEE, July 2015
21. Du, J., Zhai, J.F., Hu, J.S., Zhu, B., Wei, S., Dai, L.R.: Writer adaptive feature extraction based on convolutional neural networks for online handwritten Chinese character recognition. In: 2015 13th International Conference on Document Analysis and Recognition (ICDAR), pp. 841–845. IEEE, August 2015
22. Viloria, A., Lezama, O.B.P.: Improvements for determining the number of clusters in k-means for innovation databases in SMEs, pp. 1201–1206, ANT/EDI40 (2019)
23. Wang, H., Ding, S., Wu, D., Zhang, Y., Yang, S.: Smart connected electronic gastroscope system for gastric cancer screening using multi-column convolutional neural networks. Int. J. Prod. Res. **57**(21), 6795–6806 (2019)
24. Sharma, N., Jain, V., Mishra, A.: An analysis of convolutional neural networks for image classification. Procedia Comput. Sci. **132**, 377–384 (2018)
25. Yim, J., Sohn, K.A.: Enhancing the performance of convolutional neural networks on quality degraded datasets. In: 2017 International Conference on Digital Image Computing: Techniques and Applications (DICTA), pp. 1–8. IEEE, November 2017
26. Silva, J., Palma, H.H., Núñez, W.N., Ruiz-Lazaro, A., Varela, N.: Neural networks for tea leaf classification. J. Phys: Conf. Ser. **1432**(1), 012075 (2020)
27. Zeng, Y., Xu, X., Fang, Y., Zhao, K.: Traffic sign recognition using extreme learning classifier with deep convolutional features. In: The 2015 International Conference on Intelligence Science and Big Data Engineering (IScIDE 2015), vol. 9242, pp. 272–280, Suzhou, June 2015
28. Ciregan, D., Meier, U., Schmidhuber, J.: Multi-column deep neural networks for image classification. In: 2012 IEEE Conference on Computer Vision and Pattern Recognition, pp. 3642–3649. IEEE, June 2012
29. Ariza, P., Piñeres, M., Santiago, L., Mercado, N., De la Hoz, A.: Implementation of MOPROSOFT level I and II in software development companies in the colombian caribbean, a commitment to the software product quality region. In: 2014 IEEE Central America and Panama Convention (CONCAPAN XXXIV), pp. 1–5, Panama City (2014). https://doi.org/10.1109/CONCAPAN.2014.7000402

Asset Productivity in Organisations at the Intersection of Big Data Analytics and Supply Chain Management

Jossy P. George and K. Sagar Chandra(✉)

CHRIST (Deemed to be University), Bangalore, India
frjossy@christuniversity.in,
sagar.chandra@mba.christuniversity.in

Abstract. A close investigation is required on the fundamental instruments of an establishments' big data analytics usage. This research paper mainly addresses how is an organizations' value creation affected due to big data analytics usage, what is big data analytics, and what are its key antecedents in an organization to understand the aspects that influence the actual usage of big data analytics. Hence, the technology, organization, and environment framework are used. The review data collected from Indian founded corporations confirm that: organizational value creation is significantly affected by big data analytics usage in that organization; organizational BDA usage is indirectly influenced by environmental factors, technological factors, and organizational factors through top management support. Collectively, this research study will guide the business managers on the usage of big data analytics, and a theory-based comprehensive analysis of big data analytics usage and its key antecedents.

Keywords: Big data · Big data analytics · Survey research · Structural equation modelling · Supply chain management

1 Introduction

"Big data" often described as unstructured as well as structured information has dramatically enhanced in variety, velocity, and volume through the emergence of networked business. Managers, businesses and third parties are looking for effectively exploiting such information collected and formed inside their administrations [1]. Throughout the various levels of the organization, insights and intelligence are being converted from data to more effectively use big data. From anecdotal evidence, it can be seen that big data improves service and product development, marketing, operations, human resources (HR), and other core business purposes by the understandings derived from big data that have the potential to renovate corporate strategies and corporate models [2]. Several companies have capitalized on the promise of big data due to the developments in a collection of influential analytical techniques and information technologies [3].

To discover valuable evidence such as unseen patterns, unidentified correlation, etc. to assist producing improved results across corporate procedures of various purposes

J. I.-Z. Chen et al. (Eds.): ICIPCN 2020, AISC 1200, pp. 319–330, 2021.
https://doi.org/10.1007/978-3-030-51859-2_29

and corporations i.e., the procedure of consuming innovative technologies to study big data is the definition of the organizational use of big data analytics [4]. Current industry study specifies that most of the corporations have still not started to capitalize on big data, despite positive testimonies of big data early movers. How to brand persuasive corporate instance for essential significant funds, the risk involved, the essential information technology and analytical skills are still what organizations appear to be learning [5]. Therefore, a close investigation is required on the performance outcomes of fundamental instruments that lead to organizations' big data analytics usage.

Despite understanding the phenomenon, studies on technology acceptance behaviours at the company level remain comparatively rare though a wide body of information technology acceptance literature exists at different levels [6]. Rather than focusing on postadoption behaviours such as value creation and usage, few of organizational information technology acceptance educations focus on adoption intentions. It is also important to emphasize the potential corporate performance impacts of big data analytics competency along with understanding the drivers of big data analytics usage [7]. Also, the advancement of big data analytics and associated capabilities need a more holistic view [8].

Therefore, the value creation process and the current understanding of organizational-level usage of big data analytics seem to be limited by the above-mentioned gaps in the works of the literature. To bridge the above-mentioned gaps, this study addresses the following: the major factors of influencing organizational or company level big data analytics usage and how does the organizational or company level big data analytics impact business growth and business productivity. The domain of supply chain management is chosen to inspect the practice of big data analytics to answer the above-mentioned research questions. The supply chain management success is essential as more and more firms are increasingly dependent on knowledge creation and organizational learning. Furthermore, in today's networked economy the required information flows and learning demands can be satisfied particularly by information technology and big data analytics.

2 Literature Review

2.1 The Technological, Organizational and Environmental Framework, and Big Data Analytics Use

Dynamic and changing competencies are frequently considered as exclusive and distinguishing procedures which arise from the pathway reliant on pasts of distinct organizations [9]. The idea of pathway reliance recommends the forces driving the capabilities of the organisation are context-specific [10]. Development in dynamic competencies has dual classes of related forces which are: technical and exterior suitability [11]. Arguments provided by dynamic competences works that they employ environmental, technological, and organizational framework in the IT organization works to classify exclusive aspects explaining the movement of the organization towards big data analytics usage in the context of supply chain management [12]. The technological, organizational, and environmental framework assumes that numerous related aspects affecting the organizational procedure of acceptance and application of technological operation stern from environmental, technological, and organizational facets of an organization's context [13].

As shown in Fig. 1. Conceptual Framework the technological, organizational and environmental outline has largely been supported in the works of the literature, though the exact procedures recognized and used of the framework within the three situations vary from study to study [14–17]. Despite the adoption of the outline as it was initially developed, we find two restrictions of the environmental, technological, and organizational studies. First, technological, organizational, and environmental have preserved each of the groups of related aspects is working comparably and directly affecting an administration's intent to adopt IT. It replicates the automatic outlook of the organization [18]. And, the administrative decision making is an entirely balanced procedure for discovering an ideal choice from the given data without bearing in mind the fact that decision production is inadequate by the knowledge they have at their disposal, and by their reasoning limitation and the period they have to make a choice [19]. Another limitation of the study is that they have treated the three features distinctly by disregarding the latent difference ways in which the outcome from technological, organizational, and environmental can be influenced [20]. Moreover, it oversees the complication of different pathways through which a key-related basis will disturb the outcome of the organization [21]. For example, it has been noted that some technological, organizational, and environmental factors have a strong influence on administrative IT implementation under certain conditions.

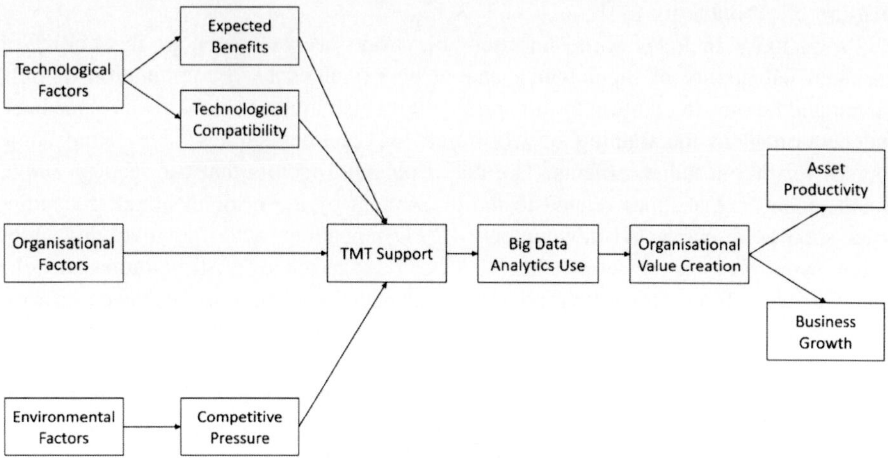

Fig. 1. Conceptual framework

To provide an all-inclusive image of the instruments which are influencing the administrative use of big data analytics in taking a decision, we suggest two extensions of the technological, organizational, and environmental framework [22]. First, incorporate the behavioural perspective of the organization like the role of the bounded rational human agency in converting external forces into managerial actions [23]. Secondly, examine the various path leading to the development of big data analytics in the organization from the technological, organizational, and environmental factors [24]. The organization specifically modifies the environmental, technological, and organizational model as per the situation of big data analytics by positioning the highest administration as an intermediary between two technological, organizational, and environmental factors. In other words, technological, organizational, and environmental can help in using big data analytics through two paths: 1) the technology will have straight influence, and 2) the impact of environmental administrative forces can be facilitated by the highest management.

H1. Technological factors will significantly influence TMT Support
H2. Organizational factors will significantly influence TMT Support
H3. Environmental factors will significantly influence TMT Support

2.2 Business Growth and Asset Productivity – Value Creation

Asset Productivity is well-defined as a result as a measure of the contributions used [30]. Several ways have been used to calculate output such as total factor productivity, capital productivity, multifactor productivity, and labour productivity [31]. Out of these, capital and labour output measures with solo aspect. Therefore, both labour and capital productivity fail to explain productivity well. Multi-factor productivity takes both these relative measures but does not explanation for aspects like organization practices and work atmosphere [32]. Total factor productivity accounts for a better measure of productivity [33].

Productivity includes some unobservable fronts as well such as the ability of manager and quality of input which cannot be explained by a combination of only capital and labour. In addition to this, productivity also includes slowdowns, speedups, and betterment in the training of workforce [34]. Earlier studies have found some sources for output enhancements. The major potential factors that can lead to output enhancements are at times related to the investment by the firms in immaterial properties such as research and development (R&D) spending, activities involving innovation, workforce capital, etc. It has also been suggested by earlier studies that the major driving force towards productivity growth is technological developments. Researchers have linked productivity in supply chain management as well, by taking into account the relationship between the market buyers and market suppliers [35–37]. Solow (1957) stated that "you can see the computer age everywhere but in the productivity statistics" [34]. Various studies have suggested that IT impacts productivity significantly. For instance, computer capital has highly contributed to productivity during 1987-91 in big U.S. firms. Asset productivity is basic to analyze the performance of the supply chain, by defining the level in which an organization efficiently uses its present resources as well as fixed resources. Major pointers of asset productivity are profitability and asset turnover rate [1].

Business growth is a component of the ability to develop a range of short-term advantages in a dynamic market. Gaining long-term advantage is tough if tried just by configuring available resources in dynamic markets [8]. The application of big data analytics can be enhanced by providing a few new insights in areas such as inventory management, customer relationship management, etc. It has been found in earlier studies that even the platform for innovation can be developed for gaining various short-term competitive advantages in supply chain management. Like, lead time can be reduced and there can be scope for product development, leading to an increase in sales by analyzing the data from inventory and shipping. Lummus, Vokurka, & Duclos (2005) suggested that analyzing the demand and inventory data with a high level of receptiveness in the supply chain is strongly associated with managers [38]. Various studies have suggested that a higher level of information availability and processing abilities produce greater receptiveness of the supply chain [39–41]. The above-mentioned capabilities can help in enhancing the managers' alertness to opportunities in driving business growth. Competitive strategy analysis is performed to direct business growth [42].

H5: Big data analytics use will significantly influence organizational value creation.

3 Research Methodology

For this empirical research keen to understand supply chain executives of Indian based companies for empirical demonstration of big data analytics' benefits. Empirical research is done to derive knowledge from actual experiences rather than from theory or belief based on observation and measured phenomena.

Figure 1 shows the relationship between the antecedents of Top Management (TMT) Support and how the TMT support influences Big data analytics use. Figure 1 also gives the relationship between big data analytics use and organizational value creation which explains the asset productivity and business growth in an organization.

A study was conducted on a sample of 468 supply chain executives from India. Web-based surveys were used to gather the data which was analyzed with structural equation modelling. The social media platform chosen to conduct the survey was Facebook. Around the world, there are 2.41 billion-plus monthly active users (MAU) as of June 2019 [43]. There has been an 8% increase of monthly active users on Facebook year on year. This is in comparison to Q1 2019 where it was 2.38 billion [44]. Facebook on average has around 1.59 billion log in each day (June 2019), who are primarily considered as daily active users (DAU). This again shows an 8% year on year increase. It shows that there is a consistency in visits to the site and there is a huge growth in the number of active users on Facebook, making them a promising prospect for marketing efforts. Around 94% of advertising revenue for Q2 2019 comes from mobile advertising revenue as reported by Facebook [45]. It is noted that there is a 23% increase year on year for mobile daily active users, which was 1.15 billion for December 2016. On average, 10 million websites are viewed daily due to the like and share button [46]. The most common age demographics is in between the age of 25 to 34, who constitute about 29.7% of the users. For many businesses' marketing efforts this is the prime target demographics, via Facebook we have a chance to engage with these key consumers. Every second five new profiles are created. Also, 74% of all internet users use social media, with men being 72% (out of 100% men) and women having a slight edge and being 76% (out of 100% women). Since this is not a large statistical difference, one can efficiently reach both genders on Facebook. There is a total of 300 million photo updates per day. It is an indication that one should marketers should target strategically as there are a large number of photos being uploaded containing substantial information, competing for the user's attention. On average, about 20 min is spent on Facebook on every visit. It implies that there is a small time frame to make your impression so one must utilize this opportunity to post unique, interesting, and relevant content and offers to get the maximum return on the marketing efforts [43]. Every 60 s on Facebook: the number of photos uploaded is 136,000, the number of comments posted is 510,000 and the number of statuses updated is 293,000 [46]. Around 50% of the users in the age group of 18–24 log onto Facebook as soon as they wake up. Hence, Facebook is important to these people and the content posted in it, provided it is done correctly. It is reported by 42% of the marketers that Facebook is critical or important to their business. As of May 2013, 16 million business pages have been created on Facebook, compared to 8 million in June 2012 which is a 100% increase [44]. How businesses will be conducted is transformed by Facebook marketing, and to extend their market the use of Facebook by local businesses continues to explode.

4 Research Instrument

For the research instrument, the researchers have used the Questionnaire method [47]. The research instrument in particular measures the following factors in detail: big data analytics usage, technological compatibility, organizational readiness, environmental factors, top management support, and organizational value creation. A total of 117 credible responses were gathered from a variety of organizations representing various industries and geographical locations. The managers (respondents) held various positions such as upper managers, middle managers, and executive managers. This justifies that the respondents had sufficient and relevant knowledge to respond to the questionnaire.

The calculation of a conventional reply rate was not likely through this method which is one of the limitations of this method. Therefore, to overcome this bias we firstly the respective industry median values were compared with the return on assets of the sampled firms using paired sample *t*-tests. The outputs show that there was no significant difference ($p > 0.05$) between industry median values and sampled firms, suggesting that there was no bias in the chosen sample.

5 Results

The exploratory factor analysis results and reliability of the constructs are shown in Table 2. It is observed that all the cumulative variance explained outstrip 60%. It is also found that Cronbach's alpha score for the selected factors ranges from 0.719 to 0.902 indicating the absolute degree of internal consistency. Besides, Joreskog's composite factor reliability score of all the selected attributes is larger than the acceptable threshold value of 0.70 suggesting dependable reliability of the retained scales structure (Table 1).

Table 1. Reliability and factorial analysis of constructs

Variables	Number of items	Mean	SD	Variance explained (%)	Cronbach alpha	Composite factor reliability
Technological factors	6	3.52	0.90	83.4	0.891	0.913
Organizational factors	5	3.25	0.82	80.6	0.856	0.889
Environmental factors	3	2.71	0.80	85.4	0.872	0.851
TMT support	3	3.42	0.92	79.8	0.902	0.924
Big data analytics use	2	3.58	0.96	80.1	0.719	0.833
Organizational value creation	3	2.68	0.99	76.6	0.842	0.933

Table 2. Multiple linear regression

Model	R	R square	Adjusted R square	Std. error of the estimate
1	.698[a]	.488	.485	.66162

a. Predictors: (Constant), Technological Factors, Organizational Factors, Environmental Factors

Big data analytics use has got the highest mean i.e., 3.58 but with the second-highest standard deviation = 0.96. Organizational Value Creation has got the least mean = 2.68, with a standard deviation = 0.99 being the highest. The least standard deviation is observed in the case of Credibility where it is equal to 0.80.

Multiple linear regression is done to establish the association between technological factors, organizational factors, environmental factors towards TMT (Top Management) Support. It is found from Table 3 that the R square value is 0.488. This means that technological factors, organizational factors, and environmental factors contribute 48.8% in understanding TMT support.

Table 3. Correlation coefficient

Correlation		Big-data analytics use	TMT support	Organizational value creation
Big-data analytics use	Pearson correlation	1		
	sig. (2-tailed)			
	N	468		
TMT support	Pearson correlation	.776**	1	
	sig. (2-tailed)	.000		
	N	468	468	
Organizational value creation	Pearson correlation	.570**	.565**	1
	sig. (2-tailed)	.000	.000	
	N	468	468	468

The correlation coefficient is performed to understand the relationship between big data analytics usage and TMT support, and also TMT support and organizational value creation. It is observed that the correlation between Big Data Analytics Use and TMT support is 77.6%, and the correlations between big data analytics use and organizational value creation are 57%. It is also found that the correlation between TMT support and organizational value creation is 56.5% (Table 4).

Table 4. Results of hypotheses testing

Hypotheses	Estimate	S.E.	C.R.	P	Accept/reject hypothesis
H1. technology factors – TMT support	.447	.035	12.667	***	Accept
H2. organizational factors – TMT support	.104	.040	2.627	.009	Accept
H3. environmental factors – TMT support	.352	.039	9.041	***	Accept
H4. TMT support – big data analytics use	.583	.044	13.119	***	Accept
H5. big data analytics use – organizational value creation	.530	.037	14.458	***	Accept

Note: ***$p < 0.0001$

6 Discussion and Managerial Implications

The paper analyzes the impact of environmental factors technological factors, and organizational factors on TMT support, big data analytics use on organizational value creation. Moreover, this paper also canvases the regulating effect of TMT Support on the association between big data analytics use and organizational value creation. The conceptual framework in this study appraised how technological factors, organizational factors, environmental factors, TMT support, and big analytics use could influence organizational value creation. This research model conveys a further step compared to the previous researches that have been consistent and has been associated with the above-mentioned variables such as technological factors to organizational value creation to a more cabalistic understanding that some variables affect the above-mentioned variables which result in varied responses from the consumers (Table 5).

Table 5. Results of structural equation modeling

Indexes	Shorthand	Acceptable fit	Proposed model	Interpretation
Absolute/predictive fit				
CMIV/DF	$\chi2$	df $\leq$ 2 or 3	2.956	Borderline Fit
Comparative fit – comparison to a baseline (independence) or other models				
Normed fit index	NFI	$\geq$.95	0.952	Perfect Fit
Incremental fit index	IFI	$\geq$.95	0.961	Good Fit
Comparative fit index	CFI	$\geq$.95	0.929	Marginal Fit
Relative fit index	RFI	$\geq$.95	0.963	Complete Fit
Parsimonious fit				
Parsimony-adjusted CFI	PCFI	Sensitive to model size	0.241	Substantial Fit
Parsimony-adjusted GFI	PGFI	Closer to 1 the better	0.832	Marginal Fit
Other fit				
Goodness-of-fit index	GFI	$\geq$.95	0.953	Good Fit
Adjusted GFI	AGFI	$\geq$.95	0.936	Bare Fit
Root mean square error of approximation	RMSEA	.06 to .08	0.069	Perfect Fit

The outcomes of this research support that the hypothesis suggesting the role of environmental factors, technological factors, and organizational factors on managers and their varied responses [48]. The results show that environmental factors, technological factors, and organizational factors contribute significantly towards TMT Support. Furthermore, it can be observed that the relationship between TMT Support and big data analytics use is towards the highest observed in this study and the correlation between just TMT Support value is less when compared to the above correlation. The outcomes of the test supported the hypothesis [40]. The second factor that was found related to big data was organizational value creation that impacts the supply chain. The study has found that technological factors, organizational factors, and environmental factors are significantly associated with big data analytics use [42]. Hence, with the inconsistency observed in previous studies, this study highlights the attention given to the environmental factors, technological factors, and organizational factors as key antecedents of big data analytics use. When it comes to the managerial implications, a manager needs to realize the effect big data analytics has on its organizations' value creation. It has become an essential part of the supply chain department of a company.

7 Conclusion

This study provides information about the key factors that ease big data analytics use, as well as a network that analyze the impression of big data analytics on organizational results. Jointly, it provides managers with guidance regarding the use of this rapidly growing technology that is big data analytics. In conclusion, this study's findings are as follows: (1) The impact that is caused on supply chain value creation due to big data analytics use; (2) Technological factors, organizational factors, and environmental factors have a direct influence on TMT support and big data analytics use; (3) TMT support and big data analytics use have an indirect influence on organizational value creation.

References

1. Chen, D.Q., Preston, D.S., Swink, M.: How the use of big data analytics affects value creation in supply chain management. J. Manag. Inf. Syst. **32**(4), 4–39 (2015)
2. Chen, H., Chiang, R.H., Storey, V.C.: Business intelligence and analytics: from big data to big impact. MIS Q. **36**(4), 1165 (2012)
3. Waller, M.A., Fawcett, S.E.: Data science, predictive analytics, and big data: a revolution that will transform supply chain design and management. J. Bus. Logist. **34**(2), 77–84 (2013)
4. Davis, F.D.: Perceived usefulness, perceived ease of use, and user acceptance of information technology. MIS Q. **13**(3), 319 (1989)
5. Goodhue, D.L., Thompson, R.L.: Task-technology fit and individual performance. MIS Q. **19**(2), 213 (1995)
6. Venkatesh, V., Morris, M.G., Davis, D.B., Davis, G.B.: User acceptance of information technology: toward a unified view. MIS Q. **27**(3), 425 (2003)
7. Zhu, K., Kraemer, K.L.: Post-adoption variations in usage and value of e-business by organizations: cross-country evidence from the retail industry. Inf. Syst. Res. **16**(1), 61–84 (2005)
8. Eisenhardt, K.M., Martin, J.A.: Dynamic capabilities: what are they? Strateg. Manag. J. **21** (10–11), 1105–1121 (2000)
9. Zieba, M., Bolisani, E., Paiola, M., Scarso, E.: Searching for innovation knowledge: insight into KIBS companies. Knowl. Manag. Res. Pract. **15**(2), 282–293 (2017)

10. Lippert, S.K.: Technology trust: an inventory of trust infrastructures for government and commercial information systems in support of electronic commerce. In: Proceedings of the 2002 InSITE Conference (2002)
11. Rai, A., Patnayakuni, R., Seth, N.: Firm performance impacts of digitally enabled supply chain integration capabilities. MIS Q. **30**(2), 225 (2006)
12. Huiskonen, J.: Maintenance spare parts logistics: special characteristics and strategic choices. Int. J. Prod. Econ. **71**(1–3), 125–133 (2001)
13. Liu, P., Huang, S.H., Mokasdar, A., Zhou, H., Hou, L.: The impact of additive manufacturing in the aircraft spare parts supply chain: supply chain operation reference (scor) model based analysis. Prod. Plan. Control **25**(13–14), 1169–1181 (2013)
14. Stecke, E.K., Talbot, F.B.: Heuristics for loading flexible manufacturing systems, flexible manufacturing systems. Flex. Manuf. Syst.: Recent Dev. Ref. Mod. Instrum. Tech. Technol. **23**, 171–182 (1995)
15. Browne, J., Sackett, P., Wortmann, J.: Future manufacturing systems—towards the extended enterprise. Comput. Ind. **25**(3), 235–254 (1995)
16. Chan, F.T., Chung, S.H., Wadhwa, S.: A heuristic methodology for order distribution in a demand driven collaborative supply chain. Int. J. Prod. Res. **42**(1), 1–19 (2004)
17. Wadhwa*, S., Rao, K.S., Chan, F.T.: Flexibility-enabled lead-time reduction in flexible systems. Int. J. Prod. Res. **43**(15), 3131–3162 (2005)
18. London, K.A., Kenley, R.: An industrial organization economic supply chain approach for the construction industry: a review. Constr. Manag. Econ. **19**(8), 777–788 (2001)
19. Sun, L., Li, X.: Comparison analysis on the bilateral efforts of farmers and the third-party organization under multiple contract modes. Kybernetes **48**(5), 818–834 (2019)
20. Kumar, R.: Study on supply chain management issues: case study of an indian fast moving electrical goods organization. Int. J. Sci. Eng. Comput. Technol. **8**(1), 18–22 (2018)
21. Opengart, R.: Supply chain management and learning organization: a merging of literatures. Int. J. Commer. Manag. **25**(2), 183–195 (2015)
22. Chen, P., Wu, M.: A modified failure mode and effects analysis method for supplier selection problems in the supply chain risk environment: a case study. Comput. Indu. Eng. **66**(4), 634–642 (2013)
23. Chi, T., Kilduff, P.P., Gargeya, V.B.: Alignment between business environment characteristics, competitive priorities, supply chain structures, and firm business performance. Int. J. Prod. Perform. Manag. **58**(7), 645–669 (2009)
24. Uluskan, M., Godfrey, A.B.: Business environment-supply chain framework and benchmarking supply chain structures. Benchmarking Int. J. **25**(8), 3021–3044 (2018)
25. Ardito, L., Scuotto, V., Del Giudice, M., Petruzzelli, A.M.: A bibliometric analysis of research on big data analytics for business and management. Manag. Decis. **57**(8), 1993–2009 (2019)
26. Grover, P., Kar, A.K.: Big data analytics: a review on theoretical contributions and tools used in literature. Glob. J. Flex. Syst. Manag. **18**(3), 203–229 (2017)
27. Hofmann, E., Rutschmann, E.: Big data analytics and demand forecasting in supply chains: a conceptual analysis. Int. J. Logist. Manag. **29**(2), 739–766 (2018)
28. Queiroz, M.M., Telles, R.: Big data analytics in supply chain and logistics: an empirical approach. Int. J. Logist. Manag. **29**(2), 767–783 (2018)
29. Mandal, S.: Exploring the influence of big data analytics management capabilities on sustainable tourism supply chain performance: the moderating role of technology orientation. J. Travel Tour. Mark. **35**(8), 1104–1118 (2018)
30. OECD. OECD Publishing, Paris (2001)

31. Böhme, T., Escribano, A., Heffernan, E.E., Beazley, S.: Causes and mitigation for declining productivity in the Australian mid-rise residential construction sector. Built Environ. Proj. Asset Manag. **8**(3), 253–266 (2018)
32. Loosemore, M.: Improving construction productivity: a subcontractor's perspective. Eng. Constr. Architect. Manag. **21**(3), 245–260 (2014)
33. Carson, C., Abbott, M.: A review of productivity analysis of the New Zealand construction industry. Constr. Econ. Build. **12**(3), 1–15 (2012)
34. Solow, R.M.: Technical change and the aggregate production function. Rev. Econ. Stat. **39** (3), 312 (1957)
35. Keller, W.: International technology diffusion (2001)
36. Werner, S.: Recent developments in international management research: a review of 20 top management journals. J. Manag. **28**(3), 277–305 (2002)
37. Blalock, G., Veloso, F.M.: Imports, productivity growth, and supply chain learning. World Dev. **35**(7), 1134–1151 (2007)
38. Lummus*, R.R., Vokurka, R.J., Duclos, L.K.: Delphi study on supply chain flexibility. Int. J. Prod. Res. **43**(13), 2687–2708 (2005)
39. Christopher, M., Lowson, R., Peck, H.: Creating agile supply chains in the fashion industry. Int. J. Retail Distrib. Manag. **32**(8), 367–376 (2004)
40. Li, X., Chung, C., Goldsby, T.J., Holsapple, C.W.: A unified model of supply chain agility: the work-design perspective. Int. J. Logist. Manag. **19**(3), 408–435 (2008)
41. Williams, B.D., Roh, J., Tokar, T., Swink, M.: Leveraging supply chain visibility for responsiveness: the moderating role of internal integration. J. Oper. Manag. **31**(7–8), 543–554 (2013)
42. Qi, Y., Zhao, X., Sheu, C.: The impact of competitive strategy and supply chain strategy on business performance: the role of environmental uncertainty*. Decis. Sci. **42**(2), 371–389 (2011)
43. Hamelin, N., Nwankwo, S., El Hadouchi, R.: Faking brands: consumer responses to counterfeiting. J. Consum. Behav. **12**(3), 159–170 (2012)
44. Brailovskaia, J., Margraf, J., Schillack, H., Köllner, V.: Comparing mental health of Facebook users and Facebook non-users in an inpatient sample in Germany. J. Affect. Disord. **259**, 376–381 (2019)
45. MacDonald, L., Rasmussen, H.: Revenue management with dynamic pricing and advertising. J. Revenue Pricing Manag. **9**(1–2), 126–136 (2009)
46. Pham, G.V., Shancer, M., Nelson, M.R.: Only other people post food photos on Facebook: third-person perception of social media behavior and effects. Comput. Hum. Behav. **93**, 129–140 (2019)
47. Zhang, Q., Hou, J., Yan, Y., Li, T.: Research on circuit board fault detection and diagnosis system based on virtual instrument. J. Electron. Meas. Instrum. **25**(2), 135–140 (2011)
48. Knights, D., Vurdubakis, T.: Information technology as organization/disorganization. Inf. Organ. **15**(3), 181–184 (2005)

Prediction of Intraday Trend Reversal in Stock Market Index Through Machine Learning Algorithms

K. S. Uma(✉) and Srinath Naidu

Department of Computer Science and Engineering, Amrita School of Engineering, Amrita Vishwa Vidyapeetham, Bengaluru, India
umaks84@gmail.com, r_srinath@blr.amrita.edu

Abstract. Research on stock price prediction has gained much importance in stock markets because of the immense economic value of being able to predict stock prices in advance. In this paper, we are trying to predict intraday trend reversals in a stock market index. Stock market indices often exhibit sharp intraday trend reversals of significant magnitude. If this reversal can be predicted or detected as soon as the move starts, it will be of immense economic value. Based on past data and statistical methods we can establish support and resistance levels for index and constituent shares. Usually, a sudden change in moves happens as a reversal at these levels or on crossing these levels with strong momentum. In general, traders use statistical methods like RSI (Relative Strength Index), MACD (Moving Average Convergence Divergence) to identify the trend reversals during day trading. One problem with all these methods is that they identify trend reversals with a time lag. We provide in this paper a novel approach to predict such sharp reversals by using machine learning algorithms. We analyze past and present data to predict intraday trend reversals in the index with precision and robustness.

Keywords: Stock market prediction · Intraday move · Machine learning

1 Introduction

During day trading in a stock market, the price of the stock sometimes moves a significant number of points swiftly within a short period. Theoretically, option buying has the potential to generate maximum returns in the stock market. Major risks faced by optional buyers are time decay, IV (implied volatility), contraction, and underlying contract moving against the predicted direction. In a market with moderate to less IV (implied volatility), being able to predict the direction of a sudden movement will help us mitigate the above risks and generate superior returns. Option trading always is riskier but is the most profitable trade in stock markets. There are two types of option trading - call options and put options [7]. Here we are trying to analyze the trend reversal in the stock market index during day trading so that we will be able to decide which trade we can take to get maximum profit. Many parameters affect the trend reversals in the stock market. We are trying to identify the parameters that will have a huge impact when the market reverses from one direction to another. There are two

J. I.-Z. Chen et al. (Eds.): ICIPCN 2020, AISC 1200, pp. 331–341, 2021.
https://doi.org/10.1007/978-3-030-51859-2_30

ways in which the trend can be reversed. The up move can suddenly move downwards, and the down move can suddenly move upwards. So according to the direction of the move, we must trade put options and call options, respectively.

Figure 1 explains how a trend reversal happens. The position marked as 'T' in the figure is the position at which the upward move in the index started moving against its direction. We aim to identify the point 'T' before it starts or the moment it starts. If we can predict this move accurately, we can take the trade accordingly and it will be of high economic value.

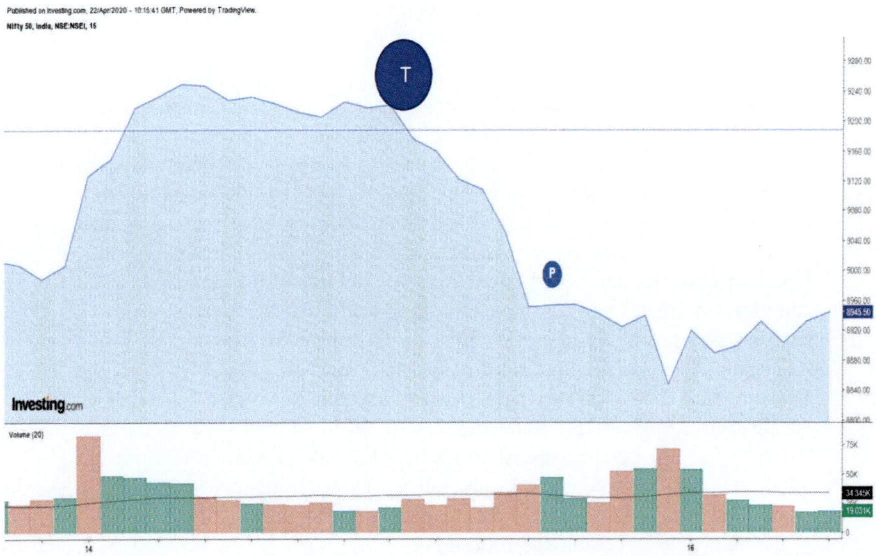

Fig. 1. An example of a trend reversal

Implied volatility (IV) [6] is a parameter that always tells the future volatility of the market. Implied volatility has a high impact on option trading [15]. Therefore, we use IV as one of the inputs to the learning machine to find the future trend reversals in the stock market index.

2 Literature Survey

Some research [1] has been carried out on stock price movement using different sources of information such as news, blogs, and numerical data in Chinese markets. The tools used in [1] were machine learning, deep learning, and sentiment analysis. The main difference between this paper and our problem is that we analyze how a trend movement in day trading happens, especially in the index. In this paper, we introduce a method that will analyze sudden movements of the index during a day.

An automated framework for intraday NIFTY [4] explains the future of the market the next day by giving positive and negative signals. This method has been able to effectively predict the market index for the next day by considering historical data. But with this framework, we cannot predict the current movement during intraday trading. Technical indicators like the Relative Strength Index (RSI) always indicate stock market movement with a time lag. It can provide the support and resistance level, but it cannot give the reversals at these levels. Therefore, there is still a need to predict trend reversals by analyzing the past and present data.

The work of [5] describes stock market prediction using news, Wikipedia hits, Google news, technical indicators to predict the future indices. This work cannot be used to predict the timely movement of index or stock price. Since our problem is completely related to intraday trading, we need to analyze current market data and establish a method to identify the trend reversals.

The work of [2] deals with predicting stock market movement with multiple sources and multiple instance learning using sentiment analysis and quantitative stock data. However, it cannot predict trend reversals during day trading. This is because capturing trend reversals requires working with the current data and analyzing current data movement.

The work of [3] on stock market volatility on time series data has used the ARIMA model to visualize the future of the stocks. But if we use this model to predict intraday trend reversal, these components always predict trend reversals with a time lag. The advantage of our work is that we predict intraday trend reversals ahead of the start of the trend reversal to get the greatest benefit in terms of economic value. To be able to do this, we must deal with current market data to produce intraday movements in the stock market.

Research has done on implementing an artificial market model [10, 12] in recent years. From our personal experience in day trading and a detailed study from Zerodha [7], we have learned the factors that are affecting trend reversal in the stock market index.

3 Proposed System

The proposed system consists of the following modules as shown in Fig. 2.

3.1 Data Collection

As we aim to predict the intraday trend reversals, we extracted data of options in day trading for the correlation analysis from true data (data provider). We have extracted the data for five days for nifty, call options, and put options. The available parameters in the above-extracted data are date, time, open, high, low, close prices, open interest, and volume. We took data at intervals of 5 min during a day to analyze the trend reversals in the system.

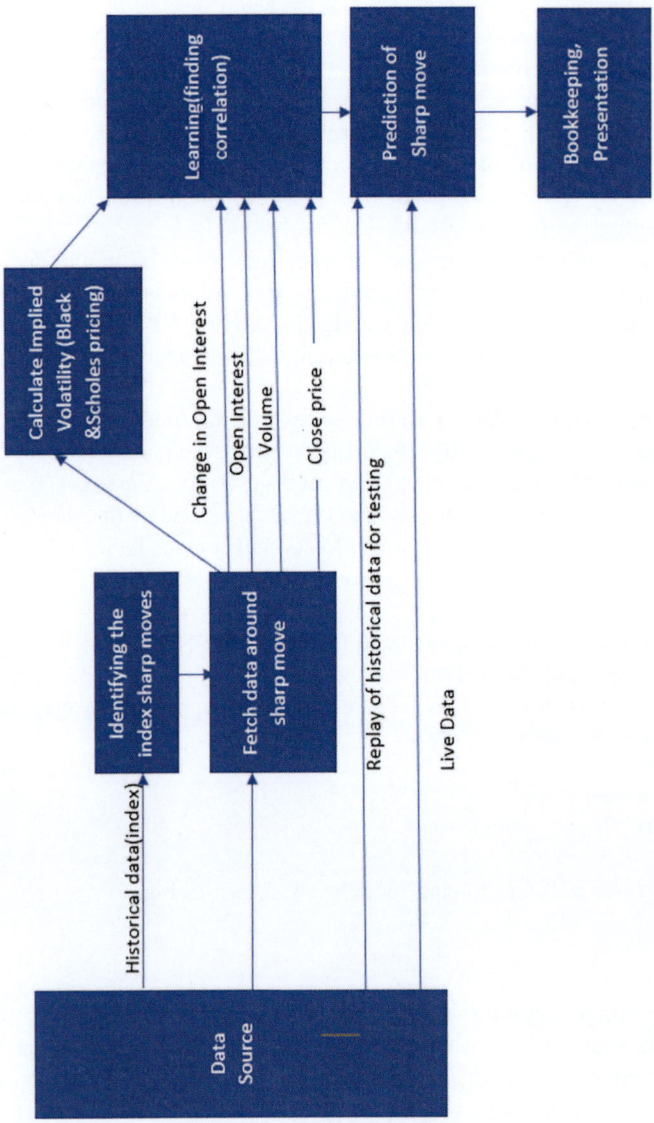

Fig. 2. A modular view of the proposed system for trend reversal

3.2 Trend Reversal Analysis from Past Data

In this module, we are trying to analyze the trend reversal from past data. Here we are plot close prices of nifty in each interval of five minutes. From this graph, we can understand the trend reversals at a point, and we can analyze the data surrounded by the time at which reversal happened.

3.3 Correlation Analysis

After identifying the trend reversals from the past data, we must analyze the data just before the trend reversal and just after the direction of the trend gets stopped. To analyze this data, we have taken past 5-min. data before the trend starts and past five-minute data before the trend ends. We must find out the correlation between data near to trend reversal and nifty index.

3.4 Implied Volatility Calculation

Implied volatility is not directly available in past data and can be calculated using the Black-Scholes pricing formula [8]. Here in past data implied volatility is not directly available. So, we must calculate implied volatility at each strike price to analyze the market move.

$$\textbf{The theoretical option price for call option} = pN(d1) - s(e^{-rt})N(d2) \quad (1)$$

$$\textbf{Where } d1 = ln\left(\frac{p}{s}\right) + \frac{\left(r + \frac{v^2}{2}\right)t}{v\sqrt{t}},$$

$$d2 = d1 - v\sqrt{t}$$

$$\textbf{The theoretical option price for put option} = N(-d2)se^{-rt} - N(-d1)p \quad (2)$$

p——>Stock Price
s——>Striking Price
t——>time remaining until expiration expressed in percent of a year
r——>current risk-free interest rate
v——>volatility measured by annual standard deviation
ln——>natural logarithm
N(x)——>cumulative normal density function

From Eq. 1 and Eq. 2 we must apply reverse engineering to find out the implied volatility of call option and put option. That is, we have to back solve volatility value by using an iterative approach with p, s, t, r as inputs and we find out the value of ‘v’. Equation 1 is used to find out the implied volatility of the call option and Eq. 2 is used to find out the implied volatility of the put option. I arbitrarily took 0.5 as the initially implied volatility.

3.5 Prediction Using Machine Learning Algorithms

This module is used to predict the future trend reversals using machine learning algorithms. Here prediction can create signals. That is whether the market will reverse from its current position. From this we can take three decisions -bull (buy call), bear (buy put), cover (exit from position).

3.6 Bookkeeping and Presentation

This module will give all the details such as total trades, profitable trades, highest profit, average profit/profitable trades, most consecutive wins, losing trades, most consecutive losses, highest loss, average loss/ losing trade, and the net returns. The presentation module will give data visualization in the form of a graph.

4 Experimental Results

4.1 Trend Reversal Analysis

We took the past 5 days of data (each five minutes data of nifty) and try to plot close prices of each day. We also try to identify the trend reversals in the past, the time at which the move starts, and the time at which the move ends. After analyzing the trend reversals from the graph, we try to identify what will happen to the data when a reversal happens and try to identify the correlation among them. Figure 3 below shows the nifty movements of each five minutes on 3rd September 2019. High peaks and troughs in this figure show that trend reversals happen on that day.

Fig. 3. Trend reversal analysis

4.2 Quantitative Models for Correlation Analysis

We aim to identify the scenario that accompanies a trend reversal. To identify how a reversal happens in intraday trading, we must identify the correlation of the data and nifty index. Initially, we have tried to analyze the correlation between the available parameters. Then we tried to calculate the implied volatility with the available parameters. After analyzing the live data, we understood that implied volatility has a high impact on intraday trading and especially on options trading. Our next step was to calculate the implied volatility of options call and option put data at each five-minute time (from the start of the market at 9.15 AM to the closing of the market at 3.30 PM) using Black Scholes pricing formula. We then plotted the implied volatility curve of a call option and put option on a day.

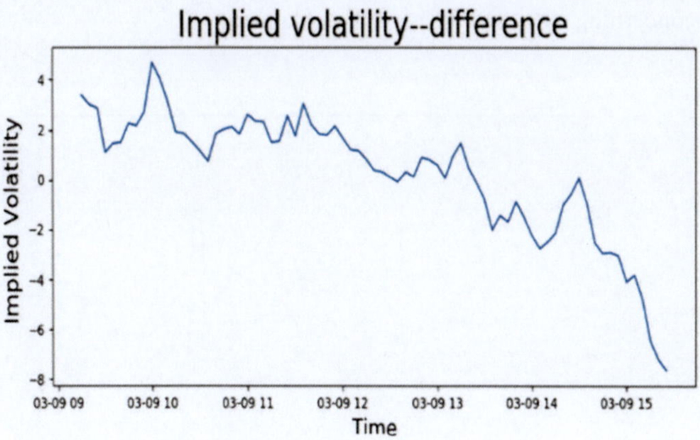

Fig. 4. Implied volatility

The following sections explain the different strategies and quantitative models we have used to analyze the correlation among data during trend reversals.

4.2.1 Implied Volatility Calculation

After noting that implied volatility has a very high impact on intraday trading, we tried to plot implied volatility (IV) of each day and every five minutes time frame. In this paper, we use the black Scholes pricing formula to calculate the implied volatility. If there is a significant change in implied volatility, we can see that market is moving up or down. An increase in IV indicates that the market is going up and a dip in IV shows that the market is going down. But IV alone cannot determine whether a sharp move happens or not. So, we added many parameters to improve accuracy. Figure 4 shows the implied volatility on a day. From this graph, we can identify the point of time at which a sudden dip in implied volatility happens. Figure 3 gives the trend reversals versus time. We can obtain from Fig. 3 the time at which a trend reversal happens and can check what happens to implied volatility at that time. Here we can see that when the reversal happens there is a dip in IV also. Therefore, IV has a high correlation with the nifty index.

4.2.2 Open Interest (OI) Variation

We analyzed the options data (both call and put) of the same strike price and try to find the change in open interest and try to save it. Then we plot the change in open interest at each time frame in both call and put. We take the change in open interest a time 5 min before a sharp move happens. We can see that in many cases when a sharp down move happens, or an up move happens the change in open interest is very high. But in many cases change in open interest alone does not dictate whether the market will reverse from the current position. So to improve the accuracy of predicting a sharp move, we add many parameters. In Fig. 5 the blue and green lines give the open interest on the put side and call side, respectively. Here we can see that whenever there is a sudden dip in open interest, the Nifty index changes sharply. The close prices of the Nifty index concerning time can be identified from Fig. 3.

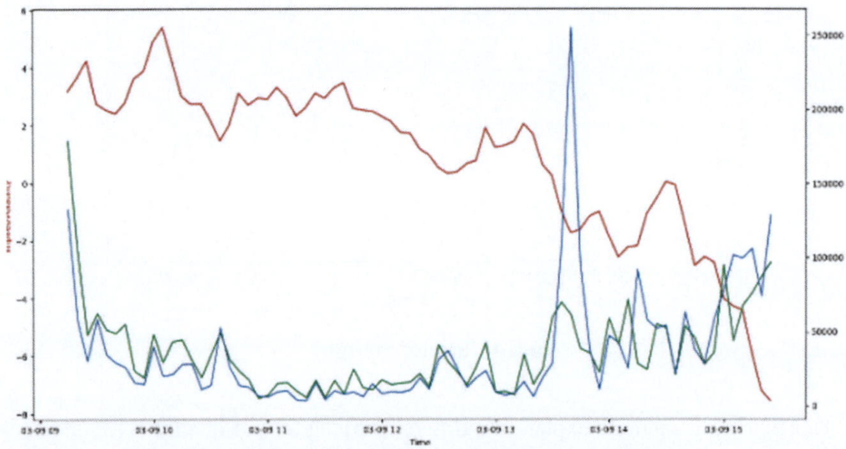

Fig. 5. Change in Open Interest from the call side and put side

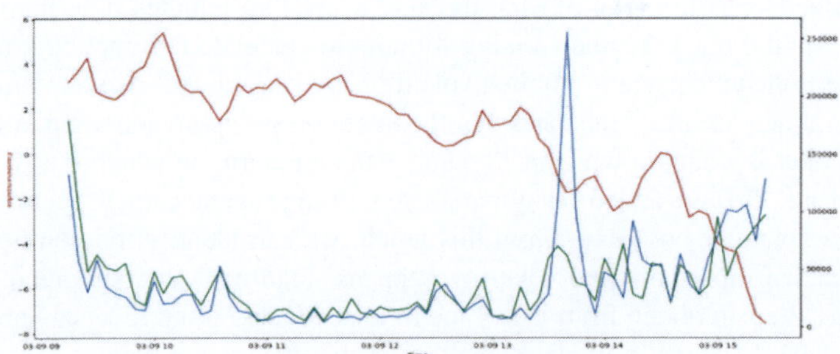

Fig. 6. Change in OI and IV

4.2.3 Open Interest Variation with Implied Volatility

From Sects. 1 and 2 we see that change in open interest and implied volatility have high impact when a trend reversal happens. To analyze this, we try to plot OI change and IV together to understand the correlation among them. In many cases, we can easily identify that if there is a sudden drop in open interest and implied volatility is going down in the call side, then a reversal occurs from the current up move. At the moment this happens we can see that on the put side there will be a slight change or no change. However, on the call side, there is a sudden dip. The red line in Fig. 6 below shows the implied volatility and the other two lines show the OI of the call side and put side.

4.2.4 Prediction Using Machine Learning Algorithms

To predict the future trend reversal, we take the raw data available from the data provider for five days and added the derived parameters implied volatility and change in open interest for a strike price. All parameters in the system are close price, open interest, volume, implied volatility, change in open interest of ATM (At the money), ITM (In the money), OTM (Out of money) of a particular strike price of both put and call options and Nifty close price. We have merged all these data through a python script and manually labeled the reversal and non-reversal points. We have defined the target as a reversal and non-reversal class. We have tried different machine learning algorithms to fit the best model. In this section I will describe all the machine learning algorithms we have tried, to implement the future trend reversal at a particular point of time.

Linear Regression

Our initial trial was with linear regression [14]. We can define the output as the function of the following parameters

$$\begin{aligned}
&\mathrm{Trend_Reversal_Value} = \\
&\quad f(CallATMClose, CallATMOI, CallATMVolume, CallATMIV, CallATMChangeOI, CallITMClose, CallITMOI, \\
&CallITMVolume, CallITMIV, CallITMChangeOI, CallOTMClose, CallOTMOI, CallOTMVolume, \\
&\qquad CallOTMIV, CallOTMChangeOI, PutATMClose, PutATMOI, PutATMVolume, \\
&\qquad PutATMIV, PutATMChangeOI, PutITMClose, PutITMOI, PutITMVolume, PutITMIV, PutITMChangeOI, \\
&PutOTMClose, PutOTMOI, PutOTMVolume, PutOTMIV, PutOTMChangeOI, NiftyClose)
\end{aligned} \tag{3}$$

We made this as a binary classification problem and if Trend_Reversal_Value >THRESHOLD we classify it as trend reversal event and otherwise it is not a reversal point. But with linear regression accuracy was very low. We have tried multiple linear regression [9] also. That also gave the same result as a linear regression. That prompted us to move on logistics regression.

Logistic Regression

We can define the output of logistic regression as

$$\text{Trend\_Reversal\_Value} = \log(CallATMClose, CallATMOI, CallATMVolume, CallATMIV, CallATMChangeOI, CallITMClose, CallITMOI, CallITMVolume, CallITMIV, CallITMChangeOI, CallOTMClose, CallOTMOI, CallOTMVolume, CallOTMIV, CallOTMChangeOI, PutATMClose, PutATMOI, PutATMVolume, PutATMIV, PutATMChangeOI, PutITMClose, PutITMOI, PutITMVolume, PutITMIV, PutITMChangeOI, PutOTMClose, PutOTMOI, PutOTMVolume, PutOTMIV, PutOTMChangeOI, NiftyClose) \quad (4)$$

We made this as a binary classification problem and if Trend_Reversal_Value >THRESHOLD we classify it as trend reversal event and otherwise it is not a reversal point. Here we have divided our dataset into 80% training data and 20% as test data and we got the accuracy as 85.3%. It can identify both the classes. But the precision of the Non-Reversal class is more than the Reversal class because the reversal point data will be very much less than the non-reversal point. So, to improve the precision of reversal class, we moved on to KNN (K-Nearest Neighbor algorithm).

KNN

In this algorithm, we have divided 50% data as a training set and 50% of the data as a test set. I have divided the data set into 50% is because to get more reversal class in the test set. We got the best accuracy at k = 3 and it is 86.2%. It is identifying both reversal and non-reversal class and the precision of reversal class is more than compared to logistics regression.

Among above all algorithms KNN gave the best accuracy score. To improve accuracy, we are trying to apply different machine learning algorithms to get the best results. To improve accuracy, we can use ANN [11] too.

5 Conclusion and Future Enhancement

After analyzing the above scenarios, we can conclude that change in Open interest and dip in implied volatility is highly correlated to nifty index reversals. That means whenever there is a sudden change in the open interest and implied volatility, the stock market index will reverse its position from the current trend. We are trying to automize the system with machine learning algorithms to predict the future trend reversals. Among three machine learning algorithms linear regression, logistics regression, and KNN, KNN is found to be more accurate.

Now we have identified the parameters that are affecting the sudden reversals in the market from past data. Our future work is to predict a reversal point more accurately using machine learning algorithms and artificial neural networks. In the initial trial, we are done with linear regression, logistic regression, and KNN. But we must come up with the best model by experimenting with different machine learning and deep learning approaches. We can add sentiment analysis also to improve the accuracy of trend reversal because the stock market will always react to sudden sentiments.

References

1. Liu, G., Wang, X.: A numerical-based attention method for stock market prediction with dual information. IEEE Access **7**, 7357–7367 (2017)
2. Zhang, X., Siyu, Q., Huang, J., Fang, B., Yu, P.: Stock market prediction via multi-source multiple instance learning. IEEE Access **6**, 50720–50728 (2018)
3. Idrees, S.M., Afshar Alam, M., Agarwal, P.: A prediction approach for stock market volatility based on time series data. IEEE Access **7**, 17287–17298 (2019)
4. Bhat, A.A., Kamath, S.S.: Automated Stock Price Prediction and Trading Framework for Nifty Intraday Trading. In: IEEE ICCCNT 2013, Department of Information Technology, National Institute of Technology, Suratkal, Karnataka (2013)
5. Weng, B.: Application of machine learning techniques for stock market prediction (2017)
6. Szakmary, A., Ors, E., Kim, J.K., Davidson, W.N.: The predictive power of implied volatility: evidence from 35 futures markets. J. Bank. Finance **27**, 2151–2175 (2003)
7. "Introduction to Stock markets" by zerodha varsity
8. https://en.wikipedia.org/wiki/Black%E2%80%93Scholes_model
9. James, G., Witten, D., Hastie, T., Tibshirani, R.: An Introduction to Statistical Learning (2013)
10. Dr. (Col) Kumar, P.N., Abhinaya, K., Vennila, S., Rupika, R.: Implementing an agent based artificial stock market model in JADE-an illustration. Int. J. Eng. Technol. **5**, 2636–2648 (2013)
11. Nair, B.B., Kumar, P.N., Prasad, S.R., Singh, L.M., Vijayalakshmi, K., Sai Ganesh, R., Reshma, J.: Forecasting short-term stock prices using sentiment analysis and Artificial Neural Networks. J. Chem. Pharmaceutical Sci. (2016)
12. Kumar, P.N., Ram, A.T., Gowtham, S.V., Krishna, S., Karthika, M.N., Mohandas, V.P.: A survey of agent based artificial stock markets (Continuous session models). Int. Res. J. Finance Econ. **64**, 126–139 (2011)
13. Müller, A.C., Guido, S.: Introduction to Machine Learning with Python (2016)
14. Murphy, K.P.: Machine Learning a Probabilistic Perspective (2012)
15. He, P., Yau, S.S.-T.: Forecasting stock market volatility using implied volatility. In: IEEE American Control Conference 2007 (2007)

Survey on Fake News Detection Techniques

Sanjeev M. Dwivedi[1(✉)] and Sunil B. Wankhade[2]

[1] Vidyalankar Institute of Technology, Wadala, Mumbai, India
sanjeev.dwivedi@vit.edu.in
[2] Rajiv Gandhi Institute of Technology, Andheri, Mumbai, India
sunil.wankhade@mctrgit.ac.in

Abstract. In recent years widespread rumors and fake news has given rise to many social and political problems. Most of the information today is acquired from digital sources. In Digital media it is difficult to assign accountability to the opinion due to which the data received cannot be authenticated. Lack of constant supervision has motivated the miscreants to spread fake information. Fake news articles that are planted over digital media shares important linguistic features such as immoderate usage of unconfirmed hyperbole and non-verified quotes. It is necessary to invent an automated mechanism to identify fake news and also to minimize its impact by restricting its spread. This survey comprehensively and systematically studies different methodologies in the detection of fake news in digital media. The survey identifies and specifies fundamental theories in Machine Learning, to facilitate and enhance the research of fake news detection. By understanding the different methodologies in fake news studies, we highlight some potential research gaps at the end of this survey.

Keywords: Fake news detection · Sentence matching · Natural language processing · Deep learning · Word embedding · tf-idf · Sentiment

1 Introduction

Fake news is a reason for many problems in society. Fake articles are written with the purpose to impose the author's political and economical inclination on the readers. Internet news and social media generate huge volumes of data that makes it difficult to identify fake news. İmpact of fake news on Twitter during the 2016 US election is still a debatable issue. Analysis of dataset of 171 million tweets from 11 million users over five weeks before the election suggests that over 25% of tweets were biased [1]. Different methodologies have been proposed to detect fake news using natural language processing (NLP) techniques to analyze the content of news articles. However, the content-based analysis limits the accuracy of detection. It is important to understand different other characteristics that can enhance the efficiency of content-based analysis. A lot of research is done on the classification of fake news using NLP technology, recently however deep learning technology like convolutional neural network (CNN) and Transformer Network is proposed to identify discrepancies between contents of the title and text [2].

Machines can solve fake news problems better than humans due to vastness and variance in the news generated in digital platforms. Machines may be more efficient in

J. I.-Z. Chen et al. (Eds.): ICIPCN 2020, AISC 1200, pp. 342–348, 2021.
https://doi.org/10.1007/978-3-030-51859-2_31

surveying a knowledge base to find all relevant articles and answering based on those many different sources. Machines can solve the fake news problem using supervised learning that extracts features of the language and content only within the source in question, without utilizing any fact-checker or knowledge base. Machine learning and NLP techniques can be utilized to create a model that can expose documents that are, with high probability, fake news articles [3].

Machine learning can be used to track sudden surge of new topics posted on digital media by connecting the articles with the origin and the spread of the articles can be curtailed until it is validated by the legitimate medium. This can automate the review process not requiring human intervention every time. It is difficult to stop the creation of fake news articles, but machine learning algorithms can be employed to detect anomaly in news articles and thus can prevent the spread of fake news stories. Using machine learning in fake news detection will not only increase the speed but also the scale of detection. It can also be used to detect anomaly in real-time news articles.

The fake news detection presents the following challenges:

i. A detailed study of the role played by linguistic features in fake news detection must be made [4].
ii. The content and context-based features extracted from different digital media posts for a given time instance cannot be used to generalize the real-world data extracted from a different context.
iii. The availability of standard datasets for fake news articles and related social media posts is still a challenge. Available resources might not help to attain innovative judgement on relevant properties of fake news and to build models that can operate properly in a real world scenario.
iv. The scarcity of characteristics information on influencers who spread misinformation on digital platforms is a significant challenge in profiling fake news.

The varying characteristics of fake news make it's detection a complicated and multifaceted job. The detection techniques exploit the news articles not only based on its headlines, body text, and publisher but also considers the feedback received and how the news propagated. We summarize and compare current studies within this framework.

2 Literature Survey

Sherry Girgis, Eslam Amer, Mahmoud Gadallah, proposes to build a classifier that not only considers the content of news articles but also suggested to apply the vector of words prepared using a word embedding to deep learning perspectives like recurrent neural network (RNN) models and gated recurrent units (GRU) and long short-term memories (LSTM) [5]. The authors then apply the result to LIAR, a publicly available dataset used to detect fake news. To improve the efficiency author seeks a hybrid model between GRU and RNN.

Tim Isbister and Fredrik Johansson, presented a transfer learning approach where a deep learning model pre-trained on specific domain dataset are used to classify news articles [6]. The authors try to classify the news articles to be hyperpartisan argumentation or not. The authors used traditional machine learning classifiers like support vector machine (SVM) and logistic regression, built upon traditional text features such as word and character n-gram term frequencies (weighted with inverse document frequency). For the final submission, they made use of transfer learning-based models as the last layer in deep learning structure. The difference here is the percentage of data used for training/validation splits when fine-tuning the models and the number of epochs for which the models were trained [7]. Despite that, only a few hundred examples were used for fine-tuning the pre-trained universal language model fine-tuning (ULMFiT)-models, accuracies, and F1-scores of approximately 0.8 were achieved on the unseen test data. The authors suggest to improve the result by utilizing the larger available by publisher training set for improving the fine-tuning of the language model on the target domain and would be to train a classifier based on Google AI's bidirectional encoder representations from transformers (BERT), which make use of a deep bidirectional transformer instead of a multilayered LSTM.

Anders E. Lillie et al., suggested on stance classification and fake news detection [8]. Stance classification is used to assess the legitimacy of news articles. Here the author's uses classification as a tool to determine if the opinion in the news articles is inclining toward a specific target. Stance classification presents challenges in regards to social and psychological aspects that appear in microblogs due to polarization of opinion [9]. Usage of the same platform in Stance classification results in the model organism problem. The author also highlights the impact of unbalanced data in the result obtained. Of the many methods, hidden markov model (HMM) is found to be the best in analyzing rumors in microblogs data. The decision tree model is found to yield good results in Stance classification.

Kyeong-hwan Kim and Chang-sung Jeong, proposes a system that verifies the input proposition against related articles to establish semantic legitimacy of input proposition [4]. To overcome the linguistic problem, it is recommended to use of repository built and regularly updated by direct human interaction after verifying its truth. The methodology creates an abstract of the news articles by continuously selecting the upper two sentences that have high cosine similarity. The input proposition is then created by model from the abstract and is assigned to Bidirectional Multi-Perspective Matching for Natural Language Sentence (BiMPM) that returns 0 if an article contains the content of proposition else returns 1 [10]. Also BiMPM, the author's suggested a hybrid method of using article abstraction along with entity matching function to overcome the problem of low performance in case of longer length and unlearned words.

Bhavika Bhutani et al., incorporates sentiment as a significant feature for enhancing the accuracy of detecting fake news [3]. Considering accuracy as the evaluation metric the authors select tf-idf of all the different text pre-processing techniques, then it enhances the quality of the merged dataset using the sentiment. Author's suggested the use of LIAR dataset and twitter data as training to predict the sentiment of a dataset. The methodology Merges Co-Sine similarity, tf-idf, and sentiment analysis [11]. Training model used here is Naïve Bayes and Random Forest.

Ye-Chan Ahn and Chang-Sung Jeong, suggested the use of Fact Data Corpus which is considered as fact to judge the extracted related sentence from input sequence [2]. Here BERT is used to construct a Korean language-specific pre-training model.

It implements a transformer network that uses self-attestation and feed-forward networks instead of LSTM and CNN. The Transformer Network captures global dependency and is capable of parallelization of multi-head Attention. Transformers were developed to solve the problem of sequence transduction, or neural machine translation [12].

Oluwaseun Ajao et al., proposes a hypothesis that relates the fake messages or rumors with the sentiments of the text posted online [13]. Basic machine learning algorithms are used to classify the labeled dataset and deep learning models like LSTM and hierarchical attention network (HAN) are also implemented to improve scale and efficiency. The result obtained consist of different variations in terms of words within text (TX), the emotional ratio (ER) and additional features (AD) that includes counts of uppercase words, exclamation marks, positive and negative emotions, hashtags and quotations.

Songqian Li, Kun Ma et al., proposes an automatic prediction model that combines text pre-processing, feature extraction, and model fusion to yield swift and accurate prediction [14]. It uses a combination of ensemble learning model using stacking and latent semantic analysis. The authors use a slice of base classifiers like grand boosting and random forest and to reduce the dimension of input data latent semantic analysis (LSA) was used. Ensemble learning is a model that combines many weak supervisory models to create a strong and all-inclusive supervisory model. The ensemble learning model helps incorrect prediction even if one of the sub-model fails as different sub-models can have different responsive abilities on different data [15].

Palagati Bhanu Prakash Reddy et al. presented a multinomial voting algorithm for fake data detection [16]. The classification algorithms here use training data as a bag of words model which is constructed using count vectorizer. Ensembled learning is found faster than deep learning. The author's concluded that ensembled algorithms can yield better results as compared to the single algorithm in a particular aspect.

Our detailed study of literature helps us to understand the approaches incorporated by different fake news detection methodologies and also to identify the probable research gaps in each.

3 Summary

Our survey of the literature has shown an inclination toward supervised classification approaches for fake news and rumors detection. This may continue for short period, however, supervised classification models based on deep learning have drawn the interest of researchers for a variety of applications. Basic NLP approaches like Naive Bayes, logistic regression, and SVM are popularly used for feature extraction from text. Complete understanding of fake news from a linguistic perspective is still a challenge. News content could suffer from concept drift problems due to the streaming nature of digital media, hence exhaustive research must be made on understanding the features of classification that will be robust to concept drift. Creating an updated and large dataset for training plays a very significant role in building detection models. The use of deep learning bidirectional transformers yields better results as compared to Multilayered LSTM. Stance Classification has the challenge of social and psychological aspects that can occur in microblogs in polarised opinion. Usage of only linguistic features can be problematic in the case of ambiguous news. In article abstraction, entity matching can be used for longer input of unlearned words. In Table 1, we have summarized the different methodologies for fake news detection, in terms of their approach and features

Table 1. Comparative study of fake news detection methodologies

Author's name	Detection methodology	Research gaps
Ye-Chan Ahn and Chang-Sung Jeong [2019]	Use Fact Data Corpus which is considered as fact to judge the extracted related sentence from input sequence. It implements the transformer network that uses self-attestation and feed-forward networks instead of LSTM and CNN	Corpus must be created in the colloquial form that is frequently used in Social Network Service. Multi-head attention comes at the cost of reduced effective resolution because of averaging attention-weighted positions
Bhutani, B., Rastogi, N., Sehgal, P., & Purwar, A. [2019]	Incorporates sentiment as a significant feature for enhancing the accuracy of detecting fake news The methodology merges Co-Sine similarity, tf-idf, and sentiment analysis	It focusses only on text news and not on images and videos. Neural Network Algorithms, may result in better accuracy
Kyeong-hwan Kim and Chang-sung Jeong [2019]	Proposes a system that verifies the input proposition against related articles to establish semantic legitimacy of input proposition. The author's suggested a hybrid method of using Article Abstraction along with Entity Matching function	Article Abstraction along with Entity Matching can be used to overcome the limitation of BiMPM for longer input sequences of unlearned words
Tim Isbister and Fredrik Johansson [2019]	The author tries to classify the news articles to be hyperpartisan argumentation or not. For the final submission, they made use of transfer learning-based models as the last layer in deep learning structure	The result can be improved by using a deep bidirectional transformer instead of a multi-layered LSTM
Anders E. Lillie and Emil R. Middelboe [2019]	Stance classification as a tool is used to determine if the opinion in the news articles is inclining toward specific target Conversations in microblogs are used for stance classification. Uses RNN to represents arbitrarily sized structured input in fixed-size vector	Stance classification can be challenging where polarised opinions take effect

(continued)

Table 1. (*continued*)

Author's name	Detection methodology	Research gaps
Ajao, O., Bhowmik, D., & Zargari, S. [2019]	Establishes a relation between fake news or rumors and sentiments of the news content posted on digital media. The Sentiment-aware classification improves the detection and prediction of fake news in digital media	Additional sources of sentiment extracted from other multimedia sources may be considered to enhance model performance
Songqian Li, Kun Ma et al. [2019]	It combines text pre-processing, feature extraction, and model fusion. It combines Ensemble learning model and latent semantic analysis	Generalization ability to the stack can be improved by cross-validation of classification models and by considering different features for features models
Palagati Bhanu Prakash Reddy et al. [2019]	This paper presented a multinomial voting algorithm and the use of Ensembled Learning is recommended	Ensembled learning cannot process very large no of features that can be done by Deep Learning

and also specified some insight on the potential research gaps of different methodologies that can facilitate further development in fake news research.

4 Conclusion

In this paper, we tried to perform a comprehensive and extensive study of different fake news detection methodologies. The literature emphasizes content and context-based features for fake news detection. Our survey highlights how the recent research trends are relying more on deep learning classifiers instead of standard supervised machine learning algorithms. Neural network algorithms may result in better accuracy than NLP algorithms. Ensemble learning models can be used to create a better and all-inclusive supervisory model. The consideration of multimedia features is yet to receive significant importance in the literature. Multimedia features of news like images and videos have lately influenced wider audiences and are susceptible to fraudulent intent, hence multimedia features to detect fake news content are growing in importance.

References

1. Bovet, A., Makse, H.A.: Influence of fake news in Twitter during the 2016 US presidential election. Nat. Commun. **10**, 7 (2019). https://doi.org/10.1038/s41467-018-07761-2
2. Ahn, Y.-C., Jeong, C.-S.: Natural language contents evaluation system for detecting fake news using deep learning. In: 2019 16th International Joint Conference on Computer Science and Software Engineering (JCSSE) (2019). https://doi.org/10.1109/jcsse.2019.8864171

3. Bhutani, B., Rastogi, N., Sehgal, P., Purwar, A.: Fake news detection using sentiment analysis. In: 2019 Twelfth International Conference on Contemporary Computing (IC3) (2019). https://doi.org/10.1109/ic3.2019.8844880
4. Kim, K.-H., Jeong, C.-S.: Fake news detection system using article abstraction. In: 2019 16th International Joint Conference on Computer Science and Software Engineering (JCSSE) (2019). https://doi.org/10.1109/jcsse.2019.8864154
5. Girgis, S., Amer, E., Gadallah, M.: Deep learning algorithms for detecting fake news in online text. IEEE (2018). 978-1-5386-5111-7/18/$31.00
6. Isbister, T., Johansson, F.: Transfer learning for hyperpartisan news detection. In: Proceedings of the 13th International Workshop on Semantic Evaluation (SemEval-2019). Minneapolis, Minnesota, USA, June 6–7, pp. 939–943. Association for Computational Linguistics (2019)
7. Seo, S., Cho, S.-B.: Offensive sentence classification using character-level CNN and transfer learning with fake sentences. Lecture Notes in Computer Science, pp. 532–539 (2017). https://doi.org/10.1007/978-3-319-70096-0_55
8. Lillie, A.E., Middelboe, E.R.: Fake news detection using stance classification: a survey. arXiv:1907.00181v1 [cs.CL]. Accessed 29 June 2019
9. Dungs, S., Aker, A., Fuhr, N., Bontcheva, K.: Can rumour stance alone predict veracity? In: Proceedings of the 27th International Conference on Computational Linguistics, Santa Fe, New Mexico, USA, pp. 3360–3370 (2018)
10. Wang, Z., Hamza, W., Florian, R.: Bilateral multi-perspective matching for natural language sentences. arXiv preprint (2017). arXiv:1702.03814
11. Tata, S., Patel, J.M.: Estimating the selectivity of tf-idf based cosine similarity predicates. ACM SIGMOD Rec. **36**(2), 7–12 (2007). https://doi.org/10.1145/1328854.1328855
12. Towards Data Science. https://towardsdatascience.com/transformers-141e32e69591
13. Ajao, O., Bhowmik, D., Zargari, S.: Sentiment aware fake news detection on online social networks. In: ICASSP 2019 - 2019 IEEE International Conference on Acoustics, Speech and Signal Processing (ICASSP) (2019). https://doi.org/10.1109/icassp.2019.8683170
14. Li, S., Ma, K., Niu, X., Wang, Y., Ji, K., Yu, Z., Chen, Z.: Stacking-based ensemble learning on low dimensional features for fake news detection. In: 2019 IEEE 21st International Conference on High Performance Computing and Communications; IEEE 17th International Conference on Smart City; IEEE 5th International Conference on Data Science and Systems (HPCC/SmartCity/DSS) (2019). https://doi.org/10.1109/hpcc/smartcity/dss.2019.00383
15. Mahabub, A.: A robust technique of fake news detection using ensemble voting classifier and comparison with other classifiers. SN Appl. Sci. **2**, 525 (2020). https://doi.org/10.1007/s42452-020-2326-y
16. Reddy, P.B.P., Reddy, M.P.K., Reddy, G.V.M., Mehata, K.M.: Fake data analysis and detection using ensembled hybrid algorithm. In: 2019 3rd International Conference on Computing Methodologies and Communication (ICCMC) (2019). https://doi.org/10.1109/iccmc.2019.8819741

Neural Network Configuration for Pollen Analysis

Amelec Viloria[1(✉)], Darwin Mercado[1], and Omar Bonerge Pineda Lezama[2]

[1] Universidad de la Costa, St. 58 #66, Barranquilla, Atlántico, Colombia
{aviloria7, dmercado}@cuc.edu.co

[2] Universidad Tecnológica Centroamericana (UNITEC), San Pedro Sula, Honduras
omarpineda@unitec.edu

Abstract. Palynology is a botanical discipline devoted to the study of pollen and spores [1], focusing mainly on the analysis of the external morphology that presents structural patterns different from those of the variations in the exine, which is the external wall of the pollen grains. The study and microscopic analysis of its symmetry, wall opening, contour, shape, size, etc., have a taxonomic value and allows distinguishing different taxa at different levels: family, genera, species. The study of pollen grains is a difficult task, in its different phases, from small microscopic samples. The analysis of these is an important source of information for many scientific and industrial applications, making palynology a valuable tool for various areas of knowledge [1]. In palynology, neural networks have been successfully applied for the classification of pollen grains. For this purpose, RPROP was selected as a neural network training algorithm for the classification of a previously reported dataset.

Keywords: Genetic algorithm · Pollen analysis · Neural network configuration

1 Introduction

Plants have undergone a complex process of evolution, which has ensured their adaptation and permanence on the planet. In the course of this process, they have developed a great variety of ways and methods of survival that rely, in a high percentage, on pollination. This is a transcendental phenomenon since life depends largely on it. According to [2], pollination is a biological coevolutionary phenomenon of a great impact on life on the planet. It consists of the mechanical transfer of the pollen grains from the anthers of a flower to the stigma of the same flower, or another flower on the same plant, or other flowers of other plants, but the same species. Subsequently, these grains can germinate to start the fertilization process, which culminates in the fertilization of the ovum, a phenomenon that triggers subsequent physiological changes, which promote the transformation of the fertilized ovule or seminal rudiments fertilized into seeds and the ovaries, and some reminiscences of the perianth of the flowers in fruits [3]. Through this process, plants ensure their success in the ecosystem. Furthermore, the study of pollen grains has allowed research to be carried out in

J. I.-Z. Chen et al. (Eds.): ICIPCN 2020, AISC 1200, pp. 349–358, 2021.
https://doi.org/10.1007/978-3-030-51859-2_32

various fields such as paleontology and medicine, as it stands out in aspects of prevention and cure of allergies, respiratory infections, memory aid and to counteract prostate problems [4]. For these reasons, botany has given it a very important place and has called it Palynology, a science that focuses on the analysis of the external morphology of the pollen grains and, therefore, in the study of the structural patterns of exine and its different variations. Among the main morphological characteristics are size, symmetries, presence and number of openings, contours, shapes, ornaments, etc. This characterization has allowed the generation of taxonomic or classification systems at different levels: family, genus, and species (see Fig. 1) [5–7].

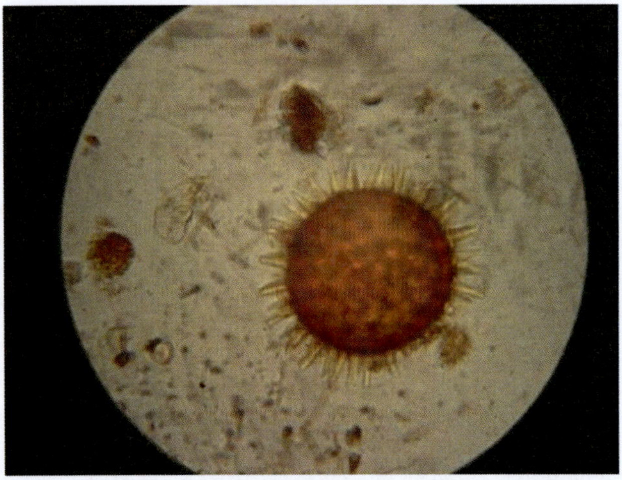

Fig. 1. Microphotograph of the species Malacra fasciata (Family Malvaceae), known as Malva peluda. Spinel ornaments are appreciated in exine [2].

Semi-automatic or automatic systems methods for pollen analysis based on image analysis and artificial intelligence techniques, in particular neural networks and vector support machines, have been proposed for a long time [4, 8–10]. In general terms, the proposed methods segment the image and extract texture characteristics both in intensity images and in some color space. [11] proposes a methodology to segment superimposed pollen grains. The method separates pollen according to the process of average change and erosion by minima based on the Fibonacci series. From the segmented image, they extract shape characteristics, and from the CIELab color space they obtain first- and second-order texture characteristics. Three classifiers are considered for the recognition of pollen grains: random tree forest, multilayer perceptron, and Bayes' net. The classification results are 96.2% and 96.1% in sensitivity and accuracy with a multilayer perceptron.

This paper analyzes the performance of three neural network topologies trained with the RPROP algorithm with and without weight regression for the classification of a pollen grain dataset, which consists of 12 palynological classes of pollen with 618 instances [12]. The above-mentioned topologies were obtained using a genetic algorithm, which determines the appropriate architecture that maximizes accuracy.

2 Theoretical Framework

2.1 Neural Networks

An artificial neural network is an information processing system that functions in a similar way to biological neural networks [13]. The basis for these models are the following assumptions [14–16]:

- Information processing takes place in a large number of simple elements called neurons.
- The signals are transmitted between the neurons along with the connecting links.
- Each connecting link is assigned a weight that multiplies the transmitted signal.
- Each neuron applies a trigger function to its network input (which is the sum of the weighted input signals) to obtain its output signal.

There are two structures in the development of neural networks, single-layer networks, and multilayer networks. An example of a neural network is shown in Fig. 2, indicating the interconnections of the network, as well as the weights between them.

2.2 RPROP

In [17], they introduced RPROP as an extension of the classical backpropagation training algorithm. RPROP replaces the constant learning rate ρ by adaptive weight sizes Δij that are controlled during learning of the wij weights. The step size Δij of the steps is updated as [4]:

$$\Delta_{ij}^{(t-1)} = \begin{cases} \Delta_{ij}^{(t-1)} \cdot \eta^{+} & if \quad \dfrac{\partial E^{(t)}}{\partial w_{ij}} \cdot \dfrac{\partial E^{(t-1)}}{\partial w_{ij}} > 0 \\ \Delta_{ij}^{(t-1)} \cdot \eta^{-} & if \quad \dfrac{\partial E^{(t)}}{\partial w_{ij}} \cdot \dfrac{\partial E^{(t-1)}}{\partial w_{ij}} > 0 \\ \Delta_{ij}^{(t-1)} & else \end{cases} \tag{1}$$

With parameters $0 < n^{-} < 1 < n^{+}$ which specifies the step size adaptation magnitude, if the gradient sign $\frac{\partial E^{(T-1)}}{\partial w_{ij}}$ changes, where E indicates the error in the grid output. The weights are updated with [18]:

$$\Delta w_{ij}^{(t)} = -\sin\left(\frac{\partial E^{(t)}}{\partial w_{ij}}\right) \cdot \Delta_{ij}^{(t)} \tag{2}$$

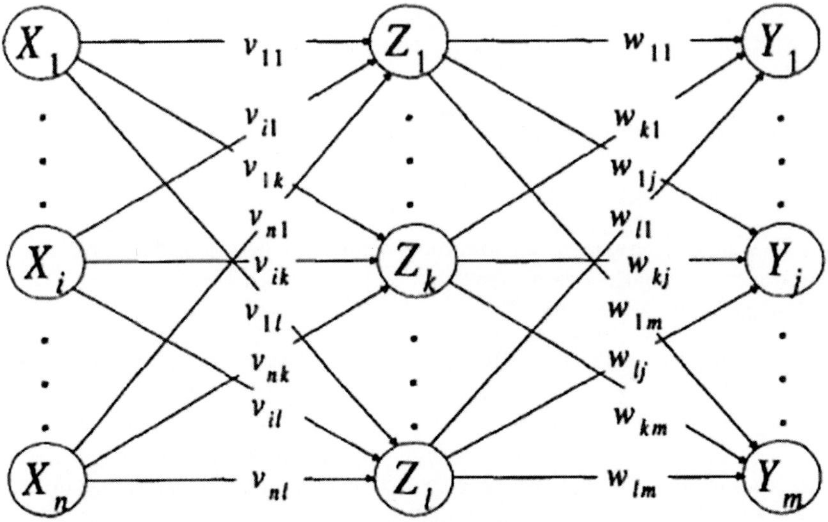

Fig. 2. Structure of a multilayer neural network [4]

The update value Δij is reduced by a factor $\eta-$. If the derivative retains its sign, the updated value increases slightly by the factor $\eta+$ to accelerate convergence in shallow regions [19, 20].

Table 1 shows a comparison between the Backpropagation training algorithm and its RPROP variant [21]. RPROP considers backward propagation with and without weight regression, as defined by the RPROP+ and RPROP− algorithms, respectively [22].

Table 1. Algorithm comparison.

Backpropagation	RPROP
● Employee two-phase propagation-adaptation cycle ● It is iterative ● Their convergence is delayed ● It does not require the standardization of input vectors ● It can generalize, correctly classifying noisy or incomplete patterns	● Its scale of time and memory requirements is linear with the number of parameters to be optimized ● Provides faster training and convergence rate ● It is robust ● It is considered one of the best learning methods among the algorithms ● It is complicated to schedule

2.3 Topology

An important aspect of neural network applications is determining an appropriate topology, i.e. the number of hidden layers and the number of neurons per hidden layer, training and test ratios, initial weights, and the number of iterations (epochs). While there are some general rules, an optimal solution to this problem cannot be determined in a reasonable time [23]. Generally, a network topology is selected by trial and error, choosing the best one. Finding out how many hidden layers should be used, how many neurons each hidden layer should have, what kind of firing function to use, what algorithm and learning rate should be used for neural network training is complex.

Genetic algorithms (GA) together with neural networks have been used in various applications [24]. They have been used to find the topology of the hidden layers, the number of connectivity and optimal training parameters to analyze biomedical data related to cardiac diseases [25]. The use of GA replaces trial-and-error methods to determine the optimal neural network topology and increases the speed, accuracy, and efficiency of the training phase for a higher degree of classification accuracy. GA is used for multivariate optimization and can solve a wide range of optimization problems, using three simple operations: selection, crossing, and mutation [26]. Consequently, by combining genetic algorithms and neural networks, optimal network architectures are obtained in less time. With this approach, besides reducing the execution time when executing the algorithm achieving optimal points. For this study, within the space of search of the genetic algorithm, it was considered to maximize the accuracy of the neural network configurations. Table 2 shows the 3 configurations which give the highest accuracy with different training and testing reasons [27, 28]. It is worth mentioning that each of these configurations has 33 input neurons and 12 output neurons.

As can be seen in Table 2, the number of neurons in the hidden layers for each of the configurations considered are too many, which is why none of the analyzed topologies are shown since the hidden layers with their respective neurons, interconnections and *wij* weights are not distinguished.

Table 2. Selected configurations.

	Layer inbound	Hidden layers			Times	Reason for training and testing
		First	Second	Third		
Configuration 1	38	44	30	21	32	72–35
Configuration 2	38	15	32	51	31	95–12
Configuration 3	38	37	50	21	41	84–22

3 Dataset

The dataset used in this document consists of 1,025 instants divided into 12 different classes of pollen [29]. Figure 3 shows a representative picture of each of the pollen classes being analyzed. The classes of the dataset, with their respective palynological name and the number of observations per class, are shown in Table 3.

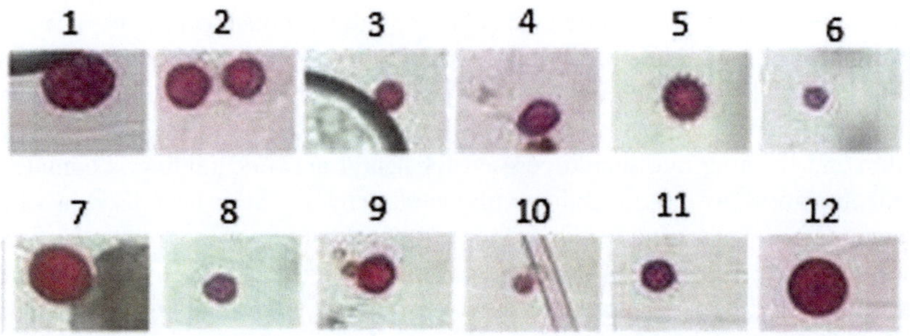

Fig. 3. Pollen grains by classes of 1–12 and are represented as follows: 1-Huizache, 2-Alpha, 3-Nettle, 4-Pasture, 5-Herb, 6-Morera, 7-Pacana, 8-Olive, 9-Mesquite, 10-Willow, 11-Pimper, 12-Sorghum.

Table 3. Palynological classification of the pollen species image dataset

Class	Pollen name	Comments
1	Huizache	47
2	Alfalfa	47
3	Nettle	55
4	Grass	38
5	Herb	39
6	Mulberry tree	60
7	Pacana	31
8	Olive	55
9	Mesquite	79
10	Sauce	66
11	Pepper	69
12	Aleppo Sorghum	35

The characteristics reported in the dataset were extracted from segmented images in the CIELab color space. The CIELab is a uniform perceptual color model in which the L component represents the luminosity of an object from black (0) to white (100), *a* represents the variation from green (−) to red (−), and *b* represents the variation from blue (−) to yellow (+). A numerical variation of these values corresponds to approximately the same amount of visually perceived change. From the segmented images, the shape characteristics were obtained, and from the CIELab color space, the first and second-order texture characteristics, described by gray level co-occurrence matrices. These data are shown in Table 4. As mentioned, there is a total of 33 characteristics, 6 of shape, 15 of first-order texture, and 12 of second-order texture.

Table 4. Characteristics of the dataset.

Features		
Form	Texture (CIE-L*a*b*)	
	First-order	Second-order
Width Height Perimeter Circularity Area Compactness	Average Medium Variance Standard deviation Entropy	Contrast Energy Correlation Homogeneity

4 Results

To test each of the topologies described in Table 1, the statistical software R and its graphic user interface RStudio were used on a laptop with i3 processor, 8 GB RAM, a 64-bit operating system.

Table 5 shows an example of the confusion matrix obtained in configuration 1 using RPROP+, with the sampling rate 70−30, i.e. 70% of the dataset for training the neural network and 32% for testing. This configuration obtained an accuracy of 99.11%. In the confusion matrix, the red color indicates that there was an error of classification, e.g., alpha was confused with both grass and mesquite.

Table 5. Confusion matrix 70–30 with the Rprop+ algorithm

	Reference													
	70%	1	2	3	4	5	6	7	8	9	10	11	12	Total
Classification	1	47	0	0	0	0	0	0	0	0	0	0	0	47
	2	0	47	0	1	0	0	0	0	1	0	0	0	47
	3	0	0	55	0	0	0	0	0	0	0	0	0	55
	4	0	0	0	38	0	0	0	0	0	0	1	0	38
	5	0	0	0	0	39	0	0	0	0	0	0	0	39
	6	0	0	0	0	0	60	0	0	0	0	0	0	60
	7	0	0	0	0	0	0	31	0	0	0	0	0	31
	8	0	0	0	0	0	0	0	55	0	0	0	0	55
	9	0	0	0	0	0	0	0	0	79	0	0	0	79
	10	0	0	0	0	0	1	0	0	0	66	0	0	66
	11	0	0	0	0	0	0	0	0	0	0	69	0	69
	12	0	0	0	0	0	0	0	0	0	0	0	35	35
	Total	47	47	55	38	39	60	31	55	79	66	69	35	621

Table 6 shows the results obtained from each of the configurations mentioned above. It is worth mentioning only the results of the proportions that obtained the best accuracy. As mentioned, the training algorithms of the topologies were Rprop+ and Rprop−.

Table 6. Results.

	Reason for training and testing	Rprop+	Rprop-	Cross-validation
Configuration 1	70–30	99.11%	98.74%	96.25
Configuration 2	90–10	99.92	99.20	94.06
Configuration 3	80–20	98.77	99.11	97.75

In configurations 1 and 2, the Rprop algorithm with weight regression (Rprop+) obtained the best accuracy result reaching values from 99.11% to 99.92%, while in configuration 3 the Rprop algorithm without weight regression (Rprop−) reached an accuracy of 99.11%, this being higher than the Rprop+ accuracy result for this configuration.

The validity of the results was checked through cross-validation, obtaining results of accuracy between 94 and 97%. Cross-validation is a statistical method used to evaluate, compare, and avoid over-training of learning algorithms, by using the subset of data labeled for validation [30–32].

In cross-validation, the information is divided into two subsets: one is used for training the model, while the other is used to validate it, its most basic form is known as k-folds where the training and validation subsets are exchanged through iterations so that each subset of data can be validated.

5 Conclusion

The current method for the classification of pollen grains is a qualitative process, by observation and discrimination of characteristics of the pollen grain. This procedure depends on the experience of the expert in analyzing the morphology of this grain and knowing how to use and interpret the palynological guidelines and keys. Despite its effectiveness, the process is slow and a large amount of time and resources must be invested, which limits the investigation processes, since these depend, to a large extent, on the availability of the expert. Furthermore, there is currently no quantitative support for the classification of pollen grains. Therefore, automating this process in a safe and precise way speeds up the recognition, identification, and power of research processes in this area.

In this study, three neural network topologies trained with the RPROP algorithm to perform pollen classification were considered. The results obtained with these networks are superior to those obtained previously [30] since in both cases the same dataset is used. The accuracy of these topologies is above 98%, while the reported results are 96% accurate, also using artificial neural networks, but the topology they used to make their respective analysis is unknown.

References

1. Rodriguez, I.F., Mégret, R., Acuna, E., Agosto-Rivera, J.L., Giray, T.: Recognition of pollen-bearing bees from video using convolutional neural network. In: 2018 IEEE Winter Conference on Applications of Computer Vision (WACV), pp. 314–322. IEEE, March 2018
2. Carpenter, G.A.: Neural-network models of learning and memory: leading questions and an emerging framework. Trends Cogn. Sci. **5**(3), 114–118 (2001)
3. Al-Saqer, S.M., Hassan, G.M.: Artificial neural networks based red palm weevil (Rynchophorus Ferrugineous, Olivier) recognition system. Am. J. Agric. Biol. Sci. **6**, 356–364 (2011)
4. Burki, C., Šikoparija, B., Thibaudon, M., Oliver, G., Magyar, D., Udvardy, O., Pauling, A.: Artificial neural networks can be used for Ambrosia pollen emission parameterization in COSMO-ART. Atmos. Environ. **218**, 116969 (2019)
5. Li, P., Treloar, W.J., Flenley, J.R., Empson, L.: vf: vf towards automation of palynology 2: the use of texture measures and neural network analysis for automated identification of optical images of pollen grains. J. Quat. Sci.: Publ. Quat. Res. Assoc. **19**(8), 755–762 (2004)
6. Li, P., Flenley, J.R.: Pollen texture identification using neural networks. Grana **38**(1), 59–64 (1999)
7. Dewan, P., Ganti, R., Srivatsa, M., Stein, S.: NN-SAR: a neural network approach for spatial autoregression. In 2019 IEEE International Conference on Pervasive Computing and Communications Workshops (PerCom Workshops), pp. 783–789. IEEE, March 2019
8. Friedman, M., Kandel, A.: Introduction to Pattern Recognition Statistical, Structural, Neural and Fuzzy Logic Approaches. World Scientific Publishing Company Inc., 2 March 1999
9. Zewdie, G.K., Lary, D.J., Levetin, E., Garuma, G.F.: Applying deep neural networks and ensemble machine learning methods to forecast airborne ambrosia pollen. Int. J. Environ. Res. Public Health **16**(11), 1992 (2019)
10. Zhao, X., Yue, S.: Analysing and forecasting the security in supply-demand management of Chinese forestry enterprises by linear weighted method and artificial neural network. Enterp. Inf. Syst. 1–18 (2020)
11. Riedmiller, M., Braun, H.: Direct adaptive method for faster backpropagation learning: the RPROP algorithm. In: Proceedings of the IEEE International Conference on Neural Networks, pp. 586–591 (1993)
12. Daood, A., Ribeiro, E., Bush, M.: Sequential recognition of pollen grain Z-stacks by combining CNN and RNN. In: The Thirty-First International Flairs Conference, May 2018
13. Raj, J.S., Ananthi, J.V.: Recurrent neural networks and nonlinear prediction in support vector machines. J. Soft Comput. Paradig. (JSCP) **1**(01), 33–40 (2019)
14. Varela, N., Silva, J., Gonzalez, F.M., Palencia, P., Palma, H.H., Pineda, O.B.: Method for the recovery of images in databases of rice grains from visual content. Procedia Comput. Sci. **170**, 983–988 (2020)
15. Khorissi, N.E., Mellit, A., Guessoum, A., Mesaouer, A.: GA-based feed-forward neural network for image classification: application for the grains of pollen. J. Appl. Comput. Sci. **17**(2), 83–96 (2009)
16. Viloria, A., Acuña, G.C., Franco, D.J.A., Hernández-Palma, H., Fuentes, J.P., Rambal, E.P.: Integration of data mining techniques to PostgreSQL database manager system. Procedia Comput. Sci. **155**, 575–580 (2019)
17. Pentoś, K., Łuczycka, D., Wróbel, R.: The identification of the relationship between chemical and electrical parameters of honeys using artificial neural networks. Comput. Biol. Med. **53**, 244–249 (2014)

18. Al-Mahasneh, M.A., Rababah, T.M., Ma'Abreh, A.S.: Evaluating the combined effect of temperature, shear rate and water content on wild-flower honey viscosity using adaptive neural fuzzy inference system and artificial neural networks. J. Food Process Eng **36**(4), 510–520 (2013)
19. Viloria, A., Lezama, O.B.P.: Improvements for determining the number of clusters in k-Means for innovation databases in SMEs. In: ANT/EDI40, pp. 1201–1206 (2019)
20. Rashidi, M.M., Galanis, N., Nazari, F., Parsa, A.B., Shamekhi, L.: Parametric analysis and optimization of regenerative Clausius and organic Rankine cycles with two feedwater heaters using artificial bees colony and artificial neural network. Energy **36**(9), 5728–5740 (2011)
21. Pentoś, K., Łuczycka, D., Kapłon, T.: The identification of relationships between selected honey parameters by extracting the contribution of independent variables in a neural network model. Eur. Food Res. Technol. **241**(6), 793–801 (2015)
22. Peyron, O., Vernal, A.D.: Application of artificial neural networks (ANN) to high-latitude dinocyst assemblages for the reconstruction of past sea-surface conditions in Arctic and sub-Arctic seas. J. Quat. Sci.: Publ. Quat. Res. Assoc. **16**(7), 699–709 (2001)
23. Mokin, V.B., Kozachko, O.M., Rodinkova, V.V., Palamarchuk, O.O., Vuzh, T.Y.: The decision support system for the classification of allergenic pollen types based on fuzzy expert data of pollen features on the microscope images. In: 2017 IEEE First Ukraine Conference on Electrical and Computer Engineering (UKRCON), pp. 850–855. IEEE, May 2017
24. Todd, G.: Fuzzy neural network interface: development and application: a thesis presented in partial fulfilment of the requirements for the degree of Master of Engineering in Information Engineering at Massey University, Palmerston North, New Zealand, (Doctoral dissertation, Massey University (2003)
25. Cho, H., Berger, B., Peng, J.: Generalizable and scalable visualization of single-cell data using neural networks. Cell Syst. **7**(2), 185–191 (2018)
26. Lehky, S.R., Sejnowski, T.J.: Neural network model of visual cortex for determining surface curvature from images of shaded surfaces. Proc. R. Soc. Lond. B Biol. Sci. **240**(1298), 251–278 (1990)
27. Dell'Anna, R., Lazzeri, P., Frisanco, M., Monti, F., Campeggi, F.M., Gottardini, E., Bersani, M.: Pollen discrimination and classification by Fourier transform infrared (FT-IR) microspectroscopy and machine learning. Anal. Bioanal. Chem. **394**(5), 1443–1452 (2009)
28. Chiroma, H., Khan, A., Abubakar, A.I., Saadi, Y., Hamza, M.F., Shuib, L., Herawan, T.: A new approach for forecasting OPEC petroleum consumption based on neural network train by using flower pollination algorithm. Appl. Soft Comput. **48**, 50–58 (2016)
29. Tomassetti, B., Lombardi, A., Cerasani, E., Di Sabatino, A., Pace, L., Ammazzalorso, D., Verdecchia, M.: Mapping of Alternaria and Pleospora concentrations in Central Italy using meteorological forecast and neural network estimator. Aerobiologia **29**(1), 55–70 (2013)
30. Guyon, V.N., Astwood, J.D., Garner, E.C., Dunker, A.K., Taylor, L.P.: Isolation and characterization of cDNAs expressed in the early stages of flavonol-induced pollen germination in petunia. Plant Physiol. **123**(2), 699–710 (2000)
31. Ramos-Pollán, R., Guevara-López, M.Á., Oliveira, E.: Introducing ROC curves as error measure functions: a new approach to train ANN-based biomedical data classifiers. In: Iberoamerican Congress on Pattern Recognition, pp. 517–524. Springer, Heidelberg, November 2010
32. Raghu, P.P., Poongodi, R., Yegnanarayana, B.: Unsupervised texture classification using vector quantization and deterministic relaxation neural network. IEEE Trans. Image Process. **6**(10), 1376–1387 (1997)

Framework to Enhance the Reachability of AI Techniques in Medical

Upasana Gaur(✉) and Monika Sharma

AIIT, Amity University, Noida, Uttar Pradesh, India
uppugaur998@gmail.com, msharma5@amity.edu

Abstract. Artificial Intelligence is being used in medical area because of its benefits as it is more capable of solving complex medical data in many ways. As per the physician says that technology help more in giving treatment to the patient because we can use the technology for getting report and information about the disease. Computer based information helps in making early decisions about the treatment. Medicine and web looks were completed utilizing the catchphrases 'Artificial intelligence reasoning' and 'neural systems (PC)'. Artificial intelligence is being used almost in every medical field. There are many techniques like fuzzy expert systems, evolutionary computation and hybrid intelligent systems and many more these are used in many different clinical areas according to their needs. In this paper, we have analysed the artificial intelligence technique uses in different medical treatments and another work of the medical area. We have also proposed a framework on how we can improve the limitations of the techniques.

Keywords: Artificial intelligent · Neural systems · Techniques · Hybrid intelligence · DXplain · Germ watcher

1 Introduction

Artificial intelligence is a trending field nowadays but it is very useful in the medical area as well. Artificial intelligence/technologies are used to diagnose and treatment of the patients who required special treatment and care for their health. Diagnosing a person using the techniques of artificial intelligence is not a single simple step instead it required many steps. Because of artificial intelligence all these steps can be done automated which leads to saving the time of professional doctors who can do such important work that can't be done with the help of artificial techniques. All the work will be done in less time. Although the professionals are also aware of the fact that we can't use the AI and automation in every work because medical is not the work which can be done by technologies without any help of the man-strength but the work in which we use the automation and technologies that can save more work done time. The one and only goal of using the artificial intelligence in medical is to reduce the time and give the best results to the patients [1, 3].

J. I.-Z. Chen et al. (Eds.): ICIPCN 2020, AISC 1200, pp. 359–365, 2021.
https://doi.org/10.1007/978-3-030-51859-2_33

2 Literature Review

In present scenario, artificial intelligence is used in various fields but medical is using its best in giving treatments to the people and analysing the reports for different diseases.

Gawad J. and Bonde C [2] also includes the various gadgets that have being used in houses for the diagnose of the diseases and they are available at the amazon propelled market from where everyone can access them for their use. They also include the term called telemedicine in which people can reach the doctor on the telephone and ask the problem and get the treatment all over the phone.

AN Ramesh, C Kambhampati, JRT Monson, PJ Drew [1] they discussed the use of artificial intelligence in neural networks in the diagnose of diseases, as image analysis in radiology and histopathology, data interpretation in intensive care setting and waveform analysis [4]. These are used in diagnosing breast, gastric, thyroid, oral epithelial cells these are focused to be analysis by the neural networks with the help of artificial intelligence [5]. And in radiology we observe the human observations and digital observation on the images like ultrasound, CT, MRIs.

Mudit verma [7], he proposed various fields where artificial intelligence has been used in various domains like education, gaming, entertainment, emergency services, and he also includes the use of AI in the medical area to its best as he includes that AI saves major time in making decisions for the treatment which is given to the patients [6]. By using machines with the AI features for the treatment we get accurate results of the tests and can get better treatment.

Yong jiang, Hui Zhi, Yi Dong [3] they all are from the Department of Neurology, China. They mentioned the various AI devices in the detail as they studied and conclude that there are two type of AI devices which falls in the medical area [7]. First category includes machine learning (ML) techniques which is used in analyse the structured date and imaging which gives the accuracy in the process of diagnosing the patients. Second category includes the Natural language processing (NLP) method [6–10] this is used to get the simple data from the complicated clinical notes. These are used to convert the normal text into machine structured data which can be analyse by the Machine learning techniques.

3 How AI Is so Suitable for Medical Area

AI is very much able to perform different and difficult, both the tasks in well fixed manner and in medical area AI is so much capable of doing text and written documents, statistics numbers and visual images and its all will done in very less time as compare to a human being.

As AI is being saving time for the doctors who are also doing many other things like researching on any other diseases and there treatments so because of AI they gets more time for this, they can spend more time studying medicines instead of managing statistic data and making report manually themselves. All this is leading increase of people getting knowledge of Artificial algorithms and machine analytical algorithms for make their work more accurate. Growth in use of analytical algorithms and machine works that also mean that AI will become more powerful and more effective in performing their assigned tasks. So the work on improving the technology for medicine is always in active stage where they test new technologies for effective medicines. There are so many applications we are using [4].

4 AI Based Techniques Works for Medical

As there is already a increasing amount of technology and automated work that plays very important role in medicines.

People can do many with the help of AI as they can book there appointments online and patients can check the availability of the hospitals and clinics just by using their cell phones.

AI is being used in every fields for increment the result of their work so as AI is adapted by the medical area and AI changed many ways in which it seeks to take care of medicine:

4.1 DXplain

It helps in asking decision more accurately, in this we get a list of possible diagnoses that can help to reach the correct diagnose to start the treatment of the patients.

4.2 Germ Watcher

This helps in getting information from laboratory, this application is used to detect the investigate the infection from which patients are suffering.

4.3 La Vinci

The La Vinci robotic surgical system is responsible for robotic arms, precise movement and magnetic vision helps the doctor to do surgery which can not be done manually.

4.4 Therapy

In AI "Therapy" is an online course which helps the patient to reduce their anxiety.

4.5 Babylon

As in medicine we have to handle big data and many documents, as we know the technology is much time saver so be use Babylon application which help working accurately for the big data and documents because of this occurring of error reduced [3, 5].

5 AI for Diagnostics

5.1 Medical Imaging

Medical imaging can be diagnoses because of AI application to improve both the time and accuracy in performing the given tasks. Development of algorithms and machine learning is the biggest research area in the AI technologies. Medical imaging is the sufficient application which is able to aggregate the image data into other data for further calculations. And working in proper manner it helps the doctor to predict the diagnose for the individual patients. Because of the difficult work of diagnose the patients on the basis of X-Ray reports can be get wrongly done by the most skilled doctors or radiologists so we are using AI application to be accurate in the results of diagnosis.

5.2 Medical and Physical Health Screening

As there is a big problem which is shortage of doctors and nurses because of this treatment of individual patients but we can solve this problem by using the AI screening application by which patient can get the diagnose through the AI application.

5.3 AI in Drugs Discovery

A relevant subset of medical discovery is to discover new drugs according to the need of patients and diseases they are facing is being the long and arduous work to be done. There are 5 stages of discovering the new drugs and making to market ready:

A. Discovery and development
B. Preclinical research
C. Clinical research
D. FDA review
E. FDA -post market safety monitoring

The presentation of AI frameworks is viewed as the answer for quicker, less expensive and progressively proficient medication disclosure, ready to crunch a lot of information, play out various emphases of compound re-enhancement and demonstrating, and perceiving designs in information that will prompt medications that are increasingly compelling. We may likewise observe drugs that are customized to every individual's natural well being profile, a key part of customized medicinal services.

AI helping in both saving time and cheaper in cost for discovering the new drugs also effect the patients as they can get the treatment on and before the time and as well as in very less cost.

Simulated intelligence will make these procedures less demanding and quicker. Medication disclosure determines its bits of knowledge into malady from a gigantic pool of information, including high-goals restorative pictures, genomic profiles, subatomic structures and organic data, to a great extent distributed in papers, diaries and medicinal preliminary reports. Human specialists regularly just read a couple of hundred distributed restorative articles for every year, out of the thousands that are accessible [4, 6].

6 Customized Health Care Is the Future

Customized medicinal services is the name of the amusement, and AI is viewed as the best way to achieve this result. The huge information investigation capacities of profound learning AI frameworks guarantee quicker and progressively exact analyses, custom treatment and clinical consideration plans, and less expensive, increasingly viable medications custom-made to every patient's infection and necessities - and these applications are only a glimpse of a larger problem. Even though it's still early days, the accomplishments officially made by AI in social insurance are energizing, and point to a sound future [1, 4, 5].

7 Limitations of AI Techniques Used in Healthcare

There are so many techniques used in the medical area but at some extent, they create some problems in using them.

Such as neural networking which has been used in diagnosing as it have the better-generalized ability which helps to predict the results but the problem which occurs in this is we can get stuck in the local minima during the training process rather than converging to the global minima.

While using the techniques we have to check them whether it is performing properly or not so for that we need someone to monitor it as AI techniques is not optimal for all the diagnoses (Fig. 1).

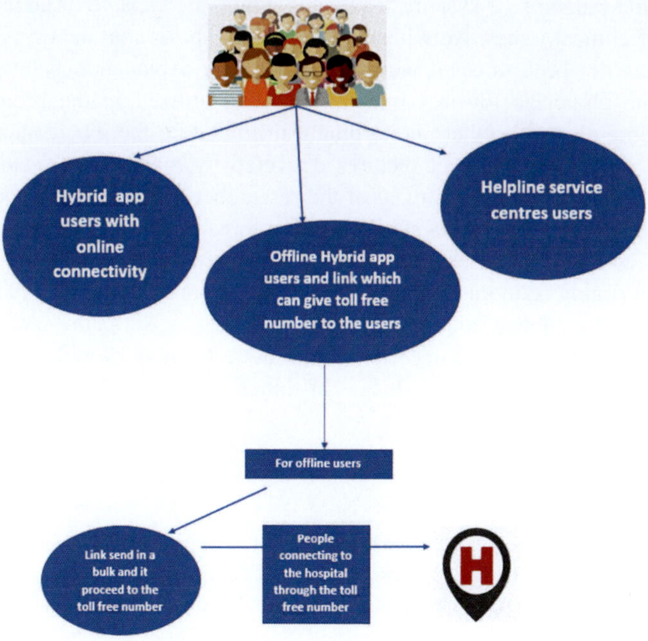

Fig. 1. Framework to improve TELEHEALTH technique.

8 Proposed Framework

As we know the telehealth technique is not reachable to all people so we have proposed the solution:

In the above framework, there are three parts where we can solve three different problems of the users.

For those who have an internet connection and have a smart phone to use the app or to use the telehealth techniques, we can also design for them a hybrid app by which they have both the elements native and web application so that the app can work on every platform. This will help the user and app will work whenever they need it.

In second condition user may have cell phones but they don't have internet connectivity and because of that, they can't use the app. So to solve this problem we can provide offline working app for registered users so that they can use the app on those places also where they are unable to connect themselves from the internet. And in the emergency user can call the toll free number to connect themselves to the hospitals.

And in the third condition, we have considered for poor and slum areas where people don't have smartphones and neither the internet connection to get rid of this problem we can build some service centers around the slum areas, where people can visit anytime and can get the facilities of Telehealth technique and get themselves the treatment from the hospital.

9 Conclusion

There is a wide range of AI systems accessible which are capable of taking care of an assortment of clinical issues. Notwithstanding, despite prior confidence, restorative AI innovation has not been grasped with eagerness. One explanation behind this is the attitude of the clinicians towards innovation being utilized in the basic leadership process. Incomprehensibly, there is no qualm in tolerating the biochemical outcomes created from an auto-analyzer or pictures delivered by magnetic resonance imaging. Notwithstanding, it is the commitment of the researcher dynamic in this field to create proof that these techniques take a shot at a down to earth level. The need to undertake more randomized controlled examinations to demonstrate the adequacy of AI frameworks in prescription is, in this manner, vital. There is convincing proof that restorative AI can assume a vital job in helping the clinician to convey wellbeing care efficiently in the 21st century. There is little uncertainty that these techniques will serve to upgrade and supplement the 'Medical knowledge' of things to come, clinician.

References

1. Ramesh, A.N., Kambhampati, C., Monson, J.R.T., Drew, P.J.: Artificial intelligence in medicine. Ann. Royal Coll. Surgeons England **86**(5), 334 (2004)
2. Gawad, J., Bonde, C.: Artificial intelligence: future of medicine and healthcare. Biochem. Ind. J. **11**(2), 113 (2017)

3. Jiang, F., Jiang, Y., Zhi, H., Dong, Y., Li, H., Ma, S., Wang, Y., Dong, Q., Shen, H., Wang, Y.: Artificial intelligence in healthcare: past, present and future. Stroke Vasc. Neurol. **2**(4), 230–243 (2017)
4. Ashizawa, K., Ishida, T., MacMahon, H., Vyborny, C.J., Katsuragawa, S., Doi, K.: Artificial neural networks in chest radiography: application to the differential diagnosis of interstitial lung disease. Acad. Radiol. **6**(1), 2–9 (1999)
5. Pandian, A.P.: Artificial intelligence application in smart warehousing environment for automated logistics. J. Artif. Intell. **1**(02), 63–72 (2019)
6. Tsang, L., Kracov, D.A., Mulryne, J., Strom, L., Perkins, N., Dickinson, R., Wallace, V.M., Jones, B.: The impact of artificial intelligence on medical innovation in the european union and United States. Intellect. Prop. Technol. Law J. **29**, 3–12 (2017)
7. Verma, M.: Artificial intelligence and its scope in different areas with special reference to the field of education. Artif. Intell. **3**(1) (2018)
8. Shapiro, S.C.: Encyclopedia of Artificial Intelligence, Second edn. John (1992)
9. https://www.google.com/search?rlz=1C1CHBF_enIN796IN797&tbm=isch&q=people+icon&chips=q:people+icon,g_1:crowd:kIgcV62XVJA%3D&usg=AI4_-kTm1bEIUg9OLEjADSLR24TvFhpnqg&sa=X&ved=0ahUKEwjP_cPxirbhAhU07nMBHe7EBuIQ4lYILCgC&biw=1280&bih=610&dpr=1.5#imgdii=wkgSPBk3i_FDsM:&imgrc=fJLji7oNELE0pM:
10. https://www.google.com/search?rlz=1C1CHBF_enIN796IN797&biw=1280&bih=610&tbm=isch&sa=1&ei=HsmlXKnWLqSEmgeIkYzoDw&q=hospital+icon&oq=hospital+icon&gs_l=img.3…1640793.1647016..1647732…0.0..0.0.0…….1….1..gws-wiz-img…..0.oAFE8VKYtE8#imgdii=hCF0m5Icg6ngDM:&imgrc=0IedD_f4dDPPQM:

Blockchain Based E-Healthcare Record System

Sharyu Kadam and Dilip Motwani(✉)

Computer Engineering Department, Vidyalankar Institute of Technology, Mumbai, India
sharyuk09@gmail.com, dilip.motwani@vit.edu.in

Abstract. Electronic Healthcare Record (EHR) is an electronic variant of patient's health history, which is sustained by the system over a period, and can incorporate the entirety of the key authoritative clinical information which is significant to the users care under a particular framework supplier, contains socioeconomics, progress notes, hitches, medications, essential signs, past clinical history, vaccinations, and radiology reports. The use of an EHR is significant to make progress, just as to ensure an elevated level of sheltered and compelling degree of patient consideration framework. Patient records are fully monitored by hospitals rather than patients, which becomes complicated to seek clinical advice from various medical clinics. Patients face stringent need to concentrate on the subtleties of their human services and reestablish the executives of their clinical information This advanced system will account for interactions with this medical data in an auditable, transparent and secure way on EHR's distributed ledger. By using a decentralized patient-centered approach users will be able to leverage their medical data. This system will use to improve care for individuals by setting the patient at the focal point of the advanced change of healthcare. The rapid advancement of blockchain innovation with smart contracts and IPFS storage improves the existing medicinal services, including clinical records just as patient-associated evidence. This advanced model provides high flexibility and EHRs availability for healthcare services.

Keywords: Blockchain · Electronic Health Record (EHR) · Security

1 Introduction

E-Healthcare Records (EHRs) serves an invaluable health record amassing organization, which bolster customary patient clinical records on paper to be electronically open on the web [1]. It is viewed as that wellbeing data innovation, especially EHR, will improve the quality and effectiveness of social insurance associations, from little practices to huge gatherings. Dependent on these potential advantages, the government-supported EHR selection under the health information technology for economic and clinical health (HITECH) Act. In recuperation, numerous emergency clinics are endeavoring to partner these frameworks and exhibit its significant use.

This framework is practiced to permit patients for securing the standard of producing, overseeing, and sharing EHRs with family, companions, social insurance suppliers, and other approved information purchasers [3]. Besides, contributed that the human services analyst and suppliers of such assistance get to these EHRs over the

J. I.-Z. Chen et al. (Eds.): ICIPCN 2020, AISC 1200, pp. 366–380, 2021.
https://doi.org/10.1007/978-3-030-51859-2_34

onboard, the progress program of medicinal services arrangement is required to be accomplished. Nonetheless, in the current circumstance, patients communicate their EHRs over the various regions during life occasions, making the EHRs move to start with one specialist co-op database then onto the next. Thusly, the patient may lose control of the present human administration data, while the pro association, for the most part, keeps up the essential stewardship.

This blockchain-based E-healthcare record system will provide deep mutual trust between each organization. Presently, EHRs are overseen by emergency clinics and suppliers, while patients denied the option to uninhibitedly regulate their own EHRs. By using blockchain innovation, principles for recording information and overseeing character are built up, and the blockchain of EHRs is developed [4]. Additionally, this innovation records the evaluating hints of all exchanges in a changeless circulated record, which ensures duty and straightforwardness in the parade of information trade. In this manner, the patient has the office to record social insurance and analytic data from specialists.

2 Literature Review

[1] "Blockchain Technology Applied in Health" Andrei Cirstea, Florentina Magda Enescu, Nicu Bizon, Cosmin Stirbu, Valeriu Manuel Ionescu, ECAI 2018 - International Conference – 10th Edition Electronics, Computers, and Artificial Intelligence, 2018. This paper acquaints an insignificant presentation with blockchain innovation, trailed by a clinical application (MedBlocks). The machine is predicated on this innovation, with the intent to shape the clinical field increasingly proficient. Blockchain innovation can upset both medication and some other fields for framework decentralization. This will change the whole world through proficiency and a secure approach to act. The target of this paper is to call attention to the remarkable capability of this innovation and how it will on a very basic level change all parts of getting, transmitting, and making sure about information.

[2] "Towards using blockchain technology for eHealth data access management"- N Rifi; Elie Rachkidi; N-Agoulmine; N C Taher. 2017 IEEE Fourth International Conference on Advances in Biomedical Engineering (ICABME) Publisher: IEEE. This work presents an innovation that is developing in significance after some time, shifting from remote access to Medical Records, as Electronic Healthcare Records (EHR), or Electronic Medical Records (EMR), to constant information trade from various on-body sensors originating from various patients. With this immense measure of basic information being traded, issues and difficulties emerge. Security and secrecy of this basic clinical information are of high worry to the patients and approved people to utilize this information. Then again, versatility and interoperability additionally are significant issues that ought to be considered inside the holocaust. This paper represents the exact issues and features the advantages of the blockchain innovation for the organization of a protected and a versatile answer for clinical information trade to have the most ideal presentation.

[3] "Securing Blockchain-based Electronic Health Record using Multilevel Authentication Radhakrishnan, A Sam Joseph, S. Sudhakar, International Conference on Advanced Computing & Communication Systems (ICACCS), 2019. Electronic health records have the patient's medication nuances and their health history. The health records pull in the thought of the aggressors' since it has noteworthy information. Loss of electronic health record prompts a misguided medication or clinical method. Social protection systems give fewer security endeavors to ensure health records. Blockchain is a coursed and decentralized record that accepts a basic activity in ensuring the data and trades. The introduction of blockchain inside the human administration systems shields the health records from the aggressors. Nevertheless, the blockchain faces phishing, word reference-based, cold wallets, and hot wallets ambushes. This paper proposes an amazing affirmation based arrangement to shield the blockchain from the attacks referenced already.

[4] "Identity and Access Management with Blockchain in Electronic Healthcare Records", Tomas Mikula and Rune Hylsberg Jacobsen, IEEE, 2018. Blockchain has demonstrated itself to alter safe and secure. It's undeniably getting consideration from organizations changing from unified to decentralized frameworks. This paper proposes a framework for character and accesses the board utilizing blockchain innovation to help confirm and approve elements in an advanced framework. A model shows the use of blockchain in character and access the board utilizing the Hyperledger Fabric structure. It gives an image of an idea upheld a utilization case concerning Electronic Health Records from the social insurance area where a changeless and auditable history is wanted for information concerning patients. Essential confirmation and approval tasks are prepared to execute in 2–3 s with an underlying size of the blockchain of about 3.8 MB covering doctors in Denmark.

[5] "Cloud-Assisted EHR Sharing with Security and Privacy Preservation Via Consortium Blockchain", Yong. Wang1, Aiqing. Zhang1, (Member, IEEE), Peiyun. Zhang2, (Member, IEEE), Huaqun. Wang3, (MEMBER, IEEE), IEEE, 2016. The sharing of electronic wellbeing records (EHRs) has incredible positive noteworthiness for research of ailment and specialists' conclusion. As of late, a cloud-based electronic clinical record sharing plan has brought huge amounts of accommodation, yet the centralization of cloud opens dangers unavoidably to information security and protection conservation. Blockchain innovation is frequently observed as a promising answer to manage these issues by its remarkable properties of decentration, secrecy, unforgeability, and evidence. In this paper, we propose a blockchain-based secure and protection safeguarding EHR sharing convention. Information requester can look through wanted watchword from information supplied to discover significant EHRs on the EHR consortium blockchain and get the re-encryption ciphertext from the cloud server after getting the information proprietor's approval. The plan, for the most part, utilizes accessible encryption and restrictive intermediary re-encryption to acknowledge information security, protection safeguarding, and get to control. Moreover, verification of approval is planned as the agreement instrument for consortium blockchain to ensure the framework's accessibility. Security investigation delineates the proposed convention can accomplish security objectives. Besides, we imitate the cryptographic natives and actualize the proposed conspire on the Ethereum stage.

Execution assessment shows that the proposed conspire has high computational productivity.

3 Problem Statement

The customary strategy which is followed in the greater part of the emergency clinics for keeping up records is the manual technique Information is kept in a centralized form in the database where hospitals can access all the data [6]. The patient needs to carry the records in the form of files and papers. The problem with the current system is that medical data is scattered and not available with the patient all the time. So, getting historical data or any vital information about the history of the patient can be difficult for the patient as they need to contact hospital management every time. Another issue related to the current system is providing confidentiality and patient medical records security which almost vital, however, clinicians/doctors will retrieve the stored data [7, 8] (Fig. 1).

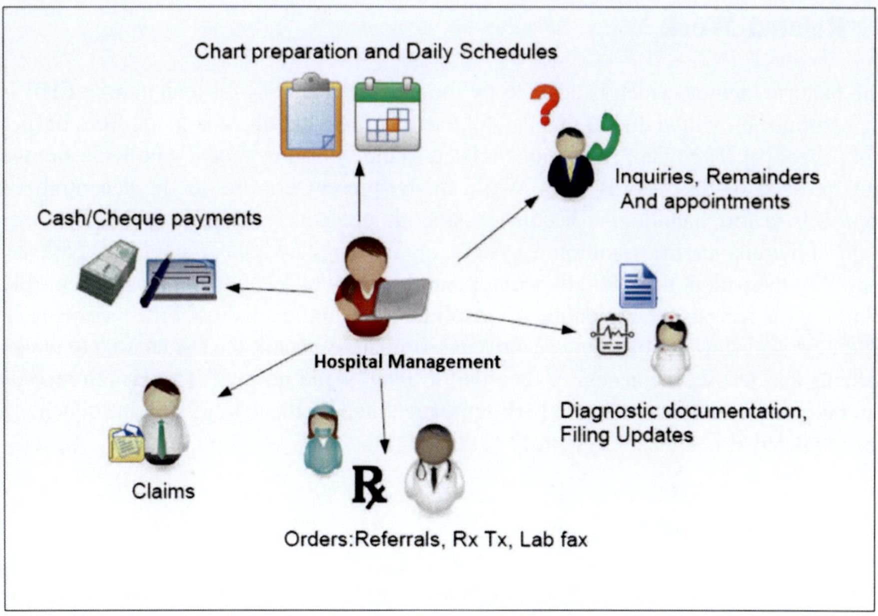

Fig. 1. Existing centralized EHR system

The above figure illustrates the framework of E-healthcare records which are completely constrained by clinics rather than patients, which confounds looking for clinical advice from various medical clinics. In this procedure, data/records are uploaded on a server by doctors or hospital authorities after treating or visiting the patient. In the above system, every time the patient needs to ask the authority of the hospital for their medical history/record which is accessed only by the hospital authority. In this system getting medical history from the hospital sometimes become a time consuming process for the patient. Due to centralized framework data loss is the major issue of this system.

3.1 Objectives

The main purpose is to supply classification to persistent clinical records exploitation "Blockchain Technology". Tolerant data mystery or classification is one of the principal important mainstays of medication. Ensuring the private subtleties of a patient is not just an issue of moral regard, it is fundamental to security trust between the specialist and the patient. The goal of this examination was to work out whether medical clinics with completely actualized EHR frameworks would be advised to quiet results contrasted with emergency clinics with halfway or no executed EHR framework after controlling for other significant patient and clinic qualities. Our investigation gives new data about the association between the execution of an EHR framework and accordingly the nature of human services conveyed inside the inpatient setting.

4 Related Work

The historic method which is adapted by most of the hospitals for maintaining EHR is the centralized method due to this Patient needs to carry the records in the form of files and papers [9]. The major restrictions related to the system is to deal with large storage and retrieval of the records [11]. While in the present era due to the decentralized approach record handling is become a smooth procedure for patients [3]. Another major dilemma in the traditional system is securing the vital records of patients, however, the patient may hold the doctor and the hospital accountable for irresponsibly handling the records, manipulating it, or splitting the confidentiality of the records [12]. At the present stage maintaining security is a primary aspect of the system due to which patients can restrict the access of records for doctors and hospitals [5]. Health records can be easily and quickly shared between patient and medical sectors by introducing a decentralized E-healthcare system [7].

5 Proposed System

The cutting-edge human services framework is utilizing computerized techniques for keeping up patient's health records. By and by human services associations utilize an incorporated technique for sparing patient's data, indicative reports, and specialist's solution. As considered, it is a concentrated framework there are odds of information getting spilled or misused for different reasons as patients don't have power over their information, and trading of the recorded information is tedious and a luxurious procedure. With the point of taking care of these issues, the system introducing a thought of changing a concentrated framework to a decentralized framework utilizing Blockchain innovation. On the Blockchain arrange, health records of the patient are frequently spared utilizing their novel IDs and open key. Social insurance associations can get to the patient's non-recognizable information through keen agreements just the interesting IDs coordinate. Whenever required patients can impart the general open key to the social insurance associations yet without the private key information would consistently remain non-recognizable to frame sure the clients will to share/recover the information, shrewd agreements are vital (Fig. 2).

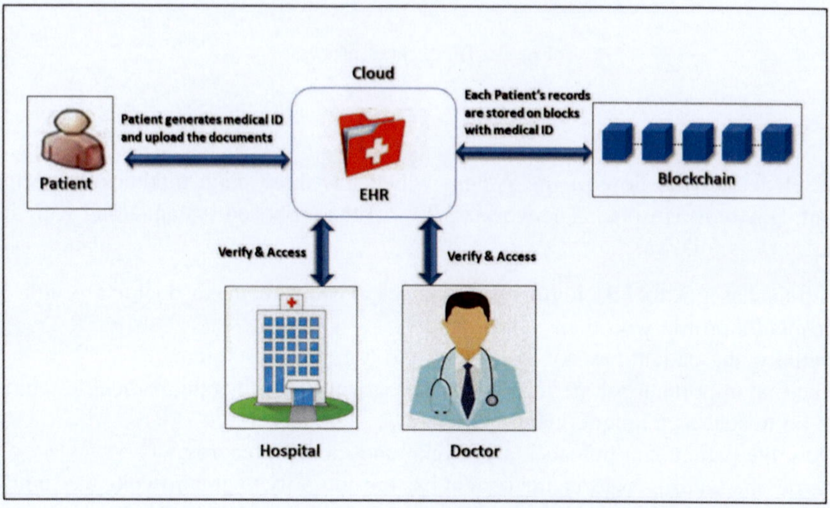

Fig. 2. Overview of the blockchain-based EHR system

5.1 Blockchain Background

This innovation was presented by Satoshi Nakamoto for his well-known work of advanced cash or digital money, i.e., bitcoin [10]. Blockchain is a chain of hinders that are associated together and are constantly developing by putting away exchanges on the squares. This stage utilizes a decentralized methodology that permits the data to be disseminated and that each bit of appropriated data or usually known as information, has shared possession [13]. Blockchains hold clusters of exchanges that are hashed hence giving them security and they are. A blockchain has certain benefits like security, anonymity, and integrity of knowledge with no outsider intercession [12]. These advantages make it an economical alternative to store patient's clinical records subsequently because the development of innovation inside the human services industry has made the security of patient's clinical information a top need [18]. An assortment of analysts has additionally distinguished that utilizing blockchain innovation in human services would be a doable arrangement.

Figure 3 illustrates the general blockchain process for transactions in which a sender can create a transaction of a block. This block of the transaction is distributed and validated via cryptographic hashing. Further, this distributed hashed transaction is committed and rewarded by miners and transferred to the receiver.

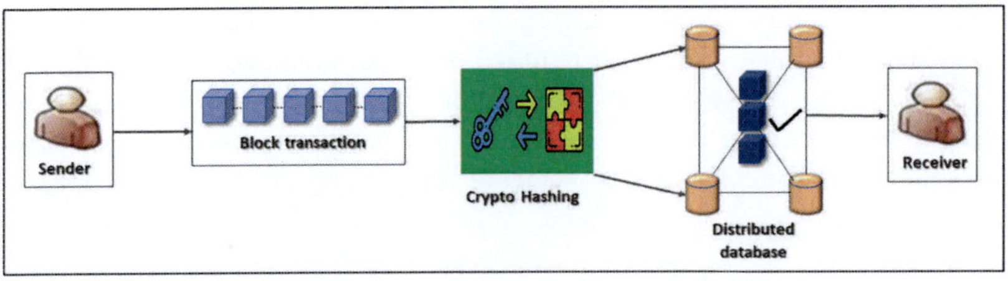

Fig. 3. Blockchain process

5.2 Flow of the System

Figure 4 defines the flow of the system which has three main modules are Admin, Patient, Doctor & Hospital. The working flow of the proposed system along with these modules is as follows:

1. Admin has got to first login itself and responsible to make doctor's profile and hospital's profile with their valid details.
2. Initially, the patient has got to register and login to the system.
3. The first important task performed by the patient is to generate medical id which is to be treated as a unique identification.
4. Now the patient can upload it's personal and medical records.
5. As per the security aspects, the patient has the authority to grant/revoke the rights on the access of records.
6. Like admin and patient, doctor and hospital has got to login to the system and may search patient by using unique medical id.
7. To access the patient's records doctor and the hospital has to send an access request to the patient.
8. If the patient grants permission for document access then only the doctor and hospital can view the documents.

5.3 System Architecture

We believe a blockchain-based E-healthcare record is a cloud stage where patient records are accumulated from a system of nearby doors and put away on an open cloud for offering to social insurance suppliers as appearing in Fig. 5. E-healthcare records may incorporate individual data and clinical history which are given by patients. Patients can generate their unique medical ID, upload the records on blockchain-based cloud, and grant the rights to access those records by doctors and hospitals.

Blockchain-based cloud plays an important role in this system which includes the following major components:

EHRs Manager: The EHRs chief assumes an important job in our information-sharing structure. It is dependable to control all client exchanges on the blockchain arrange, including information stockpiling procedures and information get to [12]. The administration capacity of EHRs administrator is empowered by shrewd agreements through exacting client strategies.

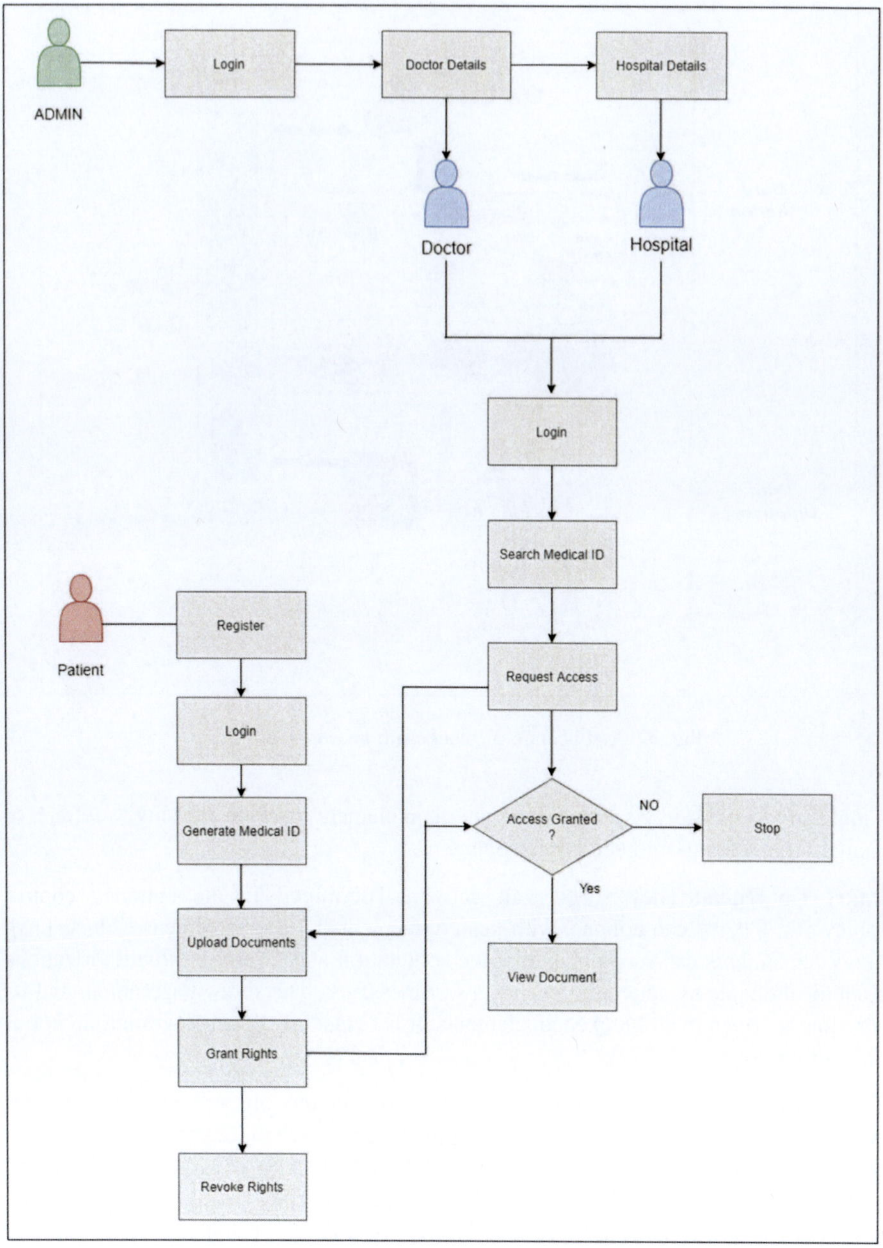

Fig. 4. The flow of the system

Admin: It is utilized to oversee exchanges and procedures on the cloud by the methods for including, changing or renouncing access consents [12]. The administrator

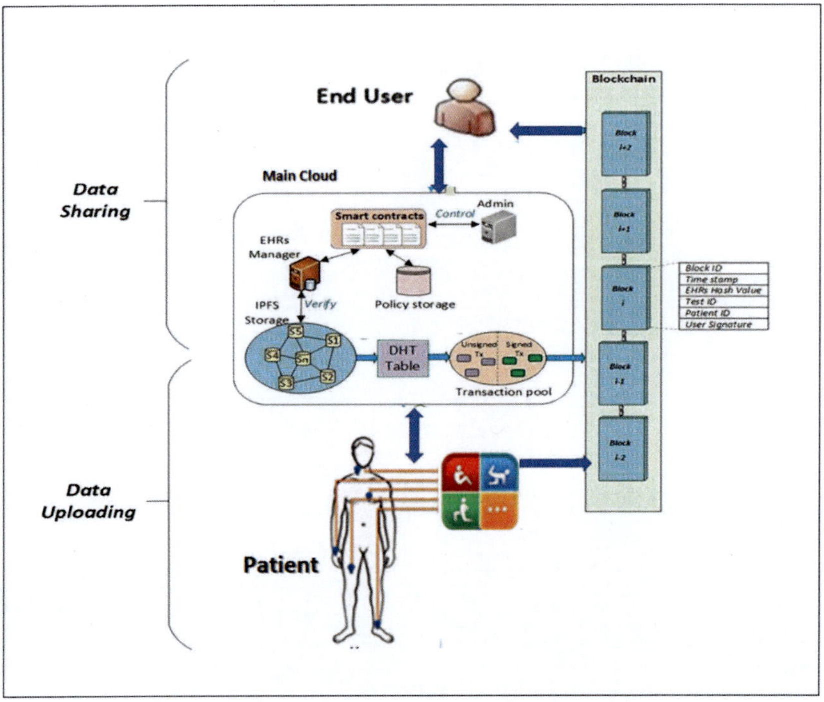

Fig. 5. Architecture of blockchain based EHR system

is mindful to send savvy contracts and the main element with the capacity to refresh or change arrangements in keen agreements.

Smart Contracts: They specify all activities permitted in the entrance control framework. Clients can connect with keen agreements by the agreement address [15]. Savvy agreements can recognize, approve solicitation and award get to authorizations for clinical clients by activating exchanges or messages. The shrewd agreement and its activities are open to all blockchain elements. It is treated as center programming in our e-health platform.

Decentralized Storage: Capacity of information on the blockchain causes fundamental issue i.e., capacity limit [15]. The information on the blockchain is obvious to everybody available on the chain this makes the information powerless which is certainly not an ideal result for a decentralized stage. The information put away on the blockchain would contain quite clinical history, records, lab results, X-beams reports, MRI results and numerous different reports, the entirety of this voluminous information is to be put away on the blockchain that would exceptionally influence the capacity limit of the blockchain. To take care of this issue Off-Chain Storage will be used.

Off-Chain Storage: Off-chain non-transactional data that is too large to be sorted in the blockchain efficiently, or requires the ability to be changed or deleted. Figure 6 shows examples of off-chain data types. To store such a substantial amount of data system attract towards interplanetary file system (IPFS) and distributed hash table (DHT) decentralized distributed file framework.

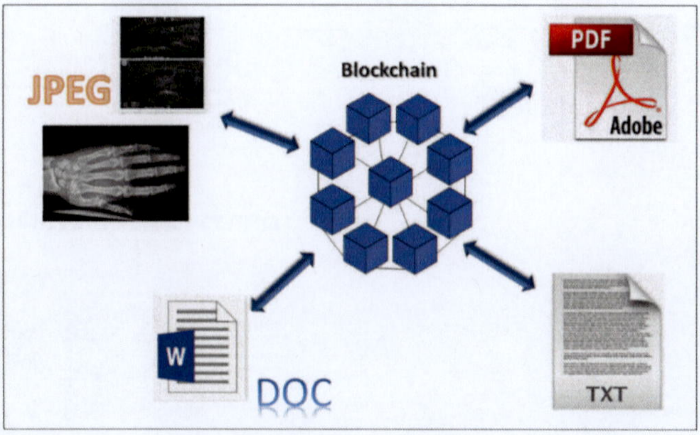

Fig. 6. Examples of off-chain storage

IPFS (Interplanetary File System)

IPFS is a convention that utilizations a distributed system for information stockpiling. It gives secure information stockpiling as information put away on IPFS is protected from any change. It utilizes a cryptographic identifier that shields the information from adjustment as any endeavor to make a change on the information put away on IPFS must be finished by changing the identifier [15].

The mechanism behind IPFS storage describes in Fig. 7. All the information records put away on IPFS contains hash esteem that is produced cryptographically. It is special and is utilized for distinguishing proof of put away information record on the IPFS. This safe stockpiling system of the IPFS convention settles on it a positive decision for putting away basic and touchy information. The cryptographic hash that is produced could be put away on the decentralized application to lessen the thorough computational activities over the blockchain. Unlike the World Wide Web, content is not stored on individual servers and instead is distributed across several nodes. To distribute this content, IPFS makes use of a Bit Torrent like DHT protocol.

Figure 8 illustrates the role of IPFS storage for the proposed system. IPFS is a peer-to-peer protocol wherever every record stores a set of hashed files verified by the EHR manager through smart contract policy. IPFS is a decentralized protocol which follows the content-based addressing. Once hash value generated for a specific record, that hashed record along with unique key is transferred to the distributed hash table.

Distributed Hash Tables (DHT)

A distributed hash table (DHT) is a peer-to-peer storage system that provides lookup and storage schemes almost sort of a hash table, with value [12]. Each node in a DHT is responsible for keys along with the mapped values. Any node can competently recover the value associated with a given key. DHTs are distributed. Each "peer" (or "node") is liable for a subset of the DHT. At the point when a friend gets a solicitation, it either

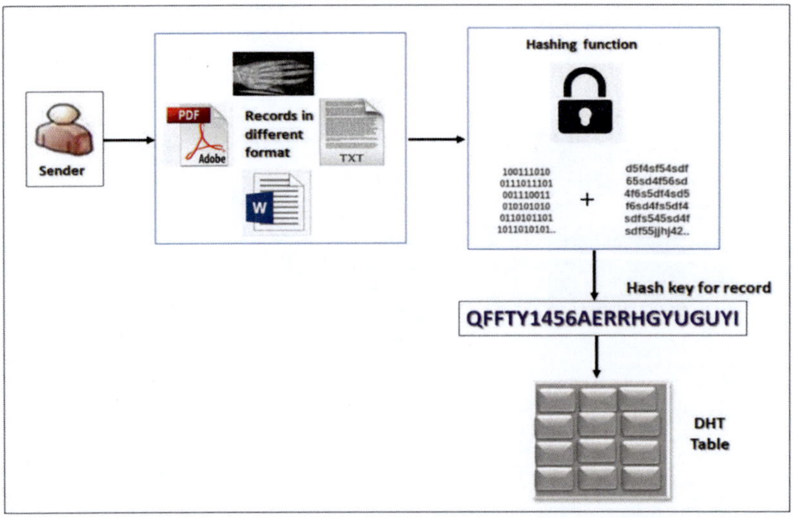

Fig. 7. IPFS process

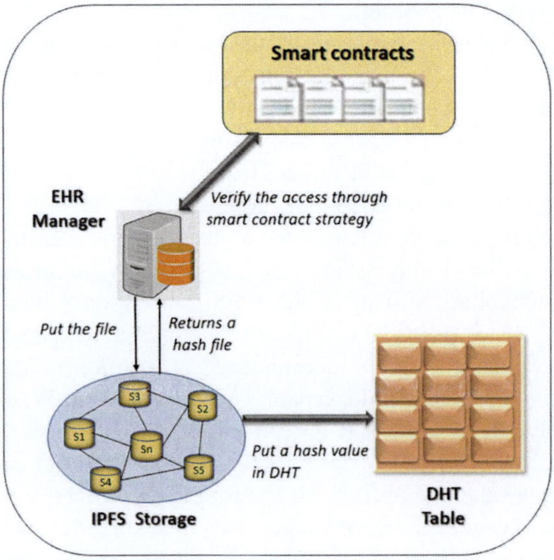

Fig. 8. IPFS storage for the system

answers it, or the solicitation is passed to another companion until a companion that can answer it is found. Contingent upon the execution, a solicitation not replied by the main hub reached can be: - sent from distributed, with the last friend reaching the mentioning peer - sent from shared, with the appropriate response sent after a similar way - replied with the contact data of a hub that has better opportunities to have the option to reply. IPFS utilizes this methodology. Figure 9 gives an example of DHT structure.

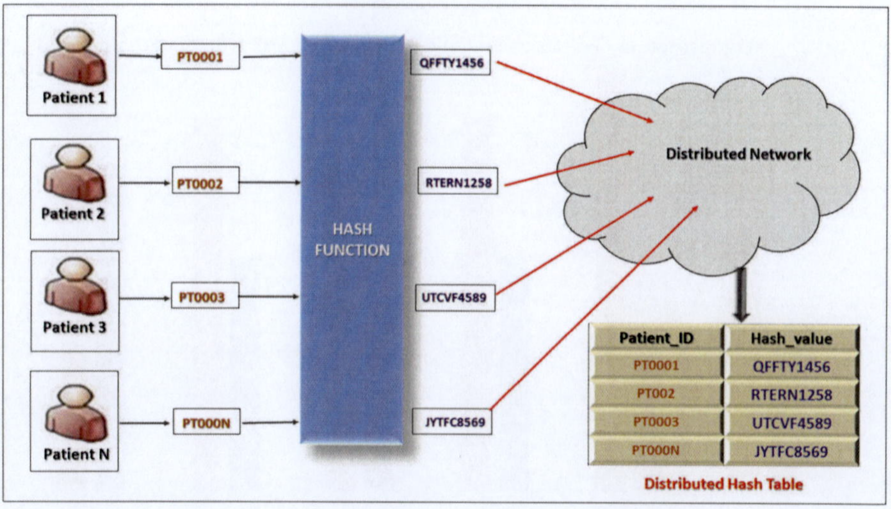

Fig. 9. The structure of a distributed hash table (DHT)

6 Result Analysis

The proposed decentralized system compares with the traditional system which follows the centralized approach under various performance characteristics to validate the viability of the model for real usability scenarios. Table 1 shows characteristics and levels for the centralized EHRs system and the proposed system. Evaluation of the proposed system executed based on comparison which is given in Fig. 10.

Table 1. Performance characteristics of centralized EHRs and decentralized EHRs

Characteristics	Centralized EHR	Decentralized HER using blockchain
Flexibility	3	7
Security	3	9
Data maintenance	5	8
Date access	3	7
Efficiency	4	8
Data redundancy	3	6

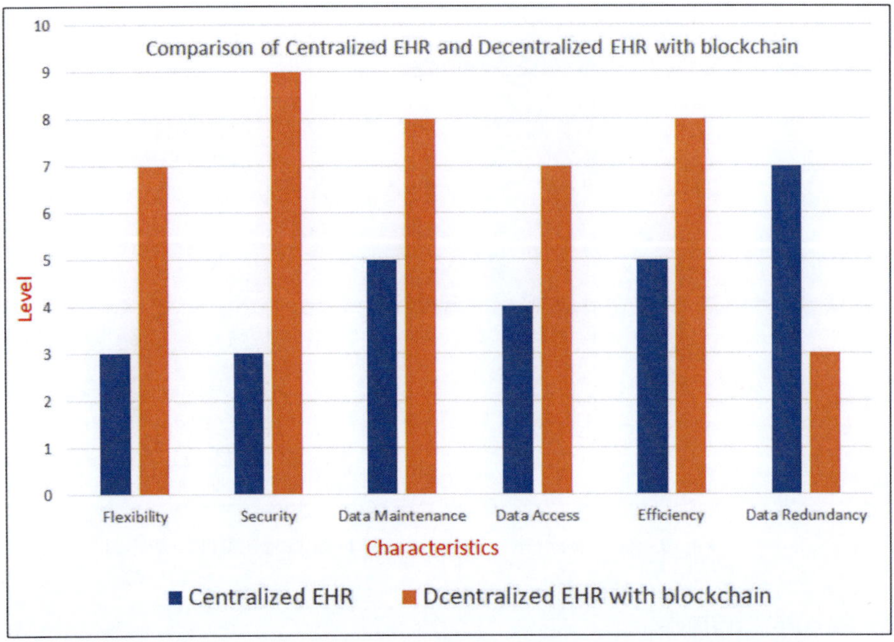

Fig. 10. Graph for comparison of centralized EHRs and decentralized EHRs

7 Conclusion

The world has commenced towards constant driven interoperability where patients give access to their health records on request. In the current system, the patient is treated as an individual owner to his health records who may select what data needs to be shared and with whom. Security is also a major concern in this case as the medical data should not be leaked or should be available only with the concern of the patient. Blockchain plays an important role in securing the data as all the medical info and transactional data would be stored on the blockchain in ledgers. The system will make sure that the data from the system is not leaked unless the authority has permission to access the data. Even when the patient visits any medical institute like a hospital, the patient will have to give out the medical id, all the data that hospital on the patient's medical id which can be accessed anywhere. This proposed structure is a mix of secure record stockpiling alongside the granular access rules for those records. It makes such a framework, that is simpler for the clients to utilize and comprehend. Additionally, the structure proposes measures to guarantee the framework handles the issue of information stockpiling as it uses the off-chain stockpiling component of IPFS. The approach additionally conveys shrewd agreements, which is a code, which executes on its own when both the gatherings concur on the arrangement of conventions. The proposed system permits the patient to upload insurance-related documents along with other records whereas verification and claim settlement process is not included in the system since it is one of the limitations of the system. The proposed system is a web-based application hence delay in granting or revoking access from the patient side can be another issue related to the system, hence the implementation of a mobile-based EHR application is one of the feasible solutions.

Acknowledgment. I want to stretch out my genuine gratitude to all who helped me with the undertaking work. I want to earnestly express gratitude towards Dr. Dilip Motwani for their input and steady direction for giving critical data concerning the commission likewise, for their help in completing this task work. I want to offer my express gratitude towards individuals and folks from the vidyalankar institute of technology [VIT] for their thoughtful co-activity and support.

References

1. Cirstea, A., Florentina, M.E., Nicu, B., Cosmin, S., Valeriu, M.I.: Blockchain technology applied in health the study of blockchain application in the health system (II). In: 2018 10th International Conference on Electronics, Computers and Artificial Intelligence (ECAI), pp. 1–4. IEEE (2018)
2. Rifi, N., Elie, R., Nazim, A., Nada, C.T.: Towards using blockchain technology for eHealth data access management. In: 2017 Fourth International Conference on Advances in Biomedical Engineering (ICABME), pp. 1–4. IEEE (2017)
3. Radhakrishnan, B.L., Joseph, A.S., Sudhakar, S.: Securing blockchain based electronic health record using multilevel authentication. In: 2019 5th International Conference on Advanced Computing & Communication Systems (ICACCS), pp. 699–703. IEEE (2019)
4. Mikula, T., Jacobsen, R.H.: Identity and access management with blockchain in electronic healthcare records. In: 2018 21st Euromicro Conference on Digital System Design (DSD), pp. 699–706. IEEE (2018)
5. Wang, Y., Zhang, A., Zhang, P., Wang, H.: Cloud-assisted EHR sharing with security and privacy preservation via consortium blockchain. IEEE Access **7**, 136704–136719 (2019)
6. Harshini, V.M., Danai, S., Usha, H.R., Kounte, M.R.: Health record management through blockchain technology. In: 2019 3rd International Conference on Trends in Electronics and Informatics (ICOEI), pp. 1411–1415. IEEE (2019)
7. Novikov, S.P., Kazakov, O.D., Kulagina, N.A., Azarenko, N.Y.: Blockchain and smart contracts in a decentralized health infrastructure. In: 2018 IEEE International Conference Quality Management, Transport and Information Security, Information Technologies (IT&QM&IS), pp. 697–703. IEEE, 2018
8. Azaria, A., Ariel, E., Thiago, V., Andrew, L.: Medrec: using blockchain for medical data access and permission management. In: 2016 2nd International Conference on Open and Big Data (OBD), pp. 25–30. IEEE (2016)
9. Kang, A.: Bitcoin's growing pains: intermediation and the need for an effective loss allocation mechanism. Mich. Bus. Entrepreneurial L. Rev. **6**, 263 (2016)
10. Pramod, P., Tripathy, P.K., Harshit, B., Manjunath, R.K.: Role of natural language processing and deep learning in intelligent machines. In: IEEE International Conference on Electrical, Communication, Electronics, Instrumentation and Computing (ICECEIC) (2019)
11. Suma, V.: Security and privacy mechanism using blockchain. J. Ubiquitous Comput. Commun. Technol. (UCCT) **1**(01), 45–54 (2019)
12. Nguyen, D.C., Pathirana, P.N., Ding, M., Seneviratne, A.: Blockchain for secure EHRs sharing of mobile cloud based e-Health systems. IEEE Access **7**, 66792–66806 (2019)

13. Zheng, Q., Yi, L., Ping, C., Dong, X.: An innovative IPFS-based storage model for blockchain. In: 2018 IEEE/WIC/ACM International Conference on Web Intelligence (WI), pp. 704–708. IEEE (2018)
14. Shan, J., Cao, J., Wu, H., Yang, Y., Ma, M., He, J.: Blochie: a blockchain-based platform for healthcare information exchange. In: 2018 IEEE International Conference on Smart Computing (Smartcomp), pp. 49–56. IEEE (2018)
15. Novikov, S.P., Oleg, D.K., Kulagina, N.A., Azarenko, N.Y.: Blockchain and smart contracts in a decentralized health infrastructure. In: 2018 IEEE International Conference Quality Management, Transport and Information Security, Information Technologies (IT&QM&IS), pp. 697–703. IEEE (2018)
16. Vidhya, R., Kumar, T., Bracken, A., Liyanage, M., Ylianttila, M.: Secure and efficient data accessibility in blockchain based healthcare systems. In: 2018 IEEE Global Communications Conference (GLOBECOM), pp. 206–212. IEEE (2018)

Differential Evolution with Different Crossover Operators for Solving Unconstrained Global Optimization Algorithms

Konjeti Harsha Saketh, Konjeti B. V. N. S. Sumanth, P. V. S. M. S. Kartik, K. S. S. Aneeswar, and G. Jeyakumar(✉)

Department of Computer Science and Engineering, Amrita School of Engineering, Amrita Vishwa Vidyapeetham, Coimbatore, India
{cb.en.u4cse17036, cb.en.u4cse17035, cb.en.u4cse17033, cb.en.u4cse17007}@cb.amrita.students.edu, g_jeyakumar@cb.amrita.edu

Abstract. The Evolutionary Algorithms (*EAs*) are the part of bio-inspired algorithms used for solving a wide variety of real-world optimization problems. Differential Evolution (*DE*) is one of the algorithms in the pool of *EA*. *DE* is designed particularly for real-valued parameter optimization problems. The mutation and crossover operators of *DE* are designed in such a way that they are apt for real-valued parameters. The native crossover operators of *DE* are Binomial Crossover and Exponential Crossover. This paper proposes to test the suitability of other Crossover operators in the *EA* community to *DE*. The empirical study presented in this paper includes four crossover operators – *DE*'s native crossover (binomial crossover), simple arithmetic crossover, single arithmetic crossover, and uniform arithmetic crossover. The *DE* with these crossover operators is implemented to solve four standard benchmarking functions with different features. The empirical results are compared based on the accuracy and speed of the algorithm. The primary observations found were the arithmetic crossover was good at achieving good optimal results than the binomial crossover. However, the binomial crossover was faster in reaching a threshold set for the optimal value in the experiment.

Keywords: Differential Evolution · Crossover · Real parameter · Recombination · Arithmetic recombination

1 Introduction

Differential Evolution (*DE*), the optimization algorithm under the group of *EAs*, proposed by [1] is popular among the researchers and practitioners for its simplicity and robustness. It is a stochastic, population-based optimization algorithm for solving nonlinear optimization problems.

Similar to other *EAs*, the first stage of *DE* decide the representation of the candidate solutions of the given problem. These steps define the genotype and the corresponding mapping strategy to map the genotype to phenotype. This helps to construct a population of candidate solutions. *DE* maintains a population of candidate solutions during

J. I.-Z. Chen et al. (Eds.): ICIPCN 2020, AISC 1200, pp. 381–388, 2021.
https://doi.org/10.1007/978-3-030-51859-2_35

its evolutionary process, which has multiple generations. At each generation, it creates new candidate solutions (called as offspring) by applying variation operators viz mutation and crossover (recombination) on a selected set of candidate solutions (called as parents). Then, the best among the parents and offspring are reserved for the next generations.

The mutation operator of *DE* is termed as *Differential Mutation*. The differential mutation process is to generate the mutated solution for each candidate in the current population. This process includes adding a difference vector of two chosen vectors from the population with another vector. The vector thus produced is called a mutant vector. Each mutant vector is recombined with its corresponding target vector, using a crossover operator, to produce the trailing vector. Then the one-to-one knockout competition between the mutant vector and the target vector decides the survivor for the next generation.

Focusing on the crossover operator of the *DE* algorithm, the objective of this paper is to observe and analyze the performance of *DE* by replacing its native crossover operators with three other crossover operators in the state-of-the-art of *EAs*.

The rest of the paper is organized as follows – Sect. 2 discusses few recent works which are related to the work presented in this paper, Sect. 3 presents the design of experiments, the Sect. 4 presents, and results and the discussion part of the work and finally, Sect. 5 concludes the paper.

2 Related Works

The recent literature related to discussing the crossover component of *DE* is summarized in this section. There are numerous works presented in the literature where the performance and working principles of *DE* crossover operators are discussed in detail, with the mathematical background and empirical observations. Few such works are presented here.

A comparative analysis of binomial and exponential crossover of *DE* is presented in [2] by Daniela Zaharie. A theoretical analysis of the role of crossover rate on deciding the number of mutant components in the target vector is elucidated. The similarity and differences between these two crossover operators were discussed with their theoretical properties and associated control parameters. The impact of the crossover operators and its control parameters on the performance of *DE* is presented in [3]. Theoretical analysis and empirical comparative study of *DE's* crossover operators are presented in [4]. Two new crossover schemes named consecutive binomial crossover and non-consecutive exponential were also designed and added in the comparative study [4]. A comparative study of binomial crossover and the contiguous binomial crossover is presented in [5]. The contiguous crossover used a fashion similar to that of the exponential crossover. An adaptive *DE* with an adaptive crossover rate scheme is proposed in [6]. The proposed *DE* used a new crossover scheme where the crossover rate is adapted based on the previous experience of the executions. A note on the poor performance of exponential crossover comparing to the binomial crossover is highlighted in [7]. This work also demonstrated that the shuffled exponential crossover is performing better than the classical exponential crossover.

[8] has presented a summarized classification of different mutations and crossover strategies for *DE*. There are studies in the literature reporting strategies for improving the performance of classical mutation and crossover operators. A crossover matrix instead of a crossover operator is introduced in [9]. A covariance matrix is designed, based on the Gaussian distribution, between the best individual and several better individuals. The matrix-based crossover operators have shown improved performance than the classical crossover operators. Researchers tried using *DE* for solving problems of smaller dimensions without the crossover operators also. As stated in [10], the *DE* with only the mutation operators itself able to solve small dimension problems.

DE in its classical form or modified form is widely applied to many real-world problems around us viz sensor placement optimization [11], video analytics [12], and object identification problem [13], etc.

On understanding the role and importance of crossover operators of *DE*, this paper focuses on providing a comparative study on the performance of *DE* with different crossover operators. This study chose three crossover operators of real-valued representation, from the literature of *EA* along with the binomial crossover of *DE*.

3 Design of Experiments

The *DE* parameters are set as follows - the population size (*NP*) = 50, the dimension (*D*) = 30, mutation factor (*F*) = 0.2, the crossover rate (C_r) = 0.6 and the number of generations = 100. The crossover operators included in the experiment were (1) Binomial Crossover (*BinCr*) (2) Simple Arithmetic Crossover (*SimAriCr*) (3) Single Arithmetic Crossover (*SinAriCr*) and (4) Uniform Arithmetic Crossover (*UniAriCr*). The benchmarking functions [14] used are (1) Sphere Function (2) Schwefel's Function (3) Generalized Restrigin's Function and (4) Ackley's Function. The detailed description of the benchmarking functions is provided in Table 1.

Table 1. Details of the benchmarking functions

f_1 - Sphere Function $f_{sp}(x) = \sum_{i=1}^{n} x_i^2,$ $-100 \leq x\_i \leq 100;$ $x^* = (0, 0, \ldots, 0); f_{sp}(x^*) = 0;$	**f_2 - Schwefel's Function** $f_{sch2}(x) = \sum_{i=1}^{n} \left(\sum_{j=1}^{i} x_j\right)^2$ $-100 \leq x_i \leq 100;$ $x^* = (0, 0, \ldots, 0); f_{sch2}(x^*) = 0;$
f_3 - Generalized Restrigin's Function $f_{grf}(x) = \sum_{i=1}^{n} \left[x_i^2 - 10cos(2\pi x_i) + 10\right]$ $5.12 \leq x_i \leq 5.12;$ $x^* = (0, 0, \ldots, 0);$ $f_{grf}(x^*) = 0;$	**f_4 - Ackley's Function** $f_{ack}(x) = 20 + e - 20exp\left(-0.2\sqrt{\frac{1}{n}\sum_{i=1}^{n} x_i^2}\right) - exp\left(\frac{1}{n}\sum_{i=1}^{n} cos(2\pi x_i)\right)$ $-5.12 \leq x_i \leq 5.12;$ $x^* = (0, 0, \ldots, 0); f_{ack}(x^*) = 0;$

The initial population is generated in such a way that no gene in a particular chromosome is beyond the boundaries of the benchmarking function, and no chromosome has a dimension more or less than the given dimension. 50 such chromosomes are generated. The fitness of a chromosome is calculated by the benchmarking functions. The calculated fitness value is appended at the end of the chromosome. Now for every chromosome in this population mutation operation is done. the process of mutation is as follows:

1. Select three chromosomes from the population leaving the candidate selected for mutation and name them as x_1, x_2, x_3, and x_t (selected candidate).
2. Find $x_2 - x_3$ for all the genes present in the chromosome.
3. Now add genes of x_1 to the difference by multiplying it with the mutation factor (*F*) this will give the donor vector $v_{mutant} = x_1 + F * (x_2 - x_3)$.

Then the crossover is performed between v_{mutant} and x_t which gives the trial vector v_{trial} vector. The fittest between the x_t and v_{trial} is found using a selection strategy and populated in the next generation. The above mutation, crossover, and selection steps are repeated for all the candidates in the current population. This is meant as the completion of one generation. Then the next generation starts with the above steps but for the population newly created in the previous generation. The fitness of the best candidate in the population after 100 generations is declared as the optimum solution of the benchmarking function used for the current run.

4 Results and Discussions

The experiment covered solving four benchmarking functions by *DE* with four different crossover operators for 10 runs. The performance metrics used for the comparison are the optimum solution attained and the speed. The optimum solution is measured as the best fitness value reached at the end of the run. The average, minimum, and maximum optimum value attained across all the ten runs are presented in Table 2.

The results show that the performance of *DE* with all the crossover operators are almost similar. However, for the f_1- Sphere function, the Binomial Crossover has given the poorest optimum value comparing to the arithmetic crossovers. Among the arithmetic crossovers, the Simple Arithmetic Crossover achieves good optimum values than the others. For f_2 - Schwefel's Function, irrespective of the crossover operators, *DE* reached the expected global optimum of zero. For f_3 - Generalized Restrigin's Function and f_4 - Ackley's Function also, the arithmetic crossover operators could produce a better result than the binomial crossover. The Single Arithmetic and Simple Arithmetic Crossovers have outperformed other arithmetic crossover operators in f_3 and f_4, respectively.

Table 2. The optimum values measured for the benchmarking functions.

	BinCr	*SimAriCr*	*SinAriCr*	*UniAriCr*
	f_1 - Sphere Function			
Average	12722.73	**1762.52**	3600.72	2777.98
Min	12086.59	**1674.39**	3420.69	2639.08
Max	13358.87	**1850.64**	3780.76	2916.88
	f_2 - Schwefel's Function			
Average	0.00	0.00	0.00	0.00
Min	0.00	0.00	0.00	0.00
Max	0.00	0.00	0.00	0.00
	f_3 - Generalized Restrigin's Function			
Average	186.20	106.15	**104.67**	121.24
Min	176.89	100.84	**99.44**	115.18
Max	195.51	111.46	**109.91**	127.30
	f_4 - Ackley's Function			
Average	5.16	**2.72**	3.78	2.92
Min	4.90	**2.58**	3.59	2.78
Max	5.41	**2.85**	3.97	3.07

Next, to measure the speed of the algorithms target values were set for the optimum value. The minimum and maximum value for the target are set as given in Eqs. (1) and (2), respectively.

$$Min\_Target = 0.95 * Agv_{ov} \tag{1}$$

$$Max\_Target = 1.05 * Agv_{ov} \tag{2}$$

where Agv_{ov} is the average optimum value, which was empirically calculated for each function-algorithm combination before the actual experimental runs.

The speed is measured in two different ways – (1) the number of generations taken by the algorithm to cross the minimum threshold of the optimum value and (2) the execution time (in seconds) taken by the algorithm to cross the same threshold. The values measured for the speed by the number of generations are presented in Table 3. The results indicate that, for f_1 and f_2, the binomial crossover is faster in reaching the threshold. Similarly, for f_3 and f_4, the Uniform Arithmetic crossover has shown faster performance in reaching the minimum threshold. The faster convergence nature of the binomial crossover is the reason for its poorer performance in reaching the global optimal solution for the f_1. For f_2, all the crossover reached the global optimum, the binomial crossover was faster. Though single arithmetic and simple arithmetic crossovers have outperformed other crossover in producing good solutions, they are outperformed by the Uniform Arithmetic Crossover by the speed.

Table 3. The speed – no of generations measured.

	BinCr	*SimAriCr*	*SinAriCr*	*UniAriCr*
	f_1 *- Sphere Function*			
Average	**29.20**	49.70	98.00	24.50
Min	**15.00**	26.00	89.00	5.00
Max	**100.00**	100.00	100.00	100.00
	f_2 *- Schwefel's Function*			
Average	**11.70**	16.30	18.40	5.60
Min	**5.00**	4.00	7.00	3.00
Max	**22.00**	35.00	40.00	8.00
	f_3 *- Generalized Restrigin's Function*			
Average	80.90	92.60	98.90	**64.10**
Min	54.00	68.00	94.00	**10.00**
Max	100.00	100.00	100.00	**100.00**
	f_4 *- Ackley's Function*			
Average	45.70	52.10	91.40	**5.90**
Min	17.00	27.00	79.00	**3.00**
Max	100.00	100.00	100.00	**8.00**

The speed measurement based on the execution time is presented in Table 4. The execution time is measured based on the system clock. It is well known that the execution time is directly proportional to the number of computational steps needed to complete a task. Hence, the crossover operators which take less execution time is regarded as the crossover operator with simple algorithmic steps. It is evident from the results, in Table 4, that the binomial crossover has taken very little execution time in all the function cases. Hence, it is understandable that the computational logic of binomial crossover is simple to compare to the arithmetic crossover operators. As the name says, the arithmetic crossover operators perform additional arithmetic operations during the crossover process.

Though *DE's* native crossover operators are binomial crossover and exponential crossover, the experimental study of this paper was successful in testing that the crossover operators in the *EA* community for other real-valued parameter optimization algorithms are also adaptable to *DE*. However, it is observed from the results (Table 2, Table 3, and Table 4) that for *DE* algorithm no single crossover operator used in this study is good in all the performance measurements. By the optimum value, it was found that the arithmetic crossover operators are better than the binomial crossover and by the speed, the binomial crossover is better than the arithmetic crossover.

Since this study has a limitation in its experimental setup, only four test functions and two performance measurements, it can be extended in the future to include more experimental analysis to bring concrete.

Table 4. The speed – the execution time (in secs) measured.

	BinCr	*SimAriCr*	*SinAriCr*	*UniAriCr*
	f_1 - Sphere Function			
Average	**0.60**	1.26	1.78	2.62
Min	**0.55**	1.20	0.90	2.40
Max	**0.68**	1.36	2.01	2.84
	f_2 - Schwefel's Function			
Average	**0.64**	1.32	2.04	2.88
Min	**0.50**	1.08	1.82	2.50
Max	**0.79**	1.61	2.28	3.21
	f_3 - Generalized Restrigin's Function			
Average	**0.50**	1.09	1.62	2.25
Min	**0.40**	0.93	1.37	1.98
Max	**0.58**	1.24	1.83	2.53
	f_4 - Ackley's Function			
Average	**0.59**	1.25	1.90	2.67
Min	**0.54**	1.15	1.75	2.42
Max	**0.72**	1.37	2.11	2.88

5 Conclusions

The compatibility of the real-valued parameters' crossover operators of *EA* literature was tested on *DE*, along with its native binomial crossover. The experimental results revealed that *DE* could accommodate to the chosen arithmetic crossover operator. An empirical comparative study of *DE* with four different crossover operators was presented in this paper. The results showed that the arithmetic operators are good at producing good optimum solutions and the binomial crossover was good in execution time. This insight indicates that the selection of the crossover operator for the *DE* algorithm depends on the expected output of the user – a good solution or faster solution.

This study can be extended further including more crossover operators, benchmarking functions, and performance measures.

References

1. Price, K.V., Storn, R.: Differential evolution: a simple evolution strategy for fast optimization. Dr. Dobb's J. **22**(4), 18–24 (1997)
2. Zaharie, D.: A comparative analysis of crossover variants in differential evolution. In: Proceedings of the International Multi-Conference on Computer Science and Information Technology, pp. 171–181 (2007)
3. Zaharie, D.: Influence of crossover on the behavior of differential evolution algorithms. Appl. Soft Comput. **9**(3), 1126–1138 (2009)

4. Lin, C., Qing, A., Feng, Q.: A comparative study of crossover in differential evolution. J. Heuristics **17**(6), 675–703 (2011). https://doi.org/10.1007/s10732-010-9151-1
5. Weber, M., Neri, F.: Contiguous binomial crossover in differential evolution. Lecture Notes in Computer Science, vol. 7269 (2012). https://doi.org/10.1007/978-3-642-29353-5_17
6. Mohamed, A.W., Almazyad, A.S.: Differential evolution with novel mutation and adaptive crossover strategies for solving large scale global optimization problems. Appl. Comput. Intell. Soft Comput. **2017**(7974218), 18 (2017)
7. Tanabe, R., Fukunaga, A.: Reevaluating exponential crossover in differential evolution. In: Parallel Problem Solving from Nature. Lecture Notes in Computer Science, vol 8672 (2014). https://doi.org/10.1007/978-3-319-10762-2_20
8. Dhanalakshmy, D.M., Pranav, P., Jeyakumar, G.: A survey on adaptation strategies for mutation and crossover rates of differential evolution algorithm. Int. J. Adv. Sci. Eng. Inf. Technol. **6**(5), 613–623 (2016)
9. Li, Y., Feng, J., Hu, J, Covariance and crossover matrix guided differential evolution for global numerical optimization. SpringerPlus **5**(1176) (2016). https://doi.org/10.1186/s40064-016-2838-5
10. Dhanalakshmy, D.M., Jeyakumar, G., Shunmuga Velayutham, C.: Crossover-free differential evolution algorithm to study the impact of mutation scale factor parameter. Int. J. Recent Technol. Eng. (IJRTE) **7**(6), 1728–1737 (2019). ISSN 2277-3878
11. Rubini, N., Prashanthi, C.V., Subanidha, S., Jeyakumar, G.: An optimization framework for solving RFID reader placement problem using differential evolution algorithm. In: Proceedings of International Conference on Communication and Signal Proceedings, pp. 1290–1294 (2017)
12. Abraham, K.T., Ashwin, M., Sundar, D., Ashoor, T., Jeyakumar, G.: Empirical comparison of different key frame extraction approaches with differential evolution based algorithms. In: Proceedings 3rd International Symposium on Intelligent System Technologies and Applications, vol. 683 (2017). https://doi.org/10.1007/978-3-319-68385-0_27
13. Vinaya Sree, K., Jeyakumar, G.: An evolutionary computing approach to solve object identification problem in image processing applications. J. Comput. Theor. Nanosci. **17**(1), 439–444 (2020)
14. Yao, D., Liu, Y., Liang, K.H., Lin, G.: Fast evolutionary algorithms. Advances in Evolutionary Computing: Theory and Applications, pp. 45–94. Springer, Heidelberg (2003). https://doi.org/10.1007/978-3-642-18965-4_2

Potential Subscriber Detection Using Machine Learning

M. Adithi Mookambal[1(✉)] and S. Gokulakrishnan[2]

[1] Computer Science and Engineering, SCSVMV University, Kanchipuram, Chennai, India
adithimookambal16@gmail.com

[2] CSE Department, SCSVMV University, Kanchipuram, Chennai, India

Abstract. The paper potential subscriber detection helps in finding who has the potential to convert from free users to the paid users. The problem statement is about finding the potential users who can get converted from free users to paid users and also finding the paid user who has potential to do another subscription. As we all know that vikatan was a paperback subscription of the articles, now it has come out with a new idea of reaching the customers with the articles in digital form. So they got introduced with the app named 'Appappo'. This app works same way like other apps at the beginning. It is used to read only the best articles and stories from tamil nadu. So, according to this app while installing in your mobile for the first time you will be given Rs 51. Also we have the articles of Rs 1, 2, 3, 5, 10 categories. Hence according to the amount /cost of the article it will be detected from the wallet. When the wallet is empty then you have to subscribe and fill the wallet with the amount to continue reading the articles. This application or the research using machine learning algorithms to predict the exact number of subscribers and giving them offers rather than giving it to everyone. We had a major drawback in the magazines where you distribute the magazines to all of them and also the offers. It just made company to spend so much and also a waste of time for them. So they decided to make it digital and release to the customers as an android app so every people in the corner will get this and benefit with it and also not a waste of time for them and also a easy task compared to before.

Keywords: Machine learning · K- fold cross validation · Subscriber · Data · App

1 Introduction

Appappo is an app which has the meaning of from time to time, which exactly means the contents will reach the customers from time to time. This is basically a paper promoted by the vikatan. Vikatan is the No 1 across the print, Tv, and Digital medium in Tamil Nadu. It is first India's first custom curated platform. Appappo is the treasure trove of evergreen, favorite, essential contents curated and personalized for readers.

J. I.-Z. Chen et al. (Eds.): ICIPCN 2020, AISC 1200, pp. 389–396, 2021.
https://doi.org/10.1007/978-3-030-51859-2_36

Great stories from eminent authors and well researched depth articles across publishers right on users phone. India's first ever subscription model where the users pays only for the content that they consume (pay – per – article).

Content well researched stories from 25 exclusive categories [9]. Quality curated content at reasonable price. Curated means the tasteful handpicked stories from popular publishers. Using Artificial Intelligence to deliver right stories at the right time to the right audience [4, 7]. Currently serving Tamil articles, English, Malayalam, Hindi, Telugu, Kannada will be added soon. Appappo is currently aimed at readers throughout India in all major languages, not to mention, we've already set our vision for our global play [11, 13] (Tables 1 and 2).

Articles are short-form or long-form contents priced in denominations of Rs 1, 3, and 5 based on the article's depths, timing and relevance to the audience. Bundles are collection of similar articles with a discount between 10% and 20% (Please note that length of the article has little weight age to the pricing)

Overall Demographics:

(See Figs. 1, 2 and 3).

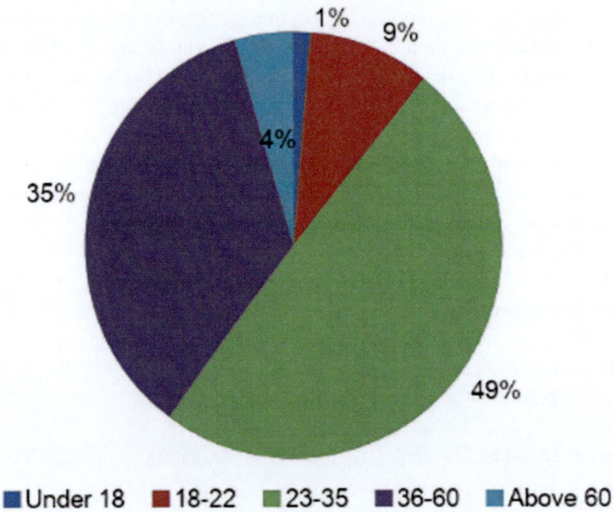

Fig. 1. Overall demographics percentage age-wise

Table 1. Overall demographics percentage

Age	Count (%)
Under 18	1%
18 - 22	9%
23 - 35	50%
36 - 60	36%
60 & Above	4%

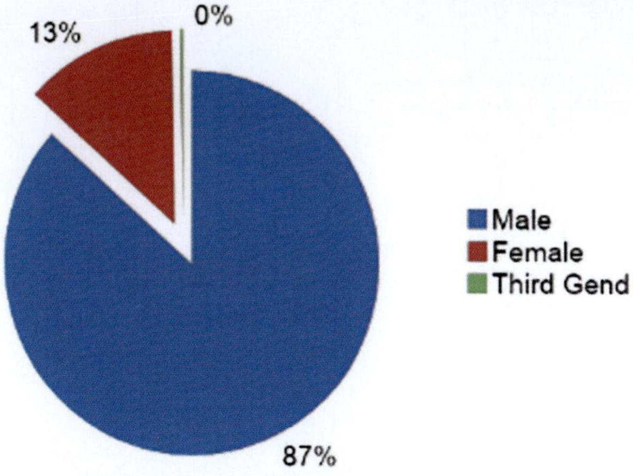

Fig. 2. Overall demographics percentage gender wise

Table 2. Overall Demographics using male, female and third gend.

Gender	Count (%)
Male	87%
Female	12%
Third Gender	1 %

Process of this app (how user uses this app):

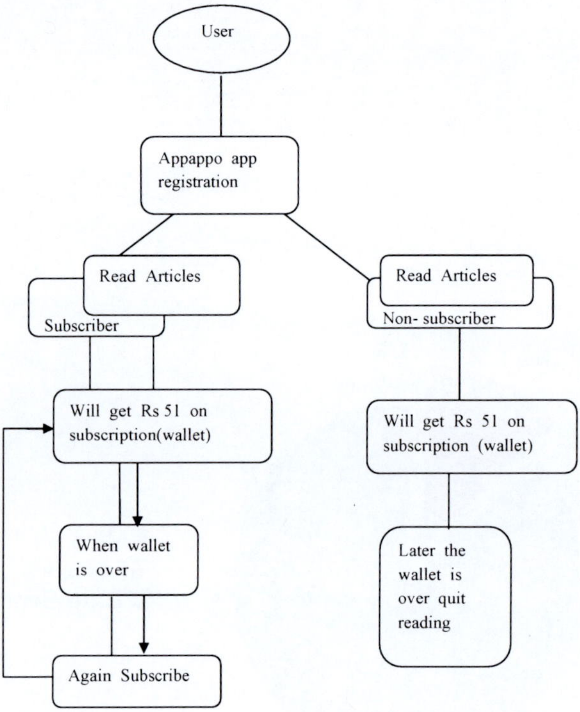

Paid Sub demographics:

Age	Count(%)
18 – 22	7%
23 – 35	47%
36 - 60	42%
60 & Above	4%

Gender	Count(%)
Male	89%
Female	11%

Article Submission: Publisher can submit the articles by logging into appappo portal or through an RSS feed. Alternatively, they can also email the articles to appappo's curation team.

Pricing: Publisher and appappo team jointly decides the pricing of the articles.

Publishing and Promotion : The articles will be published in appappo in the respective category and will be promoted to the relevant target audience. Publishers are requested to promote their article in appappo to their audience.

2 Related Work

The existing system was about to send the paperback edition to everyone and was to send the subscription message and notification to everyone [1]. Which made them spend more money and also a waste of time for them to invest more amount of money without having an prescribed idea.

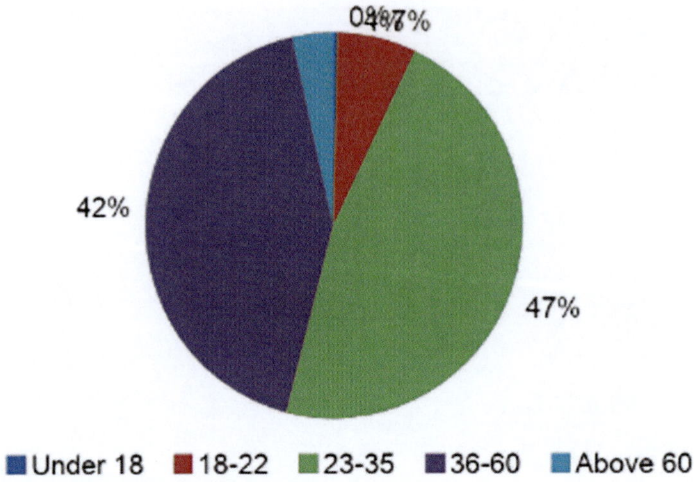

Fig. 3. Overall demographics percentage using genders age - wise

The drawbacks of the existing system is that it is waste of time to spend more money and the customers will be irritated and accordingly the support will become less [3, 14].

3 Proposed Work

The proposed system is all about finding the customers who are potential to subscribe and sending them the notification and offers. And also it is time save for them to send only to the people and there is no need to spend or waste time on the others.

The advantages of the proposed system is all about the time save and concentrate on customers to gain and concentrate on the profit.

4 Implementation Work

An initial step is the paper creator / new actions / new initiatives creator needs to set a proper target Appappo. And so to implement this, we have some basic observation data collected from the users who utilize the appappo app. If in case this app is not reached to the people then make sure it reaches everywhere and gain support.

When it comes to methodology used, we basically need to understand the original data and their parameter collected from our observation.

Here are the original data sets
Data_Users
Data_Shares
Data_Transaction.

In this data sets there are different parameters involved. So this parameter will help us solve the problem. We know our objective is to get the list of potential subscriber using algorithms. So these data will be very helpful to solve the problem.

After applying the algorithms in datasets find these accuracy, precision, recall and F1 score. The algorithms are applied and to find the best fit K Fold Cross Validation is applied. In this we take the nearest values of all the accuracies and apply them and compare them to get the best fit or the best algorithm for the dataset to find the potential subscriber.

5 Result Set

The attachment of the report generated by the proposed system is attached above according to the algorithm model and the final conclusion.

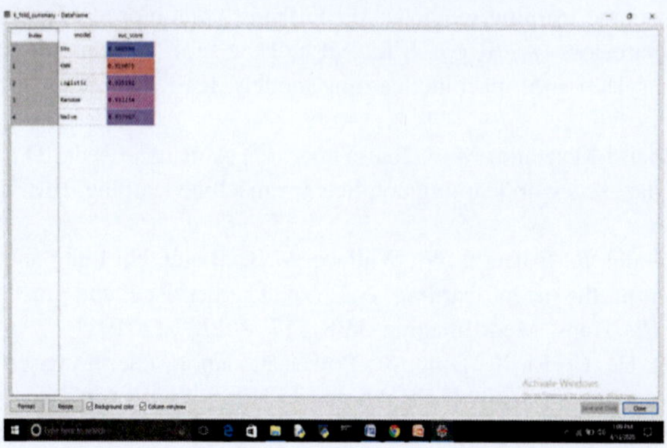

6 Conclusion

The conclusion of this proposed system is that there is the best algorithm which is suitable for the system to find the potential subscriber and then our motive is to save time and increase the profit and subscriber by providing them their favorite articles and make them utilize this app effectively.

Here I finally conclude this by, explaining in few words. I have used eight algorithms and five of them where with the similar accuracies. So I thought of applying a algorithm that compares all these five algorithm and give us the output which shows the best fit or the best algorithm for the proposed system.

So, I conclude that the support vector machine is the best suitable algorithms for the proposed system.

The scope for the further development is that we need to find the second recharge or the potential second recharge subscriber and then integrate them with live contents. This is done to automate and store the recording or tracking the subscribers and record them in the data base.

References

1. Di, C., Zhang, B., Liang, Q., Li, S., Guo, Y.: Learning automata-based access class barring scheme for massive random access in machine-to-machine communications. IEEE Internet Things J. **6**(4), 6007–6017 (2019)
2. Yu, H., Yang, X., Zheng, S., Sun, C.: Active learning from imbalanced data: a solution of online weighted extreme learning machine. IEEE Trans. Neural Netw. Learn. Syst. **30**(4), 1088–1103 (2019)
3. Krummenacher, G., Ong, C.S., Koller, S., Kobayashi, S., Buhmann, J.M.: Wheel defect detection with machine learning. IEEE Trans. Intell. Transp. Syst. **19**(4), 1176–1187 (2018)
4. Cote, D.: Using machine learning in communication networks [invited]. IEEE/OSA J. Opt. Commun. Netw. **10**(10), D100–D109 (2018)
5. Zhang, N., Cheng, K., Kang, G.: A machine-learning-based blind detection on interference modulation order in NOMA systems. IEEE Commun. Lett. **22**(12), 2463–2466 (2018)
6. Cao, W., Czarnek, N., Shan, J., Li, L.: Microaneurysm detection using principal component analysis and machine learning methods. IEEE Trans. Nanobiosci. **17**(3), 191–198 (2018)
7. Mendes, C., Barcelos, A., Rigo, S.J.: MLtool: a tool to automate the construction, evaluation, and selection of machine learning models. IEEE Lat. Am. Trans. **17**(07), 1163–1170 (2019)
8. Ledesma, S., Ibarra-Manzano, M., Cabal-Yepez, E., Almanza-Ojeda, D., Avina-Cervantes, J.: Analysis of data sets with learning conflicts for machine learning. IEEE Access **6**, 45062–45070 (2018)
9. Hussein, S., Kandel, P., Bolan, C.W., Wallace, M.B., Bagci, U.: Lung and pancreatic tumor characterization in the deep learning era: novel supervised and unsupervised learning approaches. IEEE Trans. Med. Imaging **38**(8), 1777–1787 (2019)
10. Qian, G., Li, Z., He, C., Li, X., Ding, X.: Power allocation schemes based on deep learning for distributed antenna systems. IEEE Access **8**, 31245–31253 (2020)
11. Fadlullah, Z.M., Tang, F., Mao, B., Kato, N., Akashi, O., Inoue, T., Mizutani, K.: State-of-the-Art deep learning: evolving machine intelligence toward tomorrow's intelligent network traffic control systems. IEEE Commun. Surv. Tutorials **19**(4), 2432–2455 (2017)
12. AlHajri, M.I., Ali, N.T., Shubair, R.M.: Classification of indoor environments for iot applications: a machine learning approach. IEEE Antennas Wirel. Propag. Lett. **17**(12), 2164–2168 (2018)
13. Reamaroon, N., Sjoding, M.W., Lin, K., Iwashyna, T.J., Najarian, K.: Accounting for label uncertainty in machine learning for detection of acute respiratory distress syndrome. IEEE J. Biomed. Health Inform. **23**(1), 407–415 (2019)
14. Gong, C., Liu, T., Yang, J., Tao, D.: Large-margin label-calibrated support vector machines for positive and unlabeled learning. IEEE Trans. Neural Netw. Learn. Syst. **30**(11), 3471–3483 (2019)
15. Shakya, S.: Machine learning based nonlinearity determination for optical fiber communication-review. J. Ubiquit. Comput. Commun. Technol. **1**(02), 121–127 (2019)

Efficient Deep Learning Approach for Multi-label Semantic Scene Classification

D. Senthilkumar[1], C. Akshayaa[1](✉), and D. George Washington[2]

[1] Department of Computer Science and Engineering, University College of Engineering, Anna University, Tiruchirappalli, India
desent07@gmail.com, akshayaachandrul6@gmail.com
[2] Ramanujan Computing Centre, Anna University, Chennai, India
george@annauniv.edu

Abstract. Recognition of semantic scene is a tedious task in which a single scene image is correlated with multiple classes. Also, the semantic scene classification is represented as multi instance multi label (MIML) classification, a part of multi label (ML) classification. Identifying the complex semantic relationship between the classes is the major issue in the traditional Machine Learning Technique (MLT) for MIML learning. Therefore, an efficient Deep Learning framework, Convolutional Neural Network (CNN) with a Gaussian blur filter is proposed. The proposed framework supports to identify the complex class correlations between various classes. Experimental results show that the proposed CNN for ML learning scene classification achieves a better predictive performance of 90% and a hamming loss of 0.13 when compared with the existing Machine Learning Techniques (MLT).

Keywords: Convolutional Neural Network · Deep Learning · Multi Label · Scene recognition

1 Introduction

Semantic scene classification is a problematic domain gaining renewed focus in recent days. It is a task of categorizing objects into semantic categories such as beaches, sunset, and parties [1]. Scene classification has many practical applications such as content-based indexing, automatic driving, and robot navigation [2]. A natural scene may have multiple objects meaning that multiple class labels may define the scene (for instance, a beach scene with sunset can be labeled as beach as well as sunset scene). The labeling of an image is not only based on the target it also considers the surrounding objects.

Conventional supervised learning usually implies that an entity is described by a single instance and labeled with a single class name. Classes are mutually exclusive at common classification problems. While representing in feature space, the overlapping of classes causes errors in classification. Classes in real-life scenarios are inconsistent by nature. These problems arise in the classification of natural scenes and documents and medical diagnostics [1]. Scene Classification is also represented as Multi Label (ML) learning.

J. I.-Z. Chen et al. (Eds.): ICIPCN 2020, AISC 1200, pp. 397–410, 2021.
https://doi.org/10.1007/978-3-030-51859-2_37

In Multi Label (ML) learning, an object may be simultaneously associated with multiple class labels. Most researchers either avoid such images when selecting their data sets, subjectively classify them with the most obvious base (single-label) class, or consider "beach + urban" as a separate class. In most cases, the last strategy is impractical as it would cause a dramatic increase in the number of classes to be considered. Another issue would be the availability of data for training in such a combined class category. There are drawbacks to the first two approaches too. In content-based image indexing and retrieval applications, for example, retrieving a multi-class image (e.g., beach + urban) would be more difficult for a user if we only have single labeled images of the beach or urban labels. It may require two separate queries to be conducted and the intersection of the images regained to be taken, respectively.

In a content-responsive image enhancement application, having different images of the beach, urban, beach + urban scenes settings may be desirable for the system. It is impossible to use an exclusive single label for this purpose. Identifying the semantic relationship between different classes present in the image is a tedious task. Image size, label storage, and parameters are some of the issues associated with the existing frameworks for ML scene classification. A bag-of-features (BOF) model with spatial pyramid matching (SPM) for scene recognition, was developed by Lazebnik [27]. The above-mentioned approaches make use of handcrafted features. This increases the burden of designing techniques for feature extraction. Convolutional Neural Network (CNN) approach designed especially for the classification of images does not require any separate feature extraction strategies thus reducing the complexities. Variety of classification methods have been established for multi label learning to provide different operational characteristics which include Linear Discriminant Analysis (LDA), BOOST [3], Support Vector Machines (SVM), Artificial Neural Networks (ANN), Multi- Kernel function [4], Deep Neural Networks (DNN). CNNs became able to provide a high-performance tool for scene recognition. CNN can reduce effectively image size, become more suitable for large image recognition by escalating computational efficiency. Compared to other deep learning models, CNN preserves the two-dimensional property for images and minimizes the parameter by making use of weight sharing strategy.

CNN with Dense Neural Networks based framework is proposed for Multi-Label Semantic Scene Classification to achieve a higher recognition rate with a reasonable reduction in hamming loss. The proposed approach has some major contributions:

(1) A deep learning model is proposed based on CNN with a Gaussian blur preprocessing filter for recognition of scenes.
(2) A powerful Convolutional Neural Network is used on the scene dataset for feature extraction.
(3) A solution combining the CNN and Deep Neural Network (DNN) with sigmoid activation function, yielding better results.

Section 2 reviews the works of literature based on scene classification. Section 3 discusses the proposed system and its components. Section 4 briefs about results of the experiment and discussion and Sect. 5 conclude with future work.

2 Literature Review on Multi Label Scene Classification

The minimal literature on the definition of multi labels is primarily focused on the classification of text or bioinformatics. Schapire and Singer [5] also suggested BoosTexter for document processing, expanding AdaBoost to tackle the categorization of multi-label documents. However, their model has an issue in controlling difficulty caused by overfitting. An EM-trained mixture model was suggested by McCallum [6], choosing the most likely collection of classes from the power set of possible labels and making use of heuristics to solve the complexity involved in the computations. The generative model focuses on the analysis of word frequencies in records and is therefore specific to texts. A set of binary SVM classifiers were made use by Joachims. But the technique does not address the multi-label training frameworks or specific test requirements. Clare and King [7] generalized the concept of entropy in bioinformatics to incur multi-label data (gene expression in their case), the reference strategy made use was the decision tree. As they said, because of the sparseness of the details and because they needed to learn detailed laws, they preferred a decision tree, not a complete classification.

A similar strategy for recognition of images includes the segmentation and classification of image regions (for example sky, grass) [8, 9]. A common approach to classify multi-label scenes is modeling such scenes using these labels combinations. However, there are some disadvantages to this approach. The region labeling was successfully applied only to constrained areas with minimal predetermined artifacts (for instance, landscape photographs taken from a vehicle in motion [8]). Since scenes consist of a collection of regions that are correlated, an explosion occurs in the number of the union of regions. The complexities associated with the segmentation-based strategy have made many analysts use a low-level feature, based on the exemplar approach [12–24]. Zhou's paper proposed a visual descriptor scene recognition approach that maintained the benefits of low-level object preservation as well as semantic modeling techniques [25]. The research-based on the latent variables of Latent Dirichlet Allocation, addressed image scene classification [26]. The "bag of features" method could provide effective information that explains the scene images characteristics that are independent of the effect of rotation, scale, and illumination [27]. It explains that feature extraction becomes more important in recognition of scene images to improve the accuracy of the predictive model.

These studies are based primarily on features that are handcrafted which restrict the representation of images. CNN's development has made features popular with powerful representational ability based on convolution. A convolutional method of extraction of features proposed by Gao, used on large-sized scene pictures, proposed centered factors to reduce the sources of instabilities [28]. Qi's paper used a transfer learning approach where a pre-trained CNN is used as a feature extractor [29]. This method extracts only mid-level features and does not consider any most significant level feature element. Tang proposed a GoogLenet approach for convolutional feature fusion. It involves multi-stage feature fusion known as (G-MS2F) for scene image classification [30]. Convolutional Neural Networks have also proved to be efficient in the Acoustic Scene Classification [31]. The usage of a 13 layer convolutional network for image-based fruit categorization was proposed by Zhang but it focuses only on single-label classification [32]. Real designed an image classifier that made use of an evolutionary algorithm to surpass the hand-designed models [33]. In comparison with the handcrafted features, the convolutional features consist of a more detailed representation of scene images. These features are learning-based features that contain abundant semantic information and are more robust and adaptable on scene classification. Zhou proposed machine learning classification techniques such as MIMLSVM and MIMLBOOST [3] to deal with the multi instance multi label (MIML) data. Gao, Jingyu, Jinfu Yang, Jizhao Zhang, and Mingai Li made use of Deep Neural Network (DNN) architectures to efficiently classify the natural scene data. They proposed a Convolutional Neural Network with Deep Boltzmann Machines (DBM) model for better results on fifteen scene data [2].

Boutell, Matthew R., Jiebo Luo, Xipeng Shen, and Christopher M. Brown proposed a multi-label scene classification method by using the Support Vector Machine (SVM) [1], a machine learning approach provides a reasonable recognition rate in application to scene classification. LeCun, Yann, Léon Bottou, Yoshua Bengio, and Patrick Haffner proposed a hierarchical model CNN, which extracts features by pooling and convolution layers finds application in handwritten digit recognition [10]. Techniques that are discussed so far have not been applied directly on scene recognition and none of them handles multi label problems. The objective of the proposed strategy is to consider the label correlation between the classes in the scene image thus reducing the storage space. Convolutional Neural Network doesn't require any external feature extraction thus making it simple and computationally efficient.

3 Proposed Convolutional Neural Network for Multi-lable Semantic Scene Classification

CNN is an extension to Multi-Layer Perceptron (MLP) and the structure of a basic MLP consists of an input layer with a hidden layer and finally an output layer (Fig. 1) [11]. In MLP the input layer receives the input data, hidden layers carry out computations on those data and the output layer gives the classification results. Convolutional neural networks are meant for feature extraction which in turn with dense neural networks helps in building an efficient classifier. CNN is mainly composed of the convolution layer and the pooling layer. Developing a CNN for ML scene classification comprises of two parts. The first part is to focus on constructing a CNN model with a Gaussian blur for feature extraction. The second is to focus on building a Dense Neural Network for classification.

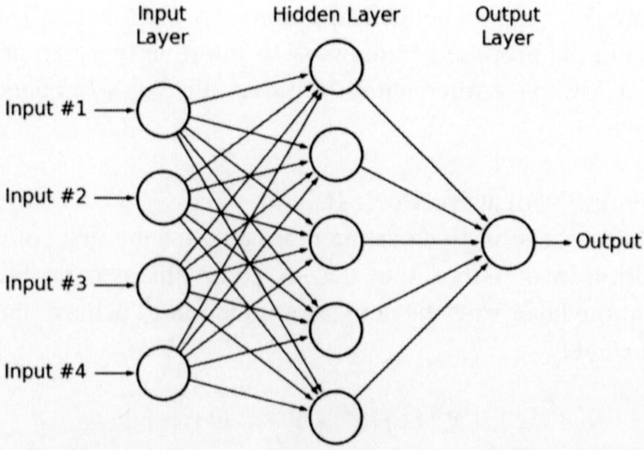

Fig. 1. Structure of Multi-Layer Perceptron.

3.1 Proposed CNN for Scene Classification

The block diagram of proposed CNN (Fig. 2) depicts various steps involved in the classification of Multi-Label (ML) scene classification.

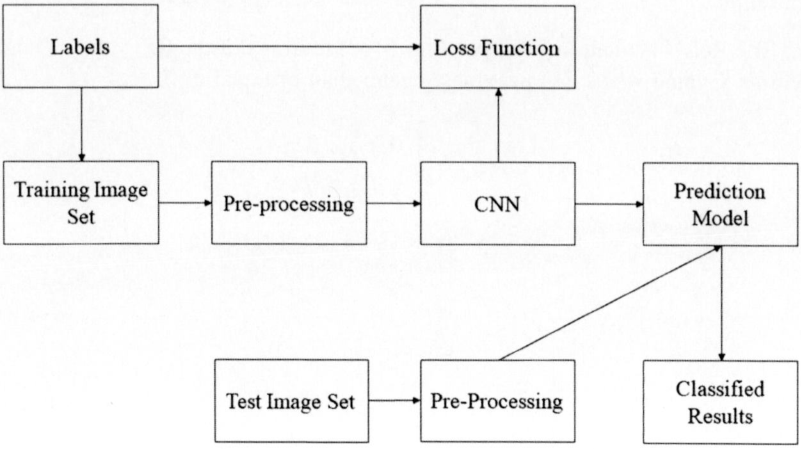

Fig. 2. Block diagram of proposed CNN model for ML scene classification

3.1.1 Preprocessing

Raw data from different sources are not suitable for analytical tasks. Preprocessing is a technique through which raw data collected from various sources are converted into a clean and analytical format. Similarly, an image in hand is probably not a flawless set of pixels, which are identical in strength values and are highly likely to be non-intensive pixels. These are called noise pixels and affect the appearance of the image that can also impair the annotation process. Noise may be mirrored, budding, dark spots, etc. Therefore, it is also beneficial to preprocess the images. The Gaussian filter has been made use in the proposed framework to improve the sharpness of the image. Gaussian filter is a low pass filter which removes the high-frequency noise signals present in the images.

3.1.2 Convolutional Neural Network (CNN)

The image of a natural scene is given as the input data for the first convolution layer on CNN. At convolution level 1 the input data is complemented by each learning convolution kernel and the layer uses the activation function to achieve the output maps. It can be given by Eq. (1)

$$X_j^1 = f(X^0 * W_j^1 + b^1) \quad (1)$$

where X_o and X_1 represents the input data and data with the convolution layer 1 respectively. j represents the number of convolutional kernels. W^1 is the layer 1 convolutional kernel. b^i represents, bias parameter. $f(\cdot)$ is the activation function which is rectilinear unit (ReLU) function.

Activation Function: The activation function determines whether or not activation of the neuron is required by measuring the weighted sum and introducing additional distortions. The purpose of the activation function is to integrate non-linearity into the neuron output.

ReLU: The ReLU activation function (Fig. 3) (2) returns 0 if the f(x) value is negative and returns x value when f(x) becomes greater than or equal to 0.

$$f(x) = \begin{cases} 0 \; if \; x < 0 \\ x \; if \; x \geq 0 \end{cases} \quad (2)$$

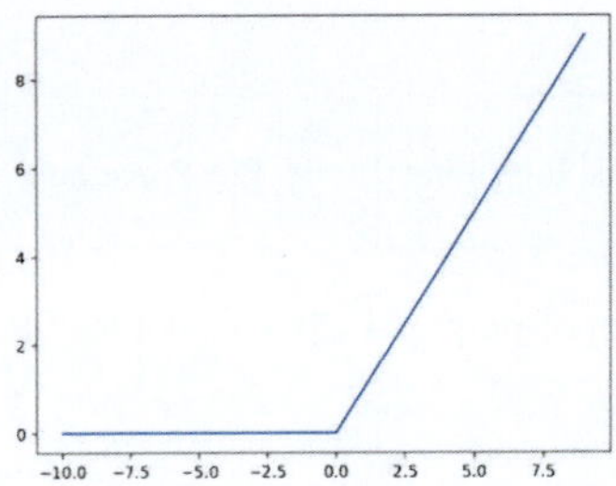

Fig. 3. ReLU activation function

The pooling layer reduces the size of the image by making use of max-pooling or means pooling technique. The output from the convolution layer 1 is given as the input to the pooling layer whose output is then forwarded to the next convolution layer or pooling layer. Finally, the output feature maps from the convolutional layer at the final stage are flattened and given as input to the Dense Neural Network (DNN). The output layer in the DNN uses the sigmoid activation function to determine the probability of each class present in the image.

3.2 Simulation Setup

CNN framework for ML scene classification is briefed with Fig. 5. The first convolution layer in the proposed system applies 32 filters on the input image (100 X 100) and the second convolution layer applies 64 filters which extract feature and resizes image (98 X 98) and then pooling is carried out which is then followed by dropout regularization to reduce overfitting. The feature vectors are further convoluted by the following third and fourth convolutional layers with filters 64, 64 respectively followed by pooling and dropout. The final feature vectors (23 X 23) are then forwarded to the dense neural network for classification. CNN doesn't require any separate feature extraction step. The Alex net architecture of the proposed model gives the feature extraction done by CNN (Fig. 4).

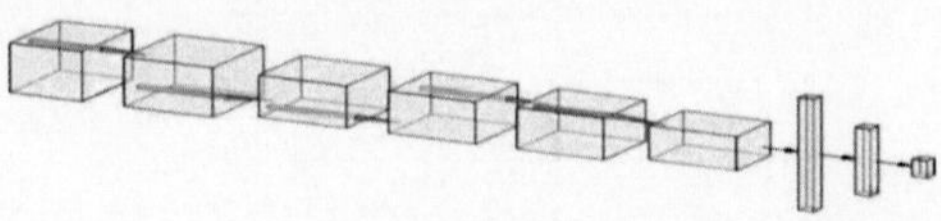

Fig. 4. Alex net architecture for various layers in CNN for ML scene classification.

3.3 Building Dense Neural Network for Classification

The feature vectors are flattened before given as input to the dense neural network. The data from the input layer is then forwarded to the hidden layers with the ReLU activation function. Then the output layer with the sigmoid activation function determines the labels associated with the image. The proposed system involves hyperparameters such as the batch size of 32, epochs of 20, and dropout regularization to prevent overfitting.

It makes use of a stochastic gradient descent optimizer for faster convergence to a globally optimal solution.

Sigmoid: The sigmoid function (Fig. 6) (3) squashes the output value between 0 and 1. It calculates the presence of one class without being affected by the presence of another class. Hence used for multi label classification. The loss function used is the Binary cross-entropy (4). Binary cross entropy tests the variance from the actual value (which is either 0 or 1) and then calculates the average class errors to determine the total loss.

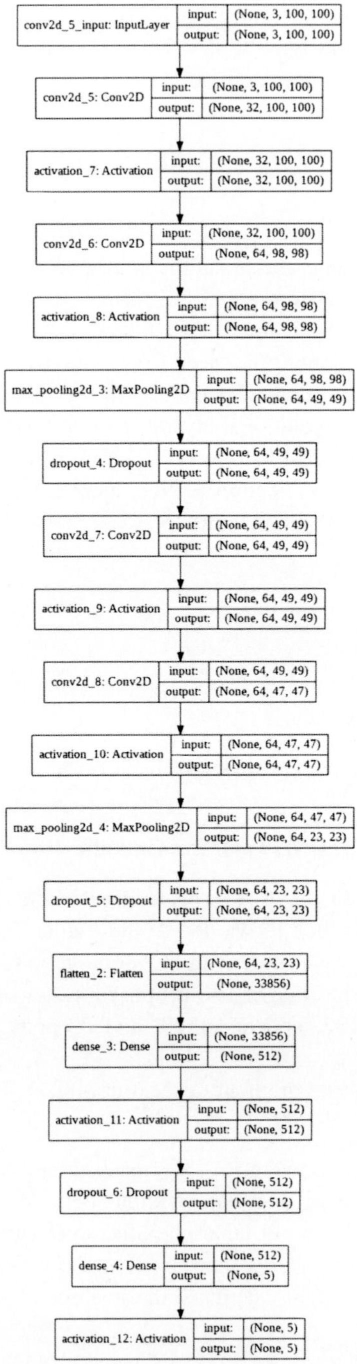

Fig. 5. Proposed CNN with the Dense Neural Network framework.

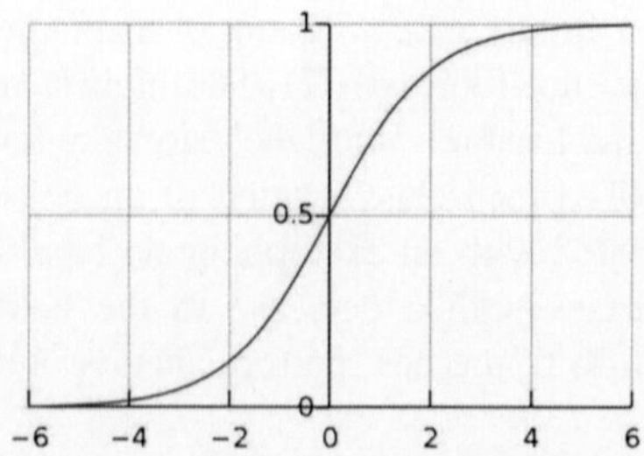

Fig. 6. Sigmoid activation function

$$f(x) = \frac{1}{1+e^{-x}} \tag{3}$$

$$L(y, \overline{y}) = -\frac{1}{N}\sum\nolimits_{i=0}^{N} (y * \log(\overline{y_i}) + (1 - y) * (1 - \overline{y_i})) \tag{4}$$

where y is the actual label and $\overline{y}$ is the predicted label. The loss is then backpropagated and the network is trained to reduce the loss.

3.4 Evaluation Parameters (Loss Function)

In MIML classification, evaluation parameters such as accuracy, hamming loss, precision [13–15], and ranking loss are considered to determine the efficiency of the model.

Accuracy: Accuracy is calculated in each situation by dividing the right labels expected to the sum of predicted and actual labels in each event (5). Total accuracy in all instances is the mean value.

$$\text{Accuracy} = \frac{1}{N}\sum_{i=1}^{N}\frac{|Y_i \cup Z_i|}{|Y_i \cap Z_i|} \tag{5}$$

where, Y_i is the actual label and Z_i is the predicted label. In the proposed model the accuracy is measured both on training as well as test dataset that is depicted with the help of an accuracy curve.

Precision: Precision can be described as the proportion of the true positive examples from all the classification model examples (6).

$$\text{Precision} = \frac{1}{N}\sum_{i=1}^{N}\frac{|Y_i \cap Z_i|}{|Z_i|} \tag{6}$$

Hamming Loss: Hamming failure takes account of estimation faults (incorrect label) and anomalies missed (label not forecast) (7). The highest rank is given to the most relevant label (1), whereas the least relevant label earns the lowest rank (M). Hamming loss tests then the probability of misclassification of an object-label combination, i.e. the incorrect label is correlated with an example or no label is estimated. The performance of the model increases with a decrease in the hamming loss value. If the hamming loss is equivalent to 0, the best performance is obtained.

$$\text{Hamming loss } = \frac{1}{N}\sum_{i=1}^{N}\frac{|Y_i \Delta Z_i|}{M} \tag{7}$$

Ranking Loss: The mean fraction of the pairs of labels that are ordered in reverse, for instance, is known as ranking loss for a particular instance. The performance of a model increases with a decrease in the ranking loss.

4 Experimental Results and Discussions

4.1 Dataset

In this paper, the multi instance multi label scene dataset is used to evaluate the proposed model. The dataset contains 2,000 photographs of the natural scene from classes such as deserts, mountains, sea, sunset, and trees. More than 22% of images belong to several labels at the same time. The dataset is preprocessed and then split into training and validation sets (Table 1). The dataset is divided as 80% for training the neural network and 20% for testing the network which means 1600 images are taken for training and 400 images are taken for testing the network.

Table 1. Distribution of classes in the MIML scene dataset.

Labels	Images	Labels	Images	Labels	Images
Desert	340	Desert + sunset	21	Sunset + trees	28
Mountains	268	Desert + trees	20	Desert + mountains + sunset	1
Trees	341	Mountains + sea	38	Desert + sunset + trees	3
Sea	216	Sea + trees	19	Mountains + sea + trees	6
Sunset	378	Mountains + sunset	106	Mountains + sunset + trees	1
Desert + mountains	19	Mountains + trees	172	Sea + sunset + trees	4
Desert + sea	5	Sea + sunset	14	Total	2000

4.2 Experimental Results

For preprocessing Gaussian filter with 3 X 3 size is used to reduce the noise present in the image and a four-layer CNN is used for feature extraction which extracts the low level and high-level features from the images while making it easy to be classified by the model (Figs. 7 and 8).

Fig. 7. Images prior sharpening by Gaussian filter.

Fig. 8. Images after sharpening by Gaussian filter.

The proposed CNN model is applied to the MIML scene dataset and the results (Table 2) are obtained. From the experimental results, it is observed that the proposed model performs well when compared with the existing frameworks. The model provides a ranking loss and precision of 0.0608 and 0.8468 respectively. The hyperparameters of the proposed CNN model are epochs (20), batch size (32), Stochastic Gradient Optimizer (learning rate = 0.01, momentum = 0.9), to prevent overfitting dropout regularization are made use.

Table 2. Evaluation parameters from existing and proposed frameworks for ML scene classification methods [3].

Frameworks	MIMLSVM	MIMLBOOST	Proposed method (CNN)
Hamming Loss	0.180 ± .017	0.189 ± .009	**0.133**
Ranking Loss	0.187 ± .018	0.181 ± .026	0.0608
Precision	0.783 ± .020	0.777 ± .025	0.8468
Accuracy	0.820 ± .024	0.811 ± .022	**0.9010**

The proposed framework makes use of the binary cross-entropy loss function. Binary cross-entropy loss increases as the predicted likelihood vary from the present tag. Predicting a probability of 0.012 would be incorrect if the real tracking tag is 1 and result in a high loss rate. A good model has a log loss of 0. The loss curve obtained from different iterations on applying the proposed CNN to the scene dataset (Fig. 9) shows that the learning rate of our model is pretty good. The test loss on the test dataset also decreases along with the increase in the number of epochs.

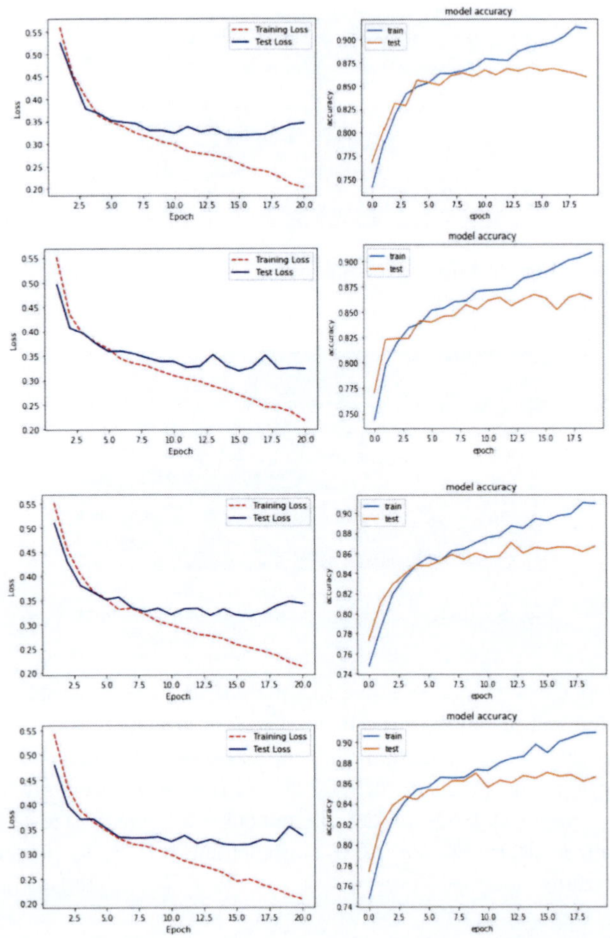

Fig. 9. Loss and accuracy curves of proposed CNN over different iterations (i, ii, iii, iv) on ML scene dataset.

5 Conclusion and Future Work

In this paper, the proposed CNN framework overcomes the problem of storage by taking into account the correlations (or dependence) among classes of objects present in an image. Preprocessing the image using a Gaussian blur filter helps in the improvement of the recognition and also feature extraction by CNN eliminates the need for external segmentation operation thus the proposed model is simple and efficient. The experimental results show that, proposed CNN achieved an accuracy rate of 90% and a hamming loss of 0.13, compared with the existing ML scene classification techniques. The backpropagation, pooling, and translational invariance involved in the Convolutional model degrade the performance of CNN. CNN does not provide any details regarding the spatial features related to an object in a scene image. Extension to future work is to consider the spatial information such as pose and orientation between the objects and reduce the loss due to the pooling layer on CNN.

References

1. Boutell, M.R., Luo, J., Shen, X., Brown, C.M.: Learning multi-label scene classification. Pattern Recogn. **37**(9), 1757–1771 (2004)
2. Gao, J., Yang, J., Zhang, J., Li, M.: Natural scene recognition based on convolutional neural networks and deep Boltzmann machines. In: 2015 IEEE International Conference on Mechatronics and Automation (ICMA), pp. 2369–2374. IEEE (2015)
3. Zhang, Z.L., Zhang, M.L.: Multi-instance multi-label learning with application to scene classification. In: Advances in Neural Information Processing Systems, pp. 1609–1616 (2007)
4. Tong-tong, C., Chan-Juan, L., Hai-Lin, Z., Shu-sen, Z., Ying, L., Xin-miao, D.: A multi-instance multi-label scene classification method based on multi-kernel fusion. In: 2015 SAI Intelligent Systems Conference (IntelliSys), pp. 782–787. IEEE (2015)
5. McCallum, A.K.: Multi-label text classification with a mixture model trained by EM. In: AAAI 1999 Workshop on Text Learning (1999)
6. Schapire, R.E., Singer, Y.: BoosTexter: a boosting-based system for text categorization. Mach. Learn. **39**(2–3), 135–168 (2000)
7. Clare, A., King, R.D.: Knowledge discovery in multi-label phenotype data. In: European Conference on Principles of Data Mining and Knowledge Discovery, pp. 42–53. Springer, Heidelberg (2001). https://doi.org/10.1007/3-540-44794-6_4
8. Campbell, N.W., Mackeown, W.P.J., Thomas, B.T., Troscianko, T.: The automatic classification of outdoor images. In: International Conference on Engineering Applications of Neural Networks, pp. 339–342 (1996)
9. Shi, X., Manduchi, R.: A study on Bayes feature fusion for image classification. In: 2003 Conference on Computer Vision and Pattern Recognition Workshop, vol. 8, p. 95. IEEE (2003)
10. LeCun, Y., Bottou, L., Bengio, Y., Haffner, P.: Gradient-based learning applied to document recognition. Proc. IEEE **86**(11), 2278–2324 (1998)
11. Botalb, A., Moinuddin, M., Al-Saggaf, U.M., Ali, S.S.: Contrasting Convolutional Neural Network (CNN) with Multi-Layer Perceptron (MLP) for big data analysis. In: 2018 International Conference on Intelligent and Advanced System (ICIAS), pp. 1–5. IEEE (2018)
12. Bautell, M., Luo, J., Gray, R.T.: Sunset scene classification using simulated image recomposition. In: Proceedings of 2003 International Conference on Multimedia and Expo, ICME 2003, (Cat. no. 03TH8698), vol. 1, pp. I-37. IEEE (2003)
13. Belongie, S., Carson, C., Greenspan, H., Malik, J.: Recognition of Images in Large Databases Using a Learning Framework. University of California at Berkeley, Berkeley (1997)
14. Hackel, T., Savinov, N., Ladicky, L., Wegner, J.D., Schindler, K., Pollefeys, M.: Semantic3D.NET: a new large-scale point cloud classification benchmark. arXiv preprint arXiv:1704.03847 (2017)
15. Lipson, P., Grimson, E., Sinha, P.: Configuration based scene classification and image indexing. In: Proceedings of IEEE Computer Society Conference on Computer Vision and Pattern Recognition, pp. 1007–1013. IEEE (1997)

16. Oliva, A., Torralba, A.: Modeling the shape of the scene: a holistic representation of the spatial envelope. Int. J. Comput. Vision **42**(3), 145–175 (2001)
17. Oliva, A., Torralba, A.: Scene-centered description from spatial envelope properties. In: International Workshop on Biologically Motivated Computer Vision, pp. 263–272. Springer, Heidelberg (2002). https://doi.org/10.1007/3-540-36181-2_26
18. Paek, S., Chang, S.F.: A knowledge engineering approach for image classification based on probabilistic reasoning systems. In: Proceedings of 2000 IEEE International Conference on Multimedia and Expo, ICME 2000. Latest Advances in the Fast Changing World of Multimedia (Cat. no. 00TH8532), vol. 2, pp. 1133–1136. IEEE (2000)
19. Serrano, N., Savakis A., Luo, A.: A computationally efficient approach to indoor/outdoor scene classification. In: Object Recognition Supported by User Interaction for Service Robots, vol. 4, pp. 146–149. IEEE (2002)
20. Smith, J.R., Li, C.-S.: Image classification and querying using composite region templates. Comput. Vis. Image Underst. **75**(1–2), 165–174 (1999)
21. Song, Y., Zhang, A.: Analyzing scenery images by monotonic tree. Multimed. Syst. **8**(6), 495–511 (2003). https://doi.org/10.1007/s00530-002-0067-y
22. Szummer, M., Picard, R.W.: Indoor-outdoor image classification. In: Proceedings 1998 IEEE International Workshop on Content-Based Access of Image and Video Database, pp. 42–51. IEEE (1998)
23. Quattoni, A., Torralba, A.: Recognizing indoor scenes. In: 2009 IEEE Conference on Computer Vision and Pattern Recognition, pp. 413–420. IEEE (2009)
24. Vailaya, A., Figueiredo, M., Jain, A., Zhang, H.J.: Content-based hierarchical classification of vacation images. In: Proceedings IEEE International Conference on Multimedia Computing and Systems, vol. 1, pp. 518–523. IEEE (1999)
25. Zhou, L., Zhou, Z., Dewen, H.: Scene classification using multi-resolution low-level feature combination. Neurocomputing **122**, 284–297 (2013)
26. Zang, M., Wen, D., Wang, K., Liu, T., Song, W.: A novel topic feature for image scene classification. Neurocomputing **148**, 467–476 (2015)
27. Zhou, L., Zhou, Z., Hu, D.: Scene classification using a multi-resolution bag-of-features model. Pattern Recogn. **46**(1), 424–433 (2013)
28. Gao, J., Yang, J., Wang, G., Li, M.: A novel feature extraction method for scene recognition based on centered convolutional restricted Boltzmann machines. Neurocomputing **214**, 708–717 (2016)
29. Qi, X., Li, C.-G., Zhao, G., Hong, X., Pietikäinen, M.: Dynamic texture and scene classification by transferring deep image features. Neurocomputing **171**, 1230–1241 (2016)
30. Tang, P., Wang, H., Kwong, S.: G-MS2F: GoogLeNet based multi-stage feature fusion of deep CNN for scene recognition. Neurocomputing **225**, 188–197 (2017)
31. Li, J., Liang, C., Zhang, B., Wang, Z., Xiang, F., Chu, X.: Neural architecture search on acoustic scene classification. arXiv preprint arXiv:1912.12825 (2019)
32. Zhang, Y.-D., Dong, Z., Chen, X., Jia, W., Sidan, D., Muhammad, K., Wang, S.-H.: Image-based fruit category classification by 13-layer deep convolutional neural network and data augmentation. Multimed. Tools Appl. **78**(3), 3613–3632 (2019)
33. Real, E., Aggarwal, A., Huang, Y., Le, Q.V.: Regularized evolution for image classifier architecture search. In: Proceedings of the AAAİ Conference on Artificial İntelligence, vol. 33, pp. 4780–4789 (2019)

Identification of Plant Diseases Using Machine Learning: A Survey

Snehal Andhare[1(✉)] and Sunil Wankhade[2]

[1] Vidyalankar Institute of Technology, Mumbai, India
snehal.andhare@vit.edu.in
[2] Rajiv Gandhi Institute of Technology, Mumbai, India
sunil.wankhade@mctrgit.ac.in

Abstract. Agriculture plays an important role across the globe. Nowadays manual plant disease identification has become difficult and time-consuming. Plant disease is an important aspect that reduces plant growth. According to the studies, plant disease has reduced the quality of plants in agriculture. Machine Learning approach helps to identify these plant diseases. Hence diseases like rice blast, stem rot, leaf scald, smut tungro, bacterial blight, brown spot, sheath blight, false, bacterial leaf streak, and fungal diseases, etc. can be identified using machine learning methods like clustering, classification, regression, etc. Different plant images that are affected by diseases are captured as well as wireless sensor network (WSN) also helps in disease identification. These images can be processed, and diseases can be identified. This practice gives rise to plant yield. This survey paper describes the machine learning approach for plant disease identification and a detailed study of various techniques is classified.

Keywords: Machine learning techniques · SVM · ANN · Plant diseases

1 Introduction

Agriculture has built up civilization. In India, the economy is dependent on crop production as India is an agrarian country. Agriculture act as the backbone of every economy. Due to the advances in agriculture, there is an increase in food production.

To have an increase in yield, up-gradation of the crop is necessary for the agriculture sector. Crop monitoring helps to identify diseases in the plant. Plant diseases are one of the factors which give rise to a reduction in the quality and quantity of crops. Machine learning has also been accounted for as the enterprise for incorporation, automation of operations, and representations for recognition. Several sensor network and machine learning tasks have been automated in computing with newer improvements. A variety of handy applications of automatic machine learning responsibilities include for example organizing data acquiring information, performing agricultural key tasks, detecting the occurred events, environmental interacting, brain-computer interface, process control, and the resembling features. These are all quite useful and supportive nowadays. Hence, the advancement of these types of variety of systems concerning the key applications is being enhanced with the extensive use of approaches

J. I.-Z. Chen et al. (Eds.): ICIPCN 2020, AISC 1200, pp. 411–421, 2021.
https://doi.org/10.1007/978-3-030-51859-2_38

of information system and technology along with computer science such as image processing, artificial intelligence, neural networks, machine learning algorithms, fuzzy logic, genetic algorithms, digital signal processing, etc.

Artificial Intelligence has always tried to improve the sensor organize correspondence. Machine Learning strategies are termed into viable methodologies for extracting information which is used to discover sensor range relations where the mathematical models are inaccessible, or very expensive. A few machine learning techniques like supervised learning, unsupervised learning, semi-supervised learning, reinforcement learning, and computational intelligent calculations utilized for bunch head (CHnode) choice, ideal way identification, finding valuable information from assembled data, limit parcel postponement and increment lifetime of WSNs. In this survey paper, different plant diseases are discussed and also various machine learning techniques are classified.

2 Related Work

Various studies have found that machine learning approaches produce more accurate results to identify plant diseases. Machine Learning provides various tools to analyze a large amount of data. To identify and detect plant disease, images are captured through the camera, then it is segmented, features are extracted and classified using different classification algorithms.

3 Plant Diseases

3.1 Rice Blast

The rice blast [1], whose causative operator is the growth Magnaporthe grisea, reporter to anamorphic stage Pyricularia grisea (Pyricularia oryzae), is the most expressive in Brazil, causing huge misfortunes in developed assortments yield. the rice impact can happen in all the plant shoots, from the underlying creating stages until the last seed creation stage (BEDENDO e PRABHU, 2005). The misfortunes brought about by the rice impact can be immediate or circuitous, while in the leaves it is roundabout and influences photosynthesis and breath (BASTIAANS e KROPFF, 1993). The impact of a sore over the fluid photosynthesis rate can be a decrease journalist to a misfortune up to multiple times the region involved by that injury (BASTIAANS e ROUMEN, 1993). In the panicles, the harms are immediate, because of its impact on various parts of the creation. BASTIAANS (1993) brings up that a serious scourge of the infection can happen during the most extreme tillering, inciting senescence of leaves. The tillering time frame would then be able to be drawn out, postponing the thriving and development. The plant development and the foliar zone arrangement diminishes definitely during the foundation of the sickness, proceeding with decreased until the development, bringing about serious loss of dry issue and grain profitability, even though the dry issue dissemination may not be influenced. The rice blast diseases can be detected using machine learning-based classification methods like k-nearest neighbor (KNN) and artificial neural network (ANN).

3.2 Bacterial Blight

The little, light green spots or streaks describe the disease which before long shows up water-drenched. The disease is extended and later on, they are identified as dry dead spots. The disease prolongs into straight streaks which may, in the long run, broaden the leaf length. Generally, illness affects the leaf sheath. In serious contaminations, a smooth dim exudate might be pressed from the cut finish of leaf displaying manifestations. Earthy colored spots on the edge of the cotyledon described plants. The microscopic organisms overwinter on crop buildup, seed, fall-planted oats, and lasting grasses. The disease is restricted by dry climate and due to bacterial contamination new leaves might be moderately liberated. The classification and identification of such disease can be done using the support vector machine technique.

3.3 Stem Rot

Stem spoil prompts the arrangement of injuries and the creation of powdery grains and unfilled panicles. The contamination of bodies or sclerotia is found in the upper soil layer. They make due in air-dry soil, covered sodden rice soil, and in faucet water. They can likewise get by on straw, which is covered in the dirt. The sclerotia coast on water system water and contaminate recently planted rice during the land arrangement. The disease is high on plants with wounds because of housing or creepy-crawly assault. The panicle dampness substance and nitrogen compost likewise impact malady advancement.

3.4 Downy Mildew

Downy Mildew affects numerous plants and appears as yellow to white fixes on the upper surfaces of more established leaves. On the undersides, these regions are detected with white to grayish, cotton-like organisms. These "downy" masses are frequently seen after a rain or substantial dew and vanish not long after bright climate resumes. As the disease advances leaves may, in the long run, turn fresh and earthy colored and tumble off even though the plant has sufficient water.

3.5 Leaf Spot

Affected plants have earthy colored or dark water-doused spots on the foliage, here and there with a yellow radiance, generally uniform in size. The spots spread and undergo wet conditions. Under dry conditions, the spots have a dotted appearance. As spots become increasingly various, whole leaves may yellow, wilt and drop. Individuals from the Prunus family (stone organic products, including cherry, plum, almond, apricot, and peach) are especially defenseless to bacterial leaf spot. The organic product is observed as spotted or has indented to have earthy colored regions. In vegetable nurseries, the bacterial leaf spot will likewise assault tomato and pepper crops.

3.6 Corn Smut

Side effects are most usually seen when the ears produce mushroom-like tumors or nerves. These “swellings” start as little, whitish-dim inconsistencies which extend and turn dark as they load up with spores. Ear nerves can develop to 4-to 5-crawls in breadth and discharge a huge number of spores as they crack. These contagious spores are passed up the breeze for significant separations to taint new plants. Irritates on leaves stay little and in the end, become hard and dry.

4 Block Diagram of Plant Disease

The block diagram of plant disease shown in Fig. 1 has various modules to detect and identify plant diseases. First, the input image is captured through the camera. The captured image is pre-processed through segmentation. Features are extracted from the segmented image. The classification algorithms compare the image from the data set and the segmented image. If the difference identified then the plant is detected as an infected plant and if no difference identified then it is termed as a normal plant.

Following is each module description of the diagram:

(a) **Image Capturing:** Input images are captured through camera. Image is in RGB (Red, Green, and Blue) form. To have color variation manufacture, an independent color space conversion is applied.
(b) **Image Pre-processing:** This module helps to remove noise in the image. Smoothing filter is used to smoothen the image. Image contrast is increased due to image enhancement.
(c) **Image Segmentation:** Segmentation means to partition images of the same skin tone into different segments. Partition of the image can be carried out using different algorithms.
(d) **Feature Extraction:** Generally, color features are extracted so that system can accurately capture the manifestation of the different diseases in the leaves. Color transformation of the image is computed. Computed result and image from the dataset is compared for disease identification
(e) **Classification:** Different classification algorithms are used such as Support Vector Machine, Neural Networks, K-Nearest Neighbor, Genetic algorithm, etc. Through these algorithms, a segmented image and the image from the dataset are compared. If the difference occurred then the plant is termed as an infected plant and if not then it is termed as a normal plant.

5 Literature Survey

The previous research work is referred to have a detailed study of this survey, previous research work is also referred. The literature survey is performed in chronological order from 2016–2019.

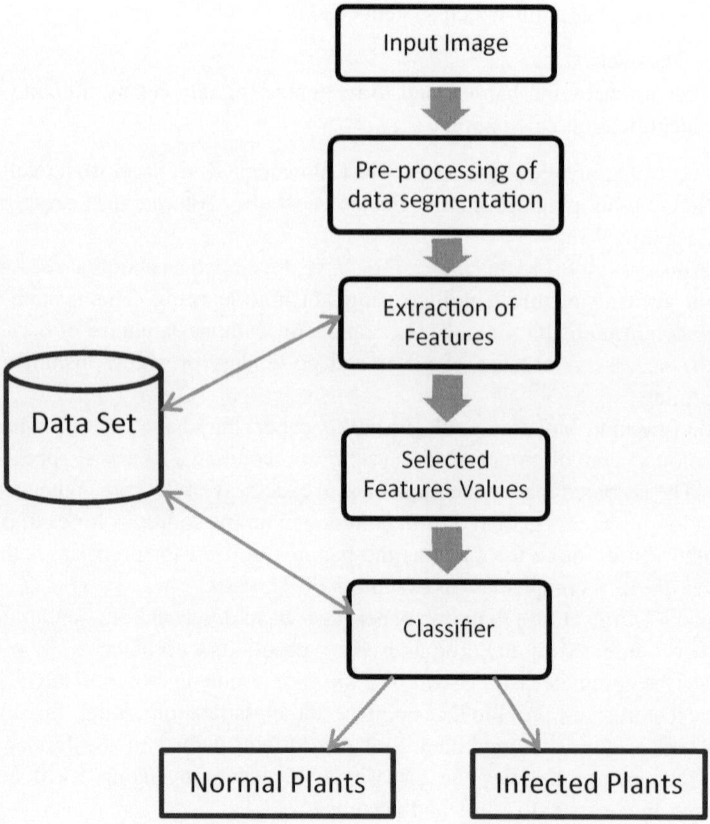

Fig. 1. Block diagram of plant disease identification

R. Poonguzhali et al. [2] in this crop condition is analyzed from leaves. Leaves are captured using a camera from a real-time environment. The captured images undergo processes like preprocessing and segmentation. Segmentation is performed using K-means clustering. After segmentation, they undergo classification using Machine learning algorithms in which healthy and diseased leaves are detected. Diseases like brown spots and leaf blasts are considered.

Michael Gomez Selvaraj et al. [3] In this research, a detection model is developed using a transfer learning approach to retrained three Convolutional Neural Network (CNN).The architecture provided the result with an accuracy of 90%. In this paper, there is an evaluation on the performance of SSD (Single Shot detector) MobileNetV1 in order to execute the detection capabilities on a mobile device.

Anuradha Badage [4] this paper has proposed framework gives the answer for consistently observing the developed territory and gives the computerized illness identification utilizing remote detecting pictures. The goal of the defined framework is to get the prior location of maladies when it starts spreading on the external layer of the leaves.

There are two phases on which system works:

(i) Trains Data-sets
(ii) Manages to check the harvest and to recognize the ailment by utilizing Canny's edge identification calculation

Identified wheat plant diseases: Leaf Rust, Powdery Mild, Stem Rust, Yellow Rust

Identified cotton plant diseases: Bacterial Blight, Alternaria, Cerespora, Grey Mildew, Fusarium Wilt.

Vijay Kumar et al. [5] in this paper, they have developed a robot that recognizes the leaf ailment utilizing picture handling using Machine learning. This system likewise screens the condition of the field, for example, soil dampness, nature of harvests, and splashes the necessary measure of water and pesticides for accomplishing the great yield in farming.

Sharada Prasanna Mohanty et al. [6] in this paper, they have built up a model that trains a profound convolutional neural system to recognize 14 harvest species and 26 infections. The prepared model accomplishes an exactness of 99.35%. When testing the model on a lot of pictures gathered from confided in online sources - for example, taken under conditions not quite the same as the pictures utilized for preparing - the model despite everything accomplishes an exactness of 31.4%.

N. Kanaka Durga et al. [7] In this paper, they have described the goal to detect the sickness that is infected by the crop and take precautions to protect the crop before it spreads over the complete crop. Use of histogram of oriented gradient (HOG) operation and predict features and provide that points to the classification model. Finally, leaves are tested, and sickness is identified, and records are passed to the farmer through message. This paper identifies the disease in the leaves by using SVM and ANN algorithms to find efficient results and accuracy.

Ronnel R. Atole et al. 2018 [8] has defined the utilization of a profound convolutional neural system in the arrangement of rice plants as indicated by wellbeing status dependent on pictures of its leaves. A three-class classifier was executed speaking to ordinary, unfortunate, and snail-swarmed plants utilizing the move gained from an AlexNet profound system. In this paper, the stochastic slope plunge was the learning calculation applied under a base learning pace of 0.0001 and a bunch size of thirty (30), which created the aftereffect of 91.23% exactness.

Jia-You Hsieh et al. 2019 [9] have worked on rice blast disease. Rice blast disease (RBD) is one of the most damaging crop diseases for rice in Taiwan. The goal is to build an early warning mechanism for the RBD using the machine learning model under current climatic conditions. With the climate data, the recursive feature elimination algorithm is conducted to select the key features that have impacts on the RBD. This paper says about which climatic factors are responsible for the aggravation of RBD.

Jayme G.A. Barbedo, 2018 [10] has received the positive outcomes that are being acquired utilizing this methodology conceal a few issues that are only from time to time considered in the individual investigations. This article presents an examination concerning the fundamental factors that influence the structure and adequacy of profound neural nets applied to plant pathology. The database utilized in the trials is unreservedly accessible and contains very nearly 50,000 pictures of 171 infections influencing 21 plant species. Be that as it may, just pictures of corn maladies were utilized with regards to this work. This subset was picked because it contains the greatest assortment of conditions and a sensible number of pictures for every one of the nine ailments which are all brought about by organisms.

Xue Wang et al. 2018 [11] To improve the adequacy of the system expectation consequence of rice protein association, the system forecast strategy for rice protein communication dependent on dark BP neural system has been proposed. Initially, a progression of key highlights about communication destinations of rice protein, for example, the arrangement range, preserved weight, entropy esteem, the open outside result of compound and grouping rate, and so on., ought to be extricated right off the bat. At that point, the dark BP neural system and their combination will be applied to the preparation and trial of these examples set. Multiple times of cross approval are applied to the preparation and test, and four gatherings of highlight blends of rice protein communication with likeness are made.

Prabira Kumar Sethy et al. 2018 [12] has spotted the side effects of rice crop sicknesses infer the reality of the malady and propose picking the best way to deal with managing the ailment. The treatment of rice crop sicknesses by applying unbalanced pesticides expands the expense and ecological contamination. Thus, the utilization of pesticides must be limited. This can be realized by focusing on the ailing territory, with a suitable amount and centralization of pesticide by evaluating illness.

Prabira Kumar Sethy et al. 2019 [13] has worked on the disease in rice crop that reduced the quality and quantity of production and mostly affects in leaf and panicle. Detection and grading of rice panicle blast are required as a prior condition for rice disease controlling. In this paper, a rice panicle sample is taken, placed individually on a white paper, and then capture the image using a smart phone camera of 12Megapixel resolution. The captured image is processed to calculate the healthy portion of the panicle (golden color) and the total area of the panicle. So ultimately, it deducts the infected portion or blast area of the panicle.

Wen-Liang Chen et al. 2019 [14] have utilized non-picture IoT gadgets to recognize rice impact. Not at all like the picture based plant ailment recognition draws near, horticulture sensors create non-picture information that can be naturally prepared and dissected by the AI instrument progressively. diminishes the stage the board cost to give ongoing preparation and expectations. To uses the climate information acquired from the Observation Data Inquire System (CODiS) of Central Weather Bureau in Taiwan and the ongoing information of the AgriTalk sensors from the miniaturized scale climate stations are set up in 4 areas.

Wei Hongyuan et al. 2018 [15] has aimed at the objective discovery of remote detecting rice field of UAV, the images of enormous size UAV is right off the bat portioned, and the sort of each image is physically recognized, and the image preparing set and confirmation set are made. At that point, the preparation model of the convolutional neural system is acknowledged by python programming. In rice region estimation, quality testing, and different angles, various calculations can be utilized to process the images took by the UAV to accomplish the objective.

D. Femi et al. 2019 [16] has developed an automated system to avoid manual method to identify plant diseases when it appears on leaf, stem, and fruits. In this paper, the author has used gradient anisotropic diffusion ımage filter for preprocessing,

Region-based segmentation algorithms for segmentation, GLCM for feature extractions, and diseases are recognized using SVM. The machine learning approach identifies whether the plant is normal or not.

Nafees Akhter Farooqui et al. 2019 [17] have used an approach for detection and identification for plant diseases by using K-means clustering, multi-class SVM, and data is processed using advanced neural network techniques. Features are extracted by using Gray Level Co-occurrence Matrix (GLCM). Plant disease is identified by SVM classifier and Advanced Neural Network is used to detect and improve accuracy.

Shabari Shedthi B et al. 2019 [18] have designed an automated framework that detects and classifies the leaf diseases accurately. In this, images are captured by a digital camera and then images undergo preprocessing, segmentation, and features extraction. Classifiers like SVM and Naïve Bayes are used. Accuracy is compared between ANN, SVM, and Naïve Bayes classifiers.

S.M. Jaisakthi et al. 2019 [19] have proposed an automatic system using image processing and machine learning techniques for detecting diseases in a grape vine. Grab Cut segmentation is used to segment grape leaf. Further, the diseased part of the leaf is again segmented using two methods: global thresholding and semi-supervised techniques. Using different Machine Learning techniques like Support Vector Machine (SVM), AdaBoost, and random forest tree different features are classified from the segmented disease part. It is found that using SVM better accuracy of 93% is achieved.

Dr. Abul Bashar, 2019 [20] has surveyed different deep learning neural network techniques to gain an accurate classification with auto-generated feature extraction. In this paper, the performance of deep learning and traditional learning is compared at the various application level. The applications are related to speech recognition, image recognition, and natural language processing. It has also presented the details of deep learning neural networks to accurately classify and predict the ordering of hierarchical data which is unlabeled and unstructured.

Dr. Abraham Candy, 2019 [21] has proposed an RGBD analysis to test various stages of fruits and identify the maturity of various stages. The color of the fruits is termed as an important factor to identify different stages of fruits. Stages are classified into three types: premature, mature, and over-mature. A Capsule network is involved. Depth Filter is used to finding image depth and a fuzzy interference system is used for image segmentation. Density-based clustering is applied to group images and classify them using Linear Support Vector Machine which is trained with various features like color, geometry, etc.

6 Comparative Analysis

Different methods employed in the recognition of plant diseases, which uses different ML approaches are summarized in Table 1. Different sorts of plant diseases like rice blast, fungal diseases, Spot on the leaves, leaf smut, and leaf diseases are identified using different classifiers. Different classification techniques utilized in various papers are surveyed and their accuracy is identified.

The performance comparison graph of different methods surveyed in Table 1 is elucidated in detail using Fig. (2) and Fig. (3). In Fig. (2) is considered, to detect plant

Table 1. Comparison of various approaches used in plant diseases diagnosis

Techniques used	Plant diseases identified	Purpose	Accuracy	Reference
Fine tuning Alex-net	Rice blast	To check whether rice plant is: (a) damaged with golden apple snails; (b) affected with diseases, or (c) normal and healthy	91.23%	8
Auto-sklearn and neural network	Rice blast	Analysis of climatic factors which is responsible for RBD (Rice Blast Disease)	72%	9
CNN	Fungal disease	Plant disease is classified and recognized	81%	10
BP and organic gray NN	Rice Protein Extraction	To extract the sequence of rice protein using gray BP neural network	63.4%	11
K-means with Fuzzy logic	Spot on leaves	Accurately quantify the disease severity	86.35%	12
SVM	Panicle blast	The calculating ratio of healthy and total area (Blast Factor)	90%	13
CNN	Rice blast	To investigate the accuracy.	89.4%	14
CNN	Leaf smut	Identification of characteristics in the rice field	94.31%	15
SVM	Leaf diseases	Solving the problem of detection and preventing crop diseases	90.50%	17
ANN, SVM, Naïve Bayes	Leaf diseases	To develop a system to detect and identify leaf diseases	90%	18
SVM	Grape Leaf Disease	To develop an automated system to classify and detect plant disease	93%	19

disease fine-tuning alexnet and fuzzy logic with k-means techniques gives better performance in accuracy than any other technique. Now consider Fig. (3), CNN and SVM give the best performance in accuracy for the identification of plant disease. But if compared to both Fig. (2) and Fig. (3), the CNN classifier is the best technique to identify and detect plant disease.

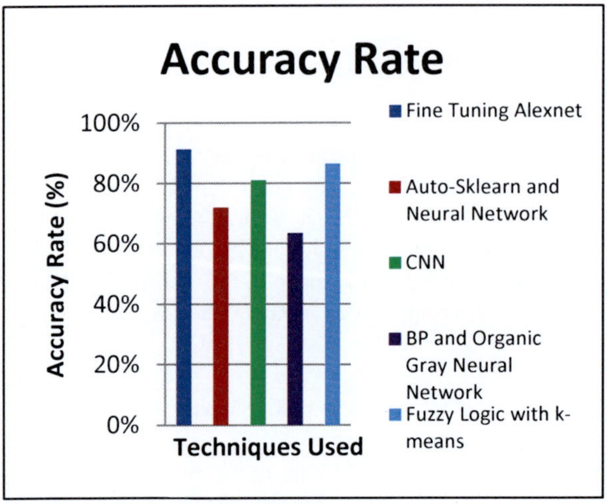

Fig. 2. Fine Tuning Alexnet, Auto-Sklearn and Neural Network, CNN, BP and Organic Gray Neural Network, Fuzzy Logic with k-means

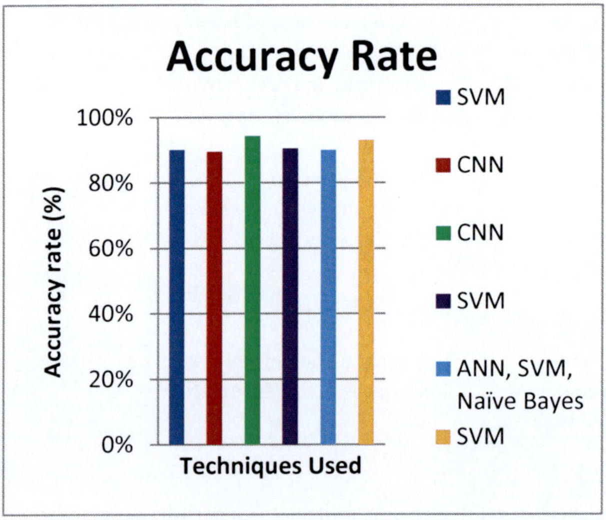

Fig. 3. SVM, CNN, ANN classifier, Naïve Bayes classifier

7 Conclusion

In this paper, the survey describes different plant diseases that occurred, also there is a study of the identification of plant diseases using machine learning. The main focus is to survey and summarize the plant disease detection using machine learning methods that will help researchers to overcome the various problems that affect crops. Further, machine learning tools can be used to predict various plant diseases.

References

1. da Soares, L.C.S., Raphael, J.P.A., Bortolotto, R.P., Nora, D.D., Gruhn, E.M.: Blast disease in rice culture. Rev. Bras. Tecnol. Apl. nas Ciências Agrárias **7**(2), 109–119 (2014)
2. Poonguzhali, R., Vijayabhanu, A.: Crop condition assessment using Machine Learning, no. 6, 897–900 (2019)
3. Selvaraj, M.G., et al.: AI powered banana diseases and pest detection. Plant Methods, 1–11 (2019)
4. Badage, A.: Crop Disease Detection using Machine Learning : Indian Agriculture, pp. 866–869 (2018)
5. Vijay Kumar, V., Vani, K.S.: Agricultural robot: leaf disease detection and monitoring the field condition using machine learning and image processing. Int. J. Comput. Intell. Res. **14** (7), 551–561 (2018)
6. Mohanty, S.P., Hughes, D., Salathé, M.: Using Deep Learning for Image-Based Plant Disease Detection (2016)
7. Durga, N.K., Anuradha, G.: Plant disease identification using SVM and ANN algorithms. Int. J. Recent Technol. Eng. (IJRTE) **5**, 471–473 (2019)
8. Atole, R.R., Park, D.: A multiclass deep convolutional neural network classifier for detection of common rice plant anomalies. Int. J. Adv. Comput. Sci. Appl. **9**(1), 67–70 (2018)
9. Hsieh, J., et al.: Building the Rice Blast Disease Prediction Model based on Machine Learning and Neural Networks Building the Rice Blast Disease Prediction Model based on Machine Learning and Neural Networks (2019)
10. Barbedo, J.G.A.: Factors influencing the use of deep learning for plant disease recognition. Biosyst. Eng. **172**, 84–91 (2018)
11. Wang, X., et al.: Gray BP neural network based prediction of rice protein interaction network. Cluster Comput. **22**(2), 4165–4171 (2018)
12. Sethy, P.K., Negi, B., Barpanda, N.K.: Measurement of Disease Severity of Rice Crop Using Machine Learning and Computational Intelligence, pp. 1–11 (2018)
13. Sethy, P.K., Sahu, S.K., Barpanda, N.K., Rath, A.K.: Rice panicle blast detection and grading based on image processing techniques, no. 5, pp. 5–8 (2019)
14. Chen, W., Lin, Y., Ng, F., Liu, C., Lin, Y.: Ricetalk: rice blast detection using internet of things and artificial intelligence technologies. IEEE Internet Things J. **7**(2), 1001–1010 (2019)
15. Hongyuan, W., Jian, M.: The Recognition of Rice Area Images by UAV Based on Deep Learning, vol. 232 (2018)
16. Femi, D., Chowdary, T.D., Anilkumar, N.: Early detection of plant disease using machine learning. Int. J. Innov. Technol. Explor. Eng. **8**(8), 950–952 (2019)
17. Farooqui, N.A., Ritika, R.: An identification and detection process for leaves disease of wheat using advance machine learning techniques. Biosci. Biotechnol. Res. Commun. **12**(4), 1081–1091 (2019)
18. Shabari Shedthi, B., Siddappa, M., Shetty, S.: Identification of plant leaf disease using machine learning techniques. Int. J. Recent Technol. Eng. **8**(3), 6077–6081 (2019)
19. Jaisakthi, S.M., Mirunalini, P., Thenmozhi, D., Vatsala.: Grape leaf disease identification using machine learning techniques. In: ICCIDS 2019 - 2nd International Conference on Computational Intelligence Data Science Proceedings, no. February, pp. 1–6 (2019)
20. Bashar, D.A.: Survey on evolving deep learning neural network architectures. J. Artif. Intell. Capsul. Netw. **2019**(2), 73–82 (2019)
21. Chandy, A.: RGBD analysis for finding the different stages of maturity of fruits in farming. J. Innov. Image Process. **1**(02), 111–121 (2019)

GigaHertz: Gesture Sensing Using Microwave Radar and IR Sensor with Machine Learning Algorithms

Misbah Ahmad(✉), Milind Ghawale, Sakshi Dubey, Ayushi Gupta, and Poonam Sonar

Rajiv Gandhi Institute of Technology, Versova, Andheri (West), Mumbai 400061, Maharashtra, India
misbahowais@hotmail.com, poonam.sonar@mctrgit.ac.in

Abstract. Intelligent systems have been important in all the industries whether its automotive, medical, automation and smart switches (in our terms replacing traditional switches). The key element of such systems is the user interface that can keep track of human gestures. Hand gesture sensing using a traditional approach i.e. camera has a serious limitation of lightning condition which triggers the use of additional infrared camera making the entire system bulky costly and less secure toward hacking problems. Short-range radar in combination with an IR sensor can sense all those gestures that a camera can by overcoming the drawbacks of the existing traditional approach. We have proposed a novel method to overcome the drawbacks of the existing systems and to use the proposed system GigaHertz for multiple applications, this method is based on the use of Doppler shift using radar and IR sensor for end to end (hardware, interface, and software) radar-IR-based system design to efficiently detect the gesture. The novelty of the proposed method is the collaboration of a radar system with an IR sensor to provide a total of eight gestures which can be configured as per user needs. The main advantage of the proposed system is in the selection of effective and efficient feature selection that enhances the classifier accuracy. A unique algorithm that we have designed to sync the various components and to configure different gestures as per user needs and also achieve high accuracy of 92% with SVM machine learning algorithm which is higher than existing gesture recognition technology. We aim at replacing the traditional switch with the proposed model device making it safer to use for multiple applications controlling a wheelchair, home automation, to control various systems in a vehicle making it safer for the driver to drive without taking eyes off the road, controlling quadcopter and many other such applications.

Keywords: IR sensor · Radar module (Yh-24g04) · Doppler shift · Real-Time Operating System (RTOS) · Gesture recognition · Machine learning classifiers

J. I.-Z. Chen et al. (Eds.): ICIPCN 2020, AISC 1200, pp. 422–434, 2021.
https://doi.org/10.1007/978-3-030-51859-2_39

1 Introduction

Gesture recognition technology is a result of one such human need to control various devices surrounding us with the help of simple gestures. This method is seen as a way for the computers to understand various human body movements interpret it and provide us a significant output. Unlike traditional devices like the keyboards or mouse [14] this method bridges the gap between the machine and humans. Gesture Control Technology is advancing at an impeccable speed and at the same time highly affecting our daily lives. The development of this technology, however, started very small with a few gestures and ranges up to which it could be applicable.

The recent availability of commercial sensors uses the camera, data-gloves, radio-frequency identification (RFID), etc. [4–11] which would be logged on to the body making the installation and cost a cumbersome and expensive process to deploy it on large scale. The use of camera compromises the security [12] as it can be easily hacked into at the same time it is sensitive to the environmental conditions while people are unwilling to carry a sensor on their body. The data-glove-based methods use sensor devices for digitizing hand and finger motions into multi-parametric data [7]. This paper addresses a device that uses a wireless system to sense and recognize the gesture. There is very little work available on gesture recognition using radar and IR sensor as in [13] where they have presented an algorithm for big gesture recognition through IR-UWB radar, but the gestures detected in that work were simply based on the position difference of the hand and may not be useful in practical applications, all the other methods have used radar and IR sensor separately for gesture recognitions.

To overcome the above problems, radar-IR-based gesture sensing device GigaHertz is designed, novelty is the proposed Gesture Recognition system of Doppler shift using radar. Doppler shift occurs whenever the source moves relative to the observer, thus whenever a gesture is performed different patterns of Doppler shift are produced.

The Doppler shift produced are difficult to detect as the change in frequency are in order of hertz and that too between 10–100 Hz, as in Wi-Fi signal [1] where a 1 m/s gesture produces 33 Hz of Doppler shift, we aim at extracting this Doppler pattern for various gesture to create our unique database with our different features, as for directions are concerned the use of infrared sensor provides a suitable choice, an infrared sensor is capable of sensing heat being emitted by the object since our body emits heat infrared sensor can sense, using of multiple infrared in a different direction can enable to detect the direction of the gesture being made.

Our proposed system GigaHertz is the first of its kind that uses a combination of radar and infrared sensor to detect the gesture with high accuracy and the first of its kind that can be used for multiple applications, as Google Soli [2] can be used in mobile phones. Some of the applications can be stated as home automation, wheelchair, drones, intelligent car system [3], and many others.

2 Proposed System of Gesture Sensing Using Microwave Radar and IR Sensor

2.1 Design of GigaHertz

Microwave radar apprehends different gesture which results from Doppler shifts and directions from the IR sensor, these signals are pre-processed using signal processing algorithm from this processed signal different features are extracted. These features are used to create a training data set. These data sets are then fed to a classifier which classifies different gestures based on their features, once classified, these training data sets are used to train the machine using a machine learning algorithm. Once trained then test samples are given for recognizing these gestures. These recognized gestures control various appliances or devices (Fig. 1).

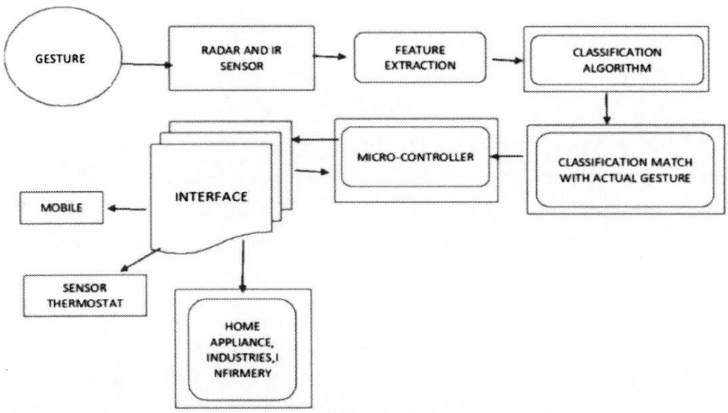

Fig. 1. Block diagram of hand gesture based automated services using radar.

Gesture. These are one out of a total of eight gestures being performed.

Radar and IR Sensor. These devices capture the Doppler shift and IR value resulting from the gesture being performed.

Feature Extraction. These are the algorithm that extracts five features from each gesture.

Classification Algorithm. These are classifiers that classify the gesture into different classes and form the database.

Classification Match with an Actual Gesture. These are some machine-learning algorithms to identify the gesture from the database.

Micro-controller. This is responsible for running the algorithm for feature extraction and maintenance of the stability of a radar module.

Interface. Various devices/appliances to be controlled come under interface.

Further explanation of the blocks is given further in detail.

Hardware Modules

Yh-24g04 Radar Module. The Yh-24g04 consist of an oscillator which generates 24 GHz signal which is emitted using transmitter antenna, the transmitted signal is reflected from the moving object in front of it, this reflected signal results in Doppler shifts which are in the range of 50–1000 Hz, this signal is then picked up by the receiving antenna which is connected to the multiplier in multiplier circuit one is the reflected signal and the other is the generated signal, one of the multipliers is multiplied with the reflected signal with a zero-degree phase shift to obtain an output IFi and the other multiplier circuit is multiplied with an input signal of 90-degree phase shift to obtain output IFq as shown in Fig. 2. The Yh-24g04 is powered by giving power supply of 5 V at Vcc and 0 V to 12 V at Vt which is responsible for generating FMCW (Frequency Multiplexed Continuous Wave), this FMCW is highly stable and immune to noise at the same time providing high resolution, the high frequency of 24 GHz is also responsible for providing a high resolution which can identify fine detained in the performed gesture and making the whole system more efficient. This forms one of the parts of our hardware.

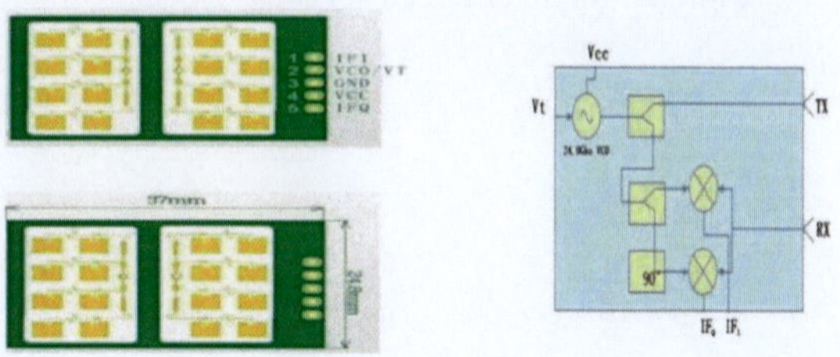

Fig. 2. Yh-24g04 Radar Module and Yh-24g04 block diagram.

Amplifier Circuit. The second most important part is the amplifier circuit (Fig. 3) which is responsible for amplifying the signal generated by the radar module in hundreds of millivolts to volt, the main function of the amplifier is to amplify the signal from the radar module by discarding various noises present in our environment, some of which are Gaussian noise, Rayleigh noise, and noises which are present due to the lead of the wire to achieve the latter problem we have designed an amplifier which can discard the noise at the same time amplify the signal, the gain of the amplifier circuit played a crucial role as the gain as to be limited so that the gestures like far and near can be detected efficiently, the gain of the amplifier is given by

$$\text{Gain} = 1 + \frac{R1}{R2} \tag{1}$$

which is approximately 10 in our case, the signal for one of the gestures is shown in Fig. 4 below as we can see that the output in absence of gesture is completely flat and we have almost removed the entire noise form the signal making it efficient and easier to extract features from the gesture.

APDS 9960. The only function of this IR sensor is to sense the directions (Fig. 5); this in collaboration with the radar system enables us to produce eight different gestures with a high level of accuracy.

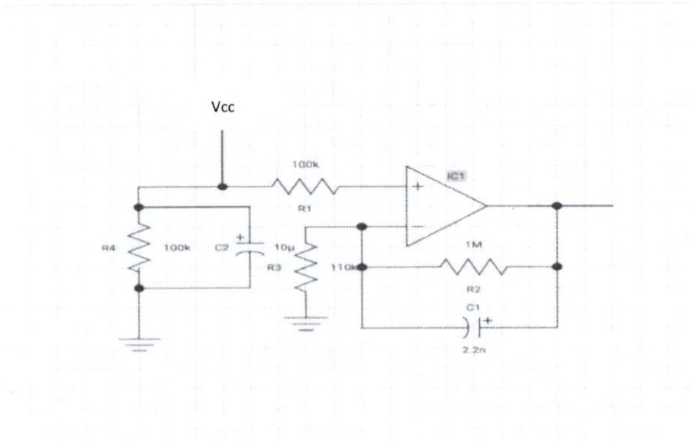

Fig. 3. Amplifier circuit for radar module.

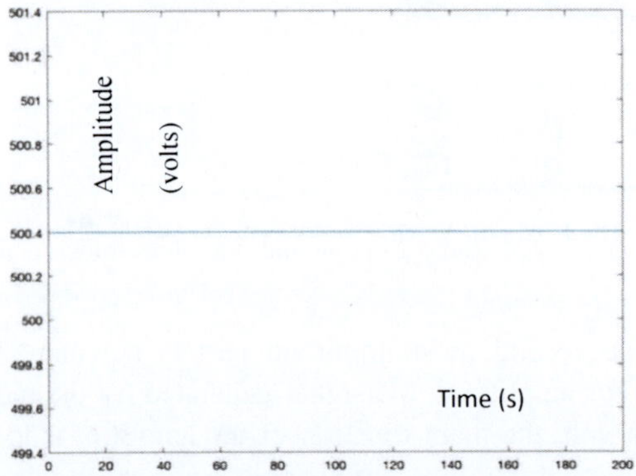

Fig. 4. Amplifier output in ideal state.

Fig. 5. APDS 9960 sensor.

2.2 Feature Extraction and Selection

The important parameters for recognizing the gesture is its feature which will differentiate it from the other gestures, initially there were total eight feature namely mean, maxima, minima, kurtosis, skewness, zero crossings, median frequency, and variance, with further analysis and training of data it became clear that the use of kurtosis and skewness were redundant as they were more suitable with the data related to a voice signal and median frequency was the factor of reducing the overall accuracy of the system, this was because when gestures were performed with different velocity then different gesture tend to have the same median frequency as Doppler shift (Fig. 6) is related to the frequency from the following formula

$$\text{Doppler Frequency} = \frac{2v}{\lambda} \tag{2}$$

Where, v = velocity
λ = wavelength of the radar

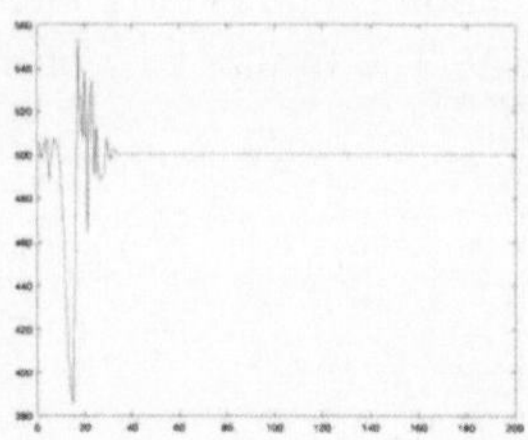

Fig. 6. Doppler shift of switch.

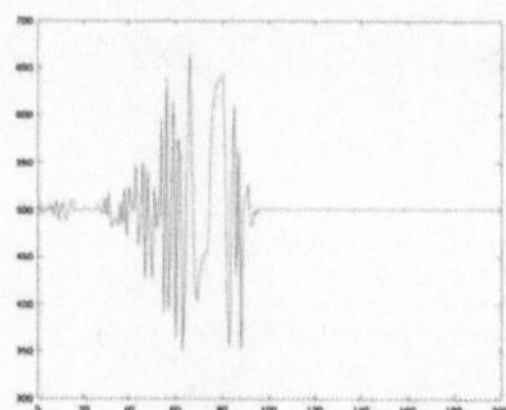

Fig. 7. Doppler shift of down.

Finally, five most significant features are selected from eight extracted features namely mean, maxima, minima, variance, zero-crossing each feature played an important role in differentiating the different gestures, as we can see from the graph, the graph of down (Fig. 7) gesture and up (Fig. 8) gesture are almost same but they are differentiated in terms of their maxima and minima value, similarly the zero crossing for on gesture is quite different from that of down and up gesture hence it is differentiated in terms of zero-crossing feature, in a very similar way these features help in differentiating the various gesture in its time domain.

(Note: zero-crossing here is given by zero-crossing – 200 where zero crossing reference is 500.4).

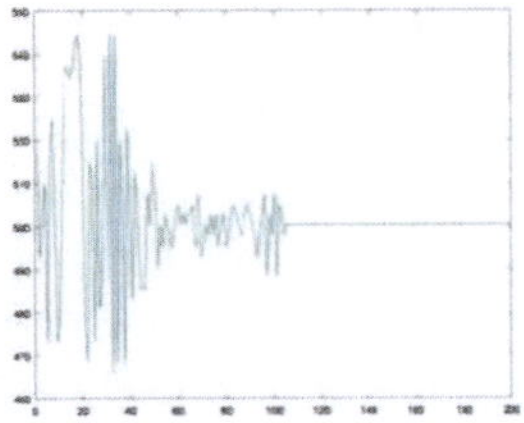

Fig. 8. Doppler shift of up.

2.3 Software

The experimental results are simulated on MATLAB 2017b and RTOS in Arduino with three stable functions (), capture (), and ML (). Due to presence of parasitic capacitance and inductance, there was a small fluctuation in the output, since some of the features like zero-crossing were dependent on the stability but these fluctuations resulted in giving the wrong value of zero-crossing periodically, hence stable () function was developed which keeps the fluctuation stable resulting in making the overall system stable it is also responsible for the invoking the function capture () to capture the data. The UML diagram for the system is shown in Fig. 9. The further gain of the signal is

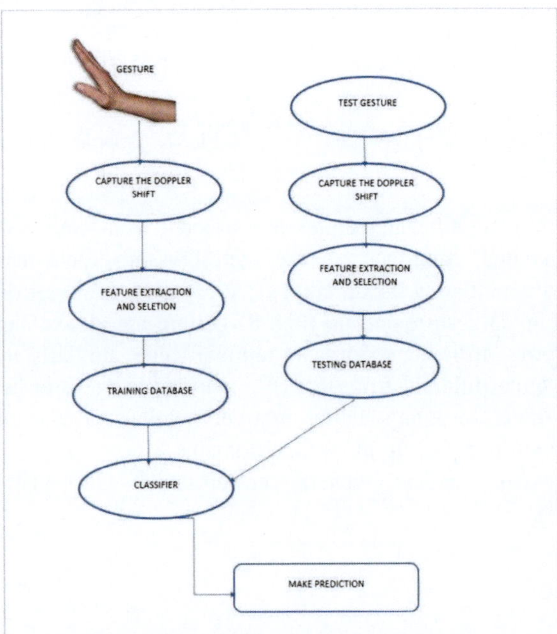

Fig. 9. Unified modelling language flow graph.

provided by the software by taking the advantage of the fact that the noise (spikes) in the system is null, so only the Doppler shifts are amplified, providing amplification in the software help to reduce the overall size of the system, the function stable() and capture() are running in RTOS in Arduino which makes the task switching fast and more reliable while the ML() function is running on MATLAB 2017b, the feature extracted from capture() function in Arduino is sent to MATLAB by serial communication which then invokes the ML() function.

2.4 Classification of Hand Gestures

As discussed in the previous section five most significant features are selected from eight extracted features namely mean, maxima, minima, variance, zero crossing. These features form the basis for classification of different gesture by using different machine learning algorithms, each feature plays an important role in the classification of each gesture, the maxima, and minima significant use is to differentiate up and down gesture, for gesture rotate and switch significant feature mean and zero-crossing were used to differentiate. IR sensor plays a key role in distributing eight gesture into two subcategories while performing gesture rotates, up, down and switch the IR sensor does not produce significant output, the changes are seen only in radar output whereas gesturing volume up, volume down, right and left result in a change in the output of IR sensor as well as the radar, hence if the five features of these four gesture are same as those of previous four gesture the IR sensor differentiates these gesture from one another hence opening the door for higher accuracy. We have used different machine learning algorithm with different parameters which are used to classify the gestures in the appropriate class. The training sets for each gesture were limited till 60+ per gesture but greater than 50 only the training data for gesture rotate was 73 hence narrowing it to the total number of training data to 640 (Table 1).

Table 1. Datasets.

Dataset	Training data	Testing data
Number of samples	640	128

The highest accuracy was achieved in Linear SVM with an accuracy of 92% (Table 2).

Table 2. Classifiers.

Classifiers	Parameters	Accuracy
Tree	i. Fine tree	89.0%
	ii. Medium tree	89.0%
	iii. Coarse tree	61.8%
SVM (kernel = 2.2)	i. Linear SVM	92.0%
	ii. Quadratic SVM	91.1%
	iii. Cubic SVM	91.0%
	iv. Fine Gaussian SVM	89.9%
KNN (k = 11)	i. Coarse KNN	55.7%
	ii. Cosine KNN	91.4%
	iii. Cubic KNN	91.2%
	iv. Weighted KNN	91.9%
Ensemble	i. Boosted tree	91.2%
	ii. Bagged tree	91.4%
	iii. Subspace discriminant	83.4%
	iv. Subspace KNN	90.1%
	v. RUSBoosted tree	91.2%

Experimental results show that SVM outperforms extensively well as compared to another machine learning classifier (Fig. 10) [15].

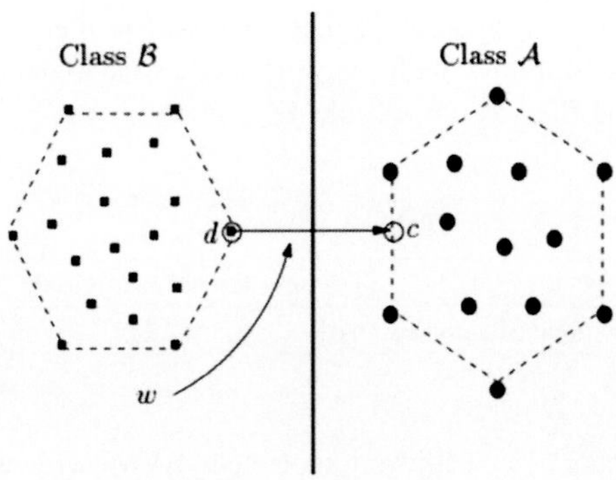

Fig. 10. SVM classification.

The gesture left, right, volume up, and volume down (Figs. 11 and 12) also produces significant changes in APDS 9960 module were according to the change in the intensity of IR these four gestures were differentiated from the other four gestures (Fig. 13).

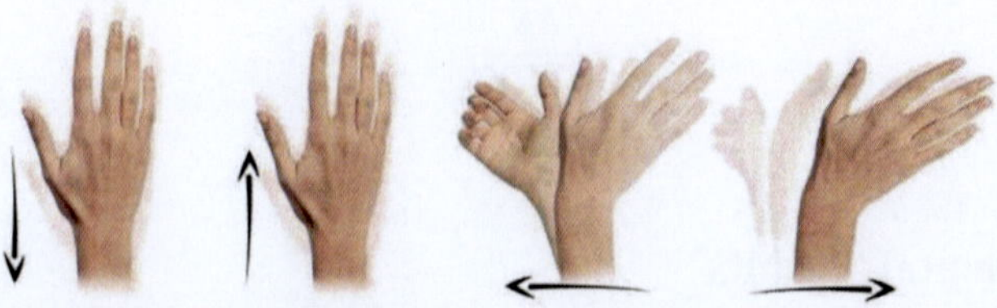

Fig. 11. Volume up & down, right & left swipe gesture.

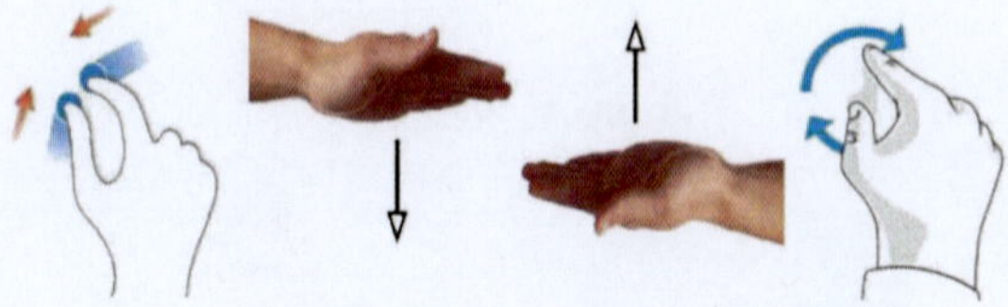

Fig. 12. Switch, down and up, rotate gesture.

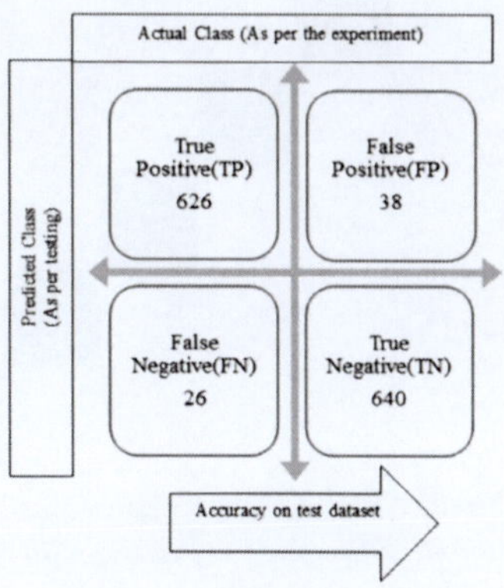

Fig. 13. Confusion matrix.

$$\text{Accuracy} = \frac{TP + TN}{\text{TP} + \text{TN} + \text{FP} + \text{FN}} = 0.9218 \tag{3}$$

$$\text{Specificity} = \frac{TN}{\text{TN} + \text{FP}} = 0.9139 \tag{4}$$

$$\text{Recall} = \frac{TP}{\text{TP} + \text{FN}} = 0.9301 \tag{5}$$

$$\text{Precision} = \frac{TP}{\text{TP} + \text{FP}} = 0.9227 \tag{6}$$

$$\mathrm{FPR} = \frac{FN}{\mathrm{TN} + \mathrm{FN}} = 0.0399 \quad (7)$$

3 Experimental Results

The following results are for the SVM machine Learning algorithm with cross-validation of 9-fold as the highest accuracy is achieved in it (Table 3).

Table 3. Results.

	Down	Rotate	Left	Button	Right	Up	Vol down	Vol up
Down	83%			5%		12%		
Rotate		95%		2%		2%		
Left			96%		2%		2%	
Button	5%	9%		79%		7%		
Right					98%			2%
Up	9%	2%		2%		88%		
Vol down					2%		98%	
Vol up					2%			98%

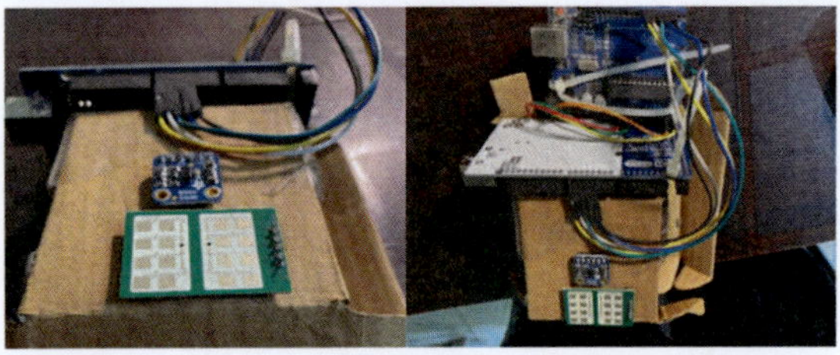

Fig. 14. Prototype of proposed model GigaHertz.

Fig. 15. Performing gesture switch on GigaHertz.

4 Conclusion

The proposed system contains microwave radar and IR sensor which transmit continuous millimeter range waveforms to capture the Doppler shifts from the performed gestures. The captured Doppler shifts amplified and processed through a designed amplifier where all the gestures produce a unique Doppler shift pattern. The microwave radar which is the core of the proposed system has many advantages over other traditional devices like camera and hand sensor also it has better power efficiency also in terms of data acquisition. The proposed system is a combination of signal processing and machine learning concept where we captured Doppler shift from the signal and extracted its various features like zero-crossing, mean, maxima, minima, variance, standard deviation, time, kurtosis out of which 5 features were selected for further classification and made the training database which includes around 2400 features of 8 gestures and made our database. We have classified all the gestures using the support vector machine model classifier and compared it with new test data of gesture. We got maximum accuracy using the SVM classifier up to 92%. SVM model gives better results and outperformed other classification algorithms and provides good recognition output. The proposed system is working on a real-time database (Figs. 14 and 15). The project is aimed to use gestures in any industry to control various application e.g. medical institution and business-level application.

References

1. Molchanov, P., Gupta, S., Kim, K., Pulli, K.: Short-range FMCW monopulse radar for hand-gesture sensing. NVIDIA Research, Santa Clara, California, USA, August 2014
2. Lien, J., Gillian, N., Karagozler, M.E., Amihood, P., Schwesig, C., Olson, E., Raja, H., Poupyrev, I.: Soli: ubiquitous gesture sensing with millimeter wave radar. ACM Trans. Graph. **35**(4), 1–19 (2016)
3. Garcia, J.M., Topa, A.L.: Gesture recognition by electromagnetic-wave reflection. IEEE, April2016
4. Zhou, R., Yuan, J., Zhang, Z.: Robust hand gesture recognition based on finger-earth mover's distance with a commodity depth camera. In: Proceedings of the 19th ACM International Conference on Multimedia, Scottsdale, AZ, USA, 28 November–1 December 2011 (2011)

5. Ying, W., Huang, T.S.: Vision-based gesture recognition: a review. In: International Gesture Workshop. Springer, Heidelberg (1999). https://doi.org/10.1007/3-540-46616-9_10
6. Chen, Q., Georganas, N.D., Petriu, E.M.: Real-time vision-based hand gesture recognition using haar-like features. In: Proceedings of the 2007 IEEE Instrumentation & Measurement Technology Conference, Warsaw, Poland, 1–3 May 2007, pp. 1–6 (2007)
7. Sturman, D.J., David, Z.: A survey of glove-based input. IEEE Comput. Graph. Appl. **14**, 30–39 (1994)
8. Pragati, G., Aggarwal, N., Sofat, S.: Vision based hand gesture recognition. World Acad. Sci. Eng. Technol. **49**, 972–977 (2009)
9. Sushmita, M., Acharya, T.: Gesture recognition: a survey. IEEE Trans. Syst. Man Cybern. Part C Appl. Rev. **37**, 311–324 (2007)
10. Parvin, A., Kulik, L., Tanin, E.: Gesture recognition using RFID technology. Pers. Ubiquit. Comput. **16**, 225–234 (2012)
11. Michael, B., Prasad, R., Philipose, M., Wetherall, D.: Recognizing daily activities with RFID-based sensors. In: Proceedings of the 11th International Conference on Ubiquitous Computing, Orlando, FL, USA, 30 September–3 October 2009 (2009)
12. Sprenger, M.E., Paul, J.G.: Radar-based gesture recognition. U.S. Patent no. 14/229,727, 1 October 2015
13. Nan, R., Quan, X., Cho, S.H.: Algorithm for gesture recognition using an IR-UWB radar sensor. J. Comput. Commun. **4**, 95–100 (2016)
14. Sultana, A., Rajapuspha, T.: A vision based hand gesture recognition. In: The Proceeding of IJCSET, vol. 3, no. 7 (2012)
15. Michahial, S.: Hand gesture recognition using support vector machine. Int. J. Eng. Sci. **4**, 42 (2015)

A Machine Learning Approach to Classify Dengue Outbreak in Tropical Conditions

G. S. S. Raj Kiran(✉), Palakodeti Rohit, Katapally Manognya, K. Likhith, N. V. Ganapathi Raju, and K. Prasanna Lakshmi

Department of Information Technology, GRIET, Hyderabad, Telangana, India
gssrajkiran@gmail.com, rohitpalakodeti@gmail.com

Abstract. A classification model was evolved to classify the severity of the dengue incidence by training on Kerala state's data. The model was developed by implementing supervised machine learning algorithms such as support vector classifier, logistic regression, and Naïve Bayes classifier to classify dengue cases. The model tries to classify a month into two class labels. The class labels are categorized based on the frequency of dengue cases recorded in that particular month. 0 ($n \leq 400$) when the number of cases recorded in that particular month is either equal or under 400, 1 ($n > 400$) when the number of cases recorded in that particular month is greater than 400 based on the meteorological data of the particular month. Among the classification algorithms, SVC (support vector classifier) with an accuracy of 92.85% and precision – 0.93, recall – 0.92, and f1 score – 0.93 was the most successful in classifying the severity of the situation.

Keywords: Dengue · Classification algorithms · Weather parameters

1 Introduction

Dengue has become a prominent viral disease in the last 2 decades universally [1]. Aedes aegypti and Aedes albopictus are two major contributing species of mosquitoes in spreading the viral fever [2]. The Dengue fever expanded into all continents and more severely near the Equatorial regions.

India has been a hotspot of Dengue fever by recording plenty of cases all over the country. Kerala state has consistently been registering very high Dengue fever cases for the past decade. The state has registered seen 1304, 4056, 7938, 2548 dengue cases in year 2011, 2012, 2013, 2014 and 4114, 7218, 21993 and 4083 in 2016, 2017, 2018 respectively.

Especially in the second half of the decade, it was the epicenter of dengue in India. Based on the observations there is a spike in cases being recorded post the monsoon season indicating the strong influence that meteorological parameters have on the dengue epidemic. Climatic factors are important determinants of dengue fever. Temperature, rainfall, and humidity are key factors for the survival, breeding, and life cycle of mosquitos, this will affect the presence and abundance of mosquitos [6]. The amount of rainfall largely affects the breeding sites for mosquitoes [9]. Various approaches were adopted to study the correlation between dengue cases and weather parameters. This plays a crucial role in classification as well as prediction.

J. I.-Z. Chen et al. (Eds.): ICIPCN 2020, AISC 1200, pp. 435–447, 2021.
https://doi.org/10.1007/978-3-030-51859-2_40

Image classification many alogorithms were utilized such as support vector, logistic regression and naïve bayes. Support vector classifier obtains 92% accuracy results, .93 precision, .921 recall and .930 f score in different situations.

2 Background Study

Nikita et al. [7] employed Naïve Bayes and Multi-regression to study and observe the relationship of meteorological parameters with the severity of different evolutionary algorithms was used to find regions with same type of spreading patterns.

Colon et al. [8] used a dataset containing dengue reports of 23 years with nine climatic regions that are used in the estimation of weather impact on dengue. Here, a Generalized Additive model was used to determine the remarkable effects of piped water and meteorological parameters on Dengue cases and discovered effects of climate were exceedingly nonlinear and variation in the climate may result in a rise in dengue incidences.

Wong Koon et al. [10] acquired the data consisting of mosquito larval recorded monthly of Thailand starting from January to December 2010. The observations pointed out that Aedes larvae were more predominantly found in rainy seasons when compared to other seasons.

Jonna et al. [11] performed geospatial analysis. The input data was Dengue cases registered from 2011 to 2014. The results pointed out that climatic conditions were influential in the dominance of the Dengue epidemic.

3 Methodology

This paper has gone through the following methodology (Fig. 1):

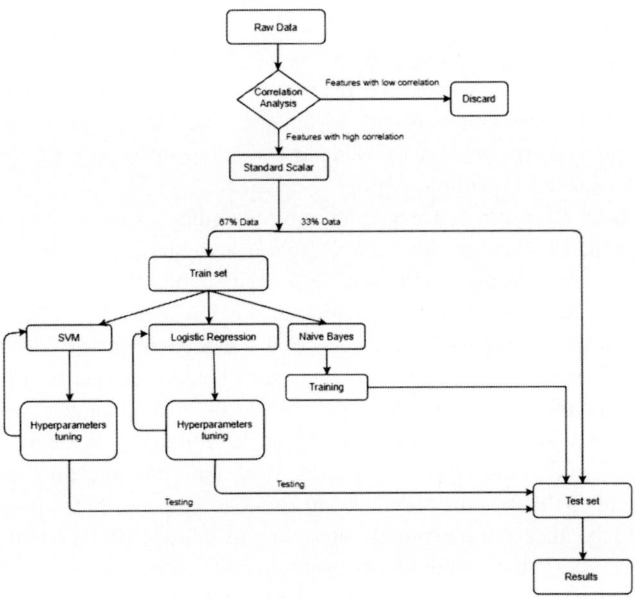

Fig. 1. Functional block diagram

3.1 Data Collection

3.1.1 Dengue Incidences Data

The Dengue cases data considered for this paper involves statistics from January 2011 to December 2018 (Excluding 2017). The Data was acquired from Health Services, Government of Kerala [7].

3.1.2 Metrological Data

The Meteorological data collected involves data belonging to the same time period of the Dengue incidences data. This data was acquired from worldweatheronline.com [12]. This Data comprises of monthly mean temperature, rainfall, and humidity.

3.1.3 Dataset Employed

The dataset consists of features such as month, year, mean temperature, mean humidity, total rainfall in that particular month, and total dengue cases (in that particular month of that year) across the state of Kerala.

3.2 Data Preprocessing

3.2.1 Data Scaling

StandardScaler belonging to the preprocessing package of sklearn was used to scale values of the training parameters of the dataset. The values were scaled in such a fashion that the mean and standard deviation of the distribution of each variable were −1 and 1.

3.2.2 Correlation Analysis

Correlation analysis reveals that there is a strong association between humidity and dengue cases [3]. The correlation between them was found to be 0.50. The dependence of dengue cases on the month was also significant as the point biserial coefficient between them is 0.32. The relationship between temperature and dengue cases was mildly inversely proportional as the correlation between them was −0.11 [7].

3.2.3 Feature Selection

Among all the meteorological variables considered Temperature was the one which varied the least periodically from one time of the year to the other throughout the State of Kerala [4]. For this reason, the temperature was excluded from the training set of considered for training the classification algorithms [5].

3.2.4 Encoding Dengue Class Labels

Since the objective of the paper is to classify the dengue severity into two different scenarios, the dengue incidents feature data were converted to categorical data by encoding each scenario into two class labels (0 or 1). 0 indicates the cases registered in that particular month across Kerala are less than or equal to 400 [$n \leq 400$], 1 indicates the cases registered in that particular month across Kerala are higher than 400 [$n > 400$]. This categorical data is considered as the target feature.

3.2.5 Cross-Validation

To validate and compute the performance index of the model the entire dataset was resampled into two parts with the aid of the function train_test_split function present in the sklearn.cross_validation package. The dataset was divided into the ratio of 67% and 33% respectively. The sample consisting of 67% data was used to train the algorithms and the other sample consisting of 33% of data was used for testing.

3.2.6 Training

In the training phase, the algorithms try to detect patterns and interesting measures in the provided training data and learn the weights of their respective activation functions. They will learn the weights based on the values of their training feature data and target column values. After learning the weights it will classify the scenario based on the values of feature parameters, in this case, the values of mean humidity, total rainfall, and month across Kerala and classify the intensity of the outbreak.

3.2.7 Model Generation for Classification

The objective of this paper is to classify the target feature which is a feature that will indicate the severity of the dengue outbreak in the state of Kerala.

- Support vector classifier, logistic regression, and naïve bayes were implemented for classifying the dengue severity.
  - The algorithms support vector classifier, logistic regression, and naïve bayes were initially trained on the training data with the values of their hyperparameters set to their default values. The results were observed. The algorithms need not necessarily perform their best when the values of the hyperparameters are set to their default values. Due to this reason, hyperparameter tuning for algorithms Support Vector Classifier and Logistic Regression was implemented.
- The hyperparameters of the algorithms fine-tuned to extract optimal performance.
  - The algorithms were retrained post extracting the ideal values of the hyperparameters of the respective algorithms. The results were observed.

3.2.8 Mechanism for Ideal Parameter Extraction

Algorithm hyperparameter tuning()

Step 1: Initialize all the hyperparameter with the list of values
Step 2: Initialize highest f1 – score with zero and ideal parameters list to an empty set
Step 3: Take a set consisting of values of all parameters
Step 4: Initialize the machine learning algorithm's hyperparameter with a set of values selected
Step 5: Train the algorithm on those values
Step 6: Evaluate the algorithm on the test set
Step 7: compute the f1 – score if is higher than the existing highest f1 score go to step 7 else go to step 8

Step 8: update the highest score and ideal set of parameters list with the current set
Step 9: repeat from step 3 to step 8 until all possible combinations of hyperparameters are implemented for training
Step 10: return the ideal set of hyperparameters for the algorithm

For extracting the best performance from any machine learning algorithm it's hyper-parameters need to be tuned. The above mechanism was implemented for classification algorithms (Excluding GaussianNB) to tune the algorithms to their respective hyper-parameters. Param1_combinations, Param2_combinations are a combination of values for a particular hyperparameter of an algorithm. The algorithm is trained on all possible sets of combination and the combination with the highest f1-score is used for training the model.

4 Results and Discussion

4.1 Support Vector Classifier (SVC)

For the generation of SVC classifiers, the Sklearn module of anaconda python-based machine learning library is used [15]. The syntax for SVC classifier is [16]

SVC(C = 1.0, decision_function_shape = 'ovr', degree = 3, gamma = 'auto', kernel = 'rbf', tol = 0.001)

On applying SVC for the non-preprocessed data, the accuracy score is 78.57%. The algorithm was re-applied post data preprocessing. Even this scenario estimator's parameters were left untouched and are set to their default values (Table 1).

Table 1. Estimator parameters of SVC were set to their default values

	Precision	Recall	F1-score	Support
0	0.79	1.00	0.88	22
1	0.00	0.00	0.00	6
Avg/total	0.62	0.79	0.69	28

As a result, we got an accuracy score of 85.71%.

SVC(C = 1.0, decision_function_shape = 'ovr', degree = 3, kernel = 'rbf', tol = 0.001, verbose = False)

In Fig. 2 Blue squares represent the scenarios where the registered dengue cases are lesser than or equal to 400 and orange triangles represent scenarios where registered dengue cases are greater than 400.

Figure 2 depicts decision curve which was fit on the default values with 'RBF' Kernel and the degree, tolerance, gamma, and the regularization parameter C were 3, 0.001,'AUTO', 1.0 respectively. Here in this case, the value of C proved to be high which resulted in slight overfitting and also the kernel 'RBF' was not suiting for the shape of the data in space.

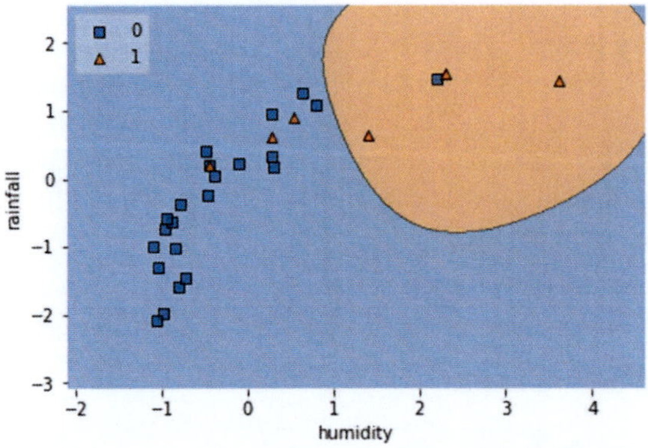

Fig. 2. Decision curve of SVC (RBF kernal) with respect to rainfall and humidity.

After that the parameters of the estimator were tuned to the values which best set to the nature and shape of data in the space (Table 2). Various combinations with parameters like C, gamma, tol, degree, and kernel.

Table 2. The parameters of SVC were set to the above values

	Precision	Recall	F1-score	Support
0	0.88	0.95	0.91	22
1	0.75	0.50	0.60	6
Avg/total	0.85	0.86	0.85	28

c = np.array([0.0001, 0.0005, 0.001, 0.005, 0.1, 0.5, 1.0, 5.0, 10.0, 50.0], dtype = 'float32') degree = np.array([1, 2, 3, 4, 5], dtype = 'float32') gamma = np.array ([1.0, 1.5, 2.0, 2.5, 3.0], dtype = 'float32') kernel = ['rbf', 'linear', 'poly'] tol = [0.0001, 0.001, 0.1, 1.0]

Different combinations of these parameters were looped through and trained the model. The best combination of tuning parameters and attained an accuracy score of 92.85% (Table 3).

Table 3. The parameters of SVC were set to the above values

	Precision	Recall	F1-score	Support
0	0.92	1.00	0.96	22
1	1.00	0.670	0.80	6
Avg/total	0.93	0.93	0.92	28

SVC(C = 0.1, decision_function_shape = 'ovr', degree = 2.0, kernel = 'poly', ol = 1.0)

In Fig. 3 Blue squares represent the scenarios where the registered dengue cases are lesser than or equal to 400 and orange triangles represent scenarios where registered dengue cases are greater than 400.

Figure 3 depicts the decision curve after adjusting the parameters such as C, gamma, degree, kernel, tolerance to values 0.1, 3.0, 2.0, 'POLY', 1.0 respectively. The transformation which took place in the decision is evident. As there is a reduction in the value of C from 1.0 to 0.1 the curve is more generic and didn't overfit this time.

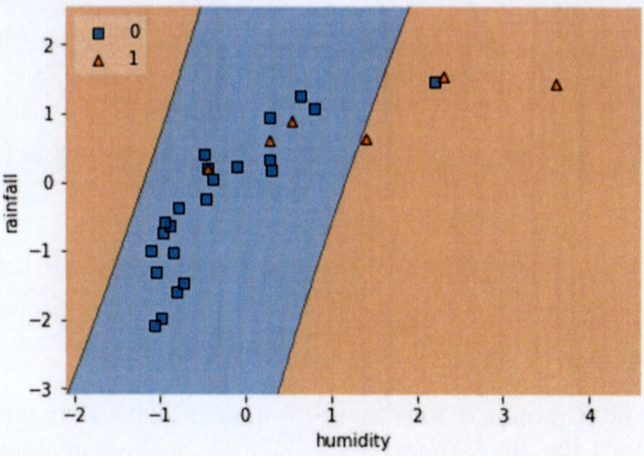

Fig. 3. Decision curve of SVC (POLY Kernal) concerning rainfall and humidity

Figure 4 plot depicts the relationship between the C (regularization parameter) and the accuracy of the SVC model. The peak accuracy is attained at 0.1 value of the C parameter.

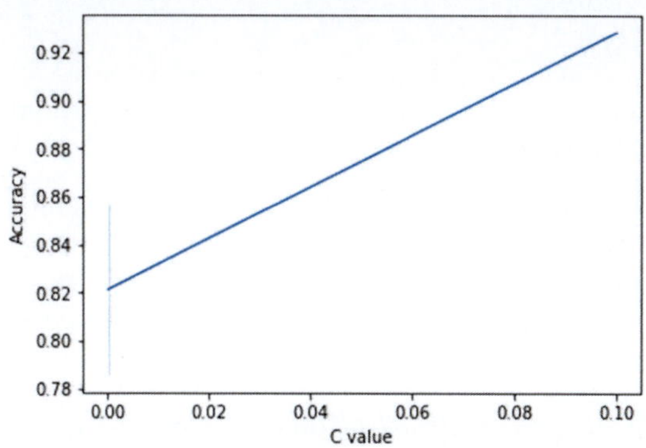

Fig. 4. Relationship between accuracy and C hyperparameter of SVC

Figure 5 plot depicts the relationship between the degree, gamma, kernel parameters, and how they affected the accuracy of the model. The blue line represents a degree, the red line represents the gamma value and the purple line represents kernel type.

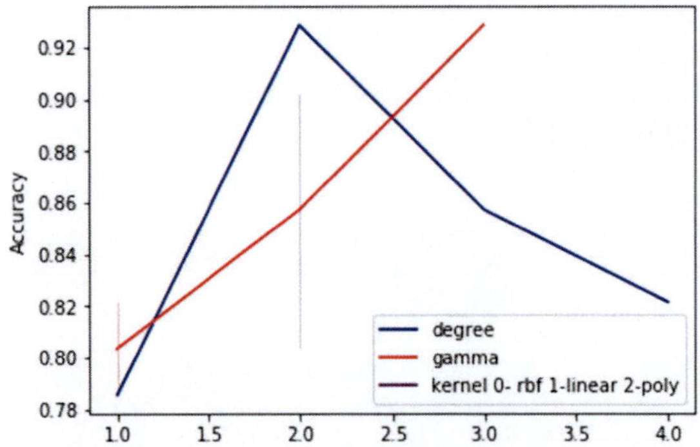

Fig. 5. Relationship between accuracy and hyperparameters degree, kernel, and gamma of SVC

4.2 Logistic Regression

On applying Logistic Regression for the un-normalized data, the algorithms scored an accuracy score of 82.14%. In the above case the estimator parameters were set to their default values which are (Table 4).

Table 4. The parameters of Logistic Regression were set to default values

	Precision	Recall	F1-score	Support
0	0.90	0.86	0.88	22
1	0.57	0.67	0.62	6
Avg/total	0.83	0.82	0.83	28

LogisticRegression(C = 1.0, fit_intercept = True, intercept_scaling = 1multi_class = 'ovr', n_jobs = 1, penalty = 'l2', random_state = None, , tol = 0.0001)

After data preprocessing, the accuracy didn't take a jump and remained at 82.14%. The coefficients of the features in the decision function were 0.02599962, 0.36452911, and 0.41128853 for month, humidity, and rainfall respectively. This shows that there is no strong relation between dengue cases and the month of the year as it was a lower-valued coefficient.

$$0.02599962(month) + 0.36452911(rainfall) + 0.41128853(humidity) - 0.75756276 < 0$$

It belongs to class label 0 i.e., to a scenario where dengue cases in a particular month are less than 400.

0.02599962(month) + 0.36452911(rainfall) + 0.41128853(humidity) − 0.75756276 > 0

It belongs to class label 0 i.e., to a scenario where dengue incidents registered in a particular month are greater than 400.

When trained with logistic regression with the following parameter values (Table 5).

Table 5. The parameters of Logistic Regression were set to the above values

	Precision	Recall	F1-score	Support
0	0.90	0.86	0.88	22
1	0.57	0.67	0.62	6
Avg/total	0.83	0.82	0.83	28

LogisticRegression(C = 1.0, fit_intercept = True, intercept_scaling = 1, multi_class = 'ovr', n_jobs = 1, penalty = 'l2', tol = .0001)

In Fig. 6 Blue squares represent the scenarios where the registered dengue cases are lesser than or equal to 400 and orange triangles represent scenarios where registered dengue cases are greater than 400.

Figure 6 depicts the curve was fit on the default values of the parameters of the LogisticRegression. The regularization parameter C needed a slight reduction, Also the parameter tolerance needed reduction.

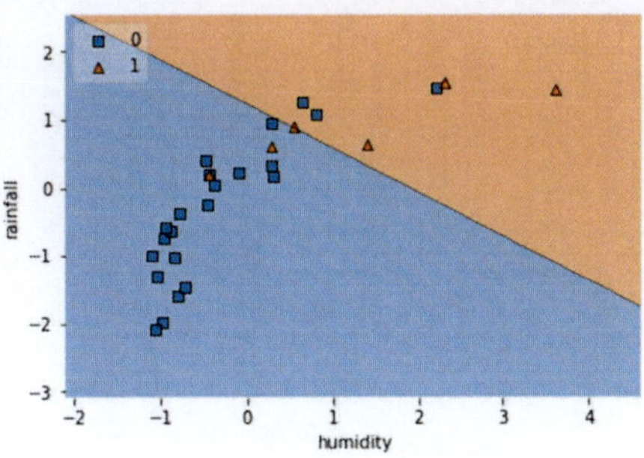

Fig. 6. Decision curve of Logistic Regression (C = 1.0)

In cases of logistic regression since the model didn't give a higher coefficient to the month on training, The model was re-trained without including the monthly data in the training data set. On testing, the model scored an accuracy of 85.71%. The coefficients of features in the decision function were 0.36389485, 0.41748337 for rainfall, and humidity respectively [13]. The reason for the slight increase in the accuracy of the model can be attributed to the fact that regression models perform better when the variables included in the training are less correlated among themselves [13]. The correlation between the humidity and month was 0.54 which is decently high.

$$0.36389485(rainfall) + 0.41748337(humidity) - 0.7558625 < 0$$

It belongs to class label 0 i.e., to a scenario where dengue incidents registered in a particular month are less than 400.

$$0.36389485(rainfall) + 0.41748337(humidity) - 0.7558625 > 0$$

It belongs to class label 0 i.e., to a scenario where dengue incidents registered in a particular month are greater than 400.

The parameters were

LogisticRegression(C = 0.10000000149011612, fit_intercept = True, intercept_scaling = 1, multi_class = 'ovr', tol = 1e− 10, verbose = 0, warm_start = False).

In Fig. 7 Blue squares represent the scenarios where the registered dengue cases are lesser than or equal to 400 and orange triangles represent scenarios where registered dengue cases are greater than 400.

Figure 7 depicts the decision curve after making changes in the parameters C and tool from 1.0 and 0.1 to 0.001 and 0.0000000001. The decision curve changed to the above form which was better than the previous one. This reduced the area of orange-coloured class as there is less number of orange labels.

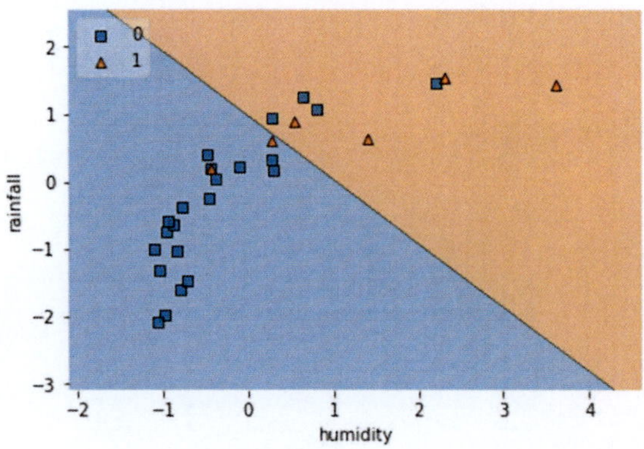

Fig. 7. Decision curve of Logistic Regression (C = 0.1, removing month feature from the training set)

4.3 Naïve Bayes (Gaussian NB)

On applying GaussianNB for the un-normalized data we got an accuracy score of 82.14%. In the above case, the estimator parameters were set to their default values which are GaussianNB (priors = None) (Table 6).

Table 6. The training data set includes temperature as a feature variable

	Precision	Recall	F1-score	Support
0	0.95	0.82	0.88	22
1	0.56	0.83	0.67	6
Avg/total	0.86	0.82	0.83	28

After data preprocessing, the accuracy score of 78.57% was obtained. This might be attributed to the fact that temperature data was removed from the training dataset as both the other models didn't benefit from the inclusion of temperature data [17]. But in the scenario of GaussianNB, it's performance took a hit because [14] GaussainNB which is also called Naïve Bayes in which Naïve means considering every feature as an independent. So, it computes prior probabilities of all the existing features concerning the class labels. So as the temperature was removed from the training data the model had one less prior probability to compute (Table 7) [14].

Table 7. The data set excludes temperature from the feature variables

	Precision	Recall	F1-score	Support
0	0.90	0.82	0.86	22
1	0.50	0.67	0.57	6
Avg/total	0.81	0.79	0.80	28

In Fig. 8 Blue squares represent the scenarios where the registered dengue cases are lesser than or equal to 400 and orange triangles represent scenarios where registered dengue cases are greater than 400.

Figure 8 depicts the curve which is the representation of the decision curve fit after the GaussianNB (Naïve Bayes) was applied to the given data.

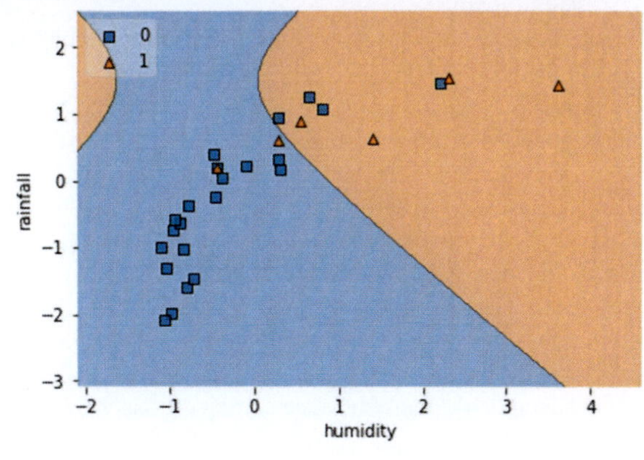

Fig. 8. Decision curve of GaussianNB

5 Conclusion

The present paper creates a predictive model for dengue incidence and its prevalence as a case study of the Kerala state data set. This research work implemented using Python programming language using Anaconda distribution on the Dengue data set collected on Kerala state of South India. In the classification phase, the model tries to classify a month into either of the two-class labels. There are two class labels represent the range in which the number of dengue cases that might occur in that particular month fall's under. The two-class labels are 0($n \leq 400$) when the number of cases occurring that particular month is either equal or under 400, 1($n > 400$) when the number of cases occurring in that particular month is greater than 400. Among the classification algorithms SVC (support vector classifier) with poly kernel performed the best by obtained accuracy of 92.850% with f− 1 measure is 0.93, precision is 0.93, recall being 0.92.

References

1. Das, S., Sarfraz, A., Jaiswal, N., Das, P.: Impediments of reporting dengue cases in India. J. Infect. Public Health **10**(5), 494–498 (2017)
2. Dengue control, WHO. https://www.who.int/denguecontrol/disease/en/. Accessed 20 Jan 2019
3. Agarwal, N., Koti, S.R., Saran, S., Senthil Kumar, A.: Data mining techniques for predicting dengue outbreak in geospatial domain using weather parameters for New Delhi, India. Curr. Sci. **114**(11), 2281–2291 (2018)
4. Geography of Kerala. https://en.wikipedia.org/wiki/Geography_of_Kerala
5. Thiruvananthapuram. https://en.wikipedia.org/wiki/Kerala
6. Dengue and Climate, Centers for Disease control and Prevention. https://www.cdc.gov/dengue/entomologyecology/climate.html
7. Ganapathi Raju, N.V., Krishna, P.G., Manognya, K., Raj Kiran, G.S.S., Rohit, P., Likhith, K.: Evolution of predictive model for dengue incidence by using machine learning algorithms. In: 2019 International Conference on Communication and Electronics Systems (ICCES), pp. 51–59. IEEE (2019)
8. Vijayakumar, T.: Neural network analysis for tumor investigation and cancer prediction. J. Electron. **1**(02), 89–98 (2019)
9. Colón-González, F.J., Fezzi, C., Lake, I.R., Hunter, P.R.: The effects of weather and climate change on dengue. PLoS Negl. Trop. Dis. **7**(11), e2503 (2013)
10. Mutheneni, S.R., Morse, A.P., Caminade, C., Upadhyayula, S.M.: Dengue burden in India: recent trends and importance of climatic parameters. Emerg. Microbes Infect. **6**(1), 1–10 (2017)
11. Wongkoon, S., Jaroensutasinee, M., Jaroensutasinee, K.: Distribution, seasonal variation & dengue transmission prediction in Sisaket, Thailand. Indian J. Med. Res. **138**(3), 347 (2013)
12. WorldWeather. https://www.worldweatheronline.com/
13. Analyticbridge, A Data Science Academy. www.analyticbridge.datasciencecentral.com/group/analyticaltechniques/forum/topics/excluding-variables-from-a-logistic-regression-model-based-on
14. Data Science. https://datascience.stackexchange.com/questions/9087/correlation-and-naive-bayes

15. Easy steps to learn Naïve Bayes Algorithm. https://www.analyticsvidhya.com/blog/2017/09/naive-bayes-explained/
16. Support Vector Machines. https://www.svm-tutorial.com/2015/06/svm-understanding-math-part-3/
17. Support Vector Machines Margin Image. https://en.wikipedia.org/wiki/Support-vector_machine

A Blockchain-Based Decentralized Framework for Crowdsourcing

Neha More and Dilip Motwani(✉)

Computer Engineering Department, Vidyalankar Institute of Technology, Mumbai, India
nehamore24@yahoo.com, dilip.motwani@vit.edu.in

Abstract. Crowdsourcing frameworks have accomplished significant concern and reception as of late. It coordinates the human insight of specific and organizations together from everywhere throughout the world to comprehend convoluted assignments. Despite, these focal frameworks subject to the shortcomings of the trust-based standard like customary monetary establishments, for example, single purpose of disappointment, high administration expense, and security revelation. Here, the concept is a blockchain-based distributed structure for crowdsourcing, in which a requester's work can be dictated by a horde of laborers with no depend on upon focal publicly supporting frameworks or expected clients to get to administrations with enlisting genuine characters. In consideration, the engineering of the proposed system where Users can enlist, post, or get an undertaking safely under this structure with high user's privacy and low administrative expenses. By expanding the versatility of crowdsourcing the purpose is illustrating convoluted crowdsourcing logic with the intelligent contract. While posting a task, the requester has to deposit the project amount which is escrowed with the system and the worker will get the project amount on the agreed task completion from the escrow. Due to this task, the requester will not pay more than what task deserves, according to a guideline disclosed when the task is posted; and each worker will surely receive a payment based on the guidelines.

1 Introduction

Crowdsourcing is becoming big waves in many industries across the globe. It tends to be profoundly incredible when utilized in the perfect spot at the opportune time. A hybrid of "crowd" and "outsourcing", Merriam-Webster defines crowdsourcing as "the practice of procuring desired services, ideas, or content by soliciting contributions from a large group of people and especially from the online community rather than from traditional workers or suppliers" [5]. People engaged in crowdsourcing sometimes work as paid freelancers, while others perform limited tasks on a charitable basis. Crowdsourcing permits organizations to subcontract work to individuals/open anyplace in the whole nation or around the globe, which lets organizations tap into an immense range of aptitudes and skills without acquiring the typical overhead expenses of inner representatives/laborers. The human-centered crowdsourcing involves three parts of roles: requesters, laborers, and a brought together crowdsourcing association.

J. I.-Z. Chen et al. (Eds.): ICIPCN 2020, AISC 1200, pp. 448–460, 2021.
https://doi.org/10.1007/978-3-030-51859-2_41

Requesters offer undertakings that are trying for PCs yet simple for a human to achieve via the crowdsourcing [8].

Crowdsourcing system as a way for all kinds of businesses to gather new ideas. The system is having some weaknesses which give challenges. First, most crowdsourcing organizations run a business on an integrated server, which endures from a single point of breakdown naturally [5]. Second crowdsourcing systems are exposed to Distributed Denial of Service attacks which makes the services unavailable [1, 8]. And third is confidentiality- when approaching individuals for their answers, an organization may need to share touchy data to get an answer that will work [5]. For Example, if an organization is seeking answers for any item bundling issue, this can be a sign that there is some insufficiency in the organization's present bundling frameworks [12]. A project may not be appropriate for crowdsourcing if it relies on commercially sensitive information. We suggest a blockchain-based distributed structure for crowdsourcing. This has benefits such as improving user security and service ease of use, improving the flexibility of crowdsourcing with low cost. Therefore, our model can interrupt the traditional model of crowdsourcing [3, 11]. Conventional crowdsourcing depends on a brought together framework to post and get errands, 3-party framework model. Astoundingly, to get agreeable outcomes, the requester just permits qualified laborers who arrive at the very least notoriety incentive to get the undertaking.

2 Literature Review

[1] **"Toward a BlockchainEnabled Crowdsourcing Platform"- Dimitrios G. Kogias and Helen C. Leligou Published by the IEEE Computer Society 11 September 2019, IEEE)**
This work studies Crowdsourcing has been sought after as how to use the office of the group for a few unique purposes in differing areas from gathering data, accumulating assets, and social event workers to execute assignments of various sizes among different targets.

[2] **"On Enabling Machine Learning Tasks atop Public Blockchains: a Crowdsourcing Approach"**
This work presents as of late developed blockchain (specifically brilliant agreement) innovation offers an appealing open door for the decentralized sharing economy. AI is regularly one significant subroutine in such a decentralized biological system. Lamentably, AI programs are typically computationally serious likewise as randomized, which fall under the natural impediments of open blockchain were mind-boggling and randomized projects can't be executed by the hidden hubs on the whole. Given likewise the limitations of existing verifiable figuring systems, we propose crowdsourcing thought from the game-theoretic viewpoint to determine the strain. We plan a simple motivation instrument all together that the execution of a decent scope of complex projects are regularly publicly supported through the blockchain, and any bogus figuring result may be deflected. particularly, our convention works inside the situations that there's no trusted third party included; Moreover, our convention not just works inside the

traditional model of non-conspiring specialist organizations, yet can likewise endure any potential alliance up to n − 1, where n is that the complete number of specialist co-ops. We additionally exhibit the best approach to utilize our convention to crowdsourcing two runs of the mill sorts of AI assignments through open blockchain. We imagine that our answer promising to dispatch decentralized applications including a decent scope of AI programs.

[3] **"Vizsafe: The Decentralized Crowdsourcing Safety Network"- Peter A. Mottur 2018 IEEE**
A smart City uses innovations to advance the conveyance of administrations to residents and increment occupants' satisfaction. While an immense exhibit of sensors and IoT gadgets will be sent to fulfill the data necessities of a reasonable way to deal with open help conveyance, any city's biggest and most complex sensor organize takes the state of its residents. since the ones who utilize the city's framework and administrations day in and day out, they're interestingly put to report any issues. What prevents a large portion of the individuals from doing so is right off the bat the lack of a stage with the satisfactory channels and the deficiency of prompt motivations—more often than not, an issue will be viewed as another person's business to manage. To tackle these problems, the Vizsafe stage has been intended to prepare the pack and gives them the motivation to present data that is to everybody's greatest advantage. Through the office of blockchain, these reports are put away on a decentralized record, and, using brilliant agreements, SPOT tokens are utilized as motivating force to people that transfer occurrence reports, regardless of whether that is a potential security danger or defective foundation. It is the slight push that makes 'another person's concern' into 'my chance'. When urban areas begin remunerating support genuinely, individuals stop to be shoppers and fix clients yet rather contribute data for their aggregate advantage, while diminishing expenses for organizations and fix suppliers. They become dynamic accomplices in keeping up the networks during which they live and work.

[4] **"ZebraLancer: Private and Anonymous Crowdsourcing System atop Open Blockchain"-Yuan Lu, Qiang Tang, Guiling Wang. 2018 IEEE 38th International Conference on Distributed Computing Systems"**
This work design two essential difficulties of decentralizing crowdsourcing, for example, information spillage and character break. To start with, our re-appropriate then-demonstrate system settles the strain between blockchain straightforwardness and information confidentiality, which is basic in crowdsourcing use-case. ZebraLancerensures: (i) a requester won't pay very what information merit, steady with an arrangement reported when her assignment is distributed through the blockchain; (ii) every specialist, in reality, gets an installment upheld the strategy, on the off chance that he submits information to the blockchain; (iii) the above properties are acknowledged without a focal referee, yet also without releasing the data to the open blockchain they execute our convention for a standard picture comment task and convey it during a test net of Ethereum. The analysis results show the appropriateness of our convention on the common genuine world blockchain.

[5] **"zkCrowd: A Hybrid Blockchain-based Crowdsourcing Platform"- S. Zhu, Z. Cai, H. Hu, Y. Li and W. Li, IEEE Transactions on Industrial Informatics. doi:10.1109**
Blockchain, a promising decentralized paradigm, are often misused not exclusively to beat the deficiencies of the ordinary crowdsourcing frameworks, yet additionally to bring specialized developments, similar to decentralization and responsibility. By and by, some basic inborn restrictions of blockchain are once in a while tended to in writing when it's consolidated into crowdsourcing, which can yield the presentation bottleneck inside the crowdsourcing frameworks. A further preferred position of the predominance of blending blockchain and crowdsourcing, during this paper, we propose an imaginative half and half blockchain crowdsourcing stage, named zkCrowd. It mixes with the upsides of mixture blockchain structure, brilliant agreement, double records, and double accord conventions to make sure about correspondences, check exchanges, and safeguard security. Both the hypothetical examination and investigations are performed to assess the advantages of zkCrowd over the state-of-the-art.

3 Problem Statement

The conventional crowdsourcing prototype supports the integrated structure, with "centralized" generally alluding to the undertaking taking care of procedure which is a basic piece of the general crowdsourcing framework [1]. The brought together crowdsourcing models acquire the benefit of straightforwardness in both buildings up an application and dealing with the framework [8]. In any case, any brought together arrangement accompanies the single purpose of disappointment helplessness, with the lone point being the focal position or administrator of the stage in the crowdsourcing worldview, as assaulting them creates the administration gets inaccessible.

Figure 1 illustrates the traditional system of crowdsourcing. In which Requesters submit errands that are trying for PCs however simple for a human to finish across the crowdsourcing framework [8]. A lot of laborers who are keen on this assignment contend and submit answers for the crowdsourcing framework, while requesters desire to choose a legitimate arrangement (normally the best or first one that comprehends the errand) and award the relating laborers the prize. In any case, despite the flourishing of the crowdsourcing frameworks, they are dependent upon the shortcomings of the customary model, which realizes some inescapable difficulties, for example, frameworks are powerless against DDoS assaults, because of brought together server odds of single point disappointment and absence of security.

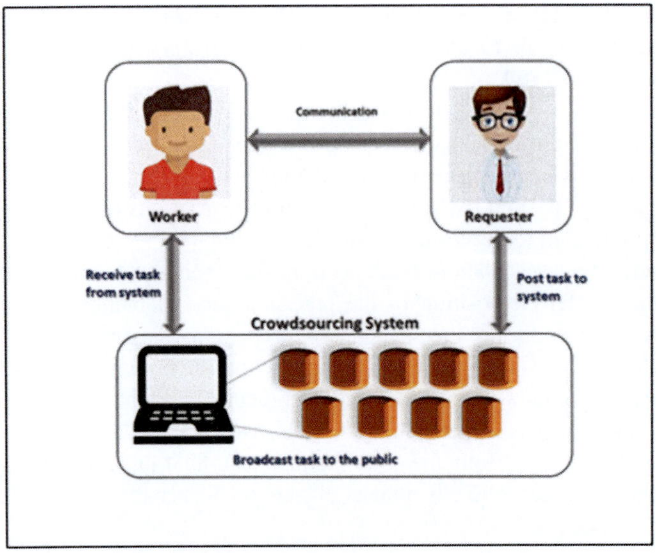

Fig. 1. Traditional crowdsourcing system on a centralized server

3.1 Objectives

The objective is to realize decentralized crowdsourcing. This will satisfy the:

a) To avert a single point of breakdown and blockage problem.
b) Reasonable trade between the information and the prizes for example on the off chance that a laborer presents the work, he will be paid with the right sum as per the predefined arrangement of assessing information.
c) Data Confidentiality i.e. the presented information will be secret to everyone except the requester means the requester is enforced to deposit the budget.
d) To maintain data security i.e. all transactions using virtual cryptocurrency are encrypted.
e) To preserve accountability.

4 Related Works

We carefully analyzed related works and shortly deliberated the insufficiencies of the new solutions [7, 14, 15]. The crowdsourcing duties, like posting a task from the requester, apply for a task selection of workers and payment structure, are obtained by the centralized crowdsourcing systems. Upwork [14] and Freelancer [7], as two well-known crowdsourcing stages, permit requesters to efficiently enlist laborers to get task arrangements, However, they required clients' definite data and put away client data, task information in the brought together stage, which may experience the ill effects of protection spillage, DDoS and the issue of the single purpose of disappointment. There are some researches on the distributed crowdsourcing system. S. Zhang, et al. [2]

presented a crowdsourcing system based on delay tolerant network in a distributed architecture and it provides less makespan. P. Yang, et al. [6] recommended a task distribution scheme by utilizing societal relationships in a crowdsourcing system While [2, 6] direct on task fulfillment in a disseminated way, they provide service through a unified system. Our design is to build a crowdsourcing system in a distributed architecture. Our new primordial can dissolve these insufficiencies by, using a blockchain-based decentralized framework with many extensive objectives which includes storing all encrypted transactions on blockchain as database and payment process is automated rite from receiving money into the escrow to reliving the payment from escrow after deducting the portal charges.

5 Proposed System

A blockchain [16] based protocol is proposed for overseeing different crowdsourcing [4, 8] advances give a promising bearing to deal with the notoriety framework. The proposed framework, a crowdsourcing stage where every single step of crowdsourcing procedure is taken care of as exchanges in Blockchain. This aides in building better confidence inside the stage clients and addresses kinds of assaults that are conceivable on a brought together crowdsourcing stage.

Figure 2. demonstrates the recommended structure of blockchain-based crowdsourcing [8]. In a given structure, operators are categorized into two types: requester and worker. There is no main server and blockchain-based crowdsourcing bolsters Turing-complete projects to delineate the procedure of crowdsourcing [8]. The requester and specialist must enlist to obtain their qualifications before getting administrations from blockchain-based crowdsourcing.

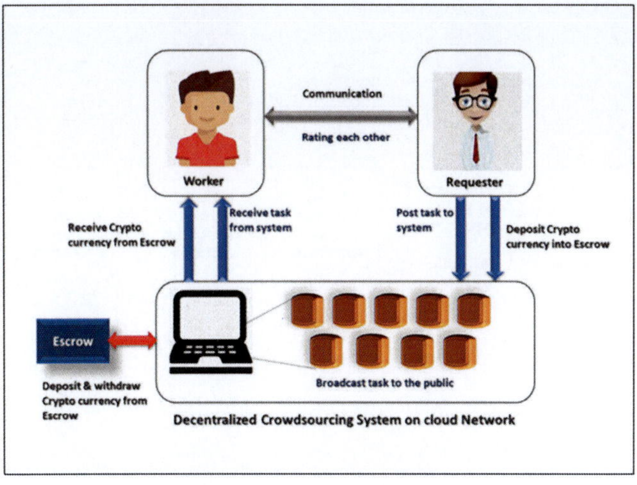

Fig. 2. Proposed system

5.1 Blockchain Technology

The blockchain technology that's seated at the top of the web is that the peer-peer network. This blockchain might be termed because of the foundational technology, resulting in the revolution of the business and therefore the government. It is a public electronic ledger built around a pear to pear system which will be openly shared among disparate users to make an unalterable record of transactions, so each time-linked to the previous one. whenever a new transaction is added, newly added data becomes another block within the chain [13]. The ledger doesn't keep track of balances, it only keeps track of each transaction. "Blocks" on the blockchain are made from digital pieces of data. Specifically, they need three parts (i) Blocks store information about transactions just like the date, time (ii) Blocks store information about who is participating in transactions (iii) Blocks store information that distinguishes them from other blocks.

Figure 3 illustrates the process of blockchain, [13] here once the exchange is mentioned, an information structure for keeping the arrangement of exchanges is conveyed to all or any hubs inside the system. all the hubs confirm the squares before adding anything to the blockchain. When the hubs do the square check, they get an award for the evidence of work. Along with these lines, each new hub in the appropriated arrangement of blockchain having a full duplicate of the blockchain. At the point when another square is shaped, it is sent to every hub inside the blockchain system. Around then, every hub affirms the square and checks whether the data communicated there is correct. If everything is fitting, the square is added to the neighborhood blockchain in each hub.

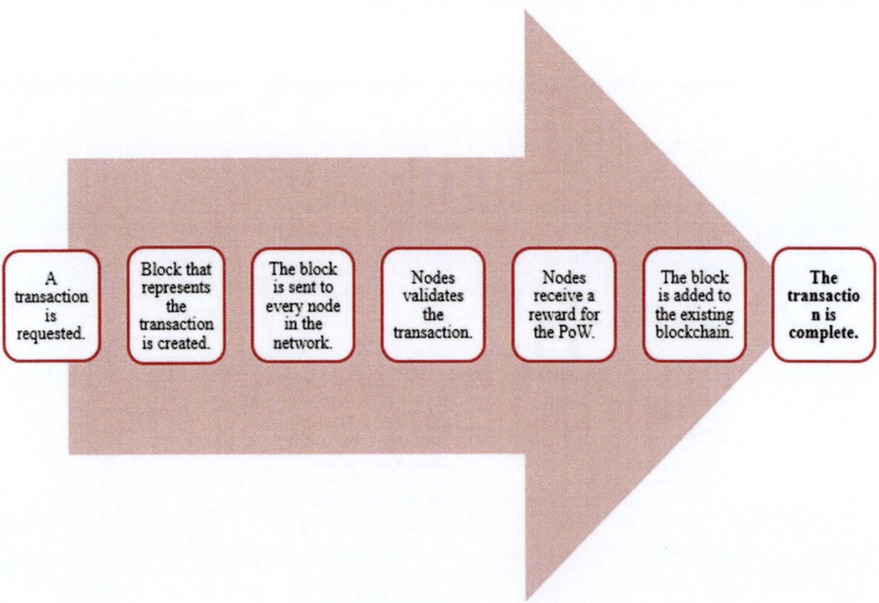

Fig. 3. Process of blockchain

5.2 SHA 256

The most adopted secure algorithms associated with the blockchain technology is SHA – 256 to get a constant hash of 256 bits every time. [10] It is having its unique quality of hash function that creates unique outputs when given different inputs This algorithm is efficient in verifying file and message integrity during the transaction, data identification, and password verification.

Figure 4 illustrates the prototype of algorithm. In this, there is an initialization vector which is of 256 bits. The input message is often of arbitrary length, so divide it in size of 512 bits. because the input will always be not an ideal multiple of 512 bits, so some a part of the input is going to be left. to the present left input, we do padding which helps to form messages in multiple of 512 bits. After these 512 bits are added with the 256 IV bits. the entire 768 bits is skilled a compression function 'c' to urge an output of 256 IV bits only. This output of 256 bits is again given to subsequent block and which are combined with 512 bits input from the next block. Again, the entire is skilled in the compression function to yield a 256-bit output. The same process is repeated until the last block. A message or data is processed by blocks of $512 = 16 \times 32$ bits, each block requiring 64 rounds.in such how all blocks are processed and eventually get 256 bits message digest.

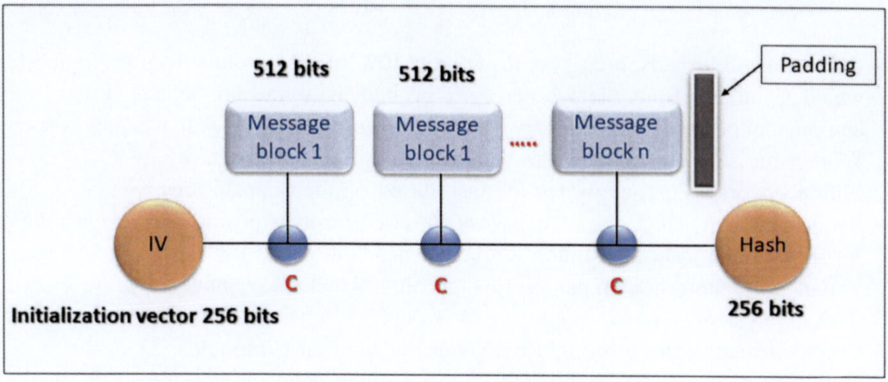

Fig. 4. Prototype algorithm of SHA 256

5.3 System Flow

Figure 5 defines the flow of the system which contains requester, worker, and admin as a module. The workflow of the proposed system along with modules are as follows:

- Requester wants to complete the task or a job, the requester will first register and login to the system, then post the task with defined amount and deadlines, on decentralized crowdsource system.
- Requester checks the balance on the wallet and deposits the fiat currency as the balance amount is less on wallet via blockchain.

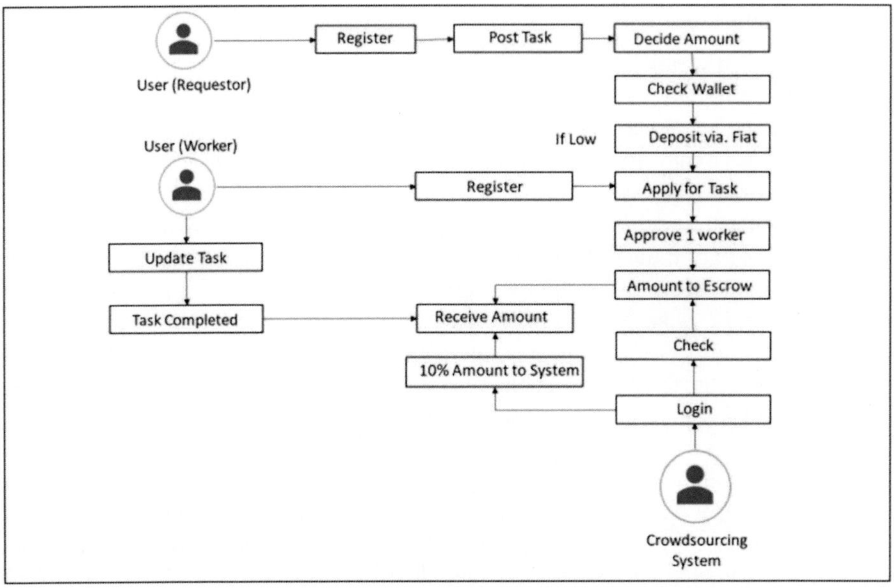

Fig. 5. The workflow of the system

- Decentralized crowdsource system will get 10% of the amount from the requester which is hidden from the worker, will be beneficial earning to the system, this amount can be used for giving reward to the new requester as well as a new worker.
- Worker register and login to the system for applying for the task.
- Multiple workers can apply for the task based on their capabilities.
- Requester, then select one of the workers based on their profile and assign a task.
- As soon as the task is assigned Requestor needs to deposit the task amount using virtual cryptocurrency in an escrow account which is managed by the crowdsourcing system.
- Cryptocurrency can be brought by Requestor via Fiat Currency.
- Cryptocurrency and all transaction details include deposits amount on the wallet, deposit amount in escrow, and deposit amount on workers' wallets are saved on the blockchain.
- The worker submits percentage or module wise tasks to the requester and the worker will receive a percentage-wise task amount via cryptocurrency from escrow as pre-decided.
- On successful completion of the task, the pre-approved escrowed amount will be deposited into the worker's account.
- Requester and worker can give ratings depending on their work and task respectively for future use.

A. The system has the following modules:

1. **Requester:** Initially, the requester needs to register and login to the system. Following are the activities to be performed by requester:
    - put the task on the crowd source system
    - decide the amount for a task
    - check the wallet amount and deposit an amount if there is an insufficient balance in the wallet
    - select worker based on their profile and assign a task
    - accept a task after completion and give a rating to the worker based on their work.

2. **Worker:** Worker needs to register and login to the system. Following are the activities to be performed by the worker
    - apply for the task
    - works on task after approval
    - completes the task and submit the task to the requester
    - receive payment on successful completion
    - giving a rating to the requester.
3. **Crowdsourcing system:** The crowdsourcing system needs to login itself. It is responsible for maintaining the escrow. The entire process is automated rite from receiving money into the escrow to reliving the payment from escrow after deducting the portal charges.

6 Result Analysis

The proposed decentralized crowdsourcing with the escrow system compares with the traditional system which follows the centralized way under various performance factors to validate the feasibility of the model for real usability scenarios. Table 1 express the factors and levels for a decentralized crowdsourcing system with the escrow and the traditional system. Evaluation of the proposed system executed based on comparison which is given in Fig. 6.

Table 1. Performance characteristics of crowdsourcing with escrow and crowdsourcing without escrow.

Factors	Crowdsourcing with ESCROW	Crowdsourcing without ESCROW
Processing Speed	70 %	40 %
Security	85 %	30 %
Duplication of task	20 %	70 %
Flexibility	75 %	50 %
Efficiency	80 %	40 %

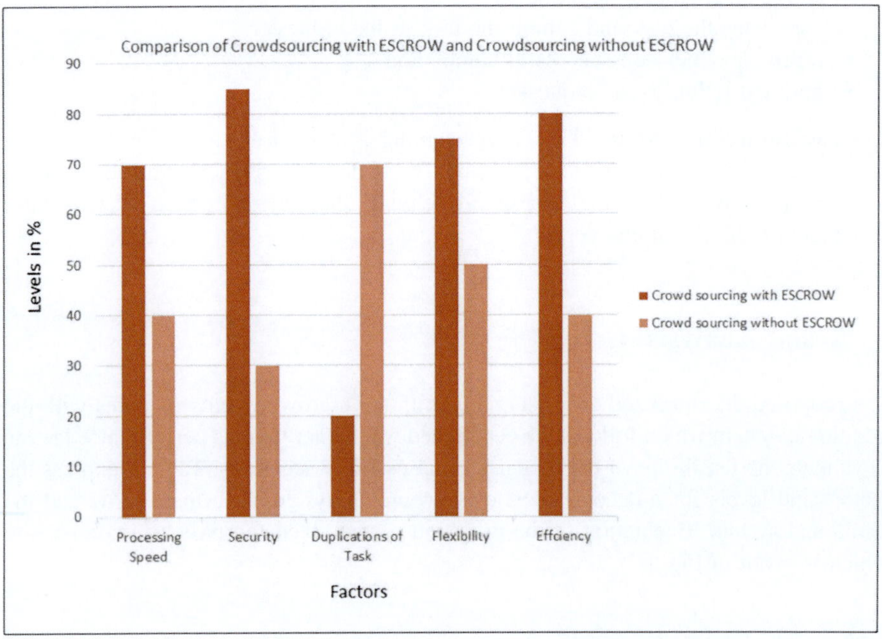

Fig. 6. Graph for comparison of crowdsourcing with escrow and crowdsourcing without escrow

7 Conclusion

This research work introduced the idea of crowdsourcing based on blockchain. Thus, it is examined that the conventional centralized crowdsourcing method suffers from disclosure of privacy, a single point of failure, high charge for services, etc. And formalized blockchain-based crowdsourcing to address crowdsourcing systems' vulnerabilities, by improving their effectiveness to solve many problems and increasing their scope for use. A Blockchain database preserves the complete, indelible, and unchanging history of all transactions, properties, and instructions executed from the very first. With this, Blockchain enables participating parties and only those parties to exchange information that is open, transparent and trusted. Adopting Blockchain technology, it is possible to keep the ledger of all transactions in a collection of nodes (which belong either to staff or to requesters), eliminating the need for a central authority / entity. The proposed system focuses on the web-based application due to its simplicity since it will slightly affect the operation speed of the crowdsourcing process. This is the limitation of the system; hence the implementation of the mobile bases application is one of the beneficial solutions.

Acknowledgment. I want to stretch out my genuine gratitude to all who helped me with the undertaking work. I want to earnestly express gratitude toward Dr. Dilip Motwani for their direction and steady supervision for giving vital data concerning the undertaking likewise, for their help in completing this task work. I want to offer my thanks to my folks and individuals from the vidyalankar institute of technology [VIT] for their thoughtful co-activity and support.

References

1. Kogias, D.G., Leligou, H.C.: Toward a blockchain enabled crowdsourcing platform. IT Prof. **21**(5), 18–25 (2019)
2. Zhang, S., Wu, J., Lu, S.: Minimum makespan workload dissemination in DTNs: making full utilization of computational surplus around. In: Proceedings of the Fourteenth ACM International Symposium on Mobile Ad Hoc Networking and Computing, pp. 293–296. ACM (2013)
3. Mottur, P.A.: Vizsafe: the decentralized crowdsourcing safety network. IEEE (2018)
4. Lu, Y., Tang, Q., Wang, G.: Zebra lancer: private and anonymous crowdsourcing system a top open blockchain. In: IEEE 38th International Conference on Distributed Computing Systems (2018)
5. Zhu, S., Cai, Z., Hu, H., Li, Y., Li, W.: zkCrowd: a hybrid blockchain-based crowdsourcing platform. IEEE Trans. Ind. Inform. **16**(6), 4196–4205 (2019)
6. Yang, P., Li, Q., Yan, Y., Li, X.Y., Xiong, Y., Wang, B., Sun, X.: Friend is treasure: exploring and exploiting mobile social contacts for efficient task offloading. IEEE Trans. Veh. Technol. **65**(7), 5485–5496 (2016)
7. Freelancer. https://www.freelancer.com
8. Xu, J., Wang, S.: A blockchain-enabled trustless crowd-intelligence ecosystem on mobile edge computing. IEEE Trans. Ind. Inform. **15**(6), 3538–3547 (2019)
9. Li, M., Weng, J., Yang, A., Lu, W., Zhang, Y., Hou, L., Liu, J.N., Xiang, Y., Deng, R.H.: CrowdBC: a blockchain-based decentralized framework for crowdsourcing. IEEE Trans. Parallel Distrib. Syst. **30**(6), 1251–1266 (2018)

10. Xiaolin, C., O' NeillCao, M.: Application-oriented SHA-256 hardware design for low-cost RFID. In: IEEE International Symposium on Circuits and Systems (2012). https://doi.org/10.1109/iscas.2012.6271509
11. Zou, J., Ye, B., Qu, L., Wang, Y., Orgun, M.A., Li, L.: A proof-of-trust consensus protocol for enhancing accountability in crowdsourcing services. IEEE Tran. Serv. Comput. **12**(3), 429–445 (2019)
12. Ding, Y., Chen, Z., Lin, F., Tang, C.: Blockchain-based credit and arbitration mechanisms in crowdsourcing. In: 2019 3rd International Symposium on Autonomous Systems (ISAS), Shanghai, pp. 490–495 (2019)
13. Rathore, H., Mohamed, A., Guizani, M.: A survey of blockchain enabled cyber-physical systems. Sensors **20**(1), 282 (2020)
14. Upwork. https://www.upwork.com
15. Truelancer. https://www.truelancer.com
16. Ethereum, pp. 777–788. ACM (2019). https://www.ethereum.org

Damping Percentage Detection Using Unconventional Methods

Tushar Bansal(✉), Sanjhi Singhal, Suprita Deswal, and S. T. Nagarajan

Department of Electrical Engineering, Delhi Technological University, Delhi, India
bansal_tushar@ymail.com, sanjhi.singhal0711@gmail.com, deswal.11@gmail.com, stnagarajan@dce.ac.in

Abstract. In a power system, inter-area modes of oscillations are the oscillations from different areas connected by tie lines which result in hazardous circumstances. In this paper, three machine learning algorithms have been used to predict the damping percentage of the distributed power system due to inter-area modes of oscillations. The data has been obtained from a simulated Kundur Two Area System along with Phasor Measurement Units (PMU) using Digsilent Power Factory software at a rate of 30 frames per second. The paper uses SVM, random forest, and XGBoost algorithms to predict the damping percentage.

Keywords: Machine learning · Hinge loss · Phasor measurement unit · Damping

1 Introduction

All the machines of a system work at a similar frequency for the operation to be stable in the power system. This is called synchronization. Therefore, the commercially chosen frequency, i.e. 50 Hz or 60 Hz is considered vital while dealing with power systems. The change in the load of one generator causes the divergence from the synchronous speed which affects all other generators in the power system. An appropriate control action is applied during this condition such as amending exciter controllers or turbines to maintain synchronization [1]. However, if the state of the network is inappropriate, low-frequency oscillations can occur. In a power system, the stability can be threatened by low-frequency oscillation that extends for a long period. These oscillations are undesirable as they increase the inefficiency during the operation of the grid. To stabilize the power system and secure the operation, it is required to ensure the stability of these oscillations. Oscillations in power systems generally contain multiple frequency components (modes), which are determined by governor control, the impedance of transmission line, generator inertia, etc. Local modes or plant modes are the oscillations associated with a single generator or single plant. The range

T. Bansal, S. Singhal and S. Deswal—These authors contributed equally to this work.

J. I.-Z. Chen et al. (Eds.): ICIPCN 2020, AISC 1200, pp. 461–471, 2021.
https://doi.org/10.1007/978-3-030-51859-2_42

of frequencies for local modes is from 0.7 Hz to 2.0 Hz [2]. Oscillations with such characteristics are well defined and analyzed. A detailed representation of the solution of the stability problem can be drawn out from the system in the vicinity of the plant. This can help in a thorough study of the factors associated with local modes. Some oscillations occur in a group of plants or a group of generators. Such oscillations are known as inter-area modes. The frequencies are in the range 0.1 Hz to 0.8 Hz. The factors influencing this mode of oscillation are not completely analyzed and understood. To study inter-area mode, a representation of the entire interconnected system is required [3–5]. In huge power systems, the inter-area modes develop instances of unstable oscillations. Such oscillations are becoming a cause of the trouble. It is observed that the inter-area modes can be perceived through the analysis of phasor measurement units. (Phasor Measurement Unit) located around the power system [6–9]. In the following paper, the Kundur-two area system has been explained in Sect. 2. Section 3 and 4 have been utilized for explaining fast Fourier transform and sliding window technique respectively. In Sect. 5, different machine learning algorithms used in the analysis have been described. Section 6 has been used for depicting the flowchart used during the analysis. The result has been presented in Sect. 7 and the conclusion in Sect. 8.

2 Kundur Two Area System

The data used in the paper has been obtained from a simulated Kundur Two Area System along with Phasor Measurement Units (PMU) using Digsilent Power Factory software at a rate of 30 frames per second.

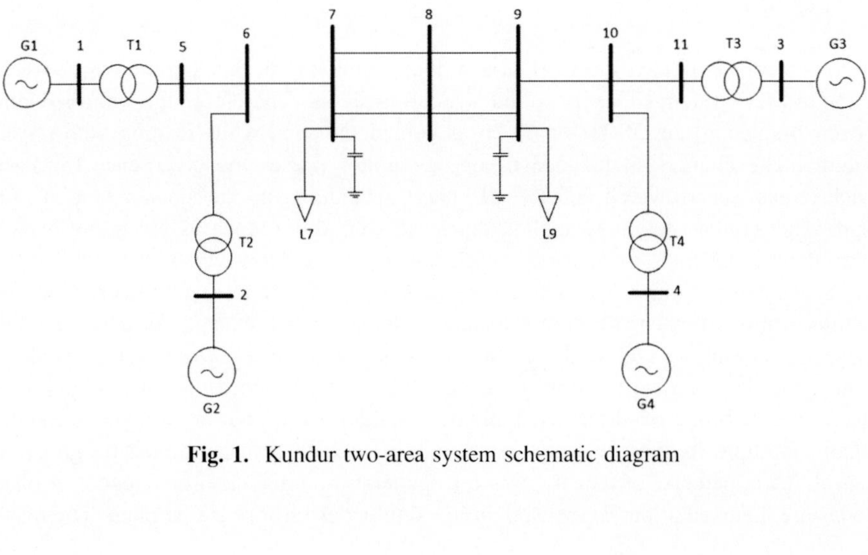

Fig. 1. Kundur two-area system schematic diagram

The simulated system consists of two areas with eleven buses (Fig. 1). Between the bus 7 and 9, the weak tie line connects the two areas. Two loads along with shunt capacitors are present at bus 7 and 9. The fundamental frequency of this system is 60 Hz. As seen from the figure, each area consists of two generators each [10, 11].

The data obtained from the PMU is categorized based on damping percentages, mainly as 2%, 4%, 6%, 8%, and 10%. Thus, the time series data is extracted along with the corresponding frequency deviation. The main external cause for the damping persistent in the system is the switching of heavy loads. There could be some internal disturbances such as loss of excitation resulting in damping of the power system as well.

The following figure (Fig. 2) shows the Fourier Transform of a set of data points, as obtained from the power system, which categorizes the system undergoing inter-area mode of oscillations.

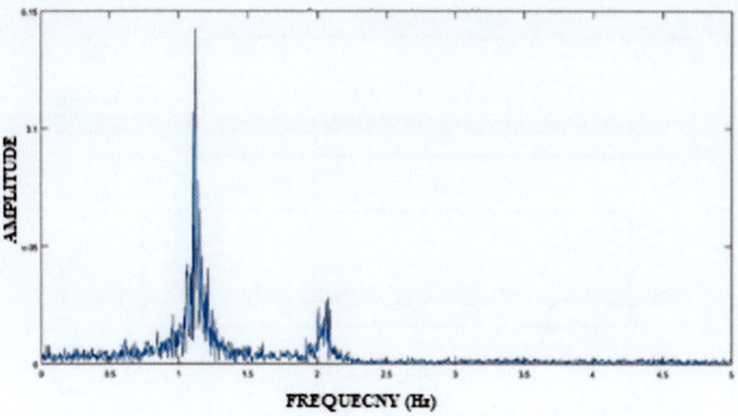

Fig. 2. Fourier Transform

3 Fast Fourier Transform (FFT)

FFT is an algorithm used in processing signals by evaluating the Discrete Fourier Transform (DFT) of a sequence. Space or time signals are converted to signals of the frequency domain. DFT matrix is factorized as the product sparse factors thus the complexity of DFT computation is decreased from $O(N^2)$ to $O(N \log N)$. Since this data is extremely large complexity is reduced significantly resulting in improved efficiency and reduced computational time. FFT has been used to derive (the peak value of the FFT) unique value to a given sample of data which is further used to classify the values into different damping percentages.

$$F(x) = \int_{-\infty}^{\infty} f(x)e^{-x}dt \tag{1}$$

4 Sliding Window Technique

The dataset that has been used is very large. To break it down into small units of the individual dataset (or samples) sliding window technique is used. Let K be the length of the window or (the number of elements in one unit of the dataset) and N be the step size by which the start and endpoints of the window move with each iteration. The window of length K moves through a distance of N repeatedly until it reaches a point where the condition $N + K - 1 \leq$ Size of the array (or dataset) [12]. This process is done to extract the FFT of various samples of a dataset rather than taking the FFT of the full dataset which outputs a single FFT. This helps in better analyses of the data set. The FFT generated provides different parameters to categorize the damping percentage, out of which the maximum peak of the graph has been chosen as the parameter for further analyses (Fig. 3).

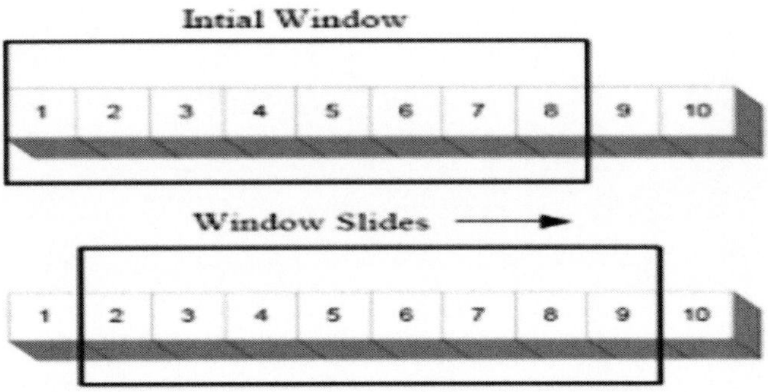

Fig. 3. Sliding window technique

5 Algorithms Used

The main idea behind the project is to predict the damping percentage in the shortest time possible. Thus, even though deep learning algorithms can predict the damping percentages much more accurately they are comparatively more computationally exhaustive than classification type supervised learning algorithms. Hence, to maintain a balance between both accuracy and processing time, SVM and tree-based algorithms have been used.

5.1 Support Vector Machine (SVM)

SVM is a type of supervised machine learning algorithm that can be used in situations relating to classification as well as regression. However, it is majorly used for classification problems.

In SVM each data point represents an N-dimensional feature vector, where the value of each coordinate represents the value of the particular feature, plotted in an N-dimensional space. The algorithm works on the principle that it creates an (N – 1) dimensional hyperplane which can separate the given data points into different groups or classes. Hyperplanes based on some conditions make a boundary to separate data points. Vectors that are close to the hyperplane are known as support vectors. These points are responsible for the orientation and position of hyperplane [13].

The SVM algorithm tries to maximize the distance between the data points and the hyperplane. Hinge loss function is used by SVM to get the maximum margin.

$$c(x, y, f(x)) = \begin{cases} 0, \mathit{if}\ y * f(x) \geq 1 \\ 1 - y * f(x),\ \mathit{else} \end{cases} \tag{2}$$

Where *c(x, y, f(x))* is the cost or loss function of the algorithm, x is the feature value, $f(x)$ is the predicted value and y is the actual output. When the actual value and the predicted value have the same signs then the value of loss function is zero. Otherwise, the value of loss function is calculated as 1 − *y*f(x)*. A regularization parameter to balance out the effects of margin maximization and loss is added.

$$min_w \lambda \|w\|^2 + \sum\nolimits_{i=1}^{n} (1 - y_i \langle x_i, w \rangle)_+ \tag{3}$$

Here w the weights associated with the feature, λ is the tradeoff parameter and $\|w\|^2$ is used to introduce a convex function to the equation to allow gradient descent to converge properly to minima.

Now, the partial derivative of the above equation concerning the weights of the feature is carried out.

$$\frac{\delta}{\delta w_k} \lambda \|w\|^2 = 2\lambda w_k \tag{4}$$

$$\frac{\delta y}{\delta w_k} (1 - y_i \langle x_i, w \rangle)_+ = \begin{cases} 0,\ \mathit{if}\ y_i \langle x_i, w \rangle \geq 1 \\ \quad -y_i x_{ik}, \mathit{else} \end{cases} \tag{5}$$

When the model correctly classifies a data point then the weights with the help of the gradient of the regularization parameter is updated.

$$w = w - \alpha \cdot (2\lambda w), \text{where } \alpha \text{ is the learning rate} \tag{6}$$

Otherwise, the weights using the gradients of the regularization parameter as well as the gradient of the loss function is updated [14].

$$w = w + \alpha \cdot (y_i \cdot x_i - 2\lambda w) \tag{7}$$

5.2 Tree Based Algorithms

Tree-based machine learning algorithms are amongst the most sought-after classification algorithms. These are easier to interpret and are much more adaptable at solving a problem. The data type for tree-based algorithms is not a concern as they can use both numerical and categorical features. The algorithms can be mapped to non-linear relationships between dependent and independent variables. They are constructed using a root node and a branch. To predict a class label, it begins from the root node and then follows the branch to the branch node by juxtaposing the features which provide the best split.

5.2.1 Decision Tree

Decision tree algorithms work through recursively dividing the whole dataset into smaller subsets whose values are as close as possible to a target class. The nodes of the tree contain R particular records that are divided due to a particular test on a feature. For example, a split on a continuous feature F can be produced by the condition $F \leq x$. This results in a left branch and right branch of the tree which contain records that are a subset of R [15].

$$R_l = \{r \in R : r(A) \leq x \tag{8}$$

$$R_r = \{r \in R : r(A) > x \tag{9}$$

Similarly, a categorical feature C can be used to produce split. For example, if $C = \{c_1, c_2, \ldots, c_k\}$ each branch j can be produced by $C = c_j$.

The split usually occurs by measuring the impurity of the parent node. The goal is to reduce the impurity of the parent node.

5.2.2 Random Forest Algorithm

Decision trees are the main driving force behind the Random forest algorithm. A decision tree algorithm works in such a way to split the dataset in a way such that the resulting groups are as unique and different from each other as possible and also the data points in each group must be as similar to each other as possible. Large numbers of decision trees form a random forest each of which is highly uncorrelated. There are two types of techniques used in the random forest algorithm: 1. Bagging (Bootstrap Aggregating) 2. Boosted trees.

In Bagging N samples are taken from the dataset. Decision tree is trained on those N samples and this process is repeated T times. T is the number of decision trees in random forest. Each iteration spews out a predicted value. The predicted output having the majority of votes is taken as the final predicted output [16].

5.3 Extreme Gradient Boosting

In XGBoost, the sequential building of trees is done to decrease the errors of the previous tree. Predecessors of trees are used to learn and update residual errors. Thus, renewed versions of the residuals are used to learn by the tree next in the sequence [17].

In boosting, base learners have high bias and hence are weak learners with predictive power approximate to random guessing. Predictive information is extracted from each weak learner and combined to generate strong learners. Both bias and variance are low in the final strong learner.

6 Flowchart

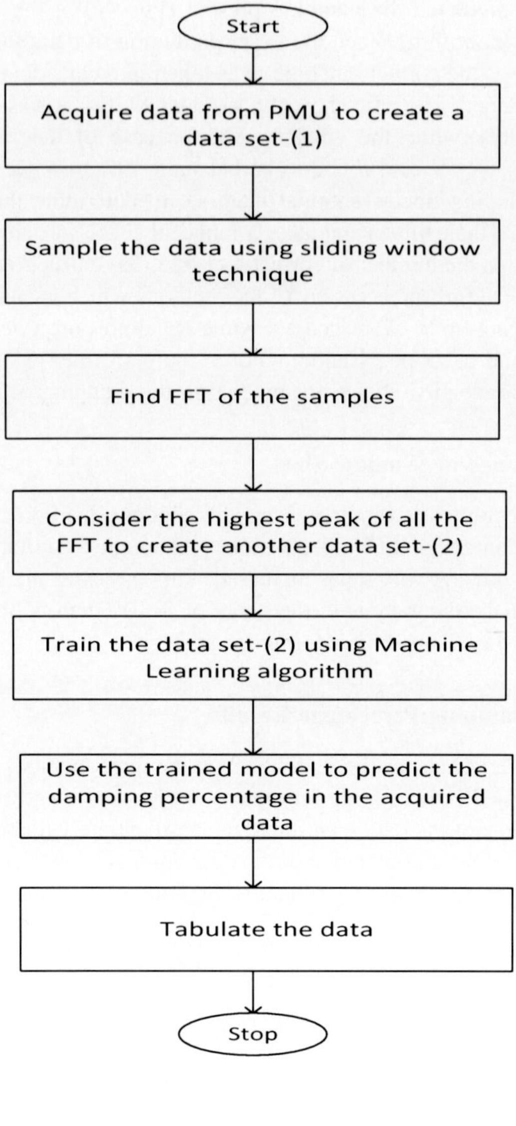

7 Results and Discussion

The three methods - support vector machine, random forest, and XGBoost - were tested with multiple windows and sliding time. The quantitative analysis for this project was performed in Python 3.7. For the analysis, Python's in-built libraries were used. Accuracy of each method was calculated based upon the total number of times the algorithm correctly predicted the damping present (Fig. 4a) in the system out of the total data fed to the algorithm. When the exact prediction of damping (i.e. taking only the precise outcome as the correct outcome) was taken into consideration the following accuracies were obtained. The efficiency of SVM and XGBoost was found to be in the range of 29% to 40% while the efficiency in the case of the random forest was increased by approx. 12.5% and found to be between 40% to 45%. Since this project was aimed at developing unconventional methods to determine the damping of the power system in less than fifteen minutes, a range of $\pm 2\%$ in damping prediction is acceptable. Due to this the number of correct hits (Fig. 5a) considered for each method increases and this increases the accuracy of the methods which is as follows. SVM and XGBoost end up being up to 77% accurate while Random Forest achieved up to 80% accuracy. In both the accounts, Random Forest outperformed SVM and XGBoost. Hence it is the recommended choice for predicting the damping of a power system.

7.1 Effect of Change in Window Size

Since the set of data points is fixed, increasing the window size from 30 s to 120 s ends up decreasing the number of FFT observations. Although quantitatively the samples have decreased and qualitatively it has increased. This increases the accuracy of all the three methods applied. Accuracy is in the range of 5–10% and is observed due to the generation of better fast Fourier transforms for analysis.

7.2 Individual Damping Percentage Result

The maximum number of correct predictions in all the three methods was seen for 10% damping followed by 2%, both lying on the end of the data spectrum. Hence, we can infer that in the middle of the data spectrum the classification boundaries are very close to each other therefore the classification of the data point becomes less accurate. And as we move towards the extremes classification boundaries become more distinct and hence results in better accuracy.

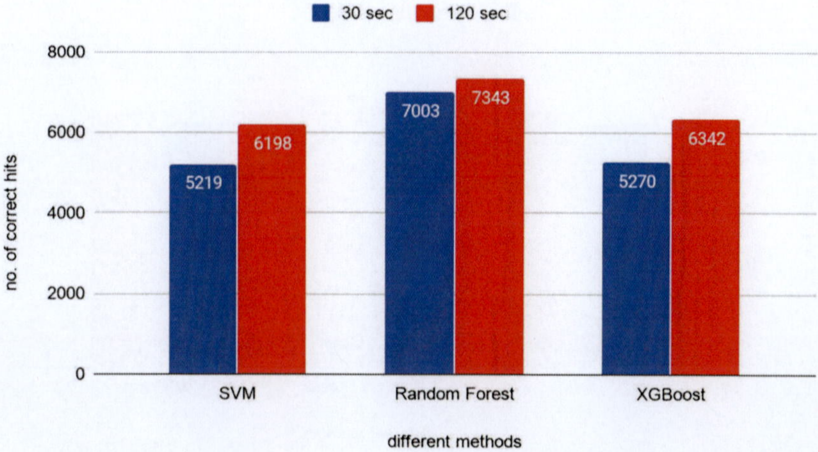

(a)

Damping Percent	Time Period in seconds	No. of correct hits		
		SVM	Random Forest	XGBoost
2	30	1337	1256	1360
	120	2107	1919	2154
4	30	668	776	670
	120	339	632	318
6	30	746	715	864
	120	980	861	1015
8	30	374	759	242
	120	318	738	261
10	30	2094	3497	2134
	120	2454	3193	2594

(b)

Fig. 4. (a) Graph for comparison of various methods. No. of correct hits (including all the damping percentages) represents the total number of times the algorithms correctly predicted the damping present in the system. (b) Table for comparison of various methods.

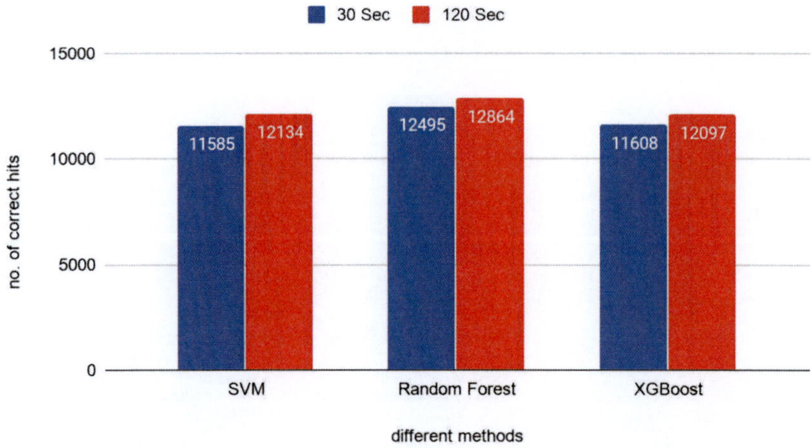

(a)

Damping Percent	Time Period in seconds	No. of correct hits		
		SVM	Random Forest	XGBoost
2	30	2129	2033	2151
	120	2859	2619	2853
4	30	2148	2324	2250
	120	2400	2463	2401
6	30	1987	2181	2000
	120	1892	2237	1780
8	30	2851	2458	2844
	120	2240	2352	2322
10	30	2470	3499	2363
	120	2743	3193	2741

(b)

Fig. 5. (a) Graph for comparison of various methods incorporating ±2% range no. of correct hits (including all the damping percentages) represents the total number of times the algorithms correctly predicted the damping present in the system within the ±2% range. (b) Table for comparison of various methods incorporating ±2% range.

8 Conclusion

Through this paper, three unconventional methods using machine learning to determine the damping of a power system with operating time, less than 15 min have been proposed. These methods were able to identify the damping in the system with up to 80% accuracy while considering a range of ±2% in damping as correct.

References

1. Jan-E-Alam, M.D.: A study on the presence of inter-area oscillation mode in Bangladesh power system network. J. Electr. Eng. Ins. Eng. **36**(2), 16–21 (2009)
2. Klein, M., Rogers, G., Kundur, P.: A fundamental study of inter-area oscillations in power systems. IEEE Trans. Power Syst. **6**(3), 914–921 (1991)
3. Paserba, J., Kundur, P., Sachez-Gasca, J., Larsen, E.: Small signal stability and power system oscillations. The Electric Power Engineering Handbook. CRC Press, Florida (2001)
4. Machowski, J., Bialek, J.W., Bumby, J.R.: Power System Dynamics: Stability and Control. Wiley, New York (2008)
5. Shim, K.S., Ahn, S.J., Choi, J.H.: Synchronization of low frequency oscillation in power systems. Energies **10**(4), 558 (2017)
6. Adamiak, M., Premerlani, W., Kasztenny, B.: Synchrophasors: definition, measurement, and application. In: Proceedings of the 59th Annual Georgia Tech Protective Relaying, pp. 27–29, Atlanta (2005)
7. Phadke, A.G., Bi, T.: Phasor measurement units, WAMS, and their applications in protection and control of power systems. Mod. Power Syst. Clean Energy **6**(4), 619–629 (2018)
8. Al-Hammouri, A.T., Nordström, L., Chenine, M., Vanfretti, L., Honeth, N., Leelaruji, R.: Virtualization of synchronized phasor measurement units within real-time simulators for smart grid applications. In: 2012 IEEE Power and Energy Society General Meeting, San Diego, pp. 1–7 (2012)
9. Konakalla, S.A.R., de Callafon, R.A.: Feature based grid event classification from synchrophasor data. In: International Conference on Computational Science (ICCS), Zurich (2017)
10. Kundur, P.: Power System Stability and Control. McGraw-Hill, New York (1994)
11. Wu, H., Wang, Q., Li, X.: PMU-based wide area damping control of power systems power system. In: 2008 Joint International Conference on Power System Technology and IEEE Power India, pp. 1–4 (2008)
12. Vafaeipour, M., Rahbari, O., Rosen, M.A., Fazelpou, F., Ansarirad, P.: Application of sliding window technique for prediction of wind velocity time series. Int. J. Energy Environ. Eng. **5**(2), 105 (2014)
13. Gomez, F.R., Rajapakse, A.D., Annakkage, U.D., Fernando, I.T.: Support vector machine-based algorithm for post fault transient stability status prediction using synchronized measurements. IEEE Trans. Power Syst. **26**, 1474–1483 (2011)
14. Hearst, M.A., Dumais, S.T., Osuna, E., Platt, J., Scholkopf, B.: Support vector machines. IEEE Intell. Syst. Appl. **13**(4), 18–28 (1998)
15. Navada, A., Ansari, A.N., Balwant, S.P., Sonkamble, A.: Overview of use of decision tree algorithms in machine learning. In: IEEE Control and System Graduate Research Colloquium (2011)
16. Livingston, F.: Implementation of Breiman's random forest machine learning algorithm. ECE591Q Mach. Learn. J. Pap. (2005)
17. Chen, T., Guestrin, C.: XGBoost: a scalable tree boosting system. In: Proceedings of the 22nd ACM SIGKDD International Conference on Data Mining, KDD 2016 (2016)

Mining Temporal Sequence Patterns Using Association Rule Mining Algorithms for Prediction of Human Activity from Surveillance Videos

D. Manju[1,2(✉)] and V. Radha[1]

[1] Department of Computer Science, Avinashilingam Institute for Home Science and Higher Education for Women, Coimbatore, India
manju.kulasekar@gmail.com, radhasrimail@gmail.com
[2] Department of Computing, Coimbatore Institute of Technology, Coimbatore, India

Abstract. One of the most interesting and important open issues in the automated video surveillance community is the analysis of human activities which requires a higher level of understanding. A framework was introduced for the prediction of human activity using temporal sequence patterns. In this framework, Sequential Pattern Mining (SPM) converted the complex symbolic sequence in the video into frequent itemsets using the Apriori algorithm. Apriori algorithm has to scan the video multiple times to find the frequent itemset that results in high execution time and memory consumption problems. To solve these problems, various association mining algorithms such as Rapid Association Rule Mining (RARM), Equivalence class clustering, and bottom-up lattice traversal (Eclat), diffset Eclat (dEclat) and Frequent Pattern-growth (FP-growth) algorithms are introduced for mining frequent itemsets. RARM speeds up the mining process at the minimum support threshold by using the Support-Ordered Trie Itemset (SOTrieIT) tree structure. The performance of frequent itemset mining is enhanced by using Eclat where depth-first approach is accommodated in it to find the frequent itemsets. However, the pruning technique in Eclat is more difficult. So, a dEclat algorithm is used where a depth-first search is performed to find frequent itemsets. But, dEclat is not more suitable for the sparse database. So, an FP-growth algorithm is introduced where a compressed data structure called FP-tree is used that solves the time and memory problem of the Apriori algorithm. The mined frequent itemsets are normal human behavior and the remaining activities are abnormal human activity. The experiments are carried out to prove the effectiveness of the RARM, Eclat, dEclat, and FP-growth for mining frequent activities in the video.

Keywords: Video surveillance system · Sequential Pattern Mining · Apriori · Rapid Association Rule Mining · Equivalence class clustering and bottom up lattice traversal · Diffset equivalence class clustering and bottom up lattice traversal · Frequent Pattern-growth

J. I.-Z. Chen et al. (Eds.): ICIPCN 2020, AISC 1200, pp. 472–483, 2021.
https://doi.org/10.1007/978-3-030-51859-2_43

1 Introduction

In recent years, the computer vision group has gained a lot of interest from the video surveillance system. The rising need for protection and safety has contributed to more insightful surveillance research. This provides a wide variety of applications such as observing people in the shopping centers, large waiting areas, health homes, traffic surveillance, clinics, etc. Presently, there is an increasing aspiration and requirement in video surveillance applications to be able to analyze human behaviors. Behavior analysis [1] includes the study and identification of motion patterns such that the behavior and relationships between objects are well defined.

Action recognition continues to be a problem, given significant research efforts during recent decades. One challenge is that in a higher-dimensional space, human gestures are depicted. Furthermore, contact between different subjects makes it difficult to scan this space. Another challenge with behavioral analysis is that different participants display significant differences with conducting equal or equivalent behaviors. Moreover, video camera visual images record real-world projective images and are prone to atmospheric conditions and lighting.

From the action recognition process, it is easy to find the abnormal behavior of a person in the video surveillance system. However, the analysis and detection of abnormal activities [2] in video surveillance is a laborious one when it is done manually. Recently, using video in machine understanding is emerging as an active research topic. Activity understanding, which involves being able to detect and classify targets of interests and analyze their actions, is one significant area. Detecting situations of special interests, like unexpected events, in particular, is one of the challenging aspects.

A general activity prediction framework [3] to predict human activity by modeling temporal sequence patterns. In this framework, sequential pattern mining (SPM) transformed object information represented as a complex symbol sequence into frequent itemsets using the Apriori algorithm. The normal activities are considered frequent itemsets and other abnormal activities are considered infrequent. However, the Apriori has problems such as high execution time and usage of memory due to the multiple scans of the video surveillance system.

So in this paper, various frequent itemset mining algorithms such as rapid association rule mining (RARM), Equivalence class clustering and bottom-up lattice traversal (Eclat) [4], Diffset Eclat (DEclat) and Frequent Pattern-growth (FP-growth) are introduced to mine the frequent activities in the video surveillance system. RARM accelerates the mining process even at low threshold values by using the Support-Ordered Trie Itemset (SOTrieIT) tree structure. RARM allows generating large 1-itemsets and 2-itemsets quickly without scanning the 2-itemset generation. To improve the performance of frequent itemset mining, Eclat algorithm is introduced. Eclat accommodates the depth first search approach to find frequent activities in the video surveillance system. However, the main disadvantage of Eclat algorithm is difficult in the pruning technique. So, an FP-growth algorithm is used to mine the frequent activities in the videos. Normal activities are frequent and abnormal activities are considered infrequent for applying the algorithm.

2 Literature Survey

A normal behavior method [5, 6] was proposed to detect abnormal event which started with building a co-occurrence matrix which through co-occurring events of learning statistics in a Spatio-temporal volume. The idea of co-occurring events was described by the common approach available in the motion label sequences. Then, Markov random field framework was applied to utilize the co-occurrence matrix as a potential function to describe the video streams concerning the probability of observing new activities. A threshold value was used in the normal behavior method which greatly influenced its performance.

A Network-Transmission-Based (NTB) algorithm [6, 7] was introduced to recognize the human activity in videos. Initially, a network was built by using an activity association between the corresponding patches and a patch of the scene. In the network, nodes signified a patch of the scene and the edges signified their corresponding activity correlation. After that, people in the video were formed as packages according to the built network and the human actions were modeled package transmissions happening in the network. Abnormal human activities were detected by analyzing the package transmission processes. However, this algorithm can perform online detection and it ends when the abnormality is detected.

A novel Gaussian Mixture Model (GMM) [8] was proposed with a particle entropy approach for detecting abnormal behavior using crowd distribution information. Crowd distance information and distribution information was used to calculate the parameters of GMM using which abnormal behaviors in the crowd could be predicted. However, this approach does not apply to complex datasets.

The context-associative hierarchical memory model [9] was introduced to recognize and predict human activities in a video surveillance system. This model was processed depends on the input activities of visual content that are previously experienced and parsed into successive sub-activities to develop context clusters to model the relation of activities on a temporal perspective. Then, subactivity of semantic attributes were arranged according to the concept hierarchy. And this was used for recognition using intervals over the contextual memory. However, this model needs improvement in terms of accuracy.

Temporally-weighted Generalized Time Wrapping (TGTW) method [10] was described. Initially in this method, Motion video was split into a sequence of short videos segmented. Each segment was signified using local Spatio-temporal statistics. At last, using the similarity derived from the TGTW integrated with the k-nearest neighbor (KNN) algorithm, an input sequence is classified among activity classes. However, this method still has prediction error as searches for the corresponding part between different segments of the input activity videos.

An approach based on graph formulation and graph kernel support vector machine [11] was proposed for abnormal activity recognition. A graph was constructed using the depicting interactions that happen of the entities that are captured in a video. After the construction of the graph, a binary support vector machine with a graph kernel was used to categorize the activities as a normal class or abnormal class. The graph was constructed based on threshold value which affects the accuracy of abnormal human activity prediction.

A Deep Neural Network (DNN) model [12] was introduced to predict human activity. It estimated the human activity through multi-view sequences made up of raw images. This model was presented as a constitute by three parts, the Convolutional Neural Network (CNN) block, the Multiple Stacked Long Short-Term Memory Residual (MSLSTRes), and the third being the dense layer. These parts enabled human activity to be differentiated by simply using sequences of raw images. Temporal evaluation of the videos in an online scenario will enhance the performance of the DNN model.

A dictionary learning algorithm [13] proposed a reference event for defining the inherent event patterns in a normal video. It was combined with sparse representation using constraints based on clustering to detect abnormal events. These normal events are represented in the learned dictionary as sparse linear combinations of bases. However, the accuracy of this algorithm is not effective as more structural information can be used for accurate abnormality detection.

3 Proposed Methodology

In this section, the frequent itemset mining algorithms such as RARM, Eclat, DEclat, and FP-growth are described in detail to mine the videos in a surveillance system to identify activities that frequently occur. The frequent itemsets are assumed as normal activities and the rest of the itemsets are treated as abnormal activities. Initially, input video is obtained from different locations. The input video scenes are split into rectangular local patches and then a normal activity pattern is learned for each patch respectively. The objects in each patch are identified and tracked based on the modified Histogram of Gradient (HOG). The activities in the local patch are considered as itemsets and the frequent itemsets are mined as a normal activity using RARM, Eclat, DEclat, and FP-growth frequent mining algorithms.

3.1 Rapid Association Rule Mining Based Frequent Activity Mining

RARM is an association rule mining algorithm which further speed up the frequent activity mining and it avoids complex candidate generation process. RARM accelerates the mining process at low support thresholds by using Support-Ordered Trie Itemset (SOTrieIT). SOTrieIT is an improved version of TrieIT which is a set of nodes in the tree. It consists of two values, the first referred to as the item label and the second referred to as item support. Here, the item represents the activities in the videos. SOTrieIT is obtained when TrieIT is sorted ordered and nodes are arranged based on support counts respectively. As a result, the lowest support node is moved to the rightmost position in the tree and similarly, the highest support item is taken to the leftmost node. From each patch of the videos, the SOTrieIT tree structure is constructed using 1-itemset and 2-itemset elements only. For example, $patch1 = \{Act_1, Act_2, Act_5\}$ possible information extracted will be $\{(Act_1), (Act_2), (Act_5), (Act_1, Act_2), (Act_1, Act_5), (Act_2, Act_5)\}$, where Act_i is the activities in the video surveillance system. Support count of those activities that already exist in the tree is increased by one, whereas a new node is created for activities that are currently not present in the tree, with activity name as label and support count marked as one.

Activities in the SOTrieIT tree are arranged according to their support count in such a way that level i of SOTrieIT generates i-Itemset. To generate 1-Itemset and 2-Itemset, depth-first search can be used to traverse the tree. Apriori algorithm is then applied to generate a large itemset from these two itemsets. This will help minimize the execution time of the process of candidate generation. Using SOTrieIT improves the performance of RARM radically. The candidate 1-Itemset and 2-Itemset can easily be extracted from the SOTrieIT. The overall process of RARM based frequent activity mining is given in the following algorithm.

RARM based frequent activity mining

Input: Local patches$L_k = \{Act_1, Act_2, \ldots, Act_n\}$

Output: Frequent activities

1. Let T be a set of SOTrieITs,N_q^y be the qth child node of parent node y, NC_p be the number of child nodes under node y and Act_n be the activity represented by node n.
2. for (i=1;i≤2;i++)
3. Get all i-itemsets in the local patches of videos intoC_i
4. for each itemset $X \in C_i$ do begin
5. Along the path X, traverse T, to find nodes
6. If the required nodes exist in T then
7. Leaf node's support count incremented
8. Use new support count to arrange updated node among siblings in decreasing order
9. else
10. Generate new nodes torepresent a path to X,and update support count as one.
11. Update T by inserting them, based on their support counts in decreasing order
12. end if
13. end for
14. end for
15. for $(h = 1; h \le NC_{Root}; h++)$
16. Let $X = N_h^{Root}$
17. if $h \ge |D| \times h \times s\%$ then begin
18. Add Act*h toL_1
19. for $(m = 1; m \le NC_h; m++)$
20. if $\sigma N_m^h \ge |D| \times H \times s\%$ then
21. Add$Act_{N_m^X}$ to L_2
22. end if
23. end for
24. end if
25. end for

26. From third iteration till n iterations, apply Apriori to find the remaining of the large itemsets from 3-itemsets onwards.

For every activity that arrives in the local patch, 1-Itemsets and 2-Itemsets are executed first. For each activity, the SOTrieIT was represented. The support counts of 1-itemsets and 2-itemsets are saved in first-level and second-level nodes correspondingly. This traversal thus needs maximum of two directions, which speeds up the process. Based on support counts of the nodes, T will then be arranged according to levels, from left to right, in decreasing order. When such a node is not found in the tree, a new node is added to T consequently. Similarly, T is then rearranged after each iteration. After the construction of SOTrieIT, T is first traversed to find a large 1-itemset and 2-itemset. Here, the depth-first search is applied to traverse the tree. The process starts with the leftmost first-level node. As T is arranged based on the support counts, on a node not satisfying the minimum support threshold, traversal is terminated. After the determination of large 1-itemsets and 2-itemsets, the Apriori algorithm is processed to find the other frequent activities in the videos. However, RARM also has a limitation. If at any point in time, the support threshold set by the user is changed, the whole process has to repeat, making it difficult to apply RARM algorithm in interactive scenarios.

3.2 Eclat Based Frequent Activity Mining

Eclat finds the frequent activities from the bottom like depth-first search. Unlike in Apriori and RARM, Eclat uses a vertical database format where activities with their list of local patches are separately maintained. TID-list [14] can be used to compute all frequent activities. A TID list is created for each activity in the first scan of the video. From k itemset, Apriori and depth-first search are applied to generate k+1 itemset. The intersection of TID-set of frequent k itemset gives (k+1)- itemset. This is repeated until it finds candidate activities. As support count information can be obtained from k itemsets, it is not necessary to scan the database to get the support of k+1 large itemset. Overhead of generating all the subsets of a local path in the video surveillance system as well as the candidate hash tree check process during support counting are avoided in this algorithm. The overall process of Eclat algorithm for mining frequent activities in the video surveillance system is given as follows.

Eclat based frequent activity mining

Input: Local patches $L_k = \{Act_1, Act_2, \ldots, Act_n\}, \min\_sup$

Output: Frequent activities

Bottom-Up (L_k) {

1. for all $Act_i \in L_k$
2. $L_k + 1 = \emptyset$
3. for all $Act_j \in L_k, i < j$
4. $N = Act_i \cap Act_j$
5. if $N.sup \geq \min\_sup$ then
6. $L_k + 1 = L_k + 1 \cup N$
7. end
8. end
9. end
10. if $L_k + 1 = \emptyset$ then
11. Bottom-Up $(L_k + 1)$
12. end}

In the Bottom-Up (L_k) process, all activities belong to the local patch L_k are processed. Check that activity exists in the local patch L_k or not. If it is present in the L_k, then the support of that activity is computed where support is the frequency of activity in that local patch L_k. After that, support of individual activity is compared with minimum decided support. Then, put those activities whose support is more than the minimum support in database $L_k + 1$. Check if $L_k + 1$ is empty or not. If $L_k + 1$ is not empty then start the same procedure for other local patches. However, Eclat requires virtual memory to perform the transformation.

3.3 Diffset Eclat Based Frequent Activity Mining

dEclat reduces memory consumption for frequent activity mining process by processing a Diffset that saves differences in the TID-list of a candidate activity from its frequent activities. From a pair of activities Act_x and Act_y having the same prefix Act in the search tree, the diffset is defined as $D(Act_{xy}) = D(Act_y) - D(Act_x)$ and $\rho(Act_{xy}) = \rho(Act_x) - |D(Act_{xy})|$, i.e., the support of activity is computed using its parent support and its own diffset. In which $D(Act_i)$ is the diffset of activity at local patch i and $\rho(Act_i)$ is the support value of an activity in the local patch i. The diffsets are generally smaller than TID-lists, so the memory consumption is reduced. From the third levels, the diffsets are calculated directly from parent diffsets. Similar to Eclat, dEclat processes are started with the creation of TID-list for each frequent 1-itemset. Then the depth-first search is delegated to the recursive function TRAVERSE. The main process of TRAVERSE is to propose a candidate Act_{xy} and check its support against min_*sup* using $\rho(Act_{xy}) = \rho(Act_x) - |D(Act_{xy})|$. This process is repeated until all frequent activities in the local patches of videos have been enumerated.

3.4 Frequent Pattern Growth Based Frequent Activity Mining

FP-tree is constructed to overcome time and memory problems of Apriori algorithm. Frequent patterns are represented in quantifiable form and stored in a prefix tree structure. Using the FP-tree, FP-growth algorithm is developed to mine the frequent activities in the video surveillance system [15]. It is a two-step approach. In the first pass of the database, support count for each activity in the local patches present in the video is calculated. Next, the remaining patterns are arranged in decreasing order, after removing infrequent patterns. FP tree is built during the second pass of the database. In the second step, frequent activities are extracted using the FP-growth algorithm. The cost of computation is decreased to a great extent, as the database is scanned only two times. Moreover, there are no candidate itemsets generated by FP-growth algorithm. It used a divide and conquer method to limit the search space for mining frequent activities in the video. The frequent activities are considered as normal activity whereas infrequent activities are considered as an abnormal activity.

4 Results and Discussion

Here, the performance of Apriori, RARM, Eclat, dEclat, and FP-growth algorithms are analyzed based on prediction accuracy, sensitivity and information gain ratio for mining frequent activities in the video. As prediction of abnormality os considered as priority, accuracy, and sensitivity are chosen over precision and specificity as a measure of performance. For the experimental purpose, Multicamera Human Action Video Data (MuHAVi)[16] dataset is used which collects a large body of human action data using 8 cameras. This dataset consists of 1904 videos, at a frame rate of 25 frames per second. The resolution of image frames is 720 × 576 pixels.

4.1 Prediction Accuracy

Prediction accuracy can be calculated as:

$$Accuracy = \frac{True\ positive + True\ negative}{True\ positive + True\ negative + False\ positive + False\ negative}$$

The prediction accuracy of the human activity prediction method using Apriori, RARM, Eclat, dEclat, and FP-growth algorithms is shown in Table 1.

Table 1. Analysis of prediction accuracy

Observation ratio (%)	Apriori	RARM	Eclat	dEclat	FP-growth
10	0.5	0.53	0.55	0.58	0.61
20	0.62	0.64	0.68	0.70	0.72
30	0.78	0.81	0.83	0.86	0.88
40	0.73	0.77	0.80	0.82	0.85
50	0.76	0.79	0.82	0.85	0.88
60	0.81	0.83	0.85	0.88	0.90
70	0.78	0.80	0.81	0.84	0.86
80	0.79	0.81	0.84	0.86	0.89
90	0.805	0.83	0.85	0.88	0.90
100	0.86	0.89	0.90	0.91	0.93

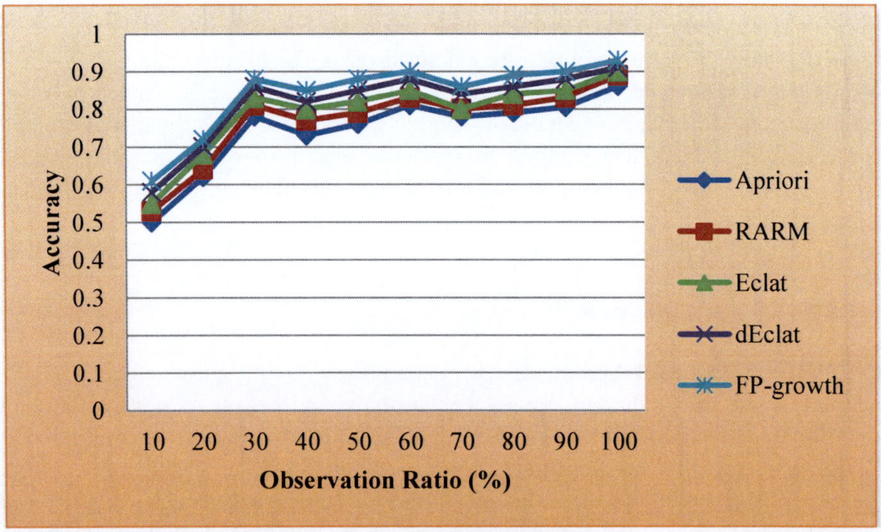

Fig. 1. Analysis of prediction accuracy

Figure 1 shows the prediction accuracy of the human activity prediction method using Apriori, RARM, Eclat, dEclat, and FP-growth for different observation ratios. X-axis denotes the observation ratio and Y-axis denotes the prediction accuracy. Prediction accuracy of the different algorithms under different observation ratios range between 0.5 to 0.93, depicted in Table 1. As the observation ratio increases, accuracy for FP-Growth has been seen to improve from around 0.61 to 0.93. The FP-Growth algorithm gives an accuracy of 0.88 at an observation ratio of 50% and 0.93 at an observation ratio of 100%. This is 15.79% greater than Apriori, 11.39% greater than RARM, 7.32% greater than Eclat, and 3.53% greater than dEclat for frequent activity mining. From this analysis, it is proved that the human activity prediction method using FP-growth algorithm has better prediction accuracy than other algorithms for frequent activity mining from video.

4.2 Information Gain Ratio

The information gain ratio of human activity prediction method using Apriori, RARM, Eclat, dEclat and FP-growth algorithms is shown in Fig. 2.

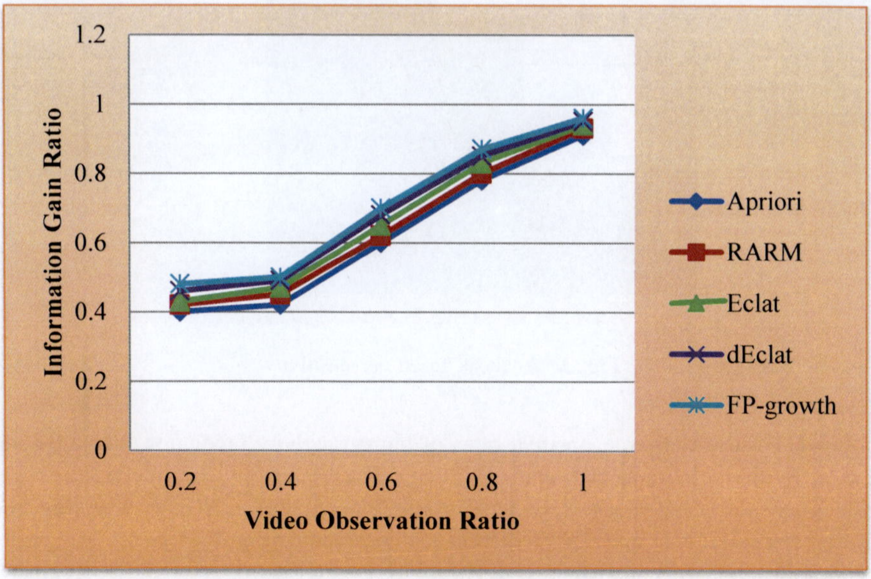

Fig. 2. Analysis based on information gain ratio

Figure 2 shows the information gain ratio of the human activity prediction method using Apriori, RARM, Eclat, dEclat, and FP-growth for different observation ratio. X-axis denotes the video observation ratio and Y-axis denotes the information gain ratio. The information gain of the human activity prediction method, at 0.6 observation ratio, using FP-growth algorithm is 16.67% greater than Apriori, 12.90% greater than RARM, 7.69% greater than Eclat and 2.94% greater than dEclat for frequent activity mining. From this analysis, it is proved that the human activity prediction method using FP-growth algorithm has a better information gain ratio than other algorithms for frequent activity mining from video.

4.3 Sensitivity

The sensitivity or true positive rate of human activity prediction method using Apriori, RARM, Eclat, dEclat, and FP-growth algorithms is shown in Fig. 3.

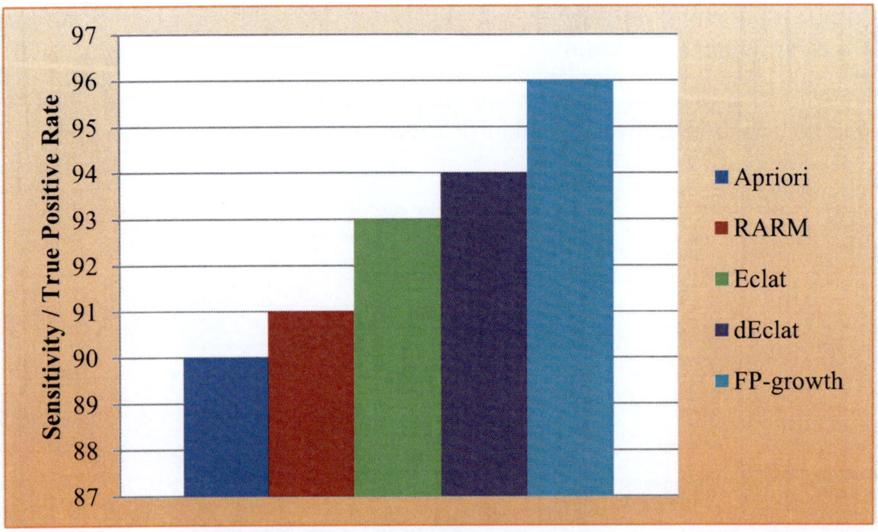

Fig. 3. Analysis based on sensitivity

Figure 3 shows the true positive rates of human activity prediction method using Apriori, RARM, Eclat, dEclat, and FP-growth. Y-axis denotes the true positive rate while X-axis denotes the association algorithms for frequent activity mining. The true positive rate of human activity prediction method using FP-growth is 6.67% greater than Apriori, 5.45% greater than RARM, 3.23% greater than Eclat and 2.13% greater than dEclat for frequent activity mining. From this analysis, it is proved that the human activity prediction method using FP-growth algorithm has a better true positive rate than other algorithms for frequent activity mining from the video.

5 Conclusion

In this paper, different association rule mining algorithms such as RARM, Eclat, dEclat, and FP-growth are introduced for predicting human abnormal activity in video surveillance systems. RARM increases the speed of the frequent activity mining process even at low threshold value by using an efficient tree structure called SOTrieIT. The performance of frequent activity mining is further enhanced by using Eclat association rule mining algorithm. It accommodates depth-first approach to find frequent activities. The pruning process of Eclat involves more complexity problems. Hence, dEclat algorithm is used where a depth-first search is employed to find the frequent activities in the video. It is not much more efficient in the spare database. So, FP-growth is introduced for frequent activity mining. The experimental results prove that the human activity prediction method using the FP-growth algorithm has better prediction accuracy, sensitivity, and information gain ratio than Apriori, RARM, Eclat, and dEclat algorithms.

References

1. Taha, A., Zayed, H.H., Khalifa, M.E., El-Horbaty, E.S.M.: Human activity recognition for surveillance applications. In: Proceedings of the 7th International Conference on Information Technology, pp. 577–586, May 2015
2. Patil, N., Biswas, P.K.: Detection of global abnormal events in crowded scenes. In: 2017 Twenty-third National Conference on Communications (NCC), pp. 1–6. IEEE, March 2017
3. Li, K., Fu, Y.: Prediction of human activity by discovering temporal sequence patterns. IEEE Trans. Pattern Anal. Mach. Intell. **36**(8), 1644–1657 (2014)
4. Shamila, N., Azam, M.A., Shehzad, K., Naeem, U., Ghazanfar, M.A.: Frequent pattern mining algorithms for finding associatedfrequent patterns for data streams: a survey. ProcediaComput. Sci. **37**, 105–116 (2014)
5. Benezeth, Y., Jodoin, P.M., Saligrama, V.: Abnormality detection using low-level co-occurring events. Pattern Recogn. Lett. **32**(3), 423–431 (2011)
6. Manju, D., Radha, V.: A survey on human activity prediction techniques. Int. J. Adv. Technol. Eng. Explor. **5**(47), 362–371 (2018)
7. Lin, W., Chen, Y., Wu, J., Wang, H., Sheng, B., Li, H.: A new network-based algorithm for human activity recognition in videos. IEEE Trans. Circuits Syst. Video Technol. **24**(5), 826–841 (2014)
8. Gu, X., Cui, J., Zhu, Q.: Abnormal crowd behavior detection by using the particle entropy. Optik **125**(14), 3428–3433 (2014)
9. Wang, L., Zhao, X., Si, Y., Cao, L., Liu, Y.: Context-associative hierarchical memory model for human activity recognition and prediction. IEEE Trans. Multimed. **19**(3), 646–659 (2016)
10. Wang, H., Yang, W., Yuan, C., Ling, H., Hu, W.: Human activity prediction using temporally-weighted generalized time warping. Neurocomputing **225**, 139–147 (2017)
11. Singh, D., Mohan, C.K.: Graph formulation of video activities for abnormal activity recognition. Pattern Recogn. **65**, 265–272 (2017)
12. Putra, P.U., Shima, K., Shimatani, K.: Markerless human activity recognition method based on deep neural network model using multiple cameras. In: 2018 5th International Conference on Control, Decision and Information Technologies (CoDIT), pp. 13–18. IEEE, April 2018
13. Yuan, Y., Feng, Y., Lu, X.: Structured dictionary learning for abnormal event detection in crowded scenes. Pattern Recogn. **73**, 99–110 (2018)
14. Zaki, M.J.: Scalable algorithms for association mining. IEEE Trans. Knowl. Data Eng. **12**(3), 372–390 (2000)
15. Manju, D., Radha, V.: An efficient spatiotemporal frequent object mining method to predict abnormal activities. J. Adv. Res. Dyn. Control Syst. **10**(2018), 1025–1029 (2018)
16. Murtaza, F., Yousaf, M.H., Velastin, S.A.: Multi-view human action recognition using 2D motion templates based on MHIs and their HOG description. IET Comput. Vis. **10**(7), 758–767 (2016)

Healthcare Analytics: Overcoming the Barriers to Health Information Using Machine Learning Algorithms

A. Veena[1,2(✉)] and S. Gowrishankar[1,2]

[1] Department of Computer Science and Engineering, Dr. Ambedkar Institute of Technology, Bengaluru 560056, India
{veenaal, gowrishankarnath}@acm.org
[2] Affiliated to VTU, Belagavi 590018, India

Abstract. In recent years, the data processed by the healthcare domain is increasing at an unprecedented rate accompanied by rich knowledge for medical research but low information has led to Healthcare Analytics. Healthcare Analytics collects the data from a myriad of areas such as clinicians, hospitals, government agencies, health insurance, pharmaceutical, and biotechnology agencies and allows for the examination of trends and patterns in various healthcare data. Based on this pattern, healthcare analytics determines how healthcare can be upgraded while constraining exorbitant spending. The retrieval of a pattern from healthcare data pertinent to the healthcare application and input to the Machine Learning algorithm is decided based on the data features. Feature selection approaches are developed to choose a subset of features that explain the details to achieve a more appropriate and compact depiction of the knowledge available. This paper summarizes Feature Selection algorithms and presents the challenges involved in healthcare data and also present an abstract architecture of data analytics in healthcare domain. Various algorithms and techniques in machine learning are compared and classified based on the applications of Healthcare Analytics.

Keywords: Healthcare analytics · Machine learning algorithms · Big data · Data analytics · Electronic health record (EHR)

1 Introduction

Formerly, conventional healthcare services are catered by healthcare service centers including healthcare record management. Some major services included are child health, the health of the geriatric population, maternal health, well-being of patients with a prolonged illness such as cancer, arthritis, cardiovascular diseases, health education, the health of critical psychiatric patients, infectious diseases and crises in public health, and data on health monitoring and governance [1]. Healthcare organizations worldwide are interested in achieving better quality and performance; it is imperative to define healthcare performance and recognize quality improvement dimensions and strategies [2]. The advancement in information collection and analytics-based

J. I.-Z. Chen et al. (Eds.): ICIPCN 2020, AISC 1200, pp. 484–496, 2021.
https://doi.org/10.1007/978-3-030-51859-2_44

technology over the past two decades has started to reform the use of data in all medicine branches [3].

Today's healthcare service ventures are moving from volume-based business into value-based business, which requires overwork from doctors and nurses to be increasingly gainful and proficient. This will improve healthcare practice, changing individual lifestyles, and driving them into longer life, prevent diseases, illness, and infections [4].

For numerous countries, healthcare is a growing segment of the economy [5]. Broad use of health information technology (IT) could theoretically improve a patient's access to their health information and promote future priorities of patient-focused consideration [6]. There has been a significant increase in electronic health data in recent years, including extensive electronic medical records (EMR) tracking patient conditions, clinical guidelines and policies, diagnostic tests, laboratory reports, imaging exams, treatments, outcomes, genomics, proteomics, claims, transaction records, and best practices [7]. A paradigm change in the healthcare sector has seen the latest trend of digitizing medical records [8].

The federal Governments change in healthcare policies, major private players, and insurance companies invested in the healthcare industry. With this change in government policies many public, private-public, and corporate health care systems emerged in many numbers. This new healthcare systems with the art of the IT technology have started creating digitized data of all healthcare population from small clinical investigation to major surgeries. Since the electronic health data is still largely underused and thus wasted [9], there is a need to turn the raw data into useable and actionable information [10]. Analytics is the way to gain knowledge by making appropriate use of data for qualitative analysis [11].

Healthcare analytics is a process of systematic data analysis that allows healthcare connoisseur to find opportunities for improvement in diagnosis, patient management, health system management, and spending. Healthcare analytics combines real-time and recorded data to predict trends, reveal actionable insights and drive long-term growth. Healthcare analytics helps to amend the quality of clinical care by increasing the healthcare organization's access to patient data and permitting health systems to crack important patterns and trends about diagnosis, treatment, and perpetuated care. It aids to access actionable perceptions that apprise further interactions with consumers, patients, and populations. Apprise population health initiatives and allow public health organizations to better manage to predict outbreaks, the spread of disease, and allocate health resources to communities in need. It also benefits to discover the root cause of consumer response, or lack of response to outreach and create tailored campaigns to improve patient engagement.

The paper is structured as detailed below: Sect. 2 addresses data forms, challenges involved in healthcare data, and public health informatics. Section 3 gives detail about the architecture of Data Analytics in Healthcare Domain, and feature selection algorithms; a survey on these feature extraction algorithms and Sect. 4 discusses the applications of the healthcare domain and literature survey on these healthcare applications.

2 Healthcare Data Types and Challenges

Formerly, medical professionals assisted by their observations dominated a significant portion of the medical procedures, and clinical trials were performed using meticulously planned and expensive investigations. The data is collected from EMG, EEG, ECG, EOG, and PPG. In the present scenario, the Internet of Things (IoT) is the largest contributor to data. These data, called EHR is the primary tool in helping the medical profession or in helping to find insights into healthcare.

In general, healthcare data can be divided into four categories [12]:

- **Clinical Data:** This data consists of surgery reports, health records of patients, medical images, and laboratory reports.
- **Patient Behavior Data:** These data are retrieved by monitoring and fitness trackers that include an accurate and detailed perspective on population well-being.
- **Pharmaceutical Research Data:** These data are obtained from high throughput test results, clinical research studies and are frequently subject to privacy issues due to business methods.
- **Health Insurance Data:** These statistics have data been accumulated and preserved for many years by different health insurance providers.

Healthcare is an extremely data-intensive industry [13]. The data from the US healthcare industry is anticipated to be alone before the Zettabyte scale and not long after the Yottabyte (10^{24} gigabytes) scale [14]. Exponential healthcare data development has been anticipated to continue to expand in different types, such as biometric data, EHR, medical imaging, clinician-reported results, biomarker data, wearable gadgets, and genomic data [15]. The recent progress, breakthroughs, challenges, gaps, and possibilities for large-scale data applications in healthcare domains like those of personal informatics, clinical informatics, public health informatics, wellness informatics, imaging informatics, urban informatics, and, bioinformatics have been described in [16].

A. Challenges Involved in Healthcare Data

Quality of the data is the basis for triumph in any company, including in healthcare systems. Specialists and analysts believe that data supremacy is of great importance to several applications in the real world and that judgments and operational performance depend on concrete data quality [17]. Along with suggestions resulting from poor organizational data quality, secure and effective healthcare depends on the admission and use of accurate, accessible, relevant, acceptable, applicable, understandable, and comprehensive knowledge [18].

Like in other application spaces, the big data transformation holds extraordinary assurance in the region of healthcare, as the accessible data about individual patients are extremely rich and hide crucial knowledge that can be exploited to improve patients' consideration while reducing its expense [19]. Some common challenges of the voluminous healthcare data are heterogeneity, inconsistency and incompleteness, scale, timeliness, privacy and data ownership, and visualization [20].

Heterogeneity
As human's process data, a ton of diversity is comfortably permitted. The wealth of the usual linguistic and its richness provide useful depth. Nevertheless, machine learning algorithms predict standard data and are slow in recognizing nuances [20].

Inconsistency and Incompleteness
Healthcare data progressively contains details of fluctuating reliability from different sources. Errors, missing and uncertain values are endemic and should be handled. On the bright side, Big Data's volume and duplication can also be used to compensate for lost data, cross-check conflicts, authorize trustworthy relationships, expose inherent classes, and expose secret relationships and models [20].

Scale
The biggest thing that everyone has in mind about Big Data is its scale. Supervising massive and rapidly expanding data volumes has long been a difficult problem. Processors have mitigated this problem in the past by getting quicker, following Moore's law. Yet a key move is now underway: data volume is rising faster than processor speeds and other computing assets [20].

Timeliness
When data grows over time, there is a need for real-time procedures to abridge and channel what to store, because it is not financially feasible to store relevant data in various instances [20].

Privacy and Data Ownership
Data security is another enormous issue and one that increases over Big Data. Severe laws are being enforced for electronic health records and data may be discovered in various circumstances. Most of the potential barriers are [21]:

1. Encryption of Data: Online transfer of clinical data happens if that information is encoded and a technique is developed to identify and validate individuals once they have been permitted to decode the information for observation and experimental use.
2. HIPAA-compliant policies: The US government's health insurance portability and accountability act (HIPAA) does not require data to be shared and used but sets policy guidelines.
3. The protocol for data transfer and exchange: Exchanging data across networks includes both EHR stakeholders and clinical databases to implement a solitary arrangement of guidelines for conveying and sharing.
4. Quality management and error detection: Just about every program that gathers reviews and uses clinical evidence from a broad variety of locations will take a severe approach to quality management and error tests.

Visualization
Numerous healthcare services, in the area of community health, patient's health-facing technologies, clinical management, and public health, leverage data visualization. Visualization methods center on data with non-spatial data features and distinct observations [22]. It is much easier to make adjustments to make the graphs and charts

as data analysis software helps users to transform one graph or chart to another. Provisional thinking on where to position the object which helps to make decisions [23].

A splendid challenge in Healthcare informatics is to find in these data relevant systemic trends and knowledge of temporal trends, which are frequently covered up and not available to the human master but would urgently need better support for decision making. Another issue is that the bulk healthcare ınformatics data sets are pitifully organized and non-standardized [24].

3 The Architecture of Data Analytics in Healthcare Domain

A defined big data system in health informatics involves a set of a process comprising complete processing of health informatics. Every phase in the process plays a very important role in providing the professional and significant results of the big data analytics [25]. This section would address in depth the various stages of Data Analytics in Healthcare.

1. Data Collection
Data collection is a huge undertaking and has a very important role to play [26]. The data are generated by human beings in the healthcare system. Each category of data contains unorganized and quasi-structured data, such as laboratory results, discharge summaries, hospital admission records, case notes, and EHR.

Data is additionally acquired from such as remote sensors, smart meters, wearable gadgets, and other gadgets that track vital signs. Data is also generated from a person's physical characteristics such as fingerprints, retinal scans, blood pressure in the heart rate, pulse, genetics, and pulse-oximetry readings, as well as x-ray and other medical images. The information is also collected from healthcare claims, billing records, social interactions, and communication tools. Data is also composed of disease records, vital health surveys, and statistical data [27].

2. Storing
Innovation and moving to EHR allow vast quantities of healthcare data with increased sophistication, from different sorts of medical images to reports from doctors, test findings, laboratory results, and so forth. Productive answers are important for financially sound storage and management [28]. In particular, there are certain simple properties for an ideal data-saving and management framework for healthcare [29].

- High Accessibility: Access to records easily, securely, and reliably anywhere, anytime is appealing for healthcare workers. Dependable and quick recovery of patient information saves doctors considerable time and empowers them to respond promptly, which can also mean the distinction between life and death.
- High Extensibility: Healthcare digital storage requirements can potentially hit thousands of terabytes, or even petabytes, the systems of the organizations should be incrementally scalable given the requirements of annual data storage and tailored to data storage.

- Cost-effective: It may be costly and difficult to store and handle a large volume. A proficient storage foundation ought to lessen the expense and complexity and secure the data without trading off performance.

Typically, there are the ranges of capability alternatives to analyze big data e.g. the computational RDBMS (relational database management systems) will scale to deal with big data applications. They are more and more suitable for organized data, however. Hadoop technology expands the ability to work with the Hadoop distributed file system (HDFS) and the Map/Reduce application system for a broader variety of content [30]. The HDFS is an integrated digital storage system utilized by Hadoop applications. Map/Reduce is a programming model and associated application for the planning and creation of large data sets using an equivalent, distributed grouping algorithm.

3. Sharing

The main challenges of recording patient and medical data are how to exchange healthcare information securely with scientists and clinicians across a healthcare group, provincial, institutional, or even national jurisdictional boundaries [31]. The research bodies of various universities and enterprises also chalk out the standard guidelines for secure, confidential data sharing on big data platforms, enabling organizations to reduce the expense of delivering tailored services to customers and deliver quality and cost-effective data services [32].

4. Analyzing

The primary step of analyzing procedures includes data processing, feature extraction and selection, and machine learning.

4.1 Data Preprocessing

Data processing is an essential phase in the process of information discovery since quality decisions must be based on the quality of data. The identification of data irregularities, resolving them early and reduction of the data to be analyzed will lead to huge benefits for decision-making [33]. Some of the most common and practical cleaning methods are rule-based data cleaning, data transformation, data deduplication, outlier detection, ML(machine learning) guided cleaning, missing value interpretation, data synchronization, data normalization, data, and human involved data cleaning [34–42].

4.2 Feature Extraction/Selection

The selection of features is otherwise called variable selection; selection of attributes or variable subsets is the method of selecting a subset of specific features (variables, predictors) for use in model creation. The major benefits of feature selection consist of pulling out important features, the reduction in data size, increased accuracy of estimates, a simple comprehension of attributes or variables, and eventually, minimizing execution time [43]. Some feature selection methods are Independent Component Analysis (ICA), Pearson's Correlation, Linear Discriminant Analysis (LDA), Tree-Based, ANOVA, Chi-Square, Correlation Feature Selection (CFS), Forward Selection, Backward Elimination, and Recursive Feature Elimination.

4.2.1 Independent Component Analysis

ICA is a computational method for separating data or signals into subcomponents. Jeanny Herault and Bernard Ans have developed the early general framework for ICA [44].

4.2.2 Pearson's Correlation

Pearson correlation is a linear correlation measure between the two variables. The coefficient has a value between +1 and -1, according to Cauchy – Schwarz inequalities [45].

4.2.3 Linear Discriminant Analysis

In DA (Discriminant Analysis) or classification, LDA is a linear discriminant analysis introduced by Fisher is the most significant and classical technique. LDA aims to identify the category of a sample object with the closest mean by group or class. The closeness is calculated by metric distance [46].

4.2.4 Correlation Feature Selection

CFS selects the attributes using the filtering method. It is introduced by Hall Correlation. The CFS uses a hypothesis that evaluates the subset features to forecast the class label along with the degree of association between them. Features that are not strongly associated with the classification are excluded. A variety of metrics such as information gain, Minimum Description Length (MDL), symmetric uncertainty can be used for measuring the correlations [47].

Table 1. Overview of feature extraction/selection algorithms for healthcare analytics

Feature extraction/Selection	Healthcare examples	Advantages	Disadvantages
ICA [49]	Schizophrenia	Reduces noise estimation, searches for distinct and uncorrelated groups, generalizations	Unsuitable for limited sample size, missing values
Pearson's Correlation [50]	fMRI – brain imaging	Finds linear dependency between pairs of elements, fast, reduction in the number of coefficients	Tedious and time-consuming to calculate
LDA [51]	ECG, Mobile Health data set	The number of features in ECG is decreased	Assumption of the normality of multi-variate
CFS [52]	Leukemia	Extract the characteristics of segmented nucleus, reduces overhead processing and space	Fail to fit the data, Less scalability
C4.5 (Tree-Based) [53]	ECG Data set	Easy, selects the attribute that has the maximum information gain, graphical illustration	Samples must belong to the same class, overfitting

4.2.5 Tree Based – C4.5

The C4.5 algorithm generates a decision tree from a training set, which is a set of instances or tuples using a technique of divide and conquer. Each attribute which is a part of an instance has either continuous or discrete values. A decision tree is a tree data structure containing leaf nodes and decision nodes that specifies a test over an attribute, which has possible outcomes. The decision is taken based on the information gain [48].

Table 1 outlines several feature extraction/selection algorithms and their advantages and disadvantages in healthcare analytics.

4.3 Machine Learning

Machine learning (ML) is the study of computer algorithms that interprets datasets and improves automatically through experience. Machine learning for health informatics has expanded as a multidisciplinary analysis in healthcare data processing using hybrid analytical methods [54]. Machine learning activities are generally classified into three groups such as supervised, unsupervised, and reinforcement learning [55].

Every named training data contains the input value as well as the ideal output value for the target. The supervised learning algorithm analyzes the sample data and generates an estimated feature that could be used to model new attributes. Through the unsupervised machine learning method, secret bits of information are extracted from unknown data sets such as cluster analysis. Reinforcement learning helps a computer to learn its actions from the input obtained from encounters with an external world [56]. From a data processing perspective, both supervised and unsupervised learning approaches favor data interpretation and reinforcement approaches favor problem-solving decisions [57]. Some well-known machine learning methods include the Artificial Neural Network (ANN), Support Vector Machine (SVM), Decision Tree, self-organization Map, and K-Means clustering, and so on. Ensemble methods also incorporate the effects of individual classification strategies and generate a better overall outcome [58].

5. Decision Making by Clinicians

Clinical Decision Support is a method to improve healthcare quality. This allows patients, doctors, and physicians to assess better [59]. These Decision Support Systems store and use knowledge to analyze the clinical data and to predict when a query comes in front of the patient dataset [60]. The decision support system can be fitting for a few potential applications in public health including analysis of observation, resource management, and the dispersal of training data.

4 Applications of Healthcare Analytics

The potential applications of this analysis are boundless, offering such possibilities as epidemic prediction, early detection, risk assessment, disease cure, preventable cure, and quality of life improvement.

Table 2 describes the above healthcare analytics and the aforementioned implementations using machine learning algorithms.

Table 2. Overview of the algorithms in Machine Learning

Applications of healthcare analytics	Technique used	Advantage	Research gap	Performance metric
Epidemic prediction	ARIMA and Gaussian Distribution Model [61]	The model can be used for policymaking and planning of Healthcare resource allocation	Forecast spatial dengue AI, and application to alert dengue patients	Root Mean Square Error, Mean Absolute Error, Mean Absolute Percentage Error
	Neural Network uses functions based on graphs [62]	The ML is used to determine parts of a solution vector to be used for replication	Crossover, local search, and strategy for multi-objective optimization	MSE, RMSE, MAE, Mean Absolute Percentage Error
Early detection	Bayesian Classification and mean-shift segmentation [63]	The mean shift technique is very efficient	Extraction of suitable descriptors	Precision, Specificity, Accuracy
	(PCA) [64]	Centered on stochastic parameters of the component space	The temporal relationship among instances, small sample space	Specificity
	K – Nearest Neighbor [65]	Defines micro-clusters and classify as harmless or cancerous	The models can be applied for other domains also	True Positives, True Negatives
	Classification (PNN Classifier) [66]	Features to enhance the accuracy in identifying breast cancer and is reliable	This technique can be used for ultrasound images as it is devoid of radiations	F measure, Recall, Precision, Accuracy
Risk assessment	Naïve Bayes Classifier [67]	Greater responsiveness	Enhances the patient tabular layout and the treatment plan appropriately	Sensitivity, Specificity, G-mean
	Bayesian Classifier and Fuzzy set theory [68]	Evaluates the harm to human health caused by pollutants in drinking water	To increase the accuracy of the model	Risk Level Distribution Curve, Reliability

(*continued*)

Table 2. (*continued*)

Applications of healthcare analytics	Technique used	Advantage	Research gap	Performance metric
Disease cure	Classification [69]	Uses the description of symptoms data	Disease forecasting	Accuracy
	Random Forest [70]	Accuracy of 98.63% is validated	Applied for model images also	Accuracy
	Neural Network [71]	Accurately predicts the disease	Applied for gene sequences	Accuracy
Quality of life improvement	ROC (Logistic Prediction) [72]	Approximation of the fall danger and is rated for usability	To provide clear instructions to geriatric persons	–

5 Conclusion

Healthcare is a very important social aspect to maintain peace and order leading to inclusive living. The main purpose of this paper is to identify the methodologies and challenges that are prevalent in the present day scenarios to solve prognostic problems using healthcare analytics. Obtaining adequate information is very crucial in healthcare analytics. Lack of information may mislead the direction of analysis. We have compared different feature extraction techniques and machine learning algorithms and have classified them accordingly to analyze the quality of data in the healthcare domain. The same machine-learning algorithm cannot be applied to a different set of data and therefore every machine learning algorithm has to be studied for its merits concerning the particular medical dataset. We have gained insight into the features of the clinical dataset that can be used for making interpretations by selecting appropriate models. Suitable machine learning algorithms are recognized for a particular clinical dataset which will help the medical practitioners to diagnose and make a well-informed decision appertaining to the patient's healthcare.

References

1. Zhao, Y., Liu, L., Qi, Y., Lou, F., Zhang, J., Ma, W.: Evaluation and design of public health information management system for primary health care units based on medical and health information. In: JIPH 2019. Elsevier (2019)
2. Khalifa, M., Zabani, I.: Utilizing health analytics in improving the performance of healthcare services. J. Infect. Public Health **9**(6), 757–765 (2016)
3. Maqbool, D., Chambers, C.: Healthcare Analytics, Essentials of Business Analytics. ISORMS, vol. 264. Springer Nature Switzerland AG (2019)
4. Alkhatib, M.A., Talaei-Khoei, A., Ghapanchi, A.H.: Analysis of research in healthcare data analytics. In: Australian Conference on Information Systems (2015)

5. Yang, J., et al.: Emerging information technologies for enhanced healthcare. Comput. Ind. **2015**(69), 3–11 (2015)
6. Baldwin, J.L., Singh, H., Sittig, D.F., Giardina, T.D.: Patient portals and health apps: pitfalls, promises, and what one might learn from the other. Healthcare **5**(3), 81–85 (2016)
7. Hu, J., Adam, P., Wang, F.: Data Driven Analytics for Personalized Healhcare, & Information Management Systems: Cases, Strategies, & Solutions, Health Informatics. Springer, Switzerland (2016)
8. Kupwade, P.H., Seshadri, R.: Big data security and privacy ıssues in healthcare. In: IEEE International Congress on Big Data (2014)
9. Belle, A., Thiagarajan, R., Soroushmehr, S.M.R., Navidi, F., Beard, D.A., Najarian, K.: Big Data Analytics in Healthcare, pp. 1–16. Hindawi Publ. Corporation, London (2016)
10. Asante-Korang, A., Jacobs, J.P.: Big Data and pediatric cardiovascular disease in the era of transparency in healthcare. Cardiol. Young **26**, 1597–1602 (2016)
11. Simpao, A.F., Ahumada, L.M., Galvez, J.A., Rehman, M.A.: A review of analytics and clinical informatics in healthcare. J. Med. Syst. **38**(4), 45 (2014)
12. Chandola, V., Sukumar, S.R., Schryver, J.: Knowledge discovery from massive healthcare claims data. In: KDD. Association for Computing Machinery (2013)
13. Kuo, M.H., Sahama, T., Kushnirul, A.W., Borycki, E.M., Grunwell, D.K.: Health big data analytics: current perspectives, challenges & potential solutions. IJBDI **1**(1–2), 114–126 (2014)
14. Raghupathi, W., Raghupathi, V.: Big data analytics in healthcare: promise and potential. Health Inf. Sci. Syst. **2**(1), 3 (2014)
15. Dhayne, H., Haque, R., Kilany, R., Taher, Y.: In Search of Big Medical Data Integration Solutions-A Comprehensive Survey, IEEE (2019)
16. Luo, J., Wu, M., Gopukumar, D., Zhao, Y.: Big Data Application in Biomedical Research and healthcare: A Literature Review, Biomedical Informatics Insights (2016)
17. Yakout, M., Elmagarmid, A.K., Neville, J.: Ranking for Data Repairs. Cyber Centre Publications (2010)
18. International Review of Data Quality. Health Information & Quality Authority. (2011). Accessed from http://www.hiqa.ie/publications/international-review-data-quality
19. Kechadi, M.-T.: Healthcare big data: challenges and opportunities. In: BDAW and Advanced Wireless Technologies, pp. l–6 (2016)
20. Jagadish, H.V., et al.: Big data and ıts technical challenges. Commun. ACM **57**(7), 86–94 (2014)
21. Dong, X.L., Naumann, F.: Data fusion. VLDB Endowment, 1654–1655 (2009)
22. Shortliffe, E.H., Sondik, E.J.: The public health informatics infrastructure: anticipation its role in cancer. Cancer Causes Control **17**, 861–869 (2006)
23. Ingram, S., Munzner, T., Irvine, V., Tory, M., Bergner, S., Moller, T.: DimStiller: workflows for dimensional analysis and reduction. In: IEEE (2010)
24. Wang, L., Wang, G., Alexander, C.A.: Big data and visualization: methods, challenges and technology progress. Dig. Technol. **1**(1), 33–38 (2015)
25. Brownson, R., Baker, E., Leet, T., Gillespie, K., True, W.: Evidence-Based Public Health. Oxford University Press, New York (2011)
26. Fang, R., Pouyanfar, S., Yang, Y., Chen, S.-C., Iyengar, S.S.: Computational health ınformatics in big data age: a survey. ACM Comput. Surv. (CSUR) **49**(1), 1–36 (2016)
27. Sofia, Batra, I., Verma, V., Malik, A.: A comprehensive analysis of data collection methods in ınternet of things. In: ICAICR 2019. ACM (2019)
28. Tippet, A.: Data capture and analytics in healthcare (2014). http://blogs.zebra.com/data-capture-analytics-in-healthcare

29. NetApp. EHR solutions: Efficient, high-availability EHR data storage & management (2011). http://www.netapp.com/us/system/pdf-reader.aspx?cc=us&m=ds-3222.pdf&pdfUri=tcm:10-61401
30. Moshe, L.: UCI machine learning repository (2013). http://archive.ics.uci.edu/ml
31. Crowne, A.: Preparing healthcare industry to capture the full potential of big data (2014). http://sparkblog.emc.com/2014/06/preparing-healthcare-industry-capture-full-potential-big-data
32. Dong, X., Li, R., He, H., Zhou, W., Xue, Z., Wu, H.: Secure sensitive data sharing on a big data platform. Tsinghua Sci. Technol. **20**(1), 72–80 (2015)
33. Written, I.H., Frank, E., Hall, M., Mark, A.: Data Mining: Practical Machine Learning Tools and Techniques, 3rd edn. Elsevier (2011)
34. Rajiv, R.S., Sailesh, C., Rahul, B.: An approach for real-time stress-trend detection using physiological signals wearable computing systems for automotive drivers. In: IEEE, pp. 1477–1482 (2011)
35. Daby, S., Deepak, S.T., Michael, S.: Mining of sensor data in healthcare: a survey. In: Managing and Mining Sensor Data, pp. 459–504. Springer (2013)
36. Anthony, S.F., et al.: Harrisons Principles of Internal Medicine, vol. 2. McGraw-Hill Medical, New York (2008)
37. Apiletti, D., Baralis, E., Bruno, G., Cerquitell, T.: Real-time analysis of physiological data to support medical applications. IEEE Trans. Inf Technol. Biomed. **13**(3), 313–321 (2009)
38. Christos A.F., et al.: On the classification of emotional biosignals evoked while viewing effective pictures: an integrated data-mining-based approach healthcare applications. IEEE, 309–318 (2010)
39. Hu, F., Jiang, M., Celentano, L., Xiao, Y.: Robust medical ad hoc sensor networks with wavelet-based ECG data mining. Ad Hoc Netw. **6**(7), 986–1012 (2008)
40. Leema, A., Hemalatha, M.: An effective and adaptive data cleaning technique for colossal RFID data sets in healthcare. WSEAS **8**, 243–252 (2011)
41. Andr'e, S.F., et al.: Data mining using clinical physiology at discharge to predict ICU readmissions. Expert Syst. Appl., 13158–13165 (2012)
42. Visalakshi, S., Radha, V.: A literature review of feature selection techniques and applications: review of feature selection in data mining. IEEE (2014)
43. Wassan, J., Wang, H., Zheng, H.: Machine learning in bioinformatics. In: Encyclopedia of Bioinformatics and Computational Biology, pp. 300–308 (2018)
44. Leach, K.N.: A survey paper on ındependent component analysis. IEEE, 239–242 (2002)
45. Van Do, L., Anh, D.T.: Some ımprovements for time series subsequence join based on pearson correlation coefficients. In: SoICT, pp. 58–65. ACM (2016)
46. Yan, H., Dai, Y.: The comparison of five discriminant methods. In: International Conference on Management and Service Science. IEEE (2011)
47. Vanaja, S., Rameshkumar, K.: Analysis of feature selection algorithms on classification: a survey. Int. J. Comput. Appl. **96**(17), 28–35 (2014)
48. Ruggieri, S.: Efficient c4.5. IEEE Trans. KDE **14**(2), 438–444 (2002)
49. Dawy, Z., Michel, S., Joachim, H., Jakob, C.M.: Fine scale genetic mapping using ındependent component analysis. IEEE ACM (TCBB) **5**(3), 448–460 (2008)
50. Taban, E., Muaaz Gul, A., Fahad, S.: A GPU based technique to compute pairwise Pearson's correlation coefficients big fMRI data. In: ACM-BCB, pp. 723–728 (2017)
51. Varatharajan, R., Gunasekaran, M., Priyan, M.: Big data classification approach using LDA with enhanced SVM method ECG signals in cloud computing, pp. 10195–10215. Springer (2017)
52. Singhal, V., Singh, P.: Correlation based feature selection for diagnosis of acute lymphoblastic leukemia. ACM, 5–9 (2015)

53. Lee, S.-J., Xu, Z., Li, T., Yang, Y.: A novel bagging C4.5 algorithm based on wrapper feature selection for supporting wise clinical decision making. JBI **78**, 1–30 (2017)
54. Athmaja, S., Hanumanthappa, M., Kavitha, V.: A survey of Machine Learning algorithms for big data analytics. In: ICIIECS (2017)
55. Bishop, C.M.: Pattern Recognition and Machine Learning. Springer, New York (2006)
56. Qui, J., Wu, Q., Ding, G., Xu, Y., Feng: A survey of machine learning for big data processing. In: EURASIP, pp. 1–16. Springer (2016)
57. Younas, K., Usman, Q., Nazish, Y., Aimal, K.: Machine learning techniques for heart disease datasets: a survey. In: ICMLC, pp. 27–35 (2019)
58. Clayton, P.D., Hripcsak, G.: Decision support in healthcare. Int. J. Biomed. Comput. **39**, 59–66 (1995)
59. Chi, C.-L., et al.: Medical Decision Support Systems based on Machine learning. University of Lowa. http://ir.uiowa.edu/cgi/viewcontent.cgi/article=1469&context=etd
60. Patsaraporn, S.: Forecasting Dengue fever epidemics using ARIMA model. ACM (2019)
61. Chris, M., Caramanis, C., Shie, M., Sanjay, S.: Detecting epidemics using highly noisy data. In: MobiHoc. Association for Computing Machinery, ACM (2013)
62. Krzysztof, M.: Informed Mutation Operator using Machine Learning for Optimization in Epidemics Prevention. Association for Computing Machinery, ACM (2019)
63. Saberian, F., Zamani, A., Gooya, M.M., Hemmanti, P., Shooredeli, M.A., Teshnehlab, M.: Prediction of seasonal Influenza epidemics in Tehran using artificial neural networks. In: ICEE (2014)
64. Bai, X., Song, W., Chen, J.: Ebola prediction with epidemic model. IEEE (2016)
65. Alam, R., et al.: Motion biomarkers for early detection of dementia-related agitation. ACM Dig. Biomarkers (2017)
66. Appukuttan, A., Sindhu, L.: Curvelet & PNN classifier based approach for early detection & classification of breast cancer in digital mammograms. In: ICICT 2016 (2016)
67. Paredes, S., Rocha, T., de Carvalho, P., Henriques, J., Morais, J., Mendes, M.: Cardiovascular disease risk assessment innovative approaches developed in Heart Cycle project. In: International Conference of the IEEE EMBC (2013)
68. Ran, W., Tian, X., Wang, Y.: Health risk assessment model of drinking water Sources based on Bayesian & Triangle Fuzzy Number. In: ICAIP. ACM (2019)
69. Shankar, M., Pahadia, M., Srivastava, D., Ashwin, T.S., Reddy, G.R.M.: A novel method for disease recognition and cure time prediction based on symptoms. In: ICCCE (2015)
70. Chen, M., Zhao, X.: Fatty liver disease prediction based on multi-layer random forest model. In: CSAI, China Association for Computing Machinery. ACM (2018)
71. Golam, M., Qasim, F., Ripon, S.H.: Tuberous Sclerosis Complex (TSC) disease prediction using optimisized convolutional neural network. In: ICCCM. ACM (2019)
72. Wu, A.Y., Munteanu, C.: Understanding older users' acceptance of wearable interfaces for sensor-based fall risk assessment. In: CHI, pp. 1–13 (2018)

A Prediction of Heart Disease Using Machine Learning Algorithms

Mohd Faisal Ansari, Bhavya AlankarKaur, and Harleen Kaur(✉)

Department of Computer Science and Engineering, School of Engineering Sciences and Technology, Jamia Hamdard, New Delhi, India
faisalrehman.2903@gmail.com,
{balankar,harleen}@jamiahamdard.ac

Abstract. Now a day's heart disease is emerging as one of the most death-dealing diseases. As per a report published by the World Health Organization [WHO], heart disease is one of the most hazardous diseases to human which causes death all over the world from the last 20 years. Approx. 12 million people are dying every year, which makes it the biggest challenge for medical professionals to develop an early diagnosis of heart disease with better accuracy. In this paper, we have applied different machine learning algorithms and compared their classification accuracies. We have proposed a modified algorithm using logistic regression with principal component analysis for predicting heart disease with more accuracy on various attributes such as age, blood pressure, chest pain, serum cholesterol levels, heart rate, and other characteristic attributes, and patients will be classified according to varying degrees of coronary artery disease.

Keywords: Machine learning · Heart disease prediction · Logistic regression · Support vector machine · Principal component analysis · Medical diagnosis

1 Introduction

In the world full of enlarged and enhanced computer technologies one of the major subfield of computer science that is Artificial Intelligence (Machine Learning and Deep Learning) is used in the medical field to pull out the predictions whether heart disease exists or not based on extracted medical records (image file or .csv file) of the patients from the medical databases called Electronic Health Record with the use of various algorithms [1, 17].

Nowadays heart disease is the most death-dealing disease. As per a report published by the World Health Organization [WHO], heart disease is one of the most hazardous diseases to human which causes death all over the world from the last 20 years. Millions of human beings around the world are suffering from heart disease. Approx. 12 million people dying every year which makes it the biggest challenge for medical professionals how important the early diagnosis of heart disease with better accuracy [2].

There are many traditional methods for predicting such illness but they are not looking sufficient, like data mining algorithms do not predict heart disease with so much accuracy like the machine learning algorithms do (support vector machine, logistic regression, naïve bayes, random forest, and decision tree). In terms of data mining, when we work with these types of algorithms the problem arises from the very

J. I.-Z. Chen et al. (Eds.): ICIPCN 2020, AISC 1200, pp. 497–504, 2021.
https://doi.org/10.1007/978-3-030-51859-2_45

first step called data extraction like incomplete data, missing values, and inconsistency, and predicted results are not so much accurate. Medical industries much needed that type of diagnosis system which can predict heart disease at an early stage and offers more and more accurate diagnosis than traditional methods [3, 18].

After the promising success of machine learning algorithms in various real-life field industries, we have also observed that it can be a promising solution with the highest accuracy for medical diagnosis and it can be seen as a key application in the healthcare industry [4, 17].

In this paper, we are applying machine learning algorithms and comparing their accuracy for classifying whether an algorithm has a more accurate percentage and on this basis, we proposed a modified algorithm for predicting heart disease on various attributes such as age, blood pressure, chest pain, serum cholesterol levels, heart rate, and other characteristic attributes, and the patient will be classified according to varying degrees of coronary artery disease. In this paper, we used the UCI machine learning dataset of 304 patients which contain 304 rows and 14 columns.

2 Related Works

Many researchers are continuously working in the field of heart disease prediction to find out better and better accuracy with the use of various algorithms [5]. From the literature survey of different numbers of researchers, various techniques have been used for heart disease prediction using large datasets to find out some trends, patterns, and associations. A short literature review is presented here. Recently integrated clustering of more than one machine learning techniques can improve model performance in the heart disease diagnosis using various algorithms.

The study shows that several researchers are using various techniques like data mining and machine learning etc., to identify the risk factors associated with heart disease. Statistical scrutiny has identified the risk factors related to heart disease to be age, blood pressure, cholesterol, smoking, high blood pressure with cholesterol levels, family history, obesity, physical inactivity, high stress. Sufficient information about the risk factors related to heart diseases helps health care professionals and doctors to identify patients at high risk of having heart disease [6].

In many types of research, the researchers also focused on security issues, when data are imported before mining. Specifically, they examine some scenarios in which data mining algorithms like association rule mining and data clustering require privacy safeguards [16]. Data mining is a hopeful approach to meet this challenging requirement [7].

Afterward, the new research area has emerged for heart disease diagnosis and prediction called machine learning. Many authors have presented the concept namely "Heart disease prediction using machine learning over data mining concepts" as a means of extracting interesting patterns using data mining concepts with the use of machine learning algorithms that focuses on expectation, based on training dataset with known factors and the data mining willidnetify the unknown data properties [8]. The machine learning concept is based on identifying unique patterns in data and extracting feasible knowledge from them. The support vector machine and logistic technique have been applied over the data for the prediction accuracy to achieve an expert system [9].

Afterward, a researcher proposed a hybrid machine learning model based on Decision Tree, Support Vector Machine, and Naïve Bayes. In this, they show how the use of three different classifiers to obtain the majority voting scheme [10]. This scheme is divided into two parts. The part says that the classifiers produce their decision separately. In the second part, these decisions are combined in a way that the new model is based on the majority voting scheme. Many research thesis mainly focuses on the task of data modeling and classification using machine learning techniques (supervised and unsupervised) from pattern recognition and data mining [11].

3 Algorithms Used

This field is originally kindled by a cardiologist who sought to develop and test computational analogy of heart muscles. In the heart disease prediction system, disease risk attributes are taken as input variables. 'Disease existing' and 'disease non-existing' are the output variables.

3.1 Logistic Regression

In logistic regression, classification method will assign distinct set of classes and it use the borrowing technique from the machine learning to obtain the field of statistics [13]. Logistic sigmoid function helps to chnage the output values in logistic regression and it returns the probability value. The advantages includes Logistic Regression uses a more intricate cost function, and that is denoted as "Sigmoid functiomn". The proposed algorithm don't obtain the weighted sum of inputs directly, but it is passed through a sigmoid function that can map any real value between 0 & 1.

The activation function is a sigmoid function $1/\left(1 - e^{-\text{Values}}\right)$. The logit function oobtain any real valued number from agives an 'S' shaped curve and plot it into a value between 0 and 1. The coefficients of the logistic regression algorithm must be calculated from your training data in Fig. 1.

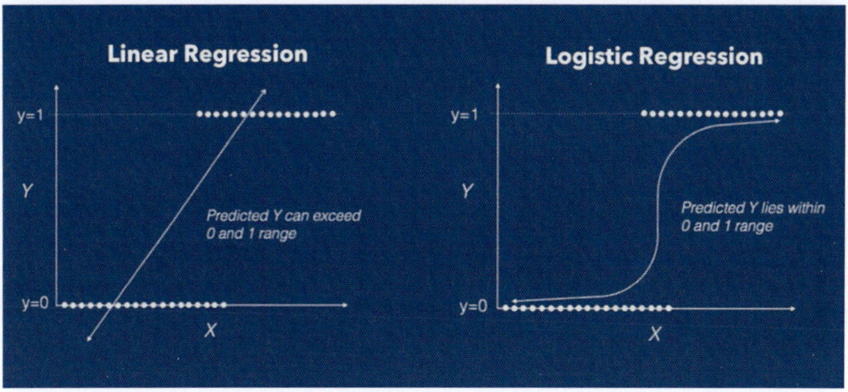

Fig. 1. Difference between graphs (Source: towardsdatascience.com)

Logistic Regression Involved to Solve the Problem
Logistic regression has been widely used in the medical research industry, for predicting disease whether it's a heart disease or a tumor (for e.g. tumor Malignant or Benign) or either cancer with the best accuracy results in the 0–1 format. It may be used to predict the risk of developing a given disease based on observed characteristics of the patients [14].

3.2 Support Vector Machine (SVM)

It is one of the machine learning technique and it obtains form the supervised category based on both classification and regression problems. In other words, support vector machines [SVM] are based on the concept of decision planes which define the boundaries of decision as shown in Fig. 2. A decision plane is one that separates a group of objects with different class memberships and can solve linear and nonlinear issues and function well for many practical issues [15].

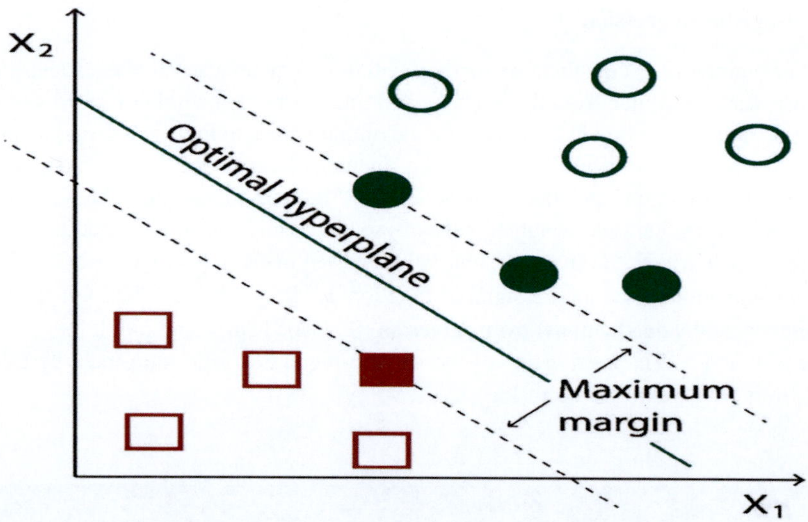

Fig. 2. Support vector machines (Source: mdpi.com)

4 Proposed Model

Steps Involved for Modelling of Dataset [as shown in Fig. 3]:

1) **Data Collection:** In this phase, we can fetch the data from the data source in the form of a CSV file through the various line of codes and processed as an input to the proposed model.

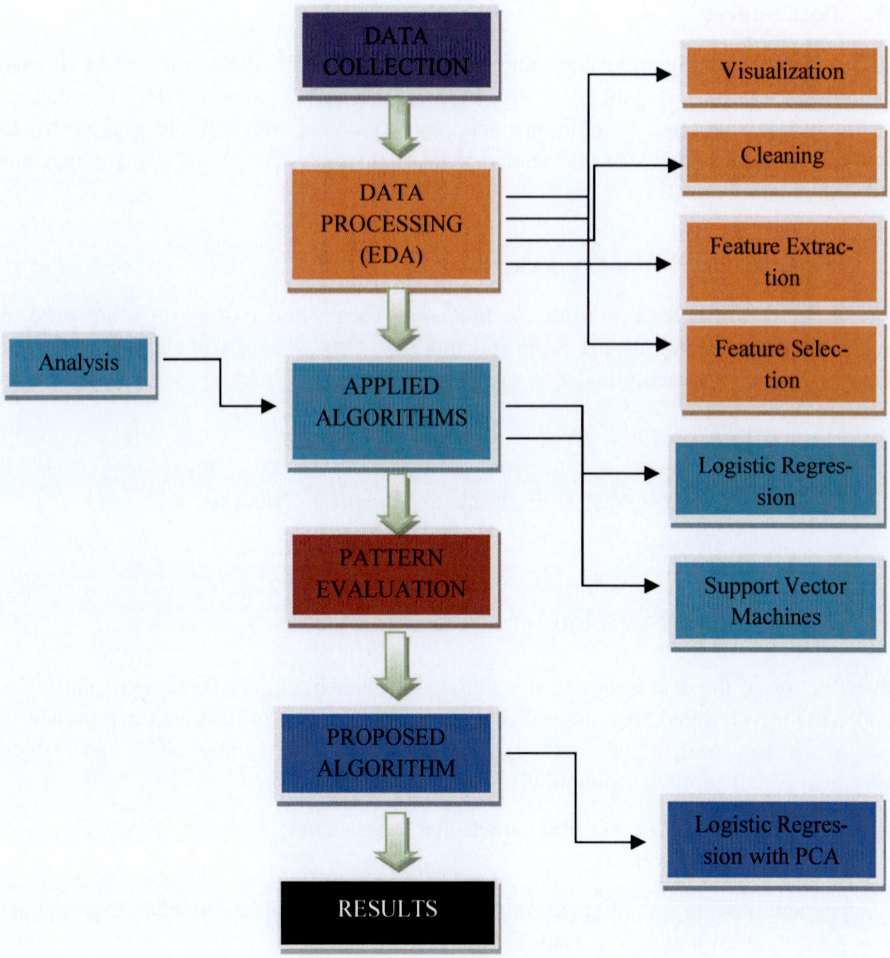

Fig. 3. Proposed Model

2) **Data Pre-processing:** This phase is also known as Exploratory Data Analysis. In this phase, we can concentrate on taking care of missing values, noise reduction, and afterward feature selection process is being carried out for training purposes.
3) **Pattern Evaluation:** In this phase, when you identify interesting patterns that represent knowledge based on some measures. In other words, we can say that performance analysis is performed to evaluate the results.
4) **Output/Results:** In this phase, we can pull out the results which we processed in the analysis part.

4.1 Data Source

UCI machine learning repository dataset is a widely used dataset for heart disease prediction system. It is generally used by machine learning researchers. The dataset contains 304 rows and 14 columns and many medical indexes, the goal is to do exploratory data analysis on the status of heart disease. The 'Num' field which has varying values from 0–1.

4.2 Performance Evaluation Criteria

Four different attributes i.e. accuracy, precision, recall, and F1-measures are used to assess the performance of the proposed models. Then, we can obtain four metrics: accuracy, precision, recall, and F1- score as follows:

$$Accuracy = TP + TN \div TP + FP + TN + FN$$
$$Precision = TP \div TP + FP \quad Recall = TP \div TP + FN$$
$$F1 - Measure = 2 \times Percision \times Recall \div Precision + Recall$$

5 Results and Discussion

Investigation of the data indicated that Oldpeak, Thalach, cp (Asymptomatic pain), CA (>1), Thal (reversible defect) are convenient useful features for forecast the presence of cardiac disease. And the age, exang, slope, trestbps, chol, gender, FBS, and restecg were also found to have a potentially slight predictive power.

Strong Predictive Power Ascribe Concluded in eda Part: Oldpeak, Thalach, cp, CA (>1), Thal (reversible defect)

i. (Patient having > 1) old peak value has more chances of having heart disease than patients with having old peak value (<1).
ii. Those who have a lower heart rate (<140) having the cardiac disease as compared to the patient not having the cardiac disease (>140).
iii. Patients who have to suffer from asymptomatic chest pain must have heart disease.

Average Predictive Power Ascribe Concluded in eda Part: Age, Exang, Slope

i. Those who have aged (>35) having more chances of heart disease.
ii. Groups of people who shared the same characteristics more chances of having chest pain after exercise.
iii. Cohort people have more chances of having a flat ST-wave slope than a non-disease cohort.

Poor Predictive Power Ascribe Concluded in eda Part: Trestbps, Chol, Gender, FBS, Restecg

Table 1. Experimental results

Models	Accuracy	Recall	Specificity	Precision	F1-score
Logistic(all attributes)	87%	.68	.65	.75	.71
Logistic(most significant attributes)	82%	.68	.69	.77	.72
Logistic(removing least significant attributes)	83%	.94	.60	.78	.85
SVM	68%	.79	.58	.65	.71
Logistic(with PCA)	86%	.68	.69	.77	.72

These ascribes have not any forecast power or can't differentiate between disease and non-disease group of people.

6 Conclusion

In this paper, we proposed a new model after applying two models. This model is evaluated on UCI machine learning repository datasets and the aim was to predict if a person has heart disease or not on attributes blood pressure, heartbeat, exang, fbs, and others with better accuracy than other models. Firstly we train logistic regression with all attributes, then we train logistic regression with strong predictive power attributes concluded in eda part, and then in last, we train logistic regression after removing the least significant attributes. Secondly, we are applying support vector machines. And then we proposed a model, logistic regression with principal component analysis. We see a logistic regression model with all the variables and logistic regression model with PCA performed best with an accuracy of 86%, recall 68%, specificity 69%, precision 77%, and f1score 72% [Table 1]. The consequences of the models are whether heart disease is existing or not with different levels of presence.

References

1. W.H Organisation: "New initiative launched to tackle cardiovascular disease, the world number one killer" Intra-Health International (2017)
2. Shen, Z., Clarke, M., Jones, R.: Detecting the risk factors of coronary heart disease by use of neural networks. In: Engineering in Medicine and Biology Society (1993)
3. Subhash, S., Patil, S.: Disease prediction using machine learning over big data. Int. J. Innov. Res. Sci. Eng. Technol. **7** (2018)
4. Ambekar, S., Phalnikar, R: Disease prediction by using machine learning. Int. J. Comput. Eng. Appl. (2018). ISSN 2321-3469
5. Wilson, P.W.F., D'Agostino, R.B., Levy, D., Belanger, A.M.: Prediction of coronary heart disease using risk factor categories. J. Am. Heart Assoc. **97**, 1837–1847 (1998)

6. Amin, S.U., Agarwal, K.: Genetic neural network based data mining in prediction of heart disease using risk factors. In: 2013 IEEE Conference on Information & Communication Technologies (ICT) (2013)
7. Kumar, B.S.: Adaptive personalized clinical decision support system using effective data mining algorithms. J. Netw. Commun. Emerg. Technol. (2018)
8. Stephen, J., Pejaver, V.: Big data in public health: terminology, mach. learning, and privacy. Annu. Rev. Public Health **39**, 95–112 (2018)
9. Raj, J.S., Ananthi, J.V.: Recurrent neural networks and nonlinear prediction in support vector machines. J. Soft Comput. Paradigm (JSCP) **1**(01), 33–40 (2019)
10. Simons, L.A., Simons, J., Friedlander, Y: Risk functions for prediction of cardiovascular disease in elderly Australians: the Dubbo study. Med. J. Aust. (2003)
11. Bashar, A.: Survey on evolving deep learning neural network architectures. J. Artif. Intell. **1** (02), 73–82 (2019)
12. Kumar, B.S.: Data mining methods and techniques for clinical decision support systems. J. Netw. Commun. Emerg. Technol. (JNCET) (2017)
13. Vapnik, V.N., Vapnik, V.: Statistical Learning Theory, vol. 2. Wiley, New York
14. Bharti, S.: Analytical study of heart disease prediction comparing with different algorithms. In: International Conference on Computing, Communication and Automation (ICCA2015) (2015)
15. Burges, J.C.: A tutorial on support vector machines for pattern recognition. Data Min. Knowl. Discov. **2**(2), 121–167 (1998)
16. Baboota, R., Kaur, H.: Predictive analysis and modelling football results using machine learning approach for English premier league. Int. J. Forecast. **35**(2), 745–755 (2019)
17. Kaur, H., Alam, M.A., Jameel, R., Mourya, A.K., Chang, V.: A proposed solution and future direction for blockchain-based heterogeneous medicare data in cloud environment. J. Med. Syst. **42**(8), 1–11 (2018). https://doi.org/10.1007/s10916-018-1007-5
18. Kaur, H., Kumari, V.: Predictive modelling and analytics for diabetes using a machine learning approach. Appl. Comput. Inf. (2018)

Multi-lingual Author Profiling: Predicting Gender and Age from Tweets!

Md. Ataur Rahman[1] and Yeasmin Ara Akter[2(✉)]

[1] Premier University, Chittagong, Bangladesh
ataur.cse@puc.ac.bd
[2] East Delta University, Chittagong, Bangladesh
yeasmin.a@eastdelta.edu.bd
http://puc.ac.bd:8082/Home/Profile?userName=ataur_cse

Abstract. This article describes how we build a multi-lingual classification system for author profiling. We have used Twitter corpus for English, Dutch, Italian and Spanish languages for building different models incorporating SVM classifier that predicts the gender and age of an author. We evaluated each model using 3-fold cross-validation on the training dataset for each of these languages. The overall maximum average accuracy for gender classification was 81.3% for Spanish while for classification of age we achieved a maximum accuracy score of 70.3% for English using the cross-validation scheme. For other languages, the results were between 64–76%.

Keywords: Author profiling · Machine learning · Data mining

1 Introduction

Authorship attribution is the task to identify the author of an anonymous text based on a closed set of candidate authors. It distinguishes between classes of authors by profiling someone based on his/her age or gender [1].

Recent research shows growing importance for author profiling mainly because of its implication in several real-life applications. For example, companies may be interested in knowing the demographics of their target group to achieve a better market segmentation [2] or a forensic analysis could be done over the text to identify a criminal.

Previous works have elucidated that the classification accuracy depends mostly on the choice of features. More fine-grained analysis in this area have shown that word and character n-grams are the best predictor for gender identification [3]. The contribution of these features was also observed by the participants of the recent author profiling shared tasks such as PAN 2018 [4], where several best-performing systems [5–7] used similar features. Rangel et el. [8] showed that tf-idf with n-gram based features, both in terms of characters and words, are very successful in gender perceptions.

J. I.-Z. Chen et al. (Eds.): ICIPCN 2020, AISC 1200, pp. 505–513, 2021.
https://doi.org/10.1007/978-3-030-51859-2_46

The choice of classical machine learning algorithms such as Logistic Regressions (LR) and Support Vector Machines (SVM) [5,9–13] are still prominent in the field of author profiling. Quite surprisingly, instead of employing a more complicated architecture such as BERT [14], a vanilla SVM decked with the character or word n-gram features and minimalist settings can often classify the gender and age attributes from the text with high accuracy [15]. This actually motivated us in order to opt for a more simplistic approach for classification using a linear SVM in our research.

In this paper, we delineate our efforts to build a multi-lingual author profiling classification system. Our endeavour was to develop two kinds of **author profiling models**, one model for **gender** identification on four different languages namely English, Dutch, Italian and Spanish and one model for **age** identification on two different languages English and Spanish.

In Sect. 2 of this study we will describe the details of the data that we have used. In Sect. 3, we start by describing the baseline model for both variants of the classification tasks and also present the approaches for model improvements. In Sect. 4 we present the results of the improved models. Finally we will discuss the results compared to the baseline model in Sect. 5.

2 Data

For this task, we used a subset of the *PAN-2015* corpus [16] comprising a collection of tweets from author files in four different languages which have been annotated for gender and age (age data is only available for English and Spanish). Gender identification is a binary classification task with labels 'F' and 'M' while age is a multi-class classification task with four age groups '18–24', '25–34', '35–49', '50-xx'. There are 24 author files for Dutch, 107 authors for English, 28 authors for Italian and 70 authors for Spanish.

Table 1 shows the distribution of author files and the number of tweets for each languages. We tried to use those data files that consisted of at least 32 tweets (with a maximum of 100 tweets) per author.

Table 1. Distribution of tweets in the training set.

Language	Files	Tweets
English	107	10007
Dutch	24	2377
Italian	28	2425
Spanish	70	6888

2.1 Preprocessing

Since the data was collected automatically, it was not entirely noise-free and we had to perform some preprocessing on the data. For every tweet in the complete dataset, we tokenize the words by using a basic split function that splits on whitespace between words. We removed "*@username*" tokens and the *URL's*. We also performed *lower-casing* of all the tokenized words and filtered out the *stop words* (such as `a`, `of`, `the`, etc.) which occurs regularly in English text and do not contribute that much in classification. These preprocessing techniques improved the overall accuracy of the models quite a bit. We have followed these preprocessing techniques from the previous PAN shared tasks, especially following [8] and [17].

For English, we filtered out *stop words* using the NLTK toolkit [18]. For Dutch, Italian and Spanish we used a python package[1] to filter out the stop words. The other pre-processing technique that we followed is the *stemming* of multiple forms of a single word into the same root, and for this, we have used the Porter's Stemmer [19] from NLTK. Another tool that we experimented with is the WordNet [20] *lemmatizer* from NLTK. Both of these did not improve the performance of our system; this may be due to the reason that for author profiling task it is essential to preserve the word as it is instead of converting it to stem/lemma. So in the end, we decided not to use these two techniques in our system.

3 Methodology

In this section, we are going to describe the author profiling models for the four different languages. First, we will describe the baseline approach that we have used for the evaluation, after that we will discuss about the features we used for model optimization.

3.1 Baseline

As baseline measure, we have used a default linear kernel SVM model without pre-processing the data and only with *word uni-grams* as features for every language. The results for baseline models are presented in Tables 2 and 3.

The formula for calculating average baseline accuracy for the joint identification of gender and age for the available languages are given below:

[1] https://pypi.python.org/pypi/stop-words.

Table 2. Baseline model for `Gender`

Language	Accuracy	F1-macro	Precision	Recall
English	0.618	0.614	0.615	0.615
Italian	0.652	0.604	0.627	0.619
Dutch	0.542	0.517	0.544	0.542
Spanish	0.754	0.744	0.759	0.741

Table 3. Baseline model for `Age`

Language	Accuracy	F1-macro	Precision	Recall
English	0.645	0.456	0.466	0.470
Spanish	0.585	0.511	0.634	0.491

$$\begin{aligned} \overline{gen} &= \frac{gen_{en} + gen_{it} + gen_{nl} + gen_{es}}{4} \\ &= \frac{0.618 + 0.652 + 0.542 + 0.754}{4} \\ &\approx 0.642 \end{aligned} \tag{1}$$

$$\begin{aligned} \overline{age} &= \frac{age_{en} + age_{es}}{2} \\ &= \frac{0.645 + 0.585}{2} \\ &\approx 0.615 \end{aligned} \tag{2}$$

For evaluation, we will compare the results of the above baseline models with the optimized model for each language as well as the average given by Eq. (1–2).

3.2 Classification Models and Features

We took the linear SVM baseline models as a starting point for model optimization. First, we pre-processed the data as described in Sect. 2.1 which noticeably improved the results (2–5% overall improvement). After that, we went to look for model optimization by tuning different parameters and experimenting with multiple features. Among the features we have tried out, `word` and `character n-grams` was prominent and showed better results. We also checked parts of speech (POS) features but it did not improve the results so we restrained ourselves from using it in our final models. For representation purpose of the features, we used `bag-of-words` (BOW) or count, `term frequency-inverse document frequency` (Tf-Idf) and `word length`; either separately or in combination (feature union). The functional block diagram for the gender and age prediction model is illustrated in Fig. 1.

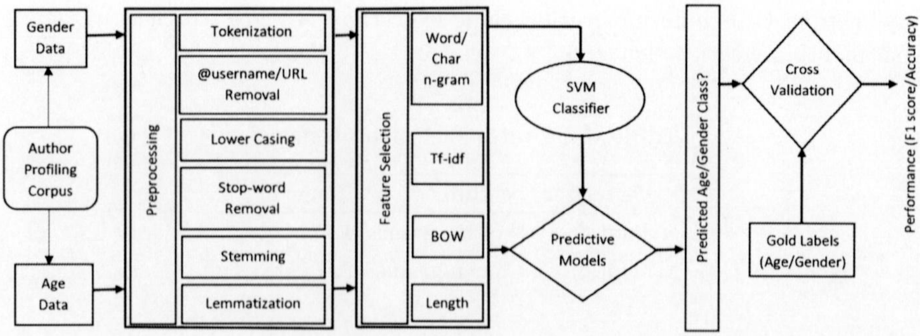

Fig. 1. Block diagram for the Gender and Age prediction models.

In this study we experimented with three different vectorizers by sklearn[2] [21] for the implementation of the aforementioned feature representations. These are commonly known as the vectorizers and we make use of the `TfidfVectorizer`, `CountVectorizer`, and `DictVectorizer` for length features. Finally, we have settled with the combination of TfidfVectorizer and CountVectorizer while we aim to find the optimal parameter settings for each language.

Table 4. Features for English - Gender and Age

Parameters	Gender	Age
Vectorizer	Tfidf+Count	Tfidf
Feature	Word + Char n-grams	Word + Char n-grams
Range	(1,2) + (3,5)	(1,2) + (3,5)

Table 5. Features for Spanish - Gender and Age

Parameters	Gender	Age
Vectorizer	Tfidf+Count	Tfidf
Feature	Word + Char n-grams	Word + Char n-grams
Range	(1,2) + (3,5)	(1,2) + (3,5)

Table 4, 5, 6 and 7 illustrates the usage of different feature combined with Tfidf and CountVectorizer. We have combined word and character n-grams with different ranges. This modification from the default word unigram settings improved the results by 5–8% for different languages regarding gender and age

[2] Source scikit-learn - http://scikit-learn.org/.

classification. Our main motivation came from the *PAN 2017* article [17] while experimenting with the features.

Table 6. Features for Dutch - Gender

Vectorizer	Feature	Range
Tfidf	Word n-grams	1, 2
Tfidf	Char n-grams	2, 7

Table 7. Features for Italian - Gender

Vectorizer	Feature	Range
Tfidf	Word n-grams	2, 2
Tfidf	Char n-grams	2, 7

4 Results and Evaluation

As mentioned earlier, we have evaluated our systems by cross-validation over the training dataset. A 3-fold cross-validation scheme was used, and the macro average of accuracy, precision, recall and f1 scores are shown as the output of our system. The results for different age and gender classification for the available languages are shown below in the Table 8 and 9.

Table 8. Cross validation scores (macro avg.) for `Gender`

Language	Accuracy	F1-macro	Precision	Recall
English	0.747	0.745	0.750	0.744
Italian	0.759	0.703	0.704	0.739
Dutch	0.667	0.657	0.700	0.667
Spanish	0.813	0.811	0.821	0.813

We calculated accuracy separately for each language regarding gender, and age class. The highest result for `Gender` classification was obtained in Spanish with an accuracy of 81.3%, and the lowest accuracy was 66.7% for Dutch.

For the class `Age`, highest was 70.3% for English and the lowest accuracy was obtained from 64.4 for Spanish. Dutch and Italian data did not consist of Age labels.

Table 9. Cross validation scores (macro avg.) for Age

Language	Accuracy	F1-macro	Precision	Recall
English	0.703	0.565	0.665	0.564
Spanish	0.644	0.535	0.809	0.504

We obtained the accuracy for the joint identification of gender and age for the available languages by averaging the results obtained per language with the following two formula (Eq. 3–4):

$$\begin{aligned}\overline{gen} &= \frac{gen_{en} + gen_{it} + gen_{nl} + gen_{es}}{4} \\ &= \frac{0.747 + 0.759 + 0.667 + 0.813}{4} \\ &\approx 0.75 \end{aligned} \tag{3}$$

$$\begin{aligned}\overline{age} &= \frac{age_{en} + age_{es}}{2} \\ &= \frac{0.703 + 0.644}{2} \\ &\approx 0.674 \end{aligned} \tag{4}$$

From Eq. (3) and Eq. (4), we have the final combined average accuracy of our system given in Eq. (5):

$$\begin{aligned} Accuracy_{combined} &= \frac{\overline{gen} + \overline{age}}{2} \\ &= \frac{0.75 + 0.674}{2} \\ &= 0.712 \end{aligned} \tag{5}$$

If we compare the baseline results with the optimized results, we can conclude that we have improved our author profiling model for **gender** identification with an average of 10.8% approximately on the four languages. Again if we compare the results for the baseline model with the optimized model for **age** identification, we could see that we have improved the model's accuracy by nearly 5.9%. Note that, in our results we also found similar patterns of improvements for each of the other metrics (*f1*, *precision* and *recall*).

After careful analysis, we noticed that while applying the pre-processor for *age* classification, the performance gets reduced by 2–5%. Thus, we did not apply the preprocessing step for age classification. We assume that younger people are more likely to use hyperlinks, symbols, a sequence of punctuation, and other kinds of informal writing style than the adult, and these could act as a feature in order to determine the age range.

5 Discussion and Conclusion

We tried to build a system that can predict the gender and age of an author based on the training data provided in the form of text for four different languages. Although we wanted to explore other machine learning classifiers such as neural nets, the intuition of selecting Linear SVM for the classification mainly came from the literature we analyzed before starting to approach the problem. The integration of different preprocessing methods and feature selections has been made by trial and error basis. We tried almost all of the preprocessing and feature selection techniques that we came across from the previous works and only kept those that contributed positively to the results. We separated four different models for four different languages, primarily because of the features and preprocessing techniques we have implemented. We build models for English and Spanish to make age and gender identification. For Dutch and Italian, we only performed gender identification.

Finally, the evaluation was done based on cross-validation, and the average accuracy was approximately 75% for gender classification and 67.4% for the task of age identification for all the languages combined. Possible (and maybe most interesting) future work could be experimenting with other parameter settings for feature extraction, use of different classification algorithms and kernels, experiment with word embedding features. It could also be interesting to look at the performance using a model that is being trained in a multilingual setting by combining all the languages.

References

1. Rangel, F., Rosso, P., Koppel, M., Stamatatos, E., Inches, G.: Overview of the author profiling task at PAN 2013. In: CLEF Conference on Multilingual and Multimodal Information Access Evaluation, pp. 352–365. CELCT (2013)
2. Rangel, F., Rosso, P., Chugur, I., Potthast, M., Trenkmann, M., Stein, B., Verhoeven, B., Daelemans, W.: Overview of the 2nd author profiling task at PAN 2014. In: CLEF 2014 Evaluation Labs and Workshop Working Notes Papers, Sheffield, UK, 2014, pp. 1–30 (2014)
3. Malmasi, S., Zampieri, M., Ljubešić, N., Nakov, P., Ali, A., Tiedemann, J.: Discriminating between similar languages and Arabic dialect identification: a report on the third DSL shared task. In: Proceedings of the Third Workshop on NLP for Similar Languages, Varieties and Dialects (VarDial3), pp. 1–14 (2016)
4. Rangel, F., Rosso, P., Montes-y Gómez, M., Potthast, M., Stein, B.: Overview of the 6th author profiling task at PAN 2018: multimodal gender identification in Twitter. Working Notes Papers of the CLEF (2018)
5. Tellez, E.S., Miranda-Jiménez, S., Moctezuma, D., Graff, M., Salgado, V., Ortiz-Bejar, J.: Gender identification through multi-modal tweet analysis using microtc and bag of visual words. In: Proceedings of the Ninth International Conference of the CLEF Association (CLEF 2018) (2018)
6. Daneshvar, S., Inkpen, D.: Gender identification in Twitter using n-grams and LSA. In: Proceedings of the Ninth International Conference of the CLEF Association (CLEF 2018) (2018)

7. Nieuwenhuis, M., Wilkens, J.: Twitter text and image gender classification with a logistic regression n-gram model. In: Proceedings of the Ninth International Conference of the CLEF Association (CLEF 2018) (2018)
8. Rangel, F., Rosso, P., Verhoeven, B., Daelemans, W., Potthast, M., Stein, B.: Overview of the 4th author profiling task at PAN 2016: cross-genre evaluations. In: Working Notes Papers of the CLEF 2016 Evaluation Labs. CEUR Workshop Proceedings/Balog, Krisztian [edit.]; et al, pp. 750–784 (2016)
9. Aragón, M.E., López-Monroy, A.P.: A straightforward multimodal approach for author profiling. In: Proceedings of the Ninth International Conference of the CLEF Association (CLEF 2018) (2018)
10. López-Santillán, R., Gonzalez-Gurrola, L., Ramfrez-Alonso, G.: Custom document embeddings via the centroids method: gender classification in an author profiling task. In: Proceedings of the Ninth International Conference of the CLEF Association (CLEF 2018), vol. 2125 (2018)
11. Ciccone, G., Sultan, A., Laporte, L., Egyed-Zsigmond, E., Alhamzeh, A.,Granitzer, M.: Stacked gender prediction from tweet texts and images note book for PAN at CLEF 2018 (2018)
12. Patra, B.G., Das, K.G., Das, D.: Multimodal author profiling for Twitter. Notebook for PAN at CLEF (2018)
13. Veenhoven, R., Snijders, S., van der Hall, D., van Noord, R.: Using translated data to improve deep learning author profiling models. In: Proceedings of the Ninth International Conference of the CLEF Association (CLEF 2018), vol. 2125 (2018)
14. Devlin, J., Chang, M.W., Lee, K., Toutanova, K.: BERT: pre-training of deep bidirectional transformers for language understanding. In: Proceedings of the 2019 Conference of the North American Chapter of the Association for Computational Linguistics: Human Language Technologies, Volume 1 (Long and Short Papers), pp. 4171–4186. Association for Computational Linguistics, Minneapolis, Minnesota (2019)
15. Basile, A., Dwyer, G., Medvedeva, M., Rawee, J., Haagsma, H., Nissim, M.: Simply the best: minimalist system trumps complex models in author profiling. In: International Conference of the Cross-Language Evaluation Forum for European Languages, pp. 143–156. Springer, Cham (2018)
16. Rangel Pardo, F.M., Celli, F., Rosso, P., Potthast, M., Stein, B., Daelemans, W.: Overview of the 3rd author profiling task at PAN 2015. In: CLEF 2015 Evaluation Labs and Workshop Working Notes Papers, pp. 1–8 (2015)
17. Rangel, F., Rosso, P., Potthast, M., Stein, B.: Overview of the 5th author profiling task at PAN 2017: gender and language variety identification in twitter. Working Notes Papers of the CLEF (2017)
18. Loper, E., Bird, S.: NLTK: the natural language toolkit. arXiv preprint cs/0205028 (2002)
19. Porter, M.F., et al.: An algorithm for suffix stripping. Program **14**(3), 130–137 (1980)
20. Miller, G.A.: Wordnet: a lexical database for english. Commun. ACM **38**(11), 39–41 (1995)
21. Pedregosa, F., Varoquaux, G., Gramfort, A., Michel, V., Thirion, B., Grisel, O., Blondel, M., Prettenhofer, P., Weiss, R., Dubourg, V., et al.: Scikit-learn: machine learning in python. J. Mach. Learn. Res. **12**(Oct), 2825–2830 (2011)

Implementation of Black Box System for Accident Analysis Using Raspberry Pi

Pravin Kumbhar[1](✉), Sudhakar R. Barbade[1,2,3,4], Utkarsh H. Jain[2], Chetan L. Chintakind[3], and Aditya Harish Barhanpurkar[4]

[1] 19/2 Navsamata Hosing Society, Solapur, India
[2] 16/302 Karanja Road, Pingle Colony, Beed, India
[3] 10/20 Bhavani Peth, Ghongade Vasti, Solapur, India
[4] C/13 Smruti Vihar Colony, Sidheshwar Nagar, Solapur, India

Abstract. The Car Black Box system is one type of data service in the vehicle. Blackbox system is used to analyze the data cause of the accident. In the black box system we are used various types of sensors named collision sensor, vibration sensor, Tilt Sensor, Temperature & humidity sensor, gas sensor, smoke sensor, Alcohol sensor, IR sensor, etc. The Raspberry Pi processor are used to control all sensors and sensing data sent to the IoT cloud after a car accident. Also, we have used Raspberry pi camera for video recording & capture photos inside the vehicles and it is stored in SD cards. The GPS is used for the location of the vehicle and we get the exact location link in cloud to analyze. Also, we have built a wireless WIFI car using NodeMcu with a motor driver and controlled on device android app and that Black box fixed with wifi.

Keywords: Raspberry PI 3 · Black box · GPS module · NodeMcu · IOT cloud computing accident analysis · Sensors

1 Introduction

As per the World Health Organization (WHO), more than one million individual peoples are losing their lives from street mishaps. A black box is an electronic device used to record instructions sent to any electronics system in a vehicle. It records specific parameters of vehicle performance using sensors. The data of all the sensors in the vehicle stored in IoT Cloud. All the cloud data displayed in the IoT app. The black box plays the job of specialists to decide the reason and ensuing investigation of the mishap, assisting with proposing measures to secure lives. Programmed fender bender discovery, or ecall frameworks, can diminish reaction time. Therefore, numerous lives could be spared, and their reality of Injuries can be diminished. Kind of mishap - Location of mishap (longitude and scope) - Time of mishap Speed. The proposed framework utilizes GPS innovation to figure the specific longitude and scope directions of the vehicle and send them through the web of things (IoT) firebase cloud for crisis administrations. Additionally uses to oversee gathering utilizing Raspberry Pi increasing speed and numerous other data Different kinds of sensors (Fig. 1).

J. I.-Z. Chen et al. (Eds.): ICIPCN 2020, AISC 1200, pp. 514–525, 2021.
https://doi.org/10.1007/978-3-030-51859-2_47

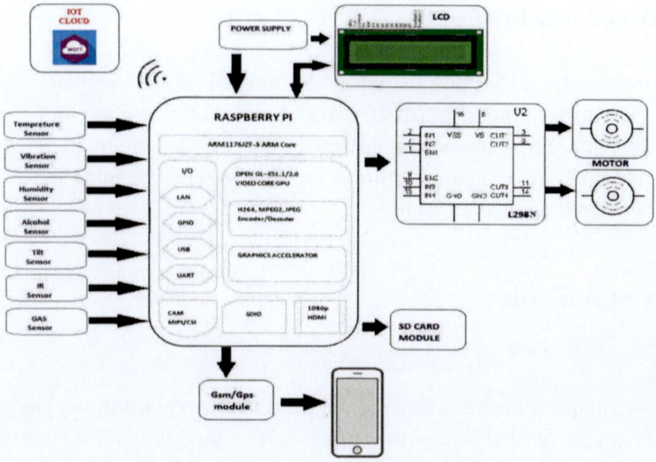

Fig. 1. Block diagram of black box

2 Proposed System

The discovery framework introduced in this paper can diminish reaction time by creating a programmed cautioning message, demonstrating the area of the vehicle to be utilized GPS gadget. This message will be sent to the firebase cloud who will settle on it Black-box crash identification and area framework depends on Raspberry Pi Followers ought to be cautioned. Since the time factor is necessary for this sort if there should be an occurrence of crisis, this procedure will spare significant time. The proposed framework design is in a rush-hour gridlock circumstance Accident, an accident detection calculation (ADA) will oversee different information sources Find the accident and request the Raspberry Pi, which thusly will deliver and send an admonition message to the firebase cloud (Fig. 2).

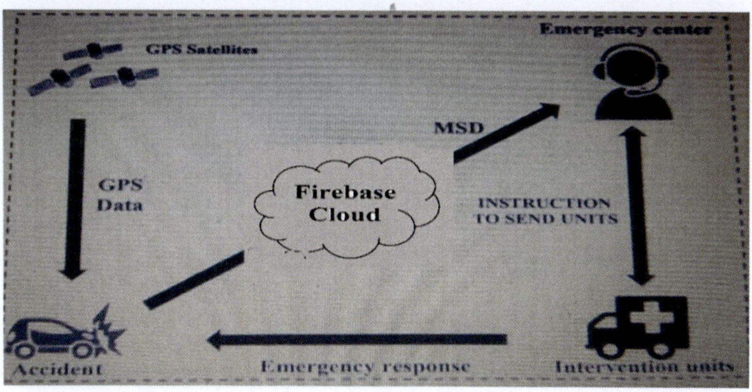

Fig. 2. Proposed system.

3 Black Box Solutions

Lamentably, most vehicles in nations where the normal time of vehicles is excessively high There is no programmed crash discovery and notice framework and the right arrangement for this situation is the utilization of the arrangement discovery eCall arrangement, Deployable gear, including individual, devoted equipment for detecting, lo-cation and correspondence capacities.

4 Figures and Tables

Flow Chart
The Fig. 3 flow diagram of the system which describes its workflow. The black box is used in car for accident analysis in this black box if an accident occurs then the first search the black box is analyze the data of cause of the accident and send data to the cloud to store the data for analysis after the accident and to inform the emergency service center. When the system starts then the black box is working automatically. İf car or vehicle is normal then no sensors are initialized. İf car is crash or accident then the siren will be activated and sensors start the sensing and send data to IoT Cloud for store the data for analysis. These data displayed on the android app and system will be stopped.

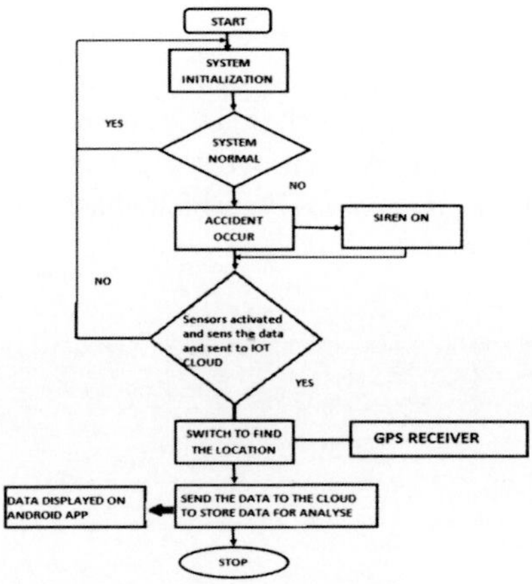

Fig. 3. Algorithm flow

Raspberry Pi OS monitors the car speed, alcohol detection and transmits those data logs through MQTT protocol. If the accident occurs then it is identified by analyzing the catalogs which we obtained from the MQTT protocol. MQTT consists of three layers Publisher, Broker, and Subscriber. If the Subscriber is connected in the cen-tralized node called Broker, data get stored in that node and is sent to the subscriber when the subscriber comes into the network. The advantage of MQTT is a lightweight protocol, even in low internet speed, it can send the data (Fig. 4).

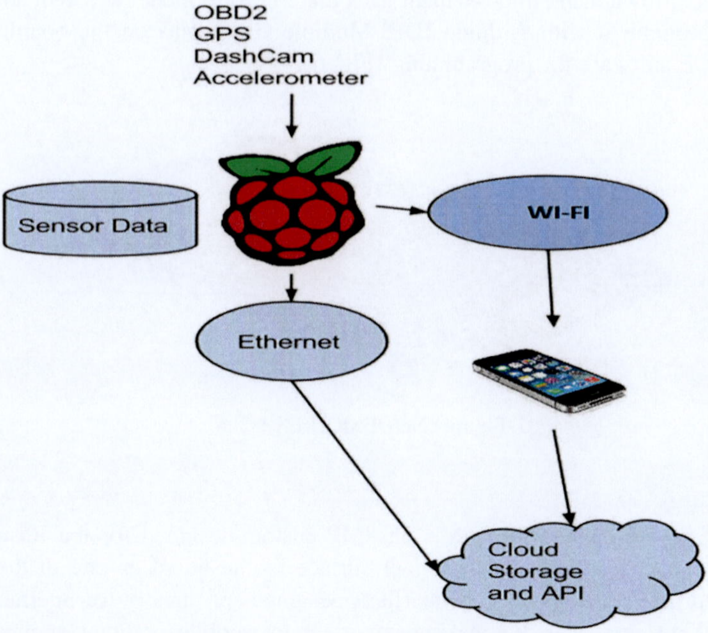

Fig. 4. General block diagram

5 Hardware Resources

Raspberry Pi 3

Raspberry PI 3 is a little PC with solitary board. The recurrence to the job is 1.2 GHz. The Raspberry PI product must be put into the SD card installed thereon. It has 4 USB ports. The source of force can be supplied by means of a 5 V Micro USB. The Raspberry PI is around 85.6 mm * 56.5 mm long. The Raspberry PI is customized using language Python (Fig. 5).

Fig. 5. Raspberry PI 3

Node MCU (ESP8266)

The NODEMCU is utilized to create an IoT application with less Integrated circuits rapidly. NODEMCU is a minimal effort Wi-Fi microchip. NodeMCU is additionally an open-source firmware and improvement pack that serves to model IoT item and we can generally program it with Arduino IDE. Multiple GPIO pins on the board. We use Arduino IDE software for programming (Fig. 6).

Fig. 6. NODEMCU ESP8266

PI CAMERA

The Raspberry Pi camera module is an 8MP custom-designed for the Raspberry Pi module. It attaches the pie to the upper surface of the board as one of the smaller sockets and uses a dedicated CSI interface, designed specifically for interfacing with cameras. It weights over 3G making it perfect for mobile or other applications. It connects to the Raspberry Pi via a short ribbon cable (Fig. 7).

Fig. 7. PI camera

GPS for Accident Detection System

GPS helps in both tracking and navigation purpose. The tracking system is used to keep track of the vehicle without the driver's intervention. But the system navigation directs the driver to reach the destination without any interference. Uses both tracking and navigation of the same architecture. Tracking as an Accident Occurs the stem detects the accident-prone vehicle and has a message Sent to the Cloud to analyze (Fig. 8).

Fig. 8. GPS receiver

5.1 Sensors Used in System

5.1.1 Alcohol Sensor

Liquor sensor is utilized to detect the nearness of liquor in air. On the off chance that the individual in the vehicle is high on liquor, at that point, it very well may be detected with the assistance of this sensor. MQ6 liquor sensor is one of the simples to utilize sensors that can be associated with the Raspberry pi. MQ6 gas sensor has high affability to liquor. It is a minimal effort sensor that can be utilized for different applications as shown in below Fig. 9.

Fig. 9. Alcohol sensor

5.1.2 Temperature and Humidity Sensor

The DHT11 is an advanced monitor of temperatures and humidity. This uses a humidity sensor and a thermistor to gage the ambient air and gives the knowledge pin an advanced symbol. It's easy to use, but it takes careful time to obtain information (Fig. 10).

Fig. 10. Temreture sensor DHT11

5.1.3 Vibration Sensor

In the systems described above, a vibration sensor is indispensable to detect the knock position of an internal combustion engine, namely, to detect engine vibration with knocking. Traditionally, a round-shaped piezoelectric vibration element has been used for this vibration sensor (Fig. 11).

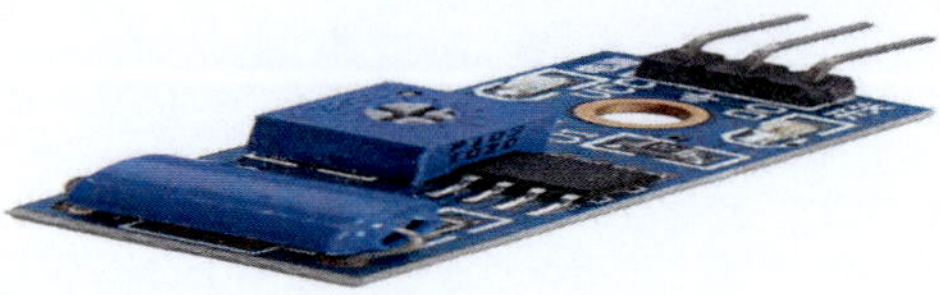

Fig. 11. Vibration sensor

5.1.4 IR Sensor

An infrared sensor is an electronic gadget that is utilized to see a few attributes of its environment. It is by emanating or distinguishing infrared radiation. Infrared sensors in estimating the warmth discharged by an item and recognizing movement (Fig. 12).

Fig. 12. IR sensor

5.1.5 Tilt Sensor

The tilt sensor is an instrument used to quantify the tilt in the various hub of the reference plane. Tilt sensors measure tilt conditions regarding gravity and are utilized in numerous applications. Like mercury switches, they can likewise be known as tilt switches or moving ball sensors (Fig. 13).

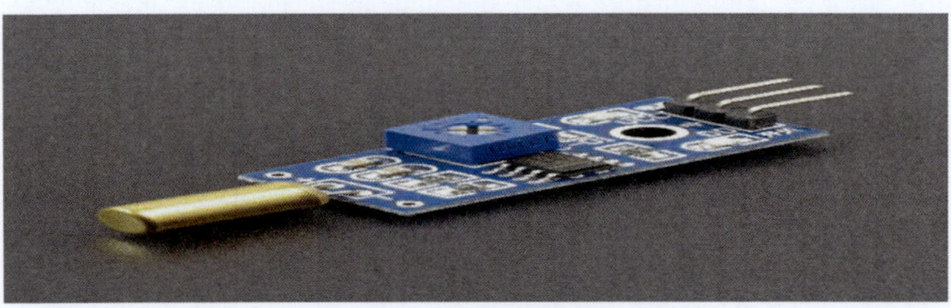

Fig. 13. TILT sensor

6 Accident Detection Algorithm

In this paper, the phase of vehicle mishap location depends on the utilization of different sensor relies upon the kind of mishap. The tilt sensor is utilized to identify in the event of vehicle impact or tilt, vehicle framework likewise has a fire sensor utilized for separately consuming. The accident area stage will at that point be actuated to get the specific position and time of a mishap utilizing GPS innovation, this data will be sent to crisis benefits just as kind of mishap through firebase cloud. A flowchart of the proposed calculation for the mishap location system appears in Fig. 3.

7 Firebase Cloud

Google Firebase is an technology development programming sponsored by Google that empowers designers to build iOS, Android, and web applications such as cloud information and cloud messaging. Firebase Cloud Messaging (FCM) is a cross-stage messaging platform that enables companies to get and distribute free-of-cost messages on iOS, Android, and the Internet (Fig. 14).

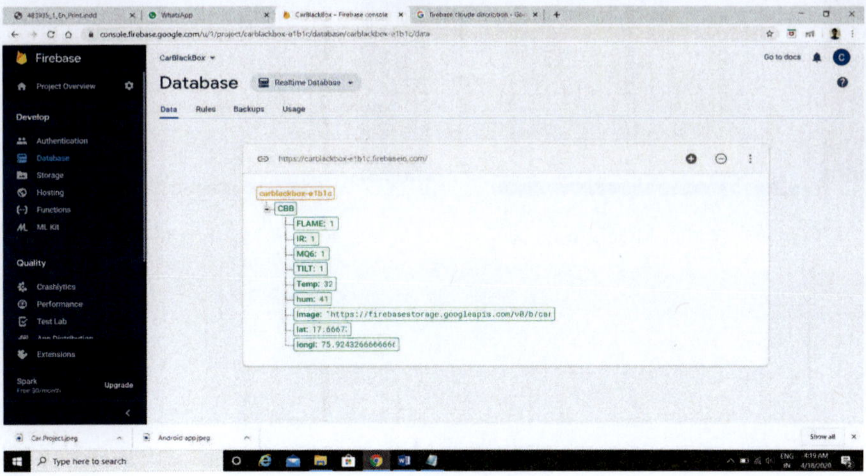

Fig. 14. Firebase cloud

8 Result and Discussion

A vigorous accident recognition and area framework has been created and presented.

This paper, an accident discovery calculation has been executed and tried all together Locate your different kinds of a car crash and send it to the area got through its GPS framework utilizing a firebase cloud. Figure 15 shows the trial of the Tilt sensor utilizing the framework model by applying solid vibration. when Fig. 16 Shows the trial of the IR sensor. Figure 17 shows the Flame sensor and Fig. 18 show the trial of the vehicle Tilt. The framework sends a short message administration to the preselected number containing significant data, for example, the mishap type, time and the specific area of the mishap, the message design is portrayed in Fig. 19.

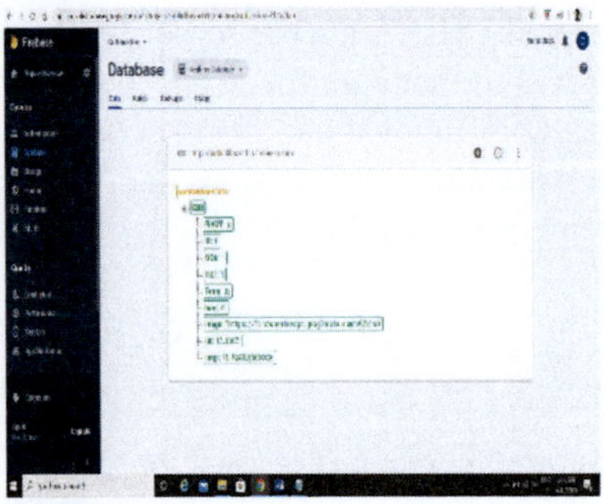

Fig. 15. Testing vibration sensor

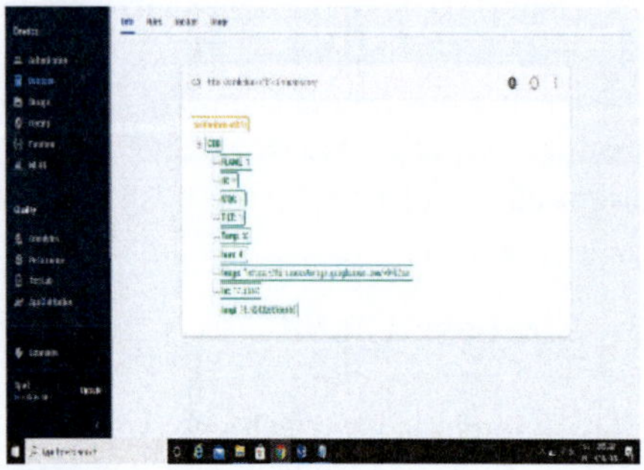

Fig. 16. Testing IR sensor

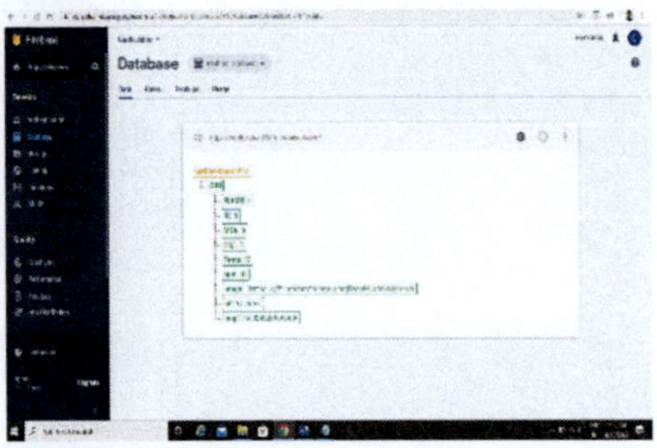

Fig. 17. Testing MQ6 sensor

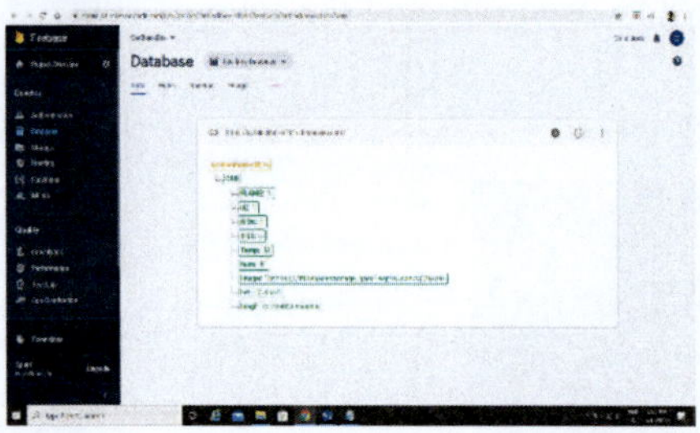

Fig. 18. Testing tilt sensor

9 Android App with GPS Location

See Fig. 20.

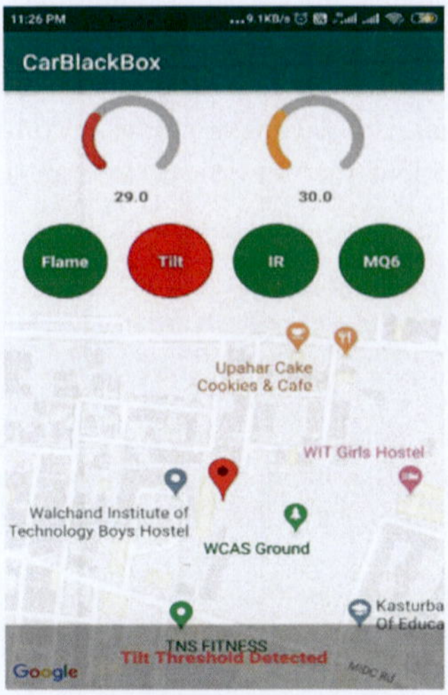

Fig. 19. Android app

Fig. 20. Output of sensors data

10 Implement of Black Box System

See Fig. 21.

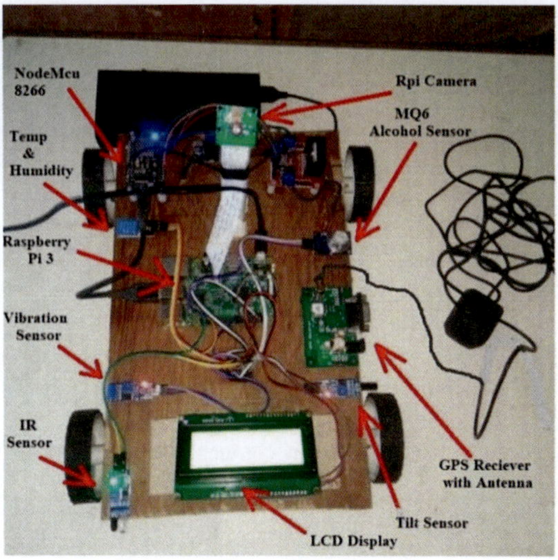

Fig. 21. Black box system

11 Conclusion

In this venture, we planned the IoT based Black Box system and executed it effectively. In this article, we presented a mishap identification and area framework, which Provides a hearty discovery system for street accident scratches utilizing a fast announcing technique using the GPS and firebase cloud advancements. The primary commitment to our framework is focused on utilizing Raspberry Pi as a principle framework controller. Reaction time the framework doesn't surpass 20 s which is a decent outcome. In future work, we are arranging Add progressively exact accident recognition procedure and test the framework to mimic genuine vehicle mishaps.

References

1. Meena, A., Iyer, S., Nimje, M., Jogjekar, S., Jagtap, S., Rahman, M.: Automatic accident detection and reporting framework for two wheelers. In: IEEE International Conference on Advanced Communication Control and Computing Technologies (ICACCCT), pp. 962–967 (2014)
2. Amin, S., Bin, M., Reaz, I., et al.: Low cost GPS/IMU integrated accident detection and location system. Indian J. Sci. Technol. **9**(10), 1–9 (2016). https://doi.org/10.17485/ijst/2016/v9i10/80221
3. Lamza, A., Wrobel, Z.: New effective strategy for computerized video adjustment for in-auto camera. In: Fifth International Conference on Multi-Media Communications Services and Security, vol. 287, pp. 180–187, June 2012
4. Kremonas, P., Pâris, J.: Everything you wanted to ask, but did not know how…. In: The European Emergency Number Association (EENA) (2015)
5. El-Rabban, A.: Introduction to GPS: The Global Positioning System, 2nd edn. Artech House, Norwood (2009)
6. World Health Organization, GLOPAL STATUS REPORT ON ROAD SAFETY, Geneva (2009)
7. Wireless Medical Technologies: A Strategic Analysis of Global Markets International Telecoms Intelligence. http://www.itireports.com
8. Hung, K., Zhang, Y.T., Tai, B.: Wearable medical devices for telehome healthcare. In: Process. 26th Annual International Conference on the IEEE EMBS (2004)
9. White, J., Thompson, C., Turner, H., Dougherty, B., Schmidt, D.C.: WreckWatch: automatic traffic accident detection and notification with smartphones. In: Springer Science + Business Media, LLC, pp. 285–303 (2011). https://doi.org/10.1007/s11036-011-0304-8
10. Wang, X., Guo, J., Hu, W., Fan, X.: Design and ımplementation of embedded web server based on ARM and Linux. Micro Comput. Inf. **23**(5–2) (2007)
11. Bonyár, A., Géczy, A., Krammer, O., Sántha, H., Illés, B., Kámán, J.: A review on current eCall systems for autonomous car accident detection. In: 40th International Spring Seminar on Electronics Technology (ISSE), pp. 1–8 (2017)
12. Sugi, T., Nakamura, M., Ide, J., Shibasaki, H.: Modeling of motor control on manual tracking for developing a hand movement compensation technique. Artif. Life Robot. **7**, 112–117 (2009)
13. Inductive proximity sensor HYP-18RL8P. http://www.hyelec.co.kr
14. Int. J. Innov. Sci. Mod. Eng. (IJISME) **2**(11) (2014). ISSN 2319-6386

15. Rahman, A.A., Natori, K., Ohnishi, K.: Disturbance decomposition of time delay system by shadow robot based on network disturbance concept. In: IEEE International Conference on Industrial Technology, pp. 1120–1125 (2009)
16. Tan, Y., Cauwenberghe, A.V.: Nonlinear neural controller with neural Smith predictor. Neural Process. Lett. **1**(2), 24–27 (1994)
17. Ajay Kumar Reddy, P., Dilip Kumar, P., Bhaskar Reddy, K.: Black box for vehicles. Int. J. Eng. Invent. **1**(7) (2012). ISSN 2278-7461

CORONA-19 NET: Transfer Learning Approach for Automatic Classification of Coronavirus Infections in Chest Radiographs

Sukkrit Sharma(✉), Smaranjit Ghose, Suhrid Datta, C. Malathy, M. Gayathri, and M. Prabhakaran

SRM Institute of Science and Technology, Chennai, Tamil Nadu, India
sukritsharma@gmail.com, smaranjitghose@protonmail.com, suhriddatta99@gmail.com, malathyc03@gmail.com, gayathrm2@srmist.edu.in, Prabhakaran_mathialagan@srmuniv.edu.in

Abstract. Coronavirus disease has been rampaging the world since its onset in the Wuhan region of China with cases skyrocketing every day. A crucial step for mitigating the havoc in this situation is the early screening of the infected patients and isolating them. Given the overwhelming number of people falling prey to this pandemic, it becomes very difficult for hospitals to provide these services within the required time. The power of Artificial Intelligence can be leveraged to solve the issue of delay in the diagnosis of coronavirus by automating the process. In this study, we propose our method, a transfer learning approach on deep convolutional neural networks to classify chest radiographs of patients as healthy or infected with Coronavirus based on the fact that Coronavirus attacks the epithelial cells of lungs resulting in pneumonia. The data used for this study has been curated from various public resources online. Our proposed model was able to achieve a validation score of 100% on the curated dataset along with high specificity and sensitivity of 1.00. This work aims to serve as a first cut solution to the automated diagnosis and doesn't claim to be the exact solution without further validation by medical professionals.

Keywords: COVID19 · Coronavirus · Deep convolutional neural networks · Transfer learning · Chest radiography

1 Introduction

Coronavirus, since it's onset, has caused a pandemic throughout the world. The ones who succumb to this are stricken with acute respiratory disorders which have seen to turn fatal in many cases. One of the measures to curb the damage caused by this is the early diagnosis of the infected individuals so that they can be given proper medical care as well as isolated from the rest for preventing further spread of the virus. The primary process for the diagnosis involves polymerase chain reaction testing. However, owing to the complexity of the procedure and lack of adequate experts it is not preferable to meet the immediate diagnostic requirements of the patients.

J. I.-Z. Chen et al. (Eds.): ICIPCN 2020, AISC 1200, pp. 526–534, 2021.
https://doi.org/10.1007/978-3-030-51859-2_48

Adding on, the limited supply of the test kits for it makes the situation even worse. A possible substitute to this is making use of chest radiography like CT Scans [2, 7] and X-Rays. However, the problem persists since there is a shortage of radiologists for the current situation. Hereby the concept of deep convolutional neural networks comes handy to build an image classifier that could extract meaningful features from the chest radiographs and differentiate between corona infected [12] and healthy ones [8]. Posteroanterior radiographs of chests are used due to their greater availability for model training and previous results obtained that support its use for the detection of pneumonia.

In our experiment, we propose Corona-19 Net, which is a model built on top of MOBILE-NET V2 [13] architecture owing to its modularity and composite design, we were able to achieve promising results in terms of validation accuracy, sensitivity, and specificity scores. The model is tuned to detect infections due to Coronavirus and thus the manifestations of the virus with dexterity. The modular design and light weightedness of the architecture further facilitates the easy deployment of it to client-side web applications to develop automatic diagnosis systems which can easily be used in hospitals (Fig. 1).

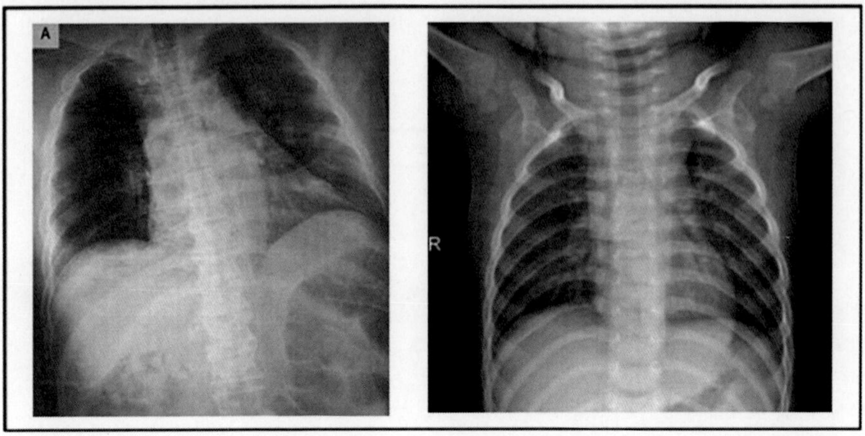

Fig. 1. Posteroanterior radiographs of Chest showing (i) Healthy patients and (ii) COVID-19 infected patients.

2 Methods

2.1 Dataset Description

The dataset used for this study comprises:

- 68 chest radiograph images of healthy patients curated from the RSNA Pneumonia Detection Challenge on Kaggle [1].
- 69 Covid-19 chest radiograph images of COVID-19 infected images curated from COVID-19 Database, SIMA, and Radiopedia [4].

Due to the small size of the dataset available various preprocessing and augmentation methods were performed to obtain a large dataset to enable the model to perform proper feature extraction and learn better from its features.

2.2 Model Architecture

To obtain an accurate image-classification model for the chest radiographs, we used a transfer-learning approach rather than training a deep convolutional neural network [14] from scratch. This is because for our study we had to compensate for the low quality and a limited number of images curated from public datasets. Transfer learning aids to leverage the previous knowledge of feature extraction that was obtained by training on large benchmark datasets like ImageNet [5] or COCO.

For our experiment, we primarily make use of MobileNetV2 architecture which com-prises two categories of blocks. There is a residual block of stride 1. While for downsizing another block of stride 2 is used. Each type of block has 3 layers. 1 × 1 convolution with ReLU6 is the first layer this time. The second layer is the depthwise convolution. Again 1 × 1 convolution is used in the third layer without non-linearity. Using ReLU again in the deep network has the power of a linear classifier on the non-zero volume part of the output domain. The depthwise separate convolution layers result in a significant decrease in model complexity. While smaller model size aids in faster training and better accuracy [10, 11] (Fig. 2).

Input	Operator	t	c	n	s
$224^2 \times 3$	conv2d	-	32	1	2
$112^2 \times 32$	bottleneck	1	16	1	1
$112^2 \times 16$	bottleneck	6	24	2	2
$56^2 \times 24$	bottleneck	6	32	3	2
$28^2 \times 32$	bottleneck	6	64	4	2
$14^2 \times 64$	bottleneck	6	96	3	1
$14^2 \times 96$	bottleneck	6	160	3	2
$7^2 \times 160$	bottleneck	6	320	1	1
$7^2 \times 320$	conv2d 1x1	-	1280	1	1
$7^2 \times 1280$	avgpool 7x7	-	-	1	-
$1 \times 1 \times 1280$	conv2d 1x1	-	k	-	

Fig. 2. Synopsis of our underlying MobileNetV2 architecture.

The table depicts the arrangement of the bottlenecks where column t shows the expansion rate of the channels, column c represents the number of input channels, column n represents the frequency of repetition of the block while s indicates if the first repetition of a block using a stride of 2 for the downsampling process. For further customization of the architecture, the output obtained is fed to an average pooling layer comprising the kernel filter of size 4 × 4 to downsample the input and then passed to the flatten layer for converting a matrix of features into vectors. The vectors are then fed to a dense layer where relu activation function is used again to introduce non-linearity. For the avoidance of overfitting, the output is fed to the dropout layer. In the end, a dense layer is added to classify the output with a softmax activation function. The learning rate of our model is set to 5e−5 with a batch size of 16 and trained for a cycle of 100 epochs.

3 Results

3.1 Model Evaluation

To evaluate the robustness of our proposed architecture, we take into consideration a thorough and versatile analysis based upon various metrics. In the confusion, the matrix comprises classes COVID and Normal where (Fig. 3):

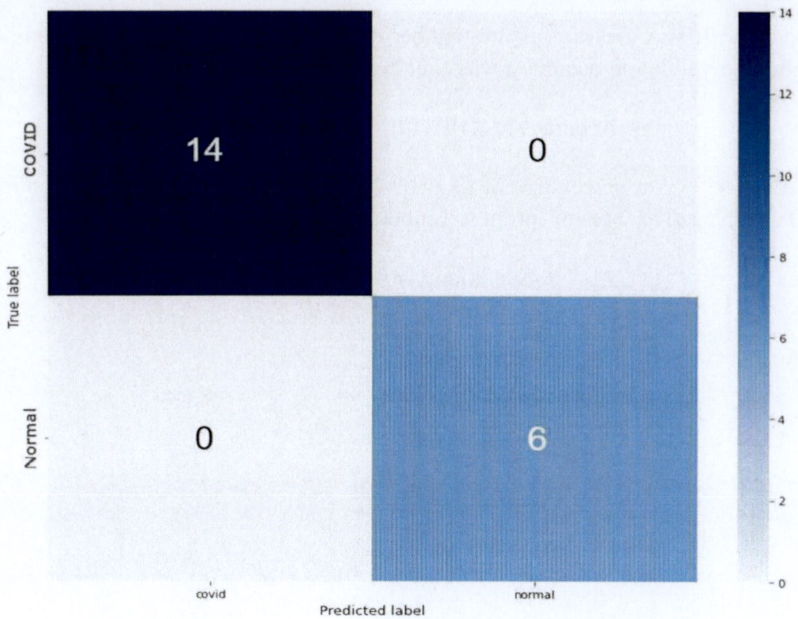

Fig. 3. Confusion Matrix of the model for (1) COVID, (2) Normal.

True positives (TP): It gives a measure of how many times the model correctly classifies COVID positive patients as COVID positive.
False Positives (FP): It gives a measure of how many times the model incorrectly classifies COVID positive patients as COVID negative.
True Negatives (TN): It gives a measure of how many times the model correctly classifies COVID negative patients as COVID negative.
False Negatives (FN): It gives a measure of how many times the model incorrectly classifies COVID negative patients as COVID positive.
Precision: It gives a measure of how many predictions of patients being COVID positive are true. For our proposed model we got a precision of: 1.

$$\text{Precision} = \text{TP}/\text{TP} + \text{FP}.$$

Recall: It gives a measure of the number of classifications of the COVID-positive patients the model can predict correctly. For our proposed model we got a recall of: 1.

$$\text{Recall} = \text{TP}/\text{TP} + \text{FN}.$$

F1 score: The F1 score being a function of precision and recall determines the number of instances our model accurately classifies without missing a significant number of instances. For our proposed model we got a score of 1.

$$\text{F1Score} = 2 \times \frac{\text{Precision} * \text{Recall}}{\text{Precision} + \text{Recall}}$$

Accuracy: It is a measure of the number of correct predictions. For our proposed model the validation accuracy was found to be 100%.

$$\text{Accuracy} = \text{TP} + \text{TN}/\text{TP} + \text{TN} + \text{FP} + \text{FN}$$

Specificity: It is a percentage of COVID Negative patients that were classified as COVID Negative, For our proposed model the score was 100% (Fig. 4).

$$\text{Specificity} = \text{TN}/\text{TN} + \text{FP}$$

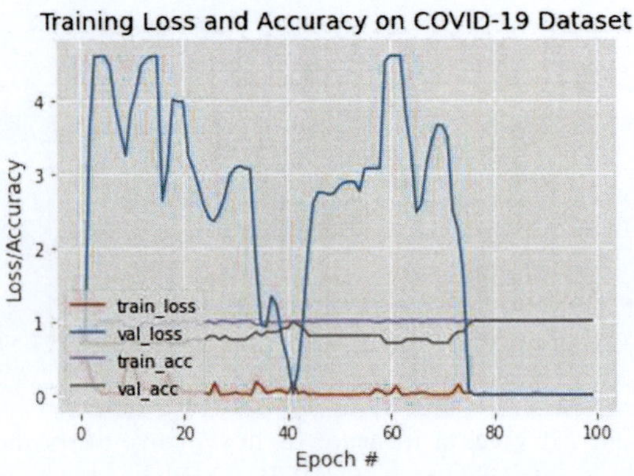

Fig. 4. Training Loss and Accuracy of our architecture.

Sensitivity: It is the percentage of COVID Positive patients that were classified as COVID Positive. For our proposed model the score was 100%.

$$\text{Sensitivity} = \text{TP}/\text{TP} + \text{FN}$$

It was also observed that our architecture strikes a great balance between both test and validation accuracy by achieving 100% in both parameters.

3.2 Performance Validation of Our Architecture

For a benchmark check of our architecture, we used other transfer learning approaches to develop our model [6]. The final performance metrics of each approach were examined to validate our proposed way.

3.2.1 VGG-16

VGG16 [9] is a 16 layer model trained on imagenet [5]. The model takes a 224 * 224 image size input image where filters with 3 * 3 receptive filters were used. Three fully connected layers follow a full stack of convolutional layers which consists of a 'soft-max' function in the last layer (Fig. 5).

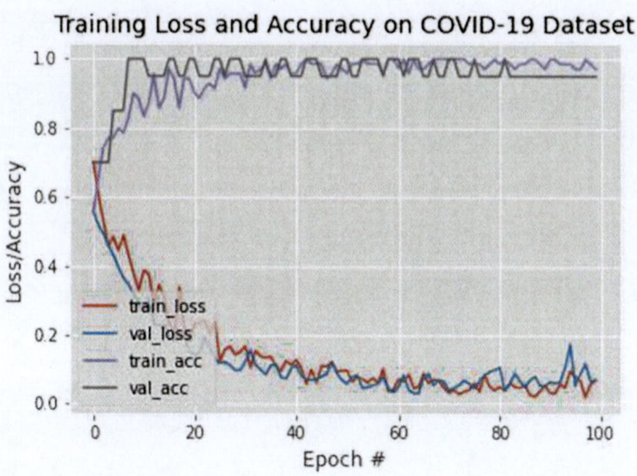

Fig. 5. Training Loss and Accuracy using VGG16 model for 100 epochs.

3.2.2 VGG-19

VGG19 is a 19 layer model also trained on imagenet. Similar to VGG16 it has fully connected layers, the first and second layers have 4096 channels each, while the third layer consists of an ILSVRC classification and consists of 1000 channels. Analogous to VGG16, it also has a soft-max layer added at the end (Fig. 6 and Table 1).

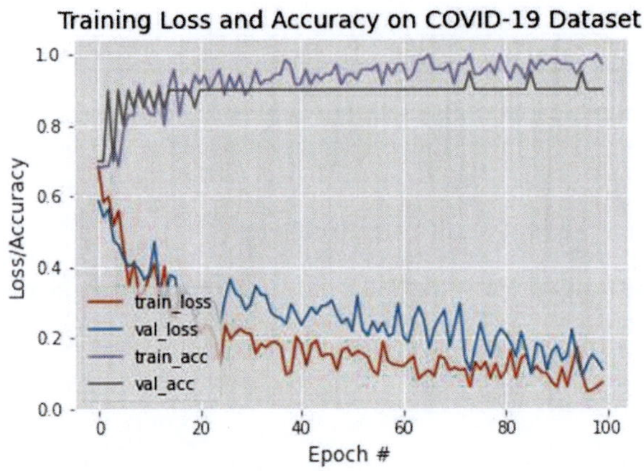

Fig. 6. Training Loss and Accuracy using VGG19 model for 100 epochs

Table 1. Performance validation comparison of our proposed work against other architectures

Model	VA	F1-Score	Precision	Recall	S	Sp
Our Model	1.00	COVID - 1.00 Normal - 1.00	COVID - 1.00 Normal - 1.00	COVID - 1.00 Normal - 1.00	1.00	1.00
VGG16	0.97	COVID - 0.96 Normal - 0.92	COVID - 1.00 Normal - 0.86	COVID - 0.93 Normal - 1.00	1.00	0.66
VGG19	0.92	COVID - 0.92 Normal - 0.86	COVID - 1.00 Normal - 0.75	COVID - 0.86 Normal - 1.00	0.84	0.66

* VA - Validation Accuracy , S - Sensitivity, Sp - Specificity.

4 Discussion

Pneumonia is a respiratory disorder caused by multiple classes of microorganisms such as [15] and bacteria. Hence in an attempt to achieve a chest radiograph classification exclusively for CoronaVirus cases, we used a similar transfer learning approach for multiclass classification in which we had 3 classes, i.e. healthy, Viral (that comprises Corona) and Bacterial. The additional data for the training was curated from the RSNA Pneumonia Detection Challenge on Kaggle and pre-processed similarly. Considering the Virus and COVID class together would lead to strong overlapping of features and reduce the feature extraction capabilities of the model. Post-training, the validation accuracy was observed to be around 92% for the multi-class case (Fig. 7).

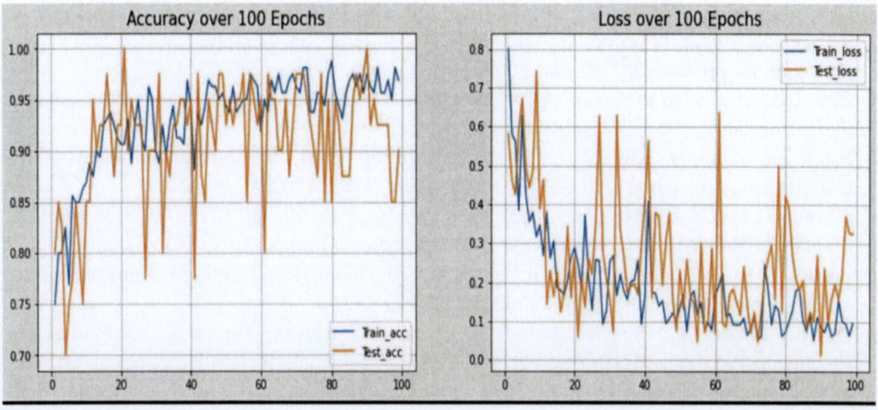

Fig. 7. Performance of model on multi-class depicting the (i) accuracy and (ii) loss over training period.

5 Conclusion

Our work proposes an automatic diagnostic system for the prediction of Corona Virus using chest radiographs of patients. The underlying model is a deep convolutional neural network designed using a transfer learning approach for achieving highly accurate predictions using an openly available dataset. To validate our work, various performance metrics were used to confirm their outcomes. Furthermore, we have discussed an additional study for a multi-class classification for the same taking into account patients suffering from pneumonia but not caused by COVID-19 as an edge case for the model. This work is meant to be a first cut solution for assisting medical professionals and requires further feedback and a larger dataset to improve and serve as a de facto solution to the early diagnosis of the pandemic.

Acknowledgment. We are sincerely thankful to Adrian Rosebrock, author of PyImageSearch for motivating us to go further with this study and Dr. Joseph Paul Cohen of Mila, University of Montreal for his efforts in accumulating various sources of the chest radiographs.

References

1. RSNA pneumonia detection challenge. Radiological Society of North America (2019)
2. Ai, T., Yang, Z., Hou, H., Zhan, C., Chen, C., Lv, W., Tao, Q., Sun, Z., Xia, L.: Correlation of chest CT and RT-PCR testing in coronavirus disease 2019 (COVİD-19) in China: a report of 1014 cases. Radiology, 200642 (2020)
3. Bashar, A.: Survey on evolving deep learning neural network architectures. J. Artif. Intell. **1** (02), 73–82 (2019)
4. Cohen, J.P., Morrison, P., Dao, L.: COVİD-19 image data collection. arXiv preprint arXiv: 2003.11597 (2020)

5. Deng, J., Dong, W., Socher, R., Li, L.-J., Li, K., Fei-Fei, L.: ImageNet: a large-scale hierarchical image database. In: 2009 IEEE Conference on Computer Vision and Pattern Recognition, pp. 248–255. IEEE (2009)
6. Lin, Z.Q., et al.: Do explanations reflect decisions? A machine centric strategy to quantify the performance of explainability algorithms. DBLP (2019)
7. Fang, Y., Zhang, H., Xie, J., Lin, M., Ying, L., Pang, P., Ji, W.: Sensitivity of chest CT for COVİD-19: comparison to RT-PCR. Radiology, 200432 (2020)
8. Gozes, O., Frid-Adar, M., Greenspan, H., Browning, P.D., Zhang, H., Ji, W., Bernheim, A., Siegel, E.: Rapid ai development cycle for the coronavirus (COVİD-19) pandemic: initial results for automated detection & patient monitoring using deep learning CT image analysis. arXiv preprint arXiv:2003.05037 (2020)
9. He, K., Zhang, X., Ren, S., Sun, J.: Deep residual learning for image recognition. In: Proceedings of the IEEE Conference on Computer Vision and Pattern Recognition, pp. 770–778 (2016)
10. Huang, C., Wang, Y., Li, X., Ren, L., Zhao, J., Yi, H., Zhang, L., Fan, G., Jiuyang, X., Xiaoying, G., et al.: Clinical features of patients infected with 2019 novel coronavirus in Wuhan, China. Lancet **395**(10223), 497–506 (2020)
11. LeCun, Y., Bengio, Y., Hinton, G.: Deep learning. Nature **521**(7553), 436–444 (2015)
12. Ng, M.-Y., Lee, E.Y.P., Yang, J., Yang, F., Li, X., Wang, H., Lui, M.M., Lo, C.S.-Y., Leung, B., Khong, P.-L., et al.: Imaging profile of the COVİD-19 infection: radiologic findings and literature review. Radiol. Cardiothor. Imaging, **2**(1), e200034 (2020)
13. Sandler, M., Howard, A., Zhu, M., Zhmoginov, A., Chen, L.-C.: MobileNetV2: inverted residuals and linear bottlenecks. In: Proceedings of the IEEE Conference on Computer Vision and Pattern Recognition, pp. 4510–4520 (2018)
14. Simonyan, K., Zisserman, A.: Very deep convolutional networks for large-scale image recognition. DBLP (2014)
15. Wang, W., Xu, Y., Gao, R., Lu, R., Han, K., Wu, G., Tan, W.: Detection of SARS-COV-2 in different types of clinical specimens. JAMA, **323**, 1843–1844 (2020)

An IoT-Based Heart Disease Detection System Using RNN

Ashif Newaz Shihab[1], Miftahul Jannat Mokarrama[2], Rezaul Karim[3], Sumi Khatun[4], and Mohammad Shamsul Arefin[1(✉)]

[1] Department of CSE, Chittagong University of Engineering and Technology, Chattogram, Bangladesh
asifnewazshihab@gmail.com, sarefin@cuet.ac.bd
[2] Department of CSE, Port City International University, Chattogram, Bangladesh
mokarramacse07@gmail.com
[3] Department of CSE, University of Chittagong, Chattogram, Bangladesh
rezaul.cse@cu.ac.bd
[4] Department of CSE, Bangladesh University of Business and Technology, Dhaka, Bangladesh
sumicuet11@gmail.com

Abstract. With the growing social pressure, most people are facing health problems and most of these are causing because of frequent heart attacks. It is very important to design an effective system that can diagnose the forthcoming happening of heart attacks or atrial fibrillations. As IoT (Internet of Things) is the most emerging and prominent technology in the modern era, a combination of the personal computer and IoT has become very efficient technology. The combined system provides various features to the user such as in emergencies to detect heart disease possibility through symptoms, sending messages to doctors according to the stage of the possibility of disease, and helps in fixing it. In case of an emergency, the system sends an emergency report to the desired doctor. In this paper, we propose an IoT based heart disease detection system using a Recurrent Neural Network (RNN). In our approach, we use Long Short Term Memory (LSTM) algorithm. The proposed system can perform the diagnosis of heart disease using RNN. The developed system helps the physician to prescribe patients without being present physically. The estimation of the result claims that the proposed system can detect heart disease efficiently.

Keywords: Recurrent Neural Network (RNN) · Long Short Term Memory (LSTM) · Internet of Things (IoT) · Atrial Fibrillation

1 Introduction

In the modern era, despite having technology in every person's reach, without the proper use of those technologies to monitor his/her health status, health problems are increasing day by day at a high pace. The death rate is also increasing. The most agitating problem now is heart failure.

J. I.-Z. Chen et al. (Eds.): ICIPCN 2020, AISC 1200, pp. 535–545, 2021.
https://doi.org/10.1007/978-3-030-51859-2_49

Heart failure, also known as Atrial Fibrillation (AF), is defined by the dysfunction of the heart's electrical system or irregular beating of a heart. Nowadays, leading a more modern life, we are being much unconscious about our health status. As heart attack may happen without any previous symptoms, it may lead to death very often. The day-by-day death rate is increasing so rapidly in this case. So it is the urge to be aware of health status so any happening may be avoided.

Therefore, a system is being proposed by designing a system that includes a wireless sensor to receive information with the respective human body heartbeat rate. Information will be transmitted on an IoT platform that is accessible by the user via the Internet. Using a continuous record of the body, those parameters are used to detect the heart-failure more efficiently. Moreover, an automatic alert generation feature is also added for emergencies.

The system architecture takes the raw signal into the input layer. After that, the hidden layer or recurrent layer using RNN and LSTM distinguishes the signal whether it is normal sinus rhythm or atrial fibrillation. Normal Sinus Rhythm (NSR) is the normal or actual sinus node impulse of heartbeat such as 72 per minute mostly ranging within 60 to 100 per minute. Atrial Fibrillation (AF), associated with an increased risk of stroke and heart failure, is a type of supraventricular tachycardia. When there is an abnormal impulse of heartbeat seen then it is assumed as a symptom of atrial fibrillation. Following this process, the possibility of heart failure can be detected through comparison.

Internet of Things defined by IoT is the more modern technology that has a vast field. Combining LSTM with IoT has become more fruitful technology. The combined system provides various features to the user such as determining heart disease possibility through symptoms, sending messages to doctors according to the stage of the possibility of disease, and helping in fixing it. In case of an emergency, the system sends an emergency report to the desired doctor.

2 Related Work

There are many methods to monitor the heart condition. As monitoring heart condition system is getting more popularity due to the easy accessibility, time-saving mechanism, and result accuracy, clinical experts are becoming dependent on these methods more than before. Therefore, it is crucial to find a robust method that can detect the problems and monitor the heart condition more accurately. Researchers have conducted a variety of researches on heart diseases adopting different approaches. Most common of them include IoT based approach for fetching data related to heart diseases from the human body [1–3] and the application of data mining and machine learning algorithms for analyzing such data and prediction of diseases [4–6]. In recent years, various deep learning algorithms are also showing promising accuracy in heart disease prediction results [7–10].

To remove the noises associated with the collected data through IoT devices, a novel approach is proposed [11] that makes use of the Sequential Recursive (SR) algorithm, Discrete Wavelet Transform (DWT) algorithm, and Fishers Linear Discriminant (FLD) algorithm for noise removal purpose. The filtered data collected this way is then matched with the existing heart disease dataset of MIT-BIH to diagnosis the problems in the heart. Though this way of diagnosis is evaluated using real-time data of different people, it still lacks in meeting some major problems related to the system like discovering and adjustment of the error in the retrieved data from sensors, the frequent updates of the knowledge-base in online, and the security issues.

To predict and diagnose heart diseases, another IoT based process is proposed in [12] with the help of cloud storage. Here, at first, data related to heart diseases are collected from the human body which is to be stored in cloud storage. The system also holds an existing dataset on heart diseases collected from the UCI (University of California, Irvine) repository and the patient's historical data from the previous experiments. Thus it tries to find out the diseases applying several machine learning algorithms on the available data. The system found that among the four classifiers used for testing, J48 outperforms others constituting about 91.50% accuracy in F1-score.

To provide instant and efficient information of heart diseases, a system is introduced in [13] that consists of Data Collection module, Data Prediction module, and User Interface module. The purpose of Data Collection module is collecting analog data by sensors from human body and converting, sampling, and grouping that data in digital format for further processing. In Data Prediction module, naive bayes algorithm is applied to the featured data extracted by the wavelet transform algorithm during data collection phase. Thus the system finally predicts whether a person has any of the three categories of heart diseases (Tachycardia, Bradycardia, Ventricular Tachycardia) with an accuracy of 95.67%.

To enhance the performance of the prediction result of the heart diseases, Khan [14] applied Modified Deep Convolutional Neural Network (MDCNN) for classifying sensor data as normal or abnormal. He also made a comparative analysis between MDCNN and other conventional classifying algorithms that revealed the feasibility of MDCNN over other algorithms he used.

To ensure the security of data in networking associated with IoT enabled artificial intelligent systems, a smart framework is proposed for the hospital atmosphere in [15]. However, the system was tested in small-scale and real-time operability is yet to be addressed.

3 System Architecture and Design

3.1 Architecture of the System Framework

The system architecture of the heart failure detection framework in Fig. 1 comprises four modules: input module, training module, testing module, and output module. Here the input module is responsible for fetching data from sensors. The training module processes data and saves weight and bias for further checking. The testing module predicts results through comparison and the output module provides the result of whether Atrial Fibrillation or Normal Sinus Rhythm.

Input Module. The first module is the input module. This module is used for fetching real-time data from sensors using Arduino. Fetched data is then passed to a personal computer for further analysis.

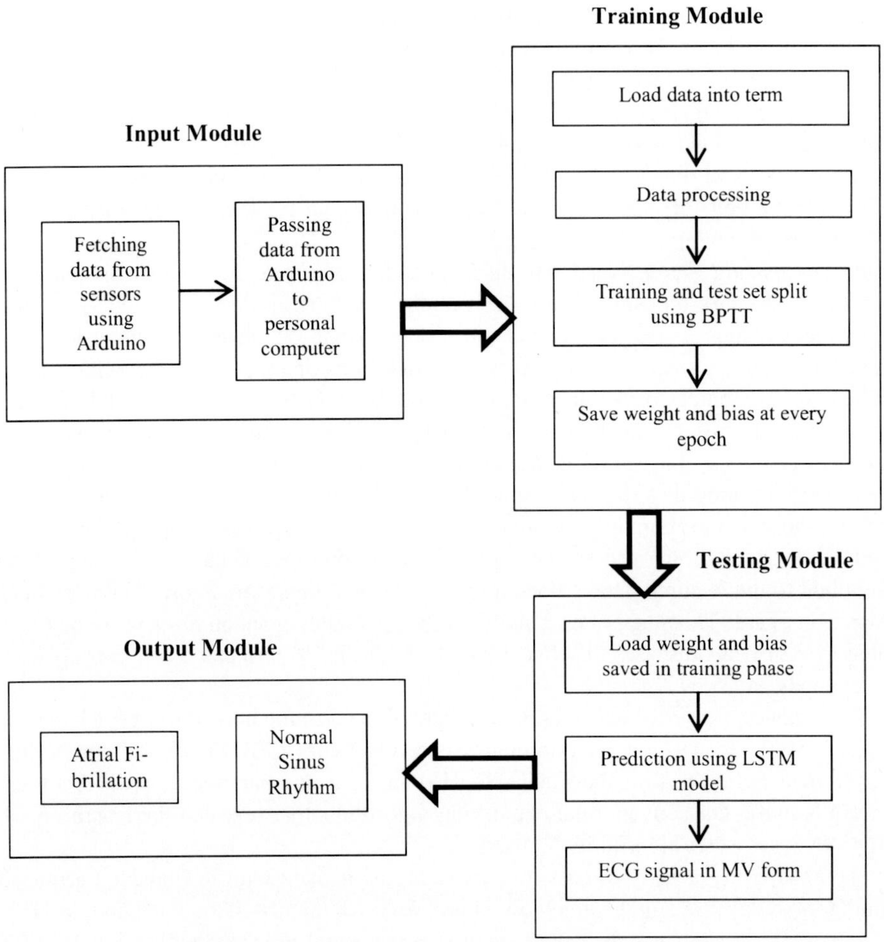

Fig. 1. The architecture of the proposed framework

Training Module. The second module is the training module. In this module after loading into data, term data is processed and split for training using Backpropagation through time (BPTT) algorithm. In each epoch weight and bias is generated. Weight and bias of each epoch are saved for further analysis. For training LSTM network sigmoid activation function, binary cross-entropy for loss calculation and Adam optimizer are used. In the training process, the Euclidian distance between two consecutive samples is calculated. Propagating through this process, the error is calculated and weight and bias are updated.

Testing Module. In the third module, the testing module test prediction is given. Here trained weight and bias of the model are loaded. AD8 single LED data is used as test data. Test data is taken using Arduino. Data was taken for 1 min. 15000 data alike are given as input. The predicted result is got based on this input.

Output Module. The fourth or final module is the output module. In this module comparison between testing dataset and training dataset is done. The final predicted result is given based on this comparison. The result prediction algorithm is used for assuming the final result. The final result may either be Atrial Fibrillation or Normal Sinus Rhythm.

3.2 Tasks of the System Framework

We divided the work of our framework into three different subtasks. These subtasks are discussed below.

Training LSTM Model. After taking the data set, training LSTM is started. Firstly dataset is split and knowing the ground truth x_train and y_train is set where x-train takes a large number of samples and the last column defines y_train. In x_train there are diverse samples whereas y_train contains only 0 or 1.

After adding weight and bias to both x_train and y_train the training process starts. We used sigmoid function as activation function containing the equation$1/(1 + e^\wedge (-x))$. Error is also measured using this activation function. The weight and bias of maximum accuracy are saved for further calculation. Training through BPTT, we got accuracy with 100%. The algorithm and an example of the LSTM training model are shown in Fig. 2 and Fig. 3 respectively.

Algorithm: Training LSTM Model
Input: Atrial fibrillation training syt.csv
1. **Begin**
2.**Set x**_train
3. **Set**y_train
4. **Add** (weight, x_train)
5. **Add** (bias, x_train)
6. **Add** (weight, y_train)
7.**Add**(bias, y_train)
8.**Sigmoid**(x_train)
9.**Sigmoid**(y_train)
10.Evaluate (x_train, y_train)
11.**If** error>0:
12.reduce(weight, bias)
13. **If** error<0
14.   increase (weight, bias)
15. Measureaccuracy
16.**If** accuracy == max
17.   save (weight, bias)
18. **End**

Fig. 2. Algorithm of training LSTM model

Testing the LSTM Model. Testing LSTM is also done after the dataset is taken. Using weight and bias to x_train, y_train, x_test, y_test test process continues. Testing provides both model accuracy and log loss. Defining Euclidian distance between samples, the error is corrected. The building of the test model completes building the LSTM

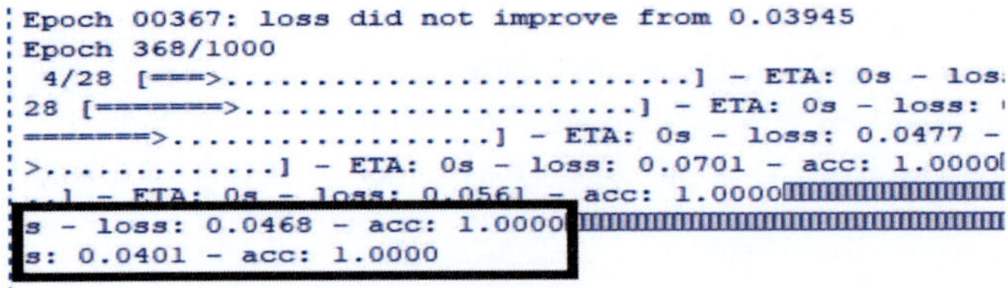

Fig. 3. Training example of LSTM network

model. The test accuracy of the model was 96%. The algorithm and an example of the LSTM testing model are shown in Fig. 4 and Fig. 5 respectively.

Algorithm: Testing LSTM Model
Input: 'Atrial fibrillation training syt.csv
1. **Begin**
2. **Set x**_train
3. **Set** y_train
4. **Add** (weight,x_train)
5. **Add** (bias,x_train)
6. **Add** (weight,y_train)
7. **Add** (bias,y_train)
8. **Sigmoid** (x_train)
9. **Sigmoid** (y_train)
10. Evaluate (x_train,x_test,y_train,y_test)
11. Measure model accuracy
12. Measure log loss
13. **Return**(model accuracy, log loss)

Fig. 4. Algorithm of testing LSTM model

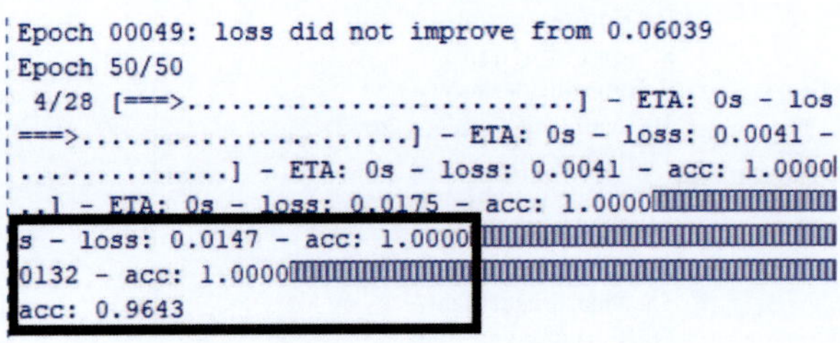

Fig. 5. Testing example of LSTM network

Prediction of Result. From the testing process, model weight and bias are taken. After fetching the raw signal, it is normalized so that the entire sample corresponds between 0 and 1. Then the raw signal is split and evaluated for prediction.

Predicted result corresponding to y_test provides whether it is Atrial Fibrillation or Normal Sinus Rhythm using 0 and 1 respectively. The algorithm for result prediction is shown in Fig. 6.

Algorithm: Result Prediction
1. **Load** (weight, bias)
2. Fetch raw signal
3. Normalize (row signal)
4. Split (raw.signal,x_test,y_test)
5. Evaluate (x_train,x_test,y_train,y_test)
6. Predict (y_test)
7. **Return** (prediction)

Fig. 6. Algorithm for result prediction

4 Implementation and Experiments

4.1 Experimental Setup

In Fig. 7 we showed the sensor connectivity. We used a single LED ECG connector to connect it with the human subject. Three electrodes: blue, red, and black wire connected to two arms and belly of a human subject. Here red and blue electrodes mainly receive electric signals and the black electrode cancels noises. To implement our system, we used python 3.6 on a personal computer, ECG sensor, Arduino, and biomedical patch.

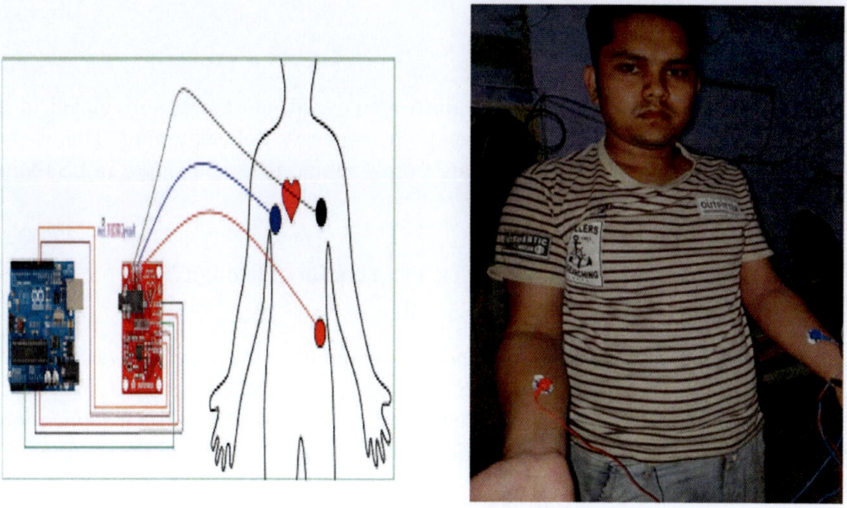

Fig. 7. Sensor connectivity

4.2 Data Collection from Sensor

Using the Py-serial library, we recorded the signal. The total number of signals recorded was 15000. The result was saved to a CSV file. We used Arduino to pass the collected data to the computer (in Fig. 8).

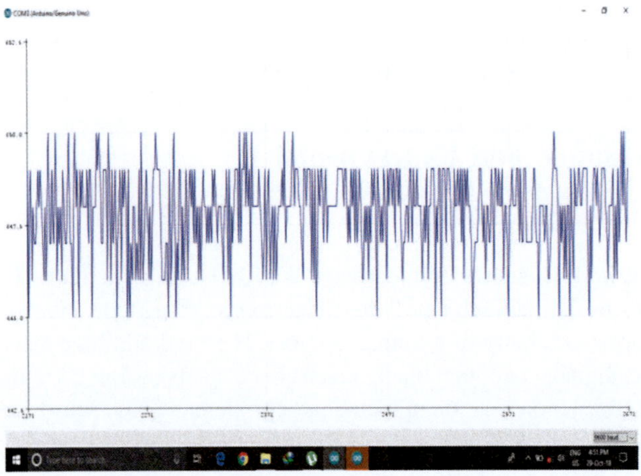

Fig. 8. Data collection using Arduino

4.3 Acquired Data Preprocessing

After recording to CSV file, the file is further analyzed and being normalized in the range 0 to 1. History of LSTM weight and bias was saved after training. This is used for prediction. Weights of LSTM and data from Arduino is used as input in LSTM.py. We used the MIT-BIH data set.

4.4 LSTM Model Training with Back Propagation Through Time and Result Calculation

We calculated several predictions. For each epoch number, a prediction was generated where backpropagation through time LSTM model propagation was used. From these annotations, the predicted result was given as either 0 or 1. 0 is for normal sinus rhythm and 1 is for atrial fibrillation. The requirements for training our system are shown in Table 1.

Table 1. The requirements for training

Hidden layer	32
Input dimension	1499
Activation function	Sigmoid
Loss calculation	Binary cross entropy
Optimizer	Adam optimizer
Accuracy check	Matrix

5 Experimental Evaluation

We evaluated the extent to which this work is successful in achieving the objectives that are defined at the beginning. In this, we have shown some real-life experimental results from our proposed system. Here, to verify the system we have talked to several patients, doctors and fulfill their requirements through our system. Later we evaluated the system.

5.1 Experimental Data

We experimented with the residence of Liver Society, Muradpur, Chattogram. We experimented this over 20 people. The result found in this experiment is given in Table 2.

Table 2. Experiment with people

Number of people	Response
19	Normal Sinus Rhythm
1	Atrial fibrillation

5.2 Result and Analysis

Here we have examined if the proposed framework is providing correct results or not. These activities can be performed based on different features for different patients. We queried whether the system was effective and responding timely in case of an emergency as our interest was helping in an emergency. We got satisfactory results by interacting with mass people.

Query with mass people provided us a better outcome. We requested them to give honest reviews about the system and got satisfactory results. In all cases, the proposed framework gained higher scores (shown in Table 3). This indicates that the framework is quite successful to remind the 25 or 30 up people to check out one of the very common diseases-heart attacks. As our key interest was to help in an emergency so that they can perform at the appropriate time, we have become able to fulfill our objectives. The bar chart in Fig. 9 illustrates it more vividly.

Table 3. List of questions after implementing the application

Question	Strongly disagree	Disagree	Neutral	Agree	Strongly agree
Q1: Do you think the system is effective?	0	0	0	7	13
Q2: Do you think the system is responding timely and correctly?	0	0	0	9	11
Q3: Do you think that the use of the emergency button will help in case of an emergency?	0	0	1	5	14
Q4: Do you think detect diseases button will helpful for 45 up ages people who will commonly face heart attack.	0	0	2	4	14
Q5: Do you think search doctor button will helpful for patients?	0	0	1	5	14

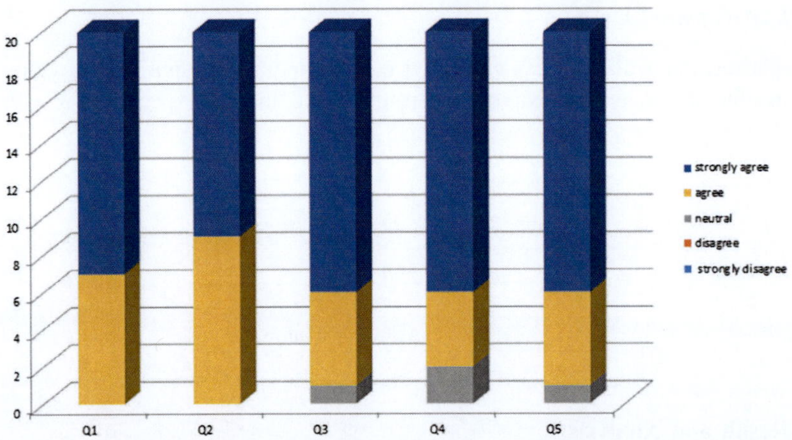

Fig. 9. User opinion after implementing the system

6 Conclusion

The combination of a personal computer and IoT has become an innovative technology in the healthcare system. In this work, a recurrent neural network model is proposed that automatically detects heart failure or atrial fibrillation from ECG signals. Here, at first, Arduino collects data from the (Heartbeat) sensor and saves in a personal computer which is then transferred wirelessly to an IoT platform using Personal Computer. Then features from ECG recording are extracted and used as the input for the recurrent neural networks (RNN) composed of recurrent layers. The recurrent neural network we considered as our base is the Long Short-Term Memory (LSTM) and for normalization Sigmoid function is used. Disease diagnosis operation using the neural network is performed based on comparison with the trained dataset. The system may help the physician to prescribe a patient without being present physically, thus developing the medical services and effective diagnosis which is based on IoT.

References

1. Sethuraman, T.V., Rathore, K.S., Amritha, G., Kanimozhi, G.: IoT based system for heart rate monitoring and heart attack detection. Int. J. Eng. Adv. Technol. (IJEAT) **8**(5), 1459–1464 (2019)
2. Ajitha, U., Aswathi, P.A., Sasidharan, A., Salman, M.V.A., Anand, V., Arvind, A.: IoT based heart attack detection and alert system. Int. J. Eng. Manag. Res. **7**(2), 285–288 (2017)
3. Lia, C., Hua, X., Zhang, L.: The IoT-Based Heart Disease Monitoring System for Pervasive Healthcare Service. In: International Conference on Knowledge Based and Intelligent Information and Engineering Systems, pp. 2328–2334. Elsevier B.V., Marsweille (2017)
4. Chitra, R., Seenivasagam, V.: Review of heart disease prediction system using data mining and hybrid intelligent techniques. Int. J. Soft Comput. **3**(4), 2229–6956 (2013)
5. Singh, P., Singh, S., Pandi-Jain, G.S.: Effective heart disease prediction system using data mining techniques. Int. J. Nanomed. **13**, 121–124 (2013)
6. Jan, M., Awan, A.A., Khalid, M.S., Nisar, S.: Ensemble approach for developing a smart heart disease prediction system using classification algorithms. Res. Rep. Clin. Cardiol. **9**, 33–45 (2018)
7. Sayad, A.T., Halkarnikar, P.P.: Diagnosis of heart disease using neural network approach. Int. J. Adv. Sci. Eng. Technol. **2**(3), 88–92 (2014)
8. Gudadhe, M., Wankhade, K., Dongre, S.: Decision support system for heart disease based on support vector machine and artificial neural network. In: International Conference on Computer and Communication Technology (ICCCT), pp. 741–745. IEEE, Allahabad (2010)
9. Sonawane, J.S., Patil, D.R.: Prediction of heart disease using multilayer perceptron neural network. In: International Conference on Information Communication and Embedded Systems (ICICES 2014), pp. 1–6. IEEE, Chennai (2014)
10. Durairaj, M., Revathi, V.: Prediction of heart disease using back propagation MLP algorithm. Int. J. Sci. Technol. Res. **4**(8), 235–239 (2015)
11. Thai, D.T., Minh, Q.T., Phung, P.H.: Toward an IoT-based expert system for heart disease diagnosis. In: The 28th Modern Artificial Intelligence and Cognitive Science Conference (MAICS 2017), pp. 157–164. IPFW, Fort Wayne (2017)
12. Ganesan, M., Sivakumar, N.: IoT based heart disease prediction and diagnosis model for healthcare using machine learning models. In: IEEE International Conference on System, Computation, Automation and Networking (ICSCAN), pp. 1–5. IEEE, Pondicherry, March 2019
13. Sorte, P., Golande, A., Yermalkar, A., Suryawanshi, V., Wanjare, U., Satpute, S.: Smart hospital for heart disease prediction using IoT. Int. J. Innov. Technol. Explor. Eng. (IJITEE) **8**(9), 321–326 (2019)
14. Khan, M.A.: An IoT framework for heart disease prediction based on MDCNN classifier. IEEE Access **8**, 3417–34727 (2020)
15. Valanarasu, M.R.: Smart and secure IoT and AI integration framework for hospital environment. J. ISMAC **1**(03), 172–179 (2019)

A Framework for Checking Plagiarized Contents in Programs Submitted at Online Judging Systems

Jannatul Ferdows[1], Sumi Khatun[2], Miftahul Jannat Mokarrama[3], and Mohammad Shamsul Arefin[1(✉)]

[1] Chittagong University of Engineering and Technology, Chittagong, Bangladesh
kfoozminus@gmail.com, sarefin@cuet.ac.bd

[2] Bangladesh University of Business and Technology, Dhaka, Bangladesh
sumicuet11@gmail.com

[3] Port City International University, Chittagong, Bangladesh
mokarramacse07@gmail.com

Abstract. Plagiarism has always been a problem in the academic area through digitalization and the widespread use of computers, copying documents has become very easy. Nowadays, all kinds of documents are available as digital content and freely accessible either legally or illegally. This development makes it easy for culprits to copy digital content found on the web without much effort. Therefore, plagiarism can be found in any field such as literature, design, scientific papers, and source code. This project focused on the detection of a particular type of plagiarism: source code plagiarism, also known as software plagiarism. Source code plagiarism is often done by computer science students. Often, they solve their assignments by copying from other students or copying it from the internet and presenting it as their own work. This project implemented source code plagiarism detection framework by detecting finder prints of programs and using them to compare each of them instead of using the whole file. We used winnowing to select fingerprints among k-gram hash values of a source code, which was generated by Rabin-Karp Algorithm. At long last, we likewise give trial results on applying enormous data sets and contrasting the runtime with Measure of Software Similarity (MOSS), the extensively used plagiarism perception facility.

Keywords: Plagiarism · Rabin Karp · Hashing · Fingerprint · Winnowing

1 Introduction

Plagiarism is the act of taking another person's work or thoughts and making them look like one's own. Students are supposed to finish their assignment on their own but many of the students tend to copy them from their colleagues

J. I.-Z. Chen et al. (Eds.): ICIPCN 2020, AISC 1200, pp. 546–560, 2021.
https://doi.org/10.1007/978-3-030-51859-2_50

or from internet. As indicated by one review, over 5% of undergrads have had plagiarism-related related experiences. Since plagiarism happens much of the time in universities, a trainer can not evaluate all understudies work reasonably. Although a long time has passed since computer science genesis, computer science educators have lagged behind in digitalizing the traditional tools and techniques for evaluating their students' assignments and homeworks. Their main tool for communicating ideas is still chalk and board. For evaluating, it is still the primitive tool – a human grader. As the number of students in computer science is increasing day by day, it has been almost impossible to check all the programs with each other by hand. Moreover source code plagiarism is more complicated than document plagiarism. That's why detecting source code plagiarism has received considerable attention.

Many useful tools have been discourse to discover plagiarism in programming assignments. The main aim of these tools is to guide the educator toward submissions that are most likely to have been plagiarized. Every source code is compared with each other and by passing them through the plagiarism detection system, a statistics of similarity measurement is published to instructors. By evaluating that, instructors decide which work is plagiarized and which work is genuine. Here, we have implemented a plagiarized contents checker for the online judging system. The output after checking the contents of the output is converted into the lightweight data-interchange format: JavaScript Object Notation (JSON) object.

2 Related Work

MOSS [13] is a free service provided by Stanford University. However, as MOSS is a client/server service with plagiarism detection processing done completely server-side, specifics of its approach are not known. The limited documentation on MOSS suggests that it relies primarily on a document fingerprinting technique called winnowing, a proficient confined fingerprinting algorithm. They likewise give experimental proceed on Web information and report involvement with MOSS, a broadly utilized plagiarism discovery service. MOSS uses string based techniques as it begins by changing over source code documents into strings. In any case, this classification is to some degree fluffy on the grounds that the engendered strings are then changed over into sets of k-grams as substrings. After begetting the k-grams, MOSS endeavors to choose a subset of the k-grams called a "fingerprint", which depicts to an archive and can be contrasted with different fingerprints to estimate the similitude between the code records. Although this is the most used free and available plagiarism detection system across higher education institutions, this is not user friendly.

A web-based detection framework that utilizes a similitude algorithm [11]. The source code is inferred and changed over into token strings in this system. Comments, name of identifiers, and white-space are disregarded. Then token strings are compared, non-overlapped and lengthiest regular substrings are selected by applying the greedy string tiling algorithm.

Sherlock [4] is an algorithm that checks for plagiarism, based on the similarity of sentences between documents. This algorithm uses the Rabin-Karp algorithm in a modified way. Sentences can have different sets of keywords means that the two sentences are not very similar and have different content. If the sentences being compared have the same set of keywords then the two sentences have the same content. Two algorithms: Knuth Morris Pratt and Rabin Karp are used to make string matches. Karnalim et. al [6] proposes a methodology for detecting plagiarism in source code using structured-based low-level approach. In another paper, Karnalim et. al [5] discussed about plagiarism attacks raised on both procedural and object-oriented programming environment for Python programming.

Kermek et. al [8] uses Sherlock which is an algorithm that checks for plagiarism, based on the similarity of sentences between documents in plagiarism detection engine. Liu et. al [10] uses mined program dependence graphs (PDGs) for plagiarism detection. A PDG is a pictorial delineation of the data and restraint conditions within a proceeding. Since PDGs are almost invariant during plagiarism, Gplag is more viable than up-to-date tools in detection of plagiarism. So as to make GPlag versatile to huge programs, a measurable lossmaking channel is utilized to prepare the plagiarism search space. This system is not available for public use.

Sim identifies similitude between programs by assessing their speciality and accuracy which is improvised by Lim et. al [9], utilizing a versatile local arrangement of keywords, and changing the changing the similarity matrix. It is notable that the matching score of keywords is inverse of the frequencies of them. That means we trait low scores for keywords of high frequencies and high scores for keywords of low frequencies. The versatile methodology is insusceptible against common plagiarizing assaults, for example, inserting or erasing unimportant or dummy keywords. Ji et. al [3] explained briefly their experience of using it in programming contests, showing that it works better than Sim. Rahal et. al [12] introduces a for detection technique of plagiarism in students' assignment to an introductory-level Visual Basic programming pro-offering course. They have used various string correlation systems, intensely used in the zone of biological sequence alignment.

Online programming judges like Codeforces [15] and Hackerrank [14] addressed the necessity of the reliable plagiarism detector for submitted codes judging systems. Regarding all of the above systems, we have indexed the importance of scrutinizing plagiarism in submitted codes in judging systems and formatting the checked output in JSON file that can be easily used in visualizing the results.

3 System Architecture and Design

The proposed system has five main phases: Tokenization with lexical analyzer (scanner) and parser, k-gram hash generation using Rabin-Karp algorithm, fingerprint engendering with the help of Winnoing method, calculation of similarity, and percentage estimation on the basis of similarity and line matches. The schematic presentation of the whole system is depicted in Fig. 1 and the phases are delineated in the next subsections.

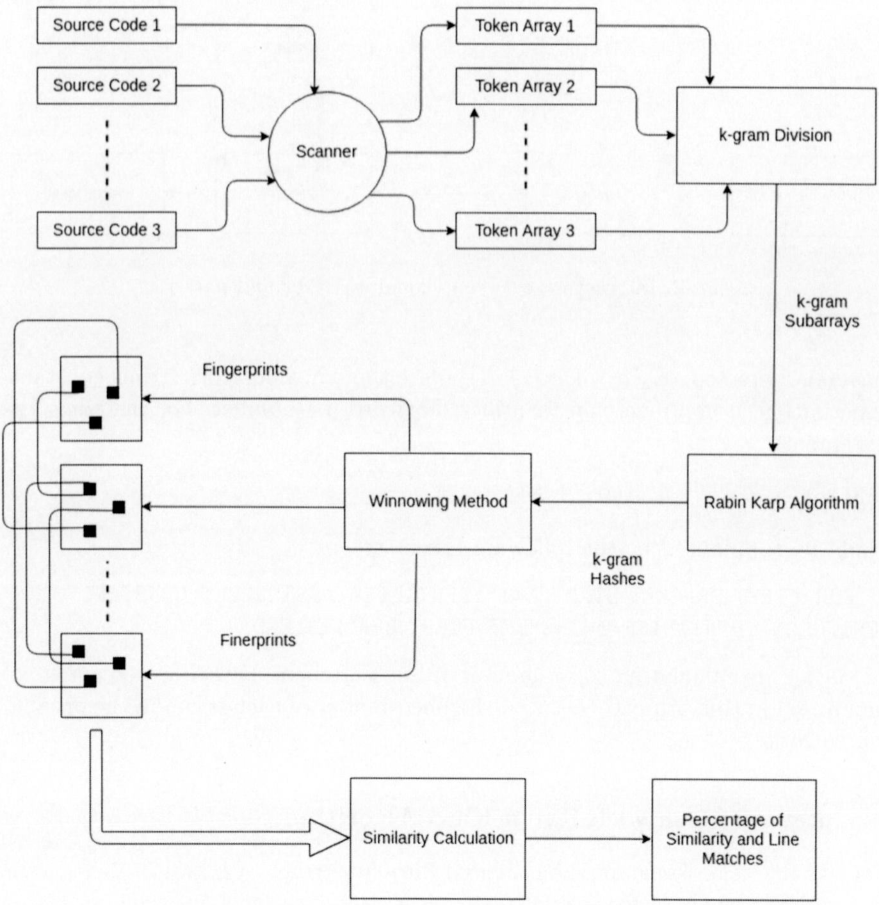

Fig. 1. Overall structure of the system.

3.1 Tokenization

Each input program is scanned where they are passed through a lexical analyzer to deliver a reduced type of its structure regarding a stream of integers named tokens. The scanner is consist of lexical analyzer and parser. The lexical analyzer is created by flex using lexical rules, so given the keywords and reserved words for a specific language, this plagiarism detection system can be easily modified to work with programs in all languages. Bison generates parser utilizing grammar rules which converts the source code to token arrays with the help of lexical analyzer. In Fig. 2 the interaction between lexical analyzer and

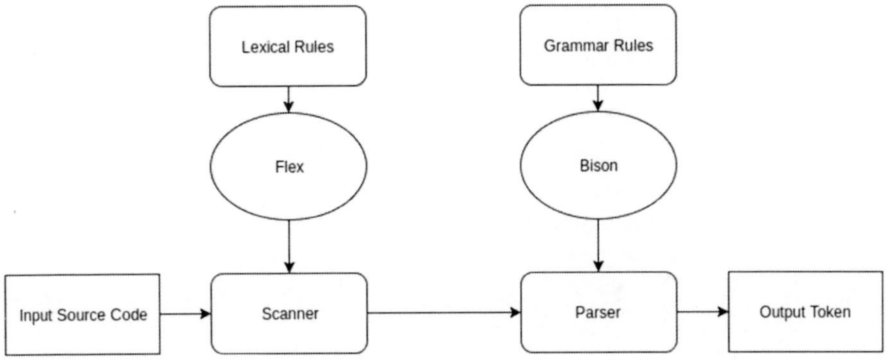

Fig. 2. Interaction between lexical analyzer and parser.

parser is depicted. Each token represents either an operator, string constant, punctuation, a comment, an identifier, keyword, or number. For instance, the statement:

```
for(int i = 0; i < 100; i ++)
```

would be supplanted by the following token stream:

```
    FOR FIRST_BRACKET_START INT IDENTIFIER ASSIGNMENT CONSTANT
SEMICOLON IDENTIFIER INCREMENT FIRST_BRACKET_END
```

As we are calculating total number of line matched (as well as percentage of similarity) in this project, we saved the line number of each token in the original source code.

3.2 k-gram Hash with Rabin-Karp Algorithm

The streams of tokens are then divided into substrings of k-gram. As we want to compare strings of fixed length, that is k, these k-grams are then converted to some hash value [7]. It's important to choose the hash function carefully to avoid hash collision. As we described before, we used Rabin-Karp Algorithm (Algorithm 1) to compute the hash value.

Algorithm 1. k-gram Hash Value Calculation

procedure Rabin-Karp(K, B)
    for $i = 0$ to $tokenSize - 1$ **do**
        if $i < k$ **then**
            $hash \leftarrow hash \cdot B + token[i]$
        else
            $hash \leftarrow (hash - token[i - K] \cdot B^{k-1}) \cdot B + token[i]$
        if $i >= k - 1$ **then**
            add *hash* and *location* to *kGramHashList*

As we saw earlier, computing hash with Rabin-Karp needs a base number. If we choose a small number like 10 as base, there'll be collisions, and if we choose a large number, the hash values will be so large that we won't be able to fit them in integers or long integers. Again, after we choose a base number, the hash values can get pretty large, depending on the value of k. So to avoid the chance of overflow, we had to choose a mod MOD. We took the Rabin-Karp value mod MOD. This number should be pretty large to increase the range of hash values. And as the base and mod should be relatively prime, we chose two prime numbers.

3.3 Winnowing Algorithm

In this step, we selected the fingerprints of a source file to which every other file is to be compared. Fingerprints are basically subset of the k-gram hash values that we computed before. We used Robust Winnowing (Algorithm 2) for this purpose, that is, selected the minimum hash of a window size W [2]. If the minimum hash of this window is selected before, it is discarded. In first sight, it might seem that we have check every value of every window to compute this. But we took advantage of the fact that we are proceeding to the end of the array by moving the window by one element, so $W - 1$ values of current window have been compared before and if minimum has of previous window is still in the current window, we just compare it to the new element. If the value is out of the current window, we computed the minimum value from scratch.

Algorithm 2. Winnowing Algorithm

```
procedure FINGERPRINT(K, W)
    minPos ← 0
    minHash ← kGramHashList[minPos]
    for i = 0 to W − 1 do
        if minHash >= kGramHashList[i] then
            minPos = i
            minHash = kGramHashList[i]
    add minHash and minPos to fingerprints
    for i = W to kGramHashListSize − 1 do
        if minPos = i − W then
            minPos = i − W + 1
            minHash ← kGramHashList[minPos]
            for j = i − W + 1 to i do
                if minHash >= kGramHashList[j] then
                    minPos = j
                    minHash = kGramHashList[j]
        else
            if minHash >= kGramHashList[i] then
                minPos = i
                minHash = kGramHashList[i]
        add minHash and minPos to fingerprints
```

3.4 Matching

After we finished selecting fingerprints, they are mapped to each of the programs in which they exist. We also saved the number of occurrence of a fingerprint in a source code. Now, for each source code, for each of its fingerprints we counted the number of occurrences of the fingerprint in other programs. So if fingerprint f_i occurs x times in program d_x and it occurs y times in program d_y then we can safely increase the count of matches of program d_x with d_y by minimum of x and y. The source code matching procedure is depicted as flowchat shown in Fig. 3

Like this, count of matches with other programs are also computed. And then we picked the program with most number of counts. This is the most important and efficient part of the program which makes it separable with other plagiarism detection systems. Because in other programs, pairwise comparison of source codes are done to find the best match. But with this fingerprinting and mapping technique, we are calculating the matching results of all source code at once.

One other improvement are also made to the system. One important fact is we are finding the best match program from which a particular program could possibly be copied. So if multiple source codes had been plagiarized from one particular program, then all the programs can be clustered and there's no point comparing all of them with the new programs. In this case, all the programs but one are flagged as plagiarized and discarded. If the percentage of similarity is greater than a specific level then the program is discarded as plagiarized if and only if the program to which it is being compared has been already checked for plagiarism.

3.5 Calculation

Total no of line matched of a program with the best matched program is counted easily as we saved the location of every token in the token array. The percentage of similarity is calculated as:

$$\%\text{ of Similarity} = \frac{\text{Total number of tokens matched}}{\text{Total number of tokens in this program}} * 100$$

4 Implementation and Experiments

4.1 Conversion from Source Code to Tokens

The tokens can be classified into the following classes [1]:

1. Only one token can be used for each keyword. A keyword pattern is analogous to the keyword itself.
2. Separate tokens are used for the operators.
3. All the identifiers are captured by one token.

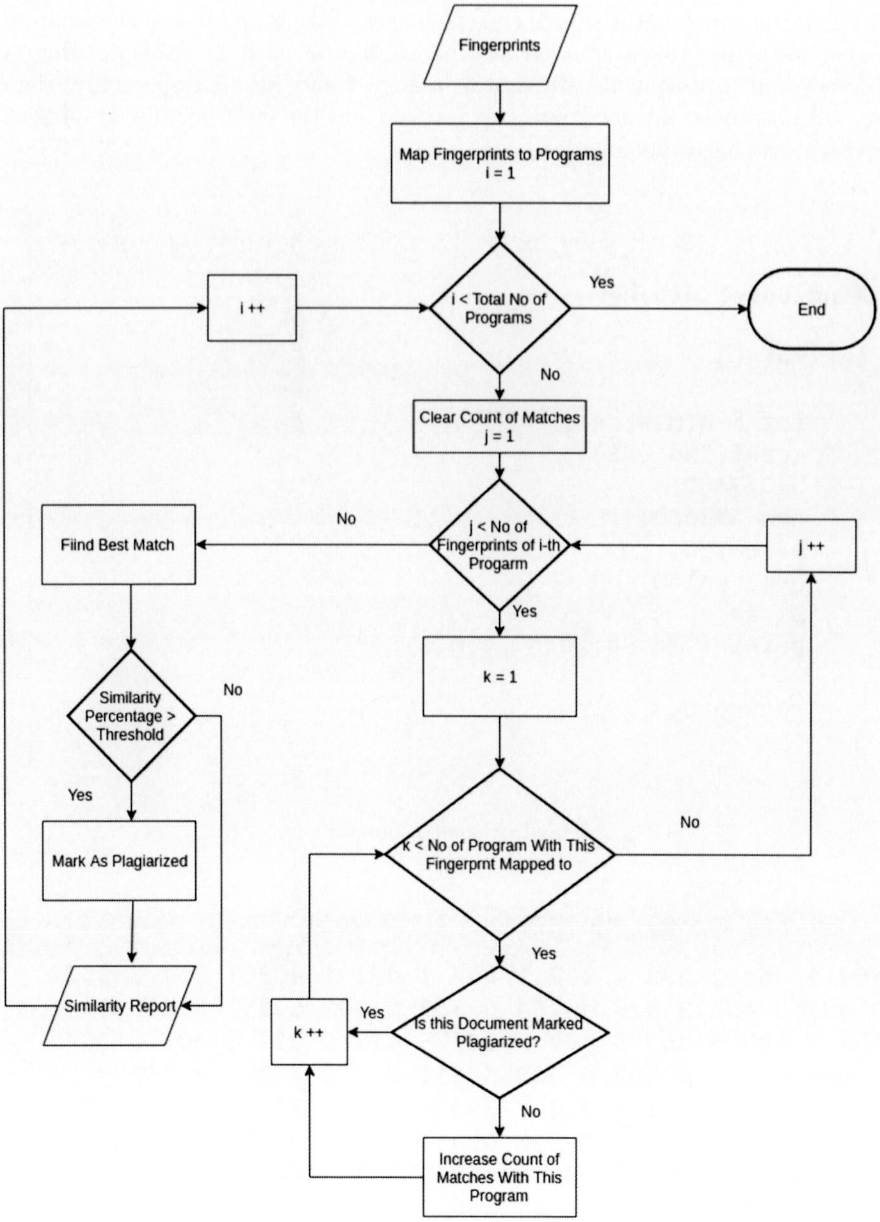

Fig. 3. Flowchart for source code matching.

4. At least one tokens representing to constants, for example, numbers and strict strings.
5. Each punctuation mark has tokens, for example, semicolon, comma, left and right parentheses, etc.

Flex & Bison converts the source code to tokens. Tokens and line of the location of that particular token is saved in separate file for each source code. Here a program and its token file are shown in Fig. 4 and Fig. 5 respectfully. Here the odd positioned integers represent a token and the even positioned integers represent the line of its location.

12752634_1704001_Ajoy_B.c × 12752634_1704001_Ajoy_B.c ×

```c
#include<stdio.h>

int main()
{
    int S,h,m,s,rem;
    scanf("%d",&S);
    h=S/3600;
    rem=S%3600;
    m=rem/60;
    rem=rem%60;
    s=rem;
    printf("%d:%d:%d\n",h,m,s);

    return 0;
}
```

Fig. 4. A sample program.

12752634_1704001_Ajoy_B.c × 12752634_1704001_Ajoy_B.c ×

```
404 1 361 1 398 1 352 1 367 1 352 1 402 1 264 3 352
3 407 3 408 3 419 4 264 5 352 5 409 5 352 5 409 5
352 5 409 5 352 5 409 5 352 5 411 5 352 6 407 6 354
6 409 6 378 6 352 6 408 6 411 6 352 7 394 7 352 7
390 7 353 7 411 7 352 8 394 8 352 8 392 8 353 8 411
8 352 9 394 9 352 9 390 9 353 9 411 9 352 10 394 10
352 10 392 10 353 10 411 10 352 11 394 11 352 11 411
11 352 12 407 12 354 12 409 12 352 12 409 12 352 12
409 12 352 12 408 12 411 12 330 14 353 14 411 14 420
15
```

Fig. 5. Token file of that program.

As the number of data is pretty high, the automation of tokenization of each program using Flex-Bison is done with Shell Scripting. The Shell program first makes a new folder name 'tokens' in the directory where the target programs are. Then it saves filename of each program in a file named 'filenames'. Again, it runs a command that passes every program file into the parser and saves the output token files in the 'tokens' folder.

4.2 Implementation with C++

The main part of this project is coded with C++. It executes the Rabin-Karp, Winnowing and Similarity Calculation. The shell script runs its last command which passes every token file as the input of the C++ file and it gives the final result as output represented in Fig. 6.

```
search_directory : .
    12752540_1704045_Robin_A.c       12752711_1704006_Fariha_A.c    84.906 percent    7 line
    12752550_1704038_Rafin_A.c        12752582_1704041_Rasel_A.c    96.721 percent    9 line
     12752555_1704001_Ajoy_A.c       12752931_1704034_Bibhas_A.c    97.959 percent    9 line
    12752562_1704051_Maruf_A.c       12752783_1704003_Tanzin_A.c    85.714 percent   10 line
     12752565_1704036_Alve_A.c       12752952_1704025_Ibadat_A.c    97.101 percent   10 line
   12752566_1704057_Shawly_A.c     12752873_1704049_Shoishob_A.c    88.710 percent   10 line
    12752582_1704041_Rasel_A.c        12752550_1704038_Rafin_A.c    96.721 percent   10 line
    12752586_1704016_Zarin_A.c        12752550_1704038_Rafin_A.c    96.721 percent    9 line
     12752593_1704044_Aman_A.c        12752550_1704038_Rafin_A.c    96.721 percent    9 line
     12752597_1704017_Anik_A.c        12752550_1704038_Rafin_A.c    96.721 percent   10 line
   12752600_1704055_Ahasan_A.c       12753418_1704023_Mahedi_A.c    97.778 percent    6 line
    12752610_1704020_Shofi_A.c        12752614_1704039_Jawad_A.c    96.825 percent    9 line
    12752614_1704039_Jawad_A.c        12752610_1704020_Shofi_A.c    96.825 percent    9 line
     12752632_1704037_MOIN_A.c        12752550_1704038_Rafin_A.c    96.721 percent    8 line
    12752641_1704047_nabin_A.c         12752565_1704036_Alve_A.c    84.932 percent   10 line
```

Fig. 6. Plagiarism report for some source codes for Task A in introductory programming class of first year students.

Value of B and MOD in Hashing. As we mentioned earlier, the value of base and prime has to be prime in order to make it uniformly distributed. For this program, we chose B as 41 and MOD as $10^9 + 7$.

4.3 Integration with a Contest Management Framework (CUET Online Judge)

We have integrated this plagiarism detection system with CUET Online Judge platform. After a contest is over, all the accepted solutions are copied on a directory. Then the server runs the plagiarism detection program with the directory as an argument to that program. After our program finishes execution, the server processes the output. It converts the output into a JSON object and sends it to the frontend. The frontend takes this JSON object and shows it in Fig. 7.

5 Experimental Result and Evaluation

We have experimented on some real source code, implemented by students. Here we concentrated on the system and time efficiency and correctness. We additionally contrasted the result of our plagiarism discovery engine with the widely used plagiarism detection system MOSS. In our experiments we have used tabular form for performance measurement of developed plagiarism detection system.

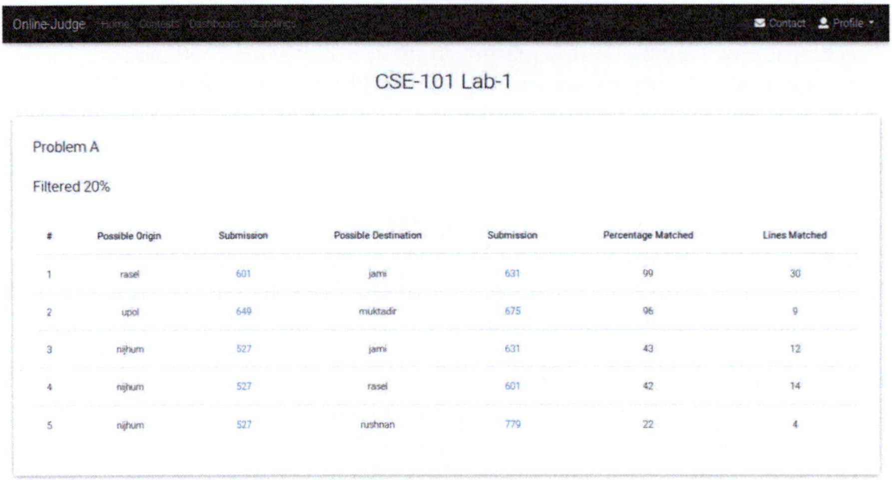

Fig. 7. Hosted on CUET online judge.

5.1 Data Set

We have collected data from two sources delinted below and also a summary is enlisted in Table 1.

1. Source Codes from 10 lab tests arranged for first and second year students. The course name of those two courses are - CSE-142 Structured Programming and CSE-242 Data Structures.
2. Source Codes from 7 practice contests, arranged for competitive programmers of CUET.

Table 1. Summary of Dataset.

Source	Number of tests/contests	Total number of tasks	Number of submissions
Lab assignments	10	24	683
Practice contests	7	103	784
Total	17	127	1467

In this table, number and type of dataset are shown. Total number of tasks defines the sum of all of the tasks for all contests. All these tasks are applied to plagiarism detection system separately. Number of Submissions defines total number of source codes submitted in these contests.

5.2 Time Efficiency

Test Run 1. For our first experiment, we used source codes from the lab test of the introductory course for first year students. The task is an easy one and the expected solution is really small shown in Table 2. We can see that the runtime is really small.

Table 2. Experiment with Data Set 1.

Total Files	63
Total Bytes	9200
Max. Size of Token Array	419
Execution Time	0.07878 s

Test Run 2. In this experiment, we used small and introductory programs but number of files is increased which is enlisted in Table 3. The data screenshot of this test is shown in Fig. 8.

Table 3. Experiment with Data Set 2.

Total Files	236
Total Bytes	56100
Max. Size of Token Array	830
Execution Time	1.4243 s

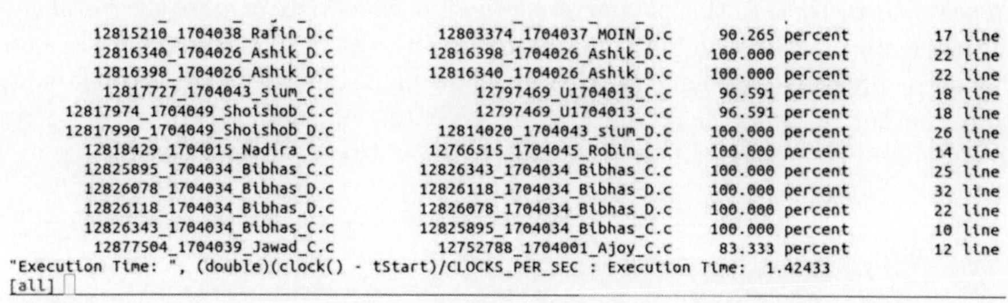
```
        12815210_1704038_Rafin_D.c          12803374_1704037_MOIN_D.c    90.265 percent     17 line
        12816340_1704026_Ashik_D.c         12816398_1704026_Ashik_D.c   100.000 percent     22 line
        12816398_1704026_Ashik_D.c         12816340_1704026_Ashik_D.c   100.000 percent     22 line
         12817727_1704043_sium_C.c            12797469_U1704013_C.c    96.591 percent     18 line
     12817974_1704049_Shoishob_C.c            12797469_U1704013_C.c    96.591 percent     18 line
     12817990_1704049_Shoishob_D.c           12814020_1704043_sium_D.c   100.000 percent     26 line
       12818429_1704015_Nadira_C.c         12766515_1704045_Robin_C.c   100.000 percent     14 line
       12825895_1704034_Bibhas_C.c        12826343_1704034_Bibhas_C.c   100.000 percent     25 line
       12826078_1704034_Bibhas_D.c        12826118_1704034_Bibhas_D.c   100.000 percent     32 line
       12826118_1704034_Bibhas_D.c        12826078_1704034_Bibhas_D.c   100.000 percent     22 line
       12826343_1704034_Bibhas_C.c        12825895_1704034_Bibhas_C.c   100.000 percent     10 line
        12877504_1704039_Jawad_C.c          12752788_1704001_Ajoy_C.c    83.333 percent     12 line
"Execution Time: ", (double)(clock() - tStart)/CLOCKS_PER_SEC : Execution Time:  1.42433
[all]
```

Fig. 8. Screenshot for Data Set 2.

Test Run 3. For experiment 3, we selected some programs designed solve advanced problem, used in competitive programming which is shown in Table 4.

Table 4. Experiment with Data Set 3.

Total Files	25
Total Bytes	15800
Max. Size of Token Array	3100
Execution Time	0.0823 s

Test Run 4. As our final experiment is enlisted in Table 5. In this regard We gathered 1368 source code from all the contests. Although they are from different contests and different tasks, as we are conducting experiment to determine if this system is efficient or not, they can be used for this purpose.

Table 5. Experiment with Data Set 4.

Total Files	1368
Total Bytes	912100
Max. Size of Token Array	4249
Execution Time	182.15 s

Because of the optimization we did in the last step of our algorithm, that is, to discard the programs that were already marked as plagiarized, we saw that the comparison process increased its speed by time. As the number of programs to compare with decreased over time, runtime of calculation of plagiarism of the new programs increased simultaneously.

5.3 Correctness

To prove correctness of this plagiarism detection system, we can use the results of our last experiment shown in Fig. 9. There are altogether 127 tasks and there's no way one source code from one task should show high similarity with another code from another program. So if our plagiarism detection method works correctly, all the high percentage of similarity reports should be from same problem.

```
       8842407_Cuet_Mystics_E.cpp        8842402_Cuet_Mystics_E.cpp     98.955 percent     35 line
       8842471_Cuet_Mystics_U.cpp             8704599_Avij33t_U.cpp     70.149 percent      9 line
        8854914_MAMUN_CUET_J.cpp        11986038_pure_amatuer_M.cpp     53.846 percent     30 line
        8855045_MAMUN_CUET_J.cpp            8757789_p_factor_J.cpp      54.266 percent     30 line
        8855940_MAMUN_CUET_U.cpp             8704599_Avij33t_U.cpp      92.157 percent      8 line
  8860377_Frustrated_coder_E.cpp      14193008_Tajir_Hasnain_A.cpp      47.107 percent     18 line
          8861238_p_factor_X.cpp       8751455_Emdadul_Hoque_X.cpp      41.418 percent     16 line
          8861343_p_factor_U.cpp            8864001_p_factor_V.cpp      56.180 percent     12 line
       8863534_Yeasin10A_E.cpp          12081368_Yeasin10A_D.cpp        49.573 percent      7 line
          8864001_p_factor_V.cpp            8726836_p_factor_D.cpp      55.932 percent     17 line
        8865958_MAMUN_CUET_I.cpp            8756251_p_factor_I.cpp      33.000 percent     38 line
  8869688_Frustrated_coder_I.cpp             8704541_Avij33t_I.cpp      34.498 percent     17 line
        8871578_nahid3334_E.cpp             8722426_p_factor_B.cpp      47.639 percent     15 line
       8873335_Yeasin10A_Q.cpp               8704652_Avij33t_Q.cpp      46.256 percent     43 line
        8873926_nahid3334_U.cpp          11913853_Naim_khan_B.cpp       47.126 percent      7 line
            8874533_Xaqir_J.cpp             8757789_p_factor_J.cpp      58.156 percent     23 line
  8874933_Frustrated_coder_V.cpp    8836620_Frustrated_coder_T.cpp      67.797 percent      8 line
            8876128_Xaqir_U.cpp           12859132_Ratul1997_C.cpp      45.283 percent     15 line
       8876196_Cuet_Mystics_I.c        11986038_pure_amatuer_M.cpp      38.721 percent     20 line
     8876254_Cuet_Mystics_V.cpp        8819156_Cuet_Mystics_W.cpp       58.108 percent      8 line
            8877028_Xaqir_V.cpp      12049596_1504054_nahid_O.cpp       39.655 percent      5 line
"Execution Time: ", (double)(clock() - tStart)/CLOCKS_PER_SEC : Execution Time:  182.15
[all]
```

Fig. 9. Screenshot for Data Set 4.

We chose around 60 pairs of code with ≥90% similarity and checked manually that all of them are from same folder or same task. Thus the correctness of this program is proved.

5.4 Evaluation

We conducted experiment 2 on MOSS too. The result is shown in Fig. 10.

```
Uploading 12818429_1704015_Nadira_C.c ...done.
Uploading 12825895_1704034_Bibhas_C.c ...done.
Uploading 12826078_1704034_Bibhas_D.c ...done.
Uploading 12826118_1704034_Bibhas_D.c ...done.
Uploading 12826343_1704034_Bibhas_C.c ...done.
Uploading 12877504_1704039_Jawad_C.c ...done.
Query submitted.  Waiting for the server's response.
http://moss.stanford.edu/results/758660385
./moss *.c  0.05s user 0.02s system 0% cpu 1:13.99 total
[all]
```

Fig. 10. MOSS Runtime for Data Set 2.

It took MOSS 1 min 13 s to provide us the result whereas our system took 1.4 s. Although MOSS is efficient and used widely in academic activities, our plagiarism detection gives us the benefit of using it from local server and without losing any time on request-response. The output representation of MOSS is not very organized. It shows similarity report of every pair of all the uploaded source code. Our system shows the best match with a specific program. Another problem with MOSS is that although it is free to use, the specifics of the making of this system is yet unknown. Schleimer et. al [13] published the base method of their system – Winnowing method, by which we implemented this system. And by doing this, we opened a door to work on this project more specific and solve limitations that MOSS has.

We can also agree that this plagiarism detection system can be used in any kind of programming event. Assuming that there are 120 students in a class, participating in a lab test or 120 teams participating in a national programming contest if number of problems is around 10 and every individual/team submits every problem this plagiarism detection will work efficiently then too.

6 Conclusion

The main objective of this project was to implement a plagiarism detection system and host it on local server. As all the public and efficient plagiarism detection service are web-based and it takes some time and effort to use or even integrate the system, not many universities has plagiarism detection service for their lab assignments and practice contests. Although this project has still some limitations, we hope those will be solved in near future. There are a few aspects that still need to be worked on. We assumed a fixed value for W and K. The efficiency of winnowing algorithm depends on these value. To find the perfect trade-off, more analysis should be done. We used a fixed threshold value for our last optimization technique. In real, this threshold should vary upon the nature of the task and solution. More analysis should be done on this threshold value. An in house database of source codes is used so, we plan to test our system with some standard databases of source codes if available.

Although winnowing technique overcomes the issues of position independence and noise suppression, it cannot detect if someone copies 100% of someone else's work and add some unnecessary statements to reduce the similarity score. A graph based solution might be helpful for this problem. The techniques that are used in this project are language independent, although the keywords vector of a language should be provided to use a language. In future, more language should be added.

References

1. Aho, A.V., Lam, M.S., Sethi, R., Ulman, J.D.: Compilers: Principles, techniques and tools (for Anna University), 2nd edn. Pearson Education India, Chennai (2003)
2. Bostan, A.S.: Winnowing algorithm for program code (2017)
3. Ji, J.H., Woo, G., Cho, H.G.: An experience of detecting plagiarized source codes in competitive programming contests. ACM SIGCSE Bull. **40**(3), 369–369 (2008)
4. Joy, M., Luck, M.: Plagiarism in programming assignments. IEEE Trans. Educ. **42**(2), 129–133 (1999)
5. Karnalim, O., Aldiansyah, A.: Python source code plagiarism attacks in object-oriented environment. Comput. Eng. Appl. J. **6**(3), 87–93 (2017)
6. Karnalim, O., Budi, S.: The effectiveness of low-level structure-based approach toward source code plagiarism level taxonomy. In: 6th International Conference on Information and Communication Technology (ICoICT), pp. 130–134. IEEE (2018)
7. Karp, R.M., Rabin, M.O.: Efficient randomized pattern-matching algorithms. IBM J. Res. Dev. **31**(2), 249–260 (1987)
8. Kermek, D., Novak, M.: Process model improvement for source code plagiarism detection in student programming assignments. Inf. Educ. **15**(1), 103–126 (2016)
9. Lim, J.S., Ji, J.H., Cho, H.G., Woo, G.: Plagiarism detection among source codes using adaptive local alignment of keywords. In: Proceedings of the 5th International Conference on Ubiquitous Information Management and Communication, p. 24. ACM (2011)
10. Liu, C., Chen, C., Han, J., Yu, P.S.: GPLAG: detection of software plagiarism by program dependence graph analysis. In: Proceedings of the 12th ACM SIGKDD International Conference on Knowledge Discovery and Data Mining, pp. 872–881. ACM (2006)
11. Prechelt, L., Malpohl, G., Philippsen, M.: Finding plagiarisms among a set of programs with JPlag. J. UCS **8**(11), 1016 (2002)
12. Rahal, I., Wielga, C.: Source code plagiarism detection using biological string similarity algorithms. J. Inf. Knowl. Manage. **13**(03), 1450028 (2014)
13. Schleimer, S., Wilkerson, D.S., Aiken, A.: Winnowing: local algorithms for document fingerprinting. In: Proceedings of the 2003 ACM SIGMOD International Conference on Management of Data, pp. 76–85. ACM (2003)
14. Sue, Y.: How plagiarism detection works at HackerRank (2018). https://blog.hackerrank.com/how-plagiarism-detection-works-at-hackerrank/
15. TooNewbie: How plagiarism checkers are implemented? Let's help CodeForces to find cheaters! (2019). https://codeforces.com/blog/entry/65874

Developing an IoT Based Water Pollution Monitoring System

Md. Mazharul Islam[1], Mohammad Shamsul Arefin[1(✉)], Sumi Khatun[2], Miftahul Jannat Mokarrama[3], and Atqiya Munawara Mahi[1]

[1] Department of CSE, Chittagong University of Engineering and Technology, Chattogram, Bangladesh
mazhar.cuetcse@gmail.com, atqiymunawaramahi@gmail.com, sarefin@cuet.ac.bd

[2] Department of CSE, Bangladesh University of Business and Technology, Dhaka, Bangladesh
sumicuet11@gmail.com

[3] Department of CSE, Port City International University, Chattogram, Bangladesh
mokarramacse07@gmail.com

Abstract. In the area of technological studies Internet of things (IoT) has become an interesting subject matter. It is the interconnection of things with each other over the internet. Average people often assume IoT in phrases of self-sustaining vehicles and smart houses; however, a number of the great programs of the IoT era are in intensely practical fields. One such example is the 'water pollution monitoring system'. With the growth of the populace and industrialization of nations at some point in the globe, water pollution has ended up a major concern for anyone. With the assist of the era, researchers have added IoT primarily based smart water pollutants tracking machine that ensures a discounted amount of time and power required to provide analytical offerings. In this task, the design of IoT based water pollutants tracking machine is proposed, which can screen the high-quality water in real-time. This device includes a few sensors which degree the water quality parameter inclusive of pH, Total Dissolved Solids (TDS), and temperature. The deliberate qualities from the sensors are prepared with the aid of a microcontroller named Arduino and transmitted to the middle controller this is NodeMcu via Serial Wire communication. Then NodeMcu passes the facts into internet-server via wireless. Finally, sensor results can view internet browser software and phone the usage of the android app. The goal is to make the machine clean for the rural location of Bangladesh, which is an absence at the back of the proper water pollutants tracking gadget. So, remote tracking and guidance focus by sharing statistics in an authenticated manner is the main aim.

Keywords: IoT · TDS · pH · Temperature · Arduino · Nodemcu

J. I.-Z. Chen et al. (Eds.): ICIPCN 2020, AISC 1200, pp. 561–573, 2021.
https://doi.org/10.1007/978-3-030-51859-2_51

1 Introduction

Globally, the presence of drinking water without contamination within the 21st century is becoming a primary assignment. Worldwide regulating our bodies made up of the United Nations (UN) and the World Health Organization (WHO) have acknowledged the human right to safe, non-stop, secure and sufficient body water and low-cost water for non-public and domestic use. According to WHO reports, 844 million human beings lack even a simple supplier of drinking water, which includes 159 million people relying on surface water [20]. Impure drinking water can lead to life-threatening diseases that include diarrhea, cholera, dysentery, typhoid, and polio. The studies report alarmingly that diarrhea alone causes about the loss of life of five lakh people every year. Hence diverse water parameters of high quality along with dissolved oxygen (DO), conductivity, pH, turbidity, and temperature must be controlled in real time. Recent times, however, water has faced challenges with the best tracking in real-time due to global warming confined water assets, growing population etc. Subsequently, higher methodologies need to grow to reveal the water quality parameters in real time.

Degrading water supplies have become normal in human distress. The traditional water quality control approaches include the guideline sequence of water trends from specific locations. Those water samples were examined using analytical techniques in the laboratory. These approaches are time consuming and not to be considered successful any more. In addition, the latest methodologies consist of an evaluation of different kinds of water-nice parameters including physical and chemical parameters. Traditional approaches for detecting water pollution have hazards such as complicated techniques, long ready time for tests, low precision measurements and high value. Therefore, in real time there is a need for the non-stop monitoring of high-quality water parameters. By focusing on the above problems, a low-price water pollution monitoring system must be expanded and designed which can monitor water in real time using the IoT environment. In this proposed method water quality parameters are measured along with pH, TDS, and temperature by means of the special water monitoring sensors.

According to the World Health Organization (WHO) by 2025, half of the inhabitants of the sector will live in water-scarce regions. This system can offer a low-priced option to water disasters for a growing country like Bangladesh, where human beings cannot have the funds for high-value water purifier. It would be very effective for them as the usage of water is vast for not only for the urban people but also for the rural people.

The leading contributions of our project are as follows:

i. For data transmission from Arduino to NodeMcu, Serial communication is used.
ii. For uploading data into the local host, the wifi module is used.
iii. The decision for water drinkability is taken by temperature, pH, and TDS sensor parameters.

The rest of the paper is organized in this way: Sect. 2 gives a big picture of relevant works about IoT based water pollution monitoring system, Sect. 3 is arranged for discussing the framework designed for IoT based water pollution observing framework.

In Sect. 4, the experimental results and assessment of the system are described. Finally, in Sect. 5 the conclusions, limitations, and suggested future improvements are presented.

2 Related Works

Regan et al. [16] structured shrewd water satisfactory tracking gadgets. In that machine, they brought water first-rate keen sensors. The sensors ship statistics to the device which gathers measurements from all the nodes via the wireless network. These insights are given to the far-flung server through the General Packet Radio Service (GPRS) system and individuals can observe measurements with the control of the remote. This framework is especially versatile, faster and user well-disposed, anyway it's so costly as a direct result of such slick sensors. Also, the size of the sensors isn't dependable for the water valve.

Rasin et al. [15] advanced a water pleasant observing device the use of ZigBee-based total wi-fi sensor community. In the introducing machine, the sensors are linked to an unpair circuit that is associated with the module named zmn2405hp of ZigBee. The recipient feature ZigBee is attached to the personal computer that shows the graphical user interface (GUI) of the circuit. Right now, high power ZigBee is utilized and it might be done to a little area of the network. Additionally, the base station is significant for the statistics garage.

Haron et. al [5] advanced a water exceptional machine for checking out cost-expending occupations of the guide track. Right now, estimated information of graded sensors is amassed through the facts package which offers information to the facts preparing unit through GSM modem for water. In the records preparing unit the records from exceptional sensors are separated and it's far persistently as contrasted and the exact parameters of the sensor esteem. On the off chance that the water isn't continually meeting its first-rate parameter, the alarm signal is there that is connected to the ringer. This contraption isn't trustworthy for significant distance additionally it will practice to the least complex single measure of water supply.

Ning et al. [2] structured an observing machine for the quality of water. This machine uses water excellent sensors for acquiring statistics, from manufacturing water and metropolitan water stockpiling, are amassed at the station where the information is handled. TCP/IP ethernet network facilitates this handled data to despatch to the main station. Again, the main station separates the facts and provides them to the surroundings division and open branch for the usage of the net. This device has improved records accuracy, reliability, and performance, also it offers powerful facts management and incorporated data structures. But, the system has the downside of not offering actual-time tracking of water attributes.

Qiao et al. [14] structured an on-line observing device of water fine essentially dependent on GPRS. The GPRS records transmission is used for sampling and sending the applicable facts to the checking center. The developed device monitors the quality of water parameters remotely which is faster and real-time than the preceding devices utilized for controlling and sensing water quality. Dong et al. [6] evolved the water monitoring device based totally on Wireless Sensor Network (WSN). This gadget primarily based on a wireless sensor community that consists of wireless water exceptional tracking network and far-flung facts center. The wireless sensor network is built on the Zigbee community protocol. WSN sample the water pleasant, and sends the facts to the internet with the help of the GPRS, having integrated TCP/IP protocol used for records transmission.

He et al. [7] designed long-distance water quality tracking gadgets using wireless sensors. In this system, all sensors transmit the statistics acquisition kit collected records sent by the wi-fi sensors.

Amruta et al. [1] created sunlight based fueled water observing machine utilizing a remote sensor arrange. Right now, WSN innovation controlled the use of solar panels. The machine comprises of base station and node. The node gathers that receivings from the exceptional wi-fi sensor. The base station clips the node through the ZigBee technology that is controlled with the guide of the solar panel. The machine is low priced but stops if the solar panel is not charged on account of nearby surroundings impact.

Le Dinh et al. [9] explored a WSN to gather real-time water parameters. Silva et al. [18] introduces WSN which traces water pollutants with the help of ZigBee and WiMAX systems via the web. The machine estimated various parameters. The device gathers and prepares real-time data from sensors. The processed data is then guided through the ZigBee gateway to the web server by utilizing the WiMAX system to screen water from a wide gap. The long-distance sensing system based on the ZigBee network is proposed by Papageorgiou et al. [12]. It is analyzed by WSN and despatched statistics to the net by utilizing the GPRS remotely.

A solar sensor network is developed by Satish et al. [17]. The system uses the ZigBee module powered by sunlight for gathering data by Base Station (BS) from distanced sensors remotely. Bhatt et al. [3] builts a real-time water quality monitoring system for the secure delivery of ingesting water. This framework is made out of various water parameters. The data is handled using a microcontroller and the ultimate records from the sensors are regarded from a web server [11]. Jing et al. [8] created a longterm wi-fi gadget for water quality monitoring using a microcontroller, that is based on GPRS. Their used microcontroller architecture is complex and expensive. Purohit et al. [13] established real-time gadgets that use liquid crystal display (LCD), GSM, sensors, and analog to digital converter (ADC). These gadgets are constrained because of their bolstered by cutting edge devoted electronic circuit boards [4, 10, 13]. Moreover, Radio-frequency identification (RFID) labels, sensor approach, cell techniques, smart technique, etc. applications are developed [19].

From, all the pointed considerations we come to understand that every single one of the devices comprised of some difficulty even though they can't address the motivation behind the cost-effective, non-stoping checking of water parameters, and real-time. Along these lines, to triumph over this downside that drives us to be progressed and format the new methodology that will be easy to use, low-cost, and real-time.

3 System Architecture and Design

Here, we have evolved a wise IoT based water quality controlling machine in which TDS sensors and pH are deployed in water tanks in the residential region which talks via the Arduino microcontroller to the NodeMcu processors wherein some parametric values are deployed closer to analyzing the facts captured for predicting the water exceptional as exact or horrific. The predicted water fine with pH and TDS for distinct types of water are updated on a webpage in the cloud for water branch personnel to do so earlier than affecting the residents. The general device supplied on this take a look at is subdivided into five modules of the device working mechanism which is shown in Fig. 1. Each purpose and working procedure of each module is described in the subsequent sections.

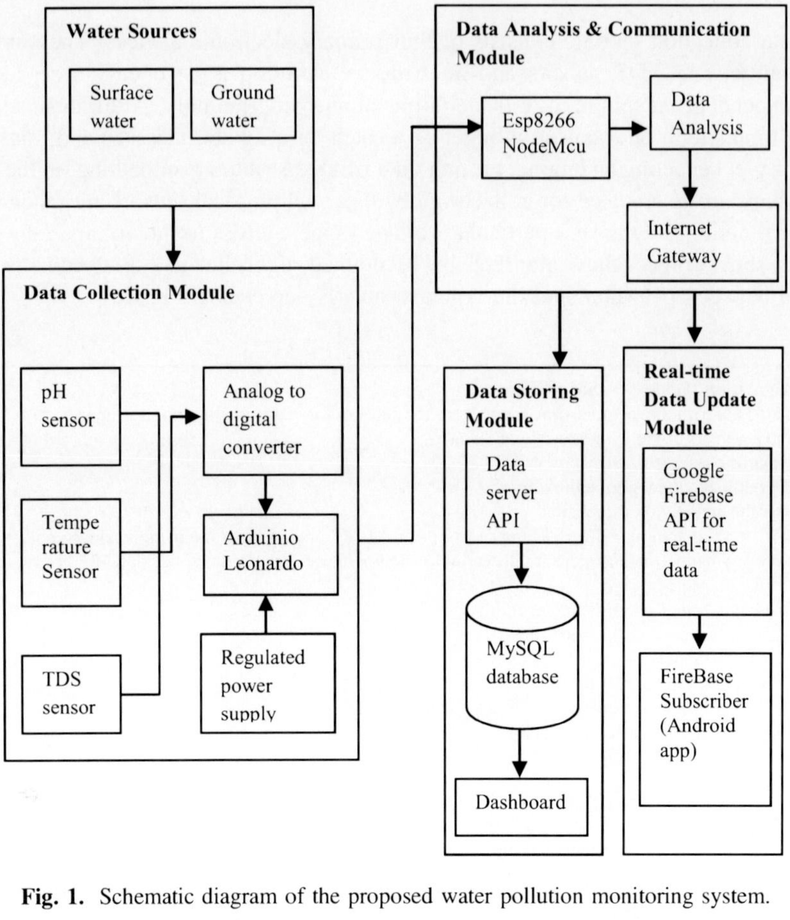

Fig. 1. Schematic diagram of the proposed water pollution monitoring system.

3.1 Sample Water Sources for Drinkable Water

There are several water sources for drinkable water. Two of them are major: Surface water and Groundwater. Humans use surface and groundwater each day for a ramification of purposes, inclusive of ingesting, cooking, and basic hygiene, further to leisure, agricultural, and commercial activities. In this project, we've used water samples from different sources, like pond water, tap water, lake water, pure drinkable water, tube well water. Pure drinkable water and tube well water is mostly used drinkable water. But in rural area, people use pond water or river water for drinking. Most of the time, the quality of this type of water is not appropriate for drinking. So, for drinking water, we've collected different types of water sources. And then, we measure the pH, TDS, and temperature of this water.

3.2 Data Collection Module

The data collection module consists of four primary electronic devices. They are pH, temperature, and TDS sensors and an Arduino. Arduino is an open-source gadgets stage dependent on simple to-utilize equipment and programming. Arduino sheets can peruse inputs light on a sensor, a finger on a catch, or a Twitter message and transform it into a yield enacting an engine, turning on a LED, distributing something on the web. The three sensors are used for collecting raw data. All physical sensors are made of an electrical node. They have a particular voltage range. Across to the voltage range, the sensors show some values modified by its defined algorithm in Fig. 2 and the calibration between pH value and the voltage value is depicted in Fig. 3.

```
Algorithm: Sensor_Value_Collection
//This algorithm takes real-time value from the sensors for water quality monitoring system
and //transform the values as a byte-stream
1.begin
2.Establish connection with Arduino
3. if the connection established
4.        Fetch the value of voltages from A0, A1, A3 analog input pin using analogRead()
          function and store it into variables 'temperature', 'pH'; 'TDS' has default gravity
          library.
5. Measure the floating value as follows
6. Temperature = (5.0*temp*1000.0)/(1024*10)
7. pH=-6.75*pH+31.0
8.if (sensor values! =0) then
          put a semicolon (;) after each sensor value for making the byte-stream and print
          values in serial.print1().
9.        Delay (500).
10.else
11.       read values again.
12.End
```

Fig. 2. The algorithm inside of the IoT-based water pollution monitoring system.

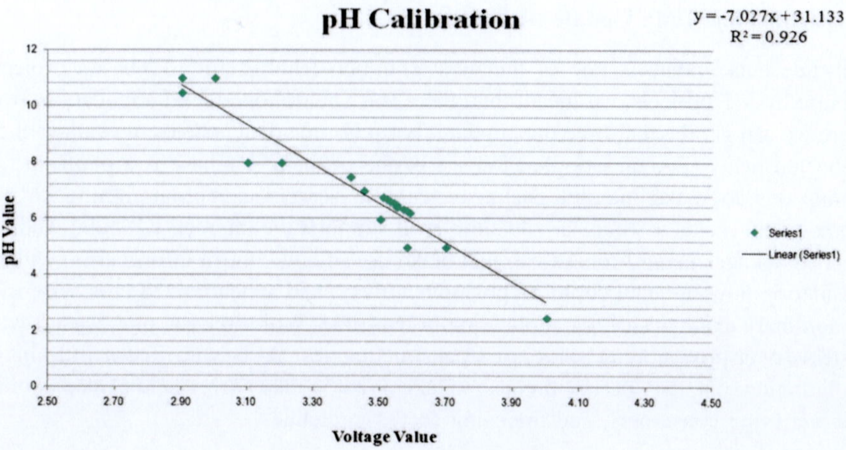

Fig. 3. pH calibration graph.

3.3 Data Analysis and Communication Module

The data analysis & communication module is constructed with mainly NodeMcu. Arduino has no WiFi module; it has only the Bluetooth module. But as we want to transmit data at a fast rate, so we need to use the WiFi module. NodeMcu has 802.11n protocol WiFi module. When the byte-stream comes from the Arduino via TX, then the NodeMcu catches it from RX pin, i.e. D7 pin. As it was a byte-stream, so we need to separate the value. Here we've used a simple C++ coding technique to separate all sensor's value. After that, we have to send the value in Hyper Text Transfer Protocol (HTTP) client and Google FireBase. So, send the data into HTTP client and Google Firebase is the communication sub-module task. Another big thing is to analyze the data. As there are several parameters to test the water, so there are several threshold values, from where we can determine the decision parameter.

3.4 Data Storing Module

Data storing module is the most important part of this project. Because by storing this huge amount of data in a server, we can do two things mainly. First of all, we can analyze the data for future work. And secondly, we can see the real-time update of this data. NodeMcu sends at least three pairs of values per five seconds. So, every day, there'll be a large quantity of data. For further analysis, this data preservation is very essential. We store the data into MySql database. MySql database is one of the best database management systems for its reliability, consistency, and security. In this project, we've used the MySql database management system. Here we've created a table with nine parameters. The ID is auto-incremented and the time is auto-generated. Without these two, all other values have come from the sensor via Arduino and NodeMcu.

3.5 Real-Time Data Update Module

Real-time data update is one of the most common features for IoT based projects. Because by IoT projects, we meant that, there is a Smartphone-based application, from where we can get the real-time data update. For real-time data updates, nowadays there are hosted many servers. Google Firebase is one of them. Firebase is a platform that permits one to make internet packages with no server-facet programming so that improvement seems to be faster and simpler. It supports the internet, IOS, and Android clients. Packages using firebase can just make use of and control information, without considering how the data could be positioned away, and synchronized crosswise over extraordinary examples of the application in real-time. With firebase, one doesn't want to stress over-provisioning servers or constructing rest APIs with just a little bit of configuration; one may permit firebase to take a risk to take each essential step: storing data, verifying consumers, and imposing access guidelines.

4 System Implementation and Experiments

4.1 Building the Setup

The system setup uses a NodeMcu for its overall controlling tasks and an Arduino-Mega for sensing tasks. Despite having NodeMcu, a microprocessor-based single-board computer (SBC), integrating Arduino with the platform may seem redundant as well as unnecessary. Even at the very beginning stage only a single piece of NodeMcu was used for all sensing and controlling tasks of the whole system. Three sensors: One digital and two analogs are selected for the connection of analog and digital input pins of Arduino-Mega. The platform also has a Wifi module, which works in the 802.11n protocol. USB ports connect the rest three items: temperature, TDS, and pH sensor. The Arduino is also connected to the NodeMcu through serial communication of TX and RX ports. Thus we have connected all the sensors to the Arduino and read them through the NodeMcu. An external battery of 11200 mAh capacity deploys the whole setup when applied outdoor.

Temperature Sensor with Arduino. DS18B20 is the temperature sensor, which we used to measure the temperature in degrees celsius. The sensor has three-pin. The first pin is for GND, which is marked by green color. The second pin is for the sensor's digital data. And the third pin is +5 V Vcc. For connecting this ds18b20, we need a digital input in Arduino. So in Arduino, we've used the 2nd digital pin for digital output. Here one important thing is that we have to connect the 4.7 kΩ pull-up resistor between the data and Vcc. The pin connection of this sensor is given in Table 1.

Table 1. Pin mapping for temperature sensors.

SL	DS18B20	Arduino
1.	+5 Vcc	+5 Vcc
2.	GND	GND
3.	DATA	PIN 2 (digital)

pH Sensor with Arduino. A pH sensor is an analog sensor. Its output is quantized by analog to digital converter. After converting the analog signal, we finally determine the digital output values. To connect the sensor, we need an analog pin in Arduino. In Arduino Leonardo, we've used the A0 pin for the pH sensor. And we've used a breadboard to connect the +5 V Vcc and GND. The pin mapping of the pH sensor is given in Table 2.

Table 2. Pin mapping for pH sensor.

SL	DS18B20	Arduino
1.	+5 Vcc	+5 Vcc
2.	GND	GND
3.	DATA	DATA A0 (Analog)

TDS Sensor with Arduino. The pin connection of this sensor is represented in Table 3. It is an analog sensor. Its output is quantized by analog to digital converter. After converting the analog signal, we finally determine the digital output values. To connect the sensor, we need an analog pin in Arduino. In Arduino Leonardo, we've used the A1 pin for the TDS sensor. And we've used a breadboard to connect the +5 V Vcc and GND.

Table 3. Pin mapping for the TDS sensor.

SL	DS18B20	Arduino
1.	+5 Vcc	+5 Vcc
2.	GND	GND
3.	DATA	DATA A0 (Analog)

Arduino and NodeMcu Connection. For transmitting data from Arduino to NodeMcu, we need serial communication. In the previous chapter, we briefly discussed the three types of serial communication between Arduino to Arduino or Arduino to NodeMcu. As we have a single byte stream for three sensors data, so we can easily use here the UART communication. Here TX0 and RX1 are used for TX, RX port in Arduino. And RX, TX is used in NodeMcu. The pin mapping is listed in Table 4. Two important things are the GND must have to connect on both sides of two microcontrollers. And another most import thing is that the RX port of NodeMcu must have to connect with the TX port on the Arduino and vice versa. Because, the data which will be transmitted from Arduino TX (T for transmission), that will be received by the RX (R for receive) of NodeMcu.

4.2 System Design for Software Implementation

In Arduino.cc IDE using C and C++ programming language programs our system's modules are written for the functioning of NodeMcu and Arduino. The main software

Table 4. Pin mapping for connecting NodeMcu & Arduino.

SL	NodeMcu	Arduino
1.	RX	TX
2.	TX	RX
3.	GND	GND

module initializes the water sensors, analog to the digital converter, WiFi module, and serial port and also checks all sensors connections at the side of the battery level. An android smartphone application has been evolved to make communication with the tool using a WiFi protocol. An 802.11n WiFi protocol has been used to set up a connection between the handheld tool and the smartphone. The android application offers a front view to show water quality parameters alongside temperature, TDS, pH, and selection parameter. Not only that, but we also store the data in Thingspeak, an online-based free IoT database platform. It provides a facility to collect location information of water sites and store them on the smartphone in an excel file format. The thingspeak also provides the data for analyzing systems by MATLAB. As we use the default thingspeak API, so when the data from NodeMcu uploads to the thingspeak, then an automatic curve will be generated for displaying the fluctuations of those parameters. The system's UI with a real-time datasheet is shown in Fig. 4.

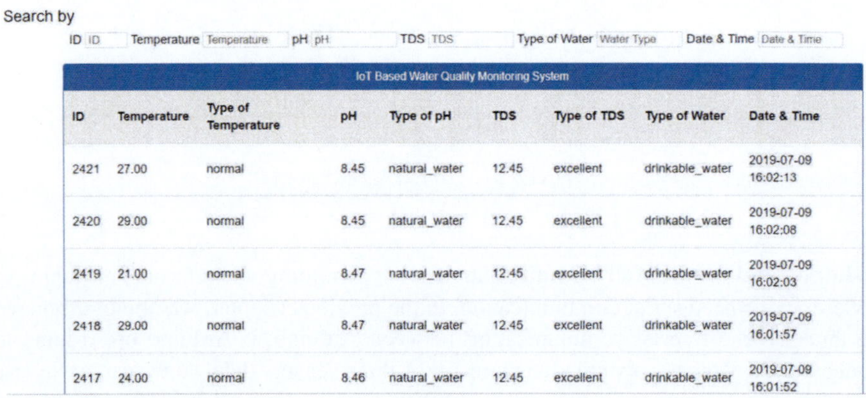
Search by
ID ID Temperature Temperature pH pH TDS TDS Type of Water Water Type Date & Time Date & Time

IoT Based Water Quality Monitoring System

ID	Temperature	Type of Temperature	pH	Type of pH	TDS	Type of TDS	Type of Water	Date & Time
2421	27.00	normal	8.45	natural_water	12.45	excellent	drinkable_water	2019-07-09 16:02:13
2420	29.00	normal	8.45	natural_water	12.45	excellent	drinkable_water	2019-07-09 16:02:08
2419	21.00	normal	8.47	natural_water	12.45	excellent	drinkable_water	2019-07-09 16:02:03
2418	29.00	normal	8.47	natural_water	12.45	excellent	drinkable_water	2019-07-09 16:01:57
2417	24.00	normal	8.46	natural_water	12.45	excellent	drinkable_water	2019-07-09 16:01:52

Fig. 4. Snapshot of datasheet for real-time sensor value and decision parameter.

4.3 Performance Evaluation

A comparison is made by the data collected from five sources of water, namely between tap water, pond water, and bottled mineral water, rainwater, tube well water. It was found that the pH levels of the water from all sources showed almost similarity with the only exceptions depending on temperature. Whereas, variations in salinity concentrations in different types of water constituted the variance of TDS values. On this project, we've tested different types of water from many water resources. They were pond water, fresh drinkable water, and regular tap water. A testing sample is shown in Fig. 5. If the value of the pH of the water is more than 8.5, then its manner the water is alkanets. If it is beneath 7, then the water is acidic. The tap water pH is 7.4. Low pH valued tap water is dangerous to human fitness because it can affect on skin inflammation and eye burn in case you clean your eyes with the aid of this water.

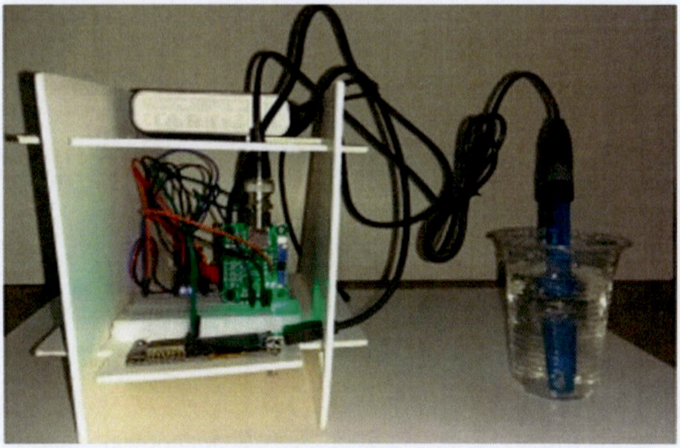

Fig. 5. Testing water quality for tap water.

Different disinfectants could be as effective for the reason of low pH, chlorine. Although people won't observe any changes while washing but the virus or bacteria might be dangerous when it touches off your skin. So, tap water pH value ought to be saved among 7.2–7.8. It was gathered three taps of water from several locations to sure our device completely okay. Three samples showed a pH average of 7. On the other hand, one showed pH studying of 6.5 that was bellowing a normal variety of pH. So, we got here that our procedure is running as we need for the pH values. Not only that, for the water, but it also should not have a temperature better than 29 °C. All 3 tap glasses of water we examined for temperature; it gave the values among 26–27 °C. After testing more than one instance all through days and nights so one can find it out any unusual outputs. Happily, we cannot find any temperature that is more than a variety targeted. Pond water is not drinkable without filtering. So, it is wastewater. It has typically pH values in the range of 6.5 to 9.5 before water special use. Our other water sample additionally gave a less pH value of 5.5. For that reason, it no longer secures for drinking or using in any domestication cause. If the pH is between 7 to 8.5 of groundwater is pure and it can be drinkable. Typically, in the night time, the pond water temperature has to be roughly 20 °C whether as day, it turns into warm for daylight and the water temperature ought to be around 35 °C. Pond water samples from unique water resources were examined to establish a reference at the parameters for every water kind. The pH accuracy graph of our system is represented in Fig. 6.

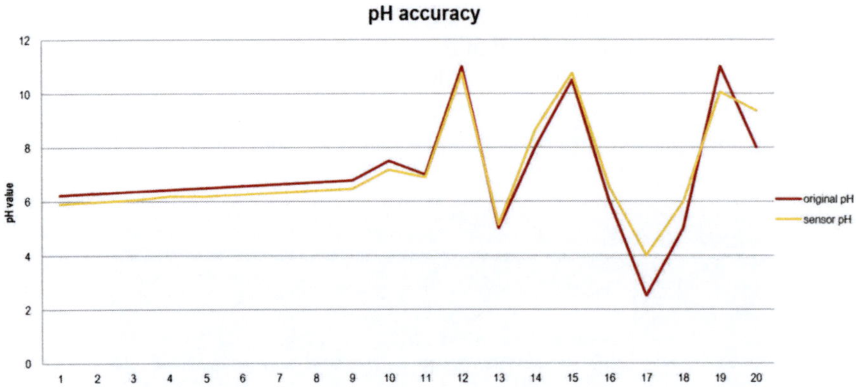

Fig. 6. pH accuracy.

5 Conclusion

Water pollution monitoring is very much essential nowadays as it's far fed on using citizens. Traditional water quality tracking and some of the era primarily based water testing were given IoT of challenges. This proves that water quality may be monitored automatically with no human involvement. In destiny, IoT based water quality tracking machines can be prolonged no longer just for storage tank however additionally for selecting ponds, rivers, and water pipes too. The same paintings can be prolonged by using looking into different water parameters rather than simply pH and TDS and for that reason control the float of water-primarily based on water quality. Ultimately the security and integrity of the records of statistics need to be secured while transmitting for analysis towards prediction and actuating the water sensors and other microcontrollers too.

Our current system for water pollution monitoring has been used for water samples collected from different sources. In the future, we plan to upgrade our system for continuous monitoring of water at different water supply systems.

References

1. Amruta, M.K., Satish, M.T.: Solar powered water quality monitoring system using wireless sensor network. In: 2013 International Multi-Conference on Automation, Computing, Communication, Control and Compressed Sensing (iMac4s), pp. 281–285. IEEE (2013)
2. An, N., An, Y.: A monitoring system for water quality. In: International Conference on Electrical and Control Engineering, pp. 4615–4617 (2010)
3. Bhatt, J., Patoliya, J.: IoT based water quality monitoring system. IRFIC, February 2016
4. Chen, S., Xu, H., Liu, D., Hu, B., Wang, H.: A vision of IoT: applications, challenges, and opportunities with China perspective. IEEE Internet Things J. **1**(4), 349–359 (2014)

5. Haron, N.S., Mahamad, M.K., Aziz, I.A., Mehat, M.: Remote water quality monitoring system using wireless sensors. In: Proceedings of the 8th WSEAS International Conference on Electronics, Hardware, Wireless and Optical Communications (EHAC 2009), Cambridge, pp. 148–154 (2009)
6. Haron, N.S., Mahamad, M.K.B., Aziz, I.A., Mehat, M.: A system architecture for water quality monitoring system using wired sensors. In: International Symposium on Information Technology, pp. 1–7. IEEE (2008)
7. He, D., Zhang, L.X.: The water quality monitoring system based on WSN. In: 2nd International Conference on Consumer Electronics, Communications and Networks (CECNet), pp. 3661–3664. IEEE (2012)
8. Jan, Jing, M.: The design of wireless remote monitoring system of water supply based on GPRS. In: International Symposium on Computer Science and Society, pp. 29–31. IEEE (2011)
9. Le Dinh, T., Hu, W., Sikka, P., Corke, P., Overs, L., Brosnan, S.: Design and deployment of a remote robust sensor network: experiences from an outdoor water quality monitoring network. In: 32nd IEEE Conference on Local Computer Networks (LCN 2007), pp. 799–806. IEEE (2007)
10. Nalajala, P., Godavarth, B., Raviteja, M.L., Simhadri, D.: Morse code generator using microcontroller with alphanumeric keypad. In: International Conference on Electrical, Electronics, and Optimization Techniques (ICEEOT), pp. 762–766. IEEE (2016)
11. Nalajala, P., Lakshmi, S.B.: A secured IoT based advanced health care system for medical field using sensor network. Int. J. Eng. Tech. **2**(20), 105–108 (2018)
12. Papageorgiou, P.: Literature survey on wireless sensor networks (2003)
13. Purohit, A., Gokhale, U.: Real time water quality measurement system based on GSM. IOSR J. Electron. Commun. Eng. (IOSR-JECE) **9**(3), 63–67 (2014)
14. Qiao, T.z., Song, L.: The Design of multi-parameter online monitoring system of water quality based on GPRS. In: International Conference on Multimedia Technology, pp. 1–3. IEEE (2010)
15. Rasin, Z., Abdullah, M.R.: Water Quality monitoring system using zigbee based wireless sensor network. Int. J. Eng. Technol. IJET **9**(10), 24–28 (2009)
16. Regan, F., Lawler, A., McCarthy, A.: SmartCoast Project - Smart Water Quality Monitoring System. Environmental Protection Agency, Ireland (2009)
17. Satish, T., Amruta, K.: Solar powered water quality monitoring system using wireless sensor network. In: IEEE Conference on Automation, Computing, Communication, Control and Compressed Sensing, pp. 281–285. IEEE (2013)
18. Silva, S., Nguyen, H.N., Tiporlini, V., Alameh, K.: Web based water quality monitoring with sensor network: employing ZigBee and WiMAX technologies. In: 8th International Conference on High-capacity Optical Networks and Emerging Technologies, pp. 138–142. IEEE (2011)
19. Stankovic, J.A.: Research directions for the Internet of Things. IEEE Internet Things J. **1**, 3–9 (2014)
20. 2.1 billion people lack safe drinking water at home, more than twice as many lack safe sanitation, 12 July 2017. https://www.who.int/news-room/

An Innovative Approach for Aerial Video Surveillance Using Video Content Analysis and Indexing

Jaimon Jacob[1(✉)], M. Sudheep Elayidom[2], and V. P. Devassia[3]

[1] Department of Computer Science and Engineering, Model Engineering College, Ernakulam, Kerala, India
jaimon@mec.ac.in

[2] Division of Computer Engineering, School of Engineering, Cochin University of Science and Technology, Kochi, Kerala, India
sudheepelayidom@gmail.com

[3] Department of Electronics and Communication Engineering, St. Joseph's College of Engineering and Technology, Palai, Kerala, India
vpdmec@yahoo.com

Abstract. There is growing interest in the use of video cameras to conduct aerial surveillance. Compared to conventional framing cameras, videos can monitor ongoing action within a scene, and adjust the camera automatically to track the action. Video documents dynamic events that are not available from aerial photos. This enables the feedback and control of dynamic event-based actions and delivers timely and important knowledge and interpretation that is not accessible elsewhere. Video recordings can be used to map and geo-locate dynamic objects or events in real-time. In this paper, we propose an innovative approach for aerial video surveillance using video content analysis and indexing, to monitor the violation of activities during the emergency lockdown period. Instead of continuously monitoring the video content, this method detects the free movement of persons and vehicles along with the location details from the geotagged videos sent by drone video cameras.

Keywords: Video aerial surveillance · Video content analysis · Video indexing

1 Introduction

Disease outbreaks and epidemics are diverse phenomena with various closely interwoven aspects: medical, social, and political. Community participation is a tool for addressing disease in social dimensions. Public participation is necessary for successful control of infectious diseases, through embracing interventions in public health. Emergency lockdowns are enforced when life threats or injury danger to humans become imminent.

Drones offer live-streaming, logically cost-effective applications for emergency services. Increase surveillance with drones and also increase the periphery of vision in the sky with the instantaneous transmission of footage to the control center. This drone

J. I.-Z. Chen et al. (Eds.): ICIPCN 2020, AISC 1200, pp. 574–583, 2021.
https://doi.org/10.1007/978-3-030-51859-2_52

operation may cover various tasks in outdoor activities, from tracking major environmental events such as flood control to pirates or refugee vessels locating or detecting them. Through programming, it can stream to a control room anywhere very quickly with very little delay on the live streaming drone with a global IP address. The geotagged video stream by drone will be received in the control center. Continuous monitoring of the video content would be difficult to track for violation of lockout policies.

Within this paper, we suggest an innovative approach to aerial video surveillance to track the free movement and crowding of people and vehicles using video content analysis and video indexing. Video sending by the drone will be interpreted as a script in this approach, which searches for the specific keywords that lead to lockdown policies being violated. At the geographical locations identified by geotagged video images, the authorities concerned can initiate actions against the violations.

The major contributions in this paper are made using the video content analysis and video indexing techniques to implement machine learning in the video aerial surveillance system to sensibly track suspicious events and activities. The methodology of video storytelling is applied to script from video footage. It avoids the task of manual video monitoring and ensures error-free threat detection.

The remaining document is set out as follows. In Sect. 2 related literature review practices are illustrated. The proposed method is set out in Sect. 3. Section 4 sets out the implementation details. Section 5 provides a result analysis, followed by a conclusion and future scope in Sect. 6.

2 Related Work

A model and overview of major components for real-time structured use of aerial video for surveillance applications are proposed in the paper [1]. The system consists of splitting an aerial video into the corresponding natural components of the scene.

The scene's three key components are the static background scenery, moving objects, and the presence of the scene's static and dynamic elements. In this technique, algorithms for the image analysis were built in jumbled scenes for 2-D/3-D frame-to-frame synchronization, shift detection, camera tracking, and moving object processing in cluttered scenes individually. By documenting the video, the geo-location of video and tracked objects is determined for surveyed reference imagery, elevation charts, and location maps. Eventually, static, interactive, and reprojected mosaics can be created for applications for compression, enhanced visualization, and mapping. Work on the detection and measurement of vehicle trajectories in congested traffic, however, is lacking.

In the paper [2], two learning-based systems for vehicle detection are proposed and compared: the human-made aggregated channel function ACF and the more rapid region-based convolutionary neural network faster R-CNN based on data-driven features. The findings show that faster R-CNN outperforms the ACF in images with large objects and those with small objects if enough data is provided. This approach collects high-precision trajectories from cars making them possible to analyze traffic dynamics based on experience with only a small number with effective samples.

The results of this study include practical guidance on how to construct a vehicle trajectory extraction system based on given conditions.

Aerial-FOV shape is quadrilateral unusual and this makes existing structural indexes ineffective to measure aerial-FOV. In paper [3] ying lu and cyrus shahabi proposed a new index structure called tetrartree that captures the geometric aerial-for property effectively.

UAVs particularly concede numerous key potential uses with the intrinsic characteristics of drones such as agility mobility and adaptive altitude in wireless systems. As aerial base stations UAVs can be used to improve wireless network penetration control reliability and energy efficiency. In contrast, UAVs can act as flying mobile terminals within a cellular network. These cell-connected UAVs can support numerous applications varying from real-time video streaming to product delivery. The paper presents a detailed tutorial on potential advantages and applications of UAVs in wireless communication [4]. Moreover, the significant threats and critical compromises in wireless networks allowed by UAV are discussed in depth.

Xu et al. [5] attempted to identify and tag events that occur concurrently and overlap in time using several sliding windows of different dimensions to produce predictions of the proposal. While this approach can work well for the task of captioning short video clips, for longer and more complex videos it fails miserably. However, the proposals expected by this approach can not be longer than the predefined sliding windows, and thus can not address the question of variation in the lengths of the proposal. Buch et al. [6] tackled this question where a new system, namely Single Stream Temporary Action Proposals (SST) was used to predict the probability of a proposal belonging to K.

Segment-CNN [7] carries out temporary action localization tasks through three segment-based 3D ConvNets (CNNs) [8], each for proposal, classification, and position. Inspired by the methods used to learn sparse dictionaries, Sparse-prop [9] extends these to describe and retrieve proposals for activities. DAPs [10] encodes a source of visual perception into discriminative states from which it generates the scores of trust for K propositions. Building on this further, the SST [6] uses a sequence encoder based on GRU to generate confidence scores for K proposals at every step of the way.

YouTube2Text [11] uses semantic hierarchies learned from the data to first pick appropriate phrase definitions and then penalize unlikely web-scale combinations of terms based on natural language corpus learning. In paper [12], the CRF learns to model the relationship between the different components of the visual input and then formulates the definition generation as a question of machine translation with the semantic relationships learned above as the source language and the sentences created as the target. The deep learning methods mostly formulate this problem in the encoder-decoder system. Venugopalan et al. [13] frames this question as a sequence-to-sequence model using a simple LSTM architecture to transform visual input into a single definition of the output of a single sentence.

In paper [14], Daniel tackles the question of counting instances of objects in images. The models can correctly estimate the number of vehicles in traffic congestion, or in a very crowded scene, they can count people. The main contribution is the idea for a novel neural network approach called Counting CNN (CNN). Essentially, the CCNN is constructed as a regression model where the network learns how to map the image patches' appearance to their corresponding density maps of objects. The second major contribution consists of a scale-aware counting model, the Hydra CNN, capable of estimating object densities in different very crowded scenarios where there is no geometric scene information available.

Paper [15], Onalenna, and J, C. Woods build a video dataset specifically intended for evaluating object counting algorithms. The main goal is to include videos that can be divided into three categories: low object density (LODV), medium object density (MODV), and high object density (HODV) videos. They describe LODV as videos for this paper which contain less than 10 objects, MODVs contain between 11 and 50 while HODVs contain more than 50 objects. They built a dataset which has different scenes for all of these types, from simple backgrounds to complex and cluttered backgrounds.

Intelligent vehicle recognition continues to be an issue due to the varying scales of vehicles that significantly impact vehicle quality counts. To answer this problem paper [17] lays out a system for vehicle recognition and counting based on a dream. Under this study, a new dataset of high-definition highway vehicles is published inside 11 129 photographs with a total of 57 290 annotated instances. The proposed dataset contains annotated tiny artifacts in the image relative to the existing public data sets and provides a full framework for deep-learning vehicle identification. Paper [19] discusses the review of the deep learning neural network architectures used in various purposes for an effective identification with automatic extraction of features.

From the above-mentioned literature study, it is noted that there are no works done in the Aerial video surveillance area to take advantage of the geoinformation of places where free movement of persons and vehicles occurs in violation of protocols during emergencies such as lockdown, using video content analysis and indexing techniques.

3 Proposed Method

Most of the video search algorithms recorded in contemporary research are evidence that there is no efficient method of collecting geo-information from the locations where such violations of the lockdown procedure occur.

However, the proposed technique allows to read and interpret the output of the geotagged video stream from the drone. A script is generated using the technique of video storytelling as an output of video content analysis. This script can be used for future visualization of the video and easy access. Using a wormhole indexing technique, the script is then used to produce a video index to ensure the minimum time complexity. The video index is similar to the page index maintain for easy of page access where each significant, relevant, and frequent keyword is kept in the sorted order. The Video index table contains information like dense and frame for locating the video shots very easily.

Figure 1 illustrates the process of the proposed Aerial video surveillance system methodology using video content analysis and video indexing. This consists of distinct modules for generating video script, video index, and video index. That video obtained from the drone is given to the module generating video script as input. The video script generation module produces a script using the technique of evaluating video content and storytelling [18]. This script is given as the input to the generator module for the video index.

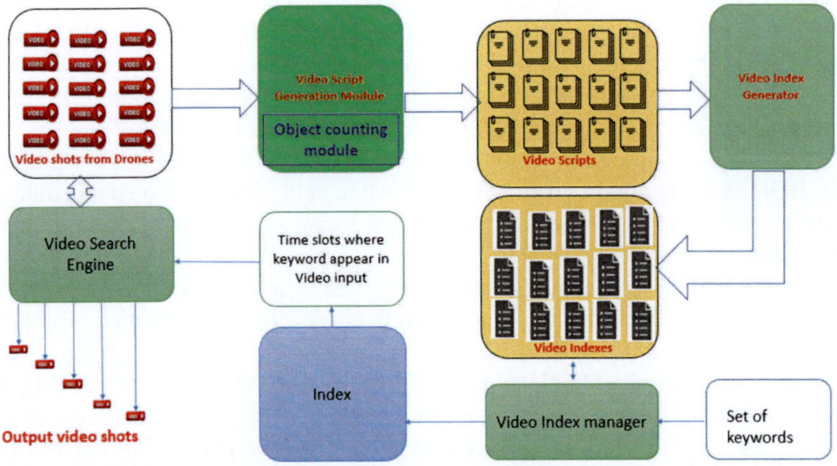

Fig. 1. Aerial video surveillance system methodology using video content analysis and video indexing- Workflow

A master index file maintains to keep the video index for multiple videos streamed by drones. The video index manager module checks the master index table with the list of keywords already stored for detecting the violation of lockdown protocols. As the rules change in the lockdown protocols, one can add or delete the keywords as suitable. Video index manager retrieves the time slot information, geotag information, and the threat, gives it to the authority who maintains the lockdown protocols. For example, the keywords crowd, many people, shopping, and streets are stored in the keyword repository to detect the violation. The video index manager checks the content of the video index table and locates the video, time slot information, and geotag information which matches with one or more keywords stored in the repository. That information is given to the end-user to proceed with.

Video files can be geotagged via metadata, audio encoding, overlay, or companion files. In the encoded video file, metadata keeps the geospatial information to be decoded for future analysis. The unmanned aerial vehicle uses MISB Standard 0601 is used by the, which enables the geocoding of corner points and horizon lines in video frames individually. The overlay includes overlaying the GPS data on the captured video as text. Companion files are separate data files that correspond to the audio/video files respectively. This metadata file is used for availing the geotag information for the corresponding locations where the threat is identified.

The data formats of companion files are typically found in the .KML and .GPX. It is plain text, and it represents in the following form:

GEO_LOCATION = (decimal latitude); (decimal longitude); ([optional]elevation in meters)

for example:

GEO_LOCATION = 32.0392; −78.3120; 210

3.1 Video Script Generation Module

The video storytelling method [18], is used for video script generation and it involves two subtasks:

(a) Discover the relevant video clips.
(b) Generate a narrative with chosen sentence-retrieving video clips.

As the first stage to tackle these problems a context-aware architecture for multi-modal semantic embedding learning is used. Embedding is taught through a two-step process locally to globally. The first process models single clip-sentence pairs for understanding an embedded locale. The second stage models the entire video as a series of images. Residual bidirectional RNN resBRNN senses the spatial structure of the video and integrates past and future contextual information into the embedding domain of multimode. It preserves temporal coherence and increases embedding complexity associated with it.

The stories are created in the second level using a Narrator model. Owing to an input recording, the narrator picks a series of different clips from it. Then a story is created by retrieving a sequence of sentences in the multimodal embedding space that best suits the clips. It is difficult to locate the right images because there is no clear idea of what visual elements are important for forming a successful narrative. To this end, we design the protagonist as a reinforcement learning agent who meets an input video sequentially and discovers a method of picking clips this optimizes the reward.

We perceive of the incentive as the literary metric between the story collected and the reference stories written in human terms. The narrator learns to discover fascinating, diverse clips that shape a strong narrative by directly optimizing the textual metrics.

3.1.1 Object Counting Module

This object counting module makes use of the methods [14–16] for counting objects. This count is used for video scripting, too. Multiple objects are tracked as described in Fig. 2, using the object box detected in algorithms called ‘Vehicle detection using YOLOv3(You Only Look Once-v3).’ The ORB (Oriented Quick and Rotated BRIEF) [19] algorithm has been used to extract the characteristics of the identified vehicles, and good findings have been achieved.

ORB is a simple, robust local feature detector implemented in 2011 by Ethan Rublee et al. Some changes were made to the Simple keypoint detector algorithm and BRIEF descriptor in this algorithm to boost performance with orientation and rotation. The FAST does not estimate the orientation. The authors change it to measure the patch’s centroid intensity with a centrally placed corner. The orientation is given by the direction of the vector from this angle to the centroid. Moments are measured to maximize the invariance of rotations. Besides, ORB authors change the BRIEF

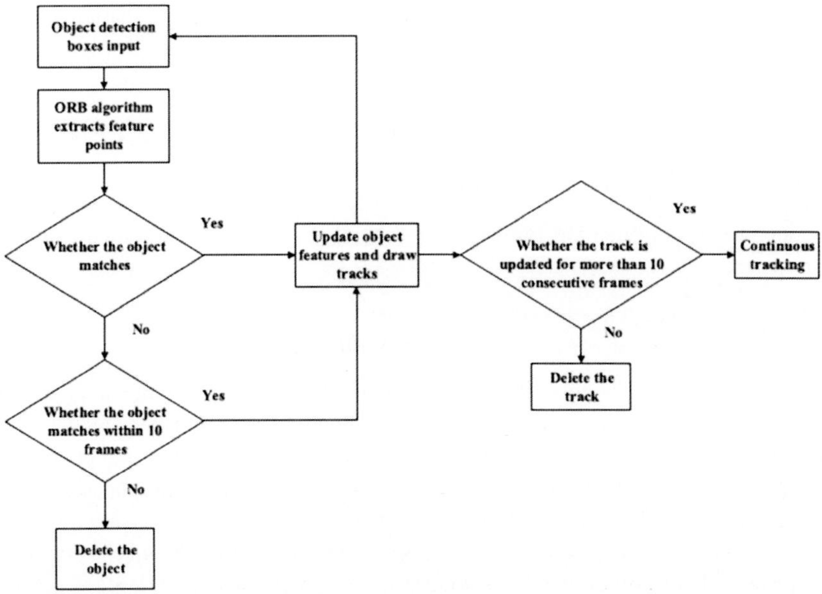

Fig. 2. Process of multi-object tracking

descriptor algorithm to better work with rotation by directing the BRIEF according to the orientation of key points.

The ORB algorithm displays superior results when it comes to the analysis of cost-effectiveness processing. This algorithm provides an outstanding alternative to the image classification algorithms SIFT and SURF. Using the Accelerated Segment Test (FAST) functionality, the ORB algorithm identifies feature points and afterward employs the Harris operator to detect corners.

3.2 Video Index Generator

The scripts generated by the generation module of video scripts provided as the input of a generator of file indexes. It uses a Wormhole [20] algorithm to look for a key with a length of L, which takes the worst-case O(log L) time. Finally, the creator of the video index produces an index that contains the keyword, the compact lower-upper frame range of the video clip in question. The video index format includes words that are stored in ordered order, along with the lower frame number and the high-density upper frame number.

4 Implementation

In the deep learning framework, the proposed solution was implemented using Python IDE and TensorFlow. A suitable VM-virtual machine identified in the google cloud platform, taking into account the size of the dataset and the model size (parameters). The code is executed in VM using the Ubuntu 18.04 VM requirements with 30 GB RAM and 8 vCPU's. It has a Regular Permanent Disk with a size of 20 GB and an NVIDIA Tesla V100 GPU is attached to fulfill the computational needs.

Video captions were generated using the video storytelling approach using the data set for video storytelling, a new dataset prepared to allow analysis. Using the wormhole algorithm, the video index generator guaranteed a minimum time complexity. Using the stories created by the ResBRNN-kNN method [17], in a forest video campfire, 86% accuracy was achieved.

5 Result Analysis

Figure 3 depicts the collection of shots used to create a script in the video clip-Campfire in a Forest. The output of the module for generating video script is as follows.

Script 1 : Many cars are on the road.

Script 2: Many people are shopping on the street.

Fig. 3. Collection of video shots and corresponding script generated by video script generator module

As shown in Fig. 3, a video is given as input to the video script generation module in this proposed methodology, and a corresponding script is generated as mentioned following. These scripts are used for generating a video index using a video index generation module. Table 1 and Table 2 show the result generated by the video index generation module corresponding to the scripts. The last column in the table describes the true positives which meant the outcome where the model correctly predicts the positive class. The keywords in the video index table are very less as the script is less. It is only because, the example used to describe the result in this paper, is merely a set of video shots instead of a long video.

Table 1. Video Index corresponding to the script 1 is shown in Fig. 3

Keyword	Frame no range	True positive
Many cars	15–28, 32–59, 78–99, 124–145	77%
Road	12–24, 35–49, 56–64	80%

Table 2. Video Index corresponding to the script 2 shown in Fig. 3

Keyword	Frame no range	True Positive
Many people	1–29, 38–59, 88–119, 124–155	78%
Shopping	12–54, 35–89, 96–134	69%
Street	1–124, 145–179	77%

6 Conclusion and Future Scope

In this paper, a creative approach is proposed and implemented for Aerial video surveillance using video content analysis and indexing through video indexing. The video storytelling approach is used to create video captions by using the video story dataset to prepare the video script and video index from the video content. Successful implementation of this technique in Aerial video surveillance would facilitate the violation of activities during civilian emergencies imposed by the authority to overcome the situation like an epidemic. Also, the relevant portion of the video streaming by Drones only needs to be viewed by the concerned authority instead of viewing the entire video. The set of keywords we keep in the keyword repository decides the relevant portion in the input video. Hence, the set of keywords can be changed to adapt to the changes in protocols. Since the wormhole indexing technique is used for indexing the script, the minimum time complexity is ensured for accessing relevant information from the video index table.

The methodology proposed and applied in this paper can be extended to various aerial video surveillance systems as well, viz

1) Manuel surveillance mission- manually deploy a drone to track incidents as they occur on the ground, during a major incident such as an oil spill, chemical leak, or explosion.
2) Robotic intrusion detection(RAID)- drones autonomously patrol areas of interest around the facility and only give alerts when it senses intruders from people and vehicles.

References

1. Kumar, R., et al.: Aerial video surveillance and exploitation. Proc. IEEE **89**(10), 1518–1539 (2001)
2. Kim, E., et al.: Extracting vehicle trajectories using unmanned aerial vehicles in congested traffic conditions. J. Adv. Transp. **2019** (2019)
3. Lu, Y., Shahabi, C.: Efficient indexing and querying of geo-tagged aerial videos. In: Proceedings of the 25th ACM SIGSPATIAL International Conference on Advances in Geographic Information Systems (2017)
4. Mozaffari, M., et al.: A tutorial on UAVs for wireless networks: applications, challenges, and open problems. IEEE Commun. Surv. Tutorials **21**(3), 2334–2360 (2019)
5. Xu, J., et al.: MSR-VTT: a large video description dataset for bridging video and language. In: 2016 IEEE Conference on Computer Vision and Pattern Recognition (CVPR) (2016)

6. Buch, S., et al.: SST: single-stream temporal action proposals. In: Proceedings of the IEEE conference on Computer Vision and Pattern Recognition (2017)
7. Tran, D., et al.: Learning spatiotemporal features with 3d convolutional networks. In: Proceedings of the IEEE International Conference on cOmputer Vision (2015)
8. Caba Heilbron, F., Niebles, J.C. Ghanem, B.: Fast temporal activity proposals for efficient detection of human actions in untrimmed videos. In: Proceedings of the IEEE Conference on Computer Vision and Pattern Recognition (2016)
9. Escorcia, V., et al.: Daps: deep action proposals for action understanding. In: European Conference on Computer Vision. Springer, Cham (2016)
10. Shou, Z., Wang, D., Chang, S.-F.: Temporal action localization in untrimmed videos via multi-stage CNNs. In: Proceedings of the IEEE Conference on Computer Vision and Pattern Recognition (2016)
11. Guadarrama, S., et al.: Youtube2text: recognizing and describing arbitrary activities using semantic hierarchies and zero-shot recognition. In: Proceedings of the IEEE International Conference on Computer Vision (2013)
12. Rohrbach, M., et al.: Translating video content to natural language descriptions. In: Proceedings of the IEEE International Conference on Computer Vision (2013)
13. Venugopalan, S., et al.: Sequence to sequence-video to text. In: Proceedings of the IEEE International Conference on Computer Vision (2015)
14. Onoro-Rubio, D., López-Sastre, R.J.: Towards perspective-free object counting with deep learning. In: European Conference on Computer Vision. Springer, Cham (2016)
15. Makhura, O.J., Woods, J.C.: Video object counting dataset. In: 2019 IEEE Conference on Multimedia Information Processing and Retrieval (MIPR). IEEE (2019)
16. Song, H., et al.: Vision-based vehicle detection and counting system using deep learning in highway scenes. Eur. Transp. Res. Rev. **11**(1), 51 (2019)
17. Bashar, A.: Survey on evolving deep learning neural network architecture. J. Artif. Intell. **1**(02), 73–82 (2019)
18. Li, J., Wong, Y., Zhao, Q., Kankanhalli, M.S.: Video Storytelling. arXiv:1807.09418v1 [cs. MM], 25 July 2018
19. Rublee, E., Rabaud, V., Konolige, K., Bradski, G.: ORB: an efficient alternative to SIFT or SURF. In: 2011 International Conference on Computer Vision, Barcelona, pp. 2564–2571 (2011)
20. Wu, X., Ni, F., Jiang, S.: Wormhole: a fast ordered ındex for ın-memory data management, arXiv:1805.02200v2 [cs.DB], May 2018

Developing a Framework to Identify Potential Firms for Job Assignment

Ariful Islam[1], Miftahul Jannat Mokarrama[2], Rezaul Karim[3], and Mohammad Shamsul Arefin[1(✉)]

[1] Department of CSE, Chittagong University of Engineering and Technology, Chattogram, Bangladesh
arifulllislam@gmail.com, sarefin@cuet.ac.bd
[2] Department of CSE, Port City International University, Chattogram, Bangladesh
mokarramacse07@gmail.com
[3] Department of CSE, University of Chittagong, Chattogram, Bangladesh
rezaul.cse@cu.ac.bd

Abstract. Nowadays, people are searching for companies via the internet to assign different types of software-related tasks. With the growing number of software firms, a problem is arising about which software firm is best suited to perform a particular job. It is not easy for a certain individual to determine the best software firm for a particular task according to his requirements manually. Therefore, by building a system to recommend the best-suited software firms for a particular job and can minimize the complexity of this important process. In this paper, a recommendation system is introduced which helps any individual or organization by suggesting them the most suited software firm according to his needs. Our system takes input job description and after processing it, suggest the user the best-suited firms according to the job description. The job description is processed by removing stop word, stemming, and extracting keywords. Porter Stemming algorithm is used to stem the job description. From the extracted keywords the job category is identified. From the software firms which have experience in working in the identified category and similar firms to these firms, some decision parameters are calculated. These decision parameters are used to identify the best-suited software firms working in the identified category.

Keywords: Recommendation system · Stemming · Pearson correlation algorithm

1 Introduction

A recommender system is a family of information filtering systems that tries to prophesy the "rating" or "preference" of a user to an item [1, 2]. Over the last few years, recommender systems have become growingly widespread and are adopted in many fields that are dependent on users' choices to expand their operability. Choosing the contents from online like books, grabbing the relevant research articles, finding the peer-fellows from social sites, searching the reasonable restaurants and hotels, etc. are

J. I.-Z. Chen et al. (Eds.): ICIPCN 2020, AISC 1200, pp. 584–593, 2021.
https://doi.org/10.1007/978-3-030-51859-2_53

all becoming more clear-cut because of the incorporation of the recommender system [3–6].

Selecting suitable organizations for assigning certain kinds of jobs is believed to be a crucial stage of HRM (Human Resource Management). The job assignment process is available in almost all organizations. In the job delegation process, supervisors have to analyze the job to identify the job expertise and obligations of the personages which is a prerequisite to the job ventures.

After the analysis supervisors will identify suitable groups to assign the job. The job assignment process is done by following the notion of "putting the right group on the right job". In this process, selected groups can accomplish their job with the utmost capability. Finally, such a job assignment process ensures efficient performance and highly productive results. To develop such a framework that assists at job delegation stages, it has to be dissected and pinpoint discernibly both the relevant attributes of the assigned jobs and the associated farms in the job assignment method [7–9]. This task belongs to Decision Support Systems (DSS). According to Sprague at el [10] DSS tends to help in the decision-making process at the underspecified problem that upper-level managers face.

In the following sections, the works in a sequence are arranged. Section 2 cites some of the groundbreaking works related to this system development process. In Sect. 3, the system architecture of our proposed approach is described. Section 4 dictates the experimental results. Finally, the future of the research directions is conveyed in Sect. 5.

2 Related Work

In a job distribution process, decision making acts as an integral part of all organizations. The rules followed by a supervisor of an organization differ in nature along the process. This depends on the nature of the business, regulations related to service and organization, and the maturity of the supervisors [11]. However, several ground rules are discussed in the context of task analysis for this purpose et al. [12, 13]. These rules include requisite task expertise, working period and knowledge, workload, etc. Those criteria are adopted by various researchers to foothold the job distribution process.

A decision support system (DSS) named Nursing Allocation System has been introduced by Trivedi and Warners [14] to assist the nurses in each ward of the hospital by formulating the timetable for the regular nurses. In this paper, they used a regression model to predict the severity of the need for nurses.

To assist in scheduling the jobs of the short-termed workers of a regional newspaper publisher in Alabama, USA, research has been conducted using DSS by Gopalakrishnan [15]. In this paper, they developed and implemented a PC-based DSS to structure and schedule the jobs of the temporary workers. A rule-based system and a heuristic search procedure have been employed to develop the requirements and schedules respectively.

Intending to support the crew scheduling to control trains at DB Schenker, Juette [16] proposed a decision support system. Here, he used a column generation solution technique. Resulted DSS can be used for strategic planning.

To schedule the shifts of works in a factory, a DSS utilizing a primo heuristic-based scheduling has been proposed by Schniederjans [17]. A real-world case-study for a US manufacturer was adopted as the standpoint for this scheduling system.

Along with these, to assign jobs, a fuzzy expert decision support system is proposed in [19]. A web-oriented DSS for supporting the task assignment process is proposed in [18] where a user's opinion on the DSS prototype is also evaluated.

3 System Architecture and Design

The system architecture of the framework for identifying potential firms for job assignment contains three basic parts. Those parts are processing the job detail, finding similar firms, and finally the recommendation generation module. The system architecture is shown in Fig. 1.

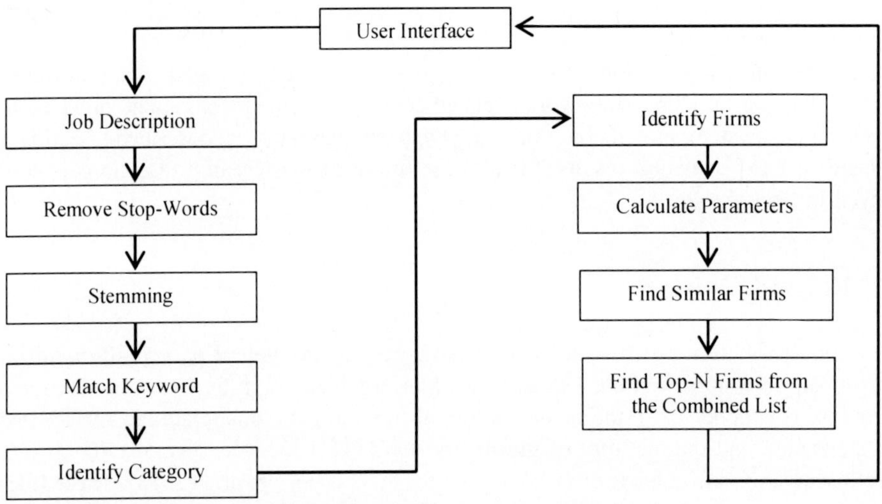

Fig. 1. The architecture of the proposed recommender system

3.1 Removing Stop Words from Job Description

From the job description, stop words are removed. Typically stop words are those words that generally do not contain any significant importance in a text document and have considered a list of stop words which is available at http://www.ranks.nl/stopwords.

Every word of the job description is matched with the list of stop words. If any word is matched that word is discarded. Stop words were removed by using Algorithm 1 as shown in Fig. 2.

```
Algorithm 1: Removing stop words
Input: Job description
Require: Remove stop words
begin
Take input the job description
Convert the job description to lower case
Take input the list of stop words from a text file
Split the list of stop words
Split the job description
Remove those words that match with the list of stop words
Join the job description in one string
End
```

Fig. 2. Algorithm for removing stop words

3.2 Stemming the Job Description

Stem or root can be defined as a portion of a word to which inflectional affixes like -ed, -ize, -s, -de, mis, etc. are added up. Stemming is the process of removing inflections from words to get the stem of that word. Stemming is used in the specialization of NLP (Natural Language Processing) to normalize a text for further processing. There are some well-known algorithms of stemming. Here, the Porter Stemming Algorithm is used to stem our job description in Algorithm 2 as shown in Fig. 3.

```
Algorithm 2: Stemming the job description
Input: Job description after removing stop words
Require: Job description
begin
split the job description
for each word from job description
   Send each word to porter stemming function
End for
End
```

Fig. 3. Algorithm for stemming job description

3.3 Searching for Keywords and Finding Software Category

A list of 72 keywords is selected. After removing the stop words and stemming the job description will search for these keywords from the remaining job description. Every keyword is related to a particular software category. If no keywords are found in the job description then will search for the synonyms of the keywords. If a synonym is found then it will be saved and it will be also used as keyword afterwards. The process is shown in Algorithm 3 of Fig. 4.

Algorithm 3: Matching Keywords and Identifying Category
Input: Job description
Require: Software Category

1. **Begin**
2. Split the job description
3. **For each word in job description do**
4. Match the word with keywords
5. **If** word is matched with a keyword
6. Select the category that the keyword is related
7. **Break**
8. **End if**
9. **End for**
10. **End**

Fig. 4. Algorithm for matching keywords and identifying category

3.4 Finding Similar Firms

From the list of top n number of firms, find out firms that are highly similar to these firms. For this, Pearson Co-relation Algorithm is used as shown in Algorithm 4 of Fig. 5.

Algorithm 4: Algorithm for similarity calculation
Input: Top n selected firms
Require: Find similar firms

1. **Begin**
2. Get id of the top n selected firms
3. Sum up the weights for each id
4. Get id for firm $\neq$ current_firm
5. **for** each firm $\neq$ current_firm**do**
6. sum up the Weights
7. calculate the similarity_weight with respect to current_firm
8. **end for**
9. sort out the similarity_weight of each firm in descending order
10. select top-N firms from sorted similar firms list
11. **End**

Fig. 5. Algorithm for similarity calculation

3.5 Removing Stop Words from Job Description

From the overall list of the top-n firms and their similar firms, again calculate some parameters such as manpower ratio, number of clients' ratio, number of installations ratio, etc. But the top n selected firms will be given more priority than similar firms. Using these values final top n number of firms will be selected and the recommended result will be sent to the user interface. An associated algorithm is shown in Algorithm 5 of Fig. 6.

Algorithm5: Recommending the result
Input: Top-N firms and their similar firms
Require: Recommending the most suitable firms

Begin
For each firm in the list of selected firms
    Calculate the man power ratio
    Calculate the number of clients ratio
    Calculate the number of installations ratio
    Total Score = Man power ratio + Number of clients ratio + Number of Installations Ratio
End for
select top-*N* firms from sorted similar firms list
End

Fig. 6. Algorithm for recommending the result

4 Experiments and Implementation

In this section, the implementation procedure is provided and performance analysis of our developed system.

4.1 Experimental Setup

The system for suggesting suitable firms for job assignment has been developed on a machine having the Windows 8, core i3 processor with 4 GB Ram. The system has been developed in Asp Dot Net. The front end was created by using asp, HTML, and CSS. And in the backend, C# and Visual Studio 2012 is used.

4.2 Implementation

Figure 7 shows the GUI of our system that takes input job descriptions and requests for recommending results. After giving the job description recommendation is generated. At first, our program will show the home page. The user has to log in to give input to the job description.

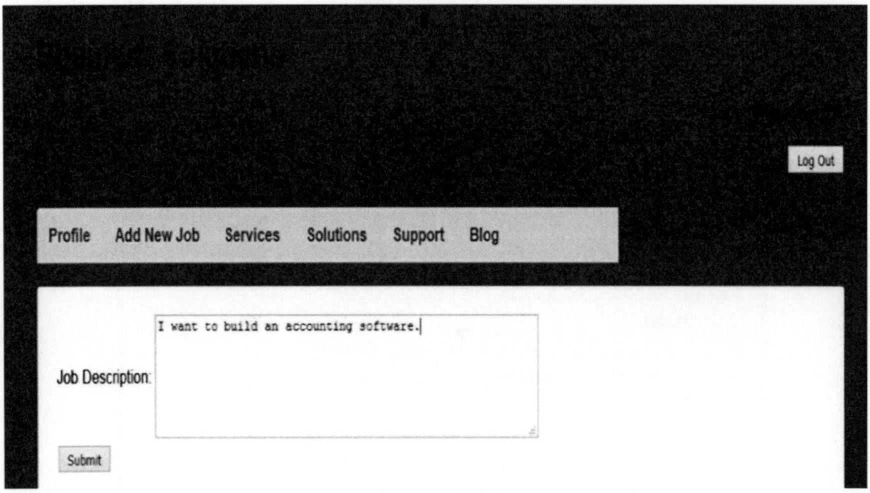

Fig. 7. Taking input from the system

4.3 Performance Evaluation

After the software category is identified, a list of recommended firms is got for each type of software. To evaluate our system, precision and recall performance evaluation metrics are used. Precision seeks for the recommended firms out of the relevant firms to the user. On the other hand, Recall seeks for the relevant firms out of the set of the recommended firms. Following formula 1 and 2 are used accordingly:

$$precision = \frac{correctly\ recommended\ firms}{total\ recommended\ firms} \tag{1}$$

$$recall = \frac{correctly\ recommended\ firms}{total\ useful\ recommended\ firms} \tag{2}$$

Worked with twenty-two categories of software firms. For each category, collected the data of manpower, number of clients, and number of installations which helps in similarity calculation. Table 1 shows the precision and recall of our system while recommending top-5 firms for each category. Figure 8 shows the corresponding graph.

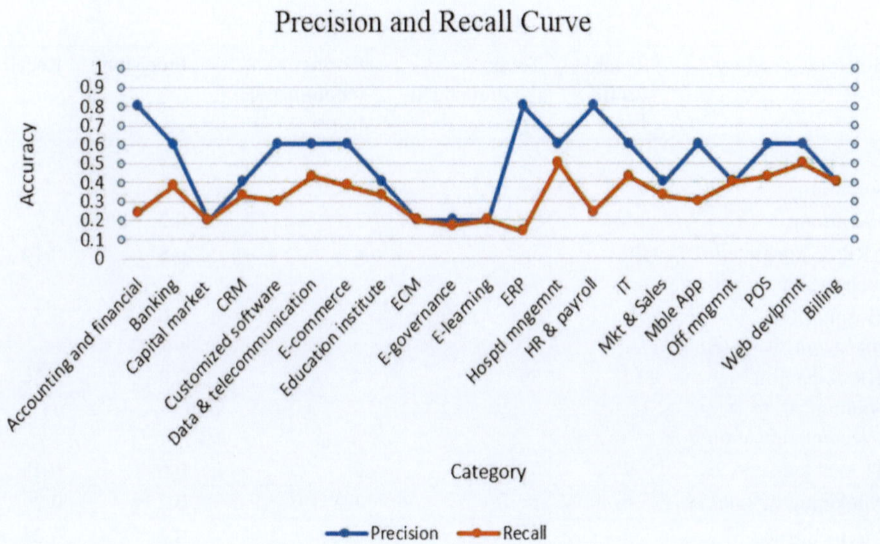

Fig. 8. Precision and Recall curve

Table 1. Precision and recall for top-5 firms.

Software category	Total useful firms	Total recommended firms	Correctly recommended firms	Precision	Recall
Accounting and financial software	17	5	4	0.8	0.24
Banking application	8	5	3	0.6	0.38
Capital market solution	5	5	1	0.2	0.2
CRM	6	5	2	0.4	0.33
Customized software development service	10	5	3	0.6	0.3
Data & telecommunication, IT infrastructure	7	5	3	0.6	0.43
E-commerce & portals	8	5	3	0.6	0.38
Education institute management application	6	5	2	0.4	0.33
Enterprise content management	5	5	1	0.2	0.2

(continued)

Table 1. (*continued*)

Software category	Total useful firms	Total recommended firms	Correctly recommended firms	Precision	Recall
E-governance solution	6	5	1	0.2	0.17
E-learning	5	5	1	0.2	0.2
ERP & Integrated business application	29	5	4	0.8	0.14
Hospital management system	6	5	3	0.6	0.5
HR & payroll solution, Data & telecommunication	17	5	4	0.8	0.24
IT enabled services	7	5	3	0.6	0.45
Marketing & sales	6	5	1	0.2	0.2
Micro finance solution	5	5	1	0.2	0.2
Mobile application	10	5	3	0.6	0.3
Office management solution	5	5	2	0.4	0.4
POS & Inventory management system	7	5	3	0.6	0.43
Web development service & hosting	6	5	3	0.6	0.5
Utility & billing	5	5	0.2	0.4	0.4

5 Conclusion

The recommendation system is an interesting research field and a branch of machine learning. It predicts for the user and acts as a proper guide. Here, a recommender system is developed to find out promising firms for job assignments. From the experiment, satisfactory performance is got from the system.

References

1. Ricci, F., Rokach, L., Shapira, B., Kantor, Paul B. (eds.): Recommender Systems Handbook. Springer, Boston (2011). https://doi.org/10.1007/978-0-387-85820-3
2. http://content.time.com/time/magazine/article/0,9171,1992403,00.html
3. Chen, H., Ororbia II, A.G., Giles, C.L.: ExpertSeer: a keyphrase based expert recommender for digital libraries, arXiv preprint (2015)
4. Chen, H., Gou, L., Zhang, X., Giles, C.: Collabseer: a search engine for collaboration discovery. In: ACM/IEEE Joint Conference on Digital Libraries (JCDL) (2011)

5. Felfernig, A., Isak, K., Szabo, K., Zachar, P.: The VITA financial services sales support environment. In: AAAI/IAAI 2007, pp. 1692–1699. Vancouver, Canada (2007)
6. Gupta, A., Goel, A., Lin, J., Sharma, A., Wang, D., Zadeh, R.B.: WTF: The Who-to-Follow system at Twitter. In: Proceedings of the 22nd International Conference on World Wide Web (2013)
7. Cascio, W.F.: Managing Human Resources (Productivity, Quality of Work Life, Profits), 6th edn. McGraw-Hill, Boston (2003)
8. Kaixuan, Z.: Some issues concerning job assignment for college graduates. Chin. Educ. Soc. **27**(3), 19–21 (1994)
9. Mondy, R.W., Noe, R.M., Premeaux, S.R.: Human Resource Management, 8th edn. Pearson Education Inc, Upper Saddle River (2002)
10. Sprague, R.: a framework for the development of decision support systems. MIS Q. **4**(4), 1–25 (1990)
11. Sauter, V.: Decision Support Systems: An Applied Managerial Approach. Wiley, New York (1997)
12. Ivancevich, J.M.: Human Resource Management, 9th edn. McGraw-hill, Boston (2004)
13. Schneier, C.E., Baird, L.S., Beatty, R.W., Shaw, D.G.: Performance, Measurement, Management, and Appraisal Sourcebook. HRD Press, Amherst (1995)
14. Trivedi, V.M., Warners, D.M.: a branch and bound algorithm for optimum allocation of flat nurses. Manage. Sci. **22**(9), 972–981 (1976)
15. Gopalakrishnan, M., Gopalakrishnan, S., Miller, D.M.: A decision support system for scheduling in newspaper publishing environment. Interfaces **23**(4), 104–115 (1993)
16. Juette, S., Albers, M., Thonemann, U.W., Haase, K.: Optimizing railway crew scheduling at DB schenker. Interfaces **41**(2), 109–122 (2011)
17. Schniederjans, M.J., Carpenter, D.A.: A heuristic job scheduling decision support system a case study. Decis. Support Syst. **18**, 159–166 (2006)
18. Vongsumedh, P.: A framework for building a web-based decision support system for job assignment. In: Proceeding of the Third International Conference on Web Information Systems and Technologies, pp. 136–141. WIA, March 2007
19. Hajiha, A., Jassbi, J.J., Khanmohammadi, S.: A fuzzy expert decision support system for job assignment. In: Fuzzy Systems Conference (2007)

Developing a Framework for Trend Prediction of Stocks Prices

Mohammad Tauhidul Islam[1], Rezaul Karim[2], Sumi Khatun[3], and Mohammad Shamsul Arefin[1](✉)

[1] Department of CSE, Chittagong University of Engineering and Technology, Chattogram, Bangladesh
sefotau@gmail.com, sarefin@cuet.ac.bd

[2] Department of CSE, University of Chittagong, Chattogram, Bangladesh
rezaul.cse@cu.ac.bd

[3] Department of CSE, Bangladesh University of Business and Technology, Dhaka, Bangladesh
sumicuet11@gmail.com

Abstract. The stock market is a very important element of a country's economy. Investment in the stock markets is a common trend among the people throughout the world. However, the investments at the stock markets may become very risky if people are not well aware of the trend of stock prices. People may gain money or lose their entire life savings by wrong investments at the stock markets. Previously the world has witnessed many stock market disasters, the crippling economic condition of an entire nation. This can be avoided to some extent if the market can be predicted in advance. Considering this fact, in this paper, a technique is proposed to predict the stock prices of different companies using the concepts of deep learning. Our system's performance is evaluated by using data from the websites of the Wall Street Journal and Alphavantage. Experimental results show that our system can perform a very good prediction of stock prices.

Keywords: Prediction system · Stock securities · Model development

1 Introduction

A stock market is one of the most vital components of today's free-market economy. The stock market is a very risky place for investment. People may gain money or lose their entire life saving by wrong investments. Recently (in the year 2014) worldwide financial crisis is seen due to a huge amount of stock price drop. This caused great damage to the international economy and many countries suffered a lot for this crisis. This situation can be avoided by some extant if the market can be predicted in advance. By this prediction, a broker may be able to find some irregularities which may alert them in advance. Also, the stock market prediction can be a useful tool for the market regulatory committee. This prediction may help them to take the necessary steps to avoid potentially harmful transactions. Any developed system that can analyse stock

J. I.-Z. Chen et al. (Eds.): ICIPCN 2020, AISC 1200, pp. 594–606, 2021.
https://doi.org/10.1007/978-3-030-51859-2_54

markets data and can provide an accurate prediction about stock prices can play a vital role to be very helpful in this regard.

In this paper, a method is presented for predicting stocks prices by breaking down information gathered from two sites: Wall Street Journal [1] and Alphavantage [2]. In our approach, the deep learning concept is used and found that our system can perform efficient prediction of stocks prices under considerations.

Our paper is composed as: In Sect. 2 a short review of related work. In Sect. 3 the methodology of our proposed strategy is clarified elaborately. The results of our experiment are presented in Sect. 4. At last, finish up and sketch future research headings in Sect. 5.

2 Related Work

The prediction problem of the stock market is not a new study, for quite a long many research have been emerged to improve the accuracy of the forecast. From that, it is found that there are several ways to predict the trend of the stock price. The studies include the areas of artificial intelligence, machine learning, artificial neural network and fuzzy set theory, big data analysis, sentimental data analysis, and many others.

Lai et al. [13] builts up a model that trains unary possibilities utilizing market value information and Gaussian Mixture Models. Graph cut based algorithm will solve the interference problem and binary markove filed helps to find the inference and learning approach effectively.

Kuttichira et al. [12] proposed a strategy for price prediction utilizing Dynamic Mode Decomposition expecting the stock market as an ever-changing framework considering the minute-wise stock price for predicting the price in the next few minutes. They worked with the time series information of the corporations indexed at the nationwide inventory exchange. This technique is the effective and computationally quick algorithm as it can anticipate the future condition of the framework with no earlier information on the hidden conditions. Even though this strategy neglects to foresee precisely when there is an unexpected change in the elements, similar to when the stock split is reported or when another product is propelled.

Golmaryami et al. [7] applied artificial neural networks that can deal with time-series data and non-linear parameters to predict the next day's stock price. It has been trained with a meta-heuristic algorithm which has a fast and powerful convergence. This algorithm adjusts the weight matrix of ANN more precisely and also it can solve problems with nonlinear variables. Though it doesn't perform that accurately when there are more complex problem spaces and unable to generate a better solution.

In the year 2015 Alostad et al. [3] worked on stock price prediction using five years monetary news corpus containing more than 50000 articles gathered from the National Association of Securities Dealers Automated Quotations (NASDAQ) site for the 30 stock symbols in Dow Jones record. They train a directional stock value prediction framework dependent on news content. They have demonstrated that breaking news based framework, in reality, yields a factually noteworthy lift in prediction accuracy contrasted with the one utilizing all news unpredictably. Although it is obscure in their method how their method will predict in case of different categories of stock.

Kara et al. [9] took a shot at predicting the direction of stock value index development utilizing artificial neural networks and support vector machines: The article utilizes specialized analysis markers to predict the bearing of the Istanbul Stock Exchange (ISE) National 100 Index. They claim great outcomes for up to 75.74% precision.

Milosevic et al. [14] have accomplished work on stock market prediction using machine learning which employs long term investment data. The paper asserts that the strategy can accurately foresee whether some organization's value will be 10% higher or not over the time of one year in 76.5% of cases.

Kim [10] discussed the stock prediction technique using support vector machine. This approach analyzed the best stock market prediction.

Schumaker et al. [16] experimented on various linguistic text formats comprising named entities approaches, noun phrases, and a bag of words for stock market prediction utilizing content investigation by breaking economic news. They utilized a few semantic printed portrayals, including a pack of words, thing phrases, and named entities approach.

Usmani et al. [19] have researched to foresee the market execution of Karachi Stock Exchange(KSE) on day closing utilizing distinctive AI methods using various properties as an input and predicts market showcase as Positive and Negative. Selvin et al. [17] use a technique that adopts a model autonomous strategy in foreseeing the stock value. But their data is not fitting enough with a particular model, they have only recognizing latent dynamic present in the data with the help of deep learning. Their proposed system can identify some interrelation within the data. Though changes occurring in the stock market don't happen in a regular pattern, their system can identify the changes in trends.

Kogan et al. [11] took a shot on text regression to predict market ventures where the content is an SEC-ordered economic paper distributed yearly by a traded on an open market organization, and the amount to be predicted is the instability of stock returns, an observational proportion of monetary hazard.

Chen et al. [5] endeavored to demonstrate and foresee the course of profit for the advertising list of the Taiwan Stock Exchange, one of the quickest developing budgetary trades in creating Asian nations and claim that PNN-based investment strategies obtain higher returns than other investment strategies.

In 2013, Nann et al. [13] gathered 2,971,381 messages relevant to the supplies of the S&P 5,00 index for six months from June 1 to November 30, 2011. At that point, they allot these reports to stocks by separating the dataset by either utilizing Twitter's "cash tag" as it is a stock's cut-up sign of dollar ($) or searching for messages in subforums involved in specific stocks on web message sheets. A tweet is not considered as spam which contains a money label alludes to the stock cost or anything identified with the budgetary estimation of the basic organization.

Chen et al. [6] made the expectation on stock costs dependent on the prominently utilized specialized indicators such as Average True Range (ATR), Relative Strength Index (RSI), Moving Average Convergence/Divergence (MACD), and Money Flow Input (MFI).

Gong et al. [8] developed a novel technique for trend prediction of the stock price using logical regression.

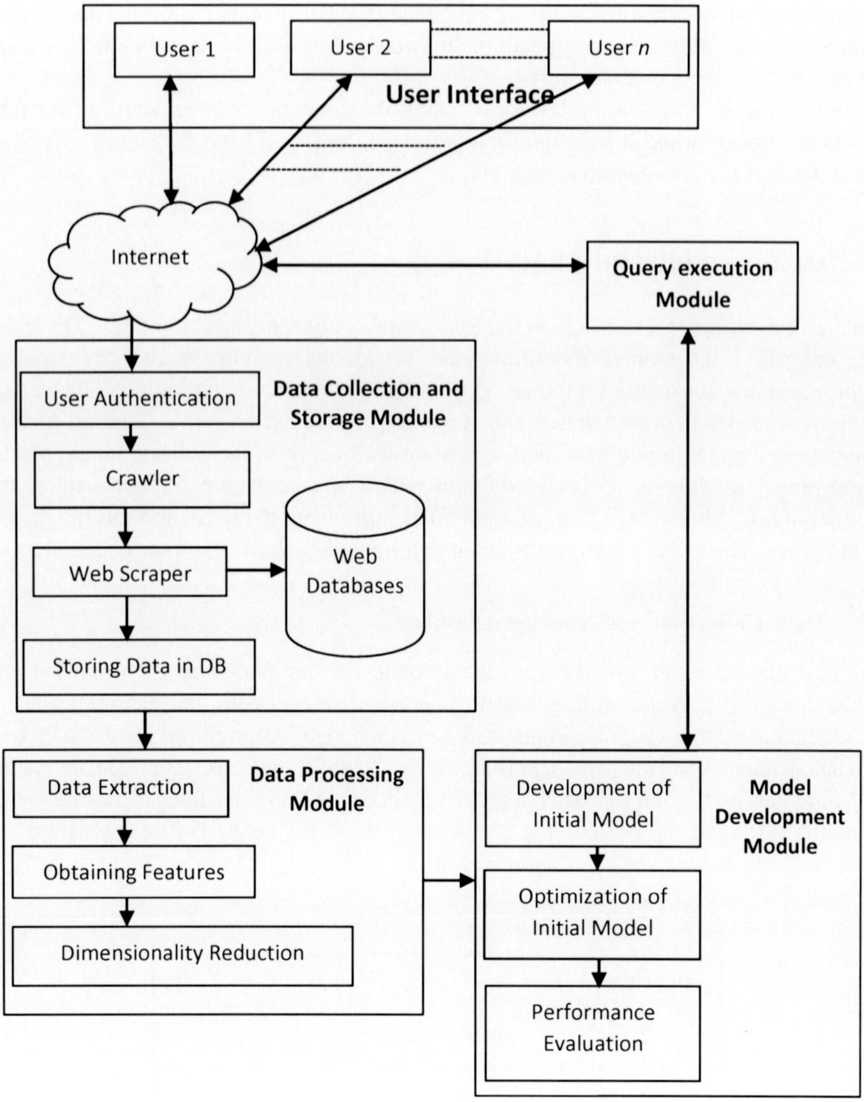

Fig. 1. The system architecture of the proposed system.

Their method has significant time-effectiveness and strong purposefulness which enable users to predict stock price trends. But some feature index variables fail in their method. Also, their system isn't capable enough to select effective variables.

Anik et al. [4] used the linear regression approach to get the prediction for individual stock securities of the Chittagong Stock Exchange (CSE) and Dhaka Stock-Exchange (DSE).

Shi et al. [18] proposed a way to deal with investigating outliers in ever-changing informational indexes. They analyzed data sets with consistent changes and follow the development of information focuses, alter the status of groups regarding their information focuses and limits, and update the information of outliers. But they fail to claim the accuracy of their methodology.

In this paper, a system is developed using the concepts of deep learning for predicting the prices of the stocks by utilizing the data acquired from the sites of the Wall Street Journal [2] and Alphavantage [1].

3 System Architecture and Design

Our framework for stocks price prediction comprises of four basic modules. These are data collection and storage module, a data processing module, model development module and query execution module. Data collection and storage module collects data from the websites in consideration and stores data in the database in a standard format. The data processing module extracts important features from the collected data. Model development module is the main module which is responsible for performing the prediction task. Query execution module takes input from the users and returns results to the users. The system architecture of our proposed system is shown in Fig. 1.

3.1 Data Collection and Storage Module

The data are collected directly from the websites of the Wall Street Journal [1] and Alphavantage [2]. Proper authentication is needed for accessing the data of the websites. Then the data using the proper API key is crawled. After giving proper API key, the data is fetched and later the data is stored into MySQL database using JSON. As the real-time data is fetched that is changed over time, the logic tier fetches any new data from the API and stores into the database without duplication. The algorithm for collecting and storing data is shown in Fig. 2.

Algorithm 1: Data collection and storage

1. **Begin**
2. Check internet connection
3. **If** no connection, **then** Step-9
4. **Else** Look for proper webpage
5. **Request** the website with API key
6. **If API** key not matched, **then** Step-9
7. **Else** get JSON data
8. **Store** in database
9. **End**

Fig. 2. Algorithm for data collection and storage

3.2 Data Processing Module

As the data from the websites is fetched through the API key, got the data as JSON format. To use this data, the data is needed to be parsed. JSON parser is used to parse the data. After that got parsed data. The parsed data contains daily information about open price *o(p)*, highest price *h(p)*, lowest price *l(p)*, and closing price *c(p)* for each record and consider each of these as the basic features. Based on these basic features, other features are computed based on the equations used in [20]. The task of dimensionality reduction is performed using the concept of [21] and then use these features as the input features for the deep neural network. The algorithm for data processing is given in Fig. 3.

Algorithm 2: Algorithm for data processing
Input: Data stored in database

1. **Begin**
2. **Convert** store data to dictionary
3. **Get** desired data
4. **For** each record **select** four basic features
5. **Compute** other features using the equations of [20]
6. Perform dimensionality reduction using the concept of [21]
7. **End**

Fig. 3. Algorithm for the data processing module

3.3 Module Development Module

For the development of the model, our crawled data are divided into two parts: training data and test data. 80% of data is considered for training and 20% data for testing. After that, the model is initialized with the features based on training data. Finally, the model is tested by our test data. Figure 4 depicts the basic approach of model development.

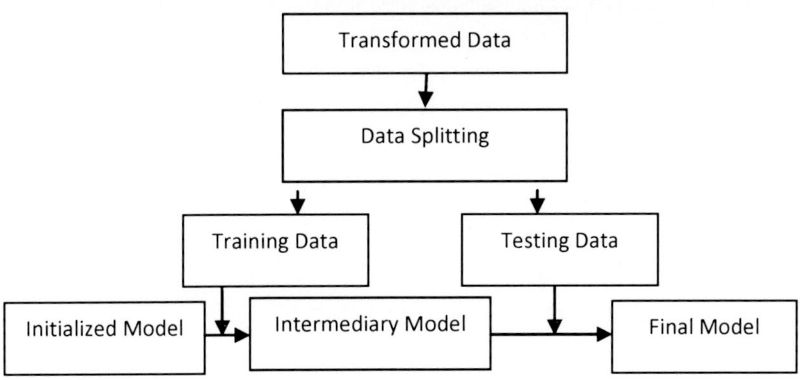

Fig. 4. Model development process

The deep neural network is used that is a multi-layer feed forward neural network and it uses supervised learning for model development. Each layer has a different role to play. Several manipulations are done in these layers. Backpropagation and forward propagation are done between these layers. When running backpropagation between the layers, calculate loss and cost between them. Using this calculation, got a new weight. This new weight is used in predicting the output. The output is generated with the help of forwarding propagation. Figure 5 shows the algorithm for prediction.

Algorithm 3: Algorithm for prediction

1. **Begin**
2. Split 80% of total data into training data set and 20% into test data set
3. **Initialize** network with regressor
4. **Compile** the network
5. Train network with training data.
6. **Validate** network with test data
7. **End**

Fig. 5. Algorithm for prediction

3.4 Query Execution Module

Query execution module is responsible to take users' queries and extracts the information in the queries. It then sends the extracted information to the system. Based on the extracted information from the queries, the system returns the related results. Finally, these retrieve results are sent to the users using the query execution module. The algorithm for query execution module is given in Fig. 6.

Algorithm 4: Algorithm for query execution

1. **Begin**
2. **Get** trading code from user input
3. **Connect** to the database.
4. **Search** for relevant data in data base
5. Show predicted price and probability of the selected company.
6. Plot data into Time vs. LTP graph
7. **End**

Fig. 6. Algorithm for query execution.

4 Implementation and Experiments

In this section, the implementation procedure is provided and performance analysis of our developed system.

4.1 Experimental Setup and Implementation

On a computer with Windows 7, 2.60 Core i5 processor with 8 GB RAM, the built prediction system was implemented. The framework was built in front end with Bootstrap CSS, HTML, JavaScript, Sublime text and in the back end with MySQL, PHP, Python. Our system is trained with the Wall Street Journal data [2] and Alpha Vantage data [1].

User can interact with the system via a well-designed interface. The user inputs the trading code of his desired security. The system then connects to the database. It looks for the desired stock and then predicts with the trained model. The output is shown in two ways. It is shown in a table and also in a graphical format. Figure 7 shows a snapshot of our system.

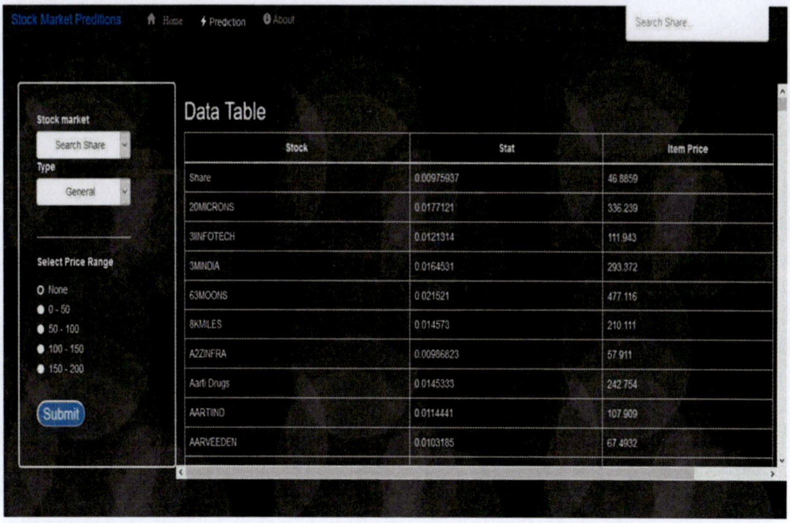

Fig. 7. A snapshot of the system showing a generalized prediction

In our system, the output can be shown in two ways. The next 10-day price of an individual security is predicted and also depicted it in a graph. User demand is also shown based on the analysis of news from wall street journal. User demand is measured based on the result of news polarity. Figure 8 shows a snapshot of the predicted price of Apple stock and its demand.

```
Prediction of Apple Stock Price in Next 10 Days :
Date                Price
10-1-2019:       27.498787
11-1-2019:       28.483566
12-1-2019:       28.483566
13-1-2019:       27.498787
14-1-2019:       27.38066
15-1-2019:       26.76038
16-1-2019:       26.514225
17-1-2019:       26.396097
18-1-2019:       26.396097
news polarity : 0.06498552489177489
User Demand Is High
```

Fig. 8. Predicted price and user demand for individual security.

4.2 Experimental Evaluation

In this section, the performance of our developed system is calculated to real-time day to day stock prices. The Last Traded Price (LTP) of the stock securities is compared manually for the predicted ones. Our system also performs analysis for the same securities. Then the performance of our system is analyzed. In our experiment, a real world change of LTP vs the predicted change of LTP predicted from data extraction from the source is used. By plotting the real time and predicted LTP on the curve and evaluated the performance of our system.

To calculate the prediction accuracy, the percentage of the difference between the predicted LTP and real world LTP is calculated using the following equation.

$$\text{Prediction Accuracy} = 100 - (|(\text{Predicted LTP} - \text{Real word LTP})|/\text{Real} - \text{worldLTP}) * 100$$

To evaluate the accuracy of the prediction module, prediction of LTP is analyzed for 49 securities. Table 1 shows the predicted LTP, real LTP and the accuracy of prediction. Figure 9 shows the corresponding graph.

The overall accuracy of our module is computed with the following equation: $\sum Prediction\,accuracy/Total\,Number\,of\,prediction$
and found that our system has an accuracy of 97.35%.

Table 1. Prediction accuracy of the system

Trading code	Predicted LTP	Real world LTP	Accuracy
ACM	33.00874	33.23	99.33
ADI	5.020361	5.82	86.25
ALGO	79.76533	80.01	99.68
ALT	2.3357584	2.54	91.73
ALX	365.0953	364.99	99.97
AMAZON	35.848064	36.02	99.5
ANTM	37.975925	37.12	97.71
APPLE	26.552622	26.83	98.95
AUTO	15.338376	14.23	92.26
BAC	16.116205	16.01	99.37
BANC	21.662254	23.03	94.05
BBC	6.3084426	6.99	90.12
BCOM	9.154940	9.56	95.71
BEAT	12.014362	11.86	98.73
BGCP	10.682398	11.96	89.29
BIDU	181.55525	183.01	99.2
BIO	178.57082	178.55	99.9
BIP	39.93315	39.65	99.29
BK	9.658916	9.54	98.84
BMI	38.576977	38.56	99.99
BRAC	23.888382	22.47	93.72
BX	14.42262	14.98	96.26
CA	26.192404	26.36	99.35
CARA	38.584026	38.64	99.84
CASC	3.9486146	4.33	91.00
CAVM	70.30226	72.01	97.62
CBMX	7.1628265	7.13	99.58
CBPX	83.40450	83.46	99.92
CBRL	149.79512	149.68	99.92
CBS	61.937267	61.54	99.36
CCA	11.175642	11.79	94.74
CCBG	19.839539	18.98	95.52
CCC	16.548199	16.78	98.56
CCO	4.552969	4.13	89.83
CDOR	11.184361	11.24	99.46
GE	30.573193	30.89	98.96
GNCMA	19.825115	20.14	98.41
GOOD	18.601177	18.46	99.24
GOOGLE	802.42377	802.90	99.29
HPQ	19.887468	19.00	95.36

(*continued*)

Table 1. (*continued*)

Trading code	Predicted LTP	Real world LTP	Accuracy
IBM	82.61967	82.69	99.99
KO	22.871609	22.76	99.21
MICROSOFT	26.29095	26.29	99.99
MO	14.750482	14.68	99.52
MRK	29.80507	29.16	97.8
NEE	101.37487	101.67	99.79
NEOG	63.992867	63.19	98.73
UTX	62.23822	62.49	99.58
XOM	56.418736	56.48	99.94

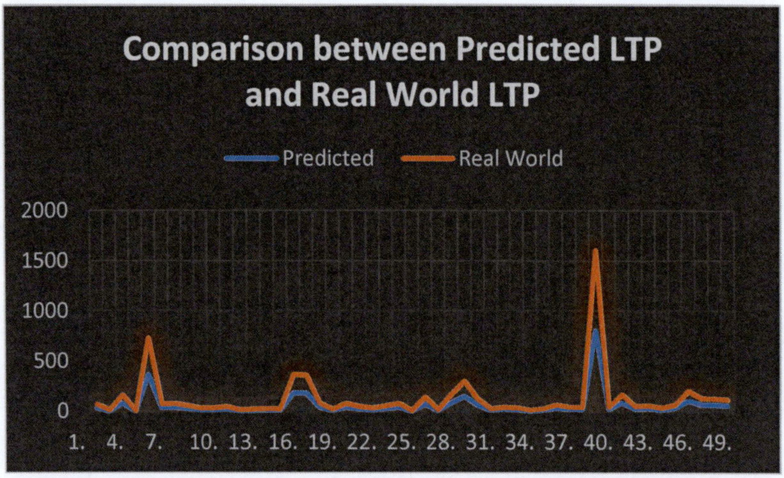

Fig. 9. Comparison between the Predicted LTP and real-world LTP.

5 Conclusion

In this paper, a stock market prediction system is proposed to help investors to invest accurately at the stock markets. In our system, the data is used from the Wall Street Journal and AlphaVantage and trained our model accordingly. Later on, experiments are conducted to test the accuracy of our system. From our experimental evaluation, the predictive accuracy of our proposed model is impressive and could be useful for predicting stocks price. In any case, it is ever pondered about what drives stock exchange—that is, the thing that components influence a stock's cost. Unfortunately, there is no clean equation that reveals to us precisely how a stock value will act. So, in future more factors are forced behind the stock market such as technical factors, market sentiment, earnings per share (EPS) etc. for predicting stock price.

References

1. Alpha Vantage. https://www.alphavantage.co/. Accessed 30 Feb 2019
2. Wall Street Journal. https://www.wsj.com/news/markets. Accessed 1 Feb 2019
3. Alostad, H., Davulcu, H.: Directional prediction of stock prices using breaking news on twitter. In: 2015 IEEE/WIC/ACM International Conference on Web Intelligence and Intelligent Agent Technology (WI-IAT), vol. 1, pp. 523–530. IEEE (2015)
4. Anik, M., Arefin, M., Dewan, M.: An intelligent technique for stock market prediction, pp. 721–733, January 2020. https://doi.org/10.1007/978-981-13-7564-460
5. Chen, A.S., Leung, M.T., Daouk, H.: Application of neural networks to an emerging financial market: forecasting and trading the Taiwan Stock Index. Comput. Oper. Res. **30**(6), 901–923 (2003)
6. Chen, H., Dyke, P: Modelling and prediction of stock price dynamics using system identification methodology based on a popularly used techniqueanalysis data. In: 2015 SAI Intelligent Systems Conference (IntelliSys), pp. 889–893. IEEE (2015)
7. Golmaryami, M., Behzadi, M., Ahmadzadeh, M.: A hybrid method based on neural networks and a meta-heuristic bat algorithm for stock price prediction. In: 2015 2nd International Conference on Knowledge-Based Engineering and Innovation (KBEI), pp. 269–275. IEEE (2015)
8. Gong, J., Sun, S.: A new approach of stock price prediction based on logistic regression model. In: 2009 International Conference on New Trends in Information and Service Science, pp. 1366–1371. IEEE (2009)
9. Kara, Y., Boyacioglu, M.A., Baykan, O.K.: Predicting direction of stock price index movement using artificial neural networks and support vector machines: The sample of the Istanbul Stock Exchange. Expert Syst. Appl. **38**(5), 5311–5319 (2011)
10. Kim, K.: Financial time series forecasting using support vector machines. Neurocomputing **55**(1–2), 307–319 (2003)
11. Kogan, S., Levin, D., Routledge, B.R., Sagi, J.S., Smith, N.A.: Predicting risk from financial reports with regression. In: Proceedings of Human Language Technologies: The 2009 Annual Conference of the North American Chapter of the Association for Computational Linguistics, pp. 272–280 (2009)
12. Kuttichira, D.P., Gopalakrishnan, E., Menon, V.K., Soman, K.: Stock price prediction using dynamic mode decomposition. In: 2017 International Conference on Advances in Computing, Communications and Informatics (ICACCI), pp. 55–60. IEEE (2017)
13. Lai, L., Li, C., Long, W.: A new method for stock price prediction based on MRFs and SSVM. In: 2017 IEEE International Conference on Data Mining Workshops (ICDMW), pp. 818–823. IEEE (2017)
14. Milosevic, N.: Equity forecast: predicting long term stock price movement using machine learning (2016). https://arxiv.org/ftp/arxiv/papers/1603/1603.00751.pdf
15. Nann, S., Krauss, J., Schoder, D.: Predictive analytics on public data the case of stock markets (2013)
16. Schumaker, R.P., Chen, H.: Textual analysis of stock market prediction using breaking financial news: the AZFin text system. ACM Trans. Inf. Syst. (TOIS) **27**(2), 1–19 (2009). https://doi.org/10.1145/1462198.1462204
17. Selvin, S., Vinayakumar, R., Gopalakrishnan, E., Menon, V.K., Soman, K.: Stock price prediction using LSTM, RNN and CNN-sliding window model. In: International Conference on Advances in Computing, Communications and Informatics (ICACCI), pp. 1643–1647. IEEE (2017)

18. Shi, Y., Japa, A.: Multi-dimensional analysis on data sets for information retrieval. In: 2017 International Conference on Intelligent Sustainable Systems (ICISS), pp. 157–162. IEEE (2017)
19. Usmani, M., Adil, S.H., Raza, K., Ali, S.S.A.: Stock market prediction using machine learning techniques. In: 3rd International Conference on Computer and Information Sciences (ICCOINS), pp. 322–327. IEEE (2016)
20. Shen, W., Guo, X., Wu, C., Wu, D.: Forecasting stock indices using radial basis function neural networks optimized by artificial fish swarm algorithm. Knowl. Based Syst. **24**, 378–385 (2011)
21. Zhang, D., Zhou, Z.: $(2D)^2$PCA : 2-Directional 2-Dimensional PCA for Efficient Face Representation and Recognition. Neurocomputing **69**, 224–231 (2005)

Machine Learning Based Diagnosis of Alzheimer's Disease

M. Karthiga[1(✉)], S. Sountharrajan[2], S. S. Nandhini[1], and B. Sathis Kumar[3]

[1] Bannari Amman Institute of Technology, Sathyamangalam, India
{karthigam,nandhiniss}@bitsathy.ac.in
[2] VIT Bhopal University, Madhya Pradesh, India
sountharrajan@gmail.com
[3] VIT University, Chennai, Tamilnadu, India
sathiskumar.b@vit.ac.in

Abstract. Alzheimer's disease is the most dreadful. Despite the various Alzheimer's treatments available nowadays, the survival rate of Alzheimer's patients is very much low. Detection of Alzheimer's disease in an earlier stage is one of the most important things to reduce the death rate. Magnetic Resonance Imaging (MRI) is an important tool in medical informatics and clinical diagnosis. This MRI image helps to diagnose and detect Alzheimer's disease at advanced stages. There are several approaches implemented to discover Alzheimer's by MRI data. Features are extracted discriminately and combinations of different classification techniques for classification is implemented in the proposed work. In this paper, the curvelet-based transform technique is utilized for extracting the features. The AdaBoost classifier is utilized for combining multiple weak classifiers into one strong classifier to improve the accuracy of the result. To improve the efficiency of the result, AdaBoost classifier with SVM was used to obtain better results than existing ones. The proposed model with 95.66% accuracy predicts the AD from the brain MRI images at an earlier stage. This accuracy value is far better than other classifiers like Decision Tree and Random Forest.

Keywords: Big data · Data mining · Data analytics · Alzheimer's disease · AdaBoost algorithm · SVM classifiers · Classification · Segmentation

1 Introduction

Today's world is driven only by data. The smooth functioning of the world cannot be imagined without data. Data is one of the important assets of every organization. Nowadays a large amount of data has been produced by different sources. Data storage is an important concern, for that where to store data and how to process data is a very big challenge. Big data will have different properties size, variety; frequent changes of such data will need a different type of storage and analytics method. The most important feature in big data is that the data should be properly analysed to extract relevant information from that.

J. I.-Z. Chen et al. (Eds.): ICIPCN 2020, AISC 1200, pp. 607–619, 2021.
https://doi.org/10.1007/978-3-030-51859-2_55

Alzheimer's disease is a common disorder for brain-related dementia that is deadliest most of the times. Identification of Alzheimer's in the earliest stage is very much important because it is 6th most dangerous means of death in the world. AD is the most dangerous brain malfunction that progressively destroys the cells of the brain and increases dementia of a person. It slowly cut-off the thinking, remembering and behavioural attitude of a person in day to day life. Finally, it leads to brain death. Cognitive impairment is a normal process of brain ageing, as long as it is not associated with AD dementia, AD does not have fixed symptoms unless it is closely looked for and it is very difficult to identify the transition of normal cognitive impairment to AD and this transition point is different for different individuals as demonstrated in [1]. MRI images help to identify the brain damage to the brain shrinking structure and extension of cerebral fluid spaces. MRI utilizes the magnetic radiations with high density to determine the defects in human organs that help the doctors to diagnose the condition effectively.

MRI of the brain gives a detailed representation of brain structure detailing about the minute divisions of cells that are damaged and to identify the changes in brain size and shape [2]. The motive of the proposed methodology is to diagnose AD with optimized features from brain MRI images followed by a prediction by utilizing various classifiers and optimiser to diagnose brain dementia priority with acute values.

2 Literature Survey

A lot of techniques are proposed in the world for AD diagnosis by utilizing brain MRI. A Three-Dimensional (3D) model of brain MRI images to detect Alzheimer's disease (AD) is done in [3]. More than 15 AD patients' 3D brain images are selected for testing purpose. The features are extracted in the form of 3D texture from the hippocampus and entorhinal cortex using a semi-automated mechanism. The classification accuracy measured in the range of 64.3% to 96.4% based on the texture analysis. The variation occurs due to dissimilarities of texture, feature extraction and feature identification as mentioned by [4]. The identified 3D texture features were compared with Mini-Mental State Examination scores. Validations imply the deviations occurred between normal patients and AD patients.

Voxel-Based Morphometry (VBM), helps to unbias the brain at the vowel level in an autonomous way [5]. This technique does not involve any prior level assumptions of the structure of the brain for accessing. The automatic tool has been designed in [6] for determining the brain tumour using Matlab. Nowadays, it has become a standard tool for examining the changes in the brain morphologically related to ageing of the human, disorders in the people psychologically and also with neurodegenerative disorders as suggested in [7].

The VBM also helps to differentiate the grey matter of the brain, the volume of the brain with normal people and with hyperthyroidism patients [8]. The brain MRI images have to be classified anatomically which is a challenging task due to the huge volume of image data. MRI based imaging schema helps for high clinical diagnosis and also in the field of medical research.

In the cingulate cortex of the brain, there is metabolic reduction posteriorly in early AD patients. A new Computer-Aided Diagnosis (CAD) tool for AD diagnosis at an early stage is utilized that collects the information related to Grey Matter along with White Matter in the brain along with the tissue distributed information. This tool aims in the diagnosis of brain dementia as proposed by [9]. Initially, segmentation of image dataset to better identify the disease using fuzzy based clustering approach is proposed in [10].

In this proposed method, Learning Vector Quantization (LVQ) algorithm is used, it helps to identify the AD images by using the tissue distribution of control and it helps to develop a set of prototypes of different class labels. The identified AD images have undergone different levels of classification process using Support Vector Machine (SVM) which will be improving the classification accuracy up to 90%. In [11], graph based matching of the features among the predicted is proposed. This predicted vector gives a detailed representation of the volume of GM or attributes of cortical thickness.

For manipulating the operations of large image dataset, cloud based architecture is utilized in [12]. The proposed techniques reveal about the various security concerns related to cloud based performance analysis. In [13] mild impairments in the brain is converted as Alzheimer by accurate predictions using hippocampal based entropy model. The dimensionality of large input data is reduced by utilizing dimensionality reduction approaches and NMSE attributes are delineated over ROI by undergoing at-test process with weighting features correlatively in [14]. To classify brain images, SVM classifier using fractional anisotropy and voxel region based mean diffusivity features are used by [15]. A fully autonomous working CAD system is proposed by [16] for the participants of ADNI.

3 Proposed Work

3.1 Database

The dataset for the proposed technique is gathered from "Alzheimer's disease Neuroimaging Initiative (ADNI) database (www.loni.ucla.edu/ADNI)" site maintained by "National Institute of Biomedical Imaging and Bioengineering (NIBIB), National Institute on Aging (NIA), Food and Drug Administration (FDA), private pharmaceutical companies and non-profit organizations" since 2003 for undergoing Alzheimer disease relevant research. The dataset contains MRI scans within the ages of 55 to 90. The MMSE (Mini Mental State Examination) scores for health are among 24 to 30. The CDR (Clinical Dementia Rate) is given as 0 for non-depressed and non-MCI people. For MCI, the MMSE scores are among 24 to 30 represents the patients with memory related complaints and CDR score, 0.5 represents the no injury as well as the absence of dementia. The number of images chosen in total is about 45 with Alzheimer disease and 60 for MCI and 45 images of healthy people. Each image is of size 256 * 256. The number of features obtained in total is 6144 (32 * 32 * 6). In this work, the extracted feature selection from MRI images based on different feature selection techniques is implemented as in Fig. 1. As extracted features are high dimensional, an optimized feature subset improved MRI brain images classification as AD, MCI and Control Subject (CS) are used.

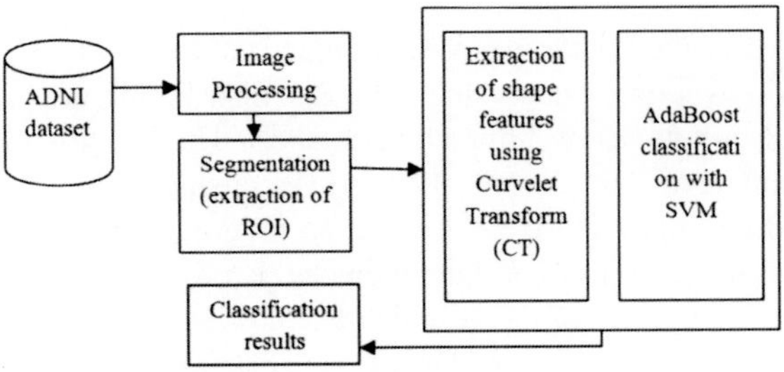

Fig. 1. Proposed methodology.

3.2 Image Pre-processing

Brain images with Alzheimer's disease contain noisier data. This noises will decrease the quality of disease identification process. Hence brain images are pre-processed to decrease the noises and to increase the classification accuracy of an image. There are so many image pre-processing techniques available that include contrast adjustment, intensity adjustment, histogram equalization, binarization and morphological operations. Here in this approach, the histogram equalization and morphological operations are used.

Histogram Normalization techniques were used for pre-processing the images. After pre-processing, it was observed that the enhancement of the images was achieved. This process will be more consistent for the dynamic data region to reduce the noise in the images. Also, this approach will distribute the intensities of the pixel so that all the range of pixel intensities are considered as shown in Fig. 2 and Fig. 3.

Fig. 2. Original image.

Fig. 3. Histogram normalized image.

The image taken for consideration will be reconstructed using morphological operations. Thresholds are utilized to extend the maxima during image transformation. The computation of regional maxima related to H-maxima transformation is the result of the extended transformation of the maxima. The resultant binary image is represented in Fig. 4.

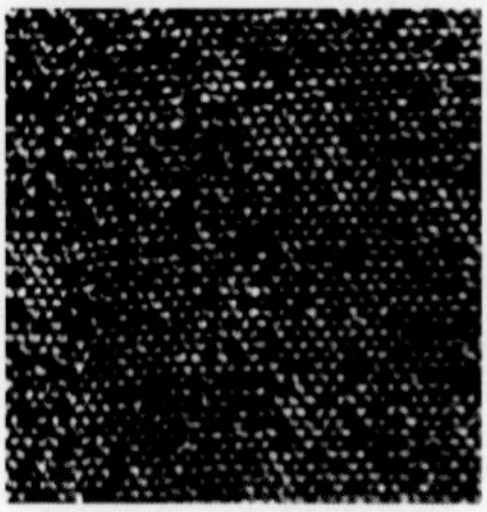

Fig. 4. Morphologically processed image.

3.3 Segmentation

Segmentation is the process of dividing the images into different partitions and the main aim of this segmentation is to make images into meaningful and easy elements to analyse. The original images taken for analysis is depicted in Fig. 5. Segmentation technique plays an important role in the Image classification process. The segmentation process can be contextual or non-contextual. The contextual segmentation concentrates on the distribution of the pixels on the greyscale images to detect the features.

To make the image meaningful and beneficial for easy analysis and understanding, the image regions should intensely correlate with the depicted features on the image which are under consideration and possess the major interest. The transformation of greyscale images into images with a high degree of feature description is the importance of having segmentation on MRI images. But the accurate segregation of greyscale images is a challenging process and only that decides the efficiency of classifying the MRI images.

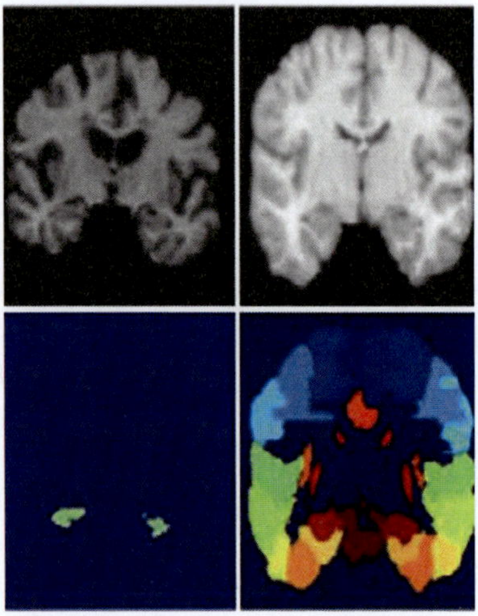

Fig. 5. Sample brain MRI images [6].

Thresholding is one of the simple types of segmentation where the greyscale images are segmented into foreground and background. To be precise, thresholding will give the binary image from the given greyscale image. Again fixing the threshold value for processing has many options. Automatic thresholding is one among the types which make the system by itself fixes the threshold values. Fixing the automatic threshold value depends on parameters like the size of the objects with high-interest, portion of the image occupied by the objects, total number of different types of such objects appearing in the image and the intensity of each object.

This segmentation process helps to identify the dead cell region from AD images are in Fig. 7. To label the images from different classes and group similar types of images in the single class region and ultimately this will help to increase the classification accuracy and the segmentation without edge detection is shown in Fig. 8. Based on the fixed threshold value, the process of image classification and class labelling is implemented over it.

3.4 Extraction of Features

Extraction of Features related to selecting the features from the segmented images which are to be used in the classification tasks. Using some mapping techniques, a new set of features are generated from the original one. The objective of selecting optimised features is to represent the original image in the reduced format to make it easier for the classification process. In our proposed work, curvelet transform based feature selection process is used. This curvelet transform was developed by [17] to represent smooth objects in an image with discontinuities.

This technique is useful to identify an important feature from the given image. These discontinuities will help to increase the classification accuracy process. The main advantage of this technique is to reconstruct the flat part of an image without blurring the edges. In this reconstruction, the curvelet transform is applied at different scale and directions. The curvelet transform is a multi-scale directional transform that allows an almost optimal non-adaptive sparse representation of objects with edges. Once the features are detected then the curvelet transform is applied for extracting the identified features from the MRI images. In the feature selection process, the curvelet transform has two stages training stage and classification stage. In the training stage, known images are given for processing with that feature extraction is performed on the new images.

In this case of multi-scale analysis, ridgelet transform is extended from the curvelet transform. For a greyscale image with function f(x, y), the transform is defined as in Eq. 1. Curvelet Transform has spatial and frequency parameters Y and Z respectively. This hologram will give a smooth surface for classification purpose.

$$R_f(a, b, \theta) = \iint \varphi_{a,b,\theta}(x, y) f(x, y) dxdy \quad (1)$$

4 Classification Technique

4.1 AdaBoost Algorithm

The AdaBoost algorithm provides improved accuracy by combining different classifiers. It works iteratively. It provides a strong and efficient classifier by gathering a lot of weak classifiers. Weight is set across each classifier and in each iteration of training samples, accurate predictions are made for unusual classifier behavior. As a base classifier, any machine learning algorithms could be utilized. In the proposed approach, SVM is used as a base classifier. AdaBoost classifier minimizes the training error during the training process in each iteration. The process of AdaBoost classification is explained in below Fig. 6.

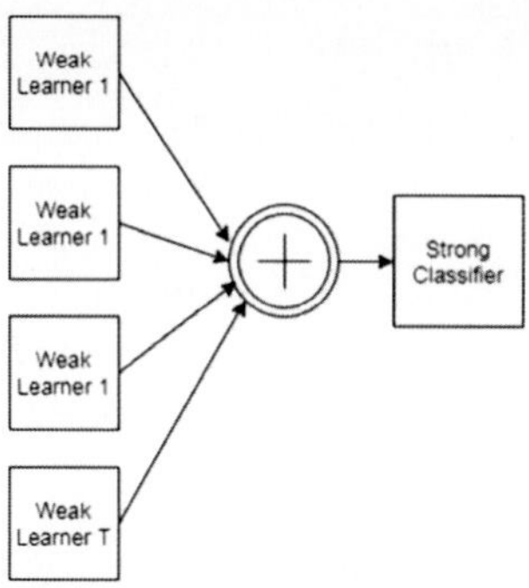

Fig. 6. The architecture of AdaBoost algorithm.

The proposed model uses the following algorithm for classifying the image dataset using Adaboost classifier.

Step 1: For every observation, X_i, assign an initial weight $W_i = 1/n$, where n relates the total number of observations.
Step 2: Train a weak model (utilizing SVM)
Step 3: For every observation do the following:
  3.1: If correctly predicted, weight W_i is decreased
  3.2: In case predicted incorrectly, weight W_i is increased
Step 4: New weak model is trained now and observations with greater values of W_i are given more priority.
Step 5: Steps 3 and 4 are repeated until the accuracy of the predictions are gained more.

4.2 Support Vector Machine (SVM)

The MRI image can be classified as normal or abnormal based on the features given by feature extraction. Features are extracted from the region where the cells are damaged and that accelerates the decision making on the MRI images. The features that got extracted are given as input to the classifiers and in turn, the classifiers assign the images to the matching category that they represent. The main aim of feature extraction is to lessen the actual data by assessing only the susceptive features that match with the input sample. The more number of features increases the computing time and also requires more storage, which may make the classification as a complex process. So dimensionality reduction helps in reducing the feature set and that is given as input to SVM classifier.

SVM is a supervised methodology for classification and it mainly classifies the data into two sets popularly known as two group classification. Also, the SVM classifier is very much work to classify limited data set with well-trained data. The AdaBoost algorithm also provides better result on binary classification and SVM associates well with AdaBoost to come up with two-group classification.

4.3 Image Classification Using SVM

Once the original MRI images of the brain are decomposed into features, the data set is ready for classification. All detailed coefficients are included in the features like horizontal level, diagonal level and vertical level. Hard and Soft threshold are the models available and that has been selected based on the values: Fixed form threshold, high penalisation, medium penalisation, low penalisation and Bal, sparsity norm (sqrt). The white in the image represents the noises in images. Unscaled white noises image structures are considered for testing.

Classification refers to the task of assigning objects to one of the various predefined categories that encompass many different applications. AdaBoost Algorithm is used to develop a strong classifier by combining the similar of the weak classifier as said above. The set of figures in Fig. 7 and Fig. 8 represents the image segmentation using threshold values as discussed before. It gives the image segmentation using threshold and without edge detection.

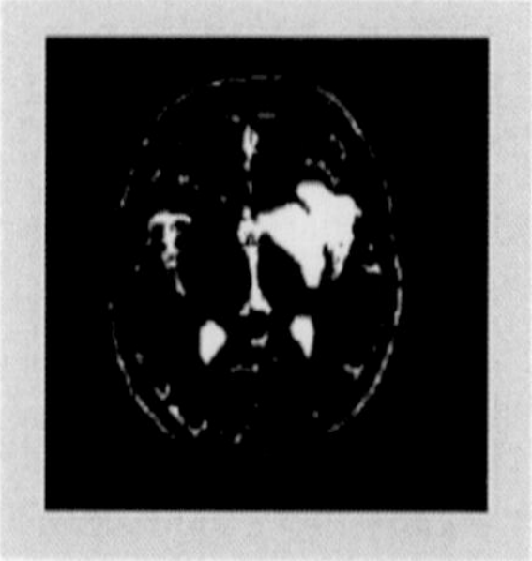

Fig. 7. Threshold segmentation.

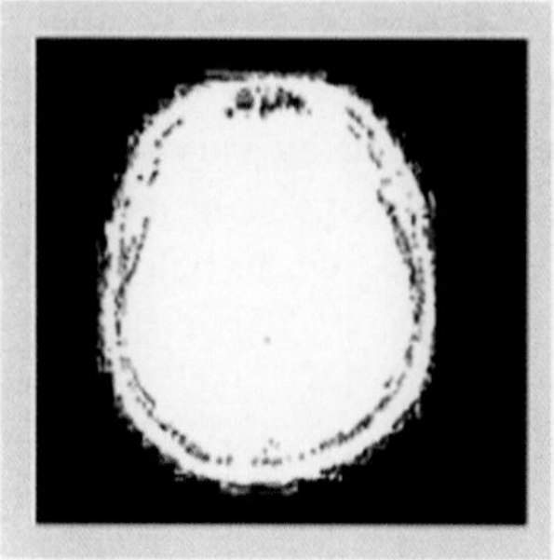

Fig. 8. Threshold segmentation without edge detection.

The holes in the above images are filled the same as the background after applying the various set of threshold values as discussed earlier. Then the image is transformed into binary format and the morphological operations are done to remove the isolated 1s and 0s that represents the isolated pixels. The operations have been applied continuously and that performs erosion followed by dilation.

This algorithm mitigates the misclassification problem by increasing the number of weak classifiers. This algorithm will consider the training pattern $(x_i, y_i) \ldots (x_n, y_n)$ for w selected features. The objective of an AdaBoost classifier runs an Algorithm T = N times.

5 Results and Discussion

The dataset comprises in total of 45 numbers of images with Alzheimer disease and 60 for MCI and 45 images of healthy people. Each image is of size 256 * 256. The number of features obtained in total is 6144 (32 * 32 * 6). For increasing the performance the dataset has to be split into training and testing dataset. The dataset is splitted using train_test_split() function in python. The test_size parameters are set to 0.3 (30% for testing and the remaining % for training). AdaBoost model is imported from scikit_learn package. AdaBoost classifier, in default, utilizes Decision Tree for classification. First, the model is trained with Decision tree and the results are observed. Three most important feature values have to be set for AdaBoost classification. One is base_estimator, which is a weak_learner technique. By default, Decision Tree is set under base_estimator. Next parameter is n_estimators, which provides information about the number of weak_classifiers utilized. In the proposed model, n_estimators are set to the value equal to 50. The last parameter is learning_rate, that contributes to the weights assigned to the weak classifiers. Initially, the learning_rate value is fixed to 1. Accuracy is obtained by comparing the actual test values with predicted results. The gained accuracy using the default classifier is about 88%.

In this proposed work the algorithm was executed T = 5 times. The weight of every training data is going to be calculated by using the αi formula. i denotes the iteration count of the weak classifier. To improve accuracy, random_forest is used as a base_estimator. But the accuracy obtained is somewhat better than using a decision tree. For further increasing the accuracies, SVM is used as base_estimators with same values for n_estimator and learning_rate. Now the classification accuracy is improved to 95.56%. The procedures are repeated by varying the number of samples taken for consideration. By using different classifiers the classification accuracy is calculated against different number features as shown in below Table 1.

Table 1. Classification accuracy for different number of features.

Number of features	Classification accuracy
30	75.43
60	78.3
90	80.55
120	84.75
150	89.45
180	92.44
200	95.56

The below Fig. 9 illustrates the classification accuracy obtained for the different number of samples by the proposed model. The results show that the proposed model provides higher accuracy for a large number of samples.

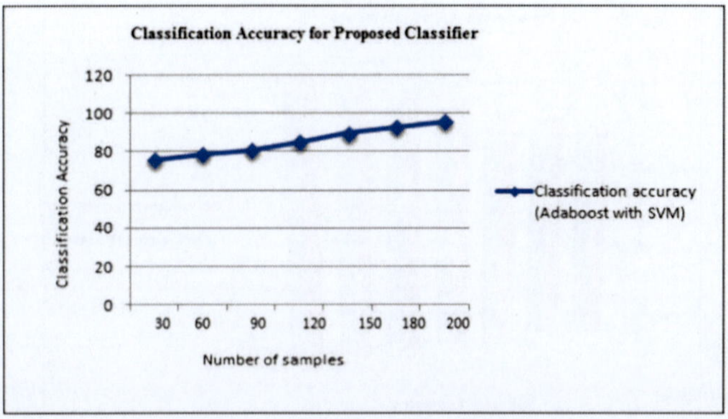

Fig. 9. Classification accuracy graph.

The results obtained for each number of features with different classifiers is represented in below Table 2. Figure 10. depicts the performance of the proposed model with other different classifiers. The results reveal that for 200 number of samples, the classifier accuracy for the proposed model is about 95.56% compared to 90.17% for Random Forest classifier and only 87.54% for the default Decision Tree classifier with AdaBoost. When the number of samples gets reduced the classification accuracy also get reduced. This proves that the classifiers work better with an increased number of training samples. A comparison of predicted accuracy with actual test values is illustrated in Fig. 10. The figure illustrates that the values are closer to each other proving that the model works better for all types of samples.

Table 2. Classification accuracy for different classifiers.

Number of samples	AdaBoost with DT	AdaBoost with RandomForest	AdaBoost with SVM
30	66.43	72.83	75.43
60	70.23	75.88	78.3
90	78.55	79.00	80.55
120	82.57	83.24	84.75
150	83.44	85.78	89.45
180	85.44	88.74	92.44
200	87.54	90.17	95.56

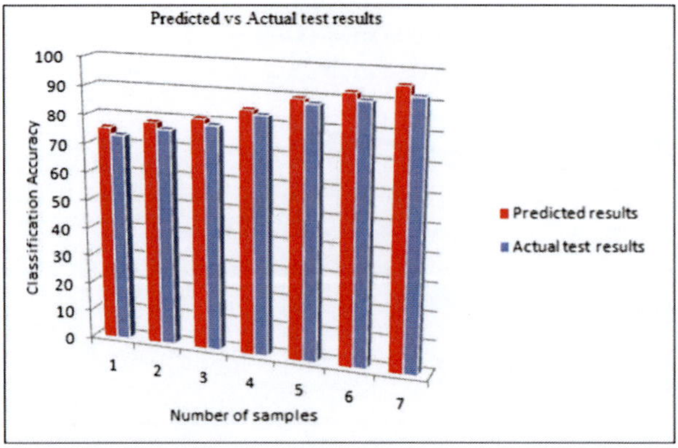

Fig. 10. Comparison of predicted and actual test results.

6 Conclusion

Alzheimer's disease is one of the most dreadful diseases nowadays. This disease increases twice the amount before 2020. In this proposed work AdaBoost Algorithm increases the classifier accuracy rate. This classifier provides fast, accurate, and cost-effective diagnosis of Alzheimer's disease and the classifier reduces the misclassification rate of small classifier and the number of weak classifiers also reduced using Boost Algorithm. The results are obtained by combining AdaBoost with different classifiers. AdaBoost with Decision Tree classifiers yields about 88% accuracy and RandomForest classifiers provide 90% accuracy. To boost the accuracy values, SVM boosting classifiers are utilized in the proposed model. The proposed model outperforms the other two classifiers and yields a result of more than 95%. Thus the proposed technique of implementing AdaBoost with SVM classifiers, correctly predicts the AD from brain MRI images with at most 95.66% accurately.

References

1. Boucheham, A., Batouche, M., Meshoul, S.: Robust hybrid wrapper/filter biomarker discovery from gene expression data based on generalised island model. Int. J. Comput. Biol. Drug Des. **8**(3), 251–274 (2015)
2. Kayalvizhi, M., Kavitha, G., Sujatha, C.M.: Analysis of ventricle regions in Alzheimer's brain MR images using level set based methods. Int. J. Biomed. Eng. Technol. **12**(3), 300–319 (2013)
3. Vapnik, V.: The Nature of Statistical Learning Theory. Springer, New York (2013)
4. Fjell, A.M., et al.: CSF biomarkers in prediction of cerebral and clinical change in mild cognitive impairment and Alzheimer's disease. J. Neurosci. **30**(6), 2088–2101 (2010)
5. Aggarwal, N., Rana, B., Agrawal, R.K., Kumaran, S.: A combination of dual-tree discrete wavelet transform and minimum redundancy maximum relevance method for diagnosis of Alzheimer's disease. Int. J. Bioinform. Res. Appl. **11**(5), 433–461 (2015)

6. Karuna, M., Joshi, A.: Automatic detection and severity analysis of brain tumors using gui in MATLAB. Int. J. Res. Eng. Technol. **2**(10), 586–594 (2013)
7. Candes, E.J., Donoho, D.L.: Edge-preserving denoising in linear inverse problems: optimality of curvelet frames. Ann. Stat. **30**(3), 784–842 (2000)
8. Saritha, M., Joseph, K.P., Mathew, A.T.: Classification of MRI brain images using combined wavelet entropy based spider web plots and probabilistic neural network. Patt. Recog. Lett. **34**(16), 2151–2156 (2013)
9. Haralick, R.M., Shanmugam, K., Dinstein, I.H.: Textural features for image classification. IEEE Tras. Syst Man Cybern. **6**, 610–621 (1973)
10. Goldgof, R.V.L.H.D.L., Clark, C.M., Silbiger, M.: MRİ segmentation using fuzzy clustering techniques. IEEE Eng. Med. Biol. Mag. **13**(5), 730–742 (1994)
11. Liu, F., Suk, H.-I., Wee, C.-Y., Chen, H., Shen, D.: High-order graph matching based feature selection for Alzheimer's Disease identification. In: Mori, K., Sakuma, I., Sato, Y., Barillot, C., Navab, N. (eds.) MICCAI 2013. LNCS, vol. 8150, pp. 311–318. Springer, Heidelberg (2013). https://doi.org/10.1007/978-3-642-40763-5_39
12. Zhuang, Y., Jiang, N., Wu, Z., Li, Q., Chiu, D.K., Hu, H.: Efficient and robust large medical image retrieval in mobile cloud computing environment. Inf. Sci. **263**, 60–86 (2014)
13. Apostolova, L., et al.: Conversion of mild cognitive impairment to Alzheimer's disease is predicted by hippocampal atrophy maps. Arch. Neurol. **63**, 693–699 (2006)
14. Alvarez, I., Górriz, J.M., Ramírez, J., Salas-Gonzalez, D., López, M., Puntonet, C.G., Segovia, F.: Alzheimer's diagnosis using eigenbrains and support vector machines. Electron. Lett. **45**(7), 342–343 (2009)
15. Fung, G., Stoeckel, J.: SVM feature selection for classification of spect images of Alzheimer's disease using spatial information. Knowl. Inf. Syst. **11**(2), 243–258 (2007)
16. Illán, I.A., et al.: 18F-FDG PET imaging analysis for computer aided Alzheimer's diagnosis. Inf. Sci. **181**(4), 903–916 (2011)
17. Apostolova, L.G., Dinov, I.D., Dutton, R.A., Hayashi, K.M., Toga, A.W., Cummings, J.L., Thompson, P.M.: 3D comparison of hippocampal atrophy in amnestic mild cognitive impairment and Alzheimer's disease. Brain **129**(11), 2867–2873 (2006)

IoT Based Vehicle (Car) Theft Detection

Rajasekhar Kommaraju[1(✉)], Rangachary Kommanduri[2], S. Rama Lingeswararao[3], Boyapati Sravanthi[3], and Cherukumalli Srivalli[3]

[1] IT Department, LakiReddy Bali Reddy College of Engineering, Mylavaram 521230, India
rajasekharkommaraju@gmail.com
[2] CSE Department, LakiReddy Bali Reddy College of Engineering, Mylavaram 521230, India
rangachary22@gmail.com
[3] LakiReddy Bali Reddy College of Engineering, Mylavaram 521230, India
ram.suvvada@gmail.com, sravanthiboyapati22@gmail.com, srivallicherukumalli@gmail.com

Abstract. It is a foregone conclusion that property crimes will hit ten million. Among such, within the steal list, the vehicle is flat-top and sometimes controls entire components of the earth. There have been some recent technical advances and new strategies are being developed to overcome this downside. The ways involved in the detection of vehicle theft are noted for interference by some or everything, as well as shields. This research helps to prevent the vehicle stealing from third parties with the help of RFID card and authorized key. Here, RFID reader connected to the car and if anyone enter into car and they requires the card authorization. Keypads were connected to the engine and the authorized person can only enter into the car by pressing passwords. If the password is incorrect, buzzer will produce noise continuously. The proposed system gives the regular updates to the car owners regarding the location of car by using GPS technology.

Keywords: RFID reader · Keypad · Buzzer · GPS · thingspeak.com · IoT · Vehicle tracking

1 Introduction

In recent years the vehicle stealing is a big problem. The protection of the vehicle is needed. There are also some limitations. Therefore, an effective security threat protection of vehicle is required. The Arduino interface is an integral a section of dc motor and GPS. Wireless fidelity module is used to search out the vehicle's location through the Global Positioning System (GPS). GPS can be a navigation system that would be used to tracking the car, and it provides entirely environmental conditions for putting the consumptive unit. Victimization the GPS antenna provides the latitude and meridian of the device. When the vehicles are left outside, currently if the vehicle is

J. I.-Z. Chen et al. (Eds.): ICIPCN 2020, AISC 1200, pp. 620–628, 2021.
https://doi.org/10.1007/978-3-030-51859-2_56

taken, the D.C. motor starts and therefore the method on top of is over and also the info is distributed through the web content. This vehicle is employed as a vehicle thievery and rescue tool for thievery hindrance and following system. Prevention of cars is often done by providing authorization for owners and by building anti-theft system in vehicles.

2 Related Work

The most usually used methods are beeps, alarms and biometrics. However, these are costly. It's easy to steal your vehicle by using a car buzzer. But when your car is away from you, you may not benefit from buzzer or alarm recognition.

Car alarm systems are used to help with various types of sensors such as pressure, tilt and shock & door sensors to prevent car robbery. However, these systems have some restrictions like peak worth, peak false warning rate and straight forward disposal. Recent developments in the PC software system and hardware for fixing this problem are to develop automobile industry reasonable authentication systems and automated biometrics primarily based identification. Several biometrics are used, like automatic face recognition, facial expression, hand pure mathematics, handwriting and voice for identity and verification of people. But biometrics is not 100% accurate and requires unification and/or additional hardware and cannot be compromised once if somebody learns it, you'll be able to continuously modification your password; however, there's no chance to modify our Iris membrane or fingerprint. If somebody features a copy of their work, you can't do a lot to remain safe.

3 Proposed Work

This paper provides an Internet Of Things based vehicle identification system. Many procedures use the date of exploitation to spot the vehicle, the planned system has exceeded the prevailing systems and system limits. This machine starts with Dc Motor, which suggests the door is unbarred exploitation, the RFID card and the car door gets locked after a few minutes. When the person enters the car, he/she have to give the authorized key to start the vehicle. During the whole process, Arduino GPS is active and entire information including car door status motor status and with the assistance of GPS location and wireless fidelity module location will be sent to the owner by thingspeak.com. The entire mechanism is supported through a switch for owner satisfaction. The projected system diagram is shown in Fig. 1 below.

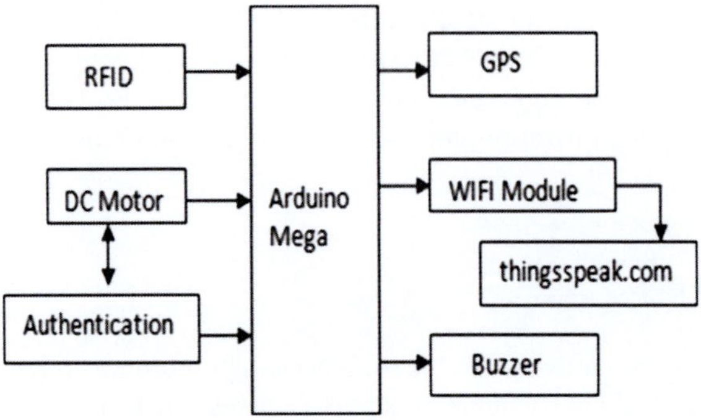

Fig. 1. Block diagram

1. *Arduino Mega*
   The microcontroller module, ATmega2560, is based on the Arduino Mega "Fig. 2". It includes fifty-four digital output/input pins (it can use fourteen PWM outputs), sixteen analog inputs, sixteen MHz crystal oscillator, four UARTs (hardware serial ports), an ICSP header, one USB association, a power jack and a reset button.

Fig. 2. Arduino mega view

2. *Buzzer*
   A sound device (Fig. 3) that uses sound and is used to sign in. The bazaar emphasizes a warning form in the form of a constantly waking sensation or beep. Initially, this device has an electroquintic system, which is surprising to an electric bell.

Fig. 3. Buzzer

3. *WiFi Module ESP8266*
ESP8266 wireless fidelity (Fig. 4) module the self- controlling system on the chip that accesses the microcontroller to your wireless fidelity network. Each module has been arranged through an instruction set instruction code associate with the engine, which is often linked to the Arduino device and suggests that Wi-Fi can gain power. The ESP8266 module is heavily and is growing anytime, the first expensive board in society.

Fig. 4. WiFi module

4. *DC Motor*
Very easy motor. D.C. motors (Fig. 5) have typically 2 leads, one is positive and the other is negative. If you connect this battery to the battery, the motor rotates. If you change the leads the motor will rotate in the opposite direction.

Fig. 5. Motor

5. *Relay*
Figure 6 - Relay can be used to activate the whole theft identity system. Switch on SMS alert and as soon as the vehicle starts, the location will be sent to the owner. If the switch is not in ON place an alert will not be sent and information of the location will not be shared. By leveraging the switch, able to order system activation and inactivity.

Fig. 6. Relay

6. *GPS (Global Positioning System)*
   Generally speaking, GPS (Fig. 7) is a predominantly satellite based navigation system. The NEO-6 M GPS module is used as a result of it's appropriate for extending GPS receivers. It is a partial ceramic antenna. The 3 V button is connected to the battery. Sometimes GPS works in anyplace within the world in weather conditions. A GPS receiver should secure a minimum of three satellite signals to estimate the second point (latitude and longitude).

Fig. 7. Keypad

7. *RFID (Radio Frequency Identification)*
   RFID is a technology which scans inside a microscope or chip remote, which is used when the RF spectrum does not use power in the affected physical connections. The associate RFID system has a reader or addresser, which releases an affiliate radio frequency signal through an associate antenna shown in Fig. 8.

Fig. 8. GPS

8. *LCD (Liquid Crystal Display)*
   Liquid crystal display as shown in Fig. 9 contains 2 layers of polar panel filters and electrodes. The alphanumeric display technology was used to display on other electronic devices such as a notebook or small computers. The layer of lighter fluid is estimated from the lens. A mixture of colorful light with crystal's grayscale image is in the form of color image. This image is then displayed on the screen (Fig. 10).

Fig. 9. RFID reader

Fig. 10. LCD

4 Flow Chart

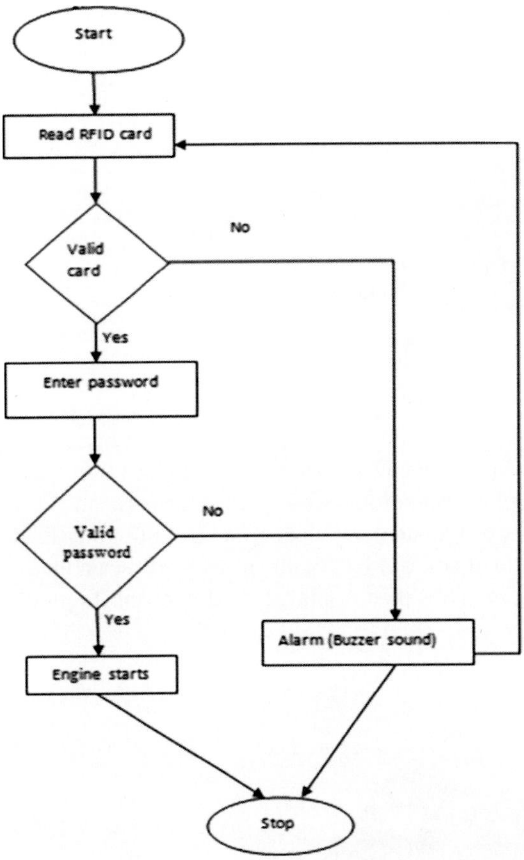

5 Result

The results are extracted by implementing software and hardware discussed in the above sections, as shown in Fig. 11. The RFID card reads the card and holds the card power, sends information to Arduino. After having the card, the owner must enter an authentic key on the keypad, then the engine starts, and the buzzer can be tracked by using the items that the owner gets a continuous sound. Information about the car and he can also track the car's location. Figure 11 and 13 represents the model using Thingspeak software kit and the location status (Fig. 12).

Fig. 11. Kit representation

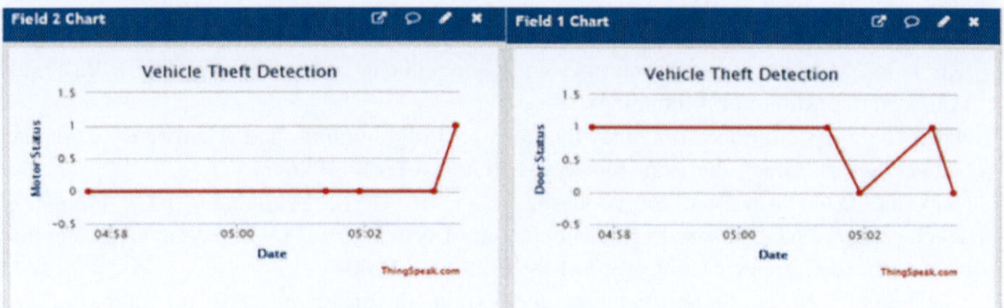

Fig. 12. Graphs representation in ThingSpeak

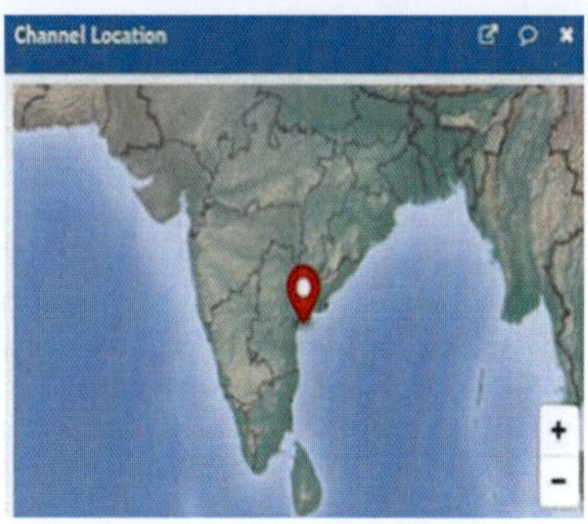

Fig. 13. Current location

6 Conclusion

Vehicle theft has inflicted more damage to its victims in monetary value and secondary financial effects. This system provides vehicle security and recognizes theft and is very cost-efficient.

7 Future Enhancement

The future scope of this paper, the biometric system can be used instead of the RFID reader. The alert message can be sent to the owner and nearby police station to track the vehicle by using the Global Positioning System and Global System for Mobile.

References

1. Ramani, R., Valarmathy, S., SuthanthiraVanitha, N., Selvaraju, S., Thiruppathi, M., Thangam, R.: Vehicle tracking and locking system based on GSM and GPS. Int. J. Intell. Syst. Appl. **9**, 86–93 (2013)
2. Poushya, M., Supritha, N., Hema, K., Tejaswini, R.: IoT based vehicle theft detection. IRE J. **1**(8), 52–55 (2018). Electronics and Communication Engineering (ECE), Vasireddy Venkatadari Institute of Technology
3. Hall, D.V.: Microprocessors and Interfacing: Programming and Hardware. Computer Science Series. Gregg division, McGraw-Hill, New York (1986)
4. Nagaraja, B.G., Rayappa, R., Mahesh, M., Patil, C.M., Manjunath, T.C.: Design & development of a GSM based vehicle theft control system. In: 2009 International Conference on Advanced Computer Control, pp. 148–152. IEEE (2009)
5. MacDonnell, M.C.: Stolen vehicles for export: a major concern for domestic and international security. Naval Postgraduate School Monterey United States (2018)
6. Manoharan, S.: An improved safety algorithm for artificial intelligence enabled processors in self driving cars. J. Artif. Intel. **1**(02), 95–104 (2019)
7. Patel, P.B.: IOT based vehicle theft detection. IJIRMPS-Int. J. Innov. Res. Eng. Multidiscip. Phys. Sci. **2**(3) (2014)
8. Predko, M.: Programming and Customizing the 8051 Microcontroller. McGraw-Hill Inc., New York (1999)
9. Nagaraja, B.G., Ravi Rayappa, M.M., Chandrasekhar, M.P., Manjunath, T.C.: Design & development of a GSM based vehicle theft control system. In: 2009 International Conference on Advanced Computer Control, pp. 148–152. IEEE (2009)
10. Singh, D.N., Tejaswi, K.: Real time vehicle theft identity and control system based on ARM 9. Int. J. Latest Trends Eng. Technol. (IJLTET) **2**(1), 240–245 (2013)
11. A Vehicle is stolen every 13 mins in Delhi; rate up 44% since last year-Times of India

EMG Signal Classification with Effective Features for Diagnosis

Abdul Wadud[1] and Md. Imran Hossain Showrov[2(✉)]

[1] Department of Computer Science, South Asian University, New Delhi, India
dewanwadud.ex@gmail.com

[2] Institute of Computer Science, Bangladesh Atomic Energy Commission, Dhaka, Bangladesh
showrov.cse@gmail.com

Abstract. Electromyography (EMG) signals are broadly used in various clinical or biomedical applications, prosthesis or rehabilitation devices, Muscle-Computer Interface (MCI), Evolvable Hardware Chip (EHW) development and many other applications. Electromyography (EMG) signal records the myopathy from nonlinear subjects in both time domain and frequency domain. It becomes very difficult to classify these various statuses. In this paper, a feature extraction and classification method of healthy and myopathy EMG signals are proposed where two features have been extracted on both healthy and myopathy EMG. Mean Squared Error (MSE) has been calculated to observe which feature will give better classification result. Then SVM is used to classify the extracted results. To evaluate the proposed model, a standard dataset collected from physionet.org is used where it shows higher accuracy than the conventional methods.

Keywords: Electromyography · Muscle computer interface · Mean Absolute Deviation · Sample Entropy

1 Introduction

Electromyography signal is a type of biomedical signal which carries properties of all conventional signals and describable in terms of their amplitude, frequency, and phase. It is originated by the movement of the muscle of the body. The movement of the muscle is known as contraction and vice versa is known as relaxation. EMG signals acquire noises while traveling through various tissues and nerves. So recorded EMG signals need to be preprocessed before use. The main purpose of EMG signal classification is in clinical diagnosis and biomedical applications. Neurological and Neuromuscular diseases can also be diagnosed with the help of different EMG signals. In terms of physiological background, there are three types of EMGs- Healthy, Myopathy and Neuropathy or Amyotrophic lateral sclerosis (ALS). Neuropathy or ALS is responsible for neurological diseases and Myopathy is responsible for neuromuscular diseases. Neuropathy or ALS is a rapidly progressive and fatal neurological disease. A statistic based on USA (ALS Association 2016), revealed that approximately 6000 peoples are diagnosed in the United States. So, research on neurological diseases is

J. I.-Z. Chen et al. (Eds.): ICIPCN 2020, AISC 1200, pp. 629–637, 2021.
https://doi.org/10.1007/978-3-030-51859-2_57

increasing day by day. Development in this research sector depends on the classification of different EMGs. Better features should be extracted for better classification result.

2 Related Works

In 1666, Francesco Redi [1] published documentation where he informed that electricity is generated by the electric ray fish by the highly specialized muscle. In 1773, Walsh demonstrated that muscle tissue of Eel fishes could generate a spark of electricity [2]. In 1792, A. Galvani illustrated that electricity could commence muscle contraction [3]. Six decades later, Dubios-Raymond found the possibility to record electrical activity throughout a voluntary muscle contraction. Marey made the first recording activity in 1890 and also introduced a new term called electromyography [3].

Gasser et al. observed the electric signals from muscles using an oscilloscope in 1922 and found only rough information due to the stochastic nature of the myoelectric signal. The detection capability EMG improved in the next 20 years. Researchers used the improved electrodes widely for experiment over muscles. In the early sixties, surface EMG was used for the treatment of particular disorders [1].

Recently, surface EMG is widely used to examine superficial muscles in clinical protocols. It is used for diagnosis of neurological and neuromuscular problems. It is also used in different research labs such as neuromuscular physiology, biomechanics and so on. When diagnoses are performed based on EMG signals, the system must classify correctly as ALS or myopathy because to handle these two disorders, many drugs and therapies are adopted [4–6]. The distinguished classification of myopathy and neuropathy is important because it is critical to delimit the treatment due to the differences in the causes of these diseases. To classify problems accurately, some feature extraction steps are followed. It is plausible to obtain excellent classification performance if extract features are extracted sufficiently well [12].

To extract features in the frequency domain, two spectral models were used - a) Fast Fourier transforms (FFT) and b) Autoregressive (AR) [3] because it can interpret the characteristics of the signal. Nevertheless, each subject carries different strength in their signals. When analyzing the EMG signals, different wavelets are being used for the time-frequency domain [7–10]. The ability to analyze various sub-bands is the main advantages of this method.

However, to analyze the EMG signals, an empirical mode decomposition technique was used for extracting signals. This technique was accepted to generate data in noisy nonstationary and nonlinear processes over an extended quantity of applications [11]. Thus, a visibility algorithm can be used to transform a nonlinear signal into a complex network. A time-series signal can be transformed into a graph using a visibility algorithm. The resulting graph bears the characteristics of the time series. For that purpose, a geometrically insignificant as well as analytically solvable horizontal visibility algorithm was used [12]. Previous works related to complex network analysis are completely theoretical having no indication of implementation in signal analysis. To classify electroencephalogram (EEG) signals, a visibility graph was used from bands with higher frequency to classify electroencephalogram (EEG) signals [5]. Contrasting with the simple entropy method, they concluded that their method was better suggested by Zhu et al. [5]. But the classification results by this method are incorrect because of using the horizontal relationship. Due to this reason, the FHVA is not fit for EMG signals (Fig. 1).

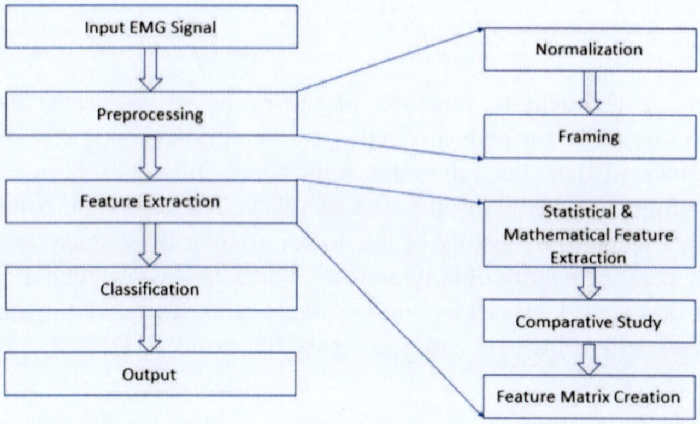

Fig. 1. A schematic representation of the proposed approach for EMG classification

3 Proposed Model

In the proposed approach, the EMG based classification starts with the preprocessing step. Then the results are calculated using two important steps - Feature extractions and classifications. The detailed description of various modules of the proposed approach illustrated in Fig.1 is defined in the following sections.

3.1 Data Preprocessing

To evaluate the proposed model, dataset first preprocessed in two steps- normalization and framing. In normalization, the signals are normalized with the bandpass filter. The filter has been designed according to needs. Finally, the normalized signals are divided into several frames.

3.2 Feature Extraction

For every frame of signal data, features are extracted and stored in feature vectors. Then, features are compared with each other to determine which feature would give more effective classification result. Finally, a feature matrix is created and classification has been performed over this matrix.

4 Results and Discussions

This section includes a detailed overview of the experiments performed, metrics for measurement, and related effects. It also gives a brief description of the datasets used for the assessment.

4.1 Dataset

To perform the experimental evaluation of the proposed approach, data has been collected from Physionet for both myopathy and healthy EMG signals. The collected data was normalized data but 6th order Butterworth filter has been designed and imposed according to needs to get the best accuracy. The data were collected by two types of patients without the history of the neuromuscular disease and with myopathy due to the longstanding history of polymyositis. The data was recorded at 50 KHz, and then sampled down to 4 kHz. Two analog filters were added during the recording cycle-a high-pass filter of 20 Hz and a low-pass filter of 5 KHz.

4.2 Experimental Setup

To evaluate the proposed model, EMG signals are divided into two levels: healthy and myopathy. The dataset is split with a ration of 50% of training and 50% of testing data respectively. Additionally, fivefold cross validation is employed in the training and testing data. Evaluation parameters are determined based on the following points.

1) Classification result using Mean Absolute Deviation (MAD) only for both subjects
2) Classification result using Sample Entropy (SE) only for both subjects
3) Classification result using both features, which is the total classification on the whole dataset

The performance obtained by the SVM classifier compared with prior works that applied different methods is observed and the total classification accuracy of the proposed method is outstanding.

4.3 Evaluation Results

In this proposed method, Mean Absolute Deviation (MAD) and Sample Entropy (SE) has been extracted from the collected dataset. Both myopathy and Healthy dataset have been divided into 20 frames, respectively. Each frame contains 630 ms data for Healthy EMG signal and 1390 ms for myopathy EMG signal. Then each of these two features has been extracted on each frame for both Healthy and Myopathy signals and stored in separate feature vectors respectively (Figs. 2 and 3).

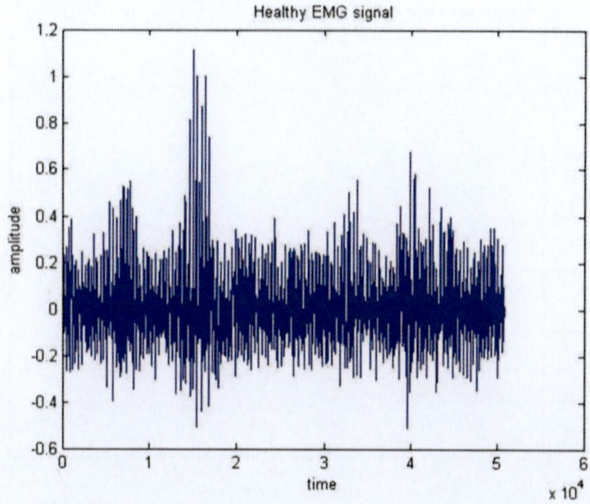

Fig. 2. Healthy EMG plotted data on MATLAB

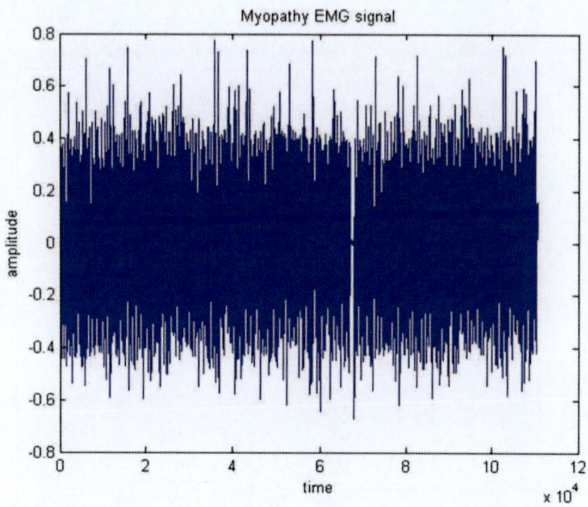

Fig. 3. Myopathy EMG plotted data on MATLAB

Figure 4 and Fig. 5 illustrate the visual differences between the two MAD feature vectors and two SE feature vectors. Mean Squared Error (MSE) also has been implemented to calculate the difference between both feature vectors theoretically. It is observed that MSE has given 0.0023 value for two MAD features and 0.1473 value for two SE features. SE features have more differences then MAD features. That is why it can be said that SE will provide better classification result than MAD.

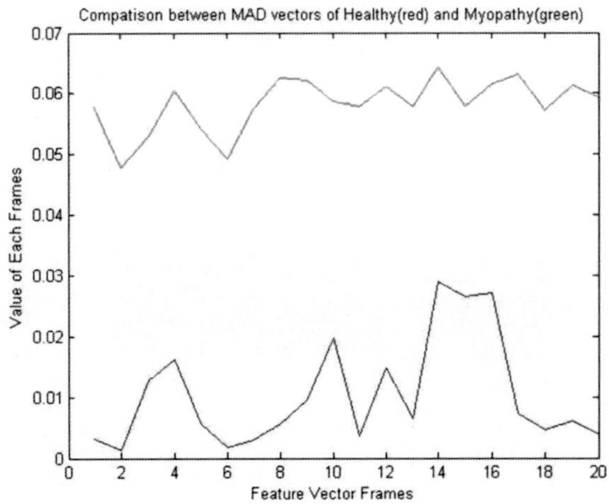

Fig. 4. Comparison between two MAD vectors of healthy (below) and myopathy (above), respectively

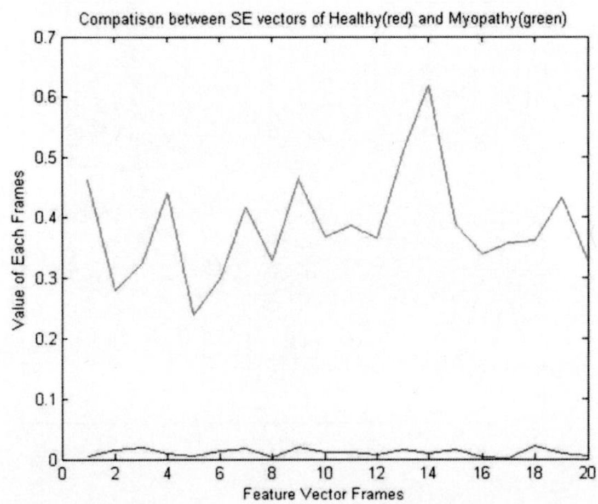

Fig. 5. Comparison between two SE vectors of healthy (above) and myopathy (below), respectively

4.4 Discussion

In this paper, a method of EMG-based feature extraction is proposed using advance effective features for healthy and myopathy detection. Because of the effectiveness of distinct features that are perfectly matched with the patterns of healthy and myopathy signals, our proposed model shows the better accuracy of the classification rather than the conventional method as given in Table 2. To develop applications in medicine in order to better human life, the outcomes should be accurate. The result obtained by the proposed method exhibits to the best accuracy that approximated 99.98%.

In this work, data has been collected from Physionet for both myopathy and healthy EMG signals. The collected data was normalized data but 6th order Butterworth filter has been designed and imposed according to needs to get the best accuracy. Mean Absolute Deviation (MAD) and Sample Entropy (SE) has been extracted from the collected dataset. Both myopathy and Healthy dataset have been divided into 20 frames, respectively. Each frame contains 630 ms data for Healthy EMG signal and 1390 ms for myopathy EMG signal. Then each of these two features has been extracted on each frame for both Healthy and Myopathy signals and stored in separate feature vectors respectively. Figure 4 and Fig. 5 show the visual differences between the two MAD feature vectors and two SE feature vectors. Mean Squared Error (MSE) also has been implemented to calculate the difference between both feature vectors theoretically. It is observed that, and 0.1473 value for two SE features. A Sample Entropy (SE) feature has more differences then Mean Absolute Deviation (MAD) features. That is why it can be said that SE will provide better classification result than MAD (Table 1).

Table 1. Feature extracted result using MAD and SE for healthy and myopathy

MAD		SE	
Healthy	Myopathy	Healthy	Myopathy
0.0032	0.0577	0.0045	0.4634
0.0012	0.0478	0.0159	0.2785
0.0128	0.0529	0.0189	0.3245
0.0164	0.0605	0.0087	0.4414
0.0055	0.0540	0.0055	0.2388
0.0017	0.0493	0.0136	0.2974
0.0030	0.0573	0.0169	0.4161
0.0057	0.0625	0.0030	0.3291
0.0095	0.0620	0.0189	0.4646
0.0197	0.0586	0.0121	0.3675
0.0035	0.0578	0.0121	0.3861
0.0148	0.0610	0.0080	0.3660
0.0065	0.0578	0.0150	0.5098
0.0289	0.0644	0.0096	0.6183
0.0266	0.0578	0.0143	0.3886
0.0272	0.0616	0.0038	0.3389
0.0073	0.0631	5.2095e−4	0.3565
0.0045	0.0572	0.0208	0.3608
0.0060	0.0613	0.0087	0.4325
0.0039	0.0593	0.0054	0.3234

SVM classifier has been used to classify the data collected from features extraction. 50% of data has been selected for training and 50% for testing using fivefold cross-validation. After this, the classifier has been trained with 50% test data using Non-Linear SVM with RBF kernel. Figure 6 shows the training figure. Non-linear SVM has been used to classify the trained data as Fig. 7. The classifier showed an unusual result for Sample Entropy and whole dataset but gives less accuracy for MAD compared with them. Table 2 shows the classification results.

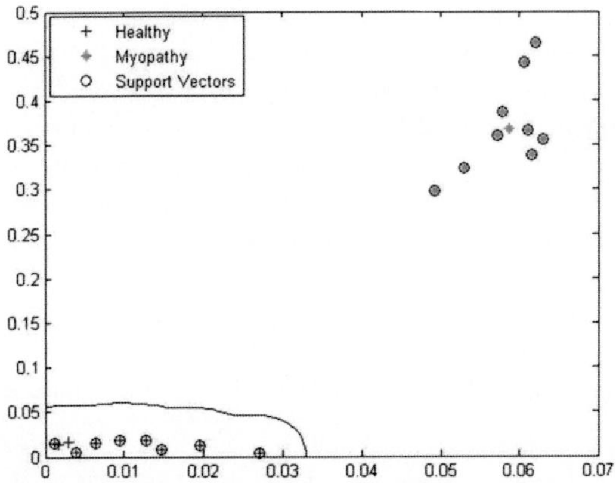

Fig. 6. Training of SVM classifier using non-linear RBF kernel

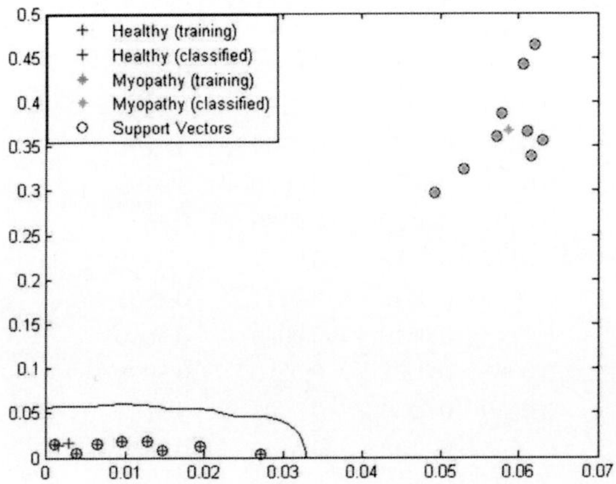

Fig. 7. SVM classification using non-linear RBF kernel

Table 2. Performance evaluation results of classification.

Subject	MAD	SE	MAD + SE
1	97%	99.98%	99.98%
2	99.98%	99.98%	99.98%

5 Conclusion

This study identifies different types of Feature Extraction techniques of EMG signal so that the proper techniques can be employed during any biomedical research, clinical diagnosis, hardware implementations, and end-user applications. In this paper, an EMG-based feature extraction model of healthy and myopathy detection is proposed where sampling of EMG signals which represent muscle responses was done based on the sampling theory for reversible discrete pulses, and two mathematical models were used for feature extraction from the obtained pulses - Sample Entropy (SE) and Mean Absolute Deviation (MAD). These mathematical features were finally classified using SVM classifiers into normal and myopathic cases. Many different adaptations, tests, and experiments have been left for the future due to the lack of a large dataset. In the future, this model can be employed in three EMGs- Healthy, Myopathy, Neuropathy and finally a system can be made for diagnosis.

References

1. Reaz, M.B.I., Hussain, M., Mohd-Yasin, F.: Techniques of EMG signal analysis: detection, processing, classification and applications. Biol. Proced. Online **8**(1), 11 (2006)
2. Sadikoglu, F., Kavalcioglu, C., Dagman, B.: Electromyogram (EMG) signal detection, classification of EMG signals and diagnosis of neuropathy muscle disease. Procedia Comput. Sci. **120**, 422–429 (2017)
3. De Luca, C.J.: The use of surface electromyography in biomechanics. J. Appl. Biomech. **13** (2), 135–163 (1997)
4. Manabe, H., Hiraiwa, A., Sugimura, T.: Unvoiced speech recognition using EMG-Mime speech recognition. In: CHI'03 Extended Abstracts on Human Factors in Computing Systems, pp. 794–795. ACM (2003)
5. Zhang, X., Zhou, P.: Sample entropy analysis of surface EMG for improved muscle activity onset detection against spurious background spikes. J. Electromyogr. Kinesiol. **22**(6), 901–907 (2012)
6. Nikolic, M.: Detailed analysis of clinical electromyography signals: EMG decomposition, findings and firing pattern analysis in controls and patients with myopathy and amytrophic lateral sclerosis, Ph.D. Dissertation (2001)
7. Khan, M., Singh, J., Tiwari, M.: A multi-classifier approach of EMG signal classification for diagnosis of neuromuscular disorders. J. Bioeng. Biomed. Sci. **3**, 2 (2016)
8. Subasi, A.: Classification of EMG signals using PSO optimized SVM for diagnosis of neuromuscular disorders. Comput. Biol. Med. **43**(5), 576–586 (2013)
9. Subasi, A.: A decision support system for diagnosis of neuromuscular disorders using dwt and evolutionary support vector machines. SIViP **9**(2), 399–408 (2015)
10. Subasi, A.: Medical decision support system for diagnosis of neuromuscular disorders using DWT and fuzzy support vector machines. Comput. Biol. Med. **42**(8), 806–815 (2012)
11. Artameeyanant, P., Sultornsanee, S., Chamnongthai, K.: An EMG based feature extraction method using a normalized weight vertical visibility algorithm for myopathy and neuropathy detection. SpringerPlus **5**(1), 2101 (2016)
12. Gokgoz, E., Subasi, A.: Comparison of decision tree algorithms for EMG signal classification using DWT. Biomed. Sig. Process. Control **18**, 138–144 (2015)

Eye Fatigue Algorithm for Driver Drowsiness Detection System

Teik Jin Lim[1(✉)], Hung Yang Leong[1], Jia Yew Pang[2], and Mohd Rizon Mohamed Juhari[1]

[1] Faculty of Engineering, Technology & Built Environment, UCSI University, Kuala Lumpur, Malaysia
lteikjin@gmail.com, {leonghy,MohdRizon}@ucsiuniversity.edu.my
[2] Heriot Watt University Malaysia, Putrajaya, Malaysia
j.pang@hw.ac.uk

Abstract. This paper proposed an algorithm that used the combination of the Viola-Jones technique, Circular Hough Transform (CHT), Histogram Equalization, Canny Edge detection and percentage of eyelid closure (PERCLOS) technique to accurately detect the eyes condition of the driver. This proposed algorithm achieved a 92.5% accuracy of eyes open images detection and 86.65% accuracy of eyes close images detection with 50 samples each. For the real-time video processing, it achieved 90% accuracy during daytime and 86% accuracy during night-time for the eyes drowsiness detection.

Keywords: Drowsiness detection · Viola-Jones · Circular Hough Transform · Image processing · Eyes detection · Eyes state analysis

1 Introduction

In every country, curbing road accidents have always been one of the utmost priorities as they contributed highly to causing high damage, even involving death accidents. Based on the data from the Traffic Investigation and Enforcement Department, Bukit Aman [1], total road accidents in Malaysia has increased from 373,071 to 533,875 in the last decade (2008–2017). In a research study, human behaviour contributes about 95% of the factors that causes road accident [3].

According to The Star newspaper, three main causes of road accidents are the poor attitude of drivers, drivers using mobile phones while driving and driver's drowsiness [2]. The poor attitude of drivers and using mobile phones while driving are considered as personal behaviour that hardly controlled or managed by a third party. The only way to prevent an accident comes from the driver's self-initiative.

To overcome driver drowsiness, a few researchers proposed an intelligent algorithm to alert the driver and prevent accidents to happen. The drowsy driver must be alert and asked to take a short nap after some key symptoms such as 'difficulty to keep the eyes open' and 'heavy eyelids' that are experienced by the driver [5]. The driver awareness is monitored using the determined key symptoms of drowsiness.

J. I.-Z. Chen et al. (Eds.): ICIPCN 2020, AISC 1200, pp. 638–652, 2021.
https://doi.org/10.1007/978-3-030-51859-2_58

2 Related Work

Some efforts have been reported in the literature on the development of driver drowsiness detection system. Various approaches and techniques were proposed. Prakash Choudhary [6] performed a series of surveys for various drowsiness detection techniques including the eyes blinking based technique, yawning based technique, PERCLOS based, template matching technique, artificial neural network based, EEG based technique and vehicular based. However, the EEG based technique and vehicular based methods were then rejected due to the high costing required and impracticality [7].

Muhammad Awais [8] designed a new approach for eye blink detection using template matching and similarity measures. The template matching is further supported using the golden ratio concept and followed by eye template creation for tracking purposes. Alshaqaqi [9] developed an algorithm to track and analyze the face and eyes of the driver using the PERCLOS method. In another research, Haq. [10] presented the detection of eyes using Viola-Jones technique. The eye-blink rate is used to determine the level of attentiveness. Alioua et al. [11] developed a novel iris edge detector algorithm by using the Circular Hough Transform to analyze the eye state. It supported that the Sobel and Prewitt method might not be able to provide the exact iris circular shape.

Sleepiness symptoms have strong associations with objective measures of sleepiness and associated impaired performance [4, 12, 13]. Caffier [14] and Schleicher [15] presented that the eye blink duration will be changed according to the increasing of drowsiness, which is expected to increase at least 50 ms compared to alert state becoming a total of 250 ms. If the eyelid is close for more than 500 ms, it is considered as microsleep, which is very dangerous because driving required high alertness.

3 Proposed Method

The flowchart of the proposed method is shown in Fig. 1, which can be breakdown into four main parts: (1) image pre-processing; (2) face and eye region detection; (3) eyes detection; (4) recognition of eye's state.

This algorithm was designed in MATLAB software platform and the support package for USB webcam is used. MATLAB is chosen because it allows matrix manipulations, plotting of functions and data, implementation of algorithms and creation of user interfaces that are fully utilized for this proposed method.

The camera, which is installed upright from the view of the driver inside the car, can capture the facial feature of the driver while he is driving on the road. The images captured from time to time will act as an input to the algorithm to detect a few symptoms that are commonly seen especially when a person is fatigue such as the longer blink duration or smaller opening of eyes. Once the driver is detected as drowsy, the alarm will ring to indicate the driver that he should take a rest.

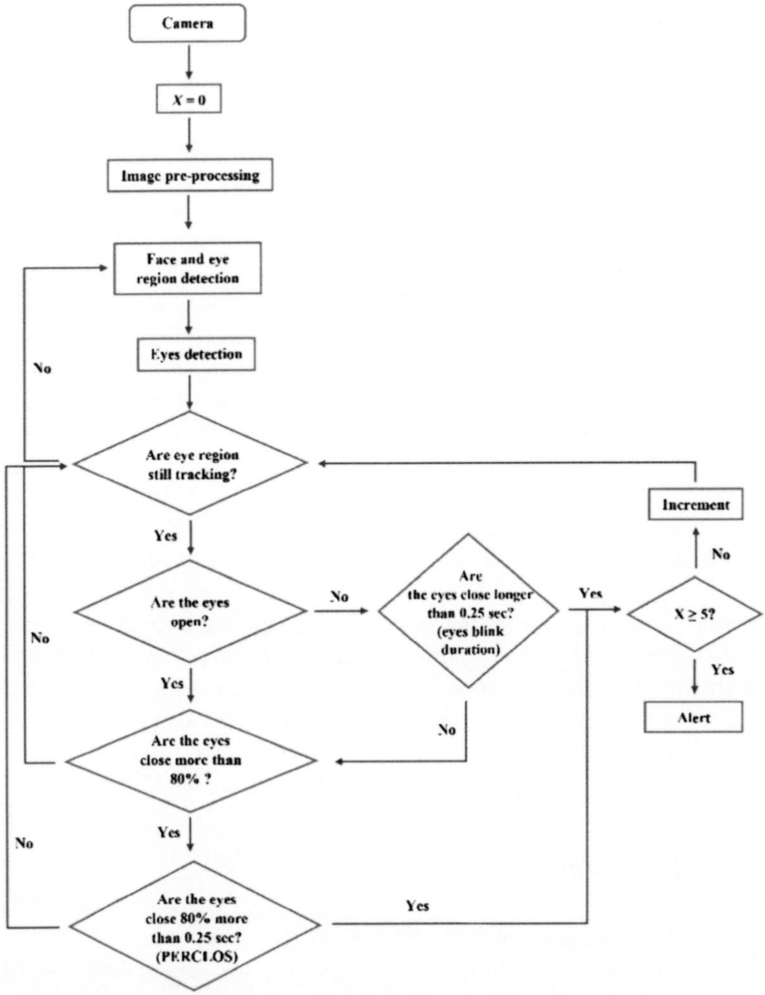

Fig. 1. Flowchart of the proposed method

3.1 Image Pre-processing

When the input images are with different backgrounds and lighting conditions, the colour intensity will play an important role in affecting the local contrast and the accuracy of the detection system. Thus, the proposed method applies histogram equalization [16] to the input image.

Applying histogram equalization increases the global contrast of the input image and overcomes issues of losing usable data of the image that is represented by close contrast values. In another word, the areas of lower local contrast are becoming higher contrast and improved the overall contrast of the whole image. Figure 2 shows the histogram before histogram equalization (left) and the histogram of the same sample image after histogram equalization (right).

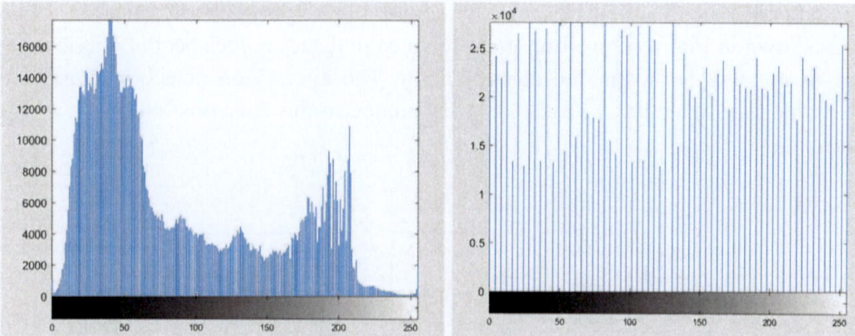

Fig. 2. Histogram before (left) and after (right) histogram equalization of a sample image

Histogram equalization does not change the type of image but only increases the contrast. Thus, RGB images are required to convert into grayscale images using rgb2gray command and im2bw command for binary images to further process in the MATLAB.

3.2 Face and Eye Region Detection

The Viola-Jones face detection method [17] applied for face and eye region detection in this proposed method is approximately 15 times faster than the Rowley-Baluja-Kanade detector [18] and about 600 times faster than the Schneiderman-Kanade detector [19]. It utilized the cascaded Adaboost classifier with Haar features for detections.

Haar features have a lot of different combinations as shown in Fig. 3. Every human being is expected to have some fixed face features in which the Haar features can be used to test on the features. Face and eyes have different patterns that portrait in Haar features and all the related Haar features are trained in the MATLAB using sample images.

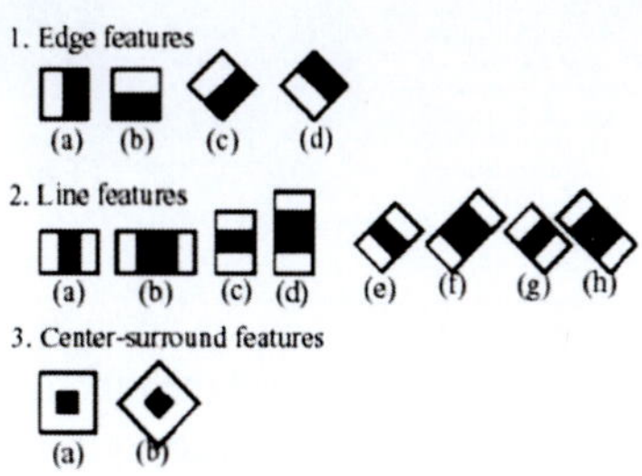

Fig. 3. Haar features such as edge features, line features and centre surrounded features

A cascaded Adaboost classifier is a strong classifier which made up of a combination of several weak classifiers while the Adaboost is a machine learning algorithm that helps in identifying the useful feature out of all the possible Haar features.

Adaboost constructs a strong classifier by combining a few weak classifiers in a linear form as shown in Fig. 4. Thus, important features in deciding the specific window have a face or not will be taken into consideration. The eye region detection worked the same as the face detection with the only difference in the Adaboost classifier of Haar features.

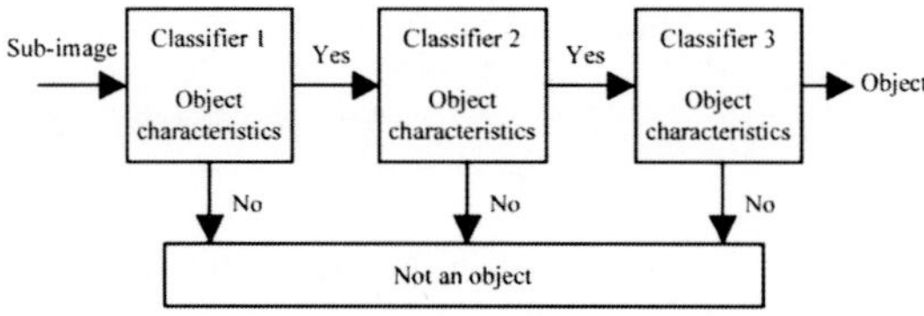

Fig. 4. Cascaded Adaboost classifier for face detection

3.3 Eyes Detection

After the successful detection of the face and eye region using Viola-Jones method, the eye region is cropped and saved as new data as shown in Fig. 5 and Fig. 6 respectively. However, the eye region is insufficient to provide accurate and precise data because eyes are the key symptoms of drowsiness [4]. Thus, the eye region that was saved previously is selected as the input source for eyes detection and further process. This proposed algorithm uses the eye region image instead of the original image because the eye region image contains fewer image pixels. It can decrease the total running time required for future process as well as decreases the possible noise detection occurred.

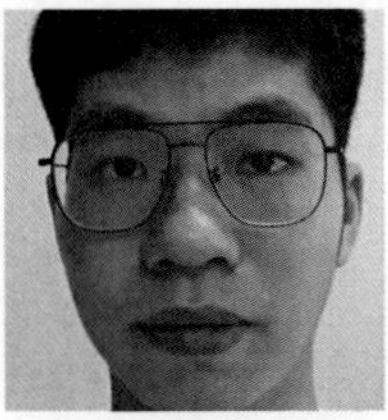

Fig. 5. Face detection

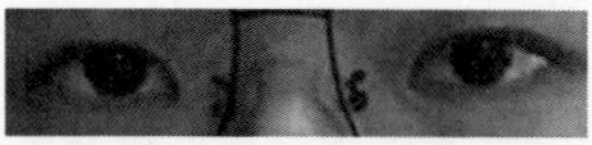

Fig. 6. Eye region detection

An edge detector is selected and applied to sketch the outline of all the edges in an eye region image because the Circular Hough Transform technique only performs on edge points in the images. Researches [20–22] supported that Canny edge detection is outperformed other edge detection methods such as the Sobel, Prewitt, Roberts and Laplacian of Gaussian method. The results obtained from other detectors such as Sobel and Prewitt provide very limited edge detected from the input image. Thus, Canny edge detector is implemented successfully in this proposed method and the image undergoes Canny edge detector is shown in Fig. 7.

After all the edges in the image are found, the Circular Hough Transform (CHT) technique [23] is used to find the circular patterns shape from the image. The Circular Hough Transform equation is as per define:

$$(x-a)^2 + (y-b)^2 = r^2 \quad (1)$$

where a and b is the abscissa and the ordinate of the circle center and r represents the radius.

The circle shapes in the image are recognized using the voting process of Circular Hough Transform. For every edge pixel detected in the processed input image, it will be voted for multiple centres (a, b) and multiple radius (r). The voting of possible circles was performed under a three dimensions accumulator where two dimensions representing the coordinates of the circle origin and the third dimension representing the radii. After voting, the positions of two highest local maxima in the accumulator matrix are selected as they gave the respective informations to the centre of circle in the input image. The two local maxima from the highest votes are usually the iris of the eyes unless more local maxima are collected.

The coordinate of the circle is visualized directly into the input image and presenting the circular shape inside the image as shown in Fig. 8.

Fig. 7. Image after Canny edge detection

Fig. 8. Eyes detection

3.4 Eye Blink Duration and PERCLOS

The successful detection of eyes using Circular Hough Transform technique provides the eyes location meanwhile it does not provide any information regarding the awareness condition for the driver. In this proposed method, two selected parameters, which are the eye blink duration and PERCLOS will determine and interpret the driver's condition.

Eye blink duration [14, 15] and the percentage of eyelids' closure (PERCLOS) [24, 25] are the parameters used in this proposed method to determine the alertness of the driver. The relationship can be concluded as: (1) The longer the duration of the eye blink, the more drowsy the driver is; (2) The greater the percentage of driver eyelids' closure (PERCLOS), the more drowsy the driver is.

According to the research [15], eyes close longer than 500 ms is considered as microsleep. The algorithm considers eyes close once it lost the detection and tracking of the iris. Then, the internal clock system performs the counting to check on the duration of the eyes close took. Any eyes close status that was longer than 250 ms will be considered as drowsy.

On the other hand, PERCLOS, the percentage of eyelid closure over the pupil over time [26], uses the P80 as the important criterion of a driver's physical state. P80 stands for the PERCLOS measurement with an eyelid closure of 80%. If the eyelid closed for more than 80%, the driver is considered as drowsiness and vice versa. To calculate the threshold pixel value for the P80, the formula is defined as:

$$\text{Threshold} = (\text{O} - \text{C}) \times 0.2 + \text{C} \quad (2)$$

O represented the pixel values during the opening of the eyes while C is the pixel value during the closing of the eyes [27]. By getting the threshold, a direct comparison can be made to decide whether the driver is under fatigue condition. The opening of eyes can have many stages in terms of opening percentage as shown in Fig. 9.

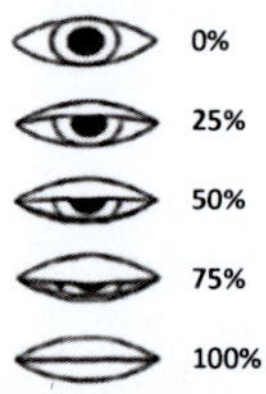

Fig. 9. The percentage of opening eyes

4 Experimental Results

In this proposed method, imaging system (offline) GUI and real-time video processing GUI were designed and tested on 50 different sets of samples images (20 sets of sample participants with glasses and 30 sets of sample participants without glasses. Each participant provided one eyes open image and one eyes close image of themselves) and six samples of real-time video processing. All the sample images are captured under different light conditions and different backgrounds as shown in Fig. 10 while the real-time video is taken with the webcam which has a resolution of 640 × 480. The software platform used is MATLAB R2018a and the whole evaluation of the proposed method is carried out on a single machine with Intel(R) Core(TM) i5-8250U CPU @ 1.60 GHz, 8 GB of RAM and Windows 10 Home as the operating system.

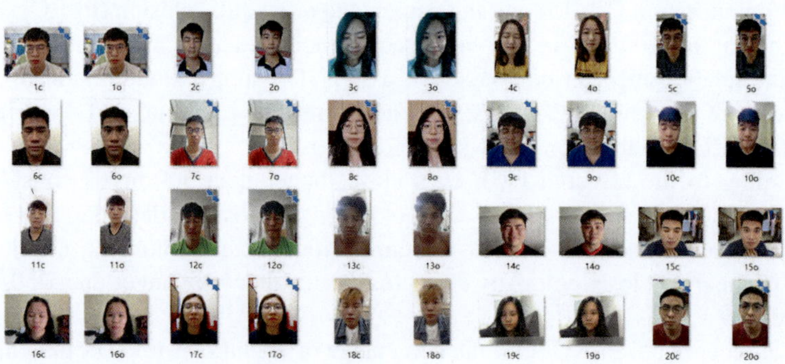

Fig. 10. 20 sets of sample images out of 50 sets of sample images

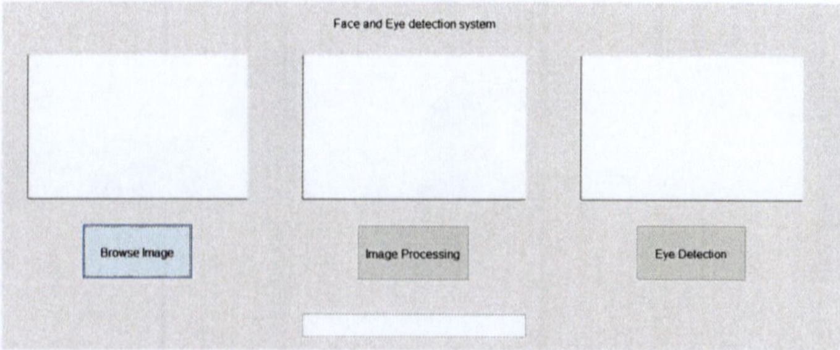

Fig. 11. GUI to test on sample images using the proposed method

The GUI as shown in Fig. 11 is used to test on the 50 sets of samples images. Each of the push buttons carries out a different function where the "Browse Image" push button allows the selection of input image, "Image Processing" push button performs the histogram equalization based on the selected input image and lastly the "Eye Detection" push button carries out eyes detection based on the cropped eyes region of the input image. The algorithm is then delivering the result as either "Eyes Open" or "Eyes Close" which is shown in Fig. 12 and Fig. 13 respectively.

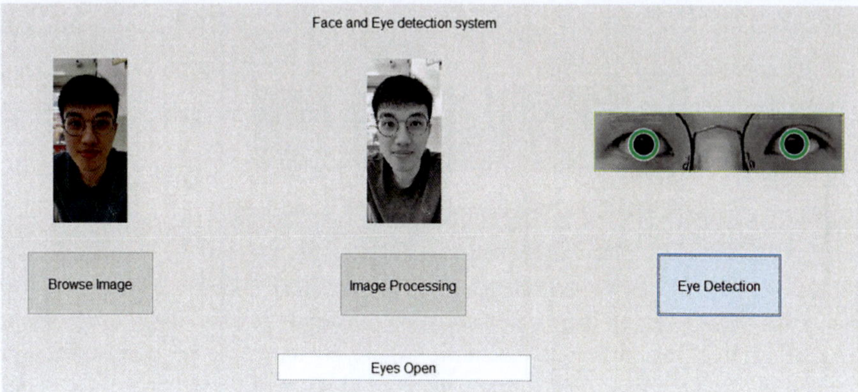

Fig. 12. Eyes open result

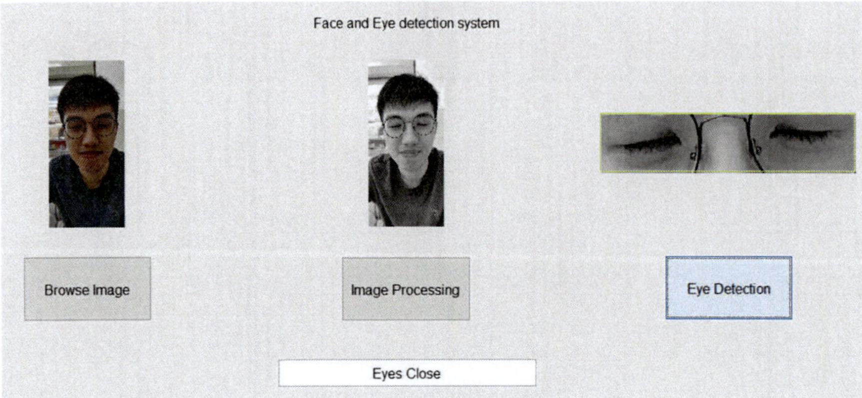

Fig. 13. Eyes close result

Although the Canny edge detection method is one of the best edge detection techniques, there is some limitation in edge detecting. During the test for 50 samples of eyes open images, all the results for the edge detection appear to be different. Some edge detected around the eyes shape might be half-circle or oval shape instead of complete circular shape. As an example, Fig. 14 shows an incomplete circle shape obtained from the edge detection of an eye region image.

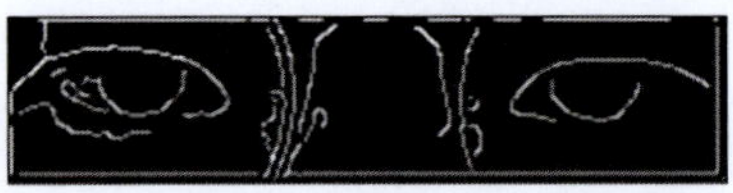

Fig. 14. Incomplete circle shape obtained

CHT technique is sensitive to detect circular shape but not for half-circle nor oval shape. Thus, in MATLAB, the parameter called sensitivity is customized to change the setting of CHT for this proposed algorithm. Figure 15 shows the relationship between the tested sensitivity varied from 0.8 to 0.92 and the successful detection of the sample images.

By default, the sensitivity for the CHT technique in MATLAB is set at 0.85. However, the eyes shape is not a perfect circle so the default value and any values lower than it performs badly. The sensitivity is increased to cope with the incomplete circular shape of eyes' edge detection. When the sensitivity is set to 0.89, 50% of success detection is achieved. From the simulation study, all 50 samples of eyes open images were successfully detected at the optimum sensitivity of 0.92.

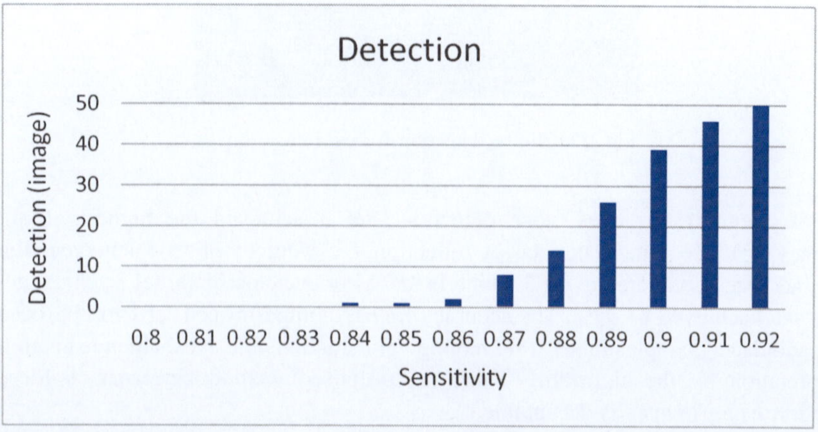

Fig. 15. Relationship between sensitivity and the total of success detection samples

Another bar chart as shown in Fig. 16 shows the accuracy of the proposed method. There are four sets of samples, which are the set 1 (samples wearing glasses with eyes open), set 2 (samples wearing glasses with eyes close), set 3 (samples without wearing glasses with eyes open) and set 4 (samples without wearing glasses with eyes close).

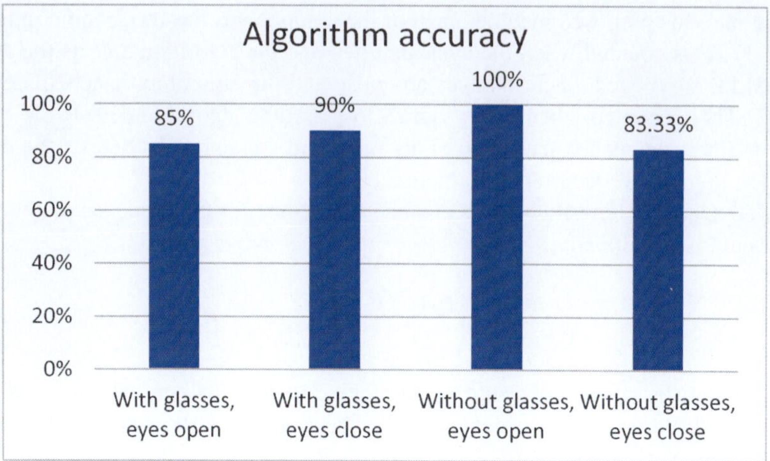

Fig. 16. Algorithm accuracy based on 50 sets of samples.

Theoretically, the detection for eyes open is more accurate than the detection for eyes close because noise detection tends to occur for eyes close images. As an example, Fig. 17 shows the noise detection for eyes close. Areas such as the edge of nose trunk and eyelid are always misinterpreted by the algorithm as a circle shape. Thus, on average, the proposed method achieved 92.5% accuracy for eyes open detection and

Fig. 17. Noise detection for eyes close image

86.67% accuracy for eyes close detection. Set 3 achieved the highest algorithm accuracy of 100%. Due to the glasses reflection, the accuracy of set 1 achieved slightly lower accuracy compared to set 3 which is 85%. For eye close data, set 2 achieved 90% while set 4 achieved 83.33%. The accuracy for set 2 outperformed set 1 for 5% because two candidates' samples in set 1 are having very small visible eyes which leads to false interpretation by the algorithm. Thus, the proposed method performance for eyes detection using images is acceptable.

The second GUI performs real-time video processing with the proposed method. The webcam is the input source of real-time video. Statistical measures such as the Confusion Matrix, Correct Classification Rate (CCR) and Kappa Statistic (k) are used to evaluate the performance of this proposed method [11, 23, 28].

Confusion Matrix is a table layout that allows the visualization of the performance of an algorithm. The column of matrix representing the number of occurrences of an estimated class, while the row of the matrix representing the number of occurrences of a real class. These measurements have a few key components which are the True Positive (TP), False Positive (FP), True Negative (TN) and False Negative (FN). By applying into this proposed method, each of the components has the specific indicators where: (1) TP is counted when the eyes are open and the algorithm detects the eyes as open; (2) FP is counted when the eyes are close and the algorithm detects the eyes as open; (3) TN is counted when the eyes are close and the algorithm detects the eyes as close; (4) FN is counted when the eyes are open and the algorithm detects the eyes as close. The sum of all occurrences is defined as T.

Correct Classification Rate (CCR) is the success rate for the proposed method [11]. The formula is as defined:

$$CCR = \frac{TP + TN}{T}. \tag{3}$$

Kappa statistic (k) is a measure of the degree of non-random agreement between observers with the expected agreement [29]. The Kappa statistic can be calculated by using the formula:

$$k = \frac{P_o - P_e}{1 - P_e} \tag{4}$$

where the P_o is the observed agreement proportion, which is similar to CCR and P_e is the random agreement proportion given by Eq. (5).

$$P_e = \frac{1}{T^2}[(P \times p) + (N \times n)] \tag{5}$$

P represents the total of estimated occurrences and p represents the total of real occurrences. N and n are the opposite of P and p respectively. The P, N, p and n variables can be calculated using the equation:

$$\mathrm{P} = \mathrm{TP} + \mathrm{FP} \tag{6}$$

$$\mathrm{N} = \mathrm{TN} + \mathrm{FN} \tag{7}$$

$$\mathrm{p} = \mathrm{TP} + \mathrm{FN} \tag{8}$$

$$\mathrm{n} = \mathrm{TN} + \mathrm{FP} \tag{9}$$

Kappa statistic is ranged from −1 to 1. By achieving a value 1 means that there is a perfect agreement between the observers with the proposed method and inversely a value −1 means total disagreement.

Table 1 shows the results of the proposed method performance during the daytime while Table 2 shows the results of the propose method performance during the night-time where the night-time is lower in light intensity compared to daytime. Therefore, the visibility and content captured by the webcam are more limited. Both the CCR and Kappa statistics show a better result during daytime eyes detection. During the daytime, the correct eyes detection achieved 90% accuracy and 0.69 for the Kappa statistic. During the night-time, the correct eyes detection obtained 86% accuracy and 0.63 for Kappa statistic. According to the interpretation table of Kappa statistic [11, 26], 0.69 and 0.63 are fall under the category of strong agreement ($0.61 < k < 0.8$). The Kappa statistic's almost perfect agreement interpretation must achieve 0.81 and above. Day-time detection accuracy is higher than night-time detection by 4%. The False Negative is occurred when the driver opens his eyes but the algorithm interprets it as close eyes. It increases significantly for night-time detection because of the limitation of visibility.

Table 1. Results of the proposed method for daytime eyes detection

Sample no.	TP	TN	FP	FN	T	CCR	Kappa
Sample video 1	72	20	5	3	100	0.92	0.78
Sample video 2	75	13	10	2	100	0.88	0.61
Sample video 3	80	11	5	4	100	0.91	0.67
Average:						0.90	0.69

Table 2. Results of the proposed method for night-time eyes detection

Sample no.	TP	TN	FP	FN	T	CCR	Kappa
Sample video 1	70	19	7	4	100	0.89	0.70
Sample video 2	65	20	8	7	100	0.85	0.62
Sample video 3	71	14	5	10	100	0.85	0.56
Average:						0.86	0.63

The GUI for real-time video processing is shown in Fig. 18 where the five separate windows are showing the real-time webcam view, cropped eye region, eyes detection, binary image of the cropped eye region and algorithm output. The counter is increased when eyes close are detected. The beeping sound from the alarm will be triggered once the pre-set threshold value is exceeded.

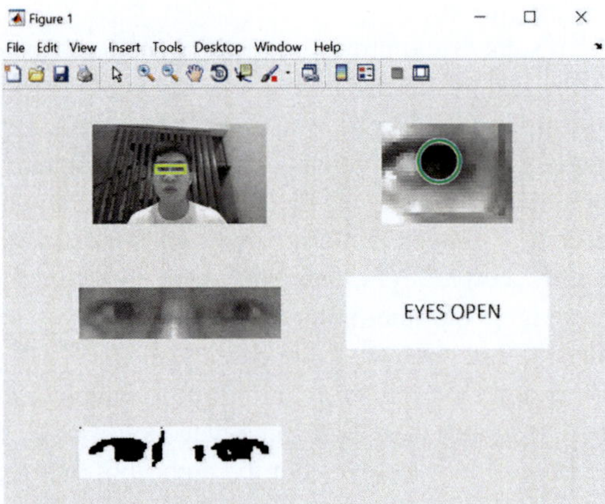

Fig. 18. GUI for real-time video processing

5 Conclusion

In this paper, a robust method for driver awareness detection and alert system is proposed. Inputs will undergo image processing and computer vision by histogram equalization, Canny edge detection, Viola-Jones algorithm and Circular Hough Transform. The sensitivity parameter for the CHT technique is fined tuned to the optimum value of 0.92 to ensure different shapes of eyes or incomplete edges detection for the eyes are still detectable. This proposed method has achieved an average of 92.5% accuracy for eyes open image detection and 86.65% accuracy for eyes close image detection. The drowsy state of the driver is not relying on a simple judgement based on the existence or non-existence of circle shape by CHT. Instead, it is added the eye blink duration and PERCLOS method to enhance the consistency and accuracy of this proposed method which gives a success rate of 90% during the daytime and 86% during the night-time. Thus, this method is extremely useful to monitor the awareness of drivers and alert the driver if fatigue happens.

In the future, the work can be extended into the additional hardware for this proposed system. Driving during the night time or midnight will be a total dark condition and it is lack of visibility inside the interior space of the car. The key symptoms of drowsiness such as the longer eyes blink frequency or smaller opening of eyes are unable to capture by a normal camera. Special hardware such as an infrared camera or infrared sensor shall be integrated for future work to overcome the limitation of visibility under total darkness conditions.

Acknowledgement. Informed consent to all the participants involved in this research work.

References

1. A Study of Road Accidents of Heavy Vehicles in Malaysia. https://people.utm.my/shamsul/wp-content/blogs.dir/949/files/2018/10/Case-study-2_MRSE2573_A-Study-of-Road-Accidents-of-Heavy-Vehicles-in-Malaysia.pdf
2. The Star Newspaper. https://www.thestar.com.my/news/nation/2018/06/08/fatigue-mobile-phone-use-among-top-causes-of-road-accidents/
3. Keall, M.D., Frith, W.J., Patterson, T.L.: The influence of alcohol. Age and number of passengers on the night-time risk of driver fatal ınjury in New Zealand. Accid. Anal. Prev. **36**, 49–61 (2004)
4. Nilsson, T., Nelson, T.M., Carlson, D.: Development of fatigue symptoms during simulated driving. Accid. Anal. Prev. **29**(4), 479–488 (1997)
5. Nordbakke, S., Sagberg, F.: Sleepy at the wheel: knowledge. Symptoms and behaviour among car drivers. Transp. Res. Part F Traffic Psychol. Behav. **10**, 1–10 (2007)
6. Choudhary, P., Sharma, R., Singh, G., Das, S.: A survey paper on drowsiness detection & alarm system for drivers. Int. Res. J. Eng. Technol. **3**(12), 1433–1437 (2016)
7. Zhang, L.M., Yang, C.: An EEG-based method for detecting drowsy driving state. In: 2010 7th International Conference on Fuzzy Systems and Knowledge Discovery, pp. 2164–2167 (2010)
8. Awais, M., Badruddin, N., Drieberg, M.: Automated eye blink detection and tracking using template matching, pp. 16–17 (2013)
9. Alshaqaqi, B., Baquhaizel, A.S., El, M., Ouis, A.: Driver drowsiness detection system. In: 8th International Workshop on Systems, Signal Processing and Their Applications (2013)
10. Haq, Z.A.: Eye-Blink Rate Detection for Fatigue Determination (2018)
11. Alioua, N., Amine, A., Rziza, M., Aboutajdine, D.: Eye state analysis using ıris detection based on Circular Hough Transform. In: International Conference on Multimedia Computing and Systems, Ouarzazate, pp. 1–5 (2011)
12. Forberg, K., Waage, S., Moen, B., Bjorvatn, B.: Subjective and objective sleep and sleepiness among tunnel workers in an extreme and ısolated environment: 10-shifts, 21-day working period, at 78 degress north. Sleep Med. **11**(2), 185–190 (2010)
13. Howard, M.E., et al.: Specific sleepiness symptoms are indicators of performance ımpairment during sleep deprivation. Accid. Anal. Prev. **62**, 1–8 (2014)
14. Caffier, P.P., Erdmann, U., Ullsperger, P.: Experimental evaluation of eye-blink parameters as a drowsiness measure. Eur. J. Appl. Physiol. **89**(34), 319–325 (2003)

15. Schleicher, R., Galley, N., Briest, S., Galley, L.: Blinks and saccades as indicators of fatigue in sleepiness warnings: looking tired? Ergonomics **51**(7), 982–1010 (2008)
16. Yeganeh, H., Ziaei, A., Rezaie, A.: A novel approach for contrast enhancement based on histogram equalization (2008)
17. Viola, P., Jones, M.: Rapid object detection using a boosted cascade of simple features (2001)
18. Mall, A., Ghosh, S.: A neural network based face detection approach. Int. J. Comput. Technol. Appl. **3**(2), 823–829 (2014)
19. Schneiderman, H., Kanade, T.: A statistical method for 3D object detection applied to faces and cars, pp. 2–7 (2000)
20. Shrivakshan, G.T.: A comparison of various edge detection techniques used in image processing. Int. J. Comput. Sci. Issues **9**(5), 269–276 (2012)
21. Chandwadkar, R., Dhole, S., Gadewar, V., Raut, D., Tiwaskar, P. S. A.: Comparison of edge detection techniques (2013)
22. Acharjya, P.P., Das, R., Ghoshal, D.: Study and comparison of different edge detectors for image segmentation (2012)
23. Hichri, S., Nakkach, H., Benzarti, F., Amiri, H.: Monitoring of driver's drowsiness based on eyes state analysis. In: 3rd International Conference on Automation, Control, Engineering and Computer Science, Hammamet, pp. 779–783 (2016)
24. Grace, R., et al.: Introduction to the Drowsy Driver Problem Fundamental Principles of a Video-based Drowsiness Detection System (1996)
25. Image, G., Yan, J., Kuo, H., Lim, Y., Liao, T.: International Symposium on Computer, Consumer and Control (2016)
26. Office of Motor Carrier Research and Standards: PERCLOS: A Valid Psychophysiological Measure of Alertness as Assessed by Psychomotor Vigilance
27. Yauri-Machaca, M., Meneses-Claudio, B., Vargas-Cuentas, N., Roman-Gonzalez, A.: Design of a vehicle driver drowsiness detection system through image processing using MATLAB. In: IEEE 38th Central America and Panama Convention (2018)
28. Ghosh, S., Nandy, T., Manna, N.: Real time eye detection and tracking method for driver assistance system. In: Gupta, S., Bag, S., Ganguly, K., Sarkar, I., Biswas, P. (eds.) Advancements of Medical Electronics. LNB, pp. 13–25. Springer, New Delhi (2015). https://doi.org/10.1007/978-81-322-2256-9_2
29. Fleissa, J.L., Cohen, J., Everitt, B.S.: Large sample standard errors of kappa and weighted kappa. Psychol. Bull. **72**(5), 323–327 (1969)

An Analytical Approach for Recognizing the Occupational Gender Bias in the Data Systems

Pranav Rai and P. N. Kumar(✉)

Department of Computer Science and Engineering,
Amrita School of Engineering, Coimbatore, Amrita Vishwa Vidyapeetham,
Coimbatore, India
rpranav75@gmail.com, pn_kumar@cb.amrita.edu

Abstract. Gender Bias is one of the most commonly seen phenomena of our society. The existing prejudices of our society are the reason for the existence of gender bias. Gender Bias causes the misrepresentation of one gender over the other gender or in simple terms one gender is considered as less important than another gender(s). In this paper, Gender Bias is studied from the perspective of engineering rather than the perspective of social sciences.

The objective of this paper is to provide an efficient way of analysing the presence of gender bias in text data systems in terms of the occupations. For the analysis, a pre-trained model which is trained on Google News articles is taken. An efficient architecture is also proposed for the detection.

Keywords: Gender Bias · word2vec · T-distributed stochastic neighbour embedding · Support Vector Machines · Principal Component Analysis

1 Introduction

Artificial Intelligence is the intelligence which is shown by the machines. These machines/systems are as good as the data on which these machines are trained, as the decision making capability of these machines is dependent on the data which is used for training it. This generates a prerequisite of good data for training. A machine which is trained with bad data can never give proper output. In other terms, the probability of making a correct decision by these machines will be very low or negligible.

Good Data can be defined as the data which is free from skewness or biases. Oppositely Bad data contains biases. The machines which will be trained on bad data will reflect these biases in its decision making or sometimes it also amplifies these biases. These biases give rise to prejudice in decision making and can make unfair decisions. Thus, it is very necessary to have unbiased data for learning which gives the foundation of detection for various type of different biases in the data. If any kind of bias is detected, the system can be saved from getting trained over the biased data.

One of the most common biases which affect the decision making capability of the machines is gender bias. These biases are results of prejudices of the society and mostly seen in the data which is taken from humans, as humans are biased by nature. Humans

J. I.-Z. Chen et al. (Eds.): ICIPCN 2020, AISC 1200, pp. 653–661, 2021.
https://doi.org/10.1007/978-3-030-51859-2_59

live in the closed proximity of the society and knowingly or unknowingly accepts the prejudices of the society in the form of the bias. Thus, any data that is generated by humans directly or directly contains and reflects the biases.

Gender Bias is been a subject of the research for a long time, but mostly the existing research on gender bias is in the perspective of social sciences. In general, gender Bias is responsible for creating various problems, but from the view of engineering, one of the areas where gender bias affects heavily is Artificial Intelligence. In this paper, the problems of Artificial Intelligence which are the result of Gender Bias are discussed and analysed. Today, Artificial Intelligence can solve the problems of multiple fields including hiring, healthcare, machine translation etc., which were used to be difficult or nearly impossible some decades ago. Artificial Intelligence is now also been used for solving the real-world problems which are quite sensitive. As the areas are sensitive, it is expected that the output of AI algorithms is fair in all means.

The study of gender bias in data systems is very important. There are cases of a biased decision made by the machine when the machine is unknowingly trained on biased human-generated data. Amazon Resume Filtering System was failed because of the biased data, where the resume was filtered based on dominant male gender instead of filtering it out based on the skill set. Google Translate also reflects gender bias while translating one language to other languages [4]. Similar other biases are also reflected from the human-generated data. COMPAS (Correctional Offender Management Profiling for Alternative Sanctions) tool which was designed to check the likelihood of a defendant becoming a recidivist. This tool was failed to make a proper decision, where it produced the results of black race offenders to be almost twice too likely as white offenders to be labelled at higher risk but not re-offended.

This paper limits the research only to the detection of gender bias. Various other demographic biases are not part of this research. A pre-trained model which is trained using Google News Articles (word2vec) is taken for the analysis. This paper proposes an end to end architecture for the detection of gender bias in terms of occupation. Machine Learning and Natural Language Processing techniques and algorithms are used in combination.

2 Related Works

In this section, some of the approaches which are related to the detection of gender bias in the data systems are discussed. This section also covers an overview of the concepts which are used for the detection of the gender bias. Gender Bias has been studied over many decades in multiple contexts, but only the relevant studies, where the context is engineering sciences are being discussed here.

Costa-jussà defines gender bias as the dominance of one gender over another gender in various contexts (ex. occupations). The consequences of gender bias in various problems are been discussed [1]. Bolukbasi et al. illustrates the gender bias, by solving the analogies, where operations on the word vectors show gender bias by showing man's vectors shows more closeness with the vector of the computer programmer while the women's vector shows more closeness with housekeeper's vector [2]. Mikolov et al. describes an efficient and improved method of generating the word vectors (embedding), which are one of the most predominant areas of studying the gender bias [1, 3, 8]. The generated vectors keep syntactic and semantic word relationships. Certain attempts are also made for reducing the gender bias by using neutral word embeddings which are generated by keeping the gender information in the dimensions of the word vectors while keeping the other dimensions free from the influence of the gender [13].

Sun et al. did a review on the studies based on the recognition and the mitigation of the gender bias. The gender debiasing approaches such as data augmentation, gender tagging, base fine-tuning are also discussed in this paper along with the advantages and disadvantages of each approach [4]. Leavy discusses an approach through which gender bias can be operationalized as text features which can be identified computationally. Susan's approaches consist of four major sections: Naming, Ordering, Biased Description, Metaphor [5].

Zhao et al. discusses a new benchmark for resolving the co-reference which is focused on gender bias. A new method is described for detecting such gender bias in the text. Certain evidence is also provided which shows the bias can be ignored [6]. Gender Bias is been studied for a wide range of datasets. Madaan et al. [9] focuses on studying gender bias in Bollywood movies. This paper also proposes an algorithm for removing gender bias from the text data. This paper is not only limited to the text data, but it also covers data such as images of the movie posters for different analysis. DecogTeller Algorithm is introduced in this paper for removing/debiasing the gender bias in the text data.

Deep learning is also employed in various works as it improves the performance of Natural Language Processing. Qian et al. employs deep learning-based approach for the debiasing the gender bias by doing direct modifications in the loss function. Gender bias is analysed as a group of 3 different categories here, co-occurrence bias, casual bias, word embedding bias [11]. Park et al. also employed deep learning-based approaches for reducing the gender bias in the abusive language text [12].

Certain researches also discover the effects of the presence of Gender Bias in the text data. Prates et al. discovers the effects of the presence of the Gender Bias in Google Translate [14].

Classification of occupations is also done as part of this study. Vijayan et al. provides a summarized view of multiple text classification algorithms [16]. Prof. Nedungadi et al. also gave views high-performance text classification algorithms [17].

3 Methodology

3.1 Data Collection

This stage primarily consists of collecting the data to do the analysis. The data that is used for the analysis is a collection of various occupations. From the internet, a collection of occupations is created. This collection of occupations consist of occupations from various industries. This collection consists of around 1300 unique job occupations. These occupations which are also referred to as jobs titles are used for finding the occupational gender bias in the data systems.

3.2 Preprocessing

The collected job titles are converted in the form of a high quality vector, primarily known as word embedding. Any operation cannot be carried out at the level of the words, thus for many operations, it is needed to convert the words into the vectors. Word2vec is one of the technique to learn word embedding vector using shallow networks. In simple terms, it can be described as a two-layer neural net which takes the word as input and returns the word in the form of vector.

The task of converting the words into the vectors is done using the word2vec embeddings which are trained on Google News Articles, which consist of around 3 million words. Word2vec is used here for the study as it is one of the commonly used techniques for converting the words into the vectors.

This word2vec embedding is developed in such a manner that the syntactical/ contextual meaning of word remain same its context as in a document if all the words are converted into vectors.

3.3 Gender Bias Detection

Cosine Similarity Measure is used for the detection of gender bias. As the occupations after converting the into vectors is a high dimensional data, Cosine Similarity is the best metric for finding the relation. As the embeddings represent the words with the same contextual meaning [15].

The word embedding represents the words with the same contextual meaning, this property can be utilized for the detection of the Gender Bias. All the word embeddings are taken and cosine similarity is measured with both male and female-centric words. The male-centric words are he, his, him, man etc. Similarly the female-centric words of the corresponding male-centric words are she, her, women etc. As the embeddings have a nature to show the more similarity with the same syntactical and contextual meaning words, the above centric words for the males and the females can cover the all the words which are related to these genders.

As embeddings are representing the same meaning as of text, so if the occupation is centric to a particular gender it represents the bias. The occupations which are closer to male-centric words are labelled as Male and others are labelled as Female (Biased).

The table here shows the cosine similarity between the occupations and male and female-centric words. The occupations here shown are just a sample of occupations from the collection of 1300 occupation. The male-centric word that is considered here is he, while the female-centric word that is considered here is she. Similar kinds of trends are seen in the case of other male and female-centric words.

Table 1 describes the similarity of the occupations with the male and female-centric words. The similarity is measured here in terms of the cosine similarity. As the word embeddings have the property of preserving the syntactical meaning of the word, the cosine similarity is used for measuring the similarity. The value of cosine similarity measured here is in the range from -1 to 1, and not been normalized from 0 to 1. The male-centric and female-centric word which are used here is he and she. Similar kind of trends was seen in the case of other male and female-centric words. The occupations which are mentioned in Table 1 is a small section of occupation from the dataset. The occupation words which are mentioned in Table 1 are also inspired by the words which are used by different authors for similar or related studies [2].

Table 1. The similarity of the occupations with male and female-centric words for a sample of occupations

Occupation	Cosine similarity with male centric word	Cosine similarity with female-centric word	Label
Maestro	0.1359914094209671	0.035805437713861465	Male
Philosopher	0.13220128417015076	0.054809048771858215	Male
Architect	0.020104922354221344	-0.0670500099658966	Male
Magician	0.1005958840250969	0.0446403473615646367	Male
Nurse	0.13776081800460815	0.2844496965408325	Female
Receptionist	0.13662581145763397	0.2935115694999695	Female
Nanny	0.07193776965141296	0.22275152802467346	Female
Interior designer	0.012551621533930302	0.11489956825971603	Female

3.4 Visualization

Data Visualization is one of the key steps for understanding the meaning of the data. Data visualization techniques give a visual summary of information which makes it easier to identify and understand the patterns and trends which are present in the data and cannot be observed when the data is present in the form of spreadsheets.

While doing any data visualization some basic information about the data must be known, otherwise, a wrong visualization technique can be employed which might result in the hiding of the information. One such basic information is the dimensionality of the data which should always be considered while doing the visualization. The pre-processed occupation data which is later converted in the form of vectors is a high dimensional data. Thus, Stochastic Neighbour Embedding is employed for visualization which has a capability of visualizing a high dimensional data without affecting the nature of data points. This technique gives each data point a location in two or three-dimensional graph. This algorithm has the benefit of preserving the similarities in high dimension and low dimension by using the cost function. Thus, it provides better visualization by reducing the tendency to crowd points together in the centre [18].

This algorithm is preferred over the other common dimensionality reduction and visualization techniques, such as Principal Component analysis which generates new dimensions of maximum variance, while t-SNE keeps the local and global structure of the data. t-SNE also have the capability of handling the nonlinear data which makes it a better choice for doing the visualization of such complex and high dimensional data.

The occupations which are converted into the form of vectors are represented into 2D graphs. t-SNE the algorithm is a randomized algorithm whose output differ every time, as it depends on multiple parameters, such as the number of iterations.

Figure 1 is a representation of a small sample of data from the dataset and has employed an iteration of 50 while representing the graph. As the algorithm is randomized, a small variation is also seen while running the algorithm at parameters.

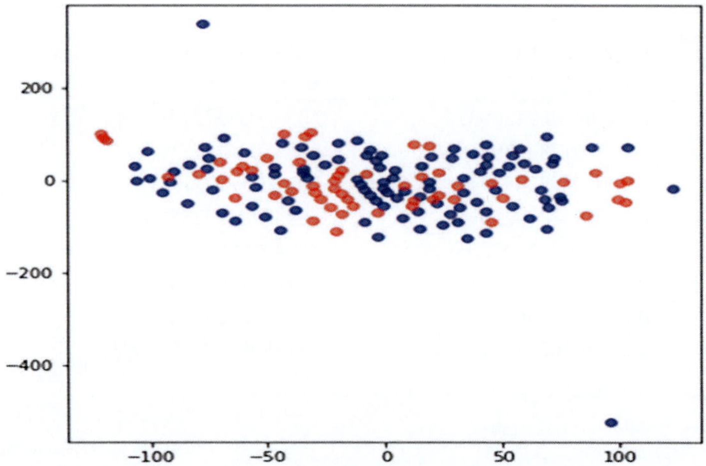

Fig. 1. t-SNE visualization of occupations

3.5 Classification

Once the Gender Bias is detected in the occupations, the occupations are tried to be get classified. Support Vector Machines are employed for the classification of the occupations. This level of classification also shows a difference between the occupation for male and females Another important prospectus of this classification is to give an approach for the detection of the gender bias at the sentence level. Sentences are the collection of words, for measuring the gender bias at the level of sentences it is necessary that, the gender bias can be measured at the level of the words. In the case of measuring the occupational gender bias, the presence of the occupation words in the sentences is assumed the most important parameter while measuring it.

For measuring the occupational gender bias, the occupation word is the most contributing feature, can consider only the words which are related to the occupations, other words of the sentences are not needed to give as high importance as compared to the words which are related to the occupations. Thus, a Support Vector Machine Classifier is used for classifying the level of the words, as the occupations are primary contributors, thus the same level of classification is expected.

There are possibilities where other words can also contribute to the occupational gender bias at the level of the sentence. In such a case, a rule can be discovered which can take the contribution from other words of the sentences, which is a separate area of research.

Support Vector Machines are employed here because of its capability of dealing with high dimensional data as it creates the hyperplane for the classification.

4 Experimental Results and Inference

The analysis shows the presence of the occupational gender bias in the google news articles on which the word2vec algorithm is trained.

Table 1 clearly shows the difference in the cosine similarity for a sample of occupations. The occupations which are closer to the male-centric words closely depicts the nature of the occupations which are expected to be done by the male. Similar nature is present for the occupations for the females. This depicts the unbalanced nature of occupations, which is resulted due to the biased text on which the algorithm is trained.

The visualization of t-Stochastic Distributed Neighbour Embedding (Fig. 1) is done on a sample of the occupations. The red points in the Figure represent the occupations which are found to be closer to male-centric words, while the blue points in the Figure represent the occupations which are found to be closer female-centric words.

The further classification of the occupations using Support Vector Machines shows the prediction of the target class for each gender bias. The support vector machine can classify with a linear kernel with an overall accuracy of 0.73.

The Classifier can classify the occupations with measures given in Table 2.

Table 2. Classifier measures

	Male	Female
Precision	0.83	0.60
Recall	0.71	0.75
F1-score	0.77	0.67

5 Conclusion

Gender bias is one of the most important aspects which is needed to detect before employing any algorithm. The above study proposes a method for the detection of the presence of the gender bias in the text data and it also shows the result which it causes if the data is trained on the gender bias data. The above methods follow an approach of finding the presence of gender bias in the text by using the context of various occupations.

The above study uses Google News article for analysis. This research works not only detects the presence of gender bias in the text but also do the classification of the occupations based on its closeness to a particular gender. The successful classification for most of the occupations again shows and verifies the presence of the gender bias in the Google News articles. Various machine learning and Natural language processing techniques are combined for the study. In this research, the presence of gender bias is limited to the scope of only two genders male and female. The scope of the study can be increased to other genders and communities such as LGBTs.

The above analysis considers gender bias as an independent phenomenon. It is quantified based on the similarity measures. Gender bias is one of the demographic bias which might come in the text because of the presence of other demographic bias. The study of the effect of other demographic bias over gender bias can also be studied. The above methodology analyzes quantifying gender bias at the level of the document, which needs a large amount of data for training. The study can be extended in the manner of analyzing at the sentence level. The above research work in limited to the detection of gender bias in the text data, this research work can be extended for other types of data formats also.

References

1. Costa-jussà, M.R.: An analysis of gender bias studies in natural language processing. Nat. Mach. Intell. 1–12 (2019)
2. Bolukbasi, T., Chang, K.-W., Zou, J., Shaligram, V., Kalai, A.: Man is to computer programmer as woman is to homemaker? Debiasing word embeddings. In: Proceedings of 30th Conference on Neural Information Processing Systems (NIPS) (2016)
3. Mikolov, T., Sutskever, I., Chen, K., Corrado, G., Dean, J.: Distributed representations of words and phrases and their compositionality. In: Proceedings of 26th Conference on Neural Information ProcessingSystems (NIPS) (2013)
4. Sun, T., Gaut, A., Tang, S., Huang, Y., ElSherief, M., Zhao, J., Mirza, D., Belding, E., Chang, K.-W., Wang, W.Y.: Mitigating gender bias in natural language processing: literature review. In: Proceedings of the 57th Annual Meeting of the Association for Computational Linguistics (2019)
5. Leavy, S.: Gender bias in artificial ıntelligence: the need for diversity and gender theory in machine learning. In: ACM/IEEE 1st International Workshop on Gender Equality in Software Engineering (2018)
6. Zhao, J., Wang, T., Yatskar, M., Ordonez, V., Chang, K.-W.: Gender bias in conference resolution: evaluation and debiasing methods. In: Proceedings of the 2018 Conference of the North American Chapter of the Association for Computational Linguistics: Human Language Technologies (2018)
7. Dixon, L., Li, J., Sorensen, J., Thain, N., Vasserman, L.: Measuring and mitigating unintended bias in text classification. In: Proceedings of the 2018 AAAI/ACM Conference on AI, Ethics, and Society (2018)
8. Mikolov, T., Chen, K., Corrado, G., Dean, J.: Efficient estimation of word representations in vector space. In: International Conference on Learning Representations (2013)
9. Madaan, N., Mehta, S., Agrawaal, T.S., Malhotra, V., Aggarwal, A., Gupta, Y., Saxena, M.: Analyze, detect and remove gender stereotyping from bollywood movies. In: Proceedings of Machine Learning Research (2018)
10. Campa, S., Davis, M., Gonzalez, D.: Deep Machine Learning Approaches to Analyzing Gender Representations in Journalism. https://web.stanford.edu/class/cs224n/reports/custom/15787612.pdf
11. Qian, Y., Muaz, U., Zhang, B., Hyun, J.W.: Reducing gender bias inword-level language models with a gender-equalizing loss function. In: Proceedings of the 57th Annual Meeting of the Association for Computational Linguistics (2019)
12. Park, J.H., Shin, J., Fung, P.: Reducing gender bias in abusive language detection. In: Proceedings of the 2018 Conference on Empirical Methods in Natural Language Processing (2018)

13. Zhao, J., Zhou, Y., Li, Z., Wang, W., Chang, K.-W.: Learning gender-neutral word embeddings. In: Proceedings of the 2018 Conference on Empirical Methods in Natural Language Processing (2018)
14. Prates, M.O.R., Avelar, P.H., Lamb, L.C.: Assessing gender bias in machine translation: a case study with Google Translate. Neural Comput. Appl. (2019)
15. Kusner, M.J., Sun, Y., Kolkin, N.I., Weinberger, K.Q.: Fromword embeddings to document distances. In: Proceedings of the 32nd International Conference on Machine Learning (2015)
16. Vijayan, V.K., Bindu, K.R., Parameswaran, L.: A comprehensive study of text classification algorithms. In: 2017 International Conference on Advances in Computing, Communications and Informatics (ICACCI) (2017)
17. Nedungadi, P., Harikumar H., Ramesh, M.V.: A high performance hybrid algorithm for text classification. In: 2014 Fifth International Conference on Applications of Digital Information and Web Technologies (ICADIWT) (2014)
18. van der Maaten, L., Hinton, G.: Visualizing Data using t-SNE. J. Mach. Learn. Res. **9**, 2579–2605 (2008)
19. https://web.cs.hacettepe.edu.tr/ ~ pinar/courses/VBM687/lectures/SVM.pdf

Highly Protective Framework for Medical Identity Theft by Combining Data Hiding with Cryptography

Babu Illuri(✉) and Deepa Jose

KCG College of Technology, Chennai 600097, Tamil Nadu, India
babucareeredeg@gmail.com, deepajosell@gmail.com

Abstract. Medical identity theft is more common since it is easy to steal the patient's record details and sell it for illegal medical trades. It happens in medical track records, personal healthcare accounts, recorded online insurance services, social security numbers, and other personally identifiable information (PII) which are disguised and used in the name of others. In this research, the proposed approach combines data hiding with hybrid encryption algorithms such as Modified Elliptic Curve Cryptography (MECC) encryption based on MIMO-OFDM framework. The hybrid encryption technique is combined with an LDPC coder for efficient information between the sender and receiver. Thereby, it improves the hiding capacity by 11.36% than SISO, 48.48% than OFDM, and 68.96% than MIMO. The sensitivity values are measured through the parameter number of pixels changes rate (NPCR) and unified average changing ıntensity (UACI) and obtained as 99.70 and 33.48 respectively. Correlation between adjacent pixels (CP) is increased compared to previous works. For SNR (Signal to Noise Ratio) at 30 dB CP is obtained as 0.9985. Simulation is carried out in MATLAB 2019a.

Keywords: ECC · Data hiding · Hybrid encryption · Low-density parity-check · BER · MIMO-OFDM

1 Introduction

Security is a major issue when the medical image is transmitted through wireless communication devices. Most of the wireless communication devices include an encryption method, to protect sensitive information but still have more probability to attack sensitive information [1]. Most of the previous research papers involve either encryption or data hiding technique [1]. In the previous study, a data hiding algorithm is implemented based on MIMO-OFDM. This paper mainly focuses on data hiding equation and capacity improvement. Its effect on OFDM parameters, guard interval, FFT lengths is analyzed [1]. The data hiding capacity based on MIMO-OFDM technologies was proposed in previous studies qualitatively, with RS Coding channel [2]. Xiaoqiang Zhang et al. used an image encryption algorithm based on Elliptical curve cryptography with the chaotic system [3]. Zhong-xun et al. implemented encryption using data hiding with LDPC codes and the information is embedded into a carrier image [4]. Rashmi.N, K.Jyothi incudes reversible data hiding steganography along

J. I.-Z. Chen et al. (Eds.): ICIPCN 2020, AISC 1200, pp. 662–671, 2021.
https://doi.org/10.1007/978-3-030-51859-2_60

with cryptography and number theory [5, 6]. The focus of this research work is to enhance the security performance efficiency of the MIMO-OFDM framework by adapting efficient hybrid encryption techniques based on their medical applications. In the proposed methodology, encryption with data hiding based on the steganography process is implemented.

The organization of the paper is as follows. In Sect. 2, the system model of the proposed technique based on MIMO-OFDM is described. In Sect. 3, the pseudo-code execution of the proposed strategy is discussed. In Sect. 4, the execution analysis of the proposed MIMO-OFDM framework is elaborated. Conclusion and Future work is given in Sects. 5 and 6.

2 System Model

The system model of the proposed technique based on the MIMO-OFDM framework is shown in Fig. 1. In the transmitter side, the transmitted data is forwarded to hybrid encryption and the secret data is generated by it. This is done by forwarding first $n/2$ blocks within MECC and next $n/2$ blocks will go through the encoder and serial to the parallel converter with modulator and finally combined with secret data based Least assignment bit process(LSB). At last, it is forwarded to LDPC coder with the embedding process to hide the data shown in Fig. 2. The MECC and data hiding outputs are merged respectively using XOR in the transmitter side. In the receiver side, following FFT, the secret data is extracted through again by XOR operation shown in Fig. 3

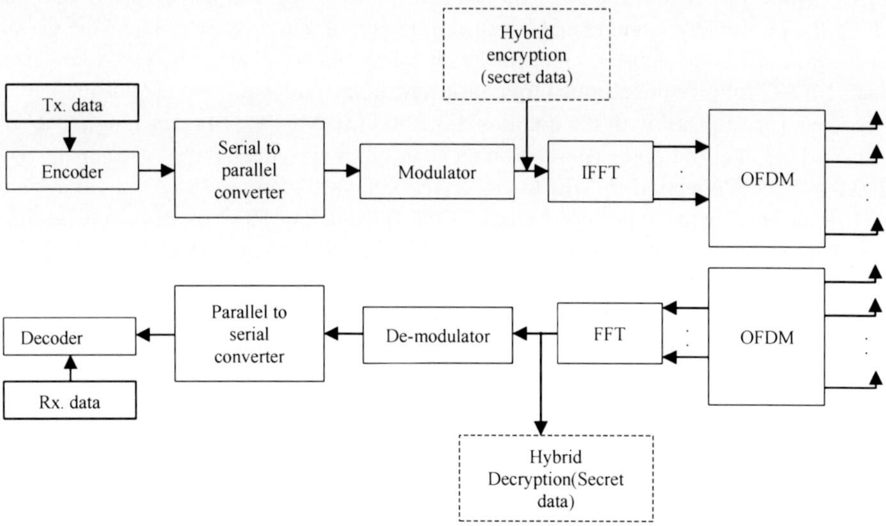

Fig. 1. System model of proposed technique based on MIMO-OFDM

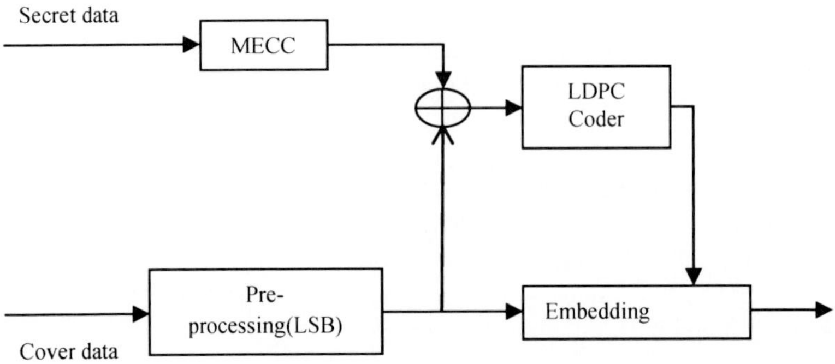

Fig. 2. A model of Encryption Process

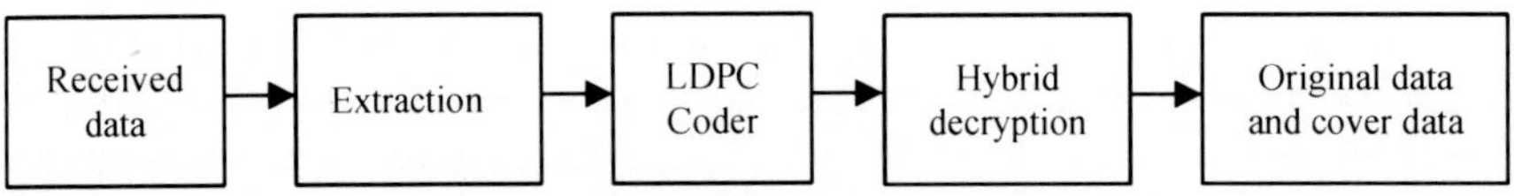

Fig. 3. A model of Decryption Process

3 Modified Elliptic Curve Cryptography (MECC) and Pseudo Code

MECC is the modified version of the standard elliptic curve cryptography (ECC) in which the public key is generated based on the elliptic curve theory. The MECC is a more efficient key generation algorithm with a fast and small size encrypted key size. Here, MECC follows the elliptic curve property; hence, the optimum size of the key is generated as compared with the traditional method [3]. MECC provides a high level of security key small bit size (164-bit) rather than other systems having a larger bit size (1024-bit) [7]. Compared to MECC and Rivest Shamir Adleman (RSA) based system [9, 10], the encrypted data of 256 bit in ECC is equivalent to 3072-bit keys in RSA. It is intended to scale down encrypted bits from 256 to 128-bit keys. The MECC includes the LSB data hiding technique to enhance security by generating hash values of each transaction for image pixels with SHA-512 [11]. So each transaction of image values generating a unique hash value, it is impossible to break hash values based on the blockchain technique [8].

3.1 Pseudo Code MECC

Key generation	
A:	Private Key: $n_A \in [1, p-1]$ and $n_B \in [1, p-1]$
B:	Public Key: $P_A = n_A.G$ and $P_B = n_B.G$
C:	Initial Key: $K_1 = n_A.P_B \approx (x, y)$ and $K_2 = n_B.P_A \approx (x, y)$
D:	$K_1 = X.G \approx (k_{11}, k_{12})$ and $K_2 = Y.G \approx (k_{21}, k_{22})$
E:	Self-invertible Key Matrix: K_m
MECC Encryption	
A:	Separate input image into blocks
B:	Block (4×1)
C:	$K_m * (P_1, P_2, P_3, \ldots) \bmod 256$
D:	$C_1 = (K_m * P_1) \bmod 256$
E:	Then, cipher image $C = (C_1, C_2, C_3, \ldots, C_n)$
MECC Decryption	
A:	Separate cipher image into blocks
B:	Block (4×1)
C:	$K_m * (C_1, C_2, C_3, \ldots) \bmod 256$
D:	$P_1 = (K_m * C_1) \bmod 256$
E:	Then, input image $P = (P_1, P_2, P_3, \ldots P_n)$

Abbreviations:

n_A, n_B = Private key generation, P_A, P_B = Public key generation, (x, y) = image pixel values, $(P_1, P_2, P_3, \ldots P_n)$ = Plain Image values, $(C_1, C_2, C_3, \ldots C_n)$ = cipher image values

4 Simulation Results and Discussion

In this section, the qualitative and quantitative evaluation of the proposed MIMO-OFDM framework method is carried out and compared with other existing methods which include RSA, AES [3]. The experimental results include two major parts. The first part discusses the MECC encryption and decryption algorithm results evaluated qualitatively. The second part presents the quantitative evaluation of the MIMO-OFDM framework after implementing hybrid encryption and decryption algorithm. The performance parameters of the MIMO-OFDM system such as BER (Bit Error Rate), PSNR (Peak to Signal Noise Ratio) and SNR (Signal to Noise Ratio) is estimated.

4.1 Qualitative Approach

In Fig. 4, the test images are taken as medical images (CT scan, MRI, etc.) are used as a carrier signal, which is transferred through the proposed MIMO-OFDM system [1]. The MECC encryption algorithm provides the secret image which is shown in Fig. 5. It depicts the carrier image and the secret image of corresponding medical images (CT scan, MRI, etc.) with an SNR value of 20 dB. The retrieved original medical Image for

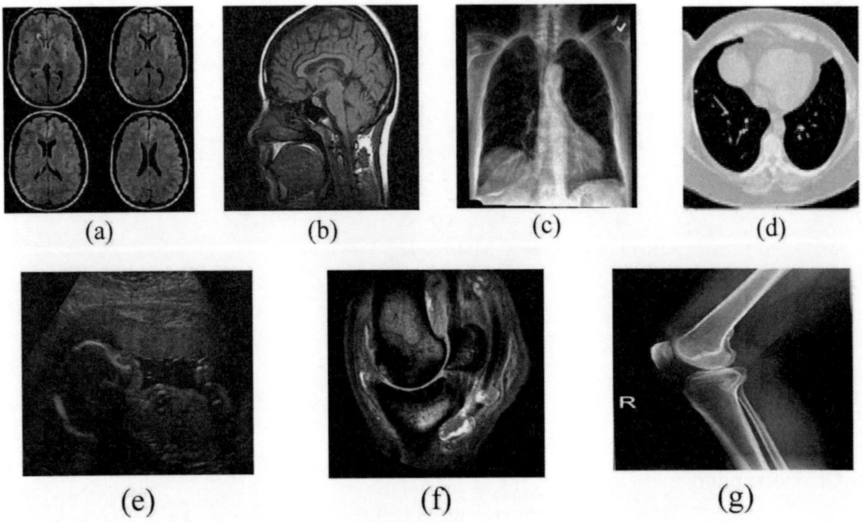

Fig. 4. Medical images (Source) (a) MRI Scan – Brain Axial (b) MRI Scan -Brain (c) Lung cancer (NSCLC) - CT (d) Chest CT Scan (e) Foetal-fetal-ultrasound-scan (f) MRI Scan (g) X-Ray

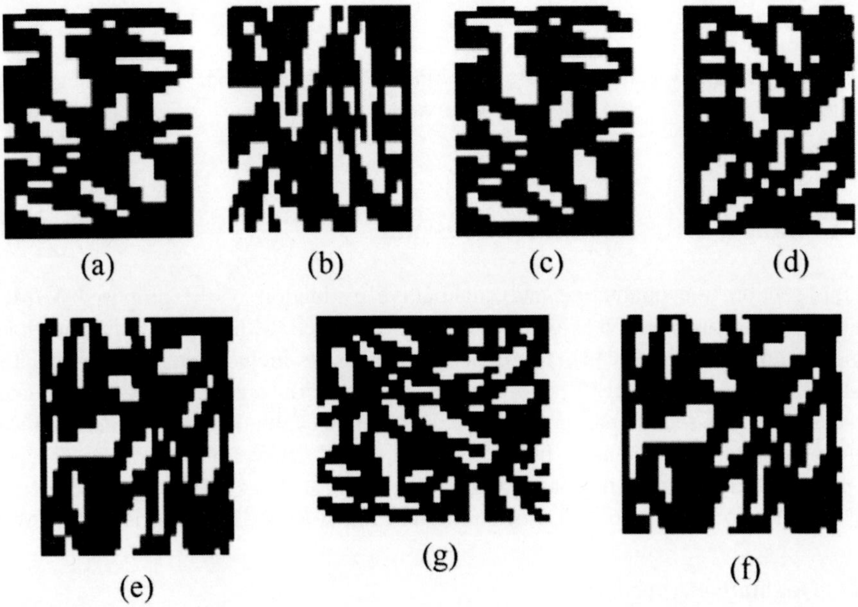

Fig. 5. Cipher images (after hybrid encryption algorithm) (a) MRI Scan – Brain Axial (b) MRI Scan - Brain (c) Lung cancer (NSCLC) - CT (d) Chest CT Scan (e) Foetal-fetal-ultrasound-scan (f) MRI Scan (g) X-Ray

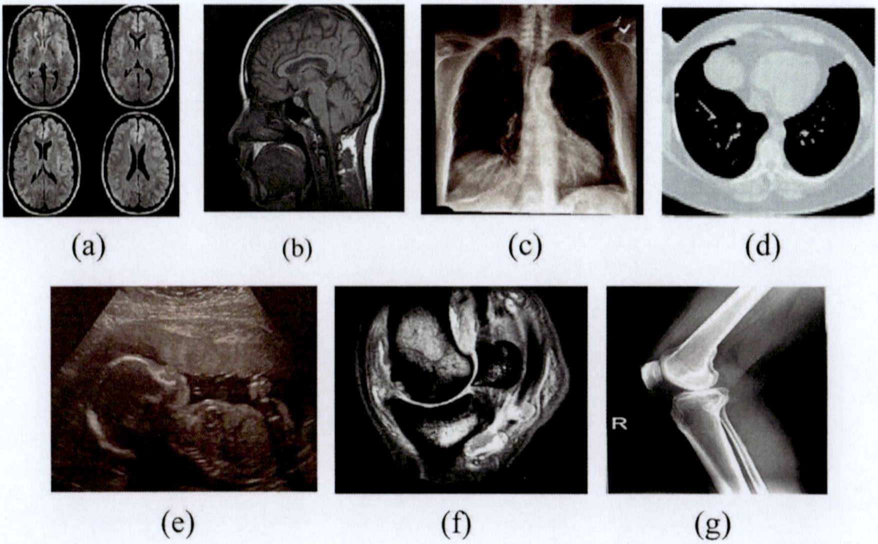

Fig. 6. Retrieve original medical image (after hybrid decryption algorithm) (a) MRI Scan – Brain Axial (b) MRI Scan - Brain (c) Lung cancer (NSCLC) - CT (d) Chest CT Scan (e) Foetal-fetal-ultrasound-scan (f) MRI Scan (g) X-Ray

different data types (CT scan, MRI, etc.) with SNR 20 dB is shown in Fig. 6. It depicts the performance of proposed hybrid encryption quality improvement based on the BER, PSNR, and SNR values. The proposed methods effectively recover the original image without any image bit reduction.

4.2 Quantitative Approach

In Fig. 7(a), [1] the plot represents the BER performance of the medical images with data covered as (carrier) and is decoded source data by proposed MIMO-OFDM system. The quality of the proposed technique is analyzed using Peak to Signal Noise

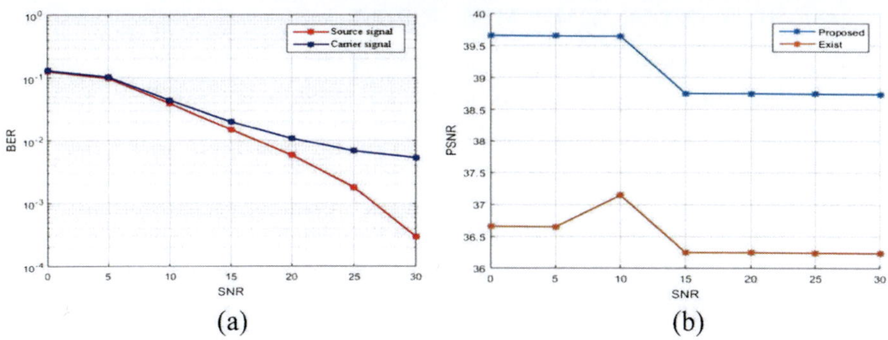

Fig. 7. Quantitative analysis of proposed and existing methods (a) BER of carrier and source data, (b) PSNR comparison

Table 1. NPCRs -UACIs for Hybrid, ECC, RSA, AES algorithms and Medical Images

Algorithms	Method (%)	MRI Scan - Brain	Lung Cancer (NSCLC) - CT	Ultrasound-Scan	MRI Scan – Brain Axial
MECC	NPCR	99.70	99.67	99.69	99.70
	UACI	33.48	33.49	33.52	33.51
	CP	99.45	96.50	93.21	92.99
Ref. [2]	NPCR	99.62	99.58	99.69	99.59
	UACI	33.63	33.36	33.61	33.52
	CP	99.75	95.00	92.65	91.21
Ref. [3]	NPCR	99.60	99.58	99.63	99.57
	UACI	33.45	33.42	33.47	33.44
	CP	99.50	96.14	94.83	93.20
Ref. [4]	NPCR	99.61	99.55	99.64	99.57
	UACI	33.37	33.31	33.35	33.33
	CP	99.21	95.20	94.68	92.81

Ratio (PSNR) parameter. Figure 7(b) shows the PSNR comparison of proposed MECC based on data hiding and existing technique [1]. In this hiding capacity ratio of the proposed MECC based MIMO-OFDM system with different SNR values 5 dB, 10 dB 15 dB, 20 dB, 25 dB, and 30 dB is obtained. The comparative analysis is shown in Table 1. From Table 1, it is seen that the hiding capacity ratio of the proposed MECCC based MIMO-OFDM system is very efficient than the existing data hiding technique [1].

The data hiding capacity is estimated by SNR value. The SNR is 30 dB with a length of the medical image as 32,768 bits and carrier data length as 40,956 bits respectively after channel coding. The secured data transmission is obtained by using the MECC encryption algorithm with LDPC error-correction codes. In this segment, the proposed MIMO-OFDM depiction is implemented using a simulation tool (MATLAB). The simulation specification of MIMO-OFDM uses the AWGN noise with six-path Rayleigh fading in the wireless communication channel [1].

4.3 Data Hiding Performance Comparison of the Proposed Method with SISO, OFDM, MIMO, MIMO-OFDM

The data hiding capability of the proposed method is evaluated by the comparative analysis made with systems like SISO, OFDM, MIMO, and MIMO-OFDM [9]. Figure 8 shows the hiding capacity of different channels

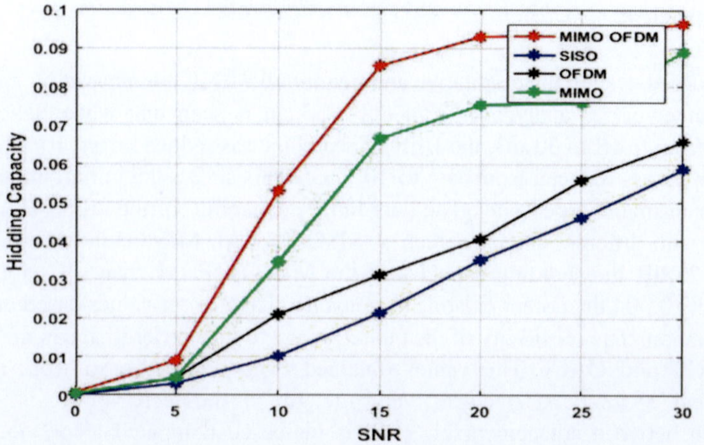

Fig. 8. Comparison of data hiding capacities SISO, MIMO, OFDM, and MIMO-OFDM

4.4 NCPR, UACI, and CP Evaluation

The cipher image will be sensitive to the plain (medical) image to reduce differential malicious attacks by measuring a slight change between plain image and cipher image. The sensitivity is measured using the number of pixels changes rate (NCPR) and the unified average changing intensity (UACI). The Correlation between adjacent pixels (CP) between cipher images and the plain image is obtained from Eq. 3. The ratio of successful detection of attacks to several malicious attacks available in the decrypted images is called several pixels changes rate. Table 3 shows the NPCR and UACI values of different medical images under varying pixel intensity and corrected the error. It is neatly observed that there variation between different methods such as MECC, Ref. [2], Ref. [3], and Ref. [4]. It is noted that the ideal NPCR and UACI values are 99.60% and 33.46% respectively obtained from Eqs. 1 and 2.

$$NPCR(\%) = \frac{\sum_{i,j} D(i,j)}{M \times N} \times 100 \tag{1}$$

$$UCAI(100\%) = \frac{1}{M \times N}\{\sum_{i,j} \frac{|C1(i,j) - C2(i,j)|}{255}\} \times 100 \tag{2}$$

$$CP = \frac{|\sum_{x=1}^{S}\sum_{y=1}^{S} I(x,y) - \mu)X(I(x,y)+1) - \mu|}{\sqrt{\sum_{x=1}^{S}\sum_{y=1}^{S}((I(x,y) - \mu))^2} \times \sqrt{\sum_{x=1}^{S}\sum_{y=1}^{S}((I(x,y)+1) - \mu)^2}} \tag{3}$$

Where $M \times N$ represents the size of the image, μ is the mean value of the pixels in an image block I, C_1 and C_2 be cipher images before and after one pixel of the plain image is modified, respectively. As a result, NPCR and UACI values are 99.62% and 33.63%. The NPCRs–UACIs values for the different medical images are evaluated which is given in Table 1.

5 Conclusion

In our proposed system the qualitative and quantitative evaluation based on the MIMO-OFDM framework is analyzed. From our work, it is seen that when SNR value is increased from 0 dB to 30 dB, the BER of secret data is reduced from 10^{-1} to$10^{-4/}$sec and carrier data is reduced from10^{-1} to $10^{-2/}$sec. This shows the minimum occurrence of an error in image processing. The data hiding capability of the proposed method is compared with different channels such as SISO, OFDM, MIMO, and MIMO-OFDM. For SNR 25 dB thresholding capacity of the MIMO-OFDM channel is found to be increased by 0.004 than other existing systems that have lesser values based on the data hiding equation. The sensitivity of the cipher image to the carrier image is measured by using NPCR and UACI. The values obtained from our proposed work are 99.70 (NPCR) and 33.48 (UACI) respectively strength of the secret image is increased. Correlation between adjacent pixels (CP) is increased compared to previous works. For SNR (Signal to Noise Ratio) at 30 dB, CP is 0.9985 which shows accurate images obtained. The proposed method utilizes the MECC algorithm with data hiding in MIMO-OFDM. The simulation results convey that the proposed method deployed a high quality of retrieval medical image with reasonable values of PSNR and MSE when compare with RSA (Rivest-Shamir-Adleman) and AES (Advanced Encryption Standard) respectively. The key size is minimized and provides more security for data using the adaptive blockchain technique, with time complexity is 1.254 s at CPU 2.6 GHz-Intel i5 processor.

References

1. Chen, L., Fan, Z., Huang, J.: Data hiding capacity of spatial domain bit replacement steganography in an MIMO-OFDM coding channel. AEU – Int. J. Electron. Commun. **70** (9), 1295–1303 (2016)
2. Ma, H.: A capacity self-adaption data hiding algorithm based on RS code. In: International Conference on Multisensor Fusion and Information Integration for Intelligent Systems, pp. 1–8 (2014)
3. Zhang, X., Wang, X.: Digital image encryption algorithm based on elliptical curve public cryptosystem. IEEE Access **6**, 70025–70034 (2018)
4. Wang, Z., Meng, X.: Digital image information hiding algorithm research based on LDPC code. EURASIP J. Image Video Process. **2018**, 1–12 (2018)
5. Rahmi, N., Jyothi, K.: An improved method for reversible data hiding steganography combined with cryptography. In: 2nd International Conference on Inventive Systems and Control (ICISC), pp. 1–12. IEEE (2018)
6. Rehman, A.: Data hiding technique in steganography for information security using number theory. J. Inf. Sci. (2018). https://doi.org/10.1177/0165551518816303
7. Liu, C.Q., Wang, Y.H.: A research on the steganography capacity of the Reed-Solomon codes. In: International Conference on Computational Intelligence and Security Workshop, CISW 2017, pp. 628–663 (2017)
8. https://blockgeeks.com/guides/what-is-blockchain-technology/

9. Kandul, A.: Steganography with cryptography in android. In: Proceedings of the 3rd International Conference on Frontiers of Intelligent Computing: Theory and Applications (FICTA), pp. 57–64 (2014). https://doi.org/10.1007/978-3-319-12012-6_7
10. l-Juaid, N., Gutub, A., Khan, E.: Enhancing PC data security via combining RSA cryptography and video based steganography. J. Inf. Secur. Cybercrimes Res. (JISCR) **1**(1) (2018)
11. Nassr, D.I.: Secure hash algorithm-2 formed on DNA. J. Egypt. Math. Soc. **27**(1), 1–20 (2019). https://doi.org/10.1186/s42787-019-0037-6

Improving the Prediction Accuracy of ASD Using Class Imbalance Mitigation Technique

S. Pushpa(✉) and K. Ulaga Priya

Department of Computer Science and Engineering,
St. Peter's Institute of Higher Education and Research, Chennai, India
pushpasangar96@gmail.com, ulagapriya@gmail.com

Abstract. The efficiency of prediction models generated by machine learning algorithms depends on the dataset used. To produce a good prediction model a well-balanced data is required. A well-balanced data set is one that has an equal distribution of samples belonging to various classes under study. Real-world datasets in various domains like health care, fraudulent behavior identification, risk management are class imbalanced. For example, in many publicly available health care data sets the number of samples having certain disease symptoms will be very few compared to healthy samples. Such a biased data set will reduce the accuracy of the prediction model generated by traditional classifiers. Hence, it becomes essential to transforming the given data set to a well-balanced one before applying any machine learning algorithms. In this paper, we analyze the autism prediction accuracy of the six machine learning algorithms such as Logistic Regression (LR), Linear Discriminant Analysis (LDA), Decision Tree (DT), K-Nearest Neighbor (KNN), Support Vector Machine (SVM) and Random Forest (RF). These algorithms are applied to the Autism Screening Adult Data Set collected from the UCI repository, the accuracy and Cohen's Kappa statistic measures are studied. As this data set is not well balanced, a prominent SMOTE algorithm is used to balance it. A comparison of the prediction accuracy shows that logistic regression performs better in a well-balanced data set.

1 Introduction

Due to the digitization of electronic health records, a tremendous amount of health care data is now made publicly available. The underlying public health records if explored to find the association between patient's disease and other socio-economic features will provide better insights. This will help public health care agencies to frame better public health policies. However, this is nearly an impossible task if carried out manually owing to the amount of data generated in the health care domain. The use of machine learning algorithms is an aid to bridge this gap with reduced human intervention and cost.

The next major problem in the healthcare domain is uneven distribution of classes in the dataset. For e.g. dataset used for analyzing a particular disease will have more healthy patient data i.e. negative samples than those affected by diseases i.e. positive samples. This class imbalance problem affects the classification accuracy of the prediction algorithm. However, the accuracy of prediction is an important metric in

J. I.-Z. Chen et al. (Eds.): ICIPCN 2020, AISC 1200, pp. 672–684, 2021.
https://doi.org/10.1007/978-3-030-51859-2_61

healthcare systems. The outcome of the prediction algorithm applied to a binary classification healthcare dataset has to correctly distinguish between positive and negative samples. The four possible outcomes of the classifier are truly positive, false positive, true negative, and false negative. In the true positive case, the algorithm classifies those personally affected by the disease correctly. In the false-positive case, the algorithm classifies the healthy patients as disease affected persons. In the case of true negative, the algorithm classifies the healthy patients correctly. In the false-negative case, the disease affected patients are classified as healthy persons. Of these four metrics, the users expect the prediction algorithms to minimize false positive and false negative values and maximize the true positive and true negative. Among the undesirable measures false positive and false negative, the false negative is dear as it may sometime cost the life of a person. False-positive involves additional medical evaluation procedures to classify them as healthy patients. However, this may cost the psychological stress in the minds of healthy individuals. These issues highlight the importance of balancing the data set before applying algorithms to classify the data. The proposed work in this paper focuses on improving the classification accuracy of autism spectrum disorder using class imbalance mitigation techniques.

Autism spectrum disorder (ASD) refers to a variety of neurodevelopment problems that begins in childhood and tend to persist throughout the lifetime of an individual. Autism related disorders are estimated to have affected 1% of the global population as of 2015 and increases every year. ASD encompasses five pervasive development disorders (PDD). They are autism, Asperger's Syndrome, Pervasive Developmental Disorder - Not Otherwise Specified (PDD-NOS), Childhood disintegrative disorder (CDD), and Rett Syndrome. Individuals affected by Autism have troubles in socializing with others and their behaviors are found to be repetitive and restricted. Asperger syndrome (AS) is a milder ASD with relatively improved communication skills and intelligence. People identified with PDD-NOS will exhibit persistent development disorders that do not fit into the diagnosis criteria of PDD.

Childhood disintegrative disorder (CDD) a.k.a as Heller's syndrome is an unusual condition where the milestones of the patients are delayed and setbacks in the communication, social, and motor abilities are observed. Rett syndrome is a genetic brain disorder that is evident in girl children after six months to 1.5 years. Apart from the symptoms of autism, the person affected by Rett syndrome is characterized by smaller head size and walking problems. Chances of an individual to develop ASD include having a sibling with ASD, older parents, persisting conditions like down syndrome, fragile X syndrome, and Rett syndrome. A very few percentages of people affected by autism can lead an independent life and join the regular workforce. While the majority of them have to be dependent on other support for their lifetime and remain unproductive. Hence, ASD has a significant economic impact on nations across the globe. ASD is not curable however early diagnosis and treatments can help to improve the living conditions. This will reduce the long term cost that results due to the delay in ASD identification.

This paper considers the open-source autism screening adult data set, taken from the UCI machine learning repository. The percentage of healthy samples in this data set constitutes around 73% and those affected with ASD are found to be around 27%. Using this unbalanced dataset will impede the prediction rate of the learning algorithms. In the proposed work attempt is made to increase the number of positive samples using SMOTE algorithm and study the prediction accuracy of the algorithms on the transformed dataset.

2 Literature Survey

Thabtah (2018) have surveyed various research works that use machine learning algorithm in ASD diagnosis. The authors have highlighted the current research in this domain lacks in conceptual, implementation, and other issues related to data. The diagnostic procedure adopted, feature selection methods, imbalance in the data set, and the performance measures used are identified as limitations in the present study. This article highlights that the use of machine learning algorithms greatly reduces the time taken and automates ASD diagnosis when incorporated in Autism diagnosis tools.

ADOS-G is a popular diagnostic tool that has four half an hour modules, Lord et al. (2000). Each unit is designed to assess individuals with varying levels of language expressiveness. Since expressive language level is a strong predictor in autism spectrum disorder the modules are designed to assess individuals without expressiveness to fluent language levels. The tasks assigned to individuals vary depending on the language level to reduce the biasing effect of varying language skills. The practitioner chooses the right module for a patient depending on age and language proficiency.

Wall et al. (2012a, b) applied a range of machine learning analysis to reduce the number of features used in ADOS-R. It is found that 8 features alone are sufficient to provide accurate results rather than the 29 features used in ADOS-R. This paper has studied the performance of the alternating decision tree algorithm on the autism genetic resource exchange (AGRE) data set. The authors have concluded that the accuracy, specificity, and sensitivity of the algorithm is almost perfect in classifying autism from normal cases.

Allison et al. (2012) developed a red flag tool to assist healthcare professionals in their decision of referring individuals including children and adults for a complete assessment of ASD. This paper has identified ten items on the Autism spectrum quotient that yields good accuracy. Participants are given the complete version of the measures used to assess individuals for ASD diagnosis. Then they are analyzed with the chosen ten best measures. The results are analyzed at a cut point of six on the AQ adult, adolescent, and children. Sensitivity, specificity, and positive predictive value are found to be 0.85 on all measures. The authors have concluded that this shorter version remains efficient for further diagnosis.

Pancerz and Derkacz (2015) proposed to adapt computational intelligence methods in diagnosing Autism Spectrum Disorders (ASDs). In this paper, the data preprocessing step is proposed for cleaning the underlying training data set used by the classifier algorithms. In the preprocessing stage, the data is split into unambiguous objects that belong to a single class. The set of unambiguous objects that are obtained are used to train classifiers. A consistency factor is suggested for rough set theory to identify objects that belong to the boundary between classes.

Duda et al. (2016) applied a range of machine intelligence algorithms to distinguish ASD from ADHD cases in minimum time. The machine learning algorithms have given accurate results by utilizing only 5 out of the 65 features from Simplex Simon Collection. This collection is based on the Social Responsiveness Scale consisting of questions answered by parents that are used to identify autistic behavior.

Preeti et al. (2016) highlighted the fact that early identification can improve outcomes in children with ASD. They have identified the average age at first consultation was two and a half years for children suspected with ASD symptoms and initiation of intervention to be carried out at around four years of age. Initiation of intervention is recommended as a good practice for all the children identified to be at high risk for ASD soon after the first consultation. Mythili and Shanavas (2014) studied the pre-diagnosis problem of ASD on a fixed dataset. Dataset chosen for this study consists of around 100 samples described with three attributes such as language skills, socializing with others, and a class label that has three values such as Mild, Moderate, and Severe. This paper has suggested machine learning as a promising research direction for the diagnosis of ASD.

Schopler et al. (1980) developed a childhood autistic rating scale (CARS) a behavior-based measure to check for ASD symptoms. Data collected from five hundred and thirty-seven autistic children is one of the largest datasets available for systematic evaluation of autism characteristics. Pratap et al. (2014) assessed the performance of unsupervised machine learning algorithms for predicting autism. Training and testing of the chosen unsupervised algorithms have identified individuals with 100% accuracy when compared against CAR values.

Hall (2012) designed the national database for autism research (NDAR) and its summary to promote research in this domain. Tenev et al. (2014) used EEG measurements to study attention hyperactivity disorder (ADHD). The sample considered for studies includes 117 adults consisting of 67 ADHD and 50 controls. EEG measurements are taken when the eyes are open and closed and two psychological tasks are carried out by the patients. The collected data is divided into four sets representing these conditions. Each set is used to train support vector machine classifiers and the outputs obtained are combined using the Karnaugh map based logical expression. The results obtained clearly distinguished between controls and ADHD and its subtypes.

Leevy et al. (2018) in the survey paper has summarized the literature for handling imbalance in data set at the level of data and algorithm. Statistical approaches to evenly distribute the data among various classes are done at the data level. The algorithmic approach modifies the conventional machine learning approaches to reduce the bias against a minority class of data. Bauder and Khoshgoftaar (2018) have analyzed the fraud that occurs in the healthcare domain. The authors have preprocessed the data and mapped the fraud labels in the LEIE dataset. RF and LR algorithms have given the best average among the various algorithms used in this paper.

He and Garcia (2009) have reviewed the imbalanced class problem based on its nature, technical solutions provided and evaluation measures used. The authors have suggested potential research directions in the class imbalance domain. The problems of imbalanced data set are analyzed in various dimensions such as imbalance that arises due to the distribution of the data. Cost-sensitive learning strategies give a cost dimension for the misclassification of data. The third dimension that has been analyzed includes sampling techniques and integration of sampling and boosting techniques.

Chawla et al. (2002) proposed an over-sampling approach (SMOTE) in which synthetic samples are generated to balance the data set. The synthetic samples are generated along the line segment joining a given class sample with its k minority neighbors. The value of k is chosen depending on the amount of oversampling required. The authors have considered five nearest neighbors of a given minority class sample.

3 Proposed Work

In the proposed approach, an attempt is made to highlight the importance of balancing the underlying data set before applying machine learning algorithms. The proposed approach analyzes the efficiency of six machine learning algorithms such as LR, LDA, DT, KNN, SVM, and RF in terms of accuracy and Cohen's Kappa Statistics. The accuracy and kappa measures of these algorithms on the imbalanced Autism Screening Adult Data Set are first studied. Later the data is balanced using the SMOTE algorithm and tested with the same machine learning algorithms. The balanced data set shows better accuracy and Kappa measures.

3.1 Logistic Regression

LR uses a logistic function f to map any real number input x to output y in the range 0 and 1. The output value of the logistic regression function is given in Eq. (1)

$$y = \frac{e^{(\beta_0 + \beta_1 x)}}{\left(1 + e^{(\beta_0 + \beta_1 x)}\right)} \tag{1}$$

Where x is the input, y is the output, β_0 is the bias and β_1 is the coefficient of x. Maximum-likelihood estimation (MLE) is used to find these coefficient values from the training data. The estimated value reduces the probability error between the actual output and the value given by this model.

3.2 Decision Tree

DT approach covers both classification and regression. This algorithm presents a descriptive representation of underlying data in a tree-like structure. Internal nodes represent attributes value of samples and the leaf nodes represent the decision taken. Splitting takes place at the internal node and the number of branches at each internal node depends on the attribute values. The split that costs the least value is chosen for further traversal down the tree. The cost of the split is computed using a Gini score given in Eq. 2.

$$\text{Gini-score(G)} = \sum pk * (1 - pk) \tag{2}$$

Where pk represents the proportion of same class inputs present in a particular group. A perfect splitting occurs when the split group of samples belongs to the same class which is either 1 or 0 and G = 0.

3.3 KNN Algorithm

KNN approach given in Algorithm 1 is a lazy machine learning algorithm that does not build a model to classify the given data. The given input is estimated to be part of a group depending on the nearest distance measure. The algorithm determines the cluster of a given data point by analyzing the data points around it. If the majority of data points around a given data point belongs to group A, then the sample is determined to belong to group A.

The outcome of the predictions of the KNN algorithm is heavily influenced by the value K. The value of the nearest neighbors of a given point K is considered as a smoothing parameter. The large variance in predictions will occur for a small value of K. Alternatively, a large value of K leads to bias. Thus, the K value has to be chosen that is maximum enough to reduce the probability of misclassification and minimum enough such that the K nearest points are closely placed near to the query point.

Algorithm 1: KNN Algorithm

```
Input:   Input sample data points and query data point, q
         Number of neighbors K
Output:  Classification of the query data point

For each data point d in the sample data
        Calculate the distance between the q and d
        Insert the distance measured and the index of the data point d in an or-
dered
        list.
End for
        Sort the ordered list from smallest to the largest distance
        Select the first K entries from the ordered list
        Get the class labels of these K entries
        Return the mode of the K labels.
```

3.4 Linear Discriminant Analysis

LDA technique is used to classify the data and reduce the dimensions. This method maximizes the separation between two given classes by reducing the similarity between-classes and increasing the similarity within-class members in a data set. LDA follows two different approaches to transform the data set and the test vector is deployed in the transformed space. A transformation that is class dependent maximizes the ratio of between-class variance to within-class variance. This approach uses two optimizing criteria for transforming the datasets independently. Another approach called class-independent transformation maximizes the proportion of overall class variance to variance within a class. LDA model makes simplifying assumptions about the underlying data that the deviation of the value of each variable is constant concerning the mean value. The mean μ of each input x belonging to class k is estimated as given in Eq. 3.

$$\mu_k = \frac{\sum x}{n_k} \tag{3}$$

Where n_k is the instances count present within class k. The variance is calculated as given in Eq. 4.

$$\sigma^2 = \frac{(x_i - \mu)^2}{n - k} \tag{4}$$

Where σ^2 is the variance. The class of a test sample is predicted by estimating the probability of each class. The test sample is classified to belong to a class that has a higher probability value. The probability of a given test sample is calculated using the following formula given in Eq. 5.

$$P(K = k|X = x) = \frac{(P_k * f_k(x))}{\sum_i P_i * f_k(i)} \tag{5}$$

Where P_k, probability of class k is obtained from the training data, referred to as prior probability is calculated as given in Eq. 6.

$$P_k = \frac{n_k}{n} \tag{6}$$

and $f_k(x)$ is the expected probability of x belonging to the class k is calculated as given in Eq. 7. Using the Gaussian distribution function and simplifying, the following equation is obtained.

$$f_k(x) = x * \left(\frac{\mu_k}{\sigma^2} - \frac{\mu_k^2}{2 * \sigma^2}\right) + \ln(P_k) \tag{7}$$

Where $f_k(x)$ is the distinguishing function for class k given x, μ_k, σ^2 and P_k.

3.5 Support Vector Machine

SVM classification algorithm aims to find an optimal hyperplane in N-Dimensional space that distinctly classifies the data points. Figure 1 shows the various hyperplanes of an SVM. The optimal hyperplane is one that is placed at a maximum margin distance from the data points of both the classes. This would ensure that new test data would be classified with more confidence. Figure 2 shows the optimal hyperplane separating the data points in two-dimensional space. The number of features determines the dimensionality of the hyperplane. For two features the hyperplane is a line and for 3 features the hyperplane is a plane and so on.

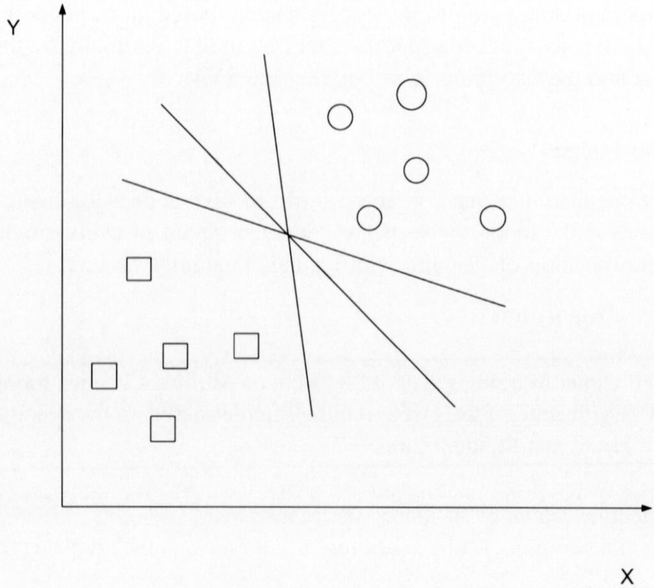

Fig. 1. Possible hyperplanes of SVM

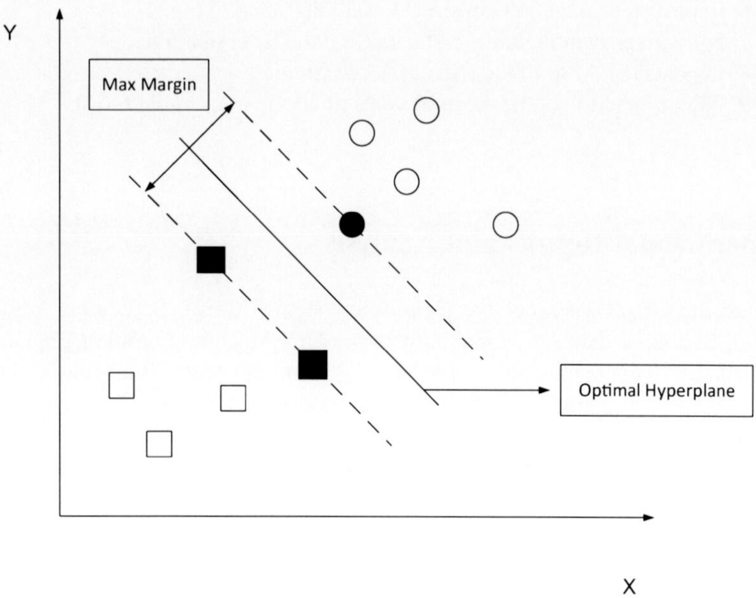

Fig. 2. Optimal hyperplane of SVM

The position and orientation of the hyperplane are based on its support vectors i.e. the data points closer to it. These support vectors are used to optimally set the margin of the hyperplane and play a vital role in building the SVM.

3.6 Random Forest

RF is an ensemble classifier that contains a forest of several decision trees. The output of this algorithm is the mode value of the predicted output of every single tree. This method is a combination of bagging with features randomly selected.

Random Forest Algorithm

Input: N- Count of training data, M – Count of Attributes in each training data.
Output: Classification of the given training samples based on the average votes of all trees in a Random forest.

Step 1: Select 'm' count of attributes, where m<<M
Step 2: Form the training set by selecting 'n' training data with replacement from N
Step 3: Form the test set from the remaining samples of step 2
Step 4: Based on 'm' attributes decide a given node in the tree
Step 5: The best split of a node is based on m attributes chosen from the training set
Step 6: Repeat steps 4&5 and construct ensemble forest of trees.
Step 7: For a given sample in the test data, find the label of a tree.
Step 8: Repeat step 7 for all the trees in the ensemble.
Step 9: Final prediction is the mean of votes of all the trees in the forest.

4 Experimental Results and Analysis

For testing the effectiveness of the proposed system "Autism Screening Adult Data Set", an open-source data set available from the UCI Machine Learning Repository is used, Thabtah (2017). The data set contains 704 samples with 20 attributes. The raw data set contains 2 numerical variables, 10 binary variables, and 8 categorical variables.

R programming language is used to implement the various machine learning algorithms and study statistical measures like accuracy, Cohen's Kappa statistic. In the initial step of the implementation, the raw imbalanced data set is used and the results obtained are given in the following Table 1 and Table 2.

Table 1. Accuracy of various machine learning algorithms before applying SMOTE algorithm

Algorithm	Min.	1st Qu.	Median	Mean	3rd Qu.	Max.
Logistic regression	0.98	1.00	1.00	0.99	1.00	1.00
Linear discriminant analysis	0.90	0.93	0.95	0.94	0.91	0.98
Decision tree	0.82	0.84	0.88	0.87	0.91	0.92
KNN	0.92	0.92	0.93	0.93	0.94	0.96
SVM	0.96	0.97	0.98	0.98	0.99	1.00
Random forest	0.93	0.95	0.96	0.96	0.96	0.99

Table 2. Cohen's Kappa statistic measures of various machine learning algorithms before applying SMOTE algorithm

Algorithm	Min.	1st Qu.	Median	Mean	3rd Qu.	Max.
Logistic regression	0.95	1.00	1.00	0.99	1.00	1.00
Linear discriminant analysis	0.78	0.84	0.88	0.87	0.93	0.95
Decision tree	0.55	0.62	0.70	0.69	0.78	0.82
KNN	0.81	0.81	0.85	0.85	0.86	0.92
SVM	0.91	0.92	0.94	0.95	0.97	1.00
Random forest	0.83	0.88	0.90	0.90	0.91	0.97

In the second step, the data is balanced by applying the SMOTE algorithm, Chawla et al. (2002) and the transformed data is then used for classification. The transformed data shows better classification accuracy and Cohen's Kappa statistics given Table 3 and Table 4 given below.

Table 3. Accuracy of various machine learning algorithms after applying SMOTE algorithm

Algorithm	Min.	1st Qu.	Median	Mean	3rd Qu.	Max.
Logistic regression	1.00	1.00	1.00	1.00	1.00	1.00
Linear discriminant analysis	0.91	0.93	0.95	0.94	0.97	0.98
Decision tree	0.84	0.87	0.89	0.89	0.91	0.99
KNN	0.93	0.94	0.95	0.95	0.96	0.99
SVM	0.95	0.97	0.97	0.98	0.99	1.00
Random forest	0.95	0.98	0.98	0.96	0.99	1.00

Table 4. Cohen's Kappa statistic measure of various machine learning algorithms after applying SMOTE algorithm

Algorithm	Min.	1st Qu.	Median	Mean	3rd Qu.	Max.
Logistic regression	1.00	1.00	1.00	1.00	1.00	1.00
Linear discriminant analysis	0.78	0.84	0.88	0.87	0.93	0.95
Decision tree	0.67	0.74	0.77	0.78	0.82	0.82
KNN	0.87	0.89	0.90	0.85	0.92	0.92
SVM	0.95	0.97	0.97	0.98	0.99	1.00
Random forest	0.95	0.98	0.98	0.98	0.99	1.00

The accuracy measure of various machine learning algorithms before and after data balancing is given in Fig. 3.

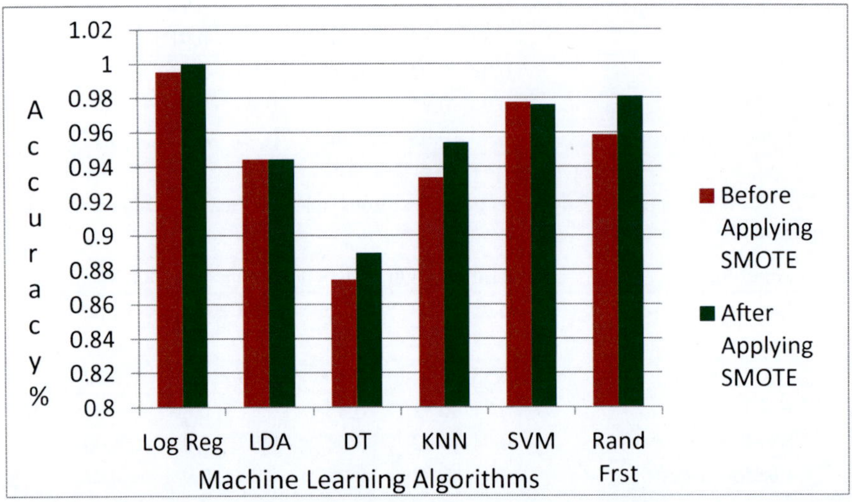

Fig. 3. Accuracy in % of various machine learning algorithms before and after data balancing using SMOTE algorithm

Cohen's Kappa statistics measure of various machine learning algorithms before and after data balancing is given in Fig. 4.

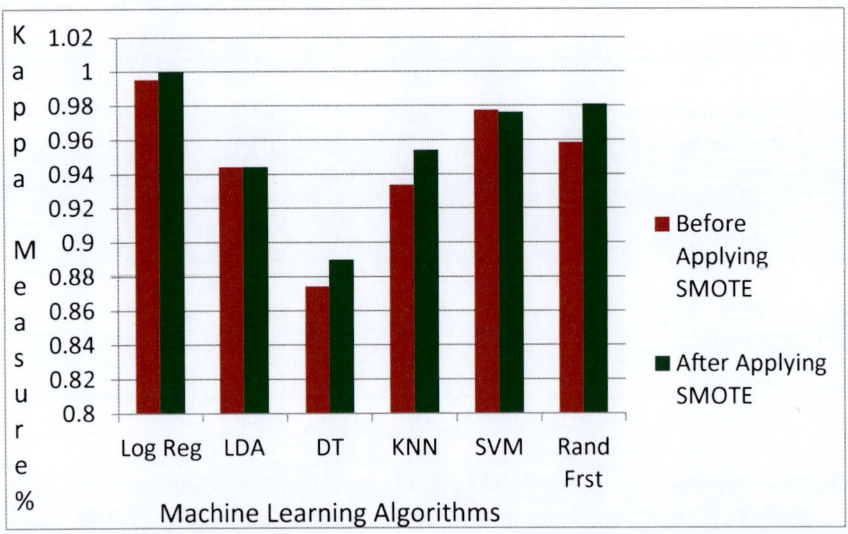

Fig. 4. Kappa measure in % of various machine learning algorithms before and after using SMOTE algorithm

5 Conclusion and Scope for Further Research

Several real-time datasets available online like credit card transactions, patient's health records are imbalanced. The efficiency of the results obtained by the machine learning algorithms using these imbalanced data sets is not reliable due to this bias. Thus, the data set has to be balanced. In recent years, machine learning algorithms that adapt automated intelligent methods are used to find interesting patterns from the given data.

This study focuses on autism prediction diagnoses using machine learning algorithms that give a fast and more accurate prediction. Autism Screening Adult Data Set collected from the UCI repository is used to study the prediction accuracy of these algorithms. An attempt is made to balance the data set using the well-known SMOTE Algorithm and the results obtained are analyzed. The prediction accuracy is found to be higher on a balanced data set. The results obtained indicate the need to balance the dataset and apply a machine-learning algorithm to obtain improved performances. Logistic regression gives better accuracy among the algorithms chosen. In future attempts will be made to integrate the existing diagnostic tools with the predictive models based on balanced data that will make them more reliable.

References

Thabtah, F.: Machine learning in autistic spectrum disorder behavioral research: a review and ways forward. Inform. Health Soc. Care (2018). https://doi.org/10.1080/17538157.2017.1399132

Lord, C., Risi, S., Lambrecht, L., Cook, E.H., Leventhal, B.L., DiLavore, P.C., Pickles, A., Rutter, M.: The autism diagnostic observation schedule-generic: a standard measure of social and communication deficits associated with the spectrum of autism. J. Autism Dev. Disord. **30**(3), 205–223 (2000). https://doi.org/10.1023/A:1005592401947

Wall, D.P., Kosmiscki, J., Deluca, T.F., Harstad, L., Fusaro, V.A.: Use of machine learning to shorten observation-based screening and diagnosis of Autism. Transl. Psychiatry **2**(2), e100 (2012a). https://doi.org/10.1038/tp.2012.10

Wall, D.P., Dally, R., Luyster, R., Jung, J.Y., DeLuca, T.F.: Use of artificial intelligence to shorten the behavioral diagnosis of autism. PLoS One **7**(8), e43855 (2012b)

Allison, C., Auyeung, B., Baron-Cohen, S.: Toward brief "red flags" for autism screening: the short Autism Spectrum Quotient and the short quantitative checklist for autism in toddlers in 1,000 cases and 3,000 controls. J. Am. Acad. Adolesc. Psychiatry **51**(2), 202–217 (2012). https://doi.org/10.1016/j.jaac.2011.11.003

Pancers, K., Derkacz, A.: Consistency-based pre-processing for classification of data coming from evaluation sheets of subjects with ASDS. In: Federated Conference on Computer Science and Information Systems, pp. 63–67 (2015)

Duda, M., Ma, R., Haber, N., Wall, D.P.: Use of machine learning for behavioral distinction of autism and ADHD. Transl. Psychiatry **9**(6), 732 (2016). https://doi.org/10.1038/tp.2015.221

Preeti, K.D.M, Srinath, S., Seshadri, S.P., Girimaji, S.C., Kommu, J.V.S.: Lost time—need for more awareness in early intervention of Autism Spectrum Disorder. Asian J. Psychiatry (2016). http://dx.doi.org/10.1016/j.ajp.2016.07.021

Mythili, M., Shanavas, M.R.: A study on Autism spectrum disorders using classification techniques. IJCSIT **5**(6), 7288–7291 (2014)

Schopler, E., Reichler, R., DeVellis, R.: Toward objective classification of childhood autism: childhood autism rating scale (CARS). J. Autism Dev. Disord. **10**, 91–103 (1980). https://doi.org/10.1007/bf02408436

Pratap, A., Kanimozhiselvi, C.S., Vijayakumar, R., Pramod, K.V.: Predictive assessment of autism using unsupervised machine learning models. Int. J. Adv. Intell. Paradig **6**(2), 113–121 (2014). https://doi.org/10.1504/ijaip.2014.062174

Hall, D., Huerta, M.F., McAuliffe, M.J., Farber, G.K.: Sharing heterogeneous data: the national database for autism research. Neuroinformatics **10**(10), 331–339 (2012). https://doi.org/10.1007/s12021-012-9151-4

Tenev, A., Markovska-Simoska, S., Kocarev, L., Pop-Jordanov, J., Müller, A., Candrian, G.: Machine learning approach for classification of ADHD adults. Int. J. Psychophys. **93**(1), 162–166 (2014). https://doi.org/10.1016/j.ijpsycho.2013.01.008

Leevy, J.L., et al.: A survey on addressing high-class imbalance in big data. J. Big Data (2018). https://doi.org/10.1186/s40537-018-0151-6

Bauder, R.A., Khoshgoftaar, T.M.: The effects of varying class distribution on learner behavior for medicare fraud detection with imbalanced big data. Health Inf. Sci. Syst. (2018). https://doi.org/10.1007/s13755-018-0051-3

He, H., Garcia, E.A.: Learning from imbalanced data. IEEE Trans. Knowl. Data Eng. **21**(9), 1263–1284 (2009)

Chawla, N.V., Bowyer, K.W., Hall, L.O., Kegelmeyer, W.P.: SMOTE: synthetic minority over-sampling technique. J. Artif. Intell. Res. **16**, 321–357 (2002)

Thabtah, F.F.: Department of Digital Technology, Manukau Institute of Technology, Auckland, New Zealand (2017). https://archive.ics.uci.edu/ml/datasets/Autism+Screening+Adult

Stock Price Prediction Based on Technical Indicators with Soft Computing Models

S. Kumar Chandar[1,2(✉)]

[1] School of Business and Management, CHRIST (Deemed to Be University), Bangalore, India
kumar.chandar@christuniversity.in
[2] Department of Management Studies, Christ University, Bangalore, India

Abstract. Stock market prediction is a very tough task in the finance world. Since stock prices are dynamic, noisy, non-scalable, non-linear, non-parametric and complicated. In recent years, soft computing techniques are used for developing stock prediction model. The main focus of this study is to develop and compare the efficiency of the three different soft computing techniques for predicting the intraday price of individual stocks. The proposed models are based on Time Delay Neural Network (TDNN), Radial Basis Function Neural Network (RBFNN) and Back Propagation Neural Network (BPNN). The predictive models are developed using technical indicators. Sixteen technical indicators were calculated from the historical price and used as inputs of the developed models. Historical prices from 01/01/2018 to 28/02/2018, where the time interval between samples is one minute, are utilized for developing models. The performance of the proposed models is evaluated by measuring some metrics. Also, this study compares the results with other existing models. The experimental result revealed that the BPNN outperforms TDNN, RBFNN as well as other existing models considered for comparison.

Keywords: Back Propagation Neural Network (BPNN) · Intraday stock price prediction · MAPE · MAE · Technical indicators · Time Delay Neural Network (TDNN)

1 Introduction

Prediction of the stock price is one of the most important issues in the financial world that has received more attention from financial analysts and researchers [1]. Stock market prediction is defined as the process of calculating the future value of stock-based on historical prices. It is regarded as a challenging task due to the nature of stock prices such as highly noisy, complicated, dynamic, irregular, nonlinear, chaotic and non-parametric [2–4]. Further to this, the financial market is also influenced by many external factors including economic conditions, political events and traders' expectations etc. This fact motivates the researchers to develop a more efficient predictive model to make a good decision and gain more profits.

Several traditional methods can be used to predict the stock market price which includes fundamental analysis, statistical analysis and technical analysis [5, 6].

J. I.-Z. Chen et al. (Eds.): ICIPCN 2020, AISC 1200, pp. 685–699, 2021.
https://doi.org/10.1007/978-3-030-51859-2_62

Fundamental analysis uses macro-economic factors, financial conditions, exchange rate and other factors to determine stock price. On the contrary, technical analysis involves historical prices. The technical analysis determines the stock value based on low, high, open and closing value. However, these traditional techniques cannot predict the stock market price accurately [20]. Recently, soft computing techniques have gained more acceptance in the financial market due to its ability to approximate any non-linear function. The key idea of using Artificial Neural Network (ANN) is to determine the mapping between input and output variables via training process [21]. Then, the trained ANN is employed to predict future values. Therefore, ANN is a promising tool for stock price prediction.

The fundamental goal of this study is to investigate the ability of soft computing techniques in predicting the intraday stock price. Three predictive models are developed by using Time Delay Neural Network (TDNN), Radial Basis Function Neural Network (RBFNN) and Back Propagation Neural Network (BPNN). Sixteen technical indicators namely Simple Moving Average (SMA) SMA_7, SMA_{14}, Exponential MA (EMA) EMA_7, EMA_{14}, Relative Strength Index (RSI), Moving Average Convergence/Divergence (MACD), Stochastic %k, Stochastic %D, Rate of Change (ROC), four-momentum, Accumulation/Distribution Oscillator (ADO) and William's R% are calculated and fed as an input to the developed models. The efficiency of the all the three models are evaluated by measuring eight metrics namely Mean Square Error (MSE), Mean Absolute Error (MAE), Root Mean Square Error (RMSE), Mean Absolute Percentage Error (MAPE), Hit Rate (HR), Error Rate (ER), Mean Absolute Deviation (MAD) and prediction accuracy and compared.

The rest of the paper is structured as follows: Sect. 2 presents a brief overview of related work in this domain. A brief overview of soft computing models used in this study is explained in Sect. 3. The proposed prediction model is presented in Sect. 4. Section 5 reports numerical results and performance comparison. Finally, the paper is concluded in Sect. 6 followed by relevant references.

2 Review of Related Work

In recent years, both financial analysts and researchers have made a tremendous effort to predict the closing price of the stock market. In the following section, this study presents the survey of earlier methods on stock market prediction. Statistical analysis, time series analysis, fundamental analysis and technical analysis are some of the conventional methods that have been applied to predict the stock market [5, 7]. However, no one methods can produce accurate prediction. Thus, it is necessary to hybrid the technical analysis, time series analysis and soft computing.

Most of the researchers have utilized one to twenty-five technical indicators to develop their model. From the review of research work done by many authors, the role technical indicators observations are used in their study. To enhance the prediction accuracy, added more technical indicators and its outcomes were investigated with three neural networks, TDNN, RBFNN and BPNN. The effectiveness of the proposed three models was demonstrated with various measures. Kara et al. [3] presented the stock market prediction model using ANN and Support Vector Machine (SVM). Ten indicators such as SMA_{10}, Weighted MA (WMA_{10}), Momentum, Stochastic %K, Stochastic %D, RSI, MACD, Larry William's R%, ADO and Commodity Channel Index (CCI) were used as inputs to the ANN and SVM. Results showed that the performance of ANN is better than SVM in terms of accuracy. Handa et al. [8] developed two prediction models: RBFNN and Error Back Propagation Network (EBPN) to predict the Indian stock market. Sixteen technical indicators were extracted from historical data and then seven technical indicators were chosen based on the ranking method. The authors proved that neural networks with seven technical indicators are the most suitable model. Further to this, results showed than the ANN technique provides better results in terms of RMSE, MAPE and MAE. Yamawaki et al. [9] presented an innovative method for predicting the intraday stock value. The authors extracted prediction strategies by using a combination of technical indicators with a Genetic Algorithm (GA). Results revealed that the moving averages are the good candidate for predicting intraday stock price. Dash et al. [10] developed a prediction model by integrating technical analysis with machine learning. Six technical indicators such as SMA_{15}, $MACD_{26}$, stochastic $\%K_{14}$, RSI_{14}, William's $\%R_{14}$, stochastic $\%D_3$ were used as the input features for the Computational Efficient Functional Link Artificial Neural Network (CEFLANN). CEFLANN is a single layer network. It is trained with the Extreme Learning Machine (ELM) algorithm for generating the stock value. Simulation results showed that the efficiency of the proposed model is superior to other existing methods such as SVM, Naïve Bayesian model, K nearest neighbor model (KNN) and Decision Tree (DT). Rout et al. [11] investigated the use of Recurrent Computationally Efficient Functional Link ANN (RCEFLANN) to predict the BSE and S&P 500. The number of input neuron is set to 8 to represents the closing price of 5 days ago and 3 technical indicators such as MA_5, BIAS and Stochastic %D. Hybrid Moderate Random Particle Swarm Optimization (HMRPSO) and Differential Evolution (DE) algorithms are applied to tune the weight of RCEFLANN. Results showed that the RCEFLANN model trained with DE was remarkably improved when compared to FLANN, RBF and Wavelet Neural Network (WNN). Niu and Wang [12] used improved RBFNN to predict time series data. Gradient Descent (GD) optimization algorithm is used to determine the parameters of RBF. The improved RBF provides highly accurate prediction. Moghaddam et al. [13] tested the ability of ANN in predicting the stock market data. The authors proved that the proposed BPNN with One Step Secant (OSS) training algorithm and Tansig activation function provides a better result.

3 Research Methodology

This section briefly reviews the basic concepts about the underlying technologies used in the study.

A. *Back Propagation Neural Network*

Perceptron is a single layer, feed-forward supervised learning algorithm-based network. It is the simplest form of ANN and can classify only the linear separable inputs. To deal with such an issue, a new network with a supervised learning algorithm named as BPNN developed by Rumehart et al. [14] in 1986. BPNN can be used for solving both linear and non-linear problems. BPNN consists of three layers: an input layer, hidden layer and output layer. The general structure of BPNN with two hidden layers is shown in Fig. 1. BPNN operates in two modes: feedforward mode and backpropagation mode. During feedforward mode, input signals are passed from the input layer to the hidden layer and then the output layer. In backpropagation mode, the error difference between target and calculated output is backpropagated. The procedure is repeated during training to minimize the error.

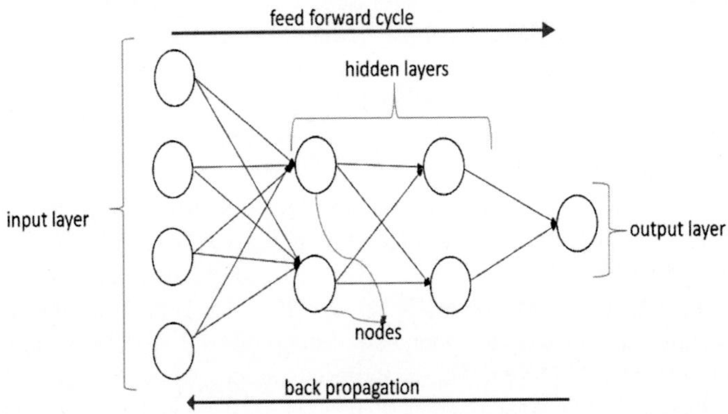

Fig. 1. Structure of BPNN.

The input layer receives the input signals from the external world. Then the signals are acts as input to the hidden layer. Each neuron is connected with the other neuron through a link named weight. Neuron sums the inputs multiplied by the weights and then added with the bias represented in Eq. (1)

$$net_j = \sum\nolimits_{i=1}^{n} x_i w_{ij} + b_j \tag{1}$$

The hidden layer calculates the output by applying the activation function

$$h_j = f\left(net_j\right) = f\left(\sum\nolimits_{i=1}^{n} x_i w_{ij} + b_j\right. \tag{2}$$

The output layer computes the solution of the problem

$$y = g(h) = g\left(\sum\nolimits_{j=1}^{p} h_j w_{jk} + b_k\right. \tag{3}$$

Where x is the input, W_{ij} is the weight between the input and hidden layer, W_{jk} is the weight between hidden and output layer and f, g is the activation function.

B. *Radial basis Function Neural Network*

RBFNN is a type feedforward artificial neural network that uses a radial basis function as the stimulation function. It has three layers: an input layer, a hidden layer and an output layer. The output of RBFNN is the linear combination of input neurons activated by activation function and neuron parameters. RBFNN finds applications in many areas including time series prediction, function approximation and classification [15, 16].

As shown in Fig. 2, the input layer receives the input and propagates to the hidden layer, h. Each neurone in the hidden layer associated with a Gaussian function.

$$\varphi_i(\|z - c_i\|) = exp\left(-\frac{1}{2}\left(\frac{\|z - c_i\|}{\sigma_i}\right)^2\right) \tag{4}$$

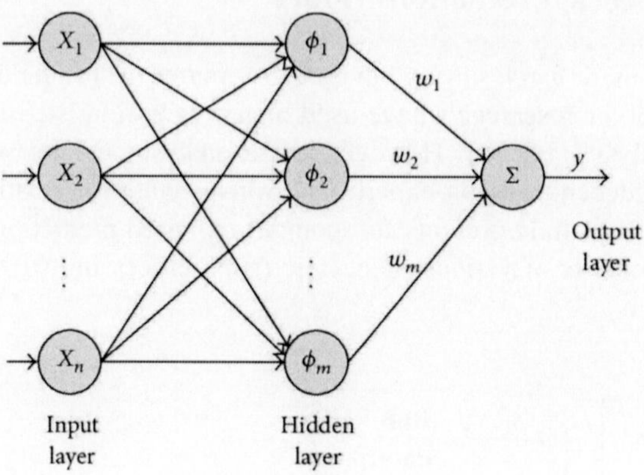

Fig. 2. Radial basis functions neural network.

The output layer function is given by,

$$f(z) = \sum\nolimits_{i=1}^{n} \lambda_i \varphi_i(\|z - c_i\|) \tag{5}$$

Where, φ is the Gaussian function, is the centre, σ is the width and λ is the weight.

C. *Time Delay Neural Network*

TDNN is a feed-forward network. TDNN can capture the dynamic relationship between consecutive variables [17]. Weights of TDNN is replaced by Finite Impulse Filter (FIR) which builds memory space for stock price prediction. Each neuron has a time-varying real-valued activation. TDNN has two layers: hidden layer and output layer [17, 18] in which the neurons time- delay neurons. Suppose a TDNN has N inputs and one output, y(t), where inputs are time series with step t.

$$I = (I_1(t), I_2(t), \ldots I_N(t)) \quad (6)$$

For each input unit $I_i(t)$, there is one bias B_i, delays p delays represented as D_i (i = 1, 2…. P) and storing the previous inputs $I_i(t - d)$. Weights of the neuron denoted by $(W_{i1}, W_{i2}, \ldots W_{iN})$. Tan sigmoid activation function F used. The output of TDNN can be expressed as,

$$y(t) = F\left(\sum\nolimits_{i=1}^{N} \left[\sum\nolimits_{D=0}^{P} I^i(t - D) * W_{iD} * B_i\right]\right) \quad (7)$$

Both the input signal at time t and previous time t − D (D = 1, 2…P) contribute to the overall response of the output neuron. Therefore, TDNN can be employed to model the dynamic behavior of the time series data.

4 Proposed Stock Prediction Model

There are numerous research efforts have been reported in the field of stock market prediction. Most of the researchers have used linear regression, statistical, fundamental and technical analysis [14, 15]. However, these analyses cannot provide consistent results due to the dependence on expertise. To investigate the relations among stock market indicators, this study presents soft computing-based prediction model as shown in Fig. 3 which consists of various processes: (i) preprocessing (ii) feature extraction and (iii) prediction.

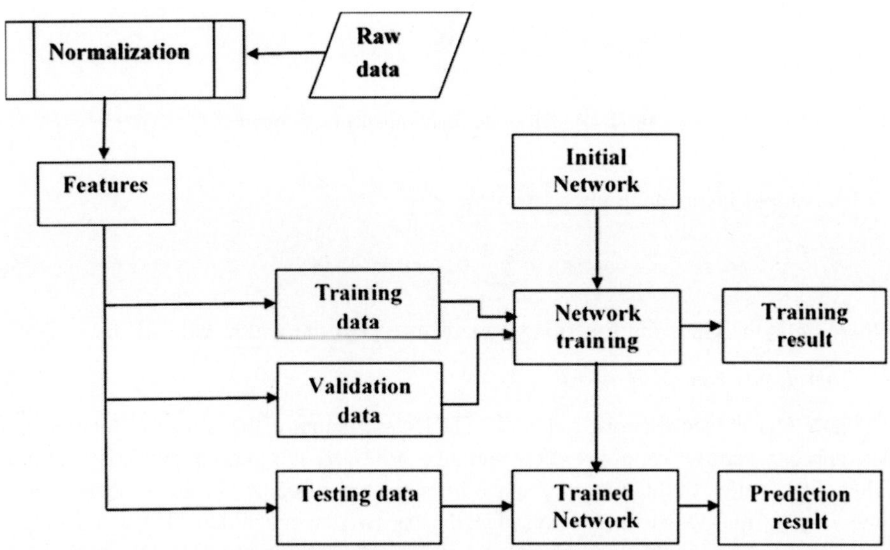

Fig. 3. Framework of the proposed stock prediction model.

A. *Preprocessing*

Data normalization plays an important role in stock prediction [19]. Historical stock data are vulnerable to noise. The quality of stock price affects the ANN results. For ANN based prediction models, historical data are normalized between 0 and 1 by using Eq. (8),

$$X_{norm} = \frac{X - X_{min}}{X_{max} - X_{min}}(h_i - l_i) \tag{8}$$

Where, X_{norm} is the normalized value of X, X is the actual value, X_{min} is the minimum value, X_{max} is the maximum value, h_i is the upper bound of the normalizing interval (in this case 1) and l_i is the lower bound of the normalizing interval (in this case 0).

B. *Feature Extraction*

Several technical indicators are available in the open literature [20]. In this study, sixteen technical indicators which include SMA_7, SMA_{14}, EMA_7, EMA_{14}, RSI, MACD, Stochastic %k, Stochastic %D, ROC, four-momentum, ADO and William's R % are derived from normalized data and employed as input to the prediction models. Technical indicators used in this study are given in Table 1.

Table 1. Technical indicators

Technical indicators	Formula
SMA	$\mathrm{SMA}_t = \frac{1}{t}\sum_{i=1}^{t} C_i$
EMA	$\mathrm{K}(C_t - EMA(t-1)) + EMA(t-1)$ $K = \frac{2}{1+n}$
RSI	$\mathrm{RSI} = 100 - \frac{100}{1+RS}$ $RS = \frac{EMA(U,n)}{EMA(D,N)}$ $U = C_t - C_{t-1} \qquad D = 0$ $D = C_{t-1} - C_t \qquad U = 0$
MACD	$MACD = EMA(12) - EMA(26)$ $Signal = EMA(MACD, 9)$
Stochastic %K	$\%K = 100\frac{C_t - C_l(n)}{C_h(n) - C_L}$
Stochastic %D	$\%D = EMA(\%K, 3)$
Momentum	$Price(t) - Price(t-n)$
ROC	$\left(\frac{Price(t)}{Price(t-n)}\right) * 100$
ADO	$\frac{H_t - C_{t-1}}{H_t - L_t}$
William's R %	$\left(\frac{H_t - C_t}{H_t - L_t}\right) * 100$

H-Highest price, C-Closing price, L-Lowest price

C. *Prediction*

Three ANNs, BPNN, TDNN and RBFNN have been used for forecasting the future price of NSE stock data, with a sample rate of one-minute. There is no standard algorithm for developing the architecture of ANNs. BPNN is designed with 16 input neurons representing technical indicators, two hidden layers with 20 hidden neurons and an output neuron corresponding to the predicted value. Tangent sigmoidal and linear activation is used at the hidden layer and output layer respectively. TDNN is designed with 16 input neurons, 20 hidden neurons and an output neuron. Tangent sigmoidal activation is applied at the hidden layer and linear activation at the output layer. Both BPNN and TDNN is trained with the Levenberg-Marquart (LM) training algorithm. RBFNN is also developed and implemented with 16 input neurons and an output neuron.

5 Simulation Results and Discussion

This section presents the numerical results of three networks such as TDNN, RBFNN and BPNN are employed for stock price prediction. All three prediction models share the common features of predicting the next minute closing price of selected stocks using technical indicators. The optimized weights of all the three networks are obtained by trial and error. Similarly, spread constant value for RBFNN is also determined by experimentation.

A. *Data set*

The examined data set comprises of minute prices form January 2018 to February 2018 of four stocks namely State Bank of India (SBI.NS), Tata steel (TATASTEEL. NS), National thermal power corporation (NTPC.NS) and Infosys (INFY.NS). Each sample consists of the opening price, lowest price, highest price and closing price of stocks traded per minute. The prediction ability of the developed models is analyzed by varying number of samples, one-day data, five days data and ten days data.

B. *Performance Indicators*

This study considered the stock data of four different sectors. The motivation behind this is to compare the prediction efficiency of the neural networks and find a suitable one. To measure the performance of the developed prediction models, eight evaluation parameters are chosen and listed in Table 2. In these parameters HR, FR, MAPE and RMSE are utilized to list whether the predicted value is close to the real value. If MAPE, MSE, MAD, MAE and RMSE are small, it means that the predicted value is equal to the original value.HR is utilized to measure the predictor accuracy. If HR is high, the predicted value is similar to real value.

Table 2. Performance indicators

Indicators	Formula
MSE	$MSE = \frac{1}{N}\sum_{k=1}^{N}(A_k - P_k)^2$
MAE	$MAE = \frac{1}{N}\sum_{k=1}^{N}\lvert A_k - P_k\rvert$
RMSE	$RMSE = \sqrt{\frac{1}{N}\sum_{k=1}^{N}(\lvert A_k - P_k\rvert)^2}$
MAPE	$MAPE = \frac{1}{N}\sum_{k=1}^{N}\left\lvert\frac{A_k - P_k}{A_k}\right\rvert \times 100$
MAD	$MAD = \frac{1}{N}\sum_{K=1}^{N}\lvert P_K - \overline{P}\rvert$
HR	$Hitrate = \frac{1}{N}\sum_{k=1}^{N} A_k = P_k$
ER	$Errorrate = \frac{1}{N}\sum_{k=1}^{N} A_k = P_k$

Where, A-Actual price, P-Predicted price

C. *Performance Evaluation*

One Day Data

In this experiment, one-day data are used to test the performance of the developed models. Thus, the total number of samples is 396. These samples are normalized to lie within the interval [0,1] using the min max method described in Sect. 4. Out of 396 samples, 276 samples used for training the models, 60 samples used for validating the models and the remaining 60 samples utilized for testing the performance of the model. After computing the output, then those values are converted back to get the original value. The obtained statistical measures are all tabulated in Table 3 for each stock and each predictive model. The higher value of HR and prediction accuracy and lower value of all statistical measures (MSE, MAE, RMSE, MAPE, MAD, ER), the better the prediction accuracy. Based on all statistical measures, it is seen from Table 3 that the performance of BPNN is better than RBFNN and TDNN in almost all the stocks.

Figure 4 graphically compares the outcome of the three predictive models against real value for all stocks. From the Fig. 4, it is observed that the predicted value obtained by BPNN is closer to the real value in almost all the stocks. Hence, it is suggested that BPNN is a good candidate for stock price prediction.

Table 3. Comparative results of statical metrics for one day data

Stock name	Predictive model	MSE	MAE	RMSE	MAPE	MAD	HR	ER	Prediction accuracy (%)
SBI.NS	TDNN	0.043	0.128	0.207	0.041	1.207	0.903	0.096	80.17
	RBFNN	0.016	0.093	0.128	0.030	0.870	0.939	0.060	90.16
	BPNN	0.014	0.037	0.119	0.012	0.249	0.983	0.016	96.28
NTPC.NS	TDNN	0.018	0.104	0.135	0.059	0.245	0.926	0.073	90.26
	RBFNN	0.363	0.284	0.603	0.161	0.353	0.718	0.281	71.50
	BPNN	0.012	0.013	0.017	0.007	0.104	0.948	0.051	98.63
TATASTEEL. NS	TDNN	0.192	0.296	0.439	0.040	1.883	0.704	0.295	61.54
	RBFNN	0.291	0.374	0.539	0.051	2.951	0.618	0.381	62.60
	BPNN	0.042	0.120	0.205	0.016	2.287	0.932	0.067	88
INFY.NS	TDNN	0.318	0.208	0.564	0.020	1.647	0.820	0.179	70.38
	RBFNN	0.120	0.245	0.346	0.023	1.925	0.742	0.257	75.44
	BPNN	0.057	0.150	0.204	0.014	1.561	0.847	0.152	84.19

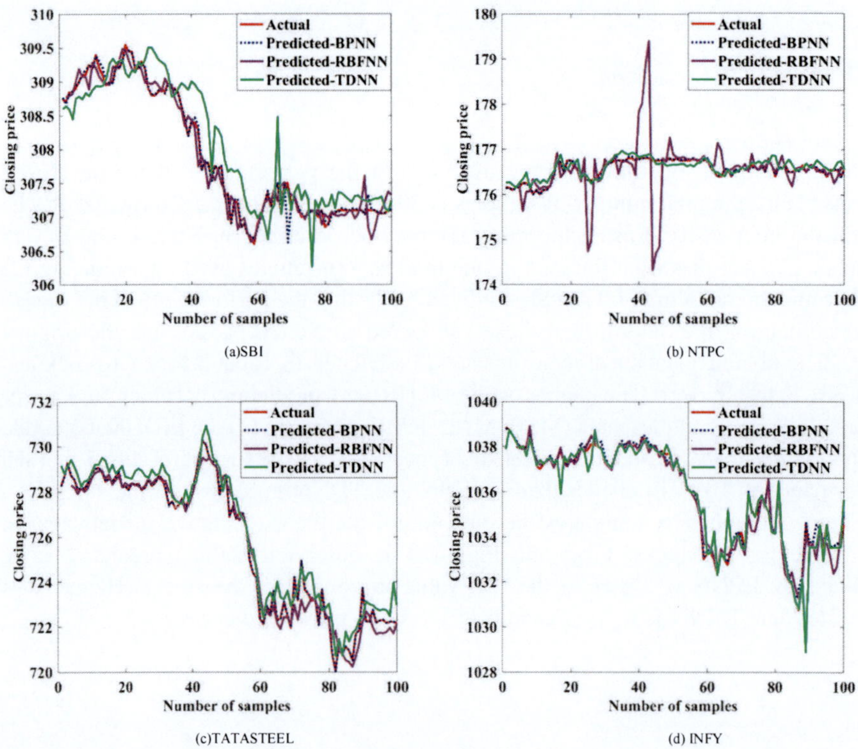

Fig. 4. Actual versus predicted stock price (one-day data).

Five Days of Data

Historical price covers five days of stock data, where the interval between samples is one minute. Therefore, a total number of samples is 1980. The entire data set is split into three groups: the first group (70%) of samples used for training the predictive models. The second group (15%) of samples utilized for validating the developed models and third group (15%) of samples employed for testing the prediction ability of the developed models. All the data are normalized in the interval [0,1]. Technical indicators are extracted and used as input to the developed models. On evolving the output, then those values are converted back to normal units to obtain predicted values. The simulation results over 5 days are summarized in Table 4 for each stock and each predictive model. From Table 4, it is inferred that the results were not satisfactory for TDNN and RBFNN. The experimental results showed that BPNN consistently performs better than other models TDNN and RBFNN in terms of all evaluation parameters in almost all the stocks. For example, in case of predicting SBI price over 5 days, MSE obtained by BPNN is very low and is almost half of that obtained by TDNN and RBFNN. It proved that BPNN with LM training algorithm can successfully forecast the intraday price of the stock market.

Table 4. Comparative results of statical metrics for five days data

Stock name	Predictive model	MSE	MAE	RMSE	MAPE	MAD	HR	ER	Prediction accuracy (%)
SBI.NS	TDNN	0.056	0.148	0.236	0.048	2.267	0.864	0.135	85.86
	RBFNN	0.025	0.106	0.158	0.034	1.325	0.846	0.153	89.36
	BPNN	0.003	0.042	0.059	0.014	0.973	0.922	0.077	95.71
NTPC.NS	TDNN	0.014	0.087	0.121	0.049	0.647	0.937	0.062	92.09
	RBFNN	0.302	0.215	0.550	0.121	0.380	0.891	0.108	78.47
	BPNN	0.001	0.027	0.035	0.015	0.165	0.986	0.013	97.24
TATASTEEL.NS	TDNN	0.376	0.411	0.613	0.055	15.76	0.630	0.369	64.19
	RBFNN	0.113	0.262	0.337	0.034	4.077	0.781	0.218	73.80
	BPNN	0.101	0.248	0.319	0.032	1.449	0.764	0.235	75.17
INFY.NS	TDNN	0.422	0.417	0.649	0.040	8.64	0.641	0.358	67.70
	RBFNN	0.144	0.272	0.380	0.026	2.96	0.778	0.221	72.75
	BPNN	0.076	0.142	0.277	0.014	1.507	0.925	0.074	85.72

To further demonstrate the performance of the developed predictive models, a graph is drawn between actual and predicted values which are shown in Fig. 4. In this graph, the predicted values are mapped and it is compared with the actual values. From Fig. 5, it is observed that the BPNN model predict the stock prices more precisely with that of the actual prices. For instance, the prediction accuracies (TDNN versus RBFNN versus BPNN) of SBI, NTPC, TATASTEEL and INFY are respectively (85.86% versus 89.36% versus 95.71%), (92.09% versus 78.74% versus 97.24%), (64.19% versus 73.80% versus 75.17%) and 67.70% versus 72.75% versus 85.72%). Hence, it is evident that BPNN outperforms the RBFNN and TDNN.

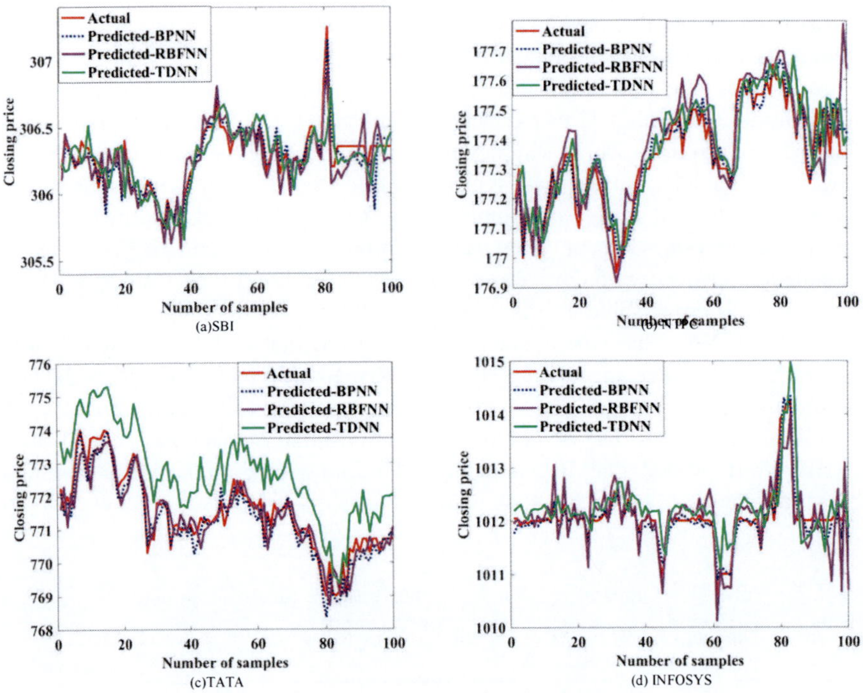

Fig. 5. Actual versus predicted stock price (Five days of data).

Ten Days of Data

This experiment considers ten trading days of data. Therefore, a total of 3960 samples. These samples are normalized between 0 and 1 and divided into three sets: training set (70% of samples), validation set (15% of samples) and test set (15% of samples).

Table 5. Comparative results of statical metrics for ten days data

Stock name	Predictive model	MSE	MAE	RMSE	MAPE	MAD	HR	ER	Prediction accuracy (%)
SBI.NS	TDNN	0.042	0.129	0.206	0.042	2.374	0.877	0.122	88.18
	RBFNN	0.766	0.235	0.875	0.078	1.763	0.851	0.148	76.45
	BPNN	0.003	0.040	0.056	0.013	0.548	0.978	0.021	95.94
NTPC.NS	TDNN	0.016	0.082	0.127	0.046	1.410	0.908	0.091	91.46
	RBFNN	0.404	0.216	0.636	0.124	0.632	0.804	0.195	78.34
	BPNN	0.002	0.035	0.052	0.020	0.317	0.838	0.161	96.41
TATASTEEL.NS	TDNN	0.369	0.406	0.607	0.053	17.67	0.638	0.361	61.43
	RBFNN	0.114	0.226	0.338	0.029	2.623	0.784	0.215	77.38
	BPNN	0.045	0.151	0.213	0.019	2.618	0.878	0.122	84.81
INFY.NS	TDNN	0.834	0.549	0.913	0.052	15.71	0.567	0.432	61.23
	RBFNN	0.679	0.343	0.789	0.088	1.725	0.832	0.167	74.22
	BPNN	0.111	0.231	0.334	0.021	1.451	0.785	0.214	76.85

Table 5 presents the statistical measures obtained by three predictive models throughout ten days of data. It can be seen from the Table 5, for SBI, BPNN possess minimal statistical value for MSE, MAE, RMSE, MAPE, MAD and ER to be 0.003,0.040,0.056,0.012,0.548 and 0.021 respectively which is minimal than RBFNN and TDNN and possess higher value for HR and prediction accuracy to be 0.978 and 95.94% which is higher compared to RBFNN and TDNN.

To demonstrate the performance of the developed predictive models, a graph is plotted between actual and predicted stock price is illustrated in Fig. 6. From the Fig. 6, it is found that the predicted curve using BPNN is closer to the actual stock curve which achieves the best accuracy. The experimental results revealed a consistently better performance of BPNN model when compared to RBFNN and TDNN over a different period of data.

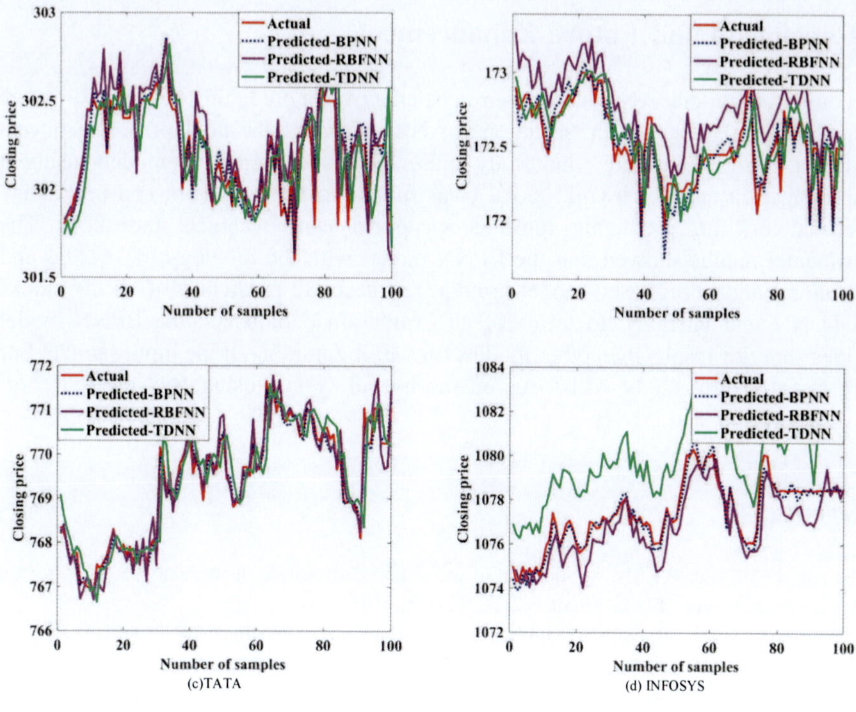

Fig. 6. Actual versus predicted stock price (Ten days of data).

Average prediction accuracy of the developed prediction models is compared with the earlier methods reported in the literature which are ANN [1], BPNN [3], Decision Tree (DT) [4] and RBFNN [8]. Figure 7 compares the performance of the developed models with the earlier models in terms of prediction accuracy. From the Fig. 7, it is found that the prediction accuracy obtained by BPNN is higher than that of RBFNN, TDNN, as well as other models, are taken for comparison.

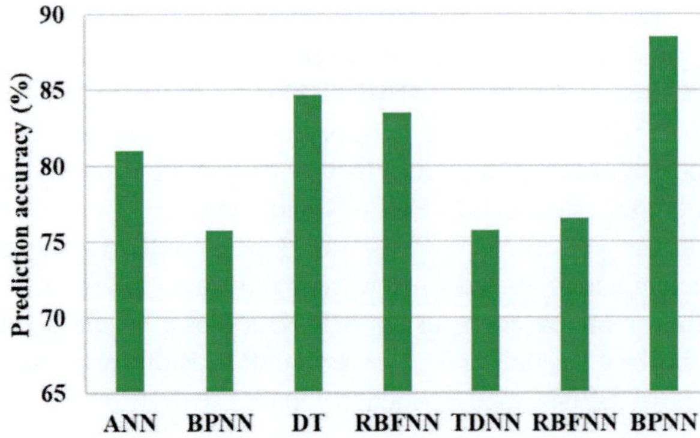

Fig. 7. Performance comparison in terms of accuracy

6 Conclusion and Future Enhancement

Soft computing model, ANN has proven to be effective in predicting the future trend of the stock market. However, the selection of ANN to forecast the stock price may have a significant effect. Therefore, this study presented three predictive models using a backpropagation neural network, radial basis function neural network and time delay neural network for predicting future stock price using technical indicators. The experimental results showed that the BPNN model with the topology of 16-20-1 and LM training algorithm-based model provides an accurate prediction of all the stocks than TDNN and RBFNN. In terms of all evaluation parameters, the BPNN model provides superior results than other models for varying number of the input sample. For future research, this study will focus on the hybrid system using Bio-inspired computing algorithms and ANNs.

References

1. Huang, C.Y., Lin, P.K.P.: Application of integrated data mining techniques in stock market forecasting. Cogent Econ. Finance **2**(1), 929505 (2014)
2. Abu-Mostafa, Y.S., Atiya, A.F.: Introduction to financial forecasting. Appl. Intell. **6**(3), 205–213 (1996)
3. Kara, Y., Boyacioglu, M.A., Baykan, Ö.K.: Predicting direction of stock price index movement using artificial neural networks and support vector machines: the sample of the Istanbul Stock Exchange. Expert Syst. Appl. **38**(5), 5311–5319 (2011)
4. Vaiz, S., Ramaswami, M.: A study on technical indicators in stock price movement prediction using decision tree algorithms. Am. J. Eng. Res. **5**(12), 207–212 (2016)
5. Adebiyi, A.A., Ayo, C.K., Adebiyi, M.O., Otokiti, S.O.: Stock price prediction using neural network with hybridized market indicators. J. Emerg. Trends Comput. Inf. Sci. **3**(1), 1–9 (2012)
6. Sureshkumar, K.K., Elango, N.M.: Performance analysis of stock price prediction using artificial neural network. Glob. J. Comput. Sci. Technol. **12**(1) (2012)

7. Ravichandran, K.S., Thirunavukarasu, P., Nallaswamy, R., Babu, R.: Estimation of return on investment in share market through ANN. J. Theor. Appl. Inf. Technol., 44–54 (2005)
8. Handa, R., Hota, H.S., Tanda, S.R.: Stock market prediction with various technical indicators using neural network techniques. Int. J. Res. Appl. Sci. Eng. Technol. **3**(VI), 604–608 (2015)
9. Yamawaki, M.T., Tokuoka, S.: Adaptive use of technical indicators for the prediction of intra-day stock prices. Physica **383**, 125–133 (2007)
10. Dash, R., Dash, P.K.: A hybrid stock trading framework integrating technical analysis with machine learning techniques. J. Finance Data Sci. **2**, 42–57 (2016)
11. Rout, A.K., Dash, P.K., Dash, R., Biso, R.: Forecasting financial time series using a low complexity recurrent neural network and evolutionary learning approach. J. King Saud Univ. Comput. Inf. Sci. **29**, 536–552 (2017)
12. Niu, H., Wang, J.: Financial time series prediction by a random data-time effective RBF neural network. Soft. Comput. **18**, 497–508 (2012)
13. Moghaddam, A.H., Moghaddam, M.H., Esfandyari, M.: Stock market index prediction using artificial neural network. J. Econ. Finance Adm. Sci. **21**, 89–93 (2016)
14. Rumelhart, D.E., Hinton, G.E., Williams, R.J.: Learning representations by back propagating errors. Nature **323**, 533–536 (1986)
15. Hota, H.S., Handa, R., Shrivas, A.K.: Time series data prediction using sliding window based RBF neural network. Int. J. Comput. Intell. Res. **13**(5), 1145–1156 (2017)
16. Wang, H., Xu, X.: Determination of spread constant in RBF neural network by genetic algorithm. Int. J. Adv. Comput. Technol. (IJACT) **5**(9), 719–726 (2013)
17. Saad, E.W., Prokhorov, D.V., Wunsch, D.C.: Comparative study of stock trend prediction using time delay, recurrent and probabilistic neural networks. IEEE Trans. Neural Netw. **9** (6), 1456–1470 (1998)
18. Meng, H., Bianchi-Berthouze, N., Deng, Y., Cheng, J., Cosmas, J.P.: Time-delay neural network for continuous emotional dimension prediction from facial expression sequences. IEEE Trans. Cybern. **46**(4), 916–929 (2015)
19. Nayak, S.C., Misra, B.B., Behera, H.S.: Impact of data normalization on stock index forecasting. Int. J. Comput. Inf. Syst. Ind. Manag. Appl. **6**(2014), 257–269 (2014)
20. Caley, J.A.: Survey of systems for predicting stock market movements, combining market indicators and machine learning classifiers. Dissertations and Theses (2013)
21. Zhang, Y.Q., Akkaladevi, S., Vachtsevanos, G., Lin, T.Y.: Granular neural web agents for stock prediction. Soft. Comput. **6**(5), 406–413 (2002)
22. Hsieh, T.J., Hsiao, H.F., Yeh, W.C.: Forecasting stock markets using wavelet transforms and recurrent neural networks: an integrated system based on artificial bee colony algorithm. Appl. Soft Comput. **11**(2), 2510–2525 (2011)

Credit Card Fraud Detection Technique Based on Hybrid Classification

Priya Gandhi(✉) and Vikas Khullar

C.T. Institute of Engineering, Management & Technology, Shahpur, India
Priya_gandhi94@icloud.com, vikas.khullar@gmail.com

Abstract. The credit card fraud detection is the prediction analysis method which can predict future possibilities based on the current information. The credit card fraud detection is the problem of prediction analysis. The credit card fraud detection has the various phases which are pre-processing, feature extraction and classification. In the previous research work, the voting classification method is designed. To reduce complexity of voting classification method and to improve accuracy ensemble classification is designed in this research work. The existing and proposed methods are implemented in Python and results are analyzed in terms of accuracy. It is analyzed that accuracy of proposed method is high as compared to existing method.

Keywords: Credit card fraud detection · Voting · KNN · Naïve Bayes

1 Introduction

Predictive Analytics encloses variety of techniques from statistics, data mining, and machine learning etc. to predict future trends based on historical and current information. A mathematical model is developed using the information related to previously occurred events such that the events that could possibly occur in future could be predicted [1]. The scores are assigned to the data based on certain predictive analytical models. The model which focuses on the behavior of an individual customer is the most commonly used predictive model in today's applications [2]. The sample data in which the known attributes are involved helps in training the model. This model can further be helpful in analyzing the new data as well as its behavior. Thus, the next actions of the customer can be predicted using this information. Credit Card Fraud Detection is one of the applications of prediction analysis. In Credit Card Fraud Detection, fraud transactions are predicted based on the historical information of credit card transactions [3]. Historical information of card transactions will be the training data for the fraud transaction prediction. It becomes easy to buy various goods and access different kinds of services instantly using the credit cards. However, the possibility of fraudulent transactions on any credit card is one of the major threats being faced today [4]. This kind of fraud is identified when the actual owner of the card has no knowledge about the transactions being made from his card. Of course there will be no contact with the card owner and the actual user will not receive any loss repayment from the fraudulent person [5]. It is very easy to conduct fraud by using very little amount of information about the card only. The fraudsters are using different and new

J. I.-Z. Chen et al. (Eds.): ICIPCN 2020, AISC 1200, pp. 700–710, 2021.
https://doi.org/10.1007/978-3-030-51859-2_63

mechanisms to perform such fraud actions so that it can be difficult to track them [6]. For reducing the fraud success it is important to introduce a fraud detection technique. The most important factor in this process is the previous transactions made by the genuine owner.

1.1 Predictive Analysis

For identifying the patterns and predicting the future possibilities, the information available in existing data sets is extracted using predictive analysis. The future events cannot be clearly known through this approach [7]. However, with certain level of reliability, the forecast of what might happen in future can be known here.

1.2 Credit Card Fraud Detection Techniques

There are two significant categorizations of the Credit Card Fraud Detection techniques [8]. In Supervised Learning Approach a database with labeled transactions are utilized. These transactions are labeled as fraudulent or normal depending on the existing historical data. The Classification models based on algorithms used to distinguish between the fraudulent or normal transactions. Every new transaction that comes in is classified on the basis of classification model. This model has two major drawbacks. Firstly, the data with labeled transactions is often unavailable, in case the labeled data is available; its accuracy is not guaranteed. Due to this, undetected fraud can be labeled as legitimate action. Secondly, the algorithms used to create these classification models are very complex and time consuming [9]. In Unsupervised Leaning Approach, the labels are not assigned to data set. A model is created on the basis of normal behavior, as it is assumed that fraudulent events are less. New transactions are compared against legitimate user behavior model, if differences in transactions are large enough; it is recognized as fraudulent activity. The activity information pertaining to account like time and location of transaction amount are included. The main advantage of this approach is, it can handle large amount of data without prior knowledge of historical data.

1.3 Classification Techniques for Credit Card Fraud Detection

Different classification techniques have been applied to detect the frauds that occur in credit card transactions.

a. Artificial neural network (ANN): It is known as a classifier that works on the concept of functioning of human brains [10]. The nodes are interlinked with each other in its architecture. The nodes that are present near to them are assigned with the weighted links of each other. It collects all the information sensed by the connecting nodes. Further, each node calculates the output values with the help of weights used with a simple function. A complex system that includes specialized tissues, cells, chemical molecules as well as organs generating an intricate network is known as the natural immune system [11]. The antibodies which have the capacity to identify and eliminate the harmful and risky antigens are known as the detectors of the immune system.

b. Artificial Immune System (AIS): On the basis of the biological metaphor of the immune system, a recent sub-field developed is the Artificial Immune System (AIS). Inspiring from the natural evolution, generating algorithms are designed lately [8]. The binary strings which denote the populations of candidate solutions are called the chromosomes. The chromosomes that have higher quality or the fitness value have higher chances of survival as per the concept of this classifier.

c. Hidden Markov model: It is known as a mechanism that generates a highly complex stochastic process with the help of a double embedded stochastic method. In the underlying system, the unobserved states are observed here [12].

d. k-NN: A non-parametric algorithm through which regression and classification are performed with high efficiency is known as k-NN. The k-nearest training samples available in the feature space are given as input for this classifier. Further, based on the classification or regression categories, the output is generated.

2 Literature Review

John O. Awoyemi et al. [14] presented a study through which the different fraudulent detection techniques designed to prevent credit card based frauds were investigated. A dataset that was used in this investigation was collected from the transactions made the European cardholders. In order to perform oversampling as well as under-sampling, the hybrid approach was proposed for the skewed data. Python was used here for applying three various techniques on both raw as well as preprocessing data. The evaluation of performances was done by calculating the performance parameters. The result outcomes showed that the proposed technique outperformed previous techniques since it provided higher accurate results. **Sharmistha Dutta et al.** [15] proposed a research that was based on identifying the crimes related to credit cards. For avoiding any identity theft few issues are faced when dealing with the non-data mining techniques. If, to prevent frauds, the non-data mining techniques are applied, few problems can be faced. To prevent such common issues, a new data mining layer was designed in this research. For generating this new layer, two algorithms known as Communal Detection and Spike Detection were applied. Large amount of time is needed to generate the results. Even with the continuous updates on algorithms, no actual evaluation is performed since the behaviors do not change depending upon the algorithms used. Thus, the adaptability concept cannot be visualized clearly here. The future work of this research can be to extend the proposed method for resolving such issues. **Kuldeep Randhawa et al.** [16] proposed a technique using machine learning to detect Credit Card Fraud Detection. The standard models were applied in the initial phase. Further, Adaboost and majority voting techniques were used for creating hybrid approaches that provided better outcomes. For evaluating the efficiency of model, the dataset available publically was utilized. The frauds were analyzed by also using another dataset generated from the financial institutions. Further, with the aim of measuring robustness of algorithms, the noise was added to data samples. The experiments were conducted on the basis of the theoretical results which show that the majority of voting methods achieve good accuracy rates in order to detect the fraud in the credit cards. Certain level of noise was added to the data sample in order to analyze the performance of hybrid approaches. For around 30% of additional noise, a score of 0.942 was achieved by few of the voting based approaches. These outcomes showed that even in the presence of noise, improved results could be achieved by proposed voting methods. **Abhimanyu Roy et al.** [17] proposed a novel approach that could detect the online transactional frauds. This approach was designed using the deep learning technology known as ANN. There were built-in components based on time and memory that provided improvements to the system. Around 80 million of the online transactions were categorized as being fraud as per the efficiency of components that detected fraud. The cloud computing scenario helped in improving the performance levels. Depending upon the performance of fraud detection, a guide for sensitivity analysis was provided related to proposed parameters. To detect the frauds, a new framework was designed by including deep learning topologies. So, by avoiding the fraud activities, the losses in financial organizations were reduced. **Shiyang Xuan et al.** [18] proposed a novel mechanism through which the behavior features of both normal and abnormal transactions were trained. Here, the two kinds of random forests were applied. Depending upon their classifiers the two random forest approaches were compared. This also affected the performance of detecting the frauds. For analyzing the performance of two kinds of random forest models, an e-commerce company of China was used. To recognize and detect the frauds from credit cards, B2C dataset was used in this paper. The random forests were applied on small dataset to evaluate the performances which showed that the improvement in outcomes. However, in comparison to other datasets few problems were still faced which could be eliminated in future.

3 Implementation of Proposed Hybrid Approach KNN & NB

In this section, we are discussing hybrid approach methodology are as following:

3.1 Working of KNN & NB

For resolving the regression and classification based issues, an easy supervised machine learning algorithm is designed known as k-NN. It was assumed that within the close proximity similar kinds of objects exist as per k-NN. Following are the steps performed in k-NN:

Step 1: Input the data.
Step 2: For selected number of neighboring values, initialize k.
Step 3: In the data, each of the examples are included.
Step 4: Among the query and current samples, distance is calculated.
Step 5: Sort the ordered collection of distances and indices from smallest to largest (in ascending order) by the distances
Step 6: Get the labels of selected K entries

To perform classification, k-NN can be applied and a class membership is given as output. Depending upon the majority vote of its neighbors, the most common class is assigned to the object. Regression can also be performed here. The values of k nearest neighbors are taken to calculate the average value.

For the recommendation systems, and in decision-making models, k-NN is applied very commonly. For recommending various kinds of services to users such as what kind of movie to watch or book to buy, this algorithm provides optimal results.

To be applicable in several classification mechanisms, the probabilistic machine learning algorithm applied is Naïve Bayes. Since the features included in the model are not based on each other, the term Naïve is added to this algorithm. The value of any other feature is not affected directly which is an altering value of the feature. Since there are important advantages with NB, this is considered to be the most powerful algorithm. It is possible to code the algorithm and make actual predictions, since this is a probabilistic model. For real-time based applications the user's requests are responded instantaneously due to which it is known to be a highly scalable and traditional algorithm.

3.2 Data Acquisition

UCI repository is use to collect the fraud transaction dataset on which experiments can be conducted. The continuous attributes that have less numbers of values are collected through this interesting dataset. However, some of the values are missing. To collect the dataset, there are few repositories available [19].

3.3 Research Methodology

This research work focuses on prediction of credit card fraud transactions using a hybrid classifier. The hybrid classifier combines KNN and Naïve Bayes algorithms. The KNN classifier will be used as the base classifier and naïve Bayes will be used as the meta-classifier. The dataset of fraud transactions will be collected from the UCI repository. The performance of the proposed approach will be analyzed in terms of certain parameters like accuracy, precision and recall and F-measure (Fig. 1).

The step wise methodology proposed in this research is explained below:

Step 1: To perform credit card fraud detection, dataset is collected from UCI repository to be given as input.
Step 2: To handle any kinds of missing or redundant values, data preprocessing is applied.
Step 3: The dataset is categorized among test and training sets by applying cross-validation method. In comparison to the test set, the training dataset is of larger size.
Step 4: The test set is predicted by applying k-NN classifier.
Step 5: The results generated from k-NN are given as input to Naïve Bayes classifier.
Step 6: Analyze performance of the hybrid model in terms of precision, recall, accuracy and f-measure

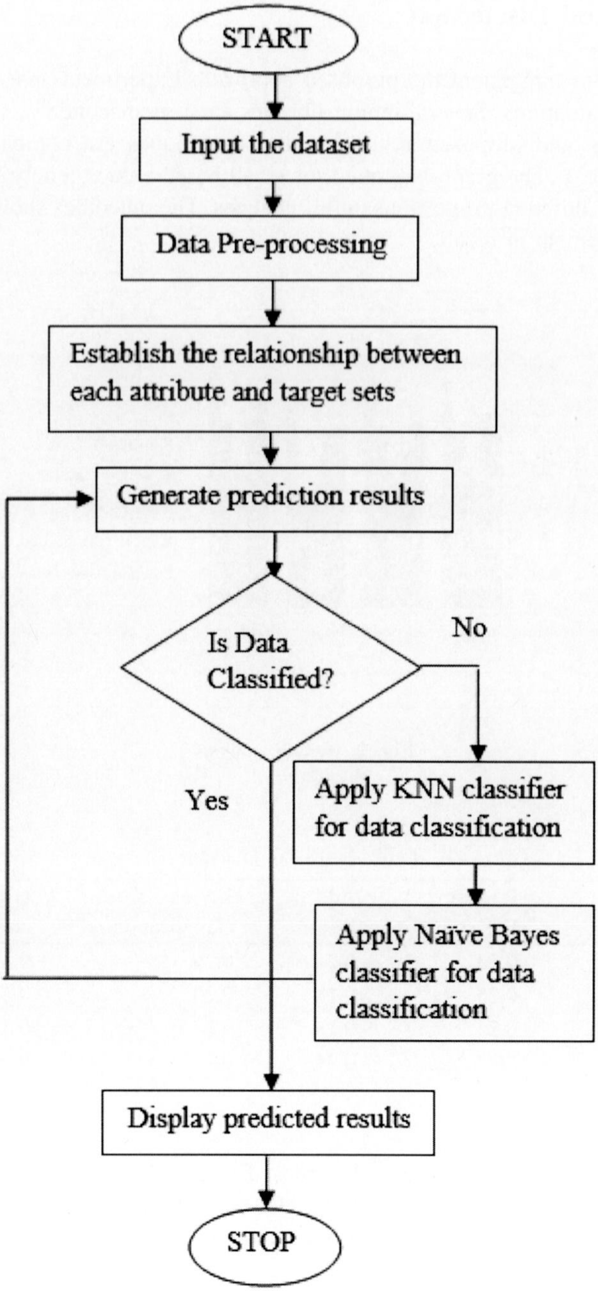

Fig. 1. Proposed flowchart

4 Results and Discussion

Python is used to implement the proposed approach. Experiments are conducted for performance evaluations. Several inbuilt libraries exist in anaconda.

The existing and proposed fraud detection techniques are compared as shown Fig. 2 and Table 1. The graph is plotted for recall based values. The training and test sets are taken at different proportions in this analysis. The outcomes show around 15 to 24% of improvement in recall.

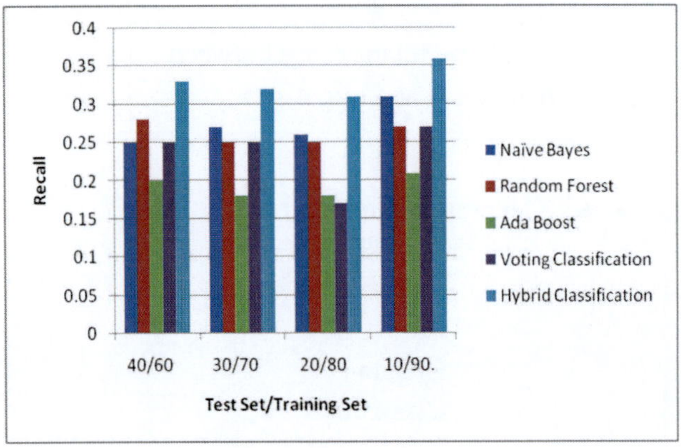

Fig. 2. Recall analysis

Table 1. Recall analysis

	Naive Bayes	Random Forest	Ada Boost	Voting classification	Hybrid classification
40/60	0.25	0.28	0.2	0.25	0.33
30/70	0.27	0.25	0.18	0.25	0.32
20/80	0.26	0.25	0.18	0.17	0.31
10/90	0.31	0.27	0.21	0.27	0.36

The performances of existing and proposed techniques in terms of precision are shown in Fig. 3 and Table 2. The training and test sets are taken at different proportions in this analysis. The outcomes show around 45% of improvement in precision for proposed technique.

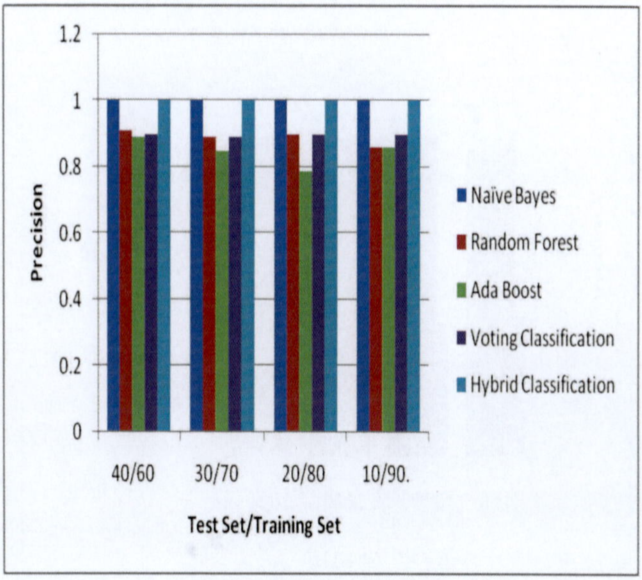

Fig. 3. Precision analysis

Table 2. Precision analysis

	Naive Bayes	Random Forest	Ada Boost	Voting classification	Hybrid classification
40/60	1	0.9	0.84	0.85	1
30/70	1	0.9	0.85	0.9	1
20/80	1	0.9	0.79	0.9	1
10/90	1	0.84	0.84	0.9	1

The performances of existing and proposed techniques in terms of accuracy are shown in Fig. 4 and Table 3. The training and test sets are taken at different proportions in this analysis. The outcomes show around 100% of improvement in accuracy for proposed technique.

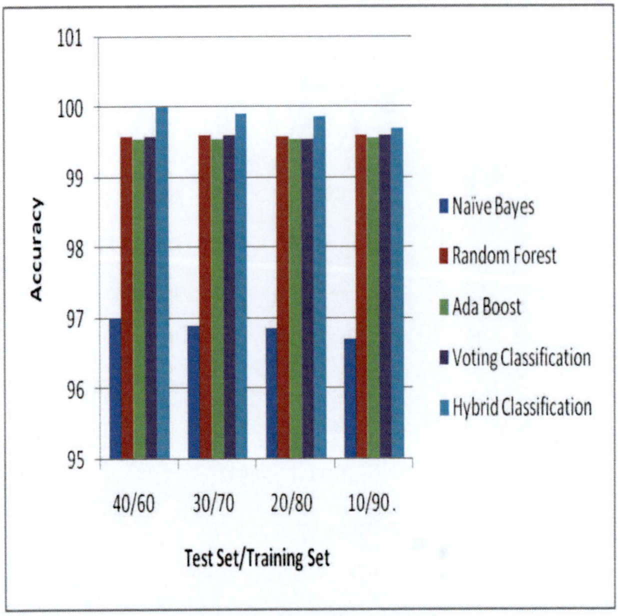

Fig. 4. Accuracy analysis

Table 3. Accuracy analysis

	Naive Bayes	Random Forest	Ada Boost	Voting classification	Hybrid classification
40/60	97	99.5	99.5	99.5	100
30/70	96.9	99.5	99.4	99.5	99.9
20/80	97.8	99.5	99.4	99.4	99.8
10/90	97.7	99.5	99.4	99.45	99.6

The performances of existing and proposed techniques in terms of execution time are shown in Fig. 5 and Table 4. The training and test sets are taken at different proportions in this analysis. The outcomes show around 2 s of reduction in execution time for proposed technique.

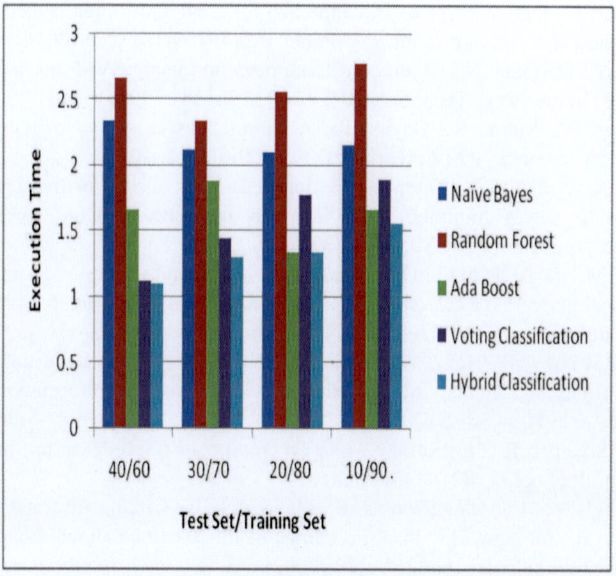

Fig. 5. Execution time

Table 4. Execution time analysis

	Naive Bayes	Random Forest	Ada Boost	Voting classification	Hybrid classification
40/60	2.3	2.6	1.6	1.2	1.1
30/70	2.1	2.5	1.9	1.4	1.3
20/80	2.1	2.55	1.3	1.8	1.3
10/90	2.1	2.8	1.7	1.9	1.6

5 Conclusion

A comparative analysis of proposed and existing techniques is presented by taking various proportions of test and training sets in this research. The results are calculated in terms of accuracy, execution time, recall and precision. The recall and precision are improved to 15–24% and 90–100% respectively when considering 40:60 proportions of test cases. The execution time is reduced to 2 s and accuracy of 100% is achieved as per this research. In the existing technique the accuracy of 99.56% which is increased to 100%. The precision, recall values of the existing methods are approx 80% which are also increased above 90%.

References

1. Raj, S.B.E., Portia, A.A.: Analysis on credit card fraud detection methods. In: International Conference on Computer, Communication and Electrical Technology, ICCCET, pp. 152–156 (2015)

2. Jain, R., Gour, B., Dubey, S.: A hybrid approach for credit card fraud detection using rough set and decision tree technique. Int. J. Comput. Appl. **139**(10), 1–6 (2016)
3. Agrawal, A.N., Dermala, N.: Credit card fraud detection using SVM and reduction of false alarms. Int. J. Innov. Eng. Technol. (IJIET) **7**(2), 176–182 (2016)
4. Phua, C., Lee, V., Smith, K., Gayler, R.: A comprehensive survey of data mining-based fraud detection research, arXiv preprint arXiv:1009.6119(2010)
5. Stojanovic, A., Aouada, D., Ottersten, B Bahnsen, A.C.: Cost-sensitive credit card fraud detection using Bayes minimum risk. In: 12th International Conference on Machine Learning and Applications, ICMLA, pp. 333–338 (2013)
6. Carneiro, E.M., Dias, L.A.V., Da Cunha, A.M., Mialaret, L.F.S.: Cluster analysis and artificial neural networks: a case study in Credit Card Fraud Detection. In: 12th International Conference on Information Technology-New Generations, pp. 122–126 (2015)
7. Aghili, S., Zavarsky, P., Hafiz, K.T.: The use of predictive analytics technology to detect credit card fraud in Canada. In: 11th Iberian Conference on Information Systems and Technologies, CISTI, pp. 1–6 (2016)
8. Bansal, M., Sonepat, H.C.E.: Survey paper on credit card fraud detection. Int. J. Adv. Res. Compu. Eng. Technol. **3**, 827–832 (2014)
9. Maes, S., Tuyls, K., Vanschoenwinkel, B., Manderick, B.: Credit card fraud detection using Bayesian and neural networks. In: Proceedings of the 1st International Naiso Congress on Neuro-Fuzzy Technologies, pp. 261–270 (2002)
10. John, S.N., Anele, C., Kennedy, O.O., Olajide, F., Kennedy, C.G.: Real-time fraud detection in the banking sector using data mining techniques/algorithm. In: International Conference on Computational Science and Computational Intelligence, CSCI, Las Vegas, NV, pp. 1186–1191 (2016)
11. Mahmud, M.S., Meesad, P., Sodsee, S.: An evaluation of computational intelligence in credit card fraud detection. In: International Computer Science and Engineering Conference, ICSEC, Chiang Mai, pp. 1–6 (2016)
12. West, J., Bhattacharya, M.: An investigation on experimental issues in financial fraud mining. In: IEEE 11th Conference on Industrial Electronics and Applications, ICIEA, Hefei, pp. 1796–180 (2016)
13. Zarrabi, H. Kazemi, Z.: Using deep networks for fraud detection in the credit card transaction. In: IEEE 4th International Conference In Knowledge-Based Engineering and Innovation, KBEI, pp. 0630–0633 (2017)
14. Awoyemi, J.O., Adetunmbi, A.O., Oluwadare, S.A.: Credit card fraud detection using machine learning techniques: a comparative analysis. In: International Conference on Computing Networking and Informatics, ICCNI, pp. 1–9 (2017)
15. Dutta, S., Gupta, A.K., Narayan, N.: Identity crime detection using data mining. In: 3rd International Conference on Computational Intelligence and Networks, CINE, Odisha, pp. 1–5 (2017)
16. Randhawa, K., Loo, C.K., Seera, M., Lim, C.P., Nandi, A.K.: Credit card fraud detection using AdaBoost and majority voting. IEEE Access **6**, 14277–14284 (2018)
17. Roy, A., Sun, J., Mahoney, R., Alonzi, L., Adams, S., Beling, P.: Deep learning detecting fraud in credit card transactions. In: Systems and Information Engineering Design Symposium, SIEDS, pp. 129–134 (2018)
18. Liu, G., Li, Z., Zheng, L., Wang, S., Jiang, C., Xuan, S.: Random forest for credit card fraud detection. In: IEEE 15th International Conference On Networking, Sensing and Control, ICNSC, pp. 1–6 (2018)

Improved Fast Block Matching Motion Estimation Using Multiple Frames

K. Priyadarshini(✉) and A. K. Thasleem Sulthana

Department of ECE, K. Ramakrishnan College of Engineering, Tiruchirapalli 621 112, Tamilnadu, India
kpriyaanilesh@gmail.com, akthasleemsulthanaece@krce.ac.in

Abstract. In video coding, the matching process for multiple reference frames is carried out and the movement across the frames in the sequence of the video is analyzed during motion estimation. An improved fast motion estimation algorithm is proposed using multiple frames for reference (MRF) as a substitute of a single frame used as reference (SRF). The results observed on simulation shows that the method put forward could speed up the steps involved and improve the PSNR value by 2.5 dB and to determine the motion vector with the reduced search space.

Keywords: Motion estimation · Multiple reference frames

1 Introduction

Multiple reference frame motion compensation is innovative technique for improving the performance of the video coding. Here, we have analyzed first that MRF provided better predictions, according to the observed empirical result. On standard test video sequences. An efficient multiframe fast search motion estimation algorithm, which can achieve nearly the same coding efficiency as the full-search approach is proposed.

2 Multiple Reference Frames

Multiple frames are considered during motion estimation processing as shown in Fig.

3 Proposed Design Steps

The proposed multiple frame analysis multi-frame motion estimation, reuses the motion information of every frame is updated to the preceding to obtain the vector of movement. Figure 1 demonstrates the optimal search among five frames. Motion vectors are generated without performing motion estimation by the "vector sum of MV".

J. I.-Z. Chen et al. (Eds.): ICIPCN 2020, AISC 1200, pp. 711–717, 2021.
https://doi.org/10.1007/978-3-030-51859-2_64

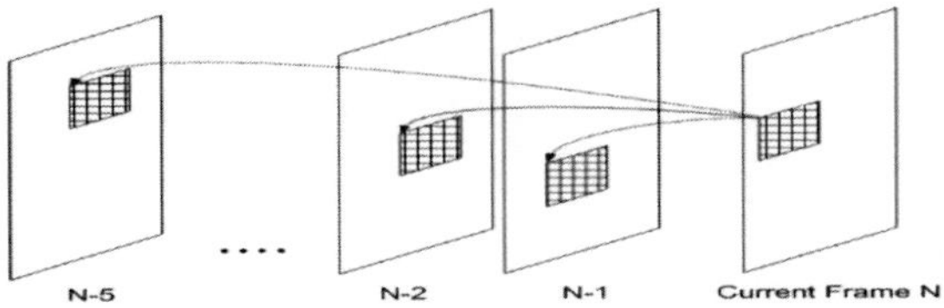

Fig. 1. Multiple reference frame ME

$$"MV^{(n-4)}_{estimated} = MV_1 + MV_2 + MV_3 + MV_4" \quad (1)$$

An efficient searching algorithm is proposed movement-vector information's of the prevailing frames of using Forward Dominant Vector Search (FDVS) approach.

The paper Su Y. et al. [2] explains a new aspect to improve "video coding performance using multiple reference frame motion estimation (MRFME) in H.264/MPEG-4 AVC. MRFME is an effective technique to improve the coding efficiency." "Several fast BMAs have been introduced to beat FS in terms of computational complexity. These include new three step search (N3SS) [4], four-step search (4SS) [5], diamond search (DS) [6], kite-cross-diamond search (KCDS) [7], modified DS (MODS) [8], and some recently proposed algorithms [9–12]. These fast motion estimation algorithms determines a solution with maximum time consumption. Hence these algorithms are carried out with modifications of multiple reference frames during processing.

4 Analysis of the Proposed Approach

This multiframe approach computational complexity is according to the number of frames used as references. Since the proposed multiple frame technique uses only Full search technique all the macro block modes are covered for comparison analysis so that the quality is not degraded during the process.

Performance analysis of this approach is measured using PSNR, and the average search points (Tables 2 and 3).

Figure 2 shows the standard sample test video sequences processed for 25, 50, 100 frames. Each frame has a macro block size of 16 × 16 with displacement inside the search range of (+7, −7) pixels in both directions.

Table 1. Standard test video sequences

Video sequence	Motion type	Frame size
Akiyo.QCIF	Medium	352 × 288
Container.QCIF	Low	352 × 288
Carphone.QCIF	Medium	352 × 288
Stefan.CIF	High	352 × 288
Foreman.QCIF	Medium	352 × 288
Bus.QCIF	High	352 × 288
News.QCIF	Low	352 × 288
Tennis.SIF	High	352 × 240
Mecid.QCIF	Medium	320 × 240
BQmall.CIF	High	160 × 120

Fig. 2. Standard sample test video sequences

5 Discussions on Results

The simulation analysis of the proposed work is carried out with standard the sequence of videos and the results are compared with FS, TSS, NTSS, FSS, and DS motion estimation algorithms. To analyse the quality performance PSNR value comparison and the average search point's computation represents the speed of the work. Table 2 indicates the PSNR of single reference frame and multiple reference frame for Full search using FDVS approach. Table 3 indicates the assessment PSNR value of SRF and MRF among block matching algorithms. Observation results shows that average PSNR values of various motion estimation algorithms are in the order as FS > TSS > DS > FSS for standard video test sequences. The cost metrics PSNR value of MRF shows that 2.5 dB improvement is achieved when compared with SRF in the FS method. The results obtained shows the average amount of search space of various motion estimation algorithm in the sequence FS > TSS > FSS > DS. The proposed method of MRF analysis using Full search method requires less number of search points when compared with SRF Full search method.

Table 2. Average PSNR value of SRF and MRF using FDVS approach

Standard video test sequence	Full Search (FS)	
	PSNR (dB)	
	MRF	SRF
Akiyo	54.66	52.77
Container	54.75	50.70
Stefan	45.75	43.97
Foreman	49.95	47.37
Bus	45.09	41.21
News	52.39	48.90
Mecid	48.07	45.94
BQmall	51.75	48.30
Mobile	43.06	43.82
Total average	49.49	46.99

Table 4 is search space of SRF as well as MRF using FDVS approach. Table 5 is the search space of SRF and MRF for BMA and Table 6 shows the comparison analysis of multiframes in terms of PSNR for Full search using FDVS approach with previous work.

Table 3. Comparison of average PSNR value of SRF and MRF among BMA

Standard video test sequence	TSS		FSS		DS	
	PSNR (dB)		PSNR (dB)		PSNR (dB)	
	MRF	SRF	MRF	SRF	MRF	SRF
Akiyo	42.61	41.25	42.62	41.26	42.63	41.32
Container	40.75	40.09	40.75	40.10	40.75	40.10
Stefan	35.09	34.37	35.00	34.24	35.00	34.26
Foreman	39.47	37.64	39.52	37.60	39.64	37.85
Bus	33.25	32.01	33.01	31.83	33.17	31.94
News	40.83	39.38	40.82	39.32	40.84	39.33
Mecid	37.43	36.27	37.27	36.20	37.14	36.25
BQ mall	40.71	37.86	40.60	37.82	40.66	38.02
Mobile	34.38	33.48	34.52	33.48	34.61	33.38
Total average	38.28	36.92	38.23	36.87	38.27	36.93

Table 4. Average search points of SRF and MRF of FS using FDVS approach

Test sequence	Full Search (FS)	
	Search points	
	SRF	MRF
Akiyo	231.95	231
Container	229.69	227
Stefan	229.79	227
Foreman	227.46	227
Bus	224.64	218
News	230.54	230.38
Mecid	226.85	222
BQmall	231.13	230.98
Mobile	229.78	226
Total average	229.02	226.59

Table 5. Comparison of average search points of SRF and MRF among BMA

Test sequence	TSS		FSS		DS	
	Search points		Search points		Search points	
	SRF	MRF	SRF	MRF	SRF	MRF
Akiyo	22.80	22.81	15.67	16.01	6.38	6.65
Container	22.80	22.84	15.66	16.42	6.43	7.05
Stefan	22.94	22.97	18.53	19.19	9.28	10.10
Foreman	22.96	23.09	18.73	21.11	9.09	12.04
Bus	20.06	23.43	20.66	21.29	11.02	11.87
News	22.79	22.79	15.76	16.21	6.51	6.97
Mecid	22.81	22.82	16.16	16.35	6.80	7.14
BQmall	22.81	22.82	15.88	16.00	6.59	6.82
Mobile	22.84	23.05	17.00	20.19	7.48	9.94
Total average	22.53	22.95	17.11	18.08	7.73	8.73

Table 6. Comparison of multi-frames in terms of PSNR for Full search using FDVS approach

References	Improvement in cost metrics
Chen et al. [1]	0.27 dB higher than SRF
Yeping et al. [2]	0.5 dB higher than SRF
Mateus Grellert etal. [3]	2.1 dB higher then SRF
Proposed method	2.5 dB higher than SRF

6 Conclusion

Our conclusion on this paper, is based on multiple reference frames instead of single frame. In the designed multi-frame motion estimation procedure, the movement vectors for the single and the multiple frames are acquired using the standard sequence of video samples. The examination is done based on the value of PSNR. Simulation results show that 2.5 dB improvements in the value of PSNR is achieved using the reference frame that is single and multiple using five frames. For the proposed method of MRF analysis using Full search, requires less number of search points when compared with SRF full search method. Experimental highlights the important enhancements of the scheme put forth.

References

1. Chen, M.J., Chiang, Y.Y., Li, H.J., Chi, M.C.: Efficient multi-frame motion estimation algorithms for MPEG-4 AVC/JVT/H. 264. In: 2004 IEEE International Symposium on Circuits and Systems (IEEE Cat. No. 04CH37512), vol. 3, pp. III-737. IEEE (2004)

2. Su, Y., Sun, M.-T.: Fast multiple reference frame motion estimation for H. 264/AVC. IEEE Trans. Circuits Syst. Video Technol. **16**(3), 447–452 (2006)
3. Grellert, M., Sampaio, F., Hecktheuer, B., de Mattos, J.C., Agostini, L.: Memory-aware multiple reference frame motion estimation for the H. 264/AVC standard. In: 2010 17th IEEE International Conference on Electronics, Circuits and Systems, pp. 575–578. IEEE (2010)
4. Li, R., Zeng, B., Liou, M.L.: A new three-step search algorithm for block motion estimation. IEEE Trans. Circuits Syst. Video Technol. **4**(4), 438–442 (1994)
5. Po, L.-M., Ma, W.-C.: A novel four-step search algorithm for fast block motion estimation. IEEE Trans. Circuits Syst. Video Technol. **6**(3), 313–317 (1996)
6. Zhu, S., Ma, K.-K.: A new diamond search algorithm for fast block-matching motion estimation. IEEE Trans. Image Process. **9**(2), 287–290 (2000)
7. Lam, C.W., Po, L.M., Cheung, C.H.: A novel kite-cross-diamond search algorithm for fast block matching motion estimation. In: 2004 IEEE International Symposium on Circuits and Systems (IEEE Cat. No. 04CH37512), vol. 3, pp. III-729. IEEE (2004)
8. Yi, X., Ling, N.: Rapid block-matching motion estimation using modified diamond search algorithm. In: 2005 IEEE International Symposium on Circuits and Systems, pp. 5489–5492. IEEE (2005)
9. Tourapis, H.Y., Tourapis, A.M.: Fast motion estimation within the H. 264 codec. In: 2003 International Conference on Multimedia and Expo. ICME 2003. Proceedings (Cat. No. 03TH8698), vol. 3, pp. III-517. IEEE (2003)
10. Kuo, C.-M., Kuan, Y.-H., Hsieh, C.-H., Lee, Y.-H.: A novel prediction-based directional asymmetric search algorithm for fast block-matching motion estimation. IEEE Trans. Circuits Syst. Video Technol. **19**(6), 893–899 (2009)
11. Lee, S.: Fast motion estimation based on adaptive search range adjustment and matching error prediction. IEEE Trans. Consum. Electron. **55**(2), 805–811 (2009)
12. Luo, J., Ahmad, I., Liang, Y., Swaminathan, V.: Motion estimation for content adaptive video compression. IEEE Trans. Circuits Syst. Video Technol. **18**(7), 900–909 (2008)

Efficient Framework for Identification of Soybean Disease Using Machine Learning Algorithms

Sachin Jadhav[1(✉)], Vishwanath Udupi[2], and Sanjay Patil[3]

[1] Bharati Vidyapeeth's College of Engineering, Kolhapur, India
Sachinbjadhav84@gmail.com
[2] Maratha Mandal Engineering College, Belagavi, India
vrudupil@gmail.com
[3] Rajgad Shri. Chattrapati Shivajiraje College of Engineering, Bhor, Pune, India
patilsbp@gmail.com

Abstract. This paper exhibits the detection and classification framework of soybean leaflet diseases. Identification and classification were performed utilizing an k-means algorithm and a multiclass support vector machine (SVM). Healthy and unhealthy leaflets infected by frogeye leaf spot, bacterial blight, and Septoria brown spot diseases were collected from a soybean field. The image database is developed by acquiring images with a constant background using a digital camera under the control environment. The image preprocessing techniques applied to the ROI of the raw image. After that, the partition of the diseased region is done using an k-means segmentation technique. The color and texture features were extracted from the segmented leaf region. The mean and standard deviation of RGB channels estimated to extract color features, and texture features were extracted using a (GLCM) method to define an image feature database. Finally, SVM was employed to identify soybean disease. The accuracy of the intended framework for classifying the defined disease was 90.20%.

Keywords: Machine learning · K-means · ROI · Segmentation · GLCM · SVM

1 Introduction

Soybean is a vital oilseed crop in India and has been reported to improve the social and economic conditions of small and marginal farmers. Soybean provides forty and twenty-five percent concerning the whole oilseed production, of India respectively. Moreover, this export of (soy meal) provides valuable foreign exchange [1]. Maharashtra, Karnataka, Rajasthan, Andhra Pradesh, Chhattisgarh, and Madhya Pradesh, are the major soybean-growing states. The contribution of these states together approximately 98% of the entire soybean production in India.

Leaf infections, such as rust, frogeye leaf spot, Septoria brown spot, downy mildew, bacterial blight, powdery mildew, and anthracnose can make immense damages in soybean production [2]. Figure 1 depicts the leaf disease samples, of soybean from a

J. I.-Z. Chen et al. (Eds.): ICIPCN 2020, AISC 1200, pp. 718–729, 2021.
https://doi.org/10.1007/978-3-030-51859-2_65

defined image database. The aforementioned leaf diseases befall at the leaflets, stem, including seed from the soybean plant. The visible indications of the diseases turn their color, shape, and size continuously while the disease grows [5].

Leaf Blight	Frogeye Spot	Brown Spot

Fig. 1. Disease samples of soybean from defined image dataset. Three different soybean fungal diseases from left to right show the leaf blight disease, frogeye spot, and brown spot.

The severity of the diseased region is evaluated by computing the number of pixels in the diseased traits region and the number of pixels in the whole leaf region. The percentage should be applied in disease evaluation methods (as severity measurement) [3]. Manual assessment methods exhibit a certain degree of subjectivity; Therefore, the result from the manual assessment may not be entirely reliable.

To devise an automatic soybean leaf disease analysis method that can be used to acquire soybean leaf images under control environment, the following challenges summarized by Barbedo (2013) must be addressed: (1) Unmanagable acquisition conditions(illumination, image quality); (2) Various infection occurrences in whole image, and (3) similarities into features within distinct disease levels; For example, Soybean brown spot and bacterial blight are difficult to distinguish during diagnosis because the affected plants exhibit similar symptoms.

Limited research has been conducted to address the challenges in soybean disease investigation. The aim of this research is to implement an automatic soybean leaf disease detection method with severity measurement and diagnosis mechanisms to identify and classify disease categories and quantify disease severity. The classified and quantified diseases can then be graded.

This article is composed as ensues:- Sect. 2 presents a concise survey of the relevant research in plant disease detection, classification, and severity measurement. Section 3 outlines the techniques used for soybean plant disease severity measurement. This section includes the following topics: as follows, leaf image acquisition, leaf image pre-processing technique, leaf disease detection, and leaf disease classification method. Section 4 involves the results and discussion, and, Sect. 5 gives the conclusion.

2 Related Studies

Many researchers have developed various image processing and computer vision-based models for identifying and classifying agriculture crop diseases. However, limited research has been conducted on severity measurement in agriculture crop disease identification. Precise but diminutive research is found in the domain of severity measurement in agriculture crop disease identification.

Meunewjinda et al. [4] used multiple artificial intelligence techniques for the automatic detection of grape leaf disease. An employed method comprised three foremost processes: (1) grape leaf color segmentation, (2) disease segmentation of grape leaf, and (3) classification of grape leaf diseases. The background information of grape leaf was segmented. Moreover, the backpropagation neural network model was employed based on color features. For grape leaf disease segmentation, a genetic algorithm was used SVM classifier to analyze disease features.

Camargo et al. [5] presented a technique for image segmentation in which the diseased regions were effectively detected. The proposed method accurately identified the target (diseased region) in images with varying intensity distribution [5].

Kim et al. [6] used the color texture applied in image smoothing for classifying grapefruit peel diseases. (SGDM) spatial gray-level dependence matrices methods were utilized for determining significant texture features, and the squared distance technique classification achieved a classification accuracy of 96.7%. Kim et al. considered various grapefruit diseases, such as cancer, copper burn, greasy spot, melanoses, and wind scar.

Al Bashish et al. [7] introduced the disease detection framework for diseases that occur on the leaves and stems of plants. The k-means segmentation method was employed for segmentation of disease, and a neural network classifier was used to classify segmented disease images, which resulted in an average accuracy of 93%.

Patil and Bodhe [8] presented a simple threshold and triangle threshold methods for segmentation. The methods proposed in the study were applied to fragment the disease area and leaf area. The severity of the disease was determined by estimating the whole leaf area and disease area.

Patil and Kumar presented a survey report on various image processing methods used to study plant diseases and traits [9]. The survey report revealed that the studied methods focused on increasing accuracy and reducing the subjective bias arising from human experts when diagnosing plant diseases.

Wu and Zhang [10] presented a novel recognition technique based on a multiclass SVM for the precise analysis of fruits. The fruit images obtained using a camera. Then, the split-and-merge method applied to eliminate the image background. The texture, color histogram, and shape features of individual fruit images were obtained. The principal component investigation was used to diminish the feature space. Lastly, various sorts of multiclass SVM were formed.

Phadikar et al. [11] presented a computerized system for detecting the blast and brown spot diseases in rice crops. The detection of infected and non-infected leaves was related to the count of peaks in the histogram technique. SVM and Bayes classification techniques were applied to classify rice leaf diseases which give accuracies of 79.5% and 68.1% respectively.

Barbados [12] conducted a survey on digital image processing methods employed to quantify, identify, and analyze diseases that occur on plants. In the survey, the identification, computation, severity, and analysis techniques of plant diseases were presented.

Ratnasari et al. [13] introduced a method for identifying and classifying sugarcane leaf disease and estimating its severity. The experimental results indicated that wide lesion results in high severity. An SVM classifier employed based on a feature database of texture and color to detect sugarcane disease in leaf images. The proposed model was highly accurate in identifying sugarcane leaf disease.

Gharge and Singh [14] used the k-means technique and neural network classifier to segment and classify leaf diseases, respectively. The diseases were graded using the region-props Matlab function. The recommended algorithm attains excellent efficiency in analysis.

Singh and Misra [15] conducted a survey on various methods employed for the detection and classification of plant leaf disease. They employed an image segmentation technique for the automated identification and classification of leaf diseases. The proposed algorithm was examined in various plant species. The proposed algorithm gives high efficiency in the identification and classification of leaf diseases. Moreover, Singh and Misra suggested numerous techniques, such as fuzzy logic, Bayes classifier and artificial neural network to enhance efficiency in future studies.

Patil et al. [16] presented an image retrieval system, which was developed to retrieving unhealthy leaflets of soybean. The texture, color and shape features did extract from an input image. The color histogram techniques utilized to obtain color features. The shape features were extracted using a scale-invariant feature transform. The combined local binary pattern (LBP) and Gabor filter techniques employed to extract texture features. With the proposed novel method, retrieval efficiencies of approximately 96%, 68%, and 76% were achieved for three soybean leaf diseases [16]. The features extracted may be used with a special classifier to enhance the achievement of the system.

Kaur et al. [17] presented a survey report on automatic detection and classification of leaf disease with computer vision and digital image processing techniques. The studies surveyed revealed various important research aspects. The authors discussed common techniques of a disease identification process. The realization of the implemented methods was investigated to diagnose the crop disease.

3 Materials and Methods

This work presents a model for disease identification, classification, and measurement of severity. The model was consist of two stages: (1) training stage and (2) testing stage. These stages had the first four of the following five stages in common:- (1) The first stage involved extraction of defined color and texture features of training and test image data of each disease class from the disease class database image set into a defined feature set; (2) The second stage included image preprocessing and image enhancement. Image preprocessing involved resizing the leaf images from a defined database into a proper form. The background of the leaf object was eliminated and the

ROI image was taken for k-means segmentation. The segmented diseased regions were considered as targets region. (3) In this stage recognized leaf area within the image was partitioned in four clusters. Amongst these, one of the clusters was likely to suit as a diseased region. (4) The fourth stage involved feature extraction. The numerous texture and color features obtained from the selected clustered sample. The extracted information was used to define the feature database. (5) The fifth stage involved classification. The SVM classifier was trained with the defined feature database and their corresponding target values. This trained classifier was then employed to analyze the test database. Figure 2 depicts a system concerning the whole approach for disease analysis.

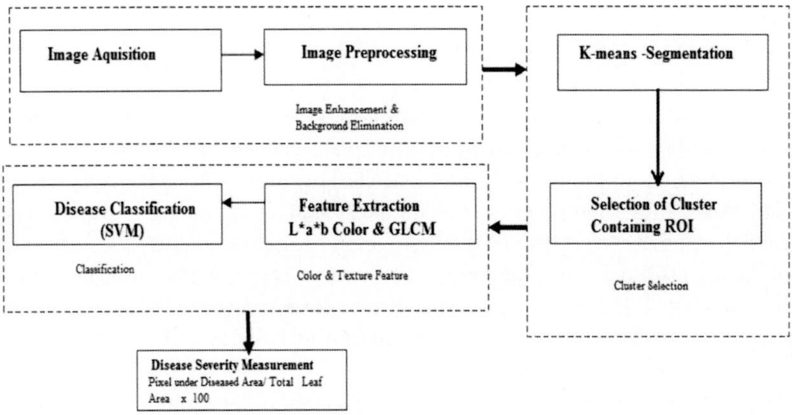

Fig. 2. Disease identification system using proposed approach.

3.1 Leaf Image Database

The leaf image dataset was developed and employed in this work, did collected of the soybean field of agriculture center Kolhapur. The images of soybean leave taken on a white background under a closed environment using a digital camera. These images consist of infected samples of Septoria brown spot, frogeye leaf spot, and bacterial blight were used in the training and test image database.

3.2 Leaf Image Preprocessing

The various digital image techniques such as contrast enhancement, background elimination, image filtering and image resizing applied to a collection of leaf images from a dataset. The input leaf image was resized to a size of 256 × 256 pixels. Figure 3 shows the output of contrast enhancement in the R, G, and B color planes of the RGB image. The irregular contrast of the image and any unexpected distortion was overcome. After that image filtering was employed. The median filter was applied to improve the highlighted diseased regions. Afterimage filtering and noise removal, background elimination was made. In this step, frequently the green-colored pixels were masked.

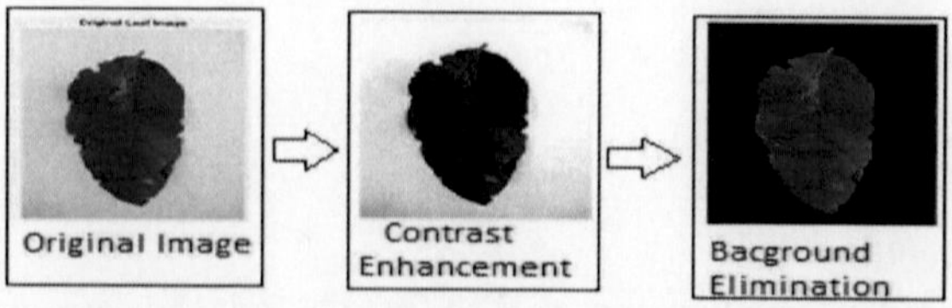

Fig. 3. Image preprocessing on set of database image

3.3 Leaf Image Segmentation

At first, the sample leaf image applied as an input to the system pre-processed. After that, the Otsu threshold-based segmentation techniques applied to eliminate the background of the leaf object of R, G, B color channel [13]. The magnitude of a threshold obtained from the masked pixels. Then the k-means segmentation technique employed to segment the ROI from the whole image.

3.3.1 K-Means Segmentation Approach

In k-means cluster analysis, data objects are grouped according to objects and their relationships. To partition the leaf object image into three clusters, k-means clustering was employed where one or more clusters comprised the disease region and the other(s) consisted of the healthy leaf object region [18, 20]. The algorithm attempts to classify pixels according to a collection of features in k number of groups. The basic operation of k-means is as follows (Zhang, 2012):

1. Clustering method based on Partition.
2. Every cluster is connected with a centroid.
3. Every point is allocated to the cluster with the nearest centroid
4. K - indicates the number of clusters, necessity is defined.

For this work the classification Squared Euclidean Distance method used; this is minimizing the sum of squares of distances among the data points and the corresponding cluster [5]; which is depicted in Eq. 1 below.

Euclidean Distance

d(p, q) indicates the distance among p^{th} and q^{th} the pixels of the similar cluster.

$$d(p, q) = \sqrt{|x_{p1} - x_{q1}|^2 + |x_{p2} - x_{q2}|^2 \ldots + |x_{pl} - x_{ql}|^2} \tag{1}$$

Calculation of Centroid (Mean) Point

The calculation of n-dimensional centroid or mean of the cluster is done by using Eq. 2 below.

$$CP(x_{1,}, x_{2,} \ldots, x_k) = \left(\frac{\sum_{i=1}^{k} x1st\, i}{k}, \frac{\sum_{i=1}^{k} x2nd\, i}{k}, \ldots.., \frac{\sum_{i=1}^{k} xnth\, i}{k} \right) \tag{2}$$

3.3.2 Disease Detection System

a) Disease detection

The system is classified into two phases: The first phase is the detection of leaf disease based on the defined test image dataset. The second phase is the measurement of severity. At first K-means segmentation technique is employed on ROI image, which partition the whole leaf object in 3 clusters. The user has to select a cluster that contains pixels to belongs diseased region only. From the selected cluster numerous features are extracted to form a feature database (knowledge database) which is further used to train SVM Classifier for disease analysis; this is shown in Fig. 4. The SVM classifier determines the disease class based on the obtained feature.

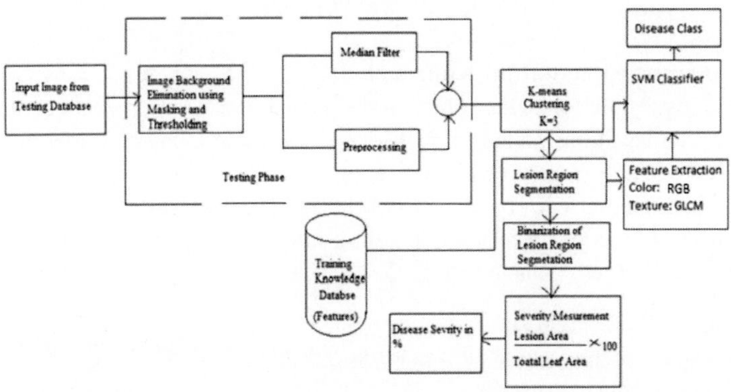

Fig. 4. Disese identification and classification system at testing phase.

3.3.3 Feature Extraction

The texture and color features are extracted to obtain better efficiency. For feature extraction, the methods computed values mean, a standard deviation of RGB color channels are employed in color and texture feature extraction GLCM method is used [19]. Based on the GLCM-kurtosis and skewness, total eight statistical parameters including color features are computed with eight feature vectors.

3.3.4 SVM-Multiclass Classification

In this work for the recognition and analysis of numerous leaf diseases of the soybean plant, the k-means clustering technique was composed of a multiclass SVM classifier. Leaf disease analysis is related to the future value of the target class. The given data pattern compared with one of the different classes. Leaf disease classification includes two stages, training and test of leaf image database using the SVM multiclass classifier. During the training stage, an SVM classifier is trained to apply feature values and their corresponding target class values. The SVM classifier applied to analyze test images from the test image dataset. In this work the segmented ROI, leaf spot images were used to train and test a multiclass SVM classifier.

4 Results and Discussions

4.1 Leaf Image Database

The soybean leaflets of distinct diseased and healthy types are collected from the Agriculture center of research Kolhapur Maharashtra. A total of 244 soybean leaf image samples considered for the training of the SVM Classifier. Out of which 80 leaf image samples analyzed for testing of image data set. Figure 1 depicts the leaflet samples soybean plant. Table 1 gives the particulars of the defined dataset.

Table 1. Classification result of multiclass SVM classifier

Leaf disease catagories	Number of leaf samples from database	Accuratly classified samples	% accuracy
Bacterial blight	55	49	89.1
Frog leaf spot	40	34	85
Septoria brown spot	49	37	75.5
(Non-desaesed) healthy	100	100	100
Composed 4 data type	244	220	90.23

4.2 K-Means Segmentation and Otsu Thresholding

4.2.1 K-Means Segmentation

The K-means method is applied to a filtered image that partitioned into three different-clusters. Figure 5 presents a segmented image into three clusters with ROI diseased cluster selection. The next step is the determination of an infected cluster from the segmented ROI image using an k-means algorithm. After that features are extracted for disease investigation.

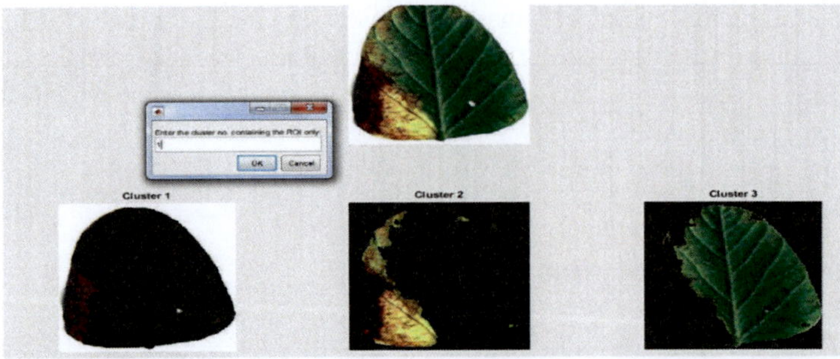

Fig. 5. Example of selection of bacterial blight disease cluster containing ROI.

4.3 Leaf Disease Classification

The concluding stage is identifying the defined soybean leaf disease class from the segmented images by employing the multiclass SVM classifier. From these segmented, leaf images the texture and color features were extracted. Further, the feature vector database is defined and, are applied as vector input to a trained multiclass SVM classifier. Then SVM analyzes the data input to four defined categories, specifically leaflet with frogeye Spot, Brown spot, Blight and healthy. A total of 244 leaf image samples analyzed to train the multiclass SVM classifier system. Out of 244 leaf image samples, 24 leaf image samples were misclassified. Figure 6 presents a Sophisticated Confusion matrix of Predicted Vs True class with TPR/FNR. So, classification efficiency for leaf type 1 is 89.1%, leaf type 2 is 85% leaf type 3 is 75.5% and leaf type 4 is 100% individually summarized in Table 1.

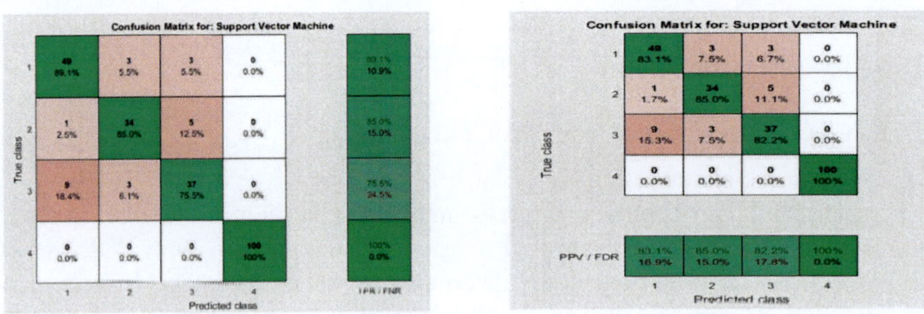

(a) (b)

Fig. 6. (a) Sophisticated confusion matrix (predicted vs true class) with TPR/FNR. (b) Sophisticated confusion matrix of predicted vs true class.

4.4 Confusion Matrix and ROC

Figure 6(a) presents the sophisticated confusion matrix of the multiclass SVM classifier. It furnishes the results of correctly and incorrectly specified disease type. It confers the percentage per true class including (TPR) true positive rates and (FNR) False-negative rates [21]. The progress rate of the multiclass SVM classification is 90.2%.

Figure 6(b) presents a sophisticated confusion matrix of the multiclass SVM Classifier, it gives (PPV) positive predictive values and (FDR) False Discovery Ratesof an overall percentage per predicted class. Figure 7 presents the sophisticated confusion matrix of the SVM classifier for the disease classification, it gives percentages over the whole confusion matrix.

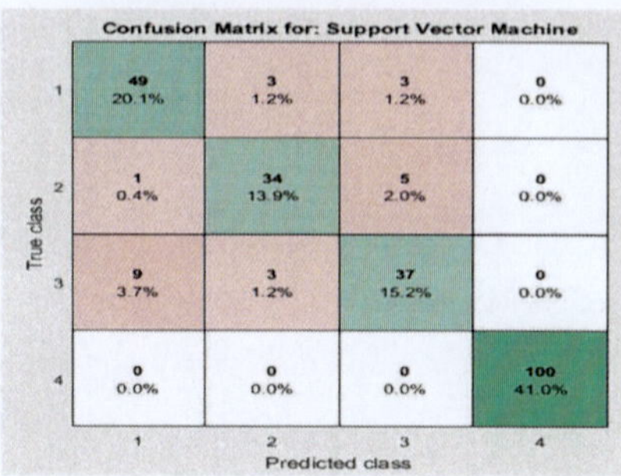

Fig. 7. Sophisticated confusion matrix

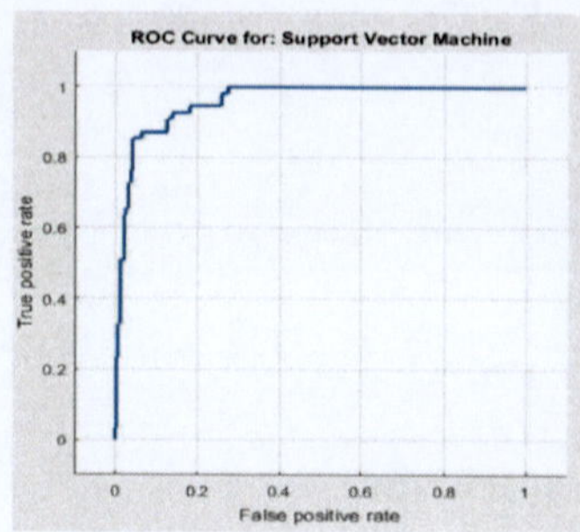

Fig. 8. Area under curve positive class 1 with TPR −0.96

ROC

The Receiver Operating Characteristics shows the Area Under Curve to specifically assigned disease class and non-diseased class. The classification accuracy of specific disease class depicted as the variation of the area under curves with the intended positive class is shown in Fig. 8 Example of Area Under Curve Positive Class 1 with TPR −0.961. Similarly, obtained for class2, class 3 and class4 respectively. The output of SVM classifier is abstracted in Fig. 9.

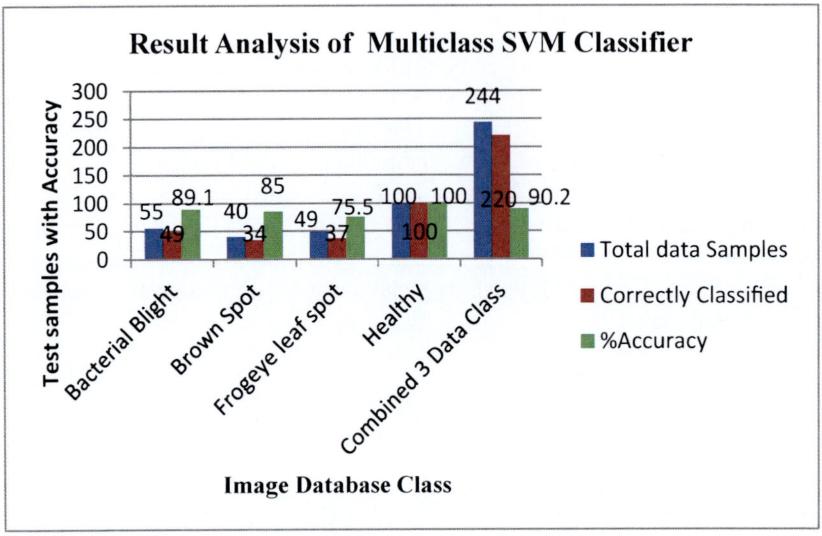

Fig. 9. Performance analysis of SVM classifier

5 Conclusion

We introduce a method for identifying and classifying soybean leaf disease and estimating its severity. The various image preprocessing methods such as image resizing, thresholding, and median filtering applied to the system. The k-means clustering technique was employed to segment the infected leaf area. Various color and texture features were extracted using the GLCM and RGB color space method. The Multi-class SVM classifier was applied to identify the class of soybean leaf. We tested the model on three distinct disease classes, specifically frogeye leaf spot, Septoria brown spot, bacterial blight, and non-diseased class. The multiclass SVM classified leaf disease by the efficiency of 90.20%. The results showed that the proposed framework method gives precise and automatic identification of soybean leaf disease. We intend to expand the size of the random and clean databases in future for covering increased disease types and achieving superior results.

Acknowledgements. This work is recommended by Dr.V.S.Patil, Assistant Professor, College of Agriculture Kolhapur, Maharashtra India.

References

1. Agarwal, D.K., Billore, S.D., Sharma, A.N., et al.: Soybean: introduction, improvement, and utilization in India—problems and prospects. Agric. Res. **2**(4), 293–300 (2013). https://doi.org/10.1007/s40003-013-0088-0
2. Godoy, C.V., Koga, L.J., Canteri, M.G.: Diagrammatic scale for assessment of soybean rust severity. Fitopatol. Bras. **31**(1), 63–68 (2006)

3. Vincelli, P., Hershman, D.E.: Assessing foliar diseases of corn, soybeans, and wheat. Principles and Practices PPFS-MISC-06 (2011)
4. Meunkaewjinda, A., Kumsawat, P., Attakitmongal, K., Srikaew, A.: Grape leaf disease detection from color imagery using hybrid intelligence system. In: IEEE Proceedings of ECTI-CON, pp. 513–516 (2008)
5. Camargo, A., Smith, J.S.: An image-processing based algorithm to automatically identify plant disease visual symptoms. Biosyst. Eng. **102**, 9–21 (2009). Research Paper: AE–Automation and Emerging Technologies
6. Kim, D.G., Burks, T.F., Qin, J., Bulanon, D.M.: Classification of grapefruit peel diseases using color texture feature analysis. Int. J. Agric. Biol. Eng. **2**(3), 41–50 (2009)
7. Al Bashish, D., Braik, M., Bani-Ahmad, S.: A framework for detection and classification of plant leaf and stem Diseases. In: IEEE International Conference on Signal and Image Processing, pp. 113–118 (2010)
8. Patil, S.B., Bodhe, S.K.: Leaf disease severity measurement using image processing. Int. J. Eng. Technol. **3**(5), 297–301 (2011)
9. Patil, J.K., Kumar, R.: Advances in image processing for detection of plant diseases. J. Adv. Bioinform. Appl. Res. **2**(2), 135–141 (2011). ISSN 0976-2604
10. Zhang, Y., Wu, L.: Classification of fruits using computer vision and a multiclass support vector machine. Open Acess Sens. **12**, 12489–12505 (2012). https://doi.org/10.3390/s120912489. ISSN 1424-8220
11. Phadikar, S., Sil, J., Das, A.K.: Classification of rice leaf diseases based on morphological changes. Int. J. Inf. Electron. Eng. **2**(3), 460–463 (2012)
12. Garcia, J., Barbedo, A.: Digital image processing techniques for detecting, quantifying and classifying plant disease. A Springer open Journal, SpringerPlus **2**, 660, 1–12 (2013). http://www.springerplus.com/content/2/1/660
13. Ratnasari, E.K., Mentari, M., Dewi, R.K., Ginardi, R.V.H.: Sugarcane leaf disease detection and severity estimation based on segmented spots image. In: IEEE International Conference on Information, Communication Technology and System, pp. 93–98 (2014)
14. Gharge, S., Singh, P.: Image processing for soybean diseaseclassification and severity estimation. In: Shetty, N., Prasad, N., Nalini, N. (eds.) Emerging Research in Computing, Information, Communication and Applications, pp. 493–500. Springer, New Delhi (2016)
15. Singh, V., Misra, A.K.: Detection of unhealthy region of plant leaves using Image processing and genetic algorithm. In: IEEE International Conference on Advances in Computer Engineering and Applications (ICACEA), pp. 1028–1032 (2015)
16. Patil, J.K., Kumar, R.: Analysis of content based image retrieval for plant leaf diseases using color, shape and texture features. Eng. Agric. Environ. Food **10**, 69–78 (2017)
17. Kaur, S., Pandey, S., Goel, S.: Plants disease identification and classification through leaf images: a survey. Arch. Comput. Methods Eng. **26**(2), 507–530 (2018). https://doi.org/10.1007/s11831-018-9255-6
18. Zhang, Z.: K-means algorithm-cluster analysis in data mining, pp. 1–16 (2012)
19. Zisserman, A.: Supervised learning support vector machines. www.inf.unibz.it/~mkacimi/svm.pdf
20. Jadhav, S., Patil, S.: Grading of soybean leaf disease based on segmented image using K-means clustering. IAES Int. J. Artif. Intell. (IJ-AI) **5**(1), 13–21 (2016). ISSN 2252-8938
21. Bishop, C.: Pattern Recognition and Machine Learning. Springer (2006)

Brain Image Visualization

Anson Antony, Sahil Bhirud, Abhishek Raj, and Niranjan Bhise(✉)

Rajarshi Shahu College of Engineering, Maharashtra 411033, India
ansonanto97@gmail.com, sahilbhirud@gmail.com,
abhishektiwarisbg@gmail.com,
niranjanbhise23@gmail.com

Abstract. This paper intends to highlight a solution that has been obtained by the amalgamation of machine learning and brain image visualization. This solution has been presented to improve the comprehension of a brain's logical proficiency to recognize and visualize objects and elements by means of noise, movements, and possible through emotions as well. This is achieved by executing the same ideas on a computer and exhibiting the objects or ideas that are produced by the brain on the screen of the computer. In relation to the tendency of humans in learning everything sequentially by amassing data and then producing a picture of it in our mind, the computers will follow the same concept by making use of datasets. Likewise, it evaluates every process consecutively by creating an image and associating it with the data presented in the dataset. After the extraction of ample data the computer further produces a pattern depending on the collected interpretations. Every time a word or a notion flashes, it tends to compare the word with the previously produced signal pattern. It also potentially exhibits the advancement of this technology for enhanced and smart robots that benefit mankind.

1 Introduction

With growing technologies each and every field is being advanced, and among them, the medical field is one. The medical field is one important field where it requires the implementation of all conceivable technologies to save the lives of the people by recognizing the diseases at early stages. Brain-Computer Interface (BCI) is one such technology in medicine field proposed in the 1970s at the University of California. Basically, the main objective of BCI was to restore the damaged sight, movement, and hearing. The BCI identifies the specific energy patterns of the brain. The machine that supports BCI is called an electroencephalogram. Electroencephalogram machine benefits to recording brain signals and do not generate an outcome that acts on user environment. In brief, BCI can be defined as the computer program that retrieves the signal from the brain, analyzes them, and translate them into commands. The commands sent from BCI performs a particular action [1].

The BCI system consists of four components they are: Signal Acquisition, Feature Extraction, Feature Translation, and Device output. Each of these components is controlled through operating control that which defines the onset and timing of operation, nature of the device commands, and details of the signal processing. The operating control allows BCI to be flexible and serve the specific requirements of a user [2].

J. I.-Z. Chen et al. (Eds.): ICIPCN 2020, AISC 1200, pp. 730–734, 2021.
https://doi.org/10.1007/978-3-030-51859-2_66

Let's understand the feature of each component in the BCI. *Signal acquisition* measures the brain signal through particular sensor modality, for instance, FMRI is used for metabolic activity and Intracranial Electrodes for Electrophysiological activity. The signals received from the brain are amplified to levels suitable for signal processing, and later those signals are transmitted to a computer. *Feature Extraction* analyzes the digital signals received from the signal acquisition are distinguished with pertinent signal characteristics from extraneous content. *Feature Translation* -the features outcome is accepted by feature translation algorithm. Feature translation converts into a command for an output device. Device output- This component helps to provide feedback to the user and hence to close the control loop [2, 7].

BCI has contributed major of its applications in the medical field that could take benefit of brain signal in association with prevention, detection, diagnosis, rehabilitation, and restoration. Human brain should have a consciousness level determination system to influence the attentiveness. If a person is addicted to smoking or alcohol that affects the brain and accelerate accidents. Hence through BCI, consciousness level monitoring of brain waves has been increased in order to guide the sick people/alcoholic people to warn while driving and thus reduces the accident rates. BCI has also contributed its function in detecting and forecasting abnormal brain structure and tumor cells (uncontrolled self-dividing cells). Rehabilitation is the process of restoring the lost function of such patients who suffered from serious injuries. In case if patients could not recover from such rehabilitation, BCI based prosthetic limbs (neuroprosthetic devices) is used in order to regain the normal functionality.

Other than medical field BCI applications have been widespread and reached many fields. BCI researchers have extended its benefits to marketing and advertising field. The researchers have considered the cognitive function in the neuromarketing field, to estimate the advertisement evaluation and hence to improve the market strategy to gain profit. BCI progression is a promising technology that benefits in enhancing the brain performance through activity modulation. By enhancing the brain activity, it benefits to invade the educational system, by utilizing the electrical signal that determines the degree of studied information. The technology elaborates on self-regulation skill through neurofeedback of functional Magnetic Resonance Imaging (fMRI).

Though the world has advent to heights in terms of technology, it is suffering from security and privacy. The efforts for improving security system by researchers is immeasurable, yet the outcome is fronting weak. Through cognitive biometric can identify the information and provides a solution for vulnerabilities. The Electroencephalogram (EEG) of BCI is used as a biometric modality that sends a secret warning when an authorized user is under the external force conditions. So, it is assumed BCI is one of the technologies that gave breakthrough in decades [1, 2].

2 Literature Review

Brain-Computer Interface (BCI) influence to improve the quality of physically weakened people. BCI measures the activity of the central nervous system and transforms into the artificial output. One greatest challenge that faced by BCI was to decode the user intent from the neural signal. To overcome the problem, the perfect solution was found, that is Electroencephalography (EEG). The EEG measures the signal at the scalp among the large population of cortical neurons. Classification and regression are the two methods used to transform into BCI from neural data [3, 4].

In order to enhance the feature of a BCI and that can make it to subject-specific BCI a machine learning approach was introduced. The evolution of BCI has driven the development of the human brain to extract the hidden patterns and to distinguish the information. In order to implicate brain to work with the particular application, an Artificial Neural Network (ANN) was used. An ANN is a machine learning tool that enables the brain and classifies actions of input. In the same way, the genetic algorithm is used to choose the number of hidden layer and neurons per layer for training data given by ANN [3].

According to [5] a BCI is based on fMRI accounts the activity of the brain through high spatial resolution. Day to Day activities carried out by human-like handling the home appliances, person identification, selecting a contact from a mobile phone, etc. all these tasks are accomplished through visual images. In the case of BCI, the daily task is achieved by controlling the BCI with the voluntary generation of the images [6]. The various part of the brain activation pattern correlate to a specific type of imaginary and perceives visual images. This research paper focuses on studying the brain image visualization using machine learning to enhance the feature of BCI.

3 Design Procedure

The significant feature of BCI is the EEG pattern classifier for image visualization. EEG pattern classifier identifies the particular pattern associated with subject's different mental states. EEG artifacts are caused by eye movements and blinking that develops the patterns and various mental states. Through experimental setup, it is reviewed the activity formed by visualization pictures is classified based on EEG recordings. The recordings are taken from Emotiv EPOC and BrainProducts ActiCap.

The experiment was conducted for four consecutive days. On the first day, a training session was designed to train the BCI classifier. On second day test was designed to give a subject accordingly which output of the BCI classifier enhance the picture imagination. On the third day the same activity is repeated as performed on the second day to improve the learning capabilities. Finally, on the last day (fourth day) the training and test session was led with the auxiliary session. The detailed description of the experiment is explained in Fig. 1.

The sub-module A in Fig. 1 represents the structure of the training. During the first three days of the experiment, EEG was recorded using Emotiv Systems Inc, EPOC 16-electrode cap. 14 electrodes were located at the different place of the head, and two were placed above the ears. The impedance of electrode communicated with the scalp was monitored visually using Emotiv Control Panel software. Finally, on the fourth day of the experiment EEG and Electrooculogram (EOG) was recorded using ActiCap (brain products). At the beginning of the training, each subject is presented with many types of pictures. For instance, when a training set was intended with the house as an input a different variety of picture of the house is presented. Based on this training session will be carried [6].

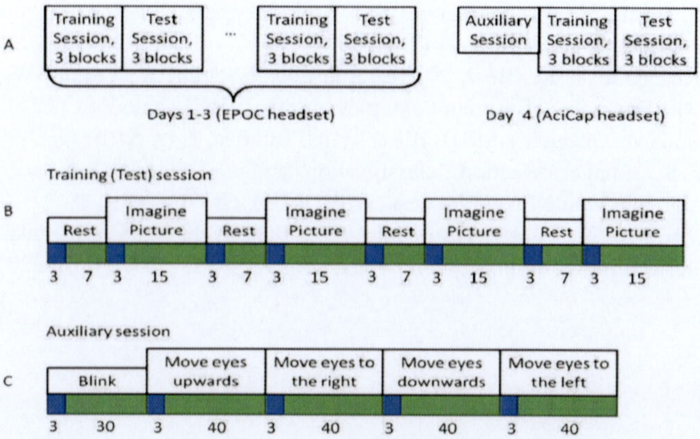

Fig. 1. Image visualization training [6].

Let's take an instance and understand how actually the visualization of the image works. When a human hears with the word 'cat', he/she imagines the image of a cat in the brain. Using the BCI approach the so formed images are collected and displayed on the system screen. In the same way, when a human hears a word which is not heard before, he/she first guess the word based on the pronunciation. Later the assumed word is transformed into a monitor. The pronunciation of the word is stored in the computer and with the help of AI search the word in an Internet and obtain the exact results. There are some cases where brain visualizes the color based on the particular sound. Different kind of sounds creates a different color spectrum in the brain.

4 Conclusion

A brain-computer interface is a computer-based system that comprises of a EEG headset that acquires the signals by the brain based on some thoughts, vision, or sound. Based on the signals collected, they are amplified (signals are purified). Thus, a feature is extracted based on the human intention by evaluating the intensity of the signal. At last, the intention of the user is defined. In order to perform all these task training sets are created which could possible only with the help of Machine Learning.

References

1. Shih, J., Krusienski, D., Wolpaw, J.: Brain-computer interfaces in medicine. Mayo Clin. Proc. **87**(3), 268–279 (2012)
2. Abdulkader, S., Atia, A., Mostafa, M.: Brain computer interfacing: applications and challenges. Egypt. Inform. J. **16**(2), 213–230 (2015)
3. Nurse, E., Karoly, P., Grayden, D., Freestone, D.: A generalizable brain-computer interface (bci) using machine learning for feature discovery. PLoS ONE **10**(6), e0131328 (2015)

4. van Erp, J., Lotte, F., Tangermann, M.: Brain-computer interfaces: beyond medical applications. Computer **45**(4), 26–34 (2012)
5. Weiskopf, N., Mathiak, K., Bock, S., Scharnowski, F., Veit, R., Grodd, W., Goebel, R., Birbaumer, N.: Principles of a brain-computer interface (BCI) based on real-time functional magnetic resonance imaging (fMRI). IEEE Trans. Biomed. Eng. **51**(6), 966–970 (2004)
6. Manoharan, S.: Image detection, classification and recognition for leak detection in automobiles. J. Innovative Image Process. (JIIP) **1**(02), 61–70 (2019)
7. Bobrov, P., Frolov, A., Cantor, C., Fedulova, I., Bakhnyan, M., Zhavoronkov, A.: Brain-computer interface based on generation of visual images. PLoS One **6**(6), e20674 (2011)

A Framework for Degraded Kannada Character Recognition

N. Sandhya[1(✉)], R. Krishnan[2], and D.R. Ramesh Babu[2]

[1] Department of Computer Science Engineering, CEC, Bengaluru, India
soni19@gmail.com

[2] Department of Computer Science Engineering, DSCE, Bengaluru, India
krishnan1234@gmail.com, bobrammysore@gmail.com

Abstract. Degraded Character Recognition (DCR) is an important area of research in the field of Document Image Analysis and Recognition (DIAR). The degradation of characters poses lot of challenges like broken characters, characters mixed with noise etc. Kannada language script has curves and complex patterns, which makes recognizing these characters very difficult. The degradation in the document can introduce gaps in these patterns, which complicates the recognition problem. In this paper, the importance of Kannada DCR system for printed scripts is addressed and also it proposes a framework consisting of four stages namely preprocessing, rebuilding, feature extraction and classification. This framework is also supported with an efficient implementation that achieved a recognition accuracy of 99% in characters with degradation.

1 Introduction

Image processing and Pattern Recognition techniques have various applications and Document Image Analysis (DIA) is one amongst them. In DIA an image of a document is analyzed, understood and represented in a form which is suitable for use. Optical Character Recognition (OCR) is an active research area in the field of Document Image Analysis. The context diagram of an OCR system is as shown in Fig. 1. Many OCR systems were developed to recognize printed and fine documents. Recognition of degraded characters is a challenging problem in the field of optical character recognition.

The deterioration of the quality in the document is meant as the degradation, this usually occurs due to the presence of the noise, The noises found in the history based documents is the categorized the as local and global, where the local is the content based and the global is the found all over the paper causing issues in the characters and skew, noisy contextual respectively. The degradation minimizes the accuracy in recognition and the causes difficulty in the recognizing the document leading to a sort out an effective method in handling the degradation.

Kannada is the official language in Karnataka state, South India. It has a rich history and the documents relating to the past history are of utmost importance and they need to be preserved and digitized. A sample Kannada degraded old document is shown in Fig. 2. The degradation of characters hinders the OCR recognition as it is mostly in the form of broken characters.

J. I.-Z. Chen et al. (Eds.): ICIPCN 2020, AISC 1200, pp. 735–745, 2021.
https://doi.org/10.1007/978-3-030-51859-2_67

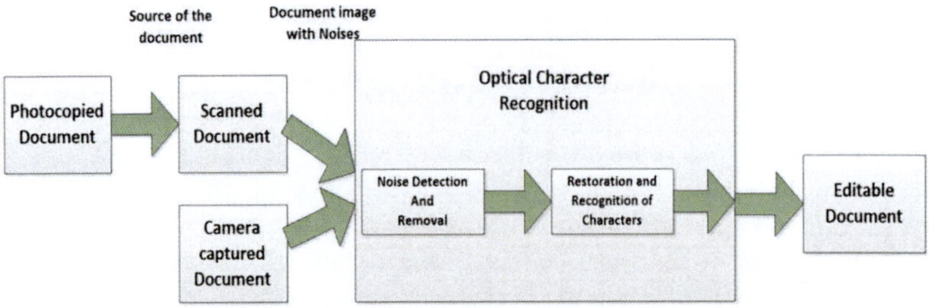

Fig. 1. Context diagram of OCR system

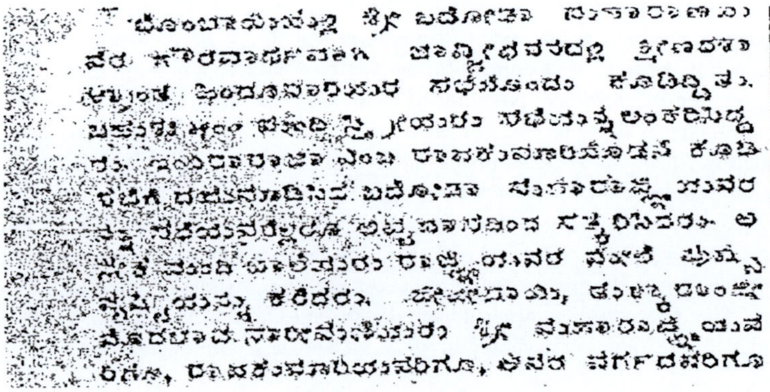

Fig. 2. Degraded Kannada document

The curves found in the Kannada characters causes complications in automatic recognition. Though the OCRs help in recognizing the degradation worsens its performance, many European Asian languages have reported work for the degraded character recognition of their languages [13–15], but very less work has been reported for Kannada degraded character recognition. This is the motivation for research on recognition of degraded Kannada characters.

In this paper, we have addressed the importance of Kannada DCR system for printed scripts and proposed a framework consisting of four stages namely preprocessing, rebuilding, feature extraction and classification. The novelty of this framework is that a new stage called rebuilding is introduced with a novel rebuilding algorithm. A recognition accuracy of 99.8% is achieved for characters with degradation.

The paper is organized as follows. Section 2 explains the end-to-end proposed system based for degraded Kannada character recognition. In Sect. 3 the preprocessing stage of the system is explained. In Sect. 4 rebuilding of Kannada character approach is explained. Section 5 details about the feature extraction and recognition. Finally Sect. 6 concludes and gives future scope.

2 Proposed System

Figure 3 representing the proposed process in recognizing degraded character is organized as five sections, [4]. Where the initial step is the loading of the input data, the next is the removal of the global noise, followed by the segmenting the characters, reconstructing and recognizing.

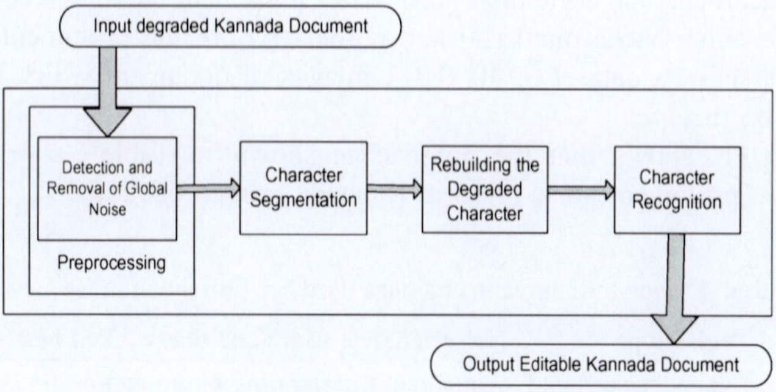

Fig. 3. Degraded character recognition framework

Figure 4 shows the input and output of each stage in our framework for degraded character recognition. As shown in the figure the input word image has degraded characters and noise present in it. After preprocessing stage the noise is removed. The characters are segmented from the image and isolated characters are obtained. The obtained isolated characters are checked for broken characters and rebuilt using end-point algorithm. These characters are classified and recognized. Finally an editable document is obtained.

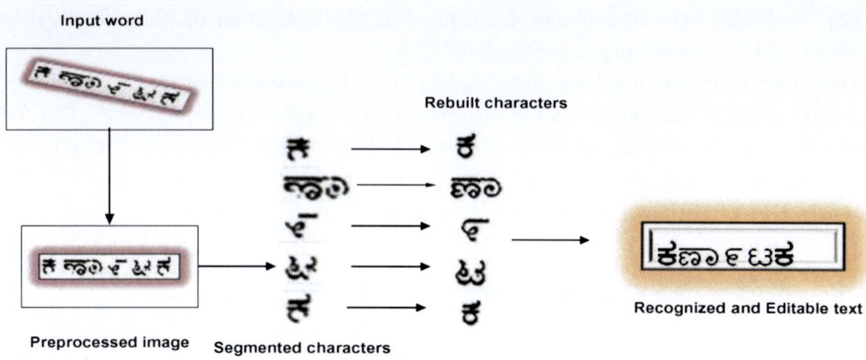

Fig. 4. Block diagram showing the steps and processing in the Framework

3 Preprocessing

The first stage in this framework is the preprocessing. The degraded documents will have various noise types present in them. In order to have better recognition accuracy these noise types have to be removed. We have identified 24 noise types and categorized them into local noises and the global [1]. Global noise establishes itself all the way through the text whereas local noise is intrinsic or content oriented noise. Most of the historic documents have multiple noise types present in them. We could establish that skew (S), noisy background (N) and border (B) are the prominent noise types based on analysis of a dataset of 4000.011 images of document which are Kannada scripts based on history.

Our study [1] shows that the approaches shown in Table 1 worked best for removing these prominent noise types in Kannada printed documents:

Table 1. Noise Removal techniques used for prominent noise types

Sl. no	Noise type	Noise removal technique that worked best
1.	Noisy background	9 quadrant thresholding based approach
2.	Skewness	Scanline based approach
3.	Border noise	Margin connected component based approach

We found that there was no published work on degradation measure. We investigated these research issues and proposed degradation measure.

The local noises were common types in documents. In [4], the brokenness and the gauging erosion is handled using the independent degradation measure. The erosion metric is calculated by the pixel difference between the perfect printed character image and degraded printed character image. The brokenness metric is calculated by number of gaps and its size in the degraded character from the perfect printed character.

We conducted experiments and determined the optimal order of noise removal to be NBS – noisy background, border, skew [2].

The order NBS has the least mean degradation rate than the other five orders of noise removal and was chosen as the right order of removal. As shown in Fig. 5a the sample degraded document before noise removal has skew and noisy background. After noise removal in the NBS order the characters are unaffected shown in Fig. 5b. The characters are highly degraded by noise removal in SNB order.

a. Before noise removal b. After noise removal in NBS order c. After noise removal in SNB order

Fig. 5. Sample degraded document for preprocessing

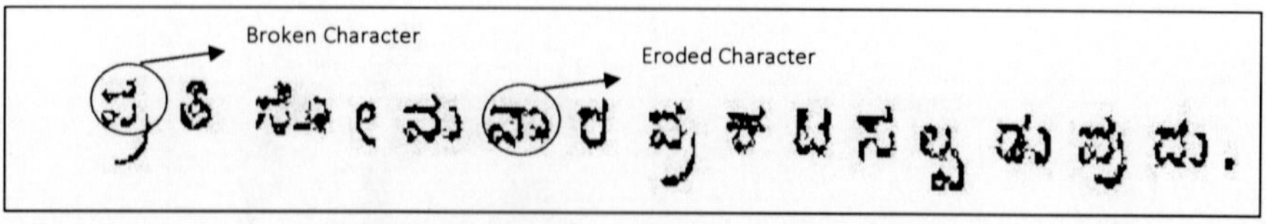

Fig. 6. Example of a degraded document

Hence as part of preprocessing stage firstly the prominent noise types of the documents are selected, and then the prominent noise types are removed with improved existing techniques in the multiple noise documents with the best order of noise removal technique.

3.1 Segmentation

Character segmentation is an important step in process of character recognition. The segmentation can be done in two ways. Character based and region based. An example for region based utilizes the vertical and horizontal forecast outline. And connected component analysis is an example for the character based segmentation. For degraded characters in Kannada the region based segmentation is used. [18] to separate the broken in the connected components [19].

In Fig. 4 the character segmentation from a word of a degraded document is clearly shown. The segmentation techniques proposed in [18, 19] were analyzed. The technique proposed in [18] for degraded printed Gurumukhi script was used for segmentation since it is region based technique. Gurumukhi script is different from Kannada script hence some modifications to the technique was done. For example the horizontal line in the Gurumukhi script is removed before initiating segmentation. But for Kannada script it is not required.

4 Rebuilding

The paper utilizes the "End point Algorithm" [3] to rebuild Kannada broken characters to improve the recognition accuracy (Fig. 6).

End point algorithm

The end point algorithm based on region growing technique [3] is used for rebuilding. The step by step output of the algorithm on the sample broken character is shown in Fig. 7. The accuracy was heightened to 89.05% and was further improved after the process of reconstructing.

This proposed scheme is most compatible for the less degradation, for the degradation where the end point could be found among the character degraded and outcome totally relies on the level of degradation. The character with the high degradation would experience an distorted reconstructing.

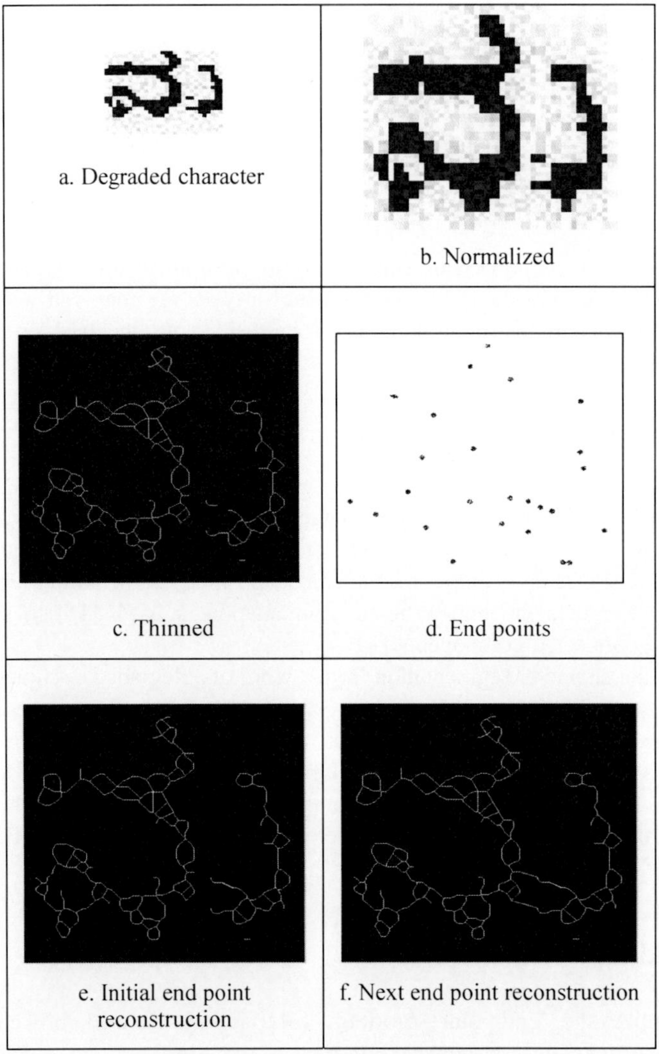

Fig. 7. Broken character rebuilding.

5 Feature Extractions and Recognition

The zonal features of the character are used and given to the FDA classifier. The following sub-sections explain these in detail. The extractions of the features are done to improve the recognition rate.

5.1 Datasets

The method utilized 7351 synthetic dataset that were created using the 150 samples of the synthetic characters that were degraded. The pixel generation were random and the with lowest = 11, moderate = 22 and the highest = 31.

5.2 Feature Extraction

Zonal Features. To enhance the recognition accuracy the, characters are segregated into zones, the space between the every pixel that is ON is estimated and reserved, the Fig. 8 shows the zoning of the characters.

a

0	0	0	0	0
0	0	0	0	0
30	25.63	0	25.63	29.41
26.17	21.09	17.02	20.12	27.58
24.41	18.60	14.03	17.20	25.23
21.54	0	0	0	20.61
17.46	11.18	0	13.34	17.46
11.05	9.05	1	9.05	11.05
0	12.80	7	12.04	0
0	14.86	11	14.86	0

b

Z.1	Z.2	Z.3	Z.4	Z.5
Z.6	Z.7	Z.8	Z.9	Z.10
Z.11	Z.12	Z.13	Z.14	Z.15
Z.16	Z.17	Z.18	Z.19	Z.20
Z.21	Z.22	Z.23	Z.24	Z.25
Z.26	Z.27	Z.28	Z.29	Z.30
Z.31	Z.32	Z.33	Z.34	Z.35
Z.36	Z.37	Z.38	Z.39	Z.40
Z.41	Z.42	Z.43	Z.44	Z.45
Z.46	Z.47	Z.48	Z.49	Z.50

c

d

0	0	0	0	0
0	0	0	0	0
28.42	25.07	25.07	25.49	31.14
26.90	20.59	20	0	28.01
25.55	16.40	12.04	16.40	23.85
6	12.52	0	13.89	22.47
16.27	12.36	3.60	10.19	19.10
19.41	9.84	4.12	10.77	20.22
20.24	11.40	7.07	13.03	22.13

e

Fig. 8. a. Character 'aa' zonal segregation b. 'aa' zonal values c. consistent zonal-numbers d. and e. the first two steps followed for 'uu'

5.3 Recognition Using FDA Approach

The FDA- Fit discriminant analysis classifier, is as follows

Confirms the type of discriminant analysis, fits the attributes of the input variable, and predicts output, with every column and row representing a variable and the observation respectively. The every output row highlights the classification accuracy of the input row the Table 2 shows the accuracy in the recognition.

Table 2. Accuracy in recognizing prior and subsequent to rebuilding

Techniques	Before rebuilding			After rebuilding		
	Low degradation	Medium degradation	High degradation	Low degradation	Medium degradation	High degradation
Proposed approach FDA	99.18%	99.18%	98.77%	99.38%	99.38%	99.38%
SVM	99.59%	98.97%	98.59%	99.79%	99.38%	95.10%
FLD	99.79%	99.26%	99.91%	100%	99.58%	99.95%

the table shows the comparative analysis between the proposed, FLD and the SVM.

6 Results and Discussion

After experimentation it was found that first Noisy background removal, then border noise removal and finally skew removal is the best order of noise removal considering only the three prominent noise types in a multi-noise document. If different techniques, other than the one proposed are chosen for removing an individual noise type like border, the related experiments in [2] may have to be repeated for determining the right order of removal.

The character rebuilding improves the character contrasts and reduces the mis-classifications [3].

FDA was chosen as the best recognition algorithm after detailed comparisons as discussed in the previous section. Table 3 shows the recognition accuracy on synthetic characters using the implementation of the framework. The recognition accuracy is the same for the three techniques i.e., 99.38% after rebuilding.

Table 3. Recognition accuracy on synthetic characters with our framework and supporting implementation

Techniques	After rebuilding		
	Low degradation	Medium degradation	High degradation
Proposed approach FDA	99.38%	99.38%	99.38%

The experimentation was conducted on a printed page without degradation. The page was preprocessed, segmented and recognized using the techniques explained in the framework. 100% accuracy was achieved on a printed page using the FDA technique.

The Table 4 highlights the recognizing percentage prior to and subsequent to the rebuilding. it was found that FDA performs was more convincing compare to the other two methods.

Table 4. Recognition accuracy achieved on real dataset

Methodology	Prior to rebuilding	Subsequent to rebuilding
Proposed approach FDA	10%	26%

7 Conclusions

Degraded character recognition is a challenging area of research. This framework with the supporting implementation provides a complete degraded Kannada character recognition system.

The prominent noise types were identified as skew, noisy background and border. The prominent noise types may vary for different datasets. For these three noise types the best in-class noise removal techniques were enhanced and implemented. A learning based approach with Zonal features and FDA technique worked best on the data sets we experimented with. We were able to achieve an accuracy of 100% on printed document without noise, 99.38% on printed documents with low levels of noise and 26% on legacy historic documents.

We have designed a framework with supporting implementation which gives best results for recognizing documents that have Kannada printed characters without noise and with low/medium degradation.

Following are the challenges in recognition that we could identify working with the historic legacy documents with very high degradation of characters: Dataset for training legacy documents is hard to find. Improved feature vectors can be identified for better recognition accuracy in legacy documents. These challenges can be addressed in future through advanced machine learning techniques.

As a future scope in research, efficient handling strategies for handwritten degraded Kannada characters can be developed since handwritten documents pose greater challenges like size variations, writing style variations etc. A Kannada document digitization system for historical documents can be developed for commercial purpose.

Competing Interest: The authors declare that they have no conflict of interest.

Informed consent: Informed consent was obtained from all individual participants included in the study.

References

1. Sandhya, N., Krishnan, R., Ramesh Babu, D.R.: A language independent characterization of document image noises in historical scripts. Int. J. Comput. Appl. **50**(9), 11–18 (2012)
2. Sandhya, N., Krishnan, R., Ramesh Babu, D.R., Das, P.: A comprehensive pre-processing approach for digital preservation of documents. In: Elsevier Proceedings of International Conference on Emerging Research in Computing, Information, Communication and Applications-(ERCICA-14). ELSEVIER (2014)
3. Sandhya, N., Krishnan, R., Ramesh Babu, D.R.: A novel local enhancement technique for rebuilding broken characters in a degraded Kannada script. In: IEEE International Advance Computing Conference (2015)
4. Sandhya, N., Krishnan, R., Ramesh Babu, D.R., Rao, B.: An efficient approach for handling degradation in character recognition. Int. J. Adv. Imaging Paradigms **14**, 14–29 (2019)
5. Sandhya, N., Madhusudan, N., Krishnan, R., Ramesh Babu, D.R.: Handwritten Kannada character recognition using zonal features and multi-class SVM. Int. J. Appl. Eng. Res. (IJAER), **9**(20) (2014). ICPRMSP'15, Special Issues
6. Website for fit discriminate analysis. http://in.mathworks.com/help/stats/fitcdiscr.html
7. Manjunath Aradhya, V.N., Hemantha Kumar, G., Noushath, S., Shivakumara, P.: Fisher linear discriminant analysis based technique useful for efficient character recognition. In: Fourth International Conference on Intelligent Sensing and Information Processing (2006)
8. Wei, C., Lin, C.-J.: A comparison of methods for multiclass support vector machines. IEEE Trans. Nueral Netw. **13**(2), 1045–1052 (2002)
9. Barekat, S., Sarrafzadeh, A., Shanbehzadeh, J.: Skew detection of scanned document images. In: Proceedings of the International MultiConference of Engineers and Computer Scientists 2013 vol I, IMECS 2013, 13–15 March 2013, Hong Kong (2013)
10. Silva, G.F.P., Lins, R.D., Silva, J.M.: Enhancing the filtering-out of the back-to-front interference in color documents with a neural classifier. IEEE (2010)
11. Website for otsu's binarization. http://en.wikipedia.org/wiki/Otsu%27s_method
12. Website for weiner filter. http://www.owlnet.rice.edu/~elec539/Projects99/BACH/proj2/wiener.html
13. Shafait, F., Breuel, T.M.: A simple and effective approach for border noise removal from document images. In: IEEE 13th International Multitopic Conference, INMIC 2009 (2009)
14. Pletschacher, S., Hu, J., Antonocopoulos, A.: A new framework for recognition of heavily degraded characters in historical typewritten documents based on semi-supervised clustering. In: ICDAR 2009 (2009)
15. Nayak, M.R., Nayak, S.: Automatic recognition of handwritten Bengali Broken Characters (BBC): simulating human pattern matching. Int. J. Comput. Appl. **59**(9), 27–32 (2012). (0975–8887)
16. Manoharan, S.: A smart image processing algorithm for text recognition information extraction and vocalization for the visually challenged. J. Innov. Image Process. (JIIP) **1**(01), 31–38 (2019)
17. Gangamma, B., Murthy, K.S., Singh, A.V.: Restoration of degraded historical document image. J. Emerg. Trends Comput. Inf. Sci. **3**(5), 792–798 (2012)
18. Website of Kaleido software & services Kannada OCR. http://kannadaocr.com/downloads/KanScanUserGuide_v10b.pdf
19. Jindal, M.K., Lehal, G.S., Sharma, R.K.: On segmentation of touching characters and overlapping lines in degraded printed Gurumukhi script. Int. J. Image Graph. **9**(3), 321–353 (2009)

20. Sachan, D., Dutta, S., Naveen, T.S., Jawahar, C.V.: Segmentation of degraded Malayalam words: methods and evaluation. In: Third National Conference on Computer Vision, Pattern Recognition, Image Processing and Graphics (2011)
21. Rajashekararadhya, S.V., Vanaja Ranjan, P.: Support vector machine based handwritten numeral recognition of Kannada script. In: IEEE International Advance Computing Conference 2009 (2009)

Classification of Soybean Diseases Using Pre-trained Deep Convolutional Neural Networks

Sachin Jadhav[1(✉)], Vishwanath Udupi[2], and Sanjay Patil[3]

[1] Bharati Vidyapeeths College of Engineering Kolhapur, Kolhapur, India
Sachinbjadhav84@gmail.com
[2] Maratha Mandal Engineering College Belagavi, Belagavi, India
vrudupil@gmail.com
[3] Rajgad Shri. Chattrapati Shivajiraje College of Engineering, Bhor, Pune, India
patilsbp@gmail.com

Abstract. In this work, a novel soybean leaf disease classification technique related to pre-trained GoogleNet deep convolutional neural networks (CNN) architecture proposed. The proposed GoogleNet architecture trained on a database of 550 image samples of unhealthy and healthy soybean leaflets with 3 symptoms of an unhealthy class particularly, septoria brown spot, bacterial blight, frogeye leaf spot, and 1 healthy class using a deep transfer learning approach. As specified 3 unhealthy class and 1 healthy class identification, we have used the 5-fold cross-validation approach, the intended pre-trained GoogleNet-CNN architecture attains an accuracy of 96.25%. It was found that the accuracy of our proposed CNN architecture is enormously more precise than the formal machine learning models. The results of performance analysis to the recognition of soybean diseases exhibit the expediency and highest success rate using the proposed GoogleNet CNN model.

Keywords: CNN · Machine learning · GoogleNet · Cassification · Diseases

1 Introduction

The recent development in convolutional neural networks (CNNs) has attained remarkable outcomes in the domain of pattern recognition [1]. This work is consisting of novel access to the improvement of plant infection diagnosis system, based on soybean diseased leaf image classification, using pre-trained GoogleNet and AlexNet deep learning architecture models. The 3 common soybean diseases include brown spot, frogeye spot and blight were considered for classification of disease task, these diseases cause significant crop yield loss and its consequence is on the livelihood and economy of a farmer. An automated system designed using recent technology like machine learning for identification of diseases related on the appearance of visible symptoms occurred on the leaf could be of numerous aid to farmers in the agriculture manner and also plant pathologists as a precise and reliable disease investigation system [2, 3].

J. I.-Z. Chen et al. (Eds.): ICIPCN 2020, AISC 1200, pp. 746–756, 2021.
https://doi.org/10.1007/978-3-030-51859-2_68

To enhance the precision and promptness of the investigation outcomes, numerous researchers have developed the computerized plant disease investigation system related to image processing and computer vision recognition systems applying support vector machine (SVM) [9], digital image processing methods [2] related on color analysis and thresholding based segmentation techniques [14] and computer vision.

Also, artificial neural networks ANNs based strategies are presently applied for recognizing plant diseases [4] these imply fused with various techniques of digital image preprocessing techniques to get more useful feature extraction from raw images applied while training the system. Recently, various research work on computerized leaf disease recognition and analysis have been carried with applying deep CNN architecture methods [10].

Jürgen Schmidhuber [1] presented a survey on relevant work in deep learning algorithms and its architectures. Finally, the author reviewed various types of deep supervised, unsupervised, and reinforcement learning techniques for short programs encoding deep and large networks.

Ferentinos [2] introduced disease identification of plant-based on CNN's architectures utilizing simplistic unhealthy and healthy leaflet images of various plants, it attains an efficiency of 99.53%.

Aravind et al. [5] implemented convolutional neural network architectures such as AlexNet and VGG16 to detect various disease which attains 97.49% and 97.29% respectively.

Mohanty et al. [6] employed a disease analysis model using the AlexNet convolutional neural network based on a transfer learning method. The proposed model was apt to identify twenty-six various diseases in fourteen crop varieties with a classification efficiency of 99.35%.

Yanming Guo et al. [7] did a survey report on numerous deep learning strategies and their contemporary advancements, with their applications in various image analysis methods, such as identification of the object, image retrieval, semantic segmentation, and image classification. Ultimately, the author has compiled future inclinations and difficulties in outlining and training deep neural networks.

Yang Lu et al. [11] introduced a convolutional neural networks model to identify popular ten various diseases of rice plant such as rice bacterial leaf blight, rice bakanae disease, rice sheath blight, rice bacterial wilt, etc. trice it achieves an accuracy of 95.45%.

The essential inspiration for improving the deep convolutional network architecture system to provide accurate information about diseases occurred on soybean plants to the farmer. Second manual feature extraction does not require for deep convolution network for identifying soybean diseases instead of that the CNNs are automatically learning from the inputs images applied during training networks in an efficient way [8]. Besides, to enhance disease analysis results, the deep CNNs are considered as dominant recognition models in image retrieval tasks [12–15]. Hence we proposed to implement CNN architecture using GoogleNet CNNN to soybean disease recognition in this research work.

The foremost contribution of this study work mainly includes (1) Retraining of pre-trained GoogleNet convolutional neural network model by modifying the output layers and then applied it to the problem of the classification of 3 common soybean diseases. Perceive that the employed technique can precisely and efficiently recognize the 3 common soybean diseases. (2) After validating the experimental results of the proposed model it shows that the CNNs method gives the highest identification accuracy than other conventional machine learning models: (a) SVM classifier, (b) PNN classifier, (c) KNN classifier.

The benefits of the proposed GoogleNet CNN is that sample images can be fed to an input layer of image dimension directly to the model. In this work, we proposed GoogleNet pre-trained CNN architecture adopted for identifying 3 distinct disease classes and a non-disease class of the soybean plant of the specified image database.

The residue of the article is composed as given below. In Sect. 2, we outline the process of retraining the GoogleNet CNNs model on new data. Section 3 offers the soybean disease classification method and presents the use of the developed GoogleNet CNN to the problem of soybean disease identification. Section 4 presents the Conclusion.

2 Materials and Methods

In the present section, we have discussed the intact procedure of employing the architecture for soybean plant disease classification using the GoogleNet convolutional neural network model [7]. The entire method is presented into various essential steps within subsections below.

2.1 Dataset

The proposed pre-trained GoogleNet deep CNN employed to identify the described soybean leaflet from the dataset. The soybean leaflets of distinct diseased and healthy types are collected from the Agriculture center of research Kolhapur Maharashtra. The training data sets which included 150 leaf image samples with leaf blight disease, 150 leaf image samples with Frogeye leaf spot disease, 100 leaf image samples with brown spot disease and 150 leaf image samples with non-disease (healthy). A total of 80 leaflets specimens used as tests dataset. Figure 1 represents the leaflet specimens from the test dataset. The details of a defined database are provided in Table 1. The total of 550 sample images fragmented within 3 disease category and 1 non-disease category.

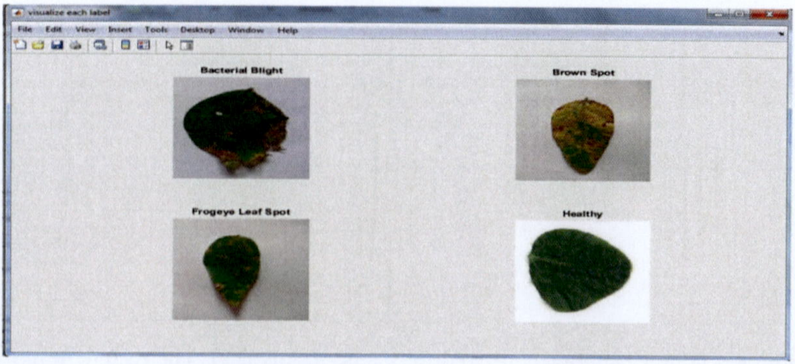

Fig. 1. Leaflet specimens of the test dataset.

Table 1. Summary of data set

Sl. No	Dataset	Training dataset	Test dataset
1	Blight	150	20
2	Brown Spot	100	20
3	Frogeye Leaf Spot	150	20
4	Healthy	150	20
Sum of 4 specimens		550	80

2.2 Image Labelling

To enhance the identification accuracy of the defined model while the extraction of features, last images meant to remain utilized as a database to proposed deep CNN classifiers were pre-processed to gain consistency. A total of 550 soybeans leaves sample images are pre-processed first to input the image dimension of 224 × 224 × 3, later the pre-processed leaflet specimens from the training dataset are applied to train, GoogleNet CNNs model. To enhance the identification efficiency of the defined model than the current machine learning models training parameters such as minibatch size, Max epoch, and the learning rate is revised..

2.3 GoogleNet Convolutional Neural Network Training

The intended strategy, illustrated in Fig. 2 comprises that four components addressed below: (1) Pre-training stage: during this stage, deep architectures train on huge datasets using high configured CPU [16]. The purpose of this stage is the network weights initialization to the subsequent stage. We have used two deep CNNs architecture models, GoogleNet. These models were pre-trained for other pattern recognition tasks using the ImageNet database. Our aim is to take the benefits of these pre-trained models to enhance classification outcomes. The method of reconfiguration of the GoogleNet architecture from the raw data image to the prophesied output probabilities of a particular disease is shown in Fig. 2. The specimen images input of the

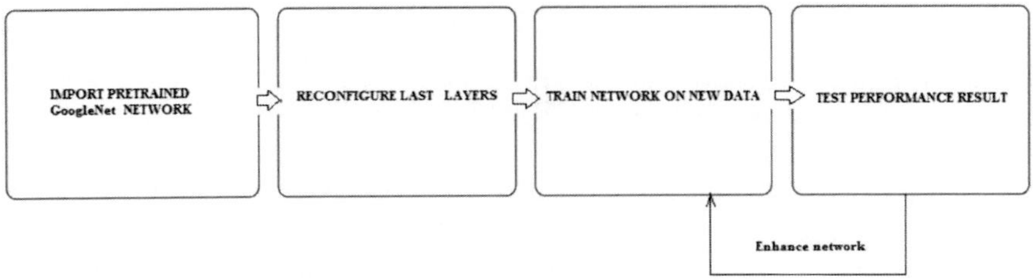

Fig. 2. Retraining GoogleNet CNN.

convolutional neural network is resized to pixels of 256*256. The results describe the probabilities of a particular disease. In the present case, we ought 3 unhealthy and 1 healthy (Fig. 1). The training of the GoogleNet CNN for making an image classification model of a database specified in Table 1. GoogleNet pre-trained architecture is reconfigured, modified and adjusted to maintain defined 4 categories. Out of 144 layer array with layers, the last few layers were modified which is presented in the study, shown in Table 2.

Table 2. GoogleNet modified layer for training the model

Layer no.	Particulars of GoogleNet layers			Modified layers
142	Loss3-Classifier	Fully Connected	1000 Fully Connected Layer	4 Fully Connected Layer (4 categories)
143	Prob	Softmax	Softmax	Softmax
144	Output	Classification Output	Crossentropyex with 'tench'	Classification Output

The layers 143 is a softmax layer and 144, is a classification output layer, that was reset in the reconfigure stage. Those layers are liable for allotting the exact categories to the raw input images. At last, the final fully connected layer set to the defined number of classes in the image dataset—four in this study.

Retraining of Google Net model

We require to reconfigure the GoogleNet network to identify 4 types of soybean diseases shown in Fig. 2 [16]. The supporting are levels for re-training the GoogleNet model.

1. Load the pre-trained GoogleNet CNN.
2. Re-configure the 141, 142 and 143 layers to implement a leaf disease identification task.
3. Retrain the model on a defined image database.
4. Validate the performance result.

3 Results and Discussion

3.1 Experimentation

The pre-trained GoogleNet CNN used for the classification of soybean disease class using a transfer learning approach.

3.2 Experimental Results

3.2.1 Training Progress Plot

One of the main objectives of this work implies to enhance the efficiency of CNN architecture. As CNN trains, the progressing plot appears [16]. Figure 3, Fig. 4, and Fig. 5 shows a plot of training progress for blight, brown spot, and frogeye leaf spot type category.

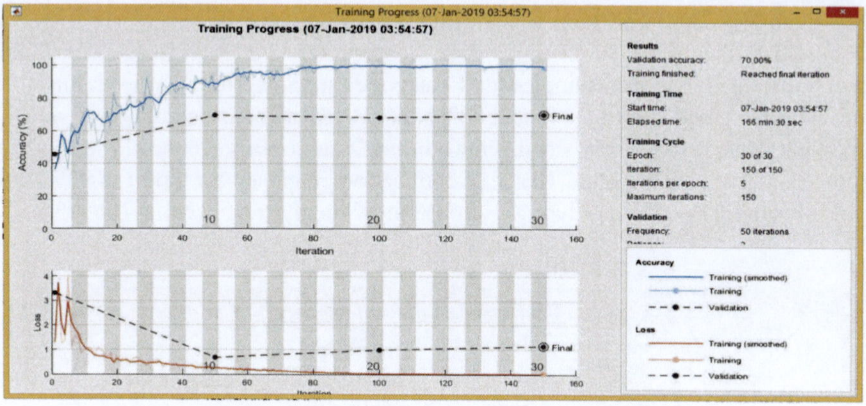

Fig. 3. Training progress plot for blight class.

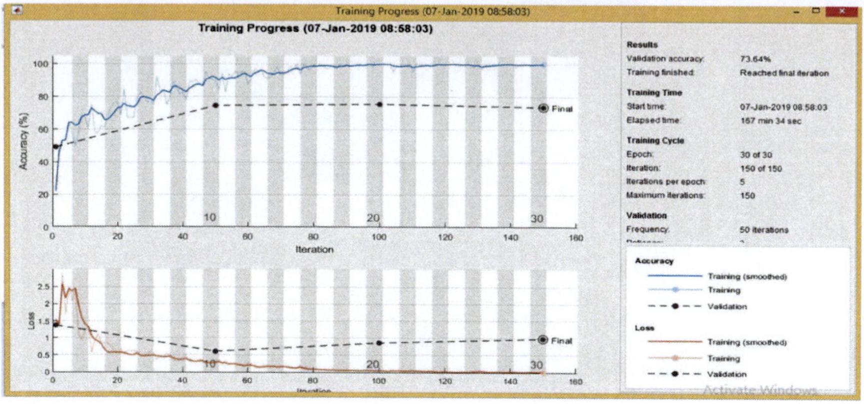

Fig. 4. Training progress plot for brown spot class.

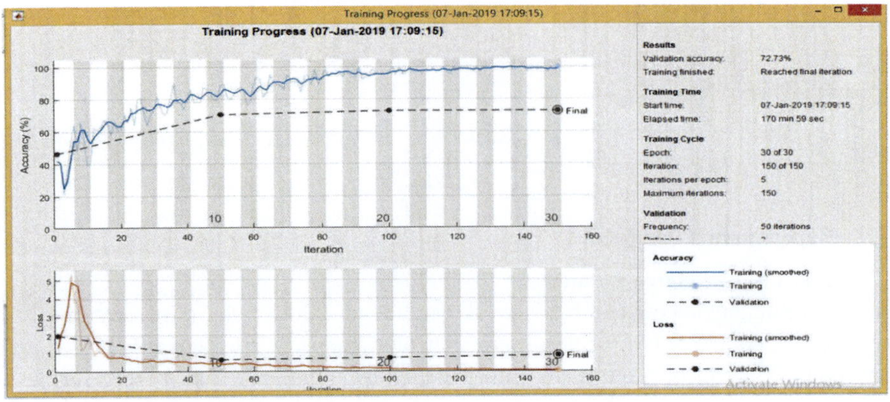

Fig. 5. Training progress plot for Frogeye leaf spot class

3.2.2 Classification and Confusion Matrix

We labeled class 1 as, blight, class 2 as, brown spot, class 3 as frogeye spots, and class 4 as non-diseased. 550 data samples for 4 categories supposed to train the architecture. To validate the execution of the model 80 leaflet images are considered. Out of which 3 leaflet specimens were misclassified, shown in Fig. 6 as confusion matrix, and Fig. 7 as sophisticated confusion matrix. Hence, identification efficiency for type 1 is 95%, type 2 is 95%, type 3 is 95% and type 4 is 100% individually; reviewed in Table 3.

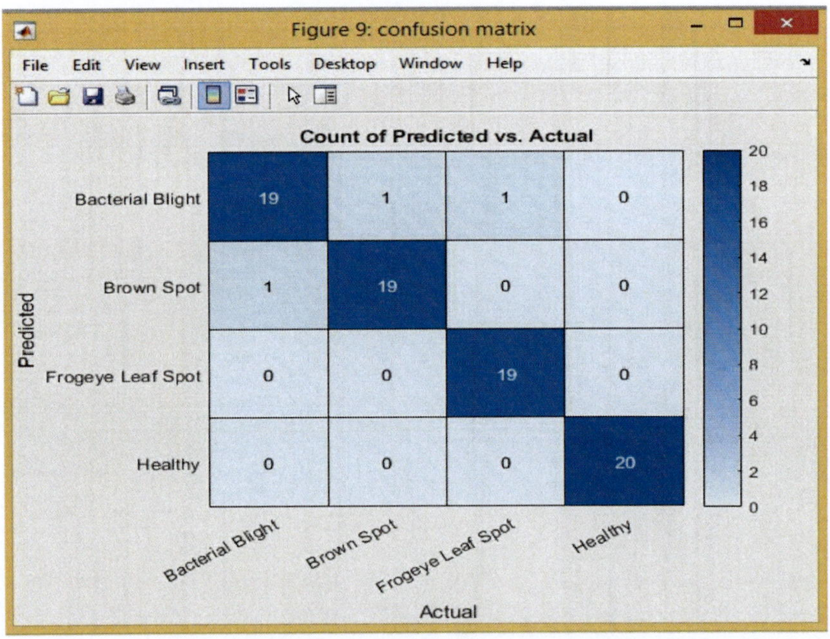

Fig. 6. Predicted vs. actual class confusion matrix.

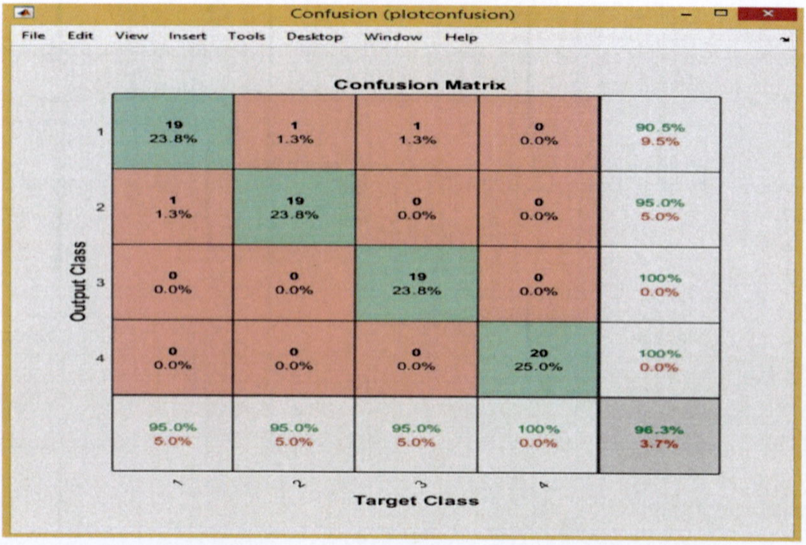

Fig. 7. Output class vs. target class sophisticated confusion matrix.

Table 3. Classification result of GoogleNet model

Leaflet database	Total test leaflets (Genuine)	Correctly classified leaflets (Predicted)	% Efficiency
Bacterial blight	20	19	95
Frog leaf spot	20	19	95
Septoria brown spot	20	19	95
Healthy	20	20	100
Composed four leaflets class	80	77	96.25

Confusion Matrix

Figure 6 and Fig. 7 show the confusion matrix and a sophisticated matrix. Figure 6 describes the confusion matrix for genuine class and predicted class classified leaflet specimens. Total of 80 disease and non-disease leaflet specimens considered to validate the performance. For the type4, every leaflet specimens were accurately identified. For the type3, type2 and type3 classes, 19 leaflet specimens from specific types were perfectly classified. It means from the aforementioned class types, 1 category in each were misclassified.

Sophisticated Confusion Matrix

Performance comparison of proposed GoogleNet convolutional neural network model with previous machine learning system implemented by System by Saylee Gharge and Priyanka Singh [13] shown in Table 4, which shows our proposed model excel over machine learning system.

Table 4. Relative study of Proposed GoogleNet CNN model

Particulars	Previous implemented system by Saylee Gharge and Priyanka Singh [15]	Proposed GoogleNet CNN classifier model
Leaf species	Soybean	Soybean
Identifed Disease	Frogeye Leaf spot, Downy mildew and Bacterial Pustule	Brown spot, Bacterial Blight, & Frogeye Leaf spot
Total data samples	30	80
Feature extracted	Manually extracted Color feature	Automatically extracted by model
Classifier	Neural Network	GoogleNet CNN Classifier
Accuracy	93.3%	96.25

Training Performance and Accuracy Result

Table 5 shows training performance with hyperparameters details. It shows the transpired time of training and overall identification efficiency later testing over distinct data.

Table 5. Proposed GoogleNet CNN Classifier Training

Epoch	Iteration	Time elapsed (hh:mm:ss)	Mini-batch accuracy	Validation accuracy	Mini-batch loss	Validation loss	Base learning rate
1	1	00:10:45	42.19%	46.36%	1.2863	1.9734	0.0010
10	50	01:05:56	89.06%	70.45%	0.3471	0.6279	0.0010
20	100	01:58:40	96.88%	72.73%	0.0878	0.7278	0.0010
30	150	02:50:59	100.00%	72.73%	0.0257	0.8870	0.0010

Elapsed time is 10571.292666 s.
Test accuracy is 0.9625

The number of sample data considered for the study is 80 in all disease class category, that is 20 in each category testing trial and accuracy of proposed GoogleNet model classification is plotted as shown in Fig. 8.

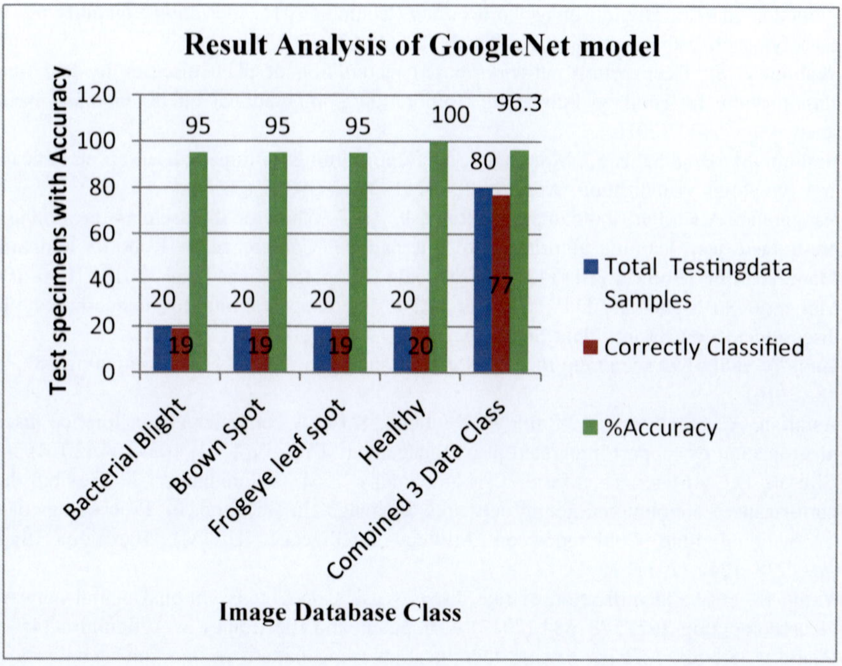

Fig. 8. Result analysis of GoogleNet model

4 Conclusion

This research work proposed a GoogleNet based deep learning method for different classifier model for disease and non-disease classes. This can be identified depends on the brown spot, bacterial infection and frogeye leaf spot. Here, the proposed techniques utilized 80 images for testing and the proposed architecture obtains 96.25% accuracy in the identification of image classification. Our experimental results show that the proposed deep convolutional GoogleNet neural network model excels in the former works in soybean disease classification. In future work, we are trying to increase more disease category and database to increase the performance rate of the model by using dense CNN models.

Acknowledgements. This research is recommended by Dr. Patil V. S, plant pathologist, Agriculture Research center Kolhapur, Maharashtra India.

References

1. Schmidhuber, J.: Deep learning in neural networks: an overview. Neural Netw. **61**, 85–117 (2015)
2. Ferentinos, K.P.: Deep learning models for plant disease detection and diagnosis. Comput. Electron. Agric. **145**, 311–318 (2018)

3. Barbedo, J.G.A.: Digital image processing techniques for detecting, quantifying and classifying plant diseases. SpringerPlus **2**(1), 660–672 (2013)
4. Sladojevic, S.: Deep neural networks based recognition of plant diseases by leaf image classification. In: Hindawi Publishing Corporation Computational Intelligence and Neuroscience, pp. 1–11 (2016)
5. Brahimi, M., Boukhalfa, K., Moussaoui, A.: Deep learning for tomato diseases: classification and symptoms visualization. Appl. Artif. Intell. **31**, 299–315 (2017)
6. Rangarajan, A.K., Purushothaman, R., Ramesh, A.: Tomato crop disease classification using pre-trained deep learning algorithm. In: International Conference on Robotics and Smart Manufacturing (RoSMa 2018) (2018). Procedia Computer Science 133 (2018) 1040–1047
7. Mohanty, S.P., Hughes, D.P., Salathe, M.: Using deep learning for image-based plant disease detection. Front. Plant Sci. **7**, Article ID 1419 (2016)
8. Guo, Y., et al.: Deep learning for visual understanding: a review. Neurocomputing **187**, 27–48 (2016)
9. Asfarian, A., Herdiyeni, Y., Rauf, A., Mutaqin, K.H.: A computer vision for rice disease identification to support integrated pest management. Crop Prot. **61**, 103–104 (2014)
10. Ciresan, D.C., Meier, U., Masci, J., Gambardella, L.M., Schmidhuber, J.: Flexible, high performance convolutional neural networks for image classification. In: Proceedings of the International Joint Conference on Artificial Intelligence (IJ-CAI), Barcelona, Spain, pp. 1238–1242 (2011)
11. Yang, L., et al.: Identification of rice diseases using deep convolutional neural networks. Neurocomputing **267**, 378–384 (2017). Computers and Electronics in Agriculture 145
12. Phadikar, S., Sil, J.: Rice disease identification using pattern recognition techniques. In: Proceedings of the IEEE International Conference on Computer and Information Technology (ICCIT), Khulna, Bangladesh, pp. 420–423 (2008)
13. Gharge, S., Singh, P.: Image processing for soybean diseaseclassification and severity estimation. In: Shetty, N., Prasad, N., Nalini, N. (eds.) Emerging Research in Computing, Information, Communication and Applications, pp. 493–500. Springer, New Delhi (2016)
14. Singh, V., Misra, A.K.: Detection of unhealthy region of plant leaves using image processing and genetic algorithm. In: IEEE International Conference on Advances in Computer Engineering and Applications (ICACEA), pp. 1028–1032 (2015)
15. Vijayakumar, T.: Neural network analysis for tumor investigation and cancer prediction. J. Electron. **1**(02), 89–98 (2019)
16. Practical deep learning examples with MATLAB Handbook. MathWorks, pp. 1–33 (2018)

Feedforward Back Propagation Neural Network (FFBPNN) Based Approach for the Identification of Handwritten Math Equations

Sagar Shinde[1(✉)], Lalit Wadhwa[2], and Daulappa Bhalke[3]

[1] JSPM Narhe Technical Campus, Pune, India
sagar.shinde5736@gmail.com
[2] NMIET, Pune, India
[3] AISSMS COE, Pune, India

Abstract. The demand for the identification of manually written mathematical equations is increasing day by day. Despite the hype, due to the increasing ambiguity in recognition, 2D and touching symbols, and complication mathematical equations, the recognition of the emerging mathematical equations has become a challenging task. The statistical, as well as complex features such as skew, kurtosis, entropy, mean, variance, standard deviation, has been considered. The classification and training have been provided using neural networks (NN) and the recognition rate has been dependent on the classifier used as well as features to be extracted. Speed of execution, efficiency and recognition rate have been enhanced by utilizing feed-forward back propagation neural network (FBBPNN) with training function gradient descent and learning rule of momentum and adaptive learning. The system can take scanned images of handwritten mathematical equations from simple through complex equations and classifies it according to the type of equations e.g. straight-line equation, the law of indices, gravity law, roots of quadratic expressions, area of a circle, convolution summation and convolution integration.

Keywords: FBBPNN · Statistical and complex features · ROC · Sensitivity · Specificity · Recognition rate

1 Introduction

Generally, the foremost common manner for the identification of the elements within the math equation is administered in either an on-line approach or offline approach. In an on-line approach, the element is identified as presently because it is a real-time entry. On the other hand with the offline approach, the equations to be identified are written on paper and scanned with the facilitate of scanner then the identification of the equation has been carried out. In both, the approaches, stylus and tablet can be utilized. It's essential to check, to develop and to validate algorithms and strategies for

J. I.-Z. Chen et al. (Eds.): ICIPCN 2020, AISC 1200, pp. 757–775, 2021.
https://doi.org/10.1007/978-3-030-51859-2_69

identification of mathematical equations, identification of texts, graphics or notations and also the extraction of data. In the identification of manually written math equations, assorted people have various handwriting styles, despite the irregularity typical for the human ability, variation in the literary genre by completely different individuals, advanced structure, touching digits or symbols, overlapping, intrinsic ambiguities, advanced linguistics as well as 2-D arrangement.

Acquiring the advancement and good identification rate of written sophisticated mathematical equations are troublesome, because of its nonlinear structure yet as several changes in written math expression dependent on the various literary genre, another noise, quality of the picture, the thickness of handwriting and missing components. The predominant piece of past work of printed ME was stress with an explanation of structures. The couples of preliminaries evade the image identification step and focus on the structure examination stage. The database in this investigation work involved math equations containing coherent images, complicated math equations, touching characters and symbols, overwrite characters, symbols in math equations. The manually written math equations data was collected from different undergraduate students, researchers and people in the society having different age group belongs to a different field. In the proposed methodology, the neural network has been used with the receiver operating characteristics, sensitivity, specificity as well as AUC (area under the Region of Convergence curve) and extracting statistical as well as complex features and evaluated the system with experimental results.

2 Literature Review

There is a lot of work has done on pattern identification for identification of printed as well as handwritten characters, symbols, special characters, math equations. Basically, Optical Character Identification (OCI) is the most crucial, essential and important part of the analysis of electronic documentation. The problem solution is achieved through pattern identification using various algorithms and natural language processing. There is only limited success in handwriting recognition, particularly for isolated, touching characters, complex equations and neatly hand-printed characters and words for limited vocabulary.

This paper demonstrates an application of Bidirectional Long Short-term Memory repetitive neural systems for the segmentation of on-line manually written English content. In addition, the acknowledgment rate of the framework additionally enhances by 3 points from 71.7% [1]. They proposed a template matching methodology for the separation of manually written and machine-printed content. The proposed methodology is training free, in the sense that training samples are not required for a classifier. A fruitful textual style detection step can lessen the number of false positives for the machine printed class and diminish the scan time for the right template from the gallery [2]. Manually written digit recognition is unique in relation to conventional digit semantics recognition, we propose particular feature definitions, including structure feature, distribution feature and projection features. In addition, meld numerous features into the profound neural systems for semantics identification. Test results show that the execution of the proposed algorithm is exceptional and shows its prevalence more than a few existing algorithms [3]. This paper proposes a way to deal with structure and execute an offline OCR framework that perceives handwritten Arabic characters; in this approach, Artificial Neural Networks (ANNs) were utilized as classifiers. The ANN was prepared dependent on the Hopfield Calculation which was planned utilizing MATLAB [4]. This paper shows a discriminative cascade model (CNN) for offline handwritten digit identification. The identification issue has been partitioned into two phases. Stage I goes for quality samples while Stage II goes for poor samples [5]. The paper represents a comparative study of classifiers on the identification of offline manually written Odia Numerals. The accuracy rate is observed to be 90.05%, 95.5% for curvature and gradient-based features respectively. Various classifiers such as support vector machine, the artificial neural network has been compared [6]. This paper represents recognizing mathematical expressions using tree transformation. They have explained a strong and effective framework for perceiving typeset and handwritten scientific documentation [7]. The authors described the classification of on-line mathematical symbols with hybrid features and recurrent neural networks [8]. This paper represents handwritten mathematical formula recognition using a statistical approach. This paper has presented a probabilistic system for a mathematical expression identification system [9]. The author has demonstrated the stroke-based performance metrics for handwritten mathematical expressions. Assessing numerical articulation identification includes a complex interaction of input primitives (e.g. pen/finger strokes), identified symbols, and identified spatial structure [10]. The author has described the Machine Learning Approach for Recognition of Mathematical Symbols. It has been discovered that the k-NN classifier has extraordinary accuracy compared with SVM as a classifier. The effectiveness of K-NN decreases with the increase in the dataset [11]. The author has described the recognition of on-line handwritten mathematical expressions using a minimum spanning tree construction and symbol dominance [12]. This paper has presented an algorithm for the recognition of offline handwritten mathematical expressions (ME) based on a predictive approach of segmentation using k-NN classification [13]. The strategy utilizes a convolutional neural network (CNN) to perform numerical image identification and acknowledgment at the same time consolidating spatial context and can deal with multi-part and touching symbols successfully [14].

3 The Flow of System with Methodology

The overall flow for the identification of complicated math expressions have explained by using Fig. 1 as shown below, including various techniques under preprocessing, segmentation and feature extraction with identification of test image.

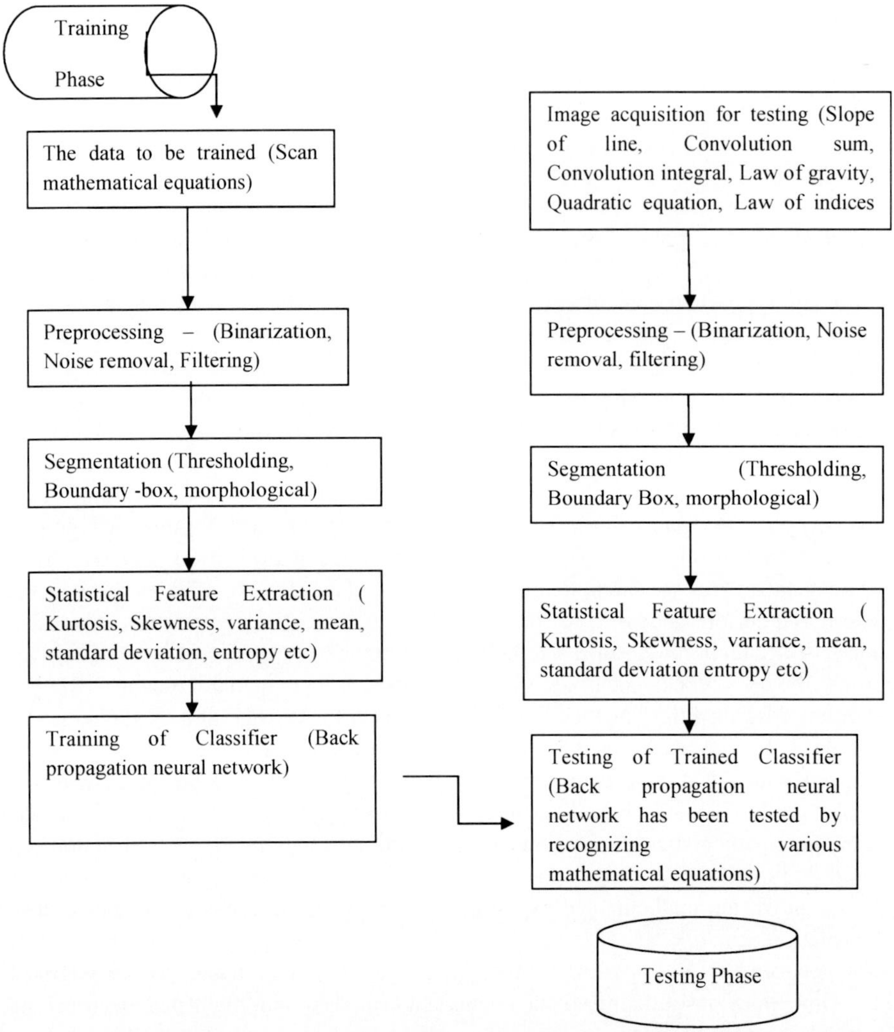

Fig. 1. Flow for identification of complicated mathematical expressions.

- Preprocessing: The important objective of this process is to enhance the quality of the image to get it available for the next process by subtracting the uncorrelated and surplus part in the background of examining the input picture. They examine pictures that are pretentious by different noises. The image filtering is used for smoothing, sharpening, removed noise and edge detection. A median filter is a nonlinear filter that is used to keep the sharpness of picture edges while subtracting the noise.
- Segmentation: As per name implies, in segmentation, each and every element of a mathematical equation is separated from each other. The input scanned image is segmented by separating all the elements. The boundary box has been created for

each and every element in mathematical equations. The most widely recognized morphological activities are widening and disintegration. The number of pixels included or expelled from the components in a picture relies upon the font as well as the shape of the structuring element used to process the picture.

- Feature extraction: The recognition rate is mainly based on the feature extracted from the input image. The features should have the same digits for the patterns within a similar class and have a significant margin in various classes. From the texture of the image, the texture features are extracted. The texture analysis is different from each other by the methodology utilized for extracting textural features. Two categories of extracting textural features are statistical features and complex features.
- Statistical feature extraction.

***a)* Mean:** The mean or average of the pixel p(i, j) values can be determined by the expression as stated below.

$$\mu = \frac{1}{NM}\sum_{i,j} p(i,j) \tag{1}$$

Where p(i, j) is the pixel value at point (i, j) of a picture of dimension M x N.

***b)* Variance:** The variance is the square of the standard deviation value.

$$\text{Variance} = \sigma^2 = \frac{1}{NM}\sum_{i,j}(p(i,j) - \mu)^2 \tag{2}$$

Where, p_{max} and p_{min} are the extreme and least estimation of a pixel in picture p(i, j)

***c)* Standard Deviation:** The standard deviation is the estimation of the average squared deviation of grey pixel estimation from its average estimation and explains the dispersion within a local area. It is determined as follows

$$\text{Standard deviation} = \sigma = \sqrt{Variance} \tag{3}$$

***d)* Entropy:** Entropy is the factual proportion of irregularity that could be utilized to describe the appearance of the acquired input picture, image or graphics. It could also be utilized to explain the distribution changes in an area. Basically, it is used to measure uniformity. Specifically, the entropy (H) of a picture is the theoretical minimum number of bits per pixel required to encode the picture without loss of information.

$$\text{Entropy} = \sum_{k=1}^{Ng} p(r,k) * \log p(r,k) \tag{4}$$

Where, $p_{r,\,k}$ is the probability of the kth grey level, could be determined as N_k/N x M, where N_k is the entire quantity of pixels with the kth grey level and N_g is the entire quantity of grey levels.

- Complex Features.

a) Contrast :

$$\sum_{n=0}^{g-1} n^2 \sum_{i=1}^{g} p(i,j) \sum_{j=1}^{g} p(i,j) \tag{5}$$

This means of contrast or local intensity changes will favorable contributions from p (i, j) away from the diagonal.

b) Correlation :

$$\frac{\sum_{i=0}^{g-1} p(i,j) \sum_{j=0}^{g-1} p(i,j)}{1+(i-j)^2} \tag{6}$$

Correlation calculates the linear dependency of grey levels of neighboring pixels. Computerized picture connection is a partner degree optical strategy that utilizes pursue associate and picture registration techniques for correct two dimensional and three-dimensional calculation of changes in pictures. This is usually accustomed to living deformation, displacement, strain and optical flow however but it is mostly used in wide areas of science and engineering technology.

c) Kurtosis :

$$\frac{\sum_{i=1}^{N} (Yi - \bar{Y})^4 / N}{s^4} \tag{7}$$

$\bar{Y}$ are stands for the mean, s stand for the standard deviation and N is stand for the number of data points. It is a determination of the flatness of data. The Kurtosis of a "Normal" distribution is 3. Kurtosis values greater than 3 denotes that the distribution is peaked relative to the normal (leptokurtic) and if it is less than three the distribution is flatter than the ideal normal curve. Skewness and kurtosis of scanned pictures of different mathematical equations have been determined and used as a feature vector. These features have been mixed with other features and applied to the classifier.

d) Skewness :

$$\frac{\sum_{i=1}^{N} (Yi - \bar{Y})^3 / N}{s^3} \tag{8}$$

Skewness could be a measure of the imbalance of the distribution around the mean. It is computed from the 3rd order moment. Skewness and kurtosis gives a statistical correlation of segment-level features and provides additional information about a scanned image.

- Classification and Recognition: Basically as we know that there is a number of classifiers are available. As per the survey carried out, it has been observed that the BPNN is a well-suited classifier to achieve the required results as an outcome in a proposed system. Neural network based on soft computing techniques. It is utilized to determine the performance of a classifier on the test samples due to such a determination, provides an unbiased estimate of its generalization error. The correctness calculated from the test set can also be utilized to compare the relative performance of a classifier on the same domain. The neural network is able to estimate the posterior probability which gives the basis for establishing the classification rule and performs the statistical evaluation. The feedforward backpropagation neural network with two hidden layers and one input layer with one output layer has been utilized with training function gradient descent and momentum adaptive learning with sum squared error for performance evaluation. In the backpropagation network, start with a training set and utilized the backpropagation method to determined the synaptic weights of a numerous layer perceptron. In this case, a system is defined to extrapolate well when the input-output mapping determined by the system is accurate for test samples never used in creating or training the network. Good extrapolation is that the system efficiency observed during training will persist with the higher possibility for patterns not used in the training.

- Backpropagation Algorithm

  a) Feedforward of the input trained ornament
  b) Backpropagation of the belonging howler
  c) Weight updates

During feedforward, each input neuron (X_i) takes the input signal and spread it to each hidden neuron, which in turn computes the stimulation and passes it on to each output unit that once more computes the activation to get total output. At the time of the training, the total output is examined with the target output and the suitable difference is measured as an error. From the error, the error factor d_k is calculated which is utilized to distribute the error back to the hidden layer. The weights are updated as per the requirement. In the same way, the error factor d_j is determined for units Z_j. Once the error factor is determined then the weights are updated at the same time.

The activation function possessed characteristics such as continuity, differentiability and nondecreasing monotony.

Usually, the activation function is expected to saturate that is approach finite extreme and least values asymptotically. The activation functions commonly used for backpropagation nets are binary sigmoidal which has a range [0, 1] and bipolar sigmoidal which has a range [+1, -1].

The terminology and training algorithm has been explained below.

∂k: part of error rectifying weight updates for wjk
∂j: part of error rectifying weight updates for vij
X: Input Training Vector (X1, X2,….Xi…Xn)
t : Output target vector (t1, t2,….ti,…tm)
a: Learning rate.
X_i: i^{th} input unit
v_{oj}: Bias on the jth unit
Z_j: j^{th} hidden unit

Net input to

$$Z_j, z_{inj} = v_{oj} + \sum x_i v_{ij} \tag{9}$$

and

$$z_j = f\left(z_{inj}\right) \tag{10}$$

w_{ok}: Bias on k^{th} output unit.
Y_k: k^{th} Output unit.
Net input to Y_k is,

$$y_{ink} = w_{ok} + \sum Z_j w_{jk} \tag{11}$$

and

$$y_k = f(y_{ink}) \tag{12}$$

Basically, in the training algorithm, the steps followed are the initialization of weights, feed-forwarding of input parameters, determination of backpropagation error and finally updating weights and biases.

4 Experimental Results with Simulations

The various input images of handwritten mathematical equations have been loaded through a graphical user interface (GUI) and processed through various stages. Finally, the inputted mathematical equation has been identified. The neural network has trained using network training function gradient descent and momentum and adaptive learning. There are one input layer, one output layer and two hidden layers with sum squared error performance. The neural network parameters have been depicted in below Fig. 2. The number of nodes 10 in each respective layer of the neural network. The total count of nodes cannot be predicted before training the network. The neural network is trained with network function gradient descent and momentum and adaptive learning. The neural network elements are as depicted in Table 1. The graphical view of performance showing the overall performance of the system with a number of epochs, training state and regression belongs to the neural network which shows training capacity of neural network with regression rate is as depicted below in Fig. 2.0, Fig. 3.0 and Fig. 4.0.

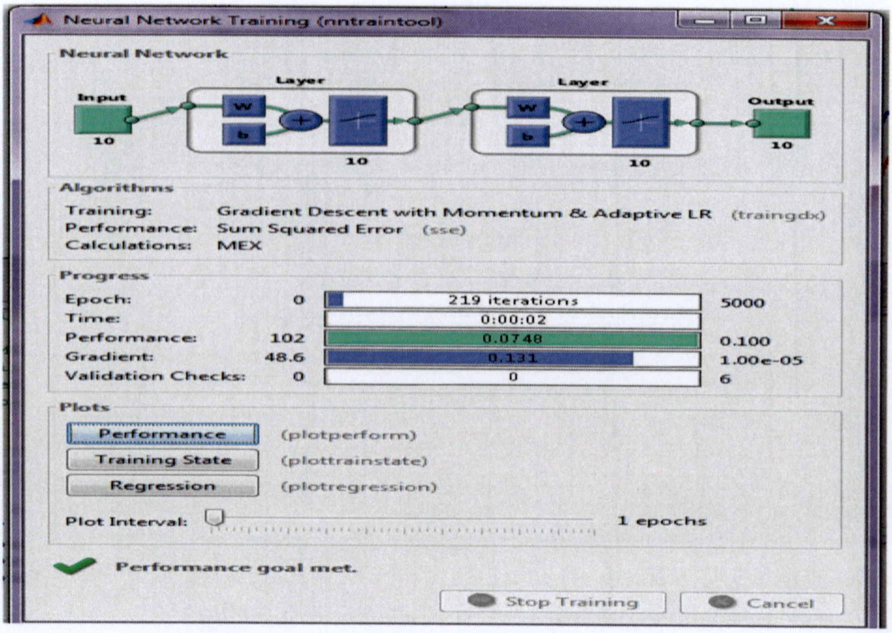

Fig. 2. Neural network training with various parameters.

The below Table 1. Shows various parameters of neural network with values.

Table 1. Parameters or elements of the neural network

Sl. No.	Parameters or Element	Type with value
1	System Instructed Function	Gradient Descent Function
2	Gaining Rule	Adaptive Learning Rule
3	Attainment Function	Mean Squared Error (MSE)
4	Instructed rate	R: 0.98918
5	Total Epochs	5500
6	Total Iteration	219 iterations
7	Attainment	0.0748
8	Gradient value	1.00e-05 i.e. 0.13126
9	Learn	6.5794 at Epoch 268
10	Computation type	MEX type

Table 1 shows the parameters in neural network architecture. It is clear that the network training function is gradient descent and adaptive LR is used. Total numbers of epochs are 5500 and training with R: 0.98918 at 219 epochs. Mean squared error (MSE) is the most extensively utilized. It is determined to evaluate the regression task.

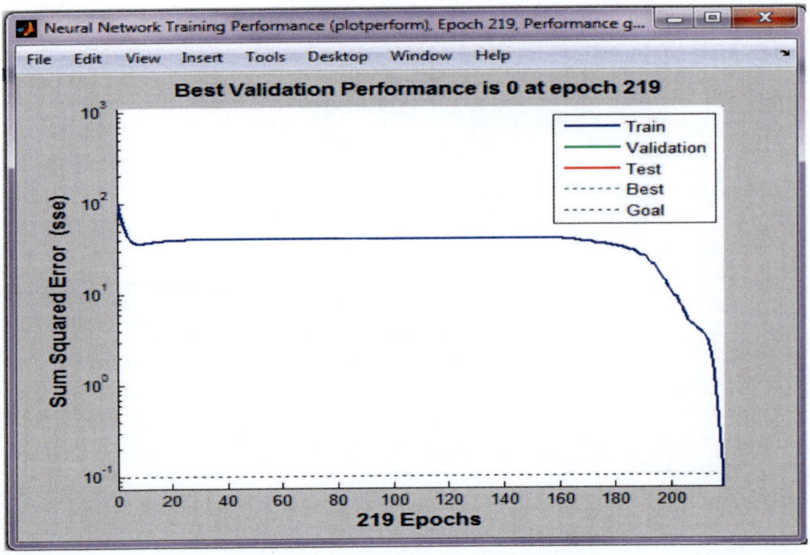

Fig. 3. Neural network training performance.

The best training performance is obtained with mean squared error (mse) 0.0000 at epoch 219. As shown in above Fig. 3, it has been observed that the mean squared error decreases with an increase in the number of epochs.

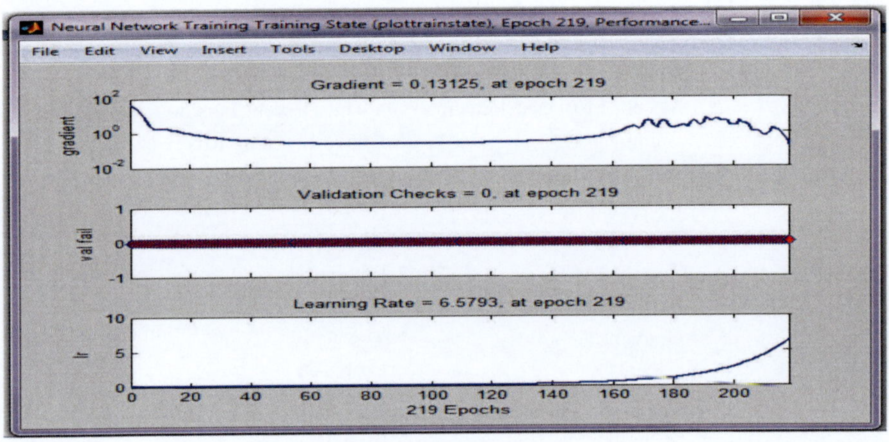

Fig. 4. Neural network training state.

The value of the gradient is 0.13125 with 219 numbers of epochs as shown in Fig. 4. The validation checks are equal to zero at epoch 219. There is no variation in validation check with the number of epochs; it remains the same through the 268 epochs. The learning rate gradually increases after 160 epochs and it is 6.5793 at epoch 268.

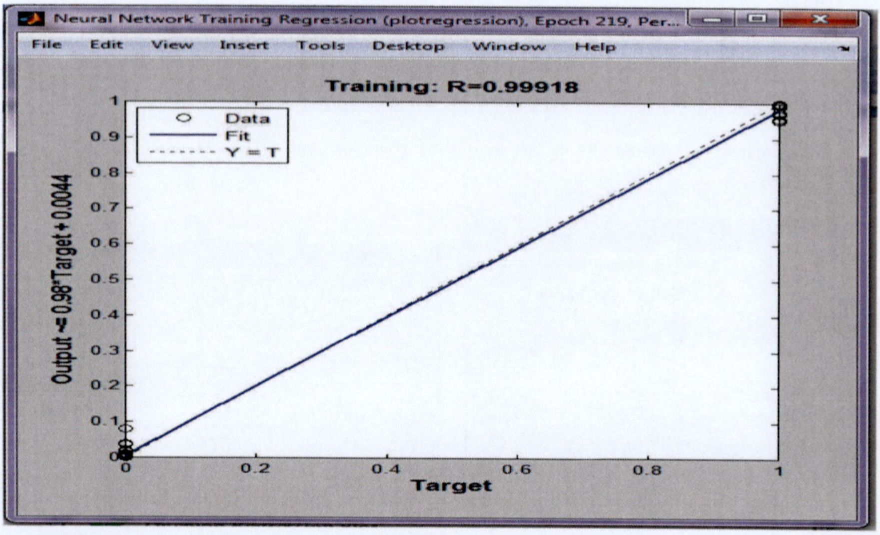

Fig. 5. Neural network training regression.

The neural network training regression with target and output with training R: 0.99918 as shown in above Fig. 5. The variation in the blue line and the dotted line indicates the difference between target and output.

In Fig. 6 as shown below, the various mathematical equations are gone through preprocessing to segmentation to get a noise-free clean and clear image of scanned input equations and are segmented to separate each and every element in a mathematical equation.

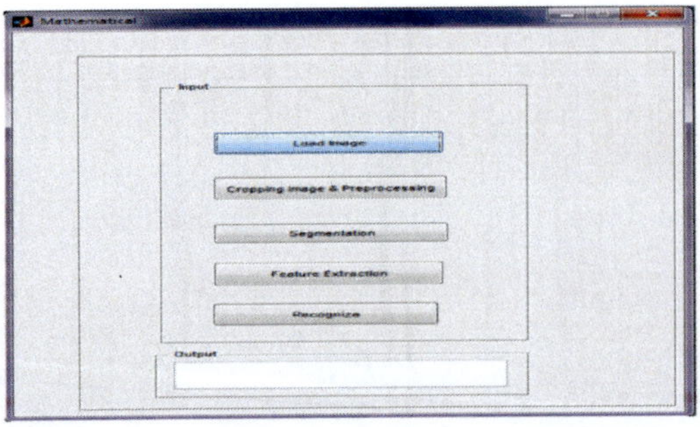

- The graphical user interface for the step by step in recognition of math equations

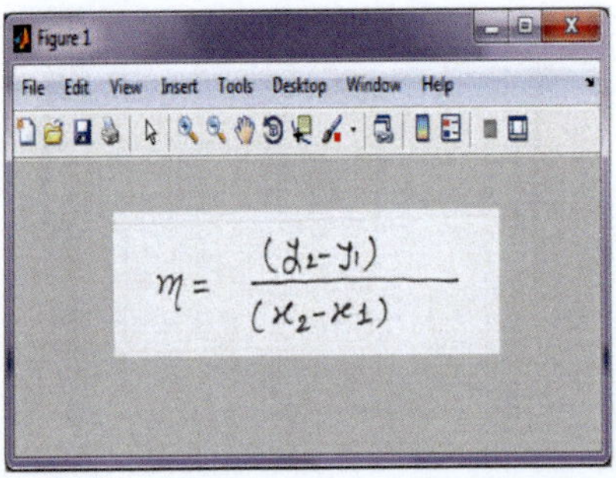

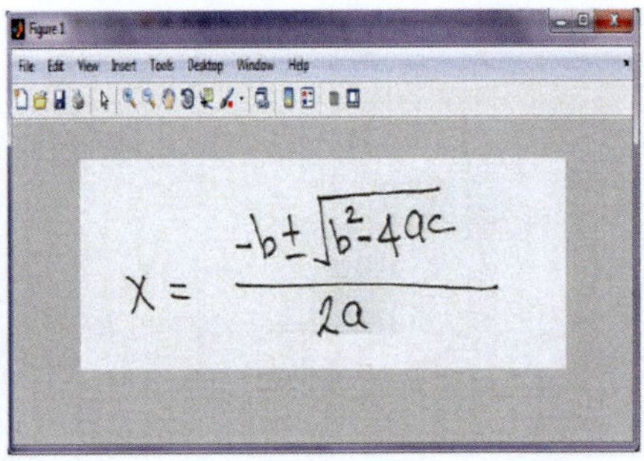

- Acquired input scanned images for the slope of line and complex conjugate equation.

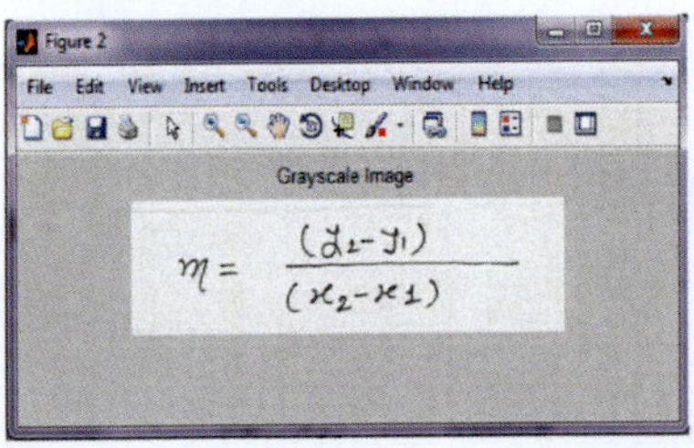

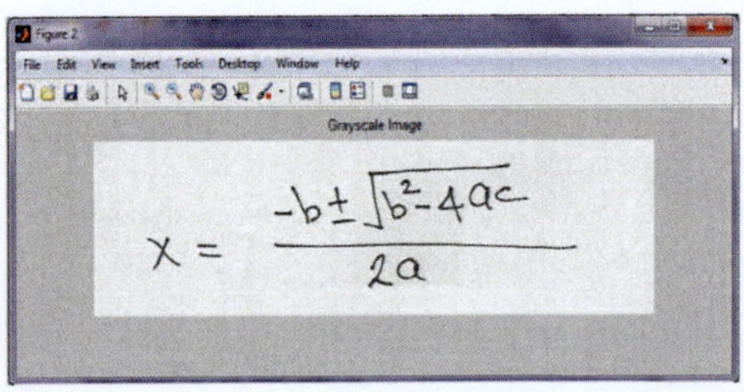

- Gray scale images

Fig. 6. Graphical user interface, preprocessed and segmented image of the equation to be recognized.

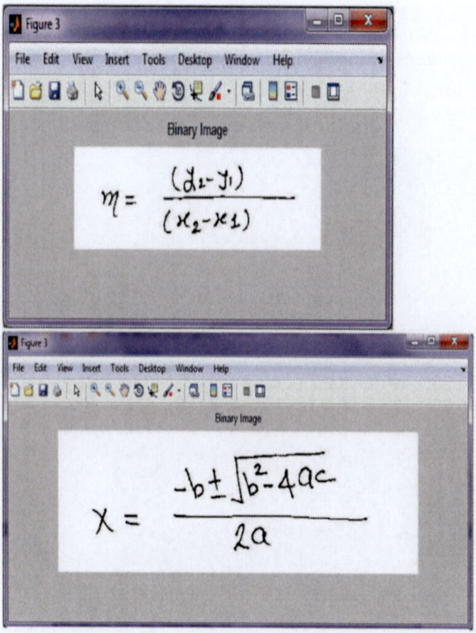

- Binary images

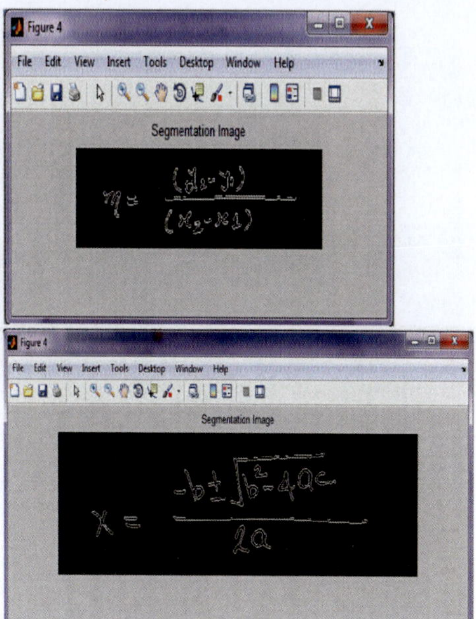

- Segmentation images

Fig. 6. (*continued*)

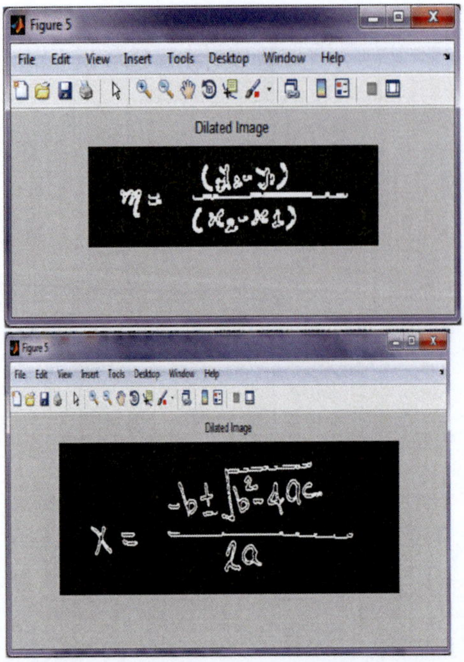

- Dialted images

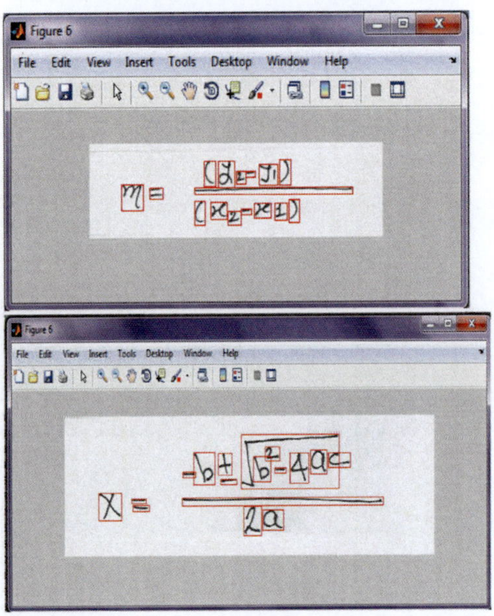

- Segmented images through bounding box.

Fig. 6. (continued)

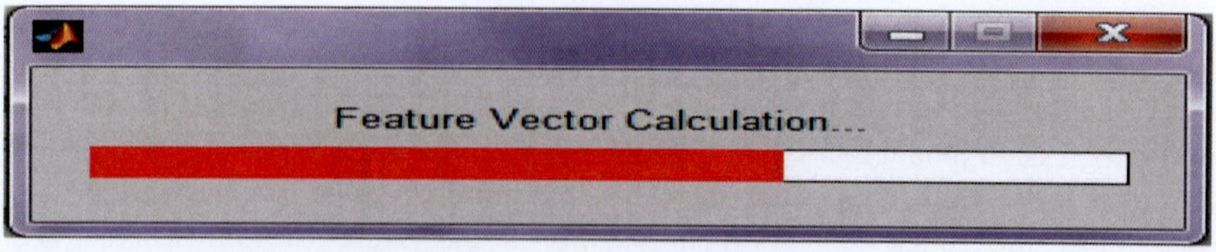

Fig. 7. Feature extracted from the equation to be recognized.

The statistical and complex features from the scanned images have been extracted for feature vector calculation as shown above in Fig. 7. The performance accuracy is depending on the extracted features. The mean, variance, standard deviation, entropy, skew and kurtosis have been extracted and BPNN trained using network function gradient descent function and momentum with adaptive LR.

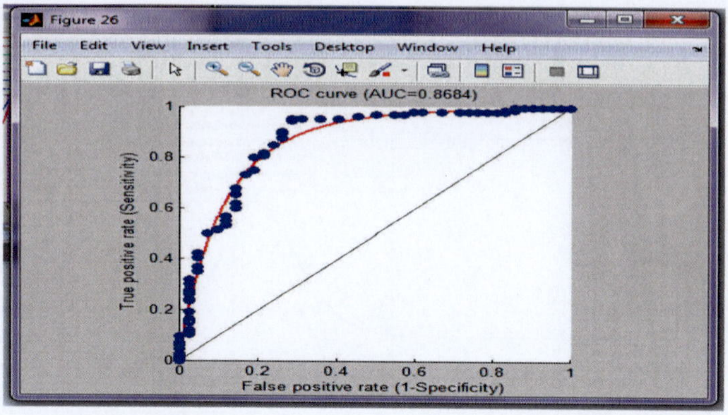

Fig. 8. Receiver operating characteristics curve with false positive rate vs true positive rate.

As shown in above Fig. 8, a ROC curve (receiver operating characteristics curve) is a graph which shows the efficiency of a classification model at all classification thresholds. This graph plots two parameters. True positive rate (TPR) also called recall and false positive rate (FPR).

True Positive accuracy $= \frac{TP}{TP + FN}$

False Positive accuracy $= \frac{FP}{FP + TN}$

Where, TP = True Positive, TN = True Negative, FP = False Positive and FN = False Negative.

A ROC curve at different classification thresholds is as shown above in Fig. 8, lowering the classification threshold more items as positive, thus enhancing both false positive and true positives. To determine the points in an ROC curve, can analyze a logistic regression model many times with various classification thresholds but it is not efficient so there is a sorting based algorithm that can provide information called AUC (Area under the ROC curve) which determines the entire two-dimensional area underneath the entire ROC curve from (0,0) to (1,1). It also gives an aggregate calculation of performance across all possible classification thresholds. AUC ranges in value from 0 to 1. A model whose predictions are 100% wrong has an AUC of 0.0, one whose predictions are 100% correct has an AUC of 1.0.

AUC is desirable as it is scale variant determines how well predictions are ranked, rather than their absolute value and it is also classification threshold invariant calculates the quality of the model's predictions regardless of what classification threshold is selected.

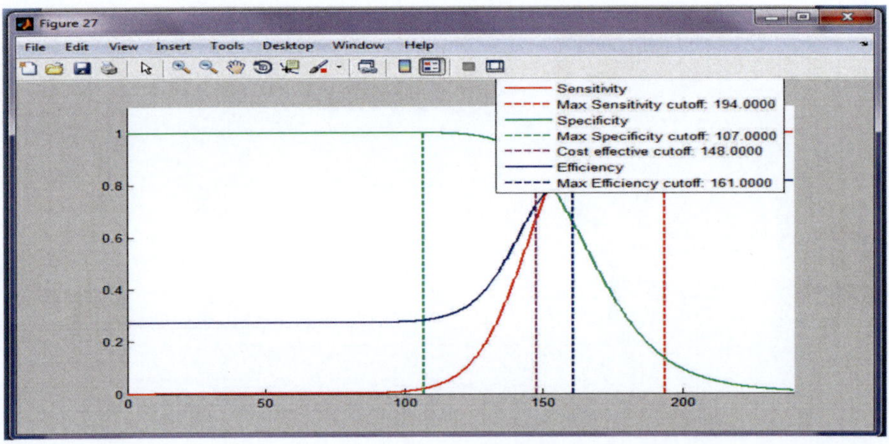

Fig. 9. Graphical representation of sensitivity, specificity and efficiency.

The sensitivity is the probability of, given a positive example, a positive test result. The specificity is the probability of, given a negative example, a negative test result. Specificity relates to the classifier's ability to identify negative results.

The below Fig. 10, has described the result of recognition through the graphical user interface. The input is a scanned image of the handwritten mathematical equation that follows the various steps such as cropping and preprocessing, segmentation, feature extraction and finally recognized that equation successfully.

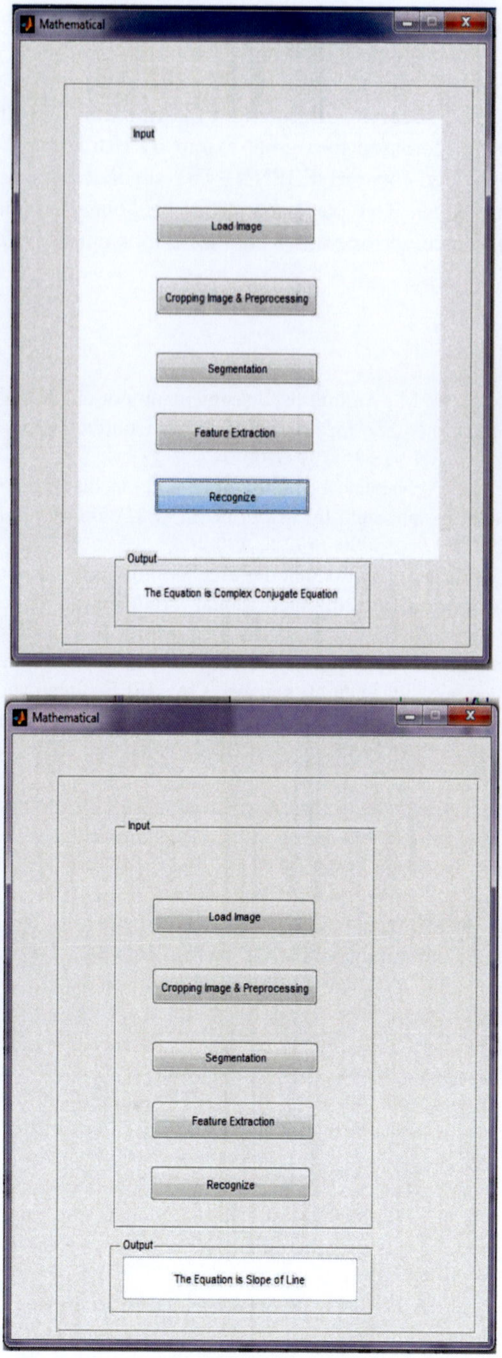

Fig. 10. Recognized handwritten mathematical equations through GUI.

5 Conclusions

The recognition of intricate handwritten mathematical equations have been implemented and demonstrated success with the help of simulation results. The recognition rate and efficiency have increased to a great extent by extracting statistical as well as complex features with feed-forward FBPNN. The success rate of segmentation and recognition is about 99.2%. The performance of the classification model has been described by using the receiver operating characteristics curve (ROC) with AUC.

References

1. Nguyen, C.T., Nakagawa, M.: An improved segmentation of online english handwritten text using recurrent neural network. In: 3rd IAPR Asian Conference on Pattern Recognition, pp. 176–180 (2015). ISBN 978-1-4799-6100-9
2. Emambakhsh, M., He, Y., Nabney, I.: Handwritten and machine-printed text discrimination using template matching approach. In: IEEE, 12th IAPR Workshop on Document Analysis Systems (2016). ISBN 978-1-5090-1792-8, 399-404
3. Ma, C., Zhang, H.: Effective handwritten digit recognition based on multi-feature extraction and deep analysis. In: IEEE, 12th International Conference on the Fuzzy System and Knowledge Discovery (FSKD) (2015). ISBN 978-1-4673-7682-2,297-301
4. Hussien, R.S., Elkhidir, A.A., Elnourani, M.G.: Optical character recognition of Arabic handwritten characters using neural network. In: IEEE, International Conference on Computing, Control, Networking, Electronics, and Embedded System Engineering (ICCNEEE), 7–9 September 2015. https://doi.org/10.1109/iccneee.2015.7381412. ISBN 978-1-4673-7869-7
5. Pan, S., Wang, Y., Liu, C., Ding, X.: A discriminative cascade CNN model for offline handwritten digit recognition. In: IEEE, 14th IAPR International Conference on Machine Vision Applications, Japan, pp. 501–504 (2015). ISBN 978-4-901122-14-6
6. Pujari, P., Majhi, B.: A comparative study of classifier on recognition of offline handwritten odia numerals. In: IEEE, International Conference on Electrical, Electronics, Signals, Communication and Optimization (EESCO) (2015). ISBN 978-1-4799-7678-2
7. Zanibbi, R., Blostein, D., Cordy, J.R.: Recognizing mathematical expressions using tree transformation. IEEE Trans. Pattern Anal. Mach. Intell. **24**(11), 1455–1467 (2002). https://doi.org/10.1109/tpami.2002.1046157. 1939-3539
8. Alvaro, F., Sanchez, J.-A., Benedi, J.-M.: Classification of on-line mathematical symbols with hybrid features and recurrent neural networks. In: IEEE 12th International Conference on Document Analysis and Recognition (2013). https://doi.org/10.1109/icdar.2013.203. ISBN 978-0-7695-4999-6
9. Celik, M., Yanikoglu, B.: Handwritten mathematical formula recognition using a statistical approach. In: 2011, IEEE 19th Signal Processing and Communications Applications Conference (SIU) (2011). https://doi.org/10.1109/siu.2011.5929696. ISBN 978-1-4577-0463-5
10. Zanibbi, R., Mouchere, H., Blostein, D.: Stroke- based performance metrics for handwritten mathematical expressions. In: IEEE, International Conference on Document Analysis and Recognition (2011). https://doi.org/10.1109/icdar.2011.75. ISBN 978-0-7695- 4520-2
11. Iffath Fathima, S., Ashoka, K.: Machine learning approach for recognition of mathematical symbols. Int. J. Sci. Res. Eng. Technol. (IJSRET) **6**(8) (2017). ISSN 2278 – 0882

12. Tapia, E., Rojas, R.: Recognition of on-line handwritten mathematical expressions using a minimum spanning tree construction and symbol dominance. In: Lladós, J., Kwon, Y.-B. (eds.) GREC 2003. LNCS, vol. 3088, pp. 329–340. Springer, Heidelberg (2004)
13. Naik, S., Metkewar, P.: Recognizing offline handwritten Mathematical Expressions (ME) based on a predictive approach of segmentation using K-NN classification. Int. J. Technol. **3**, 345–354 (2015). https://doi.org/10.14716/ijtech.v6i3.1069. ISSN 2086-9614
14. He, W., Luo, Y., Yin, F., Hu, H., Han, J., Ding, E., Liu, C.L.: Context-aware mathematical expression recognition: an end-to-end framework and a benchmark. In: 2016, IEEE 23rd International Conference on Pattern Recognition (ICPR) (2016). https://doi.org/10.1109/icpr.2016.7900135. ISBN 978-1-5090-4847-2

Ritchie's Smart Watch Data Analytics and Visualization

Nandireddy Chaitanya Nath Reddy[1], Aditya Ramesh[1], Rajkumar Rajasekaran[1](✉), and Jolly Masih[2]

[1] School of Computer Science and Engineering, Vellore Institute of Technology, Vellore, India
vitrajkumar@gmail.com
[2] Erasmus University, Rotterdam, The Netherlands

Abstract. The popularity of smartwatches has significantly risen in the past decade owing to advancements in technology, cost reduction, materials availability, production optimization and industrial automation. These factors have compounded the growth of manufacture of smartwatches and coupled with the internet-of-things boom in recent years, this field has seen rapid development and large investments of time and resources. Consequently, with a large amount of data being generated by these smartwatches, the question arises – is it feasible to obtain this data from a watch, and then visualize it in an interpretable manner? Our work answers this with a positive "yes" and demonstrates the proof-of-concept. We propose to build a platform that can extract data from a smartwatch (Mi Band 3 used in proof-of-concept) and visualize this information on a web application dashboard in order to enable the user to obtain insights regarding their physical health and data with ease and convenience.

Keywords: Data visualization · Smartwatch · Health · Analytics · Dashboard · Framework · Web application

1 Introduction

In today's age, the majority of the consumer electronics sector insists on empowering themselves with insights and actionable information. The general populace strives to be afforded personal information and the pursuit to acquire and capture data is at an all-time high. Large product firms such as Apple, Samsung, Fitbit, Google are investing large amounts of revenue into the research and development of products that allow users to gain more information about each other. This is also manifested in the consumer's side as the trend to achieve a healthier lifestyle is improving in the last few years. With the rise of wearable technology, the attention of large firms is on building smartwatches and other such wearables that equip users with the data their existence generates. Keeping these factors in mind, we propose a non-product-specific platform that was proposed to enable users to synchronize and plot the data their wearable devices generate, on a web application dashboard.

J. I.-Z. Chen et al. (Eds.): ICIPCN 2020, AISC 1200, pp. 776–784, 2021.
https://doi.org/10.1007/978-3-030-51859-2_70

The purpose of a web application is because the target end-user would hold a computer that can display and render the results generated by the visualizations. The visualization aspect of the project will be handled using Plotly and Dash due to their simplicity, versatility and usability and given they extend the user a degree of control and manoeuvrability in the visualization of their data. Bootstrap was used to rapidly prototype the proof-of-concept and the plot and dash visualizations were integrated into the front-end. The distinguishing feature from other personalized health dashboard platforms is that this proof-of-concept is product- independent and does not depend on the choice of product - the dashboard does not depend on if the watch is an Apple Watch, or a Fitbit, or another product. Furthermore, this application does not synchronize with Google and is completely client-side as the data the user possesses is local to the system. This eliminates user concerns of privacy as they have full control of their data in this manner.

2 Literature Survey

2.1 My Health Avatar: A Lifetime Visual Analytics Companion for Citizen Well-Being

The above paper considers the use cases of a software termed MyHealthAvatar, which was an undertaking funded and supported by an European commission/entity, whose intent was to design and develop a lifelong partner/companionship for the citizens of the country to track, analyze and collect healthcare, medical and lifestyle information so as to enable citizens to optimize and improve their general state of health. The paper proposed the presentation of this information using interactable components to further optimize usability.

2.2 E-Health Artificial Intelligence System Implementation

The above paper showcased modern approaches toward the electronic-health based AI/ML infrastructures and described potential ideas and techniques of implementing visual dashboards in this context, specifically for the monitoring and tracking of healthcare based data and information. The paper proposes an approach where health-based data is collected and classified from various sources, and also reiterates on the utility of dashboards in this area and field of work, by showcasing various examples.

2.3 Large Scale Mood and Stress Self-assessments on a Smartwatch

The above paper explores how in today's age, wearable technology is ever-increasing and consequently, there is a surge in available data, opening new doors into their research to predict, analyze and determine suspicious or positive trends in user's well-being and health. The paper covers a user-friendly product that is capable of storing the instant conditions of the user's emotions based on various features and development of different algorithms.

3 Proposed Model

Our proposed model is to extract data from the smartwatch on a routine basis and build a dashboard to intuitively plot out the data in order to enable the user to gain actionable insights from their own personal data. The proof-of-concept was deployed using the data extracted from a Mi Band 3 smartwatch. The smartwatch is an affordable, simple model that exhibits standard properties of smartwatches in terms of accessibility, features and tracking sensors.

The dashboard was made using Plotly and Dash for the data visualization, the dataset was stored in CSV files and was extracted raw from the smartwatch, and in order to build an elegant and user-friendly dashboard web application, Bootstrap was used (Fig. 1).

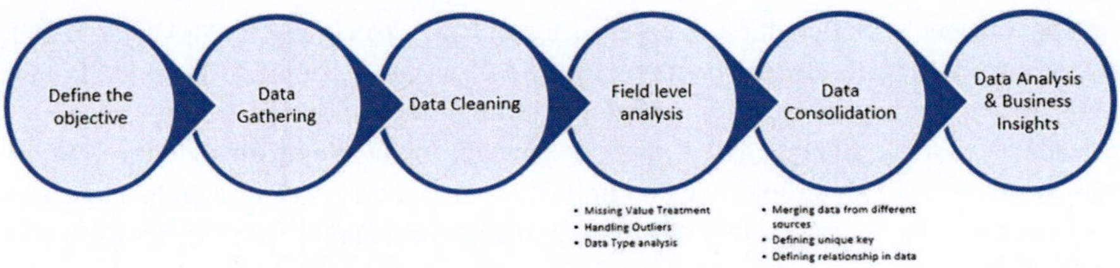

Fig. 1. Workflow for our dashboard

Our dataset consisted of the following features-

3.1 Activity

- date
- lastSyncTime
- steps
- distance
- runDistance
- calories

date	lastSyncTime	steps	distance	runDistance	calories
2018-06-05	1528224750	961	666	12	19
2018-06-06	1528330772	1072	744	81	26
2018-06-07	1528419012	966	668	40	20
2018-06-08	1528510298	4030	2860	244	88
2018-06-09	1528570894	5968	4198	420	147
2018-06-10	1528678915	3669	2575	179	81
2018-06-11	1528762320	783	542	85	16
2018-06-12	1528830233	18	12	0	1
2018-06-13	1528936536	73	50	7	2
2018-06-14	1529025742	750	519	56	14
2018-06-15	1529198983	69	48	8	2
2018-06-16	1529198986	7665	5433	152	163
2018-06-17	1529261507	1524	1054	123	35
2018-06-18	1529373269	7924	5791	167	184
2018-06-19	1529437598	7354	5321	163	162
2018-06-20	1529548894	6833	4849	196	152
2018-06-21	1529627849	7504	5487	143	172
2018-06-22	1529716704	8379	6160	289	192
2018-06-23	1529853148	3477	2415	171	77

3.2 Heartrate

- Date
- Time
- HeartRate

date	lastSyncTime	deepSleepTime	shallowSleepTime	wakeTime	start	stop
2018-06-05	1528224750	0	0	0	1528050600	1528050600
2018-06-06	1528330772	166	283	0	1528225260	1528252200
2018-06-07	1528419012	132	366	0	1528305060	1528334940
2018-06-08	1528510298	134	309	0	1528393020	1528419600
2018-06-09	1528570894	98	409	2	1528481340	1528511880
2018-06-10	1528678915	56	358	0	1528574160	1528599000
2018-06-11	1528762320	94	410	2	1528651980	1528682340
2018-06-12	1528830233	114	352	0	1528739220	1528767180
2018-06-13	1528936536	102	347	0	1528830540	1528857480
2018-06-14	1529025742	67	402	15	1528911900	1528940940
2018-06-15	1529198983	48	472	2	1528997880	1529029200
2018-06-16	1529198986	128	304	0	1529091000	1529116920
2018-06-17	1529261507	173	264	0	1529172180	1529198400
2018-06-18	1529373269	80	300	0	1529264040	1529286840
2018-06-19	1529437598	36	364	0	1529351280	1529375280
2018-06-20	1529548894	63	322	0	1529437680	1529460780
2018-06-21	1529627849	100	421	0	1529517600	1529548860
2018-06-22	1529716704	135	325	2	1529605020	1529632740
2018-06-23	1529853148	104	432	0	1529686020	1529718180

date	time	heartRate
2019-07-07	22:34	82
2019-07-08	10:05	72
2019-07-08	13:21	98
2019-07-08	13:22	78
2019-07-09	17:21	86
2019-07-09	17:59	74
2019-07-10	14:46	80
2019-07-21	20:50	84
2019-07-22	10:53	88
2019-07-22	10:54	77
2019-07-25	17:33	94
2019-07-25	18:33	75
2019-07-25	19:03	77
2019-07-25	22:33	75
2019-07-25	23:22	92
2019-07-26	00:28	88
2019-07-26	01:28	81
2019-07-26	01:51	62
2019-07-26	01:53	60
2019-07-26	01:55	60

3.3 Sleep

- date
- lastSyncTime
- deepSleepTime
- shallowSleepTime
- wakeTime
- start
- stop

The feature-engineering performed was mostly eliminating non-useful data points such as the start, stop times of sleep due to their format being entirely only machine-readable and not human-readable. Owing to the numerical, quantified nature of all the data, primarily line graphs, histograms and bar plots were used and the user was extended control on the viewing of the data by enabling them with sliders, and features such as zoom, pan, scroll, and drop down options. These features combined enabled the user to intuitively iterate over the options and access data points specific to specific dates, or above certain ranges or detect outliers in the data.

Also, worth mentioning is the fact that the data obtained above was from the Mi Band 3, and the hardware specifications of the device were as follows-

- Battery life of 110 mAh - this decided the maximal continuity of data in the dataset. Thus the dataset has no consistent information in over a 20-day period without any gaps.
- Weight of 20 gm, accordingly the data generated in regard to calories burnt, weight, etc. must be normalized.
- Operating temperature of 10–50 °C, the sensitivity of the input sensors was dependent on the ambient surrounding temperature. The data was collected in locations possessing an average temperature of 25–28 °C.
- The sensors the watch possessed were a 3-axis accelerometer to detect motion in the X, Y and Z axes, and a PPG- Heart Sensor to track pulse and sleep quality.

As the above-used smartwatch is a budget smartwatch, the results may be subject to a margin of inaccuracy, however, given the end-user would most likely possess a smartwatch of the similar price range, the proof-of-concept is far more realistic than its counterpart version wherein an expensive, hyper-sensitive smartwatch was used.

4 Demo and Results

Below are enclosed sample screens from the developed dashboard. The dashboard possesses support for activity tracking, light sleep tracking, deep sleep tracking, wakeTime tracking, and pulse monitoring. The model that we propose involves the extraction of the user-generated data live from the smartwatch using the smartwatch's exclusive APIs on a routine basis, and build a web application-based dashboard to plot out the information in order to enable the user to gain actionable insights from their own personal data. The proof-of-concept was deployed using the data extracted from a common smartwatch. The smartwatch is an affordable, simple model that features conventional properties of smartwatches in terms of accessibility, features and tracking sensors.

The dashboard was built using Plotly and Dash for the data visualization, the dataset was stored in CSV files and were extracted raw from the smartwatch, and in order to construct an elegant and user-friendly dashboard web application, Bootstrap was used.

Owing to the numerical, quantified nature of all the data, primarily line graphs, histograms and bar plots were used and the user was extended control on the viewing of the data by enabling them with sliders, and features such as zoom, pan, scroll, and drop down options. These features combined enabled the user to intuitively iterate over the options and access data points specific to specific dates, or above certain ranges or detect outliers in the data (Figs. 2, 3, 4 and 5).

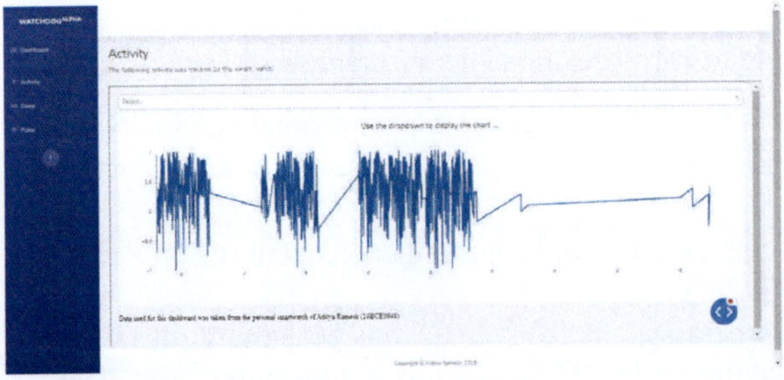

Fig. 2. Activity tracking on dashboard

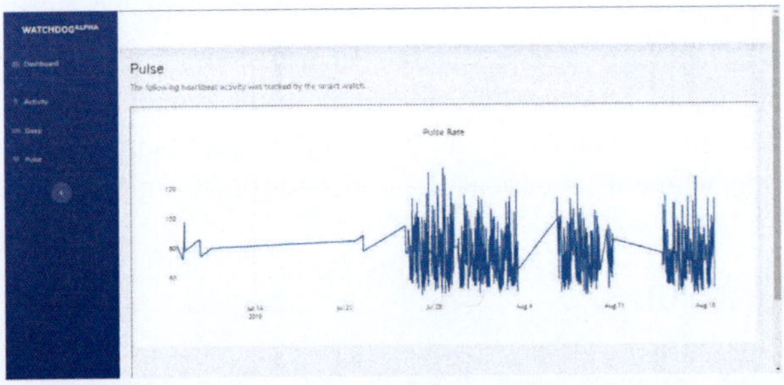

Fig. 3. Pulse tracking on dashboard

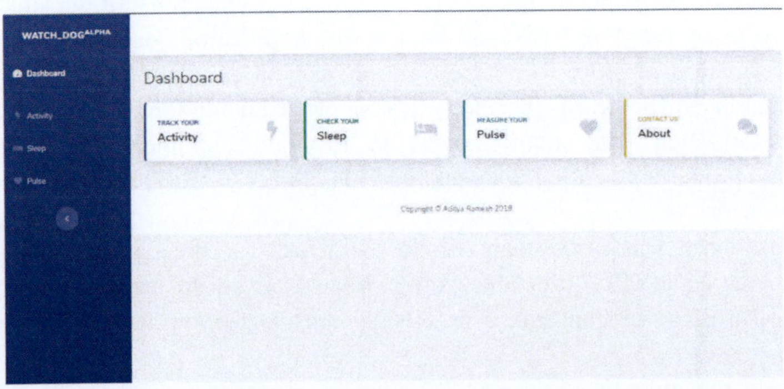

Fig. 4. Home page of dashboard

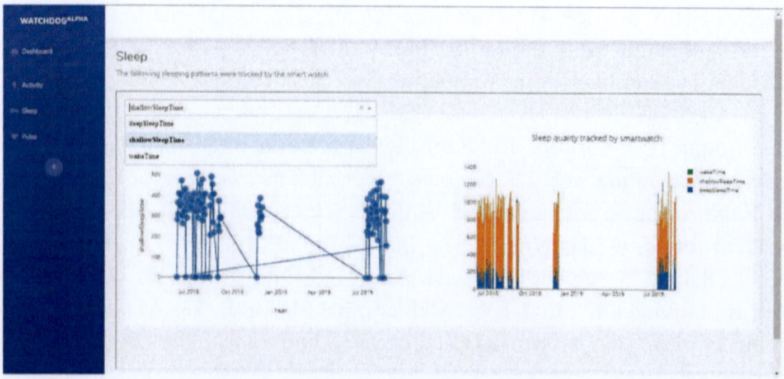

Fig. 5. Sleep tracking of dashboard

5 Conclusion

As proposed, the system developed plots out the data generated by the user's smartwatch in an intuitive and user-friendly manner leveraging human-computer interaction principles, and thus enables the user to make educated decisions based on the insights gained by this method of data visualization. Future enhancements could be further feature-engineering and providing more flexibility to modify the visualization to the users, and also extending the current visualizations to support large-scale high volume high velocity and high variety datasets, and also make the visualization smartwatch-agnostic.

References

1. Danova, T.: The wearables report: growth trends, consumer attitudes, and why smartwatches will dominate, May 2015
2. Ståhl, A., Höök, K., Svensson, M., Taylor, A.S., Combetto, M.: Experiencing the affective diary. Pers. Ubiquit. Comput. **13**(5), 365–378 (2009). https://doi.org/10.1007/s00779-008-0202-7
3. Shneiderman, B., Plaisant, C., Hesse, B.W.: Improving health and healthcare with interactive visualization methods. IEEE Comput. **46**(1), 58–66 (2013)
4. Greenhalgh, T., Stramer, K., Bratan, T., Byrne, E., Russell, J., Potts, H.W.: Adoption and non-adoption of a shared electronic summary record in England: a mixed-method case study. BMJ **340**, c3111 (2010)
5. NHS Connecting for Health. Clinical Dashboards. NHS Connecting for Health (2012)
6. Quazi, M.T., Mukhopadhyay, S.C., Suryadevara, N.K., Huang, Y.M.: Towards the smart sensors based human emotion recognition. In: 2012 IEEE International Instrumentation and Measurement Technology Conference (I2MTC 2012) (2012)
7. Ananthanarayan, S., Siek, K.A.: Persuasive wearable technology design for health and wellness. In: 2012 6th International Conference on Pervasive Computing Technologies for Healthcare (Pervasive Health), pp. 236–240. IEEE (2012)

8. Choe, E.K., Lee, N.B., Lee, B., Pratt, W., Kientz, J.A.: Understanding quantified-selfers' practices in collecting and exploring personal data. In: Proceedings of the SIGCHI Conference on Human Factors in Computing Systems, Toronto, Ontario, Canada, 26 April–01 May 2014 (2014)
9. Fernandez-Luque, L., Karlsen, R., Krogstad, T., Burkow, T., Vognild, L.: Personalized health applications in the web 2.0: the emergence of a new approach. In: Proceedings of the IEEE on Engineering in Medicine and Biology Society, pp. 1053–1056, August 2010
10. West, V.L., Borland, D., Hammond, W.: Innovative information visualization of electronic health record data: a systematic review. J. Am. Med. Inform. Assoc. **22**(2), 330–339 (2015)
11. Rajkumar, R., Govinda, K., Jindal, A., Mehtani, R., Masih, J., Sai, M.C.: Visualization effect of changing climatic trends on natural calamities: hurricanes. https://doi.org/10.36872/lepi/v50i3/201022
12. Sharma, V., Rajkumar, R., Badhrinarayanan, S.: Visualization of data mining techniques for the prediction of breast cancer with high accuracy rates. J. Comput. Sci. **15**(1) (2019). https://doi.org/10.3844/jcssp.2019.118.130

Sentiment Analysis for Product Rating Using Classification

K. Govinda(✉), K. Naveenraj, and Somula Rama Subba Reddy

SCOPE, VIT, Vellore, India
govindkm@gmail.com, naveenrajkrishnaraj@gmail.com,
svramasubbareddy1219@gmail.com

Abstract. This paper proposes a machine learning technique for implementation of Sentiment Analysis in rating the product. This system would detect hidden viewpoints in the comment section of an online marketing site and subsequently give the rating of the product. It utilizes the assessment investigation approach with the end goal to accomplish the coveted usefulness. This depends on an E-Commerce web application where clients should enroll themselves and after that they view the item and its highlights and along these lines remark about its highlights. The framework will then break down the item and dependent on the surveys given by the clients rank the item. This will utilize a database of assessment-based watchwords alongside energy or antagonism weight in database and after that dependent on these slant catchphrases mined in client remark is positioned. The remarks will be dissected by contrasting them and the catchphrases put away in the database. The framework takes in different remarks of the clients of that item in this manner indicating whether the item is great, terrible, or normal. The framework likewise holds the capacity of giving the clients a chance to see the remarks posted by different customers or clients. The job of the administrator is to add item to the framework and to include catchphrases in database. This empowers the client to effortlessly discover the right item for utilization. This is likewise valuable for those clients who need to peruse audits about the item.

Keywords: E-Commerce · Classification · NLP · ROC · AUC

1 Introduction

Sentiment Analysis is additionally called opinion mining and it makes use of computational science, content investigation, procedure linguistics and statistics to deliberately distinguish, separate, value and examine emotional states and abstract knowledge. Slant investigation is widely connected to voice of the consumer materials, for instance, audits and study reactions, on the web and online networking, and welfare materials for applications that vary from showcasing to consumer administration to clinical prescription. Emotional investigation expects to come to a decision about the state of mind of a speaker, essayist, or alternative subject as for some theme or the normally logical extremity or torrid response to a report, association, or occasion.

J. I.-Z. Chen et al. (Eds.): ICIPCN 2020, AISC 1200, pp. 785–793, 2021.
https://doi.org/10.1007/978-3-030-51859-2_71

Preprocessing of data is essential for obtaining reliable and highly accurate results from a dataset. It comprises of different processes which help to transform raw data into easily understandable data. Parameter selection is performed to remove unessential parameters and their values from the dataset to reduce the time and space complexity of the model. Missing values in the dataset are eliminated by either deleting the rows containing such values or by replacing the missing values with a constant value or an arithmetically derived value. Due to numerous complexities associated with categorical variables, they are generally converted to numerical variables by assigning fixed numerical values to each category.

Machine learning models are implemented through two modes of learning namely supervised learning and unsupervised learning. In Supervised learning, the dataset is split into the training set and the test set. The data points in the training set are marked with their outcomes while the data points in the test set are used for testing the predictive capability of the model. Initially, some data points from the training set with their respective outcomes are inputted into the classifier. The knowledge gained by the classifier model by learning about the training set data is utilized to predict the outcomes for similar data points from the test set.

Unsupervised learning also known as clustering involves utilizing unlabelled dataset. The data points in the dataset are clustered together based on similarities, distinct patterns and outliers. These clusters of data were utilized to predict the outcome. Unsupervised learning models are considered less accurate than supervised learning due to lack of human involvement. Some of the widespread applications of unsupervised models include detection of anomalous financial transactions and grouping of users and products on E-commerce and social networking sites.

Classification models are a prominent type of supervised machine learning models which utilizes single or multiple inputs to predict single or multiple outcomes. Generally, outcomes of a classifier are categorical in nature. In other words, outcomes of a classifier are well-defined and are restricted to limited number of values. Some of the prominent classifiers are Decision tree, Naïve Bayes, Random forest, Logistic regression and Gradient-boosted tree.

Classifiers have a wide range of real-world applications across different fields. Naïve Bayes classifier is widely used to detect spam emails by calculating the probability of an email belonging to spam or not spam category based on the number of spam related words and phrases in the mail. Random Forest Model is used in recommendation engines of streaming platforms like Netflix. Recommendations are provided to a user based on the viewing activities of other users with similar viewing patterns.

2 Literature Review

Sentiment Analysis is the assignment of tapping this goldmine of data. It recovers sentiments about specific items or includes and arranges them as suggested or not prescribed, that is certain or negative. The conclusion with respect to a specific item in an audit is only from time to time unequivocally constructive or antagonistic; rather individuals have a tendency to have a blended feeling about different highlights, some positive and some negative. In this manner the component particular feeling matters more than the general supposition [1].

Quick increment in web clients alongside developing intensity of online survey destinations and internet based life has brought forth Sentiment investigation or Opinion mining, which goes for figuring out what other individuals think and remark. Feelings or Opinions contain open created content about items, administrations, strategies and governmental issues. Individuals are typically intrigued to look for positive and negative assessments containing different preferences, shared by clients for highlights of specific item or administration. Consequently, item highlights or perspectives have huge part in assumption investigation [2].

Feelings towards an item through online surveys will impact buy choices of other online buyers by changing their observations on the item quality. Besides, every item viewpoint may affect purchasers' expectations in an unexpected way. In this manner, supposition investigation and econometric models are fused to look at the connection between buy expectations and angle assessment sets, which empower the weight estimation for every item viewpoint [3].

First recognize the item angles and lessen measurements to separate perspective assessment sets. Next the data pick up is figured for every angle through entropy hypothesis. In light of conclusion extremity and slant quality, we plan an econometric model by coordinating the data put on to quantify the viewpoint's weight [3].

In a regular web-based shopping website, the data with respect to the different items is put away in a RDBMS back-end storage site. As this is reasonable for online Transaction Processing, this is not the best approach for smart questioning over the items. Likewise, the appraisals for the items sold online are accessible promptly on the merchant locales or at their different purposes of offers, however these evaluations are not totally dependable. The appraisals are typically standardized to a size of 0 to 5 and either the clients don't take due thought in giving the correct evaluations or the seller themselves present fake evaluations on improve their items look. In either case, these evaluations are inclined to conclusion spamming [4].

The remarks given by client for an item is viewed as positive at one circumstance and negative at other circumstance. A few people don't express feelings similarly. Most audits will have both positive and negative remarks, which fairly sensible by breaking down sentences each one in turn. Once in a while people may give counterfeit remarks about the item, which gives the terrible audit about the item [5].

NLP is the arrangement of systems and procedures which enables PC to comprehend human discourse as it is talked. Normal NLP errands in machine learning include: sentence division, parts of discourse labelling, parsing content outcomes, profound investigation and named substance extraction. Controlled Natural Language [CNL] handling is a subset of NLP which confines the sentence structure to expand the effortlessness for labelling. Fundamental units of assessment examination are highlights which can be removed frame item surveys. These highlights can be lemmas, multi-words and valance shifters [6].

Data recovery (IR) manages the capacity, portrayal, association of, and access to data things, the portrayal and association of which furnishes the client with simple access to the data in which he is intrigued. At the end of the day, IR is discovering material of an unstructured nature that fulfils a data require from inside huge accumulations. IR frameworks distinguish the reports in a gathering which coordinates a client's question and subsequently limit the arrangement of records that are applicable to a specific issue along these lines accelerating the investigation extensively by diminishing the quantity of archives to be broke down [6].

The Apache Hadoop venture is an open-source programming work for adaptable and conveyed registering. It gives preparing systems that permits to extensive scale handling of information on groups of PCs. Hadoop works when the info/yield in the configuration of Key-Value sets [7].

Data recovery (IR) manages the capacity, portrayal, association of, and access to data things, the portrayal and association of which furnishes the client with simple access to the data in which he is intrigued. At the end of the day, IR is discovering material of an unstructured nature that fulfils a data require from inside huge accumulations. IR frameworks distinguish the reports in a gathering which coordinates a client's question and subsequently limit the arrangement of records that are applicable to a specific issue along these lines accelerating the investigation extensively by diminishing the quantity of archives to be broke down [6].

The Apache Hadoop venture is an open-source programming work for adaptable and conveyed registering. It gives preparing systems that permits to extensive scale handling of information on groups of PCs. Hadoop works when the info/yield in the configuration of Key-Value sets [7].

3 Proposed Method

The method used in this sentimental analysis is classification based on the Naïve Bayes Classification and Decision Tree based classification. The data set obtained from Amazon contains 5000 user reviews. The data set will be read using the module called pandas and data frame. The dataset is pre-processed to filter out any nonessential attributes and ensures that the dataset contains only those columns of attributes that are required for the classification methods stated above. Then the data set will be trained in order for the classifier to identify the negative, positive and neutral words in the reviews list. Then we will set up the code using Naïve Bayes Classification method that will use multinomial distribution to classify the positive and negative reviews. This will also be done for Decision Tree based classification which is done using a flowchart. The accuracy of the two classifier models are calculated using the NLTK package in python and are compared to determine the model better suited for sentiment analysis.

Naive Bayes classifiers is a "probabilistic classifier" supported relating supposition with solid (credulous) independence presumptions amongst the alternatives. Naive Bayes is a basic technique for developing classifiers: models that allot class names to issue occurrences, spoke to as vectors of highlight esteems, wherever the class marks are drawn from some restricted set. There's anything but a solitary calculation for preparing such classifiers, yet a group of calculations dependent on a typical rule: all naive Bayes classifiers assume that the approximation of a particular element is free of the approximation of some supplementary element, given the class variable. For rare kinds of probability models, Naive Bayes classifiers can be ready effectively in a focused attainment the hang of setting. Parameter estimation for naive Bayes models exploits the policy for most extreme probability. Naive Bayes is a provisional probability model given by the formula:

$$p(C_k|\mathrm{x}) = \frac{p(C_k)p(\mathrm{x}|C_k)}{p(\mathrm{x})}$$

In basic English, using Bayesian probability vocabulary, the above calculation is written as follows:

$$\text{posterior} = \frac{\text{prior} \times \text{likelihood}}{\text{evidence}}$$

We just require the numerator of that division in light of the fact that the denominator does not count on upon C and the estimations of x(i) are given so the denominator is viably steady. The numerator is identical to the joint likelihood demonstrate:

$$p(C_k, x_1, \ldots x_n)$$

Thus, comes the following chain-rule:

$$\begin{aligned}
p(C_k, x_1, \ldots x_n) &= p(x_1, \ldots x_n, C_k) \\
&= p(x_1|x_2, \ldots x_n, C_k)p(x_2, \ldots x_n, C_k) \\
&= p(x_1|x_2, \ldots x_n, C_k)p(x_2|x_3, \ldots x_n, C_k)p(x_3, \ldots x_n, C_k) \\
&= \ldots \\
&= p(x_1|x_2, \ldots x_n, C_k)p(x_2|x_3, \ldots x_n, C_k)\ldots p(x_{n-1}|x_n, C_k)p(x_n|C_k)p(C_k)
\end{aligned}$$

Thus the joint model can be written as:

$$\begin{aligned}
p(C_k|x_1, \ldots x_n) &\propto p(C_k, x_1, \ldots x_n) \\
&= p(C_k)p(x_1|C_k)p(x_2|C_k)p(x_3|C_k)\ldots \\
&= p(C_k)\prod_{i=1}^{n} p(x_i|C_k).
\end{aligned}$$

Under the above autonomy presumptions, the contingent dissemination over the class variable C is:

$$p(C_k|x_1, \ldots x_n) = \frac{1}{Z}p(C_k)\prod_{i=1}^{n} p(x_i|C_k)$$

Where Z is a scaling factor:

$$Z = p(\mathrm{x}) = \sum_k p(C_k)p(\mathrm{x}|C_k)$$

and is dependent on x (1), ….., x(n).

Decision Tree methodology of cataloging is one of the perceptive displaying ways utilized in measurements, information mining and machine learning. Tree models where the impartial variable can take a separate arrangement of abilities are referred to as characterization trees; in these tree structures, leaves express to class names and divisions speak to mixtures of highlights that prompt those class marks. choice trees where the impartial variable can take tenacious qualities (commonly genuine numbers) are referred to as relapse trees. In decision analysis, a decision tree can be used to visually and clearly represent decisions and decision making. In data mining, a decision tree labels data (but the resultant classification tree can be an input for decision making).

The aim is to form a model that predicts one of the revelatory exhibiting strategies applied in measurements, data mining and machine learning.

Choice trees wherever the target variable can take persistent qualities (commonly genuine numbers) are referred to as relapse trees.: $(\mathrm{x}, Y) = (x_1, x_2, x_3, \ldots x_k, Y)$

4 Results

The system uses the Naïve Bayes and Decision Tree method of Classification. It makes use of a data set from Amazon's products in the form of a CSV file. The data is first pre- processed to remove all the unnecessary words like stop words and all the unnecessary columns. Then the data is then trained so that the classifiers can identify the various positive, negative and neutral words in the data set. This trained data set is then processed using the Naïve Bayes classifier as well as the Decision tree so as to calculate the number of positive features as well as the negative features. The system also calculates the accuracy of the two different classification methods used so as to determine which classifier is more accurate than the other. The software's used are python 3.6, IDLE platform to run the code and also different packages to make use of the available modules in NLP.

Accuracy of Naïve Bayes Classification: **0.6875** Accuracy of Decision Tree Classification: **0.76**

Thus the classification using Decision Tree is more accurate than Naïve Bayes Classifier.

The Fig. 1 represents the Receiver Operating Characteristics curve with the Area Under. The Curve (AUC) represented as the dotted line. Since the higher the AUC, the better the distinguishable the training model would be between the two positive and negative classes and since the AUC value for the above curve is 0.98, the Decision Tree model is accurate enough in distinguishing between the positive and negative classes.

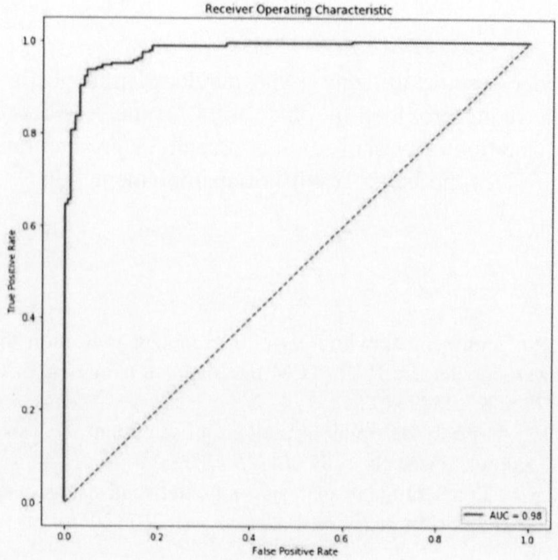

Fig. 1. ROC curve

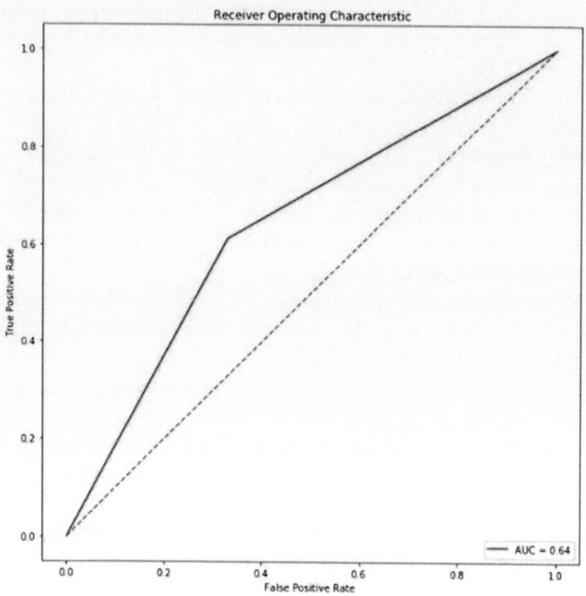

Fig. 2. AUC curve

The Fig. 2 shows AUC curve is 0.64, which details that the Naive Bayes is not capable enough to classify the products as positive or negative as compared to the training model that uses Naive Bayes.

5 Conclusion

In the todays eServices sector, utilization and purchase strategies over the internet is mainly based on the rating provided by other users for the services and products, this paper is the implementation of classification approach for product rating. The decision tree based approach gives the better result in our implementation.

References

1. Mei, J., Frank, R.: Sentiment crawling: extremist content collection through a sentiment analysis guided web-crawler. In: IEEE/ACM International Conference 2015 (2015). https://doi.org/10.1145/2808797.2809373
2. Ravi, K., Ravi, V.: A survey on opinion mining and sentiment analysis: tasks, approaches and applications. Knowl. Based Syst. **89**, 14–46 (2015)
3. Jain, A.P., Katkar, V.D.: Sentiment analysis of twitter data using data mining. In: 2015 International Conference on Information Processing (ICIP) (2015). https://doi.org/10.1109/infop.2015.7489492
4. D'Andrea, A., Ferri, F., Grifoni, P., Guzzo, T.: Approaches, tools and applications for sentiment analysis implementation. Int. J. Comput. Appl. (0975 – 8887) **125**(3), 26–33 (2015)
5. Bakshi, R.K., Kaur, N., Kaur, R., Kaur, G.: Opinion mining and sentiment analysis. In: 2016 3rd International Conference on Computing for Sustainable Global Development (INDIACom), vol. 31 October 2016
6. Pang, B., Lee, L.: Opinion mining and sentiment analysis. Found. Trends Inf. Retrieval **2**(1–2), 1–135 (2008)
7. Aciar, S.: Mining Context Information from Consumer's Reviews (2010). http://ids.csom.umn.edu/faculty/gedas/cars2010/Aciar-CARS-2010.pdf
8. Gural Vural, A., Barla Cambazoglu, B., Senkul, P.: Sentiment focused web crawling. In: CIKM 2012, ACM Transactions on the Web, Maui, HI, USA, 29 October–2 November 2012 (2012). https://doi.org/10.1145/2396761.2398564
9. Hailong, Z., Wenyan, G., Bo, J.: Machine learning and lexicon based methods for sentiment analysis: a survey. In: 2014 11th Web Information System and Application Conference (2014). https://doi.org/10.1109/wisa.2014.55
10. Kaushik, A., Kaushik, A., Naitani, S.: A study on sentiment analysis: methods and tools. Int. J. Sci. Res. (IJSR) **4**(12), 2319–7064 (2015)
11. Devika, M.D., Sunitha, C., Ganesh, A.: Sentiment analysis: a comparative study on different approaches. Procedia Comput. Sci. **87**(2016), 44–49 (2016). Fourth International Conference on Recent Trends in Computer Science & Engineering, Chennai, Tamil Nadu, India
12. Hemmantian, F., Sohrabi, M.K.: A Survey on Classification Techniques for Opinion and Sentiment Analysis (2017). https://doi.org/10.1007/s10462-017-9599-6
13. Sharma, J.: Review paper on text mining and sentimental analysis. Int. J. Eng. Comput. Sci. **6** (2017). https://doi.org/10.18535/ijecs/v6i11.05
14. Goyal, S.: Review paper on sentiment analysis of twitter data using text mining and hybrid classification approach. IJEDR **5**(2), 1–9 (2017). ISSN 2321-9939
15. Rajkumar, R., Sharma, V.: Visualization of data mining techniques for the prediction of breast cancer with high accuracy rates. J. Comput. Sci. **15**(1), January 2019. https://doi.org/10.3844/jcssp.2019.118.130

16. Shayaa, S., Jaafar, N.I., Bahri, S., Sulaiman, A.: Sentiment analysis of big data: methods, applications, and open challenges. IEEE Access **6** (2018). https://doi.org/10.1109/ACCESS.2018.2851311
17. Vohra, S.M., Teraiya, J.B.: A comparative study of sentiment analysis techniques. J. Inf. Knowl. Res. Comput. Eng. **02**(02) (2018). ISSN 0975–6760
18. Martin-Gomez, L., Caceres, M.N.: Applying data mining for sentiment analysis, trends in cyber-physical multi-agent systems. In: The PAAMS Collection - 15th International Conference, PAAMS 2017, pp. 198–205 (2018)
19. https://www.cse.iitb.ac.in/~pb/papers/cicling12-feature-specific-sa.pdf
20. http://www.statmt.org/OSMOSES/FeatureEx.pdf
21. https://www.tandfonline.com/doi/abs/10.1080/0952813X.2015.1132270?src=recsys&journalCode=teta20
22. Rajkumar, R., Masih, J., Govinda, K.: An analysis of mobile pass-codes in case of criminal investigations through social network data. Int. J. Comput. Appl. September 2019. https://doi.org/10.1080/1206212X.2019.1662169

Face Recognition System Using a Hybrid Scale Invariant Feature Transform Based on Local Binary Pattern

M. Koteswara Rao[1(✉)], K. Veera Swamy[2], and K. Anitha Sheela[3]

[1] ECE Department, JNTUCEH, Hyderabad, Telangana, India
koteswara.rao@qiscet.edu.in
[2] ECE Department, Vasavi College of Engineering, Hyderabad, Telangana, India
k.veeraswamy@staff.vce.ac.in
[3] ECE Department, College of Engineering, Hyderabad JNTUH, Hyderabad, Telangana, India
kanithasheela@jntuh.ac.in

Abstract. In this paper, the face recognition system using a hybrid scale-invariant feature transform (SIFT) based on local binary pattern (LBP) has been implemented. The face recognition system is preprocessed to extract the features based on SIFT and LBP. The modified local binary pattern is extracted the local features on the binary pattern on each subblock of the image. Histogram is implemented on each subblock of the binary number of the neighbor values. Modified scale-invariant feature transform is preprocessed to extract the features based on the scaling factors and various orientations of the images. This proposed method is classified by the support vector machine. The experimental results have been executed on the ORL, YALE databases, and based on the performance measures in terms of precision, recall, and F-Score. This paper gives better as compared to existing models.

Keywords: Modified LBP · Modified SIFT · Cosine angle based distance measure

1 Introduction

In machine learning system texture classification [4, 6, 12] is one of the most popular techniques to recognition the query from the large database. It is a popular technique for many security systems. Texture classification [5, 7, 8] has two types of phases. One is the learning phase and another one is the recognition phase. In the real world every moment has to be sorted for a better future. Behalf of that is time to give more statistical models for the machine learning phase key role in the feature extraction model. It simply exposed that learning the skills from experience. Then the recognition phase in the most powerful technique for predicting the test image from the large face dataset based on this approach can be verified as genuine or not genuine of the test image most of the recognition systems [11, 13] are preferred texture analysis over content-based image retrieval and shape-based image retrieval system. Texture analysis

J. I.-Z. Chen et al. (Eds.): ICIPCN 2020, AISC 1200, pp. 794–804, 2021.
https://doi.org/10.1007/978-3-030-51859-2_72

has different methods such as statistical, geometrical model-based and processing, etc. in big data analysis and machine learning statistical texture analysis is the key role in the research to collecting and analysis of the texture from the database. Geometrical analysis is in the area of the edge and surface of the input image. Model base machine learning is exposed to the area of latent from the different distance functions like a Gaussian mixture model. One of the robust and effective processes of the preprocessing model is the content processing. These are different types of filters such as low pass filter, high pass filter etc. texture analysis method covers the entire image of the pixel complete local binary pattern is preferred. The local binary pattern [1, 2] can be represented as different scaling operations such as rotation, orientations, etc. in the spatial system, and the pattern recognition system scale-invariant feature transform is the most important method. Generally, any image can be transformed into different features of decomposition but the image key point may not by chance at the same point in the original key point. The drawback of the existing methods is unable to convert it as the same features of the input image. So, this problem is avoiding with the help of scale-invariant feature transform (SIFT) based on the scale and orientation of the face database. SIFT is concentrated on the key point and descriptors of the input face image. This transform converted in terms of various scaling and orientations with the help of Laplacian of Gaussian. Laplacian of Gaussian is widely used for detection. LoG (Laplacian of Gaussian) is very approximation to the Difference of Gaussian (DoG). DoG is found based on the difference of blurring image of the Gaussian function with various scaling and orientations of the pyramid.

Modified LBP and modified SIFT are explained in Sect. 2, The proposed method has been described in Sect. 3, The experimental results and discussion are described in Sect. 4 and the conclusion is described in Sect. 5.

2 Modified Local Binary Pattern & Scale-Invariant Feature Transform

2.1 Modified Local Binary Pattern (MLBP)

One of the visual descriptors is the local binary pattern for the face recognition system. LBP was invented by Bouwman et al. for the texture model in the year of 1994. The local binary pattern has four parameters. The importance of radius is widely implemented in the LBP system. Neighbors of the parameters are used for identifying the number of values in the different octaves. Grid X and Grid Y are used for both horizontally and vertically respectively. In local feature extraction method for face recognition has been successfully implemented with their internal feature of the input image by using local binary pattern process. Histogram process is the one of the best methods to identify and recognition the values if the input image is continuous according to the varying along with the samples. The basic block diagram of the face recognition system has been described in detail in the following block diagram. It is represented in Fig. 1.

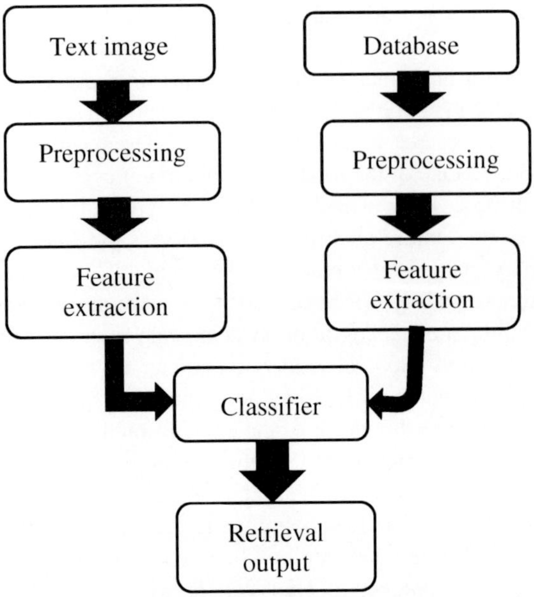

Fig. 1. Basic block diagram for face recognition system

2.2 Modified Scale Invariant Feature Transform

In machine learning, matching features are a very essential point. Generally, most of the faces are the same in terms of size but the facial expressions are different in terms of scale and orientation so scale-invariant feature transform is preferable to avoid this drawback. David Lowe was introduced SIFT. SIFT intended to generate a descriptor that was compacted to the variations consistent with characteristic viewing conditions. The descriptor is the most used part of SIFT. The applications of SIFT are face recognition, motion tracking, object recognition, navigation, etc. SIFT uses to identify the key point feature extraction. In the face recognition system, each face image is converted in terms of local features that are invariant into different scales such as rotation, translation, etc. The advantages of the SIFT are region-based, individually, quantity, effectiveness, and extended to a wide range. The region-based is concentrated on the local features from the large face database (Fig. 2).

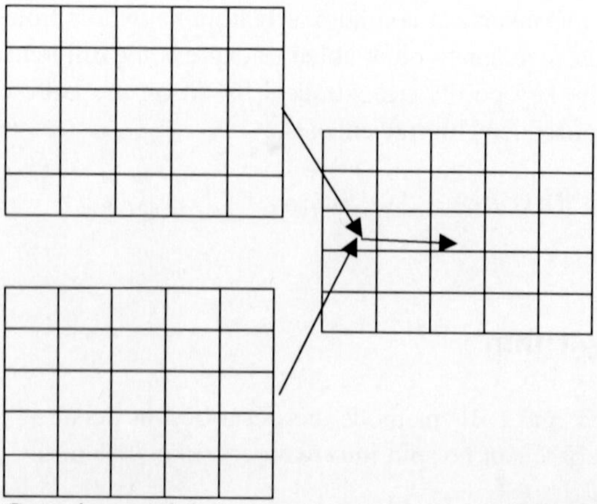

Fig. 2. Difference of Gaussian model

Individually means each local feature can be mapped to the local features of a very large face database. Quality measures such a large database faces can be represented as small objects. Effectiveness measures the performance of the retrieval images in terms of precision and recall. Due to different scaling and orientation techniques if can be easily implemented on a large face database to extract the feature for robust. The procedure of the scale-invariant feature transform has the following steps to extract the local features. These can be represented as a blurred image with a Gaussian operator. Where blur image is the convolution operation of the input image and Gaussian function.

$$L(x, y, z) = I(X, Y) \oplus G(x, y, \sigma)$$

Where

x, y = Location coordinates
I = Input image
G = Gaussian blur image
L = Blur image
$\oplus$ = Convolution operation
σ = Scale parameter = $k\sigma$

Where k is $\sqrt{2}$ is preferable

$$G(x, y, \sigma) = \frac{1}{2\pi\sigma^2} exp^{-\frac{(x^2 - y^2)}{2\sigma^2}}$$

SIFT is not only applicable for invariant of the image but also having good properties in terms of illumination and viewpoint. SIFT features are octaves and scale. These values may vary based on the size of the original face image. To recognitions better efficiency has to be taken as maximum as feature keywords from the input image. So initially input image size has to be doubled to express the different octaves in terms of different scales. The key points are extracted based on the Difference of Gaussian image between the consequent blurred images.

$$D(x, y, \sigma) = L(x, y, k_i, \sigma) - L(x, y, k_j, \sigma)$$

3 Proposed Algorithm

Hybrid modified SIFT and LBP methods are described in detail as mentioned in the below steps. The database can be split into two parts like 70% of the training and 30% of the testing images.

3.1 Modified Local Binary Pattern Algorithm

Step 1: The input image can be represented as various subbands based on a square matrix. Each sub band center pixel is compared with all its neighbors.
Step 2: After subtracting its center pixel value, 1 is replaced for greater value pixel and 0 is replaced for less than of the pixel value.
Step 3: Then, the binary number can be converted into decimal value based on the clockwise direction.
Step 4: Similarly, the same procedure for all individual pixels with no overlapping procedure. The feature extraction process of the text image is chosen from the untrained dataset (Fig. 3).

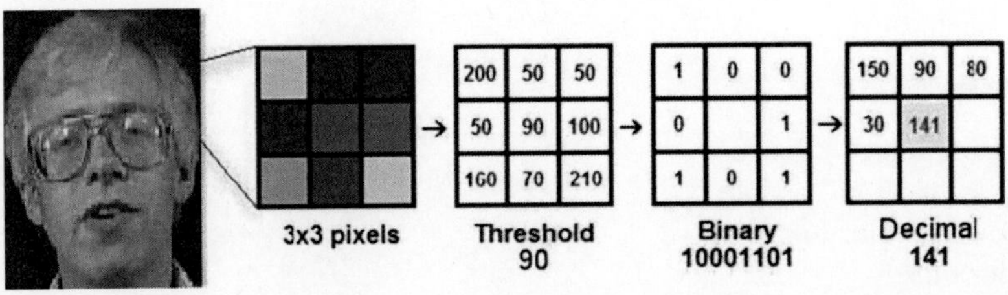

Fig. 3. Identification for decimal value in local binary pattern

Step 5: Compute the histogram of each subband. Then each histogram of the subband gives the face features of the input images.

3.2 Modified Scale Invariant Feature Transform

Step 1: **Scale space extrema detection:** scale space extreme
Detection is composed and gives

x, y, k_j, σ For Gaussian function

$$G(x, y, \sigma) = \frac{1}{2\pi\sigma^2} exp^{-\frac{(x^2 - y^2)}{2\sigma^2}}$$

Different Gaussian images has been illuminated in Fig. 4.

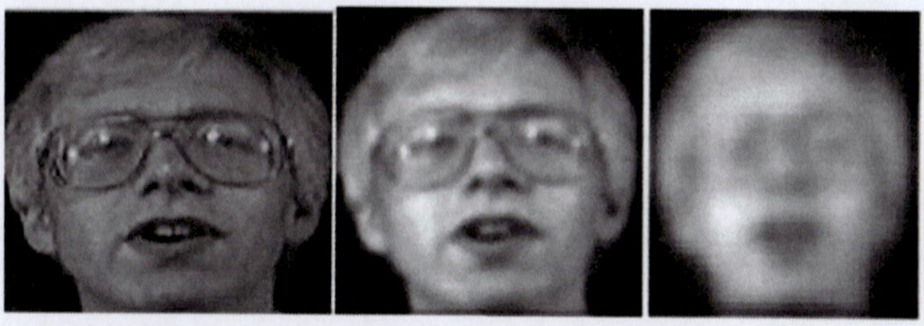

Fig. 4. Model for scale-space extrema detection [5]

Step 2: **localization:** To find the key point localization and it can be extracted the local features of the input image like blob detection. The localization is also illuminated in Fig. 5.

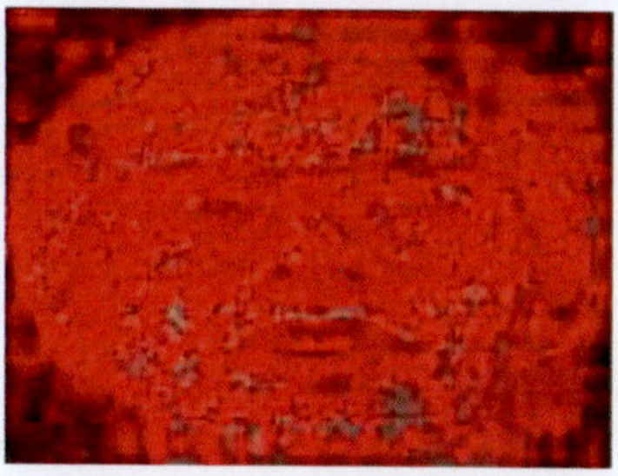

Fig. 5. Key point localization model for modified SIFT [6]

Step 3: **Assignment with various orientations:** The feature in terms of direct orientation in Fig. 6.

Fig. 6. Orientation model for SIFT

Step 4: **Features points:** To extract content based on the histogram of the subblocks. Then it can be converted as maximum number possibles (Fig. 7).

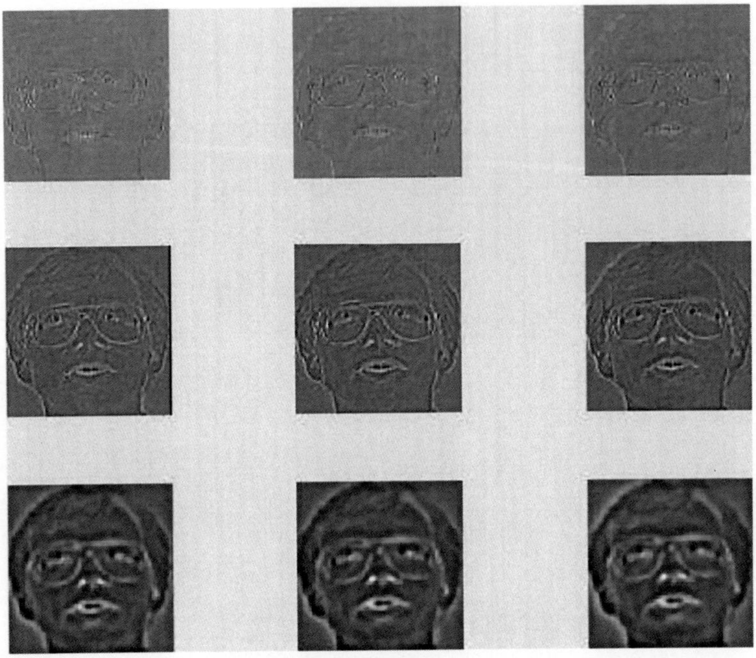

Fig. 7. Scale-invariant feature transform decomposition for three octaves and three orientations.

Step 5: **Similarity features:** To find the similar feature content based on the training model.

4 Experimental Results

The face recognition system has been executed by Yale and ORL standard databases (Fig. 8, 9, and 10 and Table 1).

Table 1. Details of Yale and ORL databases.

Parameters	Yale	ORL
Classes	15	40
Images	11	10
Total faces	165	400
Size	100 × 80	112 × 92
Type	Grey-scale	Grey-scale
Expressions	Centre-light, happy, normal, sad, Surprise etc.	Varying the lights, smiling, closed eyes, glasses, etc.
Designed	Yale University	Cambridge University

Fig. 8. ORL database images for face recognition system.

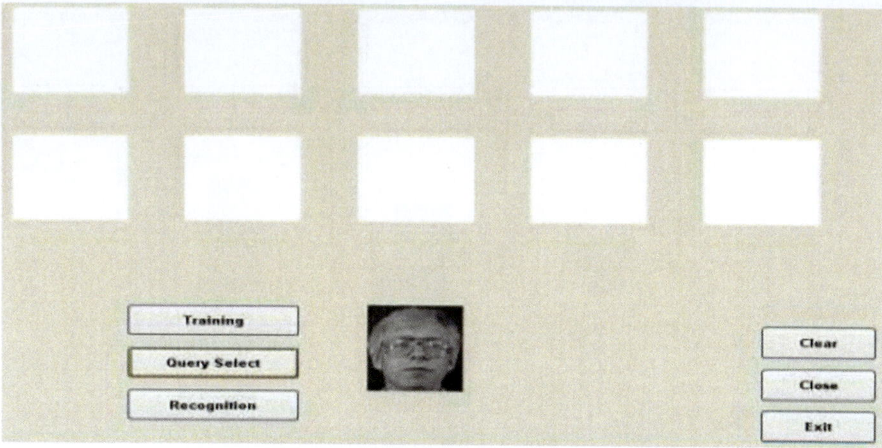

Fig. 9. Testing image for the proposed method

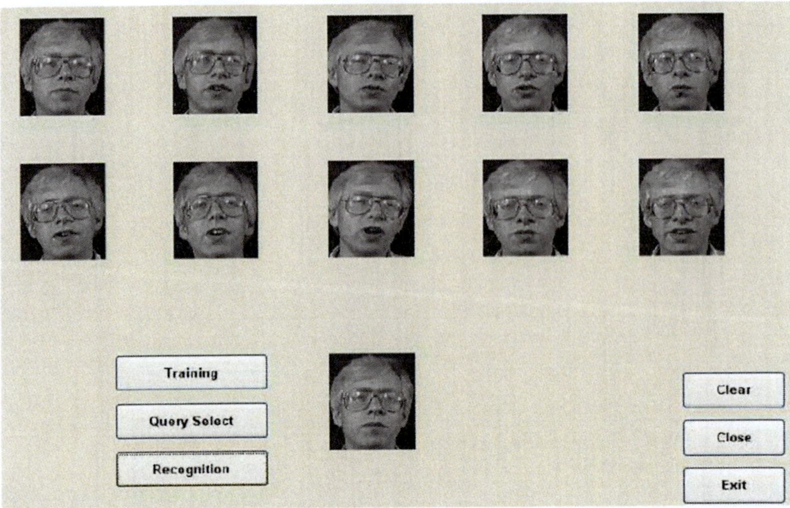

Fig. 10. Retrieval images for the proposed method

The proposed method for modified LBP and modified SIFT experimental results have been described in Table 2.

Table 2. Face recognition using performance measures

		Recognition efficiency (%)		
S.No	Methods	Precision	Recall	F-Score
1	Enhanced GWT	85	83	83.98
2	Modified LBP	91	89	89.98
3	Modified SIFT (Proposed)	95.5	93.5	94.48

The proposed comparison graph for the precision is shown in Fig. 11.

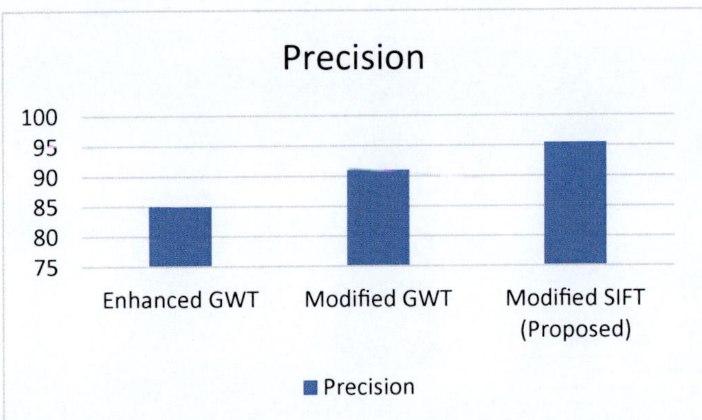

Fig. 11. Comparison graph for precision for face recognition methods

Performance measure is calculated by using precision, recall, and F-Score based on retrieval images (Figs. 12 and 13).

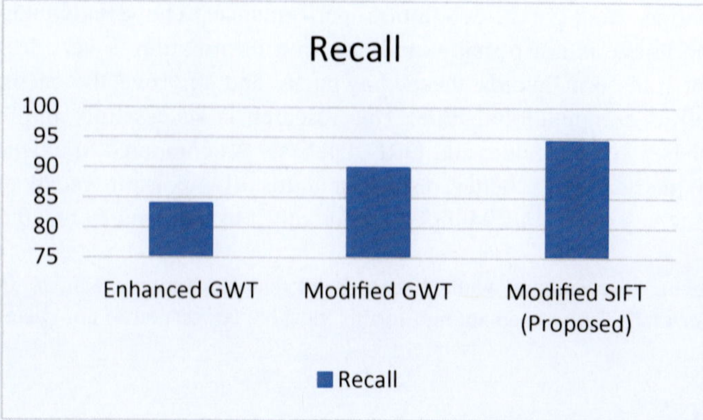

Fig. 12. Comparison graph for recall for face recognition methods

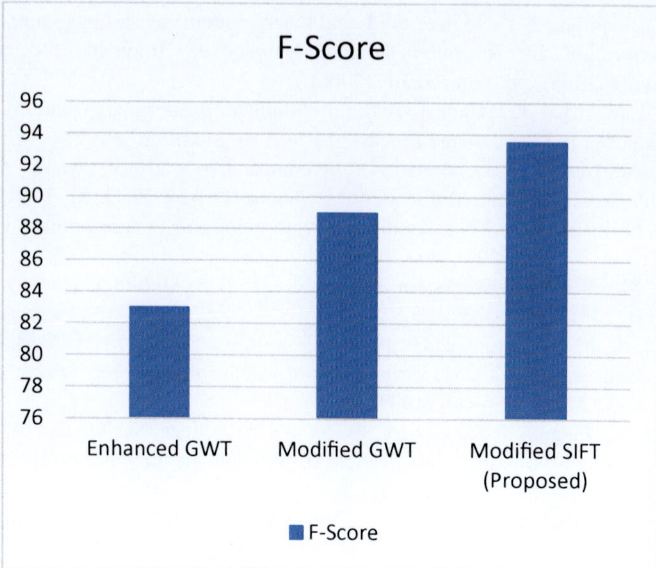

Fig. 13. Comparison graph for F-Score for face recognition methods

5 Conclusion

Modified local binary pattern and scale-invariant transform for face recognition has been implemented in this paper. Modified local binary pattern advantages are less computation time, and good recognition performance. The disadvantages are the rotation of the image is not possible and captured information is very less. Modified scale-invariant transform avoids these drawbacks and improve the recognition efficiency as well as computational time. This research is successfully implemented on different databases such as Yale, and ORL database. As compared to existing methods the proposed method gives better results in terms of precision, recall and F-Score values are 95.5%, 93.5%, and 94.48% respectively as compared to existing methods.

Acknowledgment. The authors would like to thank Dr. S. Srinivasa Kumar, Professor and Vice-Chancellor to JNTUA, Anathapuram for his valuable suggestion to complete this work.

References

1. Ahonen, T., Hadid, A., Pietikainen, M.: Face description with local binary patterns: application to face recognition. IEEE Trans. Pattern Anal. Mach. Intell. **28**(12), 2037–2041 (2006)
2. Heusch, G., Rodriguez, Y., Marcel, S.: Local binary patterns as an image preprocessing for face authentication. In: 7th International Conference on Automatic Face and Gesture Recognition (FGR06), pp. 6–pp. IEEE (2006)
3. Liao, S., Law, M.W.K., Chung, A.C.S.: Dominant local binary patterns for texture classification. IEEE Trans. Image Process. **18**(5), 1107–1118 (2009)
4. Ojala, T., Pietikäinen, M., Harwood, D.: A comparative study of texture measures with classification based on featured distributions. Pattern Recogn. **29**(1), 51–59 (1996)
5. Mäenpää, T., Pietikäinen, M.: Classification with color and texture: jointly or separately? Pattern Recogn. **37**(8), 1629–1640 (2004)
6. Bishop, C.M.: Neural Networks for Pattern Recognition. Oxford University Press, New York (1995)
7. Duda, R.O., Hart, P.E., Stork, D.G.: Pattern Classification. Wiley, New York (2012)
8. Ojala, T., Pietikäinen, M.: Unsupervised texture segmentation using feature distributions. Pattern Recogn. **32**(3), 477–486 (1999)
9. Liu, L., Lao, S., Fieguth, P.W., Guo, Y., Wang, X., Pietikäinen, M.: Median robust extended local binary pattern for texture classification. IEEE Trans. Image Process. **25**(3), 1368–1381 (2016)
10. Liu, L., Fieguth, P., Guo, Y., Wang, X., Pietikäinen, M.: Local binary features for texture classification: taxonomy and experimental study. Pattern Recogn. **62**, 135–160 (2017)
11. Ojala, T., Pietikainen, M., Maenpaa, T.: Multiresolution gray-scale and rotation invariant texture classification with local binary patterns. IEEE Trans. Pattern Anal. Mach. Intell. **24** (7), 971–987 (2002)
12. Kumar, T.S.: A novel method for HDR video encoding, compression and quality evaluation. J. Innov. Image Process. (JIIP) **1**(02), 71–80 (2019)
13. Heikkila, M., Pietikainen, M.: A texture-based method for modeling the background and detecting moving objects. IEEE Trans. Pattern Anal. Mach. Intell. **28**(4), 657–662 (2006)

Author Index

J. I.-Z. Chen et al. (Eds.): ICIPCN 2020, AISC 1200, pp. 805–807, 2021.
https://doi.org/10.1007/978-3-030-51859-2

MIX
Papier aus verantwortungsvollen Quellen
Paper from responsible sources
FSC® C105338

Printed by Books on Demand, Germany